CAMPBELL
essential
biology

CAMPBELL

essential

biology

7e

GLOBAL EDITION

Eric J. Simon • **Jean L. Dickey** • **Jane B. Reece**

New England College   Clemson, South Carolina   Berkeley, California

with contributions from
Rebecca S. Burton
Alverno College

P Pearson

Courseware Portfolio Management, Director: *Beth Wilbur*
Courseware Portfolio Management, Specialist:
  Alison Rodal
Courseware Director, Content Development:
  Ginnie Simione Jutson
Courseware Sr. Analyst: *John Burner*
Developmental Editor: *Susan Teahan*
Associate Editor, Global Edition: *Sulagna Dasgupta*
Courseware Editorial Assistant: *Alison Candlin*
Managing Producer: *Mike Early*
Content Producer: *Lori Newman*
Senior Content Developer: *Sarah Jensen*
Rich Media Content Producers: *Tod Regan, Ziki Dekel*
Media Production Manager, Global Edition:
  Vikram Kumar
Full-Service Vendor: *Integra Software Services, Inc.*

Copyeditor: *Joanna Dinsmore*
Compositor: *Integra Software Services, Inc.*
Design Manager: *Mark Ong*
Cover Designer, Global Edition: *Lumina Datamatics Ltd.*
Interior Design: *TT Eye*
Illustrators: *Lachina*
Rights & Permissions Project Manager: *Ben Ferrini*
Rights & Permissions Management: *Cenveo*
Photo Researcher: *Kristin Piljay*
Manufacturing Buyer: *Stacey Weinberger*
Senior Manufacturing Controller, Global Edition:
  Kay Holman
Product Marketing Manager: *Christa Pelaez*
Field Marketing Manager: *Kelly Galli*
Cover Photo Credit: *robdimagery/Shutterstock*

Acknowledgements of third party content appear on page A-5, which constitutes an extension of this copyright page.

Pearson Education Limited
KAO Two
KAO Park
Harlow
CM17 9NA
United Kingdom

and Associated Companies throughout the world

Visit us on the World Wide Web at: www.pearsonglobaleditions.com

British Library Cataloguing-in-Publication Data
A catalogue record for this book is available from the British Library

ISBN 10: 1-292-30709-9
ISBN 13: 978-1-292-30709-1

10 9 8 7 6 5 4 3 2 1
20 19

Typeset by Integra Software Services, Inc
Printed in Malaysia (CTP-VVP)

About the Authors

ERIC J. SIMON

is a professor in the Department of Biology and Health Science at New England College (Henniker, New Hampshire). He teaches introductory biology to science majors and nonscience majors, as well as upper-level courses in tropical marine biology and careers in science. Dr. Simon received a B.A. in biology and computer science, an M.A. in biology from Wesleyan University, and a Ph.D. in biochemistry from Harvard University. His research focuses on innovative ways to use technology to increase active learning in the science classroom, particularly for nonscience majors. Dr. Simon is also the author of the introductory biology textbook *Biology: The Core*, 2nd Edition, and a coauthor of *Campbell Biology: Concepts & Connections*, 9th Edition.

To my lifelong friends BZ, SR, and SR, who have taught me the value of loyalty and trust during decades of unwavering friendship

JANE B. REECE

was Neil Campbell's longtime collaborator and a founding author of *Campbell Essential Biology* and *Campbell Essential Biology with Physiology*. Her education includes an A.B. in biology from Harvard University (where she was initially a philosophy major), an M.S. in microbiology from Rutgers University, and a Ph.D. in bacteriology from the University of California, Berkeley. At UC Berkeley, and later as a postdoctoral fellow in genetics at Stanford University, her research focused on genetic recombination in bacteria. Dr. Reece taught biology at Middlesex County College (New Jersey) and Queensborough Community College (New York). Dr. Reece's publishing career began in 1978 when she joined the editorial staff of Benjamin Cummings, and since then, she played a major role in a number of successful textbooks. She was the lead author of *Campbell Biology* Editions 8–10 and a founding author of *Campbell Biology: Concepts & Connections*.

To my wonderful coauthors, who have made working on our books a pleasure

JEAN L. DICKEY

is Professor Emerita of Biological Sciences at Clemson University (Clemson, South Carolina). After receiving her B.S. in biology from Kent State University, she went on to earn a Ph.D. in ecology and evolution from Purdue University. In 1984, Dr. Dickey joined the faculty at Clemson, where she devoted her career to teaching biology to nonscience majors in a variety of courses. In addition to creating content-based instructional materials, she developed many activities to engage lecture and laboratory students in discussion, critical thinking, and writing, and implemented an investigative laboratory curriculum in general biology. Dr. Dickey is the author of *Laboratory Investigations for Biology*, 2nd Edition, and is a coauthor of *Campbell Biology: Concepts & Connections*, 9th Edition.

To my mother, who taught me to love learning, and to my daughters, Katherine and Jessie, the twin delights of my life

NEIL A. CAMPBELL

(1946–2004) combined the inquiring nature of a research scientist with the soul of a caring teacher. Over his 30 years of teaching introductory biology to both science majors and nonscience majors, many thousands of students had the opportunity to learn from him and be stimulated by his enthusiasm for the study of life. He is greatly missed by his many friends in the biology community. His coauthors remain inspired by his visionary dedication to education and are committed to searching for ever-better ways to engage students in the wonders of biology.

Preface

Biology education has been transformed in the last decade. The non-majors introductory biology course was (in most cases) originally conceived as a slightly less deep and broad version of the general biology course. But a growing recognition of the importance of this course—one that is often the most widely enrolled within the department, and one that serves as the sole source of science education for many students—has prompted a reevaluation of priorities and a reformulation of pedagogy. Many instructors have narrowed the focus of the course from a detailed compendium of facts to an exploration of broader themes within the discipline—themes such as the central role of evolution and an understanding of the process of science. For many educators, the goals have shifted from communicating a great number of bits of information toward providing a deep understanding of fewer, but broader, principles. Luckily for anyone teaching or learning biology, opportunities to marvel at the natural world and the life within it abound. Furthermore, nearly everyone realizes that the subject of biology has a significant impact on his or her own life through its connections to medicine, biotechnology, agriculture, environmental issues, forensics, and many other areas. Our primary goal in writing *Campbell Essential Biology* is to help teachers motivate and educate the next generation of citizens by communicating the broad themes that course through our innate curiosity about life.

Goals of the Book

Although our world is rich with "teachable moments" and learning opportunities, an explosion of knowledge threatens to bury a curious person under an avalanche of information. "So much biology, so little time" is the universal lament of biology educators. Neil Campbell conceived of *Campbell Essential Biology* as a tool to help teachers and students focus on the most important areas of biology. To that end, the book is organized into four core areas: cells, genes, evolution, and ecology. Dr. Campbell's vision, which we carry on and extend in this edition, has enabled us to keep *Campbell Essential Biology* manageable in size and thoughtful in the development of the concepts that are most fundamental to understanding life. We've aligned this new edition with today's "less is more" approach in biology education for nonscience majors—where the emphasis is on fewer topics but broader themes—while never allowing the important content to be diluted.

We formulated our approach after countless conversations with teachers and students in which we noticed some important trends in how biology is taught. In particular, many instructors identify three goals: (1) to engage students by relating biology content to their lives and the greater society; (2) to help students understand the process of science by teaching critical thinking skills that can be used in everyday life; and (3) to demonstrate how biology's broader themes—such as evolution and the relationship of structure to function—serve to unify the entire subject. To help achieve these goals, every chapter of this book includes several important

features. First, a chapter-opening essay called Biology and Society highlights a connection between the chapter's core content and students' lives. Second, an essay called The Process of Science (in the body of the chapter) describes how the scientific process has illuminated the topic at hand, using a classic or modern experiment as an example. Third, a chapter-closing Evolution Connection essay relates the chapter to biology's unifying theme of evolution. Fourth, the broad themes that unify all subjects within biology are explicitly called out (in blue) multiple times within each chapter. Finally, to maintain a cohesive narrative throughout each chapter, the content is tied together with a unifying chapter thread, a relevant high-interest topic that is touched on several times in the chapter and woven throughout the three feature essays. Thus, this unifying chapter thread ties together the pedagogical goals of the course, using a topic that is compelling and relevant to students.

New to This Edition

This latest edition of *Campbell Essential Biology* goes even further than previous editions to help students relate the material to their lives, understand the process of science, and appreciate how broad themes unify all aspects of biology. To this end, we've added significant new features and content to this edition:

- **A new approach to teaching the process of science.** Conveying the process of science to nonscience-major undergraduate students is one of the most important goals of this course. Traditionally, we taught the scientific method as a predefined series of steps to be followed in an exact order (observation, hypothesis, experiment, and so forth). Many instructors have shifted away from such a specific flow chart to a more nuanced approach that involves multiple pathways, frequent restarts, and other features that more accurately reflect how science is actually undertaken. Accordingly, we have revised the way that the process of science is discussed within our text, both in Chapter 1 (where the process is discussed in detail) and in The Process of Science essay in every chapter of the textbook. Rather than using specific terms in a specific order to describe the process, we now divide it into three broad interrelated areas: background, method, and results. We believe that this new approach better conveys how science actually proceeds and demystifies the topic for non-scientists. Chapter 1 also contains important information that promotes critical thinking, such as discussion of control groups, pseudoscience, and recognizing reliable sources of information. We believe that providing students with such critical-thinking tools is one of the most important outcomes of the nonscience-major introductory course.
- **Major themes in biology incorporated throughout the book.** In 2009, the American Association for the Advancement of Science published a document that served as a call to action in undergraduate biology education. The principles of this document, which

is titled "Vision and Change," are becoming widely accepted throughout the biology education community. "Vision and Change" presents five core concepts that serve as the foundation of undergraduate biology. In this edition of *Campbell Essential Biology,* we repeatedly and explicitly link book content to themes multiple times in each chapter, calling out such instances with boldfaced blue text. For example, in Chapter 4 (A Tour of the Cell), the interrelationships of cellular structures are used to illustrate the theme of interactions within biological systems. The plasma membrane is presented as an example of the relationship between structure and function. The cellular structures in the pathway from DNA to protein are used to illustrate the importance of information flow. The chloroplasts and mitochondria serve as an example of the transformations of energy and matter. The DNA within these structures is also used to illustrate biology's overarching theme of evolution. Students will find three to five examples of themes called out in each chapter, which will help them see the connections between these major themes and the course content. To reinforce these connections, this edition of *Campbell Essential Biology* includes new end-of-chapter questions and Mastering Biology activities that promote critical thinking relating to these themes. Additionally, PowerPoint© lecture slides have been updated to incorporate chapter examples and offer guidance to faculty on how to include in these themes within classroom lectures.

■ **Updated connections to students' lives.** In every edition of *Campbell Essential Biology*, we seek to improve and extend the ways that we connect the course content to students' lives. Accordingly, every chapter begins with an improved feature called Why It Matters showing the relevance of the chapter content from the very start. Additionally, with every edition, we introduce some new unifying chapter threads intended to improve student relevance. For example, this edition includes new threads that discuss evolution in a human-dominated world (Chapter 14) and the importance of biodiversity to human affairs (Chapter 20). As always, we include some updated Biology and Society chapter-opening essays (such as "A Solar Revolution" in Chapter 7), The Process of Science sections (such as a recent experiment investigating the efficacy of radiation therapy to treat prostate cancer, in Chapter 2), and Evolution Connection chapter-closing essays (such as an updated discussion of biodiversity hot spots in Chapter 20). As we always do, this edition includes many content updates that connect to students' lives, such as information on cutting-edge cancer therapies (Chapter 8) and recent examples of DNA profiling (Chapter 12).

■ **Developing data literacy through infographics.** Many nonscience-major students express anxiety when faced with numerical data, yet the ability to interpret data can help with many important decisions we all face. Increasingly, the general public encounters information in the form of infographics, visual images used to represent data. Consistent with our goal of preparing students to approach important issues critically, this edition includes a series of new infographics, or Visualizing the Data figures. Examples include the elemental composition of the human body (Chapter 2), a comparison of calories burned through exercise versus calories consumed in common foods (Chapter 5), and ecological footprints (Chapter 19). In addition to the printed form, these infographics are available as assignable tutorial questions within Mastering Biology.

■ **Helping students to understand key figures.** For this new edition, a key figure in each chapter is supplemented by a short video explaining the concept to the student. These Figure Walkthrough videos will be assignable in Mastering Biology. The animations are written and narrated by authors Eric Simon and Jean Dickey, as well as teacher and contributor Rebecca Burton.

Attitudes about science and scientists are often shaped by a single, required science class—*this* class. We hope to nurture an appreciation of nature into a genuine love of biology. In this spirit, we hope that this textbook and its supplements will encourage all readers to make biological perspectives a part of their personal worldviews. Please let us know how we are doing and how we can improve the next edition of *Campbell Essential Biology.*

ERIC SIMON
Department of Biology and
Health Science
New England College
Henniker, NH 03242
SimonBiology@gmail.com

JEAN DICKEY
Clemson, SC
dickeyj@clemson.edu

JANE B. REECE
Berkeley, California

The following Visual Walkthrough highlights key features of *Campbell Essential Biology 7e.*

Develop and practice science literacy skills

Learn how to view your world using scientific reasoning with *Campbell Essential Biology*. See how concepts from class and an understanding of how science works can apply to your everyday life. Engage with the concepts and practice science literacy skills with Mastering Biology and Pearson eText.

NEW! New and updated Process of Science essays present scientific discovery as a flexible and non-linear process.

Each essay summarizes the **background, method,** and **results** from a scientific study.

New Thinking Like a Scientist questions appear at the end of each Process of Science essay and involve applying a scientific reasoning skill.

Examples of new Process of Science topics include:

- Chapter 4: How Was the First 21st-Century Antibiotic Discovered? p. 95
- Chapter 9: What Is the Genetic Basis of Short Legs in Dogs? p. 190
- Chapter 11: Can Avatars Improve Cancer Treatment? p. 244
- Chapter 16: What Killed the Pines? p. 364
- Chapter 20: Does Biodiversity Protect Human Health? p. 480

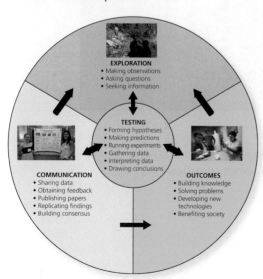

EXPLORATION
- Making observations
- Asking questions
- Seeking information

TESTING
- Forming hypotheses
- Making predictions
- Running experiments
- Gathering data
- Interpreting data
- Drawing conclusions

COMMUNICATION
- Sharing data
- Obtaining feedback
- Publishing papers
- Replicating findings
- Building consensus

OUTCOMES
- Building knowledge
- Solving problems
- Developing new technologies
- Benefiting society

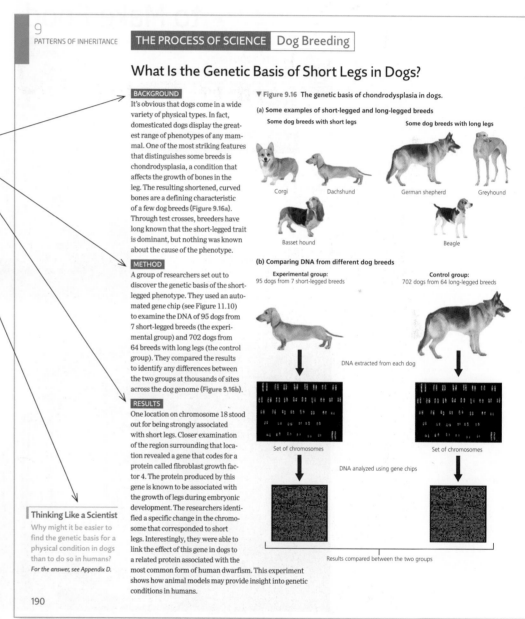

9 PATTERNS OF INHERITANCE

THE PROCESS OF SCIENCE Dog Breeding

What Is the Genetic Basis of Short Legs in Dogs?

BACKGROUND

It's obvious that dogs come in a wide variety of physical types. In fact, domesticated dogs display the greatest range of phenotypes of any mammal. One of the most striking features that distinguishes some breeds is chondrodysplasia, a condition that affects the growth of bones in the leg. The resulting shortened, curved bones are a defining characteristic of a few dog breeds (Figure 9.16a). Through test crosses, breeders have long known that the short-legged trait is dominant, but nothing was known about the cause of the phenotype.

METHOD

A group of researchers set out to discover the genetic basis of the short-legged phenotype. They used an automated gene chip (see Figure 11.10) to examine the DNA of 95 dogs from 7 short-legged breeds (the experimental group) and 702 dogs from 64 breeds with long legs (the control group). They compared the results to identify any differences between the two groups at thousands of sites across the dog genome (Figure 9.16b).

RESULTS

One location on chromosome 18 stood out for being strongly associated with short legs. Closer examination of the region surrounding that location revealed a gene that codes for a protein called fibroblast growth factor 4. The protein produced by this gene is known to be associated with the growth of legs during embryonic development. The researchers identified a specific change in the chromosome that corresponded to short legs. Interestingly, they were able to link the effect of this gene in dogs to a related protein associated with the most common form of human dwarfism. This experiment shows how animal models may provide insight into genetic conditions in humans.

▼ Figure 9.16 The genetic basis of chondrodysplasia in dogs.

(a) Some examples of short-legged and long-legged breeds

Some dog breeds with short legs

Corgi

Dachshund

Basset hound

Some dog breeds with long legs

German shepherd

Greyhound

Beagle

(b) Comparing DNA from different dog breeds

Experimental group: 95 dogs from 7 short-legged breeds

Control group: 702 dogs from 64 long-legged breeds

DNA extracted from each dog

Set of chromosomes

Set of chromosomes

DNA analyzed using gene chips

Results compared between the two groups

| **Thinking Like a Scientist**
Why might it be easier to find the genetic basis for a physical condition in dogs than to do so in humans?
For the answer, see Appendix D.

190

NEW! A new organization and new content in Chapter 1 focus on science literacy skills to introduce the process of science right from the start.

Explore biology with . . .

7 Photosynthesis: Using Light to Make Food

Why Photosynthesis Matters

Do you like to eat? We humans can trace every morsel of our food back to plants. By capturing the energy of sunlight and using it to create organic materials, plants performing photosynthesis feed the world.

NEARLY ALL LIFE ON EARTH—INCLUDING YOU—CAN TRACE ITS SOURCE OF ENERGY BACK TO THE SUN.

COVER UP! PROTECTING YOURSELF FROM SHORT WAVELENGTHS OF LIGHT CAN BE LIFESAVING.

WANT TO DO SOMETHING SIMPLE TO COMBAT GLOBAL CLIMATE CHANGE? PLANT A TREE—YOU'LL BE GLAD YOU DID!

140

Why It Matters Photo Collages have been updated to give real-world examples to convey why abstract concepts like cellular respiration or photosynthesis matter.

... the most relevant, real-world examples

New and Updated Chapter Threads weave a compelling topic throughout each chapter, highlighted in the Biology and Society, The Process of Science, and Evolution Connection essays.

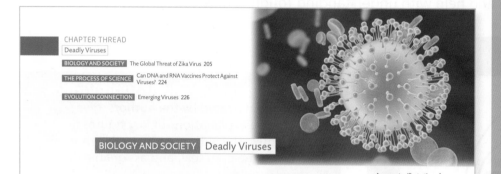

CHAPTER THREAD
Deadly Viruses
BIOLOGY AND SOCIETY The Global Threat of Zika Virus 205
THE PROCESS OF SCIENCE Can DNA and RNA Vaccines Protect Against Viruses? 224
EVOLUTION CONNECTION Emerging Viruses 226

A computer illustration of the Zika virus. Spikes made of protein enable the virus to recognize a host cell.

BIOLOGY AND SOCIETY Deadly Viruses

The Global Threat of Zika Virus

In 2015, an alarming number of babies were born in Brazil with severe damage to their central nervous systems and sensory organs. The affected babies had neurological problems (such as underdeveloped brains and seizures), slow growth, difficulty feeding, and joint and muscle problems. After a frantic search, health officials discovered a link between these abnormalities and exposure to a little-known pathogen: the Zika virus. By 2016, when the United Nations World Health Organization (WHO) issued a worldwide health emergency, Zika virus and Zika-related health problems in newborns began appearing in warm, humid regions of the United States and many other countries.

The Zika virus was first discovered to infect humans in 1952 and had been identified in African monkeys a few years earlier. Zika virus can be transmitted to humans by one species of mosquito. It can also be spread between sexual partners. But Zika virus is not dangerous to most healthy adults. In fact, some people feel just fine after being infected, while others have mild symptoms like aches or a fever. However, Zika virus can be spread from mother to fetus. Unfortunately, developing babies are particularly vulnerable to the virus's effects.

Health agencies have few weapons against Zika virus. There is no vaccine, and medicines can only treat symptoms. Nighttime mosquito netting and staying indoors after dusk can offer protection against many mosquito-borne diseases, but the mosquitoes that carry Zika virus bite both night and day. Public awareness campaigns aimed at avoiding mosquito bites and eliminating mosquito breeding grounds (such as stagnant water) have been implemented in Zika-prone areas. In November of 2016, WHO declared that the Zika global health emergency was over, not because Zika is gone, but because it is expected to be a long-term problem, the "new normal" rather than an emergency.

The Zika virus, like all viruses, consists of a relatively simple structure of nucleic acid (RNA in this case) and protein. Viruses operate by hijacking our own cells and turning them into virus factories. Combating any virus therefore requires a detailed understanding of life at the molecular level. In this chapter, we will explore the structure of life's most important molecule—DNA—to learn how it replicates, mutates, and controls the cell by directing the synthesis of RNA and protein.

NEW!
New Chapter Threads include:

- Chapter 1: Swimming with the Turtles
- Chapter 2: Helpful Radiation
- Chapter 7: Solar Energy
- Chapter 13: Evolution in Action
- Chapter 14: Evolution in the Human-Dominated World
- Chapter 20: Importance of Biodiversity

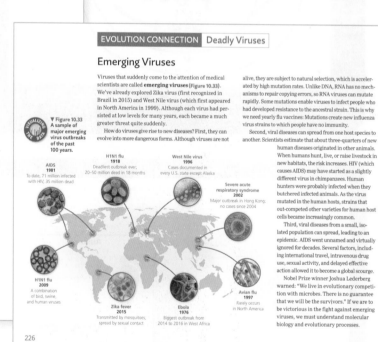

EVOLUTION CONNECTION Deadly Viruses

Emerging Viruses

Viruses that suddenly come to the attention of medical scientists are called **emerging viruses** (Figure 10.33). We've already explored Zika virus (first recognized in Brazil in 2015) and West Nile virus (which first appeared in North America in 1999). Although each virus had persisted at low levels for many years, each became a much greater threat quite suddenly.

How do viruses give rise to new diseases? First, they can evolve into more dangerous forms. Although viruses are not alive, they are subject to natural selection, which is accelerated by high mutation rates. Unlike DNA, RNA has no mechanisms to repair copying errors, so RNA viruses can mutate rapidly. Some mutations enable viruses to infect people who had developed resistance to the ancestral strain. This is why we need yearly flu vaccines: Mutations create new influenza virus strains to which people have no immunity.

Second, viral diseases can spread from one host species to another. Scientists estimate that about three-quarters of new human diseases originated in other animals. When humans hunt, live, or raise livestock in new habitats, the risk increases. HIV (which causes AIDS) may have started as a slightly different virus in chimpanzees. Human hunters were probably infected when they butchered infected animals. As the virus mutated in the human hosts, strains that out-competed other varieties for human host cells became increasingly common.

Third, viral diseases from a small, isolated population can spread, leading to an epidemic. AIDS went unnamed and virtually ignored for decades. Several factors, including international travel, intravenous drug use, sexual activity, and delayed effective action allowed it to become a global scourge.

Nobel Prize winner Joshua Lederberg warned: "We live in evolutionary competition with microbes. There is no guarantee that we will be the survivors." If we are to be victorious in the fight against emerging viruses, we must understand molecular biology and evolutionary processes.

▼ Figure 10.33 A sample of major emerging virus outbreaks of the past 100 years.

AIDS 1981
To date, 71 million infected with HIV, 35 million dead

H1N1 flu 1918
Deadliest outbreak ever; 20–50 million dead in 18 months

West Nile virus 1996
Cases documented in every U.S. state except Alaska

Severe acute respiratory syndrome 2002
Major outbreak in Hong Kong; no cases since 2004

H1N1 flu 2009
A combination of bird, swine, and human viruses

Zika fever 2015
Transmitted by mosquitoes, spread by sexual contact

Ebola 1976
Biggest outbreak from 2014 to 2016 in West Africa

Avian flu 1997
Rarely occurs in North America

226

Biology and Society essays
relating biology to everyday life are either new or updated. Some new topics:

- Chapter 7: A Solar Revolution p. 141
- Chapter 10: The Global Threat of Zika Virus p. 205
- Chapter 14: Humanity's Footprint p. 303
- Chapter 17: Evolving Adaptability p. 371

Evolution Connection essays
demonstrate the importance of evolution as a theme throughout biology, by appearing in every chapter. Some new topics:

- Chapter 1 Turtles in the Tree of Life p. 52
- Chapter 10 Emerging Viruses   p. 226
- Chapter 20 Saving the Hot Spots p. 483

Complex biological processes are explained . . .

Mastering™ Biology is an online homework, tutorial, and assessment platform that improves results by helping students quickly master concepts.

A wide range of interactive, engaging, and assignable activities, many of them contributed by *Campbell Essential Biology* authors, encourage active learning and help with understanding tough course concepts.

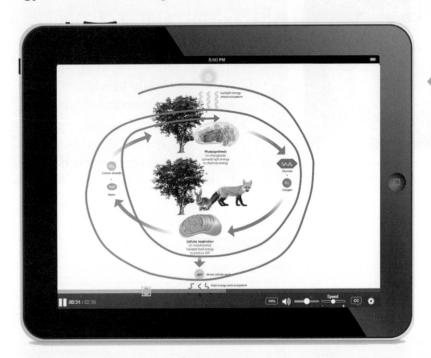

NEW! 20 Figure Walkthrough Videos, created and narrated by the authors, give clear, concise explanations of key figures in each chapter. The videos are accessible through QR codes in the print text, and assignable in Mastering Biology.

NEW! Visualizing the Data coaching activities bring the infographic figures in the text to life and are assignable in Mastering Biology.

...with engaging visuals and narrated examples in Mastering Biology

12 Topic Overview videos, created by the authors, introduce key concepts and vocabulary. These brief, engaging videos introduce topics that will be explored in greater depth in class.

Topics include:

– Macromolecules

– Ecological Organization

– Mechanisms of Evolution

– An Introduction to Structure and Function

– Interactions Between the Respiratory and Circulatory Systems

– DNA Structure and Function

. . . And more!

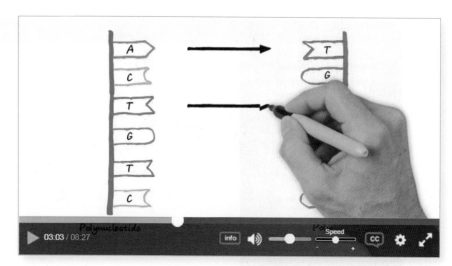

BioInteractive Short Films from HHMI, Video Tutors, BioFlix® 3D animations, and MP3 Audio Tutors support key concept areas covered in the text and provide coaching by using personalized feedback on common wrong answers.

Part A

Can you match the terms to their definitions?

Drag the terms on the left to the appropriate blanks on the right to complete the sentences.

Reset | Help

RNA

replication

base

translation

DNA

transcription

____ serves as the molecular basis for life.

DNA copies itself via the process of ____ .

RNA is produced from DNA via the process of ____ .

Proteins are produced from RNA via the process of ____ .

There are five examples of a ____ : A, G, C, T, and U.

One way that ____ is different from DNA is that it contains Us instead of Ts.

Ready-to-Go Teaching Modules make use of teaching tools for before, during, and after class, including new ideas for in-class activities. These modules incorporate the best that the text, Mastering Biology, and Learning Catalytics have to offer and can be accessed through the Instructor Resources area of Mastering Biology.

Learning Catalytics™ helps generate class discussion, customize lectures, and promote peer-to-peer learning with real-time analytics. Learning Catalytics acts as a student response tool that uses students' smartphones, tablets, or laptops to engage them in more interactive tasks and thinking.

- Help your students develop critical thinking skills
- Monitor responses to find out where your students are struggling
- Rely on real-time data to adjust your teaching strategy

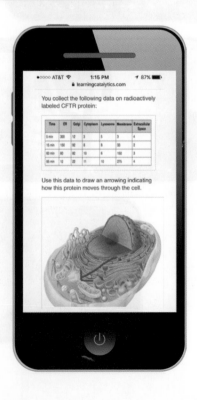

. . . and the resources to accomplish them

Extensive resources save instructors valuable time both in course preparation and during class. Instructor materials can be accessed and downloaded from the Instructor Resources area of Mastering Biology.
www.pearson.com/mastering/biology

New! Identifying Major Themes end-of-chapter questions in the text and coaching activities in Mastering Biology give instructors resources to integrate Vision and Change biological themes into their course.

Revised Guided Reading Activities in the Mastering Biology Study Area and Instructor Resources offer a simple resource that encourages students to get the most out of each text chapter. These worksheets accompany each chapter of the text and are downloadable from Mastering Biology.

Identifying Major Themes--Chapter 18

Part A

Can you identify the major theme illustrated by each of the following examples? If necessary, you can review the themes in Chapter 1 of your book.

Match the themes on the left with the examples on the right. Not all themes will be used.

Reset | Help

Solar energy from sunlight, captured by chlorophyll during the process of photosynthesis, powers most ecosystems. Pathways that transform energy and matter

After a period of lower-than-average rainfall, drought-resistant individuals may be more prevalent in a plant population. Evolution

Information flow

Reptilian scales and the waxy coating on many leaves reduce water loss.
Relationship of structure to f

Other organisms may compete
its physical and chemical enviro

Submit | My Answers | Give Up

Correct

IDENTIFYING MAJOR THEMES

For each statement, identify which major theme is evident (the relationship of structure to function, information flow, pathways that transform energy and matter, interactions within biological systems, or evolution) and explain how the statement relates to the theme. If necessary, review the themes (see Chapter 1) and review the examples highlighted in blue in this chapter.

11. The highly folded membranes of the mitochondria make these organelles well suited to carry out the huge number of chemical reactions required for cellular respiration to proceed.

12. Cellular respiration and photosynthesis are linked, with each process using inputs created by the other.

13. Your body uses many different intersecting chemical pathways that, all together, constitute your metabolism.

For answers to Identifying Major Themes, see Appendix D.

Complete the following questions as you read the chapter content—Cellular Respiration: Aerobic Harvest of Food Energy:

1. The majority of a cell's ATP is produced within which of the following organelles?

   a. mitochondria

   b. nucleus

   c. ribosome

   d. Golgi apparatus

2. Students frequently have the misconception that plant cells don't perform cellular respiration. Briefly explain the basis of this misconception.

3. Briefly explain why the overall equation for cellular respiration has multiple arrows. Use the following figure, which illustrates the equation for cellular respiration, to help you answer.

$C_6H_{12}O_6$ + 6 O_2 → → → 6 CO_2 + 6 H_2O + approx. 32 ATP

The **Instructor Exchange** in the Instructor Resources area of Mastering Biology provides successful, class-tested active learning techniques and analogies from biology instructors around the world, offering a springboard for quick ideas to create more compelling lectures. Contributor Kelly Hogan moderates contributions to the exchange.

Engage with biology concepts anytime, anywhere with Pearson eText

New to *Campbell Essential Biology* 7th edition/*Campbell Essential Biology with Physiology* 6th edition, the Pearson eText includes videos, interactives, animations, and audio tutors that bring the text to life and help you understand key concepts. Get all the help you need in one integrated digital experience.

NEW! Over 100 rich media resources, many of them created by the author team, are included in the Pearson eText and accessible on smartphones, tablets, and computers. Examples of the rich media include: Figure Walkthrough videos, Topic Overview videos, MP3 Audio Tutors, Video Tutors, and BioFlix Tutorials.

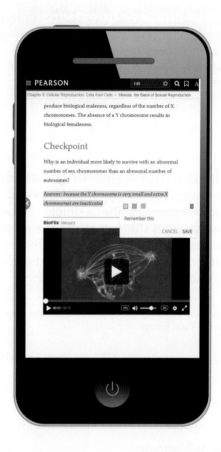

Pearson eText Mobile App offers offline access and can be downloaded for most iOS and Android phones/tablets from the Apple App Store or Google Play.

Acknowledgments

Throughout the process of planning and writing *Campbell Essential Biology*, the author team has had the great fortune of collaborating with an extremely talented group of publishing professionals and educators. We are all truly humbled to be part of one of the most experienced and successful publishing teams in biology education. Although the responsibility for any shortcomings lies solely with the authors, the merits of the book and its supplements reflect the contributions of a great many dedicated colleagues.

First and foremost, we must acknowledge our huge debt to Neil Campbell, the founding author of this book and a source of ongoing inspiration for each of us. Although this edition has been carefully and thoroughly revised—to update its science, its connections to students' lives, its pedagogy, and its currency—it remains infused with Neil's original vision and his commitment to share biology with introductory students.

This edition benefited significantly from the efforts of contributor Rebecca S. Burton from Alverno College. Using her years of teaching expertise, Becky made substantial improvements to two chapters, contributed to the development of new and revised Chapter Thread essays, and helped shape the emphasis on the unifying themes throughout the text and in Mastering Biology. We thank Becky for bringing her considerable talents to bear on this edition!

This book could not have been completed without the efforts of the *Campbell Essential Biology* team at Pearson Education. Leading the team is courseware portfolio management specialist Alison Rodal, who is tireless in her pursuit of educational excellence and who inspires all of us to constantly seek better ways to help teachers and students. Alison stands at the interface between the book development team and the educational community of professors and students. Her insights and contributions are invaluable. We also thank the Pearson Science executive team for their supportive leadership, in particular, senior vice president of portfolio management Adam Jaworski, director of portfolio management Beth Wilbur, and directors of courseware content development Barbara Yien and Ginnie Simione Jutson.

It is no exaggeration to say that the talents of the best editorial team in the industry are evident on every page of this book. The authors were continuously guided with great patience and skill by courseware senior analyst John Burner and senior developmental editor Susan Teahan. We owe this editorial team—which also includes the wonderfully capable and friendly editorial assistant Alison Candlin—a deep debt of gratitude for their talents and hard work.

Once we formulated our words and images, the production and manufacturing teams transformed them into the final book. Senior content producer Lori Newman oversaw the production process and kept everyone and everything on track. We also thank the managing content producer Mike Early for his careful oversight. Every edition of *Campbell Essential Biology* is distinguished by continuously updated and beautiful photography. For that we thank photo researcher Kristin Piljay, who constantly dazzles us with her keen ability to locate memorable images.

For the production and composition of the book, we thank senior project editor Margaret McConnell of Integra Software Services, whose professionalism and commitment to the quality of the finished product is visible throughout. The authors owe much to copyeditor Joanna Dinsmore and proofreader Pete Shanks for their keen eyes and attention to detail. We thank design manager Mark Ong and designer tani hasegawa of TT Eye for the beautiful interior and [U.S.-edition] cover designs; and we are grateful to Rebecca Marshall and Courtney Coffman and the artists at Lachina for rendering clear and compelling illustrations. We also thank rights and permissions project manager Matt Perry at Cenveo and the manager of rights and permissions Ben Ferrini. In the final stages of production, the talents of manufacturing buyer Stacy Weinberger shone.

Most instructors view the textbook as just one piece of the learning puzzle, with the book's supplements and media completing the picture. We are lucky to have a *Campbell Essential Biology* supplements team that is fully committed to the core goals of accuracy and readability. Content producer Lori Newman expertly coordinated the supplements, a difficult task given their number and variety. We also thank media project manager Ziki Dekel for his work on the excellent Instructor Resources and eText that accompanies the text. We owe particular gratitude to the supplements authors, especially the indefatigable and eagle-eyed Ed Zalisko of Blackburn College, who wrote the Instructor Guide and the PowerPoint© Lectures; the highly skilled and multitalented Doug Darnowski of Indiana University Southeast, who revised the Quiz Shows and Clicker Questions; and Jean DeSaix of the University of North Carolina at Chapel Hill, Justin Shaffer of the University of California, Irvine, Kristen Miller of the University of Georgia, and Suann Yang of SUNY Geneseo, our collaborative team of Test Bank authors for ensuring excellence in our assessment program. In addition, the authors thank Reading Quiz authors Amaya Garcia Costas of Montana State University and Cindy Klevickis of James Madison University; Reading Quiz accuracy reviewer Veronica Menendez; Practice Test author Chris Romero of Front Range Community College; and Practice Test accuracy reviewer Justin Walgaurnery of the University of Hawaii.

We wish to thank the talented group of publishing professionals who worked on the comprehensive media program that accompanies *Campbell Essential Biology*. The team members dedicated to Mastering Biology are true "game changers" in the field of biology education. We thank rich media content producers Ziki Dekel and Tod Regan for coordinating our multimedia plan. Vital contributions were also made by associate Mastering media producer Kaitlin Smith and web developer Barry Offringa. We also thank Sarah Jensen, senior content developer, for her efforts to make our media products the best in the industry.

As educators and writers, we are very lucky to have a crack marketing team. Product marketing manager Christa Pelaez and field marketing manager Kelli Galli seemed to be everywhere at once as they helped us achieve our authorial goals by keeping us constantly focused on the needs of students and instructors.

We also thank the Pearson Science sales representatives, district and regional managers, and learning technology specialists for representing *Campbell Essential Biology* on campuses. These representatives are our lifeline to the greater educational community, telling us what you like (and don't like) about this book and the accompanying supplements and media. Their enthusiasm for helping students makes them not only ideal ambassadors but also our partners in education. We urge all educators to take full advantage of the wonderful resource offered by the Pearson sales team.

Eric Simon would like to thank his colleagues at New England College for their support and for providing a model of excellence in education, in particular, Lori Koziol, Deb Dunlop, Mark Mitch, Bryan Partridge, and Wayne Lesperance. Eric would also like to acknowledge the contributions of Jim Newcomb of New England College for lending his keen eye for accuracy and for always being available to discuss teaching innovations; Jay Withgott for sharing his expertise; Elyse Carter Vosen for providing much-needed social context; Jamey Barone for her sage sensitivity; and Amanda Marsh for her expert eye, sharp attention to detail, tireless commitment, constant support, compassion, and seemingly endless wisdom.

At the end of these acknowledgments, you'll find a list of the many instructors who provided valuable information about their courses, reviewed chapters, and/or conducted class tests of *Campbell Essential Biology* with their students. All of our best ideas spring from the classroom, so we thank them for their efforts and support.

Most of all, we thank our families, friends, and colleagues, who continue to tolerate our obsession with doing our best for science education. And finally, we all wish to welcome budding superstar Leo to our *Campbell Essential Biology* family.

ERIC SIMON, JEAN DICKEY, JANE REECE

Reviewers of this Edition

Lois Bartsch
Metropolitan Community College

Allison Beck
Black Hawk College

Lisa Boggs
Southwestern Oklahoma State University

Steven Brumbaugh
Green River College

Ryan Caesar
Schriener University

Alexander Cheroske
Moorpark College

Gregory Dahlem
Northern Kentucky University

Richard Gardner
South Virginia University

Thomas Hinckley
Landmark College

Sue Hum-Musser
Western Illinois University

Brian Kram
Prince George's Community College

Tangela Marsh
Ivy Tech Community College East Central Region

Roy Mason
Mt. San Jacinto College

Mary Miller
Baton Rouge Community College

Michele Nash
Springfield Technical Community College

Mary Poffenroth
San Jose State University

Michelle Rogers
Austin Peay State University

Troy Rohn
Boise State University

Sanghamitra Saha
University of Houston Downtown

Mark Smith
Santiago Canyon College

Anna Sorin
University of Memphis

Jennifer Stueckle
West Virginia University

Alice Tarun
Alfred State SUNY College of Technology

Ron Tavernier
SUNY Canton

Anotia Wijte
Irvine Valey College

Edwin Wong
Western Connecticut State University

Calvin Young
Fullerton College

Reviewers of Previous Editions

Marilyn Abbott
Lindenwood College

Tammy Adair
Baylor University

Shazia Ahmed
Texas Woman's University

Felix O. Akojie
Paducah Community College

Shireen Alemadi
Minnesota State University, Moorhead

William Sylvester Allred, Jr.
Northern Arizona University

Megan E. Anduri
California State University, Fullerton

Estrella Z. Ang
University of Pittsburgh

David Arieti
Oakton Community College

C. Warren Arnold
Allan Hancock Community College

Mohammad Ashraf
Olive-Harvey College

Heather Ashworth
Utah Valley University

Tami Asplin
North Dakota State

Bert Atsma
Union County College

Yael Avissar
Rhode Island College

Barbara J. Backley
Elgin Community College

Gail F. Baker
LaGuardia Community College

Neil Baker
Ohio State University

Kristel K. Bakker
Dakota State University

Andrew Baldwin
Mesa Community College

Linda Barham
Meridian Community College

Charlotte Barker
Angelina College

Verona Barr
Heartland Community College

Lois Bartsch
Metropolitan Community College

S. Rose Bast
Mount Mary College

Erin Baumgartner
Western Oregon University

Sam Beattie
California State University, Chico

Allison Beck
Black Hawk College

Rudi Berkelhamer
University of California, Irvine

Penny Bernstein
Kent State University, Stark Campus

Suchi Bhardwaj
Winthrop University

Donna H. Bivans
East Carolina University

Andrea Bixler
Clarke College

Brian Black
Bay de Noc Community College

Allan Blake
Seton Hall University

Karyn Bledsoe
Western Oregon University

Judy Bluemer
Morton College

Sonal Blumenthal
University of Texas at Austin

Lisa Boggs
Southwestern Oklahoma State University

Dennis Bogyo
Valdosta State University

David Boose
Gonzaga University

Virginia M. Borden
University of Minnesota, Duluth

James Botsford
New Mexico State University

Cynthia Bottrell
Scott Community College

Richard Bounds
Mount Olive College

Cynthia Boyd
Hawkeye Community College

Robert Boyd
Auburn University

B. J. Boyer
Suffolk County Community College

TJ Boyle
Blinn College, Bryan Campus

Mimi Bres
Prince George's Community College

Patricia Brewer
University of Texas at San Antonio

Jerald S. Bricker
Cameron University

Carol A. Britson
University of Mississippi

George M. Brooks
Ohio University, Zanesville

Janie Sue Brooks
Brevard College

Steve Browder
Franklin College

Evert Brown
Casper College

Mary H. Brown
Lansing Community College

Richard D. Brown
Brunswick Community College

Steven Brumbaugh
Green River Community College

Joseph C. Bundy
University of North Carolina at Greensboro

Carol T. Burton
Bellevue Community College

Rebecca Burton
Alverno College

Warren R. Buss
University of Northern Colorado

Wilbert Butler
Tallahassee Community College

Ryan Caesar
Schreiner University

Miguel Cervantes-Cervantes
Lehman College, City University of New York

Maitreyee Chandra
Diablo Valley College

Miriam Chavez
University of New Mexico, Valencia

Bane Cheek
Polk Community College

Alexander Cheroske
Moorpark College

Thomas F. Chubb
Villanova University

Reggie Cobb
Nash Community College

Pamela Cole
Shelton State Community College

William H. Coleman
University of Hartford

Jay L. Comeaux
McNeese State University

James Conkey
Truckee Meadows Community College

Joe W. Conner
Pasadena City College

Karen A. Conzelman
Glendale Community College

Ann Coopersmith
Maui Community College

Erica Corbett
Southeastern Oklahoma State University

James T. Costa
Western Carolina University

Pat Cox
University of Tennessee, Knoxville

Laurie-Ann Crawford
Hawkeye Community College

Michael Cullen
University of Evansville

Gregory Dahlem
Northern Kentucky University

Pradeep M. Dass
Appalachian State University

Paul Decelles
Johnson County Community College

Galen DeHay
Tri County Technical College

Cynthia L. Delaney
University of South Alabama

Terry Derting
Murray State University

Jean DeSaix
University of North Carolina at Chapel Hill

Elizabeth Desy
Southwest State University

Edward Devine
Moraine Valley Community College

Dwight Dimaculangan
Winthrop University

Danielle Dodenhoff
California State University, Bakersfield

Deborah Dodson
Vincennes Community College

Diane Doidge
Grand View College

Don Dorfman
Monmouth University

Richard Driskill
Delaware State University

Lianne Drysdale
Ozarks Technical Community College

Terese Dudek
Kishwaukee College

Shannon Dullea
North Dakota State College of Science

David A. Eakin
Eastern Kentucky University

Brian Earle
Cedar Valley College

Ade Ejire
Johnston Community College

Dennis G. Emery
Iowa State University

Hilary Engebretson
Whatcom Community College

Renee L. Engle-Goodner
Merritt College

Virginia Erickson
Highline Community College

Carl Estrella
Merced College

Marirose T. Ethington
Genesee Community College

Paul R. Evans
Brigham Young University

Zenephia E. Evans
Purdue University

Jean Everett
College of Charleston

Holly Swain Ewald
University of Louisville

Dianne M. Fair
Florida Community College at Jacksonville

Joseph Faryniarz
Naugatuck Valley Community College

Phillip Fawley
Westminster College

Lynn Fireston
Ricks College

Jennifer Floyd
Leeward Community College

Dennis M. Forsythe
The Citadel

Angela M. Foster
Wake Technical Community College

Brandon Lee Foster
Wake Technical Community College

Carl F. Friese
University of Dayton

Suzanne S. Frucht
Northwest Missouri State University

Edward G. Gabriel
Lycoming College

Anne M. Galbraith
University of Wisconsin, La Crosse

Kathleen Gallucci
Elon University

J. Yvette Gardner
Clayton State University

Richard Gardner
South Virginia University

Gregory R. Garman
Centralia College

Wendy Jean Garrison
University of Mississippi

Gail Gasparich
Towson University

Kathy Gifford
Butler County Community College

Sharon L. Gilman
Coastal Carolina University

Mac Given
Neumann College

Patricia Glas
The Citadel

Ralph C. Goff
Mansfield University

Marian R. Goldsmith
University of Rhode Island

Andrew Goliszek
North Carolina Agricultural and Technical State University

Tamar Liberman Goulet
University of Mississippi

Curt Gravis
Western State College of Colorado

Larry Gray
Utah Valley State College

Tom Green
West Valley College

Robert S. Greene
Niagara University

Ken Griffin
Tarrant County Junior College

Denise Guerin
Santa Fe Community College

Paul Gurn
Naugatuck Valley Community College

Peggy J. Guthrie
University of Central Oklahoma

Henry H. Hagedorn
University of Arizona

Blanche C. Haning
Vance-Granville Community College

Laszlo Hanzely
Northern Illinois University

Sig Harden
Troy University

Sherry Harrel
Eastern Kentucky University

Reba Harrell
Hinds Community College

Frankie Harris
Independence Community College

Lysa Marie Hartley
Methodist College

Janet Haynes
Long Island University

Michael Held
St. Peter's College

Consetta Helmick
University of Idaho

J. L. Henriksen
Bellevue University

Michael Henry
Contra Costa College

Linda Hensel
Mercer University

Jana Henson
Georgetown College

James Hewlett
Finger Lakes Community College

Richard Hilton
Towson University

Thomas Hinckley
Landmark College

Juliana Hinton
McNeese State University

Phyllis C. Hirsch
East Los Angeles College

W. Wyatt Hoback
University of Nebraska at Kearney

Elizabeth Hodgson
York College of Pennsylvania

Jay Hodgson
Armstrong Atlantic State University

A. Scott Holaday
Texas Tech University

Robert A. Holmes
Hutchinson Community College

R. Dwain Horrocks
Brigham Young University

Howard L. Hosick
Washington State University

Carl Huether
University of Cincinnati

Sue Hum-Musser
Western Illinois University

Celene Jackson
Western Michigan University

John Jahoda
Bridgewater State College

Dianne Jennings
Virginia Commonwealth University

Richard J. Jensen
Saint Mary's College

Corey Johnson
University of North Carolina

Scott Johnson
Wake Technical Community College

Tari Johnson
Normandale Community College

Tia Johnson
Mitchell Community College

Gregory Jones
Santa Fe College, Gainesville, Florida

John Jorstad
Kirkwood Community College

Tracy L. Kahn
University of California, Riverside

Robert Kalbach
Finger Lakes Community College

Mary K. Kananen
Pennsylvania State University, Altoona

Thomas C. Kane
University of Cincinnati

Arnold J. Karpoff
University of Louisville

John M. Kasmer
Northeastern Illinois University

Valentine Kefeli
Slippery Rock University

Dawn Keller
Hawkeye College

John Kelly
Northeastern University

Tom Kennedy
Central New Mexico Community College

Cheryl Kerfeld
University of California, Los Angeles

Henrik Kibak
*California State University,
Monterey Bay*

Kerry Kilburn
Old Dominion University

Joyce Kille-Marino
College of Charleston

Peter King
Francis Marion University

Peter Kish
*Oklahoma School of Science and
Mathematics*

Robert Kitchin
University of Wyoming

Cindy Klevickis
James Madison University

Richard Koblin
Oakland Community College

H. Roberta Koepfer
Queens College

Michael E. Kovach
Baldwin-Wallace College

Brian Kram
Prince George's Community College

Jocelyn E. Krebs
University of Alaska, Anchorage

Ruhul H. Kuddus
Utah Valley State College

Nuran Kumbaraci
Stevens Institute of Technology

Holly Kupfer
Central Piedmont Community College

Gary Kwiecinski
The University of Scranton

Roya Lahijani
Palomar College

James V. Landrum
Washburn University

Erica Lannan
Prairie State College

Lynn Larsen
Portland Community College

Grace Lasker
*Lake Washington Institute
of Technology*

Brenda Leady
University of Toledo

Siu-Lam Lee
University of Massachusetts, Lowell

Thomas P. Lehman
Morgan Community College

William Leonard
Central Alabama Community College

Shawn Lester
Montgomery College

Leslie Lichtenstein
Massasoit Community College

Barbara Liedl
Central College

Harvey Liftin
Broward Community College

David Loring
Johnson County Community College

Eric Lovely
Arkansas Tech University

Lewis M. Lutton
Mercyhurst College

Bill Mackay
Edinboro University

Maria P. MacWilliams
Seton Hall University

Mark Manteuffel
St. Louis Community College

Lisa Maranto
Prince George's Community College

Michael Howard Marcovitz
Midland Lutheran College

Tangela Marsh
*Ivy Tech Community College East Central
Region*

Angela M. Mason
Beaufort County Community College

Roy B. Mason
Mt. San Jacinto College

John Mathwig
College of Lake County

Lance D. McBrayer
Georgia Southern University

Bonnie McCormick
University of the Incarnate Word

Katrina McCrae
Abraham Baldwin Agricultural College

Tonya McKinley
Concord College

Mary Anne McMurray
Henderson Community College

Diane Melroy
*University of North Carolina
Wilmington*

Maryanne Menvielle
California State University, Fullerton

Ed Mercurio
Hartnell College

Timothy D. Metz
Campbell University

Andrew Miller
Thomas University

Mary Miller
Baton Rouge Community College

David Mirman
Mt. San Antonio College

Kiran Misra
Edinboro University

Nancy Garnett Morris
Volunteer State Community College

Angela C. Morrow
University of Northern Colorado

Susan Mounce
Eastern Illinois University

Patricia S. Muir
Oregon State University

James Newcomb
New England College

Jon R. Nickles
University of Alaska, Anchorage

Zia Nisani
Antelope Valley College

Jane Noble-Harvey
University of Delaware

Michael Nosek
Fitchburg State College

Jeanette C. Oliver
Flathead Valley Community College

David O'Neill
*Community College of
Baltimore County*

Sandra M. Pace
Rappahannock Community College

Lois H. Peck
University of the Sciences, Philadelphia

Kathleen E. Pelkki
Saginaw Valley State University

Jennifer Penrod
Lincoln University

Rhoda E. Perozzi
Virginia Commonwealth University

John S. Peters
College of Charleston

Pamela Petrequin
Mount Mary College

Paula A. Piehl
Potomac State College of West Virginia University

Bill Pietraface
State University of New York Oneonta

Gregory Podgorski
Utah State University

Mary Poffenroth
San Jose State University

Rosamond V. Potter
University of Chicago

Karen Powell
Western Kentucky University

Martha Powell
University of Alabama

Elena Pravosudova
Sierra College

Hallie Ray
Rappahannock Community College

Jill Raymond
Rock Valley College

Dorothy Read
University of Massachusetts, Dartmouth

Nathan S. Reyna
Howard Payne University

Philip Ricker
South Plains College

Todd Rimkus
Marymount University

Lynn Rivers
Henry Ford Community College

Jennifer Roberts
Lewis University

Laurel Roberts
University of Pittsburgh

Michelle Rogers
Austin Peay State University

Troy Rohn
Boise State University

April Rottman
Rock Valley College

Maxine Losoff Rusche
Northern Arizona University

Michael L. Rutledge
Middle Tennessee State University

Mike Runyan
Lander University

Travis Ryan
Furman University

Tyson Sacco
Cornell University

Sanghamitra Saha
University of Houston Downtown

Bassam M. Salameh
Antelope Valley College

Sarmad Saman
Quinsigamond Community College

Carsten Sanders
Kutztown University

Pamela Sandstrom
University of Nevada, Reno

Leba Sarkis
Aims Community College

Walter Saviuk
Daytona Beach Community College

Neil Schanker
College of the Siskiyous

Robert Schoch
Boston University

John Richard Schrock
Emporia State University

Julie Schroer
Bismarck State College

Karen Schuster
Florida Community College at Jacksonville

Brian W. Schwartz
Columbus State University

Michael Scott
Lincoln University

Eric Scully
Towson State University

Lois Sealy
Valencia Community College

Sandra S. Seidel
Elon University

Wayne Seifert
Brookhaven College

Susmita Sengupta
City College of San Francisco

Justin Shaffer
University of California, Irvine

Patty Shields
George Mason University

Cara Shillington
Eastern Michigan University

Brian Shmaefsky
Kingwood College

Rainy Inman Shorey
Ferris State University

Cahleen Shrier
Azusa Pacific University

Jed Shumsky
Drexel University

Greg Sievert
Emporia State University

Jeffrey Simmons
West Virginia Wesleyan College

Frederick D. Singer
Radford University

Anu Singh-Cundy
Western Washington University

Kerri Skinner
University of Nebraska at Kearney

Sandra Slivka
Miramar College

Jennifer Smith
Triton College

Margaret W. Smith
Butler University

Mark Smith
Santiago Canyon College

Thomas Smith
Armstrong Atlantic State University

Anna Sorin
University of Memphis

Deena K. Spielman
Rock Valley College

Minou D. Spradley
San Diego City College

Ashley Spring
Eastern Florida State College

Robert Stamatis
Daytona Beach Community College

Joyce Stamm
University of Evansville

Eric Stavney
Highline Community College

Michael Stevens
Utah Valley University

Bethany Stone
University of Missouri, Columbia

Jennifer Stueckle
West Virginia University

Mark T. Sugalski
New England College

Marshall D. Sundberg
Emporia State University

Adelaide Svoboda
Nazareth College

Alice Tarun
Alfred State SUNY College of Technology

Ron Tavernier
SUNY Canton

Sharon Thoma
Edgewood College

Kenneth Thomas
Hillsborough Community College

Sumesh Thomas
Baltimore City Community College

Betty Thompson
Baptist University

Chad Thompson
Westchester Community College

Paula Thompson
Florida Community College

Michael Anthony Thornton
Florida Agriculture and Mechanical University

Linda Tichenor
University of Arkansas, Fort Smith

John Tjepkema
University of Maine, Orono

Bruce L. Tomlinson
State University of New York, Fredonia

Leslie R. Towill
Arizona State University

Bert Tribbey
California State University, Fresno

Nathan Trueblood
California State University, Sacramento

Robert Turner
Western Oregon University

Michael Twaddle
University of Toledo

Virginia Vandergon
California State University, Northridge

William A. Velhagen, Jr.
Longwood College

Melinda Verdone
Rock Valley College

Leonard Vincent
Fullerton College

Jonathan Visick
North Central College

Michael Vitale
Daytona Beach Community College

Lisa Volk
*Fayetteville Technical
Community College*

Daryle Waechter-Brulla
University of Wisconsin, Whitewater

Stephen M. Wagener
Western Connecticut State University

Sean E. Walker
California State University, Fullerton

James A. Wallis
St. Petersburg Community College

Eileen Walsh
Westchester Community College

Helen Walter
Diablo Valley College

Kristen Walton
Missouri Western State University

Jennifer Warner
University of North Carolina at Charlotte

Arthur C. Washington
*Florida Agriculture and Mechanical
University*

Kathy Watkins
Central Piedmont Community College

Dave Webb
St. Clair County Community College

Harold Webster
Pennsylvania State University, DuBois

Ted Weinheimer
California State University, Bakersfield

Lisa A. Werner
Pima Community College

Joanne Westin
Case Western Reserve University

Wayne Whaley
Utah Valley State College

Joseph D. White
Baylor University

Quinton White
Jacksonville University

Leslie Y. Whiteman
Virginia Union University

Rick Wiedenmann
New Mexico State University at Carlsbad

Anotia Wijte
Irvine Valley College

Peter J. Wilkin
Purdue University North Central

Bethany Williams
California State University, Fullerton

Daniel Williams
Winston-Salem University

Judy A. Williams
*Southeastern Oklahoma
State University*

Dwina Willis
Freed Hardeman University

David Wilson
University of Miami

Mala S. Wingerd
San Diego State University

E. William Wischusen
Louisiana State University

Darla J. Wise
Concord College

Michael Womack
Macon State College

Edwin Wong
Western Connecticut State University

Bonnie Wood
University of Maine at Presque Isle

Holly Woodruff (Kupfer)
Central Piedmont Community College

Jo Wen Wu
Fullerton College

Mark L. Wygoda
McNeese State University

Calvin Young
Fullerton College

Shirley Zajdel
Housatonic Community College

Samuel J. Zeakes
Radford University

Uko Zylstra
Calvin College

Acknowledgments for the Global Edition

Pearson would like to thank the following for contributing to the Global Edition:

Alan Feest
University of Bristol

Clemens Kiecker
King's College London

Sarah Taylor
Keele University

Pearson would like to thank the following for reviewing the Global Edition:

Mohamad Faiz Foong Abdullah
Universiti Teknologi MARA System

Adriaan Engelbrecht
University of the Western Cape

Paul Broady
University of Canterbury

Alan Feest
University of Bristol

Detailed Contents

4 A Tour of the Cell 88

5 The Working Cell 108

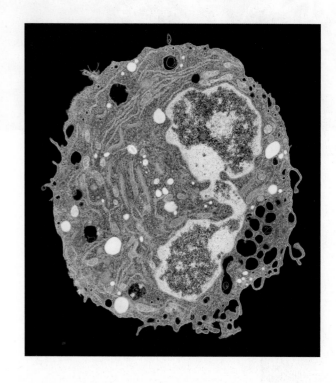

6 Cellular Respiration: Obtaining Energy from Food 124

7 Photosynthesis: Using Light to Make Food 140

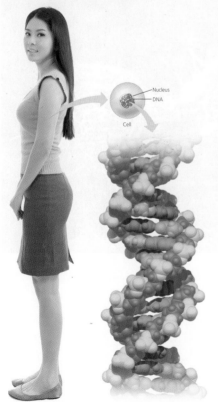

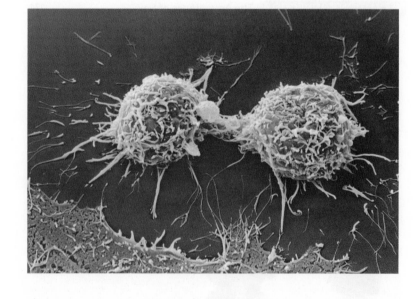

12 DNA Technology 250

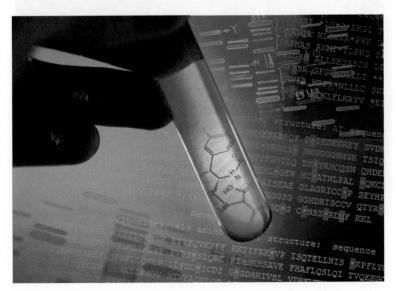

Unit 3 Evolution and Diversity 275

13 How Populations Evolve 276

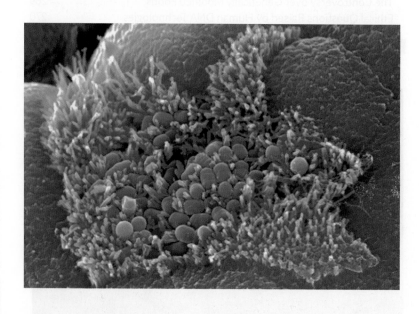

20 Communities and Ecosystems 458

Appendices

CAMPBELL

essential
biology

1 Learning About Life

Why Biology Matters

Nearly everyone has an inborn curiosity about the natural world. Whatever your connection to nature—perhaps you have pets; enjoy visiting parks, zoos, or aquariums; or watch TV shows about interesting creatures—this book will help demonstrate how the study of biology connects to your life.

YOU'RE A SCIENTIST! ALTHOUGH YOU MAY NOT REALIZE IT, YOU USE THE PROCESS OF SCIENCE EVERY DAY.

WHAT THE HECK IS THAT? IF YOU'VE WONDERED WHAT AN UNUSUAL OR ESPECIALLY BEAUTIFUL ANIMAL IS CALLED, YOU'RE CURIOUS ABOUT BIOLOGY.

IS THERE LIFE ON MARS? ONE OF THE MISSIONS OF THE MARS ROVER IS TO SEARCH FOR SIGNS OF LIFE.

BIOLOGY AND SOCIETY   Swimming with the Turtles

A Passion for Life

Imagine yourself floating gently in a warm, calm ocean. Through the blue expanse, you spy a green sea turtle gliding toward you. You watch intently as it grazes on seagrass. It's easy to be captivated by this serene sea creature, with its paddle-shaped flippers and large eyes. As you follow it, you can't help but wonder about its life—how old it is, where it is traveling, whether it has a mate.

It's very human to be curious about the world around us. Nearly all of us have an inherent interest in life, an inborn fascination with the natural world. Do you have a pet? Are you concerned with fitness or healthy eating? Have you ever visited a zoo or an aquarium for fun, taken a nature hike through a forest, grown some plants, or gathered shells on the beach? Would you like to swim with a turtle? If you answered yes to any of these questions, then you share an interest in biology.

We wrote *Essential Biology* to help you harness your innate enthusiasm for life, no matter how much experience you've had with college-level science (even if it's none!). We'll use this passion to help you develop an understanding of the subject of biology, an understanding that you can apply to your own life and to the society in which you live. Whatever your reasons for taking this course—even if only to fulfill your school's science requirement—you'll soon discover that exploring life is relevant and important to you.

To reinforce the fact that biology does indeed affect you personally, every chapter of *Essential Biology* opens with an essay—called Biology and Society—where you will see the relevance of that chapter's material. Topics as varied as green energy (Chapter 7), pet genetics (Chapter 9), and the importance of biodiversity (Chapter 20) help to illustrate biology's scope and show how the subject of biology is woven into the fabric of society. Throughout *Essential Biology*, we'll continuously emphasize these connections, pointing out many examples of how each topic can be applied to your life and the lives of those you care about.

An inborn urge to learn about life. A college student swims with a green sea turtle off the coast of Belize, Central America.

The Scientific Study of Life

Now that we've established our goal—to examine how biology affects your life—a good place to start is with a basic definition: **Biology** is the scientific study of life. But have you ever looked up a word in the dictionary, only to find that you need to look up some of the words within that definition to make sense of the original word? The definition of *biology*, although seemingly simple, raises questions such as "What is a scientific study?" and "What does it mean to be alive?"

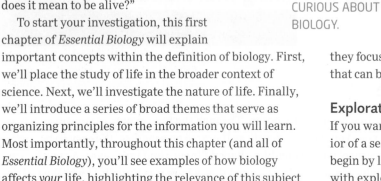

IF YOU'VE WONDERED WHAT AN UNUSUAL OR ESPECIALLY BEAUTIFUL ANIMAL IS CALLED, YOU'RE CURIOUS ABOUT BIOLOGY.

To start your investigation, this first chapter of *Essential Biology* will explain important concepts within the definition of biology. First, we'll place the study of life in the broader context of science. Next, we'll investigate the nature of life. Finally, we'll introduce a series of broad themes that serve as organizing principles for the information you will learn. Most importantly, throughout this chapter (and all of *Essential Biology*), you'll see examples of how biology affects *your* life, highlighting the relevance of this subject to society and everyone in it. ☑

✅ **CHECKPOINT**

Define biology.

■ *Answer: Biology is the scientific study of life.*

▼ **Figure 1.1 Scientific exploration.** Dr. Jane Goodall spent decades recording her observations of chimpanzee behavior during field research in the jungles of Tanzania.

An Overview of the Process of Science

The definition of *biology* as the scientific study of life leads to an obvious first question: What does it mean to study something scientifically? Notice that biology is not defined simply as "the study of life" because there are many nonscientific ways that life can be studied. For example, meditation is a valid way of contemplating life, but it is not a *scientific* means of studying life, and therefore it does not fall within the scope of biology. How, then, do we tell the difference between science and other ways of trying to make sense of the world?

Science is an approach to understanding the natural world that is based on inquiry—a search for information, evidence, explanations, and answers to specific questions. Scientists seek natural causes for natural phenomena. Therefore, they focus solely on the study of structures and processes that can be verifiably observed and measured.

Exploration

If you wanted to understand something—say, the behavior of a sea turtle—how would you start? You'd probably begin by looking at it. Biology, like other sciences, begins with exploration **(Figure 1.1)**. During this initial phase of inquiry, you may simply watch the subject and record your observations. A more intense exploration may involve extending your senses using tools such as microscopes or precision instruments to allow for careful measurement. Whatever the source, recorded observations are called **data**—the evidence on which scientific inquiry is based. In addition to gathering your own data, you may read books or articles on the subject to learn about previously collected data.

As you proceed with your exploration, your curiosity will lead to questions, such as "Why is it this way?" "How does it work?" "Can I change it?" Such questions are the launching point for the next step in the process of science: testing.

EXPLORATION
• Making observations
• Asking questions
• Seeking information

Testing

After making observations and asking questions, you may wish to conduct tests. But where do you start? You could probably think of many possible ways to investigate your subject. But you can't possibly test them all at once. To organize your thinking, you will likely begin by selecting one possible explanation and testing it. In other words, you would make a hypothesis. A **hypothesis** is a proposed explanation for a set of observations. A valid hypothesis must be testable and falsifiable—that is, it must be capable of being demonstrated to be false. A good hypothesis thus immediately leads to predictions that can be tested. Some hypotheses (such as ones involving conditions that can be easily controlled) lend themselves to **experiments**, or scientific tests. Other hypotheses (such as ones involving aspects of the world that cannot be controlled, such as ecological issues) can be tested by making further observations. The results of an experiment will either support or not support the hypothesis.

We all use hypotheses in solving everyday problems, although we don't think of it in those terms. Imagine that you press the power button on your TV remote, but the TV fails to turn on. That the TV does not turn on is an observation. The question that arises is obvious: Why didn't the remote turn on the TV? You probably would not just throw your hands up in the air and say "There's just no way to figure this out!" Instead, you might imagine several possible explanations, but you couldn't investigate them all simultaneously. Instead, you would focus on just one explanation, perhaps the most likely one based on past experience, and test it. That initial explanation is your hypothesis. For example, in this case, a reasonable hypothesis is that the batteries in the remote are dead.

After you've formed a hypothesis, you would make further observations or conduct experiments to investigate this initial idea. In this case, you can predict that if you replace the batteries, the TV will work. Let's say that you conduct this experiment, and the remote still doesn't work. You conclude that this observation does not support your hypothesis. You would then formulate a second hypothesis and test it. Perhaps you hypothesize that the TV is unplugged. You could continue to conduct additional experiments and formulate additional hypotheses until you reach a satisfactory answer to your initial question. As you do this, you are following a series of steps that provide a loose guideline for scientific investigations. These steps are shown in **Figure 1.2** and are sometimes called "the scientific method." They are a rough "recipe" for discovering new explanations, a set of procedures that, if followed, may provide insight into the subject at hand.

The steps are simply a way of formalizing how we usually try to solve problems. If you pay attention, you'll find that you often formulate hypotheses, test them, and draw conclusions. In other words, the process of science is probably your "go-to" method for solving problems. Although the process of science is often presented as a series of linear steps (such as those in Figure 1.2), in reality investigations are almost never this rigid. Different questions will require different paths through the steps. There is no single formula for successfully discovering something new; instead, the process of science suggests a broad outline for how an investigation might proceed. ☑

ALTHOUGH YOU MAY NOT REALIZE IT, YOU USE THE PROCESS OF SCIENCE EVERY DAY.

Communication and Outcomes

The process of science is typically repetitive and nonlinear. For example, scientists often work through several rounds of making observations and asking questions, with each round informing the next, before settling on hypotheses that they wish to test. In fine-tuning their questions, they rely heavily on scientific literature, the published contributions of fellow scientists. By reading about and understanding past studies, they can build on the foundation of existing knowledge.

☑ **CHECKPOINT**

Do all scientific investigations follow the steps in Figure 1.2 in that precise order? Explain.

■ *Answer: No. Different scientific investigations may proceed through the process of science in different ways.*

▶ **Figure 1.2** **Testing a common problem using the process of science.**

TESTING
- Forming hypotheses
- Making predictions
- Running experiments
- Gathering data
- Interpreting data
- Drawing conclusions

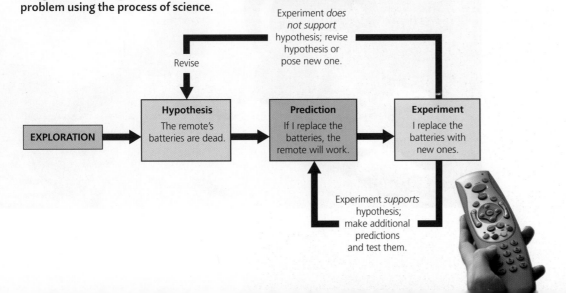

Experiment *does not support* hypothesis; revise hypothesis or pose new one.

Revise

EXPLORATION

Hypothesis
The remote's batteries are dead.

Prediction
If I replace the batteries, the remote will work.

Experiment
I replace the batteries with new ones.

Experiment *supports* hypothesis; make additional predictions and test them.

COMMUNICATION
- Sharing data
- Obtaining feedback
- Publishing papers
- Replicating findings
- Building consensus

▶ **Figure 1.3 Scientific communication.** Like these college students, scientists often communicate results to colleagues at meetings.

Additionally, scientists communicate with each other through seminars, meetings, personal communication, and scientific publications **(Figure 1.3)**. Before experimental results are published in a scientific journal, the research is evaluated by qualified, impartial, often anonymous experts who were not involved in the study. This process, intended to provide quality control, is called **peer review**. Reviewers often require authors to revise their paper or perform additional experiments in order to provide more lines of evidence. It is not uncommon for a scientific journal to reject a paper entirely if it doesn't meet the rigorous standards set by fellow scientists. After a study is published, scientists often check each other's claims by attempting to confirm observations or repeat experiments.

Science does not exist just for its own sake. In fact, it is interwoven with the fabric of society **(Figure 1.4)**. Much of scientific research is focused on solving problems that influence our quality of life, such as the push to cure cancer or to understand and slow the process of climate change. Societal needs often determine which research projects are funded. Scientific studies may involve basic research (largely concerned with building knowledge) or they may be more applied (largely concerned with developing new technologies). The ultimate aim of most scientific investigations is to benefit society. This focus on outcomes highlights the connections between biology, your own life, and our larger society.

OUTCOMES
- Building knowledge
- Solving problems
- Developing new technologies
- Benefiting society

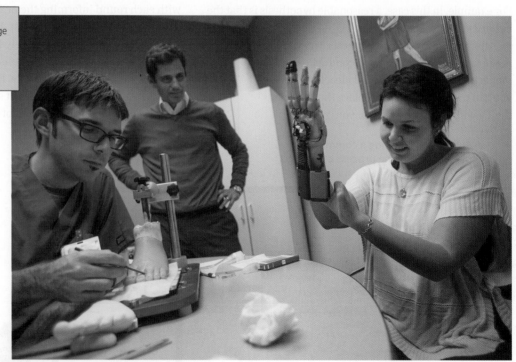

▶ **Figure 1.4 Scientific outcomes.** A 22-year-old woman tries on her new prosthetic hand with individually movable bionic fingers.

Putting all these steps together, **Figure 1.5** presents a more comprehensive model of the process of science. You can see that forming and testing hypotheses (represented in blue) are at the center of science. This core set of activities is the reason that science explains natural phenomena so well. These activities, however, are shaped by exploration (orange) and influenced by communication with other scientists (yellow) and by outcomes (green). Notice that many of these activities connect to others, illustrating that the components of the process of science interact. As in all quests, science includes elements of challenge, adventure, and luck, along with careful planning, reasoning, creativity, patience, and persistence in overcoming setbacks. Such diverse elements of inquiry allow the process of science to be flexible, molded by the needs of each particular challenge.

In every chapter of *Essential Biology*, we include examples of how the process of science was used to study the content presented in that chapter. Some of the questions that will be addressed are Do baby turtles swim (this chapter)? Can avatars improve cancer treatment (Chapter 11)? What can lice teach us about ancient humans (Chapter 17)?

As you become increasingly scientifically literate, you will arm yourself with the tools you need to evaluate claims that you hear. We are all bombarded by information every day—through commercials, social media, websites, magazine articles, and so on—and it can be hard to filter out the bogus from the truly worthwhile. Having a firm grasp of science as a process of inquiry can therefore help you in many ways outside the classroom. ✓

Hypotheses, Theories, and Facts

Since scientists focus on natural phenomena that can be reliably observed and measured, let's explore how the terms *hypothesis, theory,* and *fact* are related.

As previously noted, a hypothesis is a proposed explanation for an observation. In contrast, a scientific **theory** is much broader in scope than a hypothesis. A theory is a comprehensive and well-substantiated explanation. Theories only become widely accepted by scientists if they are supported by a large, varied, and growing body of evidence. A theory can be used to explain many observations. Indeed, theories can be used to devise many new and testable hypotheses. It is important to note that scientists use the word *theory* differently than many people tend to use it in everyday speech, which implies untested speculation ("It's just a theory!"). In fact, the word *theory* is commonly used in everyday speech in the way a scientist uses the word *hypothesis*. It is therefore improper to say that a scientific theory, such as the theory of

▼ **Figure 1.5** **An overview of the process of science.** Notice that performing scientific tests lies at the heart of the entire process.

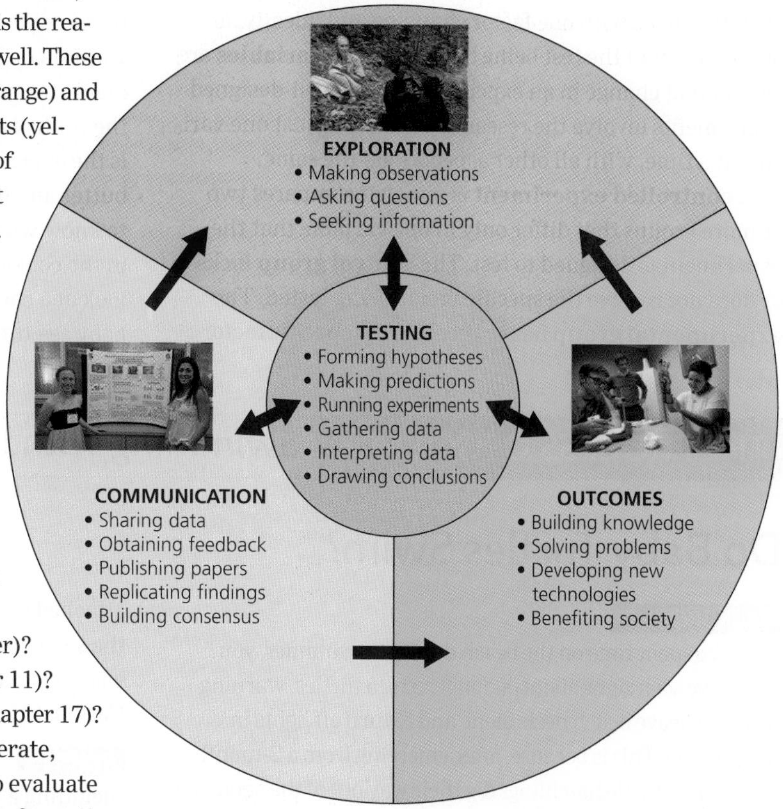

Figure Walkthrough

Mastering **Biology**
goo.gl/6bRdg9

EXPLORATION
- Making observations
- Asking questions
- Seeking information

TESTING
- Forming hypotheses
- Making predictions
- Running experiments
- Gathering data
- Interpreting data
- Drawing conclusions

COMMUNICATION
- Sharing data
- Obtaining feedback
- Publishing papers
- Replicating findings
- Building consensus

OUTCOMES
- Building knowledge
- Solving problems
- Developing new technologies
- Benefiting society

evolution, is "just" a theory to imply that it is untested or lacking in evidence. In reality, every scientific theory is backed up by a wealth of supporting evidence, or else it wouldn't be referred to as a theory. However, a theory, like any scientific idea, must be refined or even abandoned if new, contradictory evidence is discovered.

A **fact** is a piece of information considered to be objectively true based on all current evidence. A fact can be verified and is therefore distinct from opinions (beliefs that can vary from person to person), matters of taste, speculation, or inference. However, science is self-correcting: New evidence may lead to reconsideration of information previously regarded as a fact.

Many people associate facts with science, but accumulating facts is not the primary goal of science. A dictionary is an impressive catalog of facts, but it has little to do with science. It is true that facts, in the form of verifiable observations and repeatable experimental results, are the prerequisites of science. What advances science, however, are new theories that tie together a number of observations that previously seemed unrelated. The cornerstones of science are the explanations that apply to the greatest variety of phenomena. People like Isaac Newton, Charles Darwin, and Albert Einstein stand out in the history of science not because they discovered a great many facts but because their theories had such broad explanatory power. ✓

✓ **CHECKPOINT**

Why does peer review improve the reliability of a scientific paper?

■ *Answer: A peer-reviewed paper carries a "seal of approval" from impartial experts on the subject.*

✓ **CHECKPOINT**

You arrange to meet a friend for dinner at 6 P.M., but when the appointed hour comes, she is not there. You wonder why. Another friend says, "My theory is that she forgot." If your friend were speaking like a scientist, what would she have said?

■ *Answer: "My hypothesis is that she forgot."*

Controlled Experiments

To investigate a hypothesis, a researcher often runs a test multiple times with one factor changing and, ideally, all other factors of the test being held constant. **Variables** are factors that change in an experiment. Most well-designed experiments involve the researcher changing just one variable at a time, with all other aspects held the same.

A **controlled experiment** is one that compares two or more groups that differ only in one variable that the experiment is designed to test. The **control group** lacks or does not receive the specific factor being tested. The **experimental group** has or receives the specific factor

being tested. The use of a controlled experiment allows a scientist to draw conclusions about the effect of the one variable that did change. For example, you might compare cookie recipes by altering the amount of butter (the variable in this experiment) while keeping all other ingredients the same. In this case, the original cookie recipe is the control group, while the new recipe with more butter is the experimental group. If you were to vary both the butter and the flour at the same time, it would be difficult to know which variable was responsible for any changes in the cookies. To further illustrate this principle, let's look at a controlled experiment that investigated whether baby sea turtles swim or just drift in the water.

THE PROCESS OF SCIENCE | Swimming with the Turtles

Do Baby Turtles Swim?

BACKGROUND

If you've spent time on the beach during the summer, you may have seen signs about endangered sea turtles, warning people to leave beach nests alone and to turn off lights in the evening. This is because, after emerging from a 2-month incubation, turtle hatchlings dig their way out of the sand and then use moonlight to navigate to the sea **(Figure 1.6a)**. What happens next has long been a mystery to marine biologists. Can the juvenile turtles swim in ocean currents? Or do they just passively drift? No one knows how baby sea turtles travel during their first several years. In fact, some marine biologists refer to this time as "lost years" in the turtle life cycle.

METHOD

In 2015, researchers from the University of Central Florida investigated the question of whether baby green sea turtles swim or drift. They attached tiny satellite trackers to 24 green sea turtles, each between 1 and 2 years old, in the Gulf of Mexico **(Figure 1.6b)**. The researchers also

attached trackers to floating buckets and released them at the same time and in the same locations. The experiment was conducted under a scientific research permit from the National Marine Fisheries Service (NMFS).

RESULTS

Including a control group (the floating buckets) allowed the researchers to draw conclusions about the experimental group (the baby turtles). Comparing data on the paths taken by each group revealed that the turtles moved slowly (averaging only 0.4 miles per hour). However, the turtles moved faster and along different tracks than the floating buckets **(Figure 1.6c)**. These data suggest that, despite longstanding assumptions by marine biologists, very young sea turtles travel by swimming, and not just drifting. Such information may help efforts to protect endangered species of sea turtles.

Thinking Like a Scientist

What was the purpose of attaching transmitters to floating buckets?

For the answer, see Appendix D.

▼ **Figure 1.6 Tracking baby sea turtles.**

(a) Green sea turtle hatchlings scrambling to the sea

NMFS Research Permit 16733

(b) Satellite tracker on the back of a baby turtle

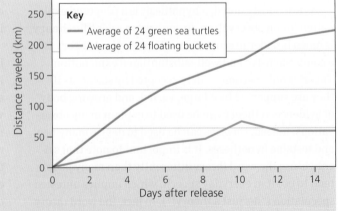

(c) Graph showing the distance traveled by the average turtle (red line) versus the average floating bucket (blue line)

NMFS Research Permit 16733

In this experiment, the independent variable (what is being changed) was the type of object being tracked: the sea turtles or the floating buckets.

The independent variable was tested for the effect upon the dependent variable.

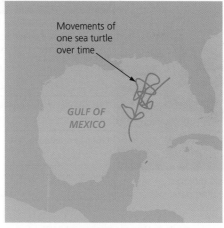

Movements of one sea turtle over time

GULF OF MEXICO

In this experiment, the dependent variable (the effect under investigation) was the speed of travel.

◀ **Figure 1.7 Independent versus dependent variables.** These hypothetical data on green sea turtle migration show the relationship between these two types of variables.

The study on whether baby sea turtles swim is a good example of a controlled experiment. The variable was the identity of the object followed: turtles versus buckets. Other factors in the experiment—such as the type of satellite tracker used, when and where they were released, how often data were collected, and how speed was calculated—were purposefully kept the same. The control group was the floating buckets, and the experimental group was the baby sea turtles. The buckets were the control group because they lacked the factor being tested: the ability to move on their own. By comparing the movements of the floating buckets with the movements of the baby sea turtles, the experimenters could be confident that any observed differences were due to the turtles being able to swim.

In a controlled experiment, like the one just described, the **independent variable** is what is being manipulated by the researchers as a potential cause—in this case, the object under investigation (either turtles or buckets). The **dependent variable** is the response, output, or effect under investigation that is used to judge the outcome of the experiment—in this case, the speed of movement. The dependent (measured) variable is affected by the independent (manipulated) variable **(Figure 1.7)**. Well-designed experiments often test just one independent variable at a time.

A controlled experiment can sometimes be a blind experiment, in which some information about the experiment is withheld from participants **(Figure 1.8)**. For example, the turtle researchers may have analyzed the trajectory data without knowing whether each track was a turtle or a bucket. The identities of the blinded components are revealed only after the experiment is complete. Performing the study blind removes bias on the part of the investigators. This type of study is called a **single-blind experiment**.

Many medical drug trials include a **placebo**, a medically ineffective treatment that allows the placebo group to serve as a control group. Typically, the placebo group does not know that they are receiving an ineffective substitute. An experiment in which neither the participant nor the experimenter knows which group is the control group is called a **double-blind experiment**. The "gold standard" for a medical trial is a "double-blind placebo-controlled study," meaning that neither the patients nor the doctors know which patients received the real treatment and which received a placebo. Such a design prevents bias on the part of the researchers and also takes into account the placebo effect, a well-documented phenomenon whereby giving patients a fake treatment nonetheless causes them to improve due to their belief that they are receiving an effective treatment. ☑

Evaluating Scientific Claims

The process of science involves evaluating scientific claims. Sometimes claims are made using scientific jargon with the intention of appearing to conform to scientific standards without actually doing so. **Pseudoscience** is any field of study that is falsely presented as having a scientific basis. Given our access to huge amounts of information, much of it unreliable, the ability to recognize pseudoscience is a very important thinking skill. Although the difference between valid science and pseudoscience can at times

☑ **CHECKPOINT**

You bake two recipes of cookies and label them "A" and "B." You ask a group of friends to rate the recipes. Design a double-blind experiment to determine which recipe is superior.

Answer: A third party should label the cookies and collect the data so that neither the investigator (you) nor the subjects (your friends) know which recipe is which.

▼ **Figure 1.8 How to recognize blind studies.**

TYPE OF STUDY	TEST SUBJECTS KNOW WHICH GROUP IS WHICH?	RESEARCHERS KNOW WHICH GROUP IS WHICH?
Not blind	Yes	Yes
Single blind	No	Yes
Double blind	No	No

FEATURES OF SCIENCE	FEATURES OF PSEUDOSCIENCE
Adheres to an established and well-recognized scientific method	Does not adhere to generally accepted processes of science
Repeatable results	Results that cannot be duplicated by others; results that rely on a single person or are solely opinion
Testable claims that can be disproven	Unprovable or untestable claims; reliance on assumptions or beliefs that are not testable
Open to outside review	Rejection of external review or refusal to accept contradictory evidence
Multiple lines of evidence	Overreliance on a small amount of data; underlying causes are not investigated

A field biologist collecting data

A pyramid that is claimed to channel energy

▶ **Figure 1.9 Features of science versus pseudoscience.**

be confusing, there are several indicators that you can use to recognize pseudoscience (**Figure 1.9**). For example, a pseudoscientific study may be based soley or largely on **anecdotal evidence**, an assertion based on a single or a few examples that do not support a generalized conclusion—for example, "Today was unusually cold, so global warming must be a hoax!" A proper scientific investigation is open to outside review (the communication step in Figure 1.5), while pseudoscientific claims often reject external review or refuse to accept contradictory evidence. Often, pseudoscientific claims are based on results that cannot be duplicated by others because they rely on a single person or are solely opinion. A proper scientific study, on the other hand, has repeatable results that stand up to external scrutiny.

One of the best ways to evaluate scientific claims is to consider the source of the information (**Figure 1.10**). Science depends upon peer review, the evaluation of work by impartial, qualified, often anonymous experts who are not involved in that work. Publishing a study in a peer-reviewed journal is often the best way to ensure that

✓ **CHECKPOINT**

If someone says, "It rained yesterday, so I don't believe that there is a drought," this is an example of what kind of improper thinking?

■ *Answer: a conclusion based on anecdotal evidence*

▼ **Figure 1.10 Recognizing a reliable source.** The more criteria a given source meets, the more reliable it is.

Souce reliability checklist

☐ Is the information current?
☐ Is the source primary (and not secondary)?
☐ Is/are the author(s) indentifiable and well qualified?
☐ Does the author lack potential conflicts of interest?
☐ Are references cited?
☐ Are any experiments described in enough detail that they could be reproduced?
☐ Was the information peer reviewed?
☐ Is the information unbiased?
☐ Is the intent of the source known and valid?

it will be considered scientifically valid. No matter the source, reliable scientific information can be recognized by being up to date, drawing from known sources of information, having been authored by a reputable expert, and being free of bias.

Now that we have explored the process of science, keep in mind that it has proven to be the most effective method for investigating the natural world. In fact, nearly everything we know about nature was learned through the process of science. ✓

The Properties of Life

Recall once again the definition at the heart of this chapter: Biology is the scientific study of life. Now that we understand what constitutes a scientific study, we can turn to the next question raised by this definition: What is life? Or, to put it another way, what distinguishes living things from nonliving things? The phenomenon of life seems to defy a simple, one-sentence definition. Yet even a small child instinctively knows that a bug or a plant is alive but a rock is not.

If someone placed an object in front of you and asked whether it was alive, what would you do? Would you poke it to see if it reacts? Would you watch it closely to see if it moves or breathes? Would you dissect it to look at its parts? Each of these ideas is closely related to how biologists actually define

life: We recognize life mainly by what living things do. Using a green sea turtle as an example, **Figure 1.11** highlights the major properties we associate with life: order, cells, growth and development, energy processing, regulation, response to the environment, reproduction, and evolution.

An object is generally considered to be alive if it displays all of these characteristics simultaneously. On the other hand, a nonliving object may display some of these properties, but not all of them. For example, a virus has an ordered structure, but it cannot process energy, nor is it composed of cells. Viruses, therefore, are generally not considered to be living organisms (see Chapter 10 for more information on viruses).

▼ **Figure 1.11 A green sea turtle displays the properties of life.**
An object is considered alive only if it displays all of these properties simultaneously.

Order is apparent in many of the turtle's structures, such as the regular arrangement of plates in the turtle shell.

Like any large organism, the body of a sea turtle is made of trillions of **cells**.

Growth and development into a mature adult sea turtle takes decades.

Scientists who study turtle **evolution** believe that they first appeared nearly 250 million years ago, making their lineage older than the dinosaurs.

As part of the turtle **reproduction** cycle, a female will lay 100-200 eggs into a hole dug in a sandy beach.

The sex of sea turtle hatchlings varies in **response to the environment:** Warmer temperatures favor the development of females, while cooler temperatures favor the development of males.

Energy processing in adult sea turtles depends upon a diet of algae and sea grass.

Although surrounded by salt water, sea turtles carry out **regulation** of the salt level in their body by excreting excess salt through their eyes.

▼ **Figure 1.12 A sample of the diversity of life in a national park in Namibia.**

Even as life on Earth shares recognizable properties, it also exists in tremendous diversity (**Figure 1.12**; see also Chapters 13–17). But must we limit our discussion to life on this planet? Although we have no proof that life has ever existed anywhere other than Earth, biologists speculate that extraterrestrial life, if it exists, could be recognized by the same properties described in Figure 1.11. The Mars rover *Curiosity*, which has been exploring the surface of the red planet since 2012, contains several instruments designed to identify substances that provide evidence of past or present life. For example, *Curiosity* is using a suite of onboard instruments to detect chemicals that could provide evidence of energy processing by microscopic organisms. In 2017, NASA announced that one of Saturn's moons, Enceladus, is the most likely place in our solar system to find extraterrestrial life, due to its abundant water and geothermal activity. NASA hopes to launch a probe there soon. The search continues. ☑

ONE OF THE MISSIONS OF THE MARS ROVER IS TO SEARCH FOR SIGNS OF LIFE.

☑ **CHECKPOINT**

Which properties of life apply to a car? Which do not?

■ *Answer: A car demonstrates order, regulation, energy processing, and response to the environment. But a car does not grow, reproduce, or evolve, and it is not composed of cells.*

Major Themes in Biology

As new discoveries unfold, biology grows in breadth and depth. However, major themes continue to run throughout the subject. These overarching principles unify all aspects of biology, from the microscopic world of cells to the global environment. Focusing on a few big-picture ideas that cut across many topics within biology can help organize and make sense of all the information you will learn.

This section describes five unifying themes that recur throughout our investigation of biology: the relationship of structure to function, information flow, pathways that transform energy and matter, interactions within biological systems, and evolution. You'll encounter these themes throughout subsequent chapters, with some key examples **highlighted in blue in the text**.

The Relationship of Structure to Function

When considering useful objects in your home, you may realize that form and function are related. A chair, for example, cannot have just any shape; it must have a stable base to hold it up and a flat area to support weight. The function of the chair constrains the possible shapes it may have. Similarly, within biological systems, structure (the shape of something) and function (what it does) are often related, with each providing insight into the other.

The correlation of structure and function can be seen at different levels, or scales, within biological systems, such as molecules, cells, tissues, and organs. Consider your lungs, which function to exchange gases with the environment: Your lungs bring in oxygen (O_2) and take out carbon dioxide (CO_2). The structure of your lungs correlates with this function (**Figure 1.13**). The smallest branches of your lungs end in millions of tiny sacs in which the gases cross from the air to your blood, and vice versa. This branched structure (the form of the lungs) provides a tremendous surface area over which a very high volume of air may pass (the function of the lungs). At quite another level, the correlation of structure and function can be seen in your cells. For example, oxygen diffuses into red blood cells as it enters the blood in the lungs. The indentations of red blood cells (**Figure 1.14**) increase the surface area through which oxygen can diffuse; without such indentations, there would be less surface area through which gases could move.

Throughout your study of life, you will see the structure and function principle apply to all levels of biological organization, such as the structures of molecules and of entire organisms. Some specific examples of the correlation

☑ CHECKPOINT

Explain how the correlation of structure to function applies to a tennis racket.

■ Answer: A tennis racket must have a large, flat surface for striking the ball, an open mesh so that air can flow through, and a handle for grasping and controlling the racket.

▼ **Figure 1.13 Structure and function: human lungs.**
The structure of your lungs correlates with their function.

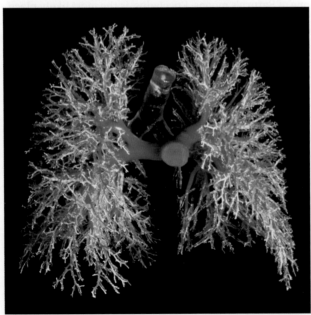

▼ **Figure 1.14 Structure and function: red blood cells.**
As oxygen enters the blood in the lungs, it diffuses into red blood cells.

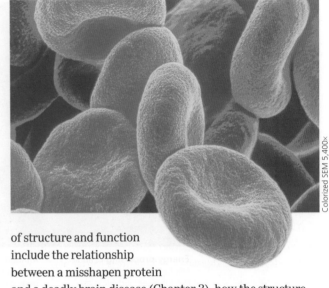

Colorized SEM 5,400×

of structure and function include the relationship between a misshapen protein and a deadly brain disease (Chapter 3), how the structure of DNA allows it to act as the molecule of heredity (Chapter 10) and to serve as the basis of forensic investigations (Chapter 12), and structural adaptations within the bodies of plants (Chapter 16). ☑

Information Flow

For life's functions to proceed in an orderly manner, information must be received, transmitted, and used. Just as our society depends upon communication between its members, a "society" of biological components cannot function as a living system without the flow of information. Such information flow is apparent at all levels of biological organization. For example, information about the amount of glucose in the bloodstream is received by organs such as your pancreas. The pancreas acts on that information by releasing hormones (including insulin) that regulate the levels of glucose in the blood. As insulin takes effect and glucose levels change, the pancreas processes this information, making continuous adjustments.

At the microscopic level, every cell contains **genes**, hereditary units of information consisting of specific sequences of DNA passed on from the previous generation. At the organismal level, as every multicellular organism develops from an embryo, information exchanged between cells enables the overall body plan to take shape in an organized fashion (as you'll see in Chapter 11). When the organism is mature, information about the internal conditions of its body is used to keep those conditions within a range that allows for life. Although bacteria and humans inherit different genes, that information is encoded in an identical chemical language common to all organisms. In fact, the language of life has an alphabet of just four letters. The chemical names of DNA's four molecular building blocks are abbreviated as A, G, C, and T (Figure 1.15). A typical gene is hundreds to thousands of chemical "letters" in length. A gene's meaning to a cell is encoded in its specific sequence of these letters, just as the message of this sentence is encoded in its arrangement of the 26 letters of the English alphabet.

How is all this information used within your body? At any given moment, your genes are coding for the production of thousands of different proteins that control your body's processes. (You'll learn the details of how proteins are produced using information from DNA in Chapter 10.) Food is broken down, new body tissues are built, cells divide, signals are sent—all controlled by information stored in your DNA. For example, information in one of your genes translates to "Make insulin," helping your body regulate blood sugar.

People with type 1 diabetes often have a mutation (error) in a different gene that causes the body's immune cells to attack and destroy the insulin-producing pancreas cells. These cells are then unable to properly respond to information about glucose levels in the blood. The

▼ **Figure 1.16 Information flow and diabetes.** In some people with diabetes, a mutation causes a disruption in the normal flow of genetic information. For such people, insulin produced by genetically engineered bacteria can be lifesaving.

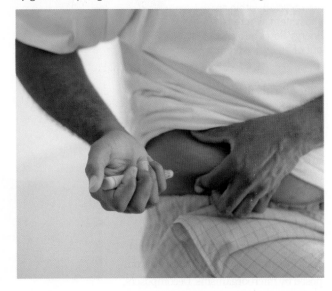

breakdown of the normal flow of information within the body leads to disease. Some people with diabetes regulate their sugar levels by injecting themselves with insulin (Figure 1.16) produced by genetically engineered bacteria. These bacteria can make insulin because the human gene has been transplanted into them. ☑

☑ CHECKPOINT
What is the term for the genetic information that encodes for a protein?

■ *Answer: gene*

Pathways That Transform Energy and Matter

Movement, growth, reproduction, and the various cellular activities of life are work, and work requires energy. The input of energy, primarily from the sun, and the transformation of energy from one form to another make life possible. Most ecosystems are solar powered at their source. The energy that enters an ecosystem as sunlight is captured by plants and other photosynthetic organisms (producers) that absorb the sun's energy and convert it into chemical energy, storing it as chemical bonds within sugars and other complex molecules. Food molecules provide energy and matter for a series of consumers, such as animals, that feed on producers. Organisms use food as a source of energy by breaking chemical bonds to release energy stored in the molecules or as building blocks for making new molecules needed by the organism. In other words, the molecules consumed can be used as both a source of energy and a source of matter. In these energy conversions between and within organisms, some energy is converted to heat, which is then lost from the ecosystem. Thus, energy flows through an ecosystem, entering as light energy and exiting as heat

▶ **Figure 1.15 Information stored in DNA.** Every molecule of DNA is constructed from four kinds of chemical building blocks that are chained together, shown here as simple shapes and letters. Genetic information is encoded in the sequence of these building blocks.

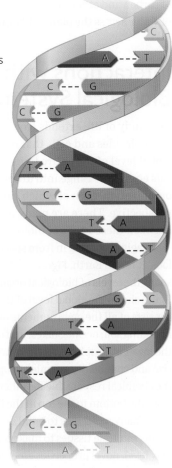

The four chemical building blocks of DNA

A DNA molecule

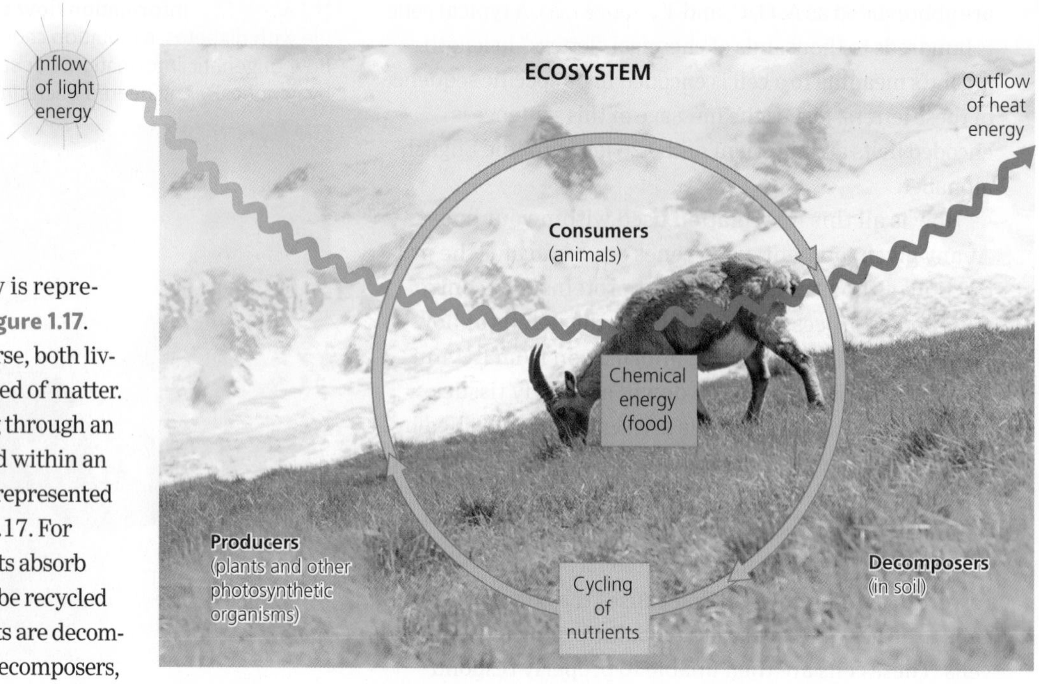

▶ **Figure 1.17**
Transformations of energy and matter in an ecosystem. Nutrients are recycled within an ecosystem, whereas energy flows into and out of an ecosystem.

ECOSYSTEM

Inflow of light energy

Outflow of heat energy

Consumers (animals)

Chemical energy (food)

Producers (plants and other photosynthetic organisms)

Decomposers (in soil)

Cycling of nutrients

energy. This flow of energy is represented by wavy lines in **Figure 1.17**.

Every object in the universe, both living and nonliving, is composed of matter. In contrast to energy flowing through an ecosystem, matter is recycled within an ecosystem. This recycling is represented by the blue circle in Figure 1.17. For example, minerals that plants absorb from the soil can eventually be recycled back into the soil when plants are decomposed by microorganisms. Decomposers, such as fungi and many bacteria, break down waste products and the remains of dead organisms, changing complex molecules into simple nutrients. The action of decomposers makes nutrients available to be taken up from the soil by plants again, thereby completing the cycle.

Within all living cells, a vast network of interconnected chemical reactions (collectively referred to as metabolism) continually converts energy from one form to another as matter is recycled. For example, as food molecules are broken down into simpler molecules, energy stored in the chemical bonds is released. This energy can be captured and used by the body (to power muscle contractions, for example). The atoms that made up the food can then be recycled (to build new muscle tissue, for example). Within all living organisms, there is a never-ending "chemical square dance" in which molecules swap chemical partners as they receive, convert, and release matter and energy. Within the ocean, for example, sunlight filtering through shallow water enables seagrass to grow. When a sea turtle grazes on seagrass, it obtains energy that it uses to swim and uses the molecules of matter as building blocks for cells within its own body.

Energy transformations can be disrupted, often with dire consequences. Consider what happens if you consume cyanide, one of the deadliest known poisons. Ingesting just 200 milligrams (about half the size of an aspirin tablet) causes death in humans. Cyanide is so toxic because it blocks an essential step within the metabolic pathway that harvests energy from glucose. When even a single protein within this pathway becomes inhibited, cells lose the ability to extract the energy stored in the chemical bonds of glucose. The rapid death that follows is a gruesome illustration of the importance of energy and matter transformations to life. Throughout your study of biology, you will see more examples of how living organisms regulate the transformation of energy and matter, from microscopic cellular processes such as photosynthesis (Chapter 7) and cellular respiration (Chapter 8), to ecosystem-wide cycles of carbon and other nutrients (Chapter 20), to global cycles of water across the planet (Chapter 18). ✓

✓ **CHECKPOINT**

1. What is the key difference between how energy and matter move in ecosystems?

2. What is the primary way by which energy leaves your body?

■ *Answers: 1. Energy moves through an ecosystem (entering and exiting), whereas matter is recycled within an ecosystem. 2. as heat*

Interactions within Biological Systems

The study of life extends from the microscopic level of the molecules and cells that make up organisms to the global level of the entire living planet. We can divide this enormous range into different levels of biological organization, each of which can be viewed from a system perspective. There are many interactions within and between these levels of biological systems.

Imagine zooming in from space to take a closer and closer look at life on Earth. **Figure 1.18** takes you on a tour that spans the levels of biological organization. The top of the figure shows the global level of the entire **biosphere**, which consists of all the environments on Earth that support life—including soil; oceans, lakes, and other bodies of water; and the lower atmosphere. At the other extreme of biological size and complexity are microscopic molecules such as DNA, the chemical responsible for inheritance. Zooming outward from the bottom to the top in the figure, you can see that it takes many molecules to build a cell, many cells to make a tissue, multiple tissues to make an organ, and so on. At each new level, novel properties emerge that are absent from the preceding one. These emergent properties are due to the specific arrangement and interactions of many parts into

▼ **Figure 1.18 Zooming in on life.**

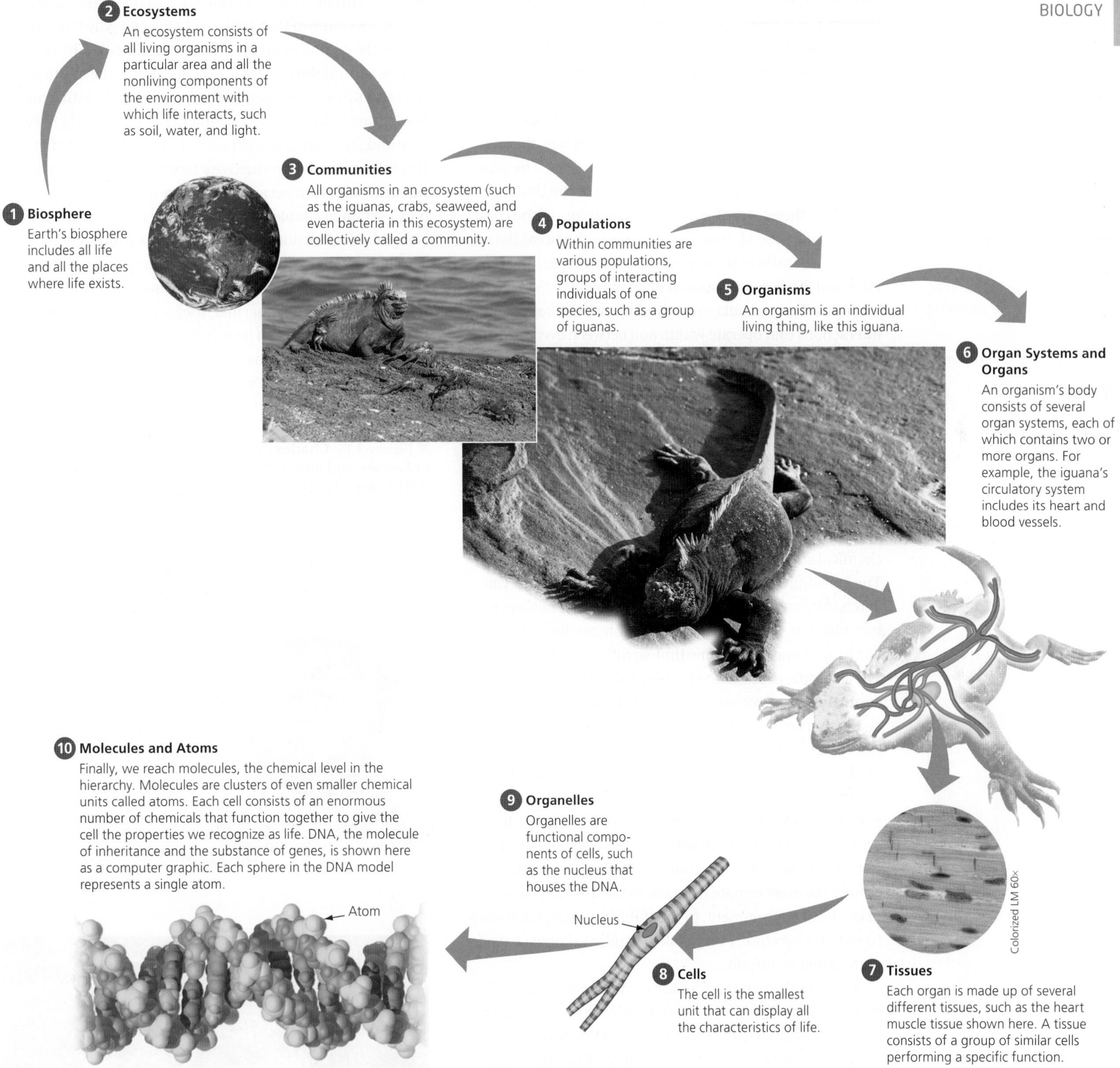

2 Ecosystems
An ecosystem consists of all living organisms in a particular area and all the nonliving components of the environment with which life interacts, such as soil, water, and light.

1 Biosphere
Earth's biosphere includes all life and all the places where life exists.

3 Communities
All organisms in an ecosystem (such as the iguanas, crabs, seaweed, and even bacteria in this ecosystem) are collectively called a community.

4 Populations
Within communities are various populations, groups of interacting individuals of one species, such as a group of iguanas.

5 Organisms
An organism is an individual living thing, like this iguana.

6 Organ Systems and Organs
An organism's body consists of several organ systems, each of which contains two or more organs. For example, the iguana's circulatory system includes its heart and blood vessels.

10 Molecules and Atoms
Finally, we reach molecules, the chemical level in the hierarchy. Molecules are clusters of even smaller chemical units called atoms. Each cell consists of an enormous number of chemicals that function together to give the cell the properties we recognize as life. DNA, the molecule of inheritance and the substance of genes, is shown here as a computer graphic. Each sphere in the DNA model represents a single atom.

← Atom

9 Organelles
Organelles are functional components of cells, such as the nucleus that houses the DNA.

Nucleus

8 Cells
The cell is the smallest unit that can display all the characteristics of life.

7 Tissues
Each organ is made up of several different tissues, such as the heart muscle tissue shown here. A tissue consists of a group of similar cells performing a specific function.

Colorized LM 60x

an increasingly complex system. Such properties are called emergent because they emerge as complexity increases. For example, life emerges at the level of the cell; a test tube full of molecules is not alive. The saying "the whole is greater than the sum of its parts" captures this idea. Emergent properties are not unique to life. A box of camera parts won't do anything, but if the parts are arranged and interact in a certain way, you can capture photographs. Add structures from a phone, and your camera and phone can interact to gain the ability to send photos to friends. New properties emerge as the complexity increases. Compared with such nonliving examples, however, the unrivaled complexity of biological systems makes the emergent properties of life especially fascinating to study.

Consider another example of interactions within biological systems, one that operates on a much larger level: global climate. For example, as the temperature of Earth's atmosphere rises, the oceans are becoming warmer. When the water is too warm, coral animals expel algae that live within them. As a result, the corals lose their color, a phenomenon called coral bleaching. The bleached coral fails to support other life that grows on and around the coral reef, thereby reducing the food supply available to sea turtles. The gases emitted from a coal plant in the Midwest of the United States can therefore affect a turtle swimming off the coast of Australia. Indeed, nearly two-thirds of the Great Barrier Reef of Australia experienced significant bleaching in 2016 and 2017, a development of great concern to marine biologists. Throughout our study of life, we will see countless interactions that operate within and between every level of the biological hierarchy shown in Figure 1.18. ☑

☑ **CHECKPOINT**

What is the smallest level of biological organization that can display all the characteristics of life?

■ *Answer: a cell*

Evolution

Life is distinguished by both its unity and its diversity. Multiple lines of evidence point to life's unity, from the similarities seen among and between fossil and living organisms, to common cellular processes, to the universal chemical structure of DNA, the molecule of inheritance. The amazing diversity of life is on display all around you and is documented in zoos, nature shows, and natural history museums. It is remarkable how life can be both so similar (unity) and at the same time so different (diversity). The scientific explanation for this unity and diversity is **evolution**, the process of change that has transformed life on Earth from its earliest forms to the vast array of organisms living today. Evolution is the fundamental principle of life and the core theme that unifies all of biology. The theory of evolution by natural selection is the one principle that makes sense of everything we know about living organisms. Evolution can help us investigate and understand every aspect of life, from the tiny organisms that occupy the most remote habitats, to the diversity of species in our local environment, to the stability of the global environment. Therefore, every biology student should strive to understand evolution.

The evolutionary view of life came into focus in 1859 when Charles Darwin published one of the most influential books ever written: *On the Origin of Species by Means of Natural Selection* (Figure 1.19). The first of two main points that Darwin presented in *The Origin of Species* was that species living today arose from a succession of ancestors that were different from them. Darwin called this process "descent with modification." This insightful phrase captures both the unity of life (descent from a common ancestor) and the diversity of life (modifications that evolved as species diverged from their ancestors).

Although other scholars had proposed similar ideas of descent with modification, Darwin was the first to propose a valid mechanism that explained how and why it occurs. This is the second main point of Darwin's book: The process of natural selection is the driving force of evolution. In the struggle for existence, those individuals with traits best suited to the local environment are more likely to survive and leave the greatest number of healthy offspring. It is this unequal reproductive success that Darwin called **natural selection** because the environment "selects" only certain heritable traits from those already existing. Natural selection does not promote or somehow encourage changes. Rather, mutations occur randomly. Natural selection "edits" those changes that have already occurred. If those traits can be inherited, they will be more common in the next generation. The results of natural selection are evolutionary adaptations, inherited traits that enhance survival in an organism's specific environment.

▼ **Figure 1.19** Charles Darwin (1809–1882), *The Origin of Species*, and blue-footed boobies he observed on the Galápagos Islands.

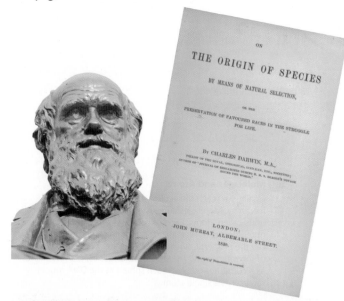

▼ Figure 1.20 Natural selection in action.

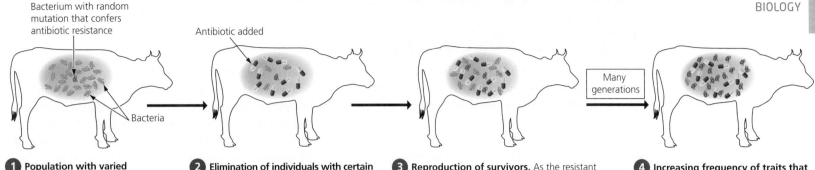

Bacterium with random
mutation that confers
antibiotic resistance

Antibiotic added

Many
generations

Bacteria

1 Population with varied inherited traits. The bacteria vary in ability to resist the antibiotic. By chance, a few bacteria are resistant.

2 Elimination of individuals with certain traits. The majority of bacteria are killed by the antibiotic. The few resistant bacteria will tend to survive.

3 Reproduction of survivors. As the resistant bacteria reproduce, genes for antibiotic resistance are passed along to the next generation in greater frequency.

4 Increasing frequency of traits that enhance survival and reproductive success. Generation after generation, the bacteria population adapts to the environment by natural selection.

The world is rich with examples of natural selection. Consider the development of antibiotic-resistant bacteria (Figure 1.20). Dairy and cattle farmers often add antibiotics to feed because doing so results in larger, more profitable animals. **1** The members of the bacteria population already, due to random mutation, vary in their susceptibility to an antibiotic. **2** Once the environment has been changed with the addition of antibiotics, some bacteria will succumb quickly and die, whereas others will survive. **3** Those that do survive will have the potential to multiply, producing offspring that will likely inherit the traits that enhance survival. **4** Over many generations, bacteria that are resistant to antibiotics will thrive in greater and greater numbers. Thus, feeding antibiotics to cows may promote the evolution of antibiotic-resistant bacterial populations.

As you'll see in Chapter 13, the theory of evolution by natural selection is supported by multiple lines of evidence—the fossil record, experiments, observations of natural selection in action, and genetic data. Evolution is the central theme that makes sense of everything we know and learn about biology. Throughout this text, we'll see many more examples of both the process and products of evolution.

To review the five unifying themes of biology, let's return to the green sea turtle (Figure 1.21). Every topic in biology can be related to these big ideas. To emphasize evolution as the central theme of biology, we end each chapter with an Evolution Connection section. Let's end the chapter the way it began, by imagining yourself floating in the ocean. ✓

☑ CHECKPOINT

1. What is the modern term for what Darwin called "descent with modification"?

2. What mechanism did Darwin propose for evolution? What three-word phrase summarizes this mechanism?

■ Answers: 1. evolution 2. natural selection; unequal reproductive success

▼ Figure 1.21 Applying the major themes of biology to the study of the green sea turtle.

The relationship of structure to function. Flippers are adapted to cruising through the sea.

Information flow. DNA comparisons with other types of turtles reveal how genes account for the unique traits of this species.

Pathways that transform energy and matter. A diet of mainly low-nutrient foods results in slow growth.

Evolution. Fossils indicate the shell evolved from expanding backbones and ribs. Natural selection favored better protection, leading to the shell we see today.

Interactions within biological systems. Changing global climate patterns affect migration routes and the growth of the turtles' prey.

EVOLUTION CONNECTION | Swimming with the Turtles

Turtles in the Tree of Life

Just as you have a family history, each species on Earth today represents one twig on a branching tree of life that extends back in time through ancestral species more and more remote. Analyzing DNA can determine relationships in a human family (settling questions of paternity, for example). Similarly, comparing DNA sequences from different species provides evidence of evolutionary relationships. The underlying assumption is that the more closely the DNA sequences of two species match, the more closely they are related. Such evidence can be used to generate an evolutionary tree, such as the one shown in **Figure 1.22**. Notice that this figure is highly abbreviated, showing just a few examples. Diagrams of evolutionary relationships generally take the form of branching trees, usually turned sideways and read from left (most ancient time) to right (most recent time).

Species that are very similar, such as the loggerhead and hawksbill turtles, share a common ancestor at a relatively recent branch point on the tree of life. In addition, all turtles can be traced back much further in time to an ancestor common to turtles, crocodiles, snakes, and all other reptiles. All reptiles have hard-shelled eggs, and such similarities are what we would expect if all reptiles descended from a common ancestor, a first reptile. And reptiles, mammals, and all other animals share a common ancestor—the first animal—even more ancient. Going further back still, at the cellular level, all life displays striking similarities. For example, all living cells are surrounded by an outer membrane of similar makeup and use structures called ribosomes to produce proteins. Such evolutionary analyses based on the structure of DNA provide insight into life both current and ancient. In this spirit, we will begin our investigation of biology by studying the chemistry of life (Chapter 2).

▼ **Figure 1.22 A partial family tree of animals.** This tree represents a hypothesis (a tentative model) based on both the fossil record and a comparison of DNA sequences in present-day animals. As new evidence about the evolutionary history of turtles emerges, this tree will inevitably change.

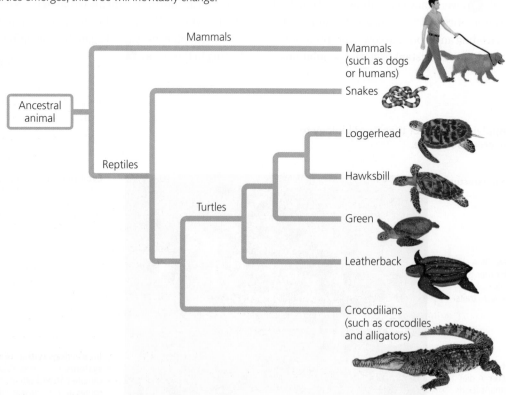

Chapter Review

SUMMARY OF KEY CONCEPTS

The Scientific Study of Life

Biology is the scientific study of life.

An Overview of the Process of Science

It is important to distinguish scientific investigations from other ways of thinking because only scientific means of studying life qualify as biology. Science usually begins with exploration, either through observations recorded as data or by gathering information from reliable sources. Exploration often raises questions that can be tested by forming a hypothesis (a proposed explanation for your observations). Hypotheses can be investigated by further observations or by conducting experiments. This process sometimes takes the form of a series of steps that can lead to discovery:

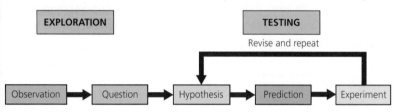

Scientists often communicate their results through peer-reviewed publications. There is no fixed process of science. Each study proceeds through exploration, testing, communication, and outcomes in different ways.

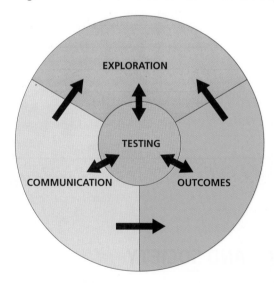

Hypotheses, Theories, and Facts

A theory is a broad and comprehensive statement about the world that is supported by the accumulation of a great deal of verifiable evidence. A fact is a piece of information that can be verified by any independent observer.

Controlled Experiments

A controlled experiment involves running the same tests on two or more groups that differ in one variable. The control group does not receive the change under investigation, while the experimental group does. A blind experiment is one where information about the experiments is withheld from the participants and/or the experimenters.

Evaluating Scientific Claims

Be wary of pseudoscience that is falsely presented as following a process of science when it does not. You can learn to recognize reliable resources by looking for indicators.

The Properties of Life

All life displays a common set of characteristics:

- order
- regulation
- growth and development
- energy processing
- response to the environment
- reproduction
- evolution
- cells

Major Themes in Biology

Throughout your study of biology, you will frequently come upon examples of five unifying themes:

MAJOR THEMES IN BIOLOGY				
Structure/ Function	Information Flow	Energy Transformations	Interconnections within Systems	Evolution

The Relationship of Structure to Function At all levels of biology, structure and function are related. Changing structure often results in an altered function, and learning about a component's function will often give insight into its structure.

Information Flow Throughout living systems, information is stored, transmitted, and used. Within your body, genes provide instructions for building proteins, which perform many of life's tasks.

Pathways That Transform Energy and Matter Within ecosystems, nutrients are recycled, but energy flows through.

Interactions within Biological Systems Life can be studied on many levels, from molecules to the entire biosphere. As complexity increases, novel properties emerge. For example, the cell is the smallest unit that can possibly display all of the characteristics of life.

Evolution Charles Darwin established the ideas of evolution ("descent with modification") through natural selection (unequal reproductive success) in his 1859 publication *The Origin of Species*. Natural selection leads to adaptations to the environment, which—when passed from generation to generation—is the mechanism of evolution.

Mastering Biology

For practice quizzes, BioFlix animations, MP3 tutorials, video tutors, and more study tools designed for this textbook, go to Mastering Biology™

SELF-QUIZ

1. Which is *not* a characteristic of all living organisms?
   a. growth and development
   b. composed of multiple cells
   c. complex yet organized
   d. uses energy

2. Place the following levels of biological organization in order from smallest to largest: atom, biosphere, cell, ecosystem, molecule, organ, organism, population, tissue. Which is the smallest level capable of demonstrating all of the characteristics of life?

3. Plants use the process of photosynthesis to convert the energy in sunlight to chemical energy in the form of sugar. While doing so, they consume carbon dioxide and water and release oxygen. Explain how this process functions in both the cycling of chemical nutrients and the flow of energy through an ecosystem.

4. How does natural selection cause a population to become adapted to its environment over time?

5. Which of the following are the proper components of the scientific method?
   a. experiment, conclusion, application.
   b. question, observation, experiment, analysis, prediction.
   c. observation, question, hypothesis, prediction, experiment, conclusion.
   d. observation, question, opinion, conclusion, hypothesis.

6. Which statement best distinguishes hypotheses from theories in science?
   a. Theories are hypotheses that have been proven.
   b. Hypotheses are tentative guesses; theories are correct answers to questions about nature.
   c. Hypotheses usually are narrow in scope; theories have broad explanatory power and are supported by a lot of evidence.
   d. Hypotheses and theories mean essentially the same thing in science.

7. _____ is the core theme that unifies all areas of biology.

8. What distinguishes a fact from an opinion?

9. Match each of the following terms to the phrase that best describes it.
   a. Natural selection
   b. Evolution
   c. Hypothesis
   d. Biosphere
   1. A testable idea
   2. Descent with modification
   3. Unequal reproductive success
   4. All life-supporting environments on Earth

For answers to the Self Quiz, see Appendix D.

IDENTIFYING MAJOR THEMES

For each statement, identify which major theme is evident (the relationship of structure to function, information flow, pathways that transform energy and matter, interactions within biological systems, or evolution) and explain how the statement relates to the theme. If necessary, review the theme descriptions in this chapter.

10. By comparing genes between green sea turtles and humans, insight can be gained into how those genes encode specific physical traits.

11. Although green sea turtles consume a lot of vegetation, they get few nutrients from each mouthful, requiring them to graze frequently.

12. As global climate changes, green sea turtles alter many aspects of their behavior.

For answers to Identifying Major Themes, see Appendix D.

THE PROCESS OF SCIENCE

13. Figure 1.20 depicts the selection of a trait—the resistance to an antibiotic—in a population of bacteria. Let us assume that this resistance is conferred by a gene that helps to neutralize or inactivate the antibiotic and is only carried by the resistant bacteria. What would happen to the bacterial population if this antibiotic is banned? Develop a hypothesis that is consistent with Darwin's theory of natural selection.

14. **Interpreting Data** The Kemp's ridley sea turtle (*Lepidochelys kempii*) is a critically endangered species. The graph presents the number of Kemp's ridley sea turtle nests found in the Padre Island National Seashore in Texas over a span of years. Write a one-sentence summary of the results presented in the graph.

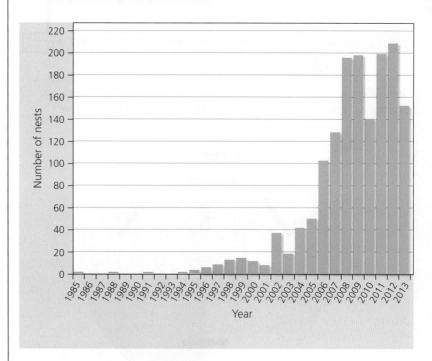

BIOLOGY AND SOCIETY

15. The development of both drugs and household items typically involves biomedical and/or biochemical research. Select three drugs and three household items, and devise hypothetical flowcharts, like the one depicted in Figure 1.2, which delineate the process of development of each of these.

16. Check the presence of the word "biological" (or "bio") on the containers of food and cleaning products. How does its use on commercial products relate to the scientific definition of biology?

Cells

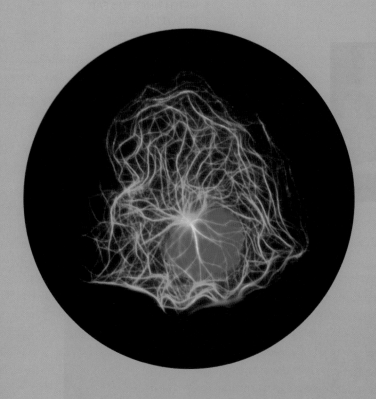

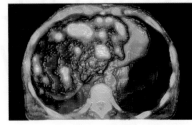

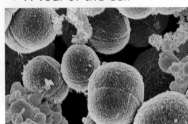
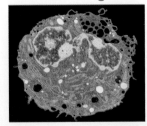

2 Essential Chemistry for Biology

Why Chemistry Matters

Every biological system can be viewed as a vast collection of chemicals. In fact, everything you do—digesting food, moving muscles, even the act of thinking—is a consequence of countless chemical reactions occurring within your body. Understanding life therefore requires learning some basic chemistry.

ADDING SALT TO YOUR FRIES? CONSIDER THIS: TABLE SALT IS A COMBINATION OF AN EXPLOSIVE SOLID AND A POISONOUS GAS.

WATCH WHAT YOU EAT! TOO LITTLE OF THE ELEMENT COPPER IN YOUR DIET CAUSES BRAIN DAMAGE, BUT TOO MUCH CAN HARM VITAL ORGANS.

THE EXHAUST RELEASED BY A CAR IN BOSTON CONTRIBUTES TO THE DEATH OF CORAL IN THE GREAT BARRIER REEF OF AUSTRALIA.

BIOLOGY AND SOCIETY   Helpful Radiation

Radioactive vision. This PET scan of a human liver highlights areas of diseased tissue.

Nuclear Medicine

The word *radioactive* probably sets off alarm bells in your mind: "Danger! Hazardous!" It is true that radiation, high-energy particles emitted by radioactive substances, can penetrate living tissues and kill cells by damaging DNA. But radiation can also be medically beneficial by, for example, helping to treat cancer. What determines whether radiation is harmful or helpful? Radiation is most dangerous when exposure is uncontrolled and covers most or all of the body, as happens when a person is exposed to radioactive fallout from a nuclear detonation or accident. In contrast, radiation applied in a carefully controlled manner can be helpful.

Perhaps the most beneficial application of controlled radiation is nuclear medicine, the use of radioactive substances to diagnose and treat diseases. Cells use radioactive atoms as if they were the normal form. Once a cell takes up a radioactive atom, its location and concentration can be detected because of the radiation it emits. This makes radioactive isotopes useful as tracers—biological spies, in effect. For example, a medical diagnostic tool called a PET scan works by detecting small amounts of radiation emitted by radioactive materials that were purposefully introduced into the body.

Radiation therapy exposes only a small part of the body to a precise dosage of radiation. When treating cancer, for example, carefully calibrated radiation beams are aimed from several angles, intersecting only at the tumor. This provides a deadly dose to cancerous cells but mostly spares surrounding healthy tissues. Radiation therapy is also used to treat Graves' disease, a condition in which an overactive thyroid gland (located in the neck) causes a variety of physical symptoms. People with Graves' disease may be treated with radioactive iodine. Because the thyroid produces a variety of hormones that use iodine, the radioactive iodine accumulates in this gland, where it then provides a steady low dose of radiation that can, over time, destroy enough thyroid tissue to reduce symptoms.

What makes something radioactive? To understand this question, we have to look to the most basic level of all living things: the atoms that make up all matter. Many questions about life, such as why radiation harms cells, can be reduced to questions about chemicals and their interactions, such as how radiation affects atoms in living issues. Therefore, knowledge of chemistry is essential to understanding biology. In this chapter, we'll review some basic chemistry that you can apply throughout your study of biology. We'll start with an examination of molecules, atoms, and their components. Next, we'll explore water, one of life's most important molecules, and its crucial role in sustaining life on Earth.

Some Basic Chemistry

Why would a biology book include a chemistry chapter? If you take any biological system apart, you eventually end up at the chemical level. In fact, you can think of your body as a big watery container of chemicals undergoing a continuous series of chemical reactions. Viewed this way, your metabolism—the total of all the chemical reactions in your body—is like a giant square dance, with chemical partners constantly swapping atoms as they move. Beginning at this basic biological level, let's explore the chemistry of life.

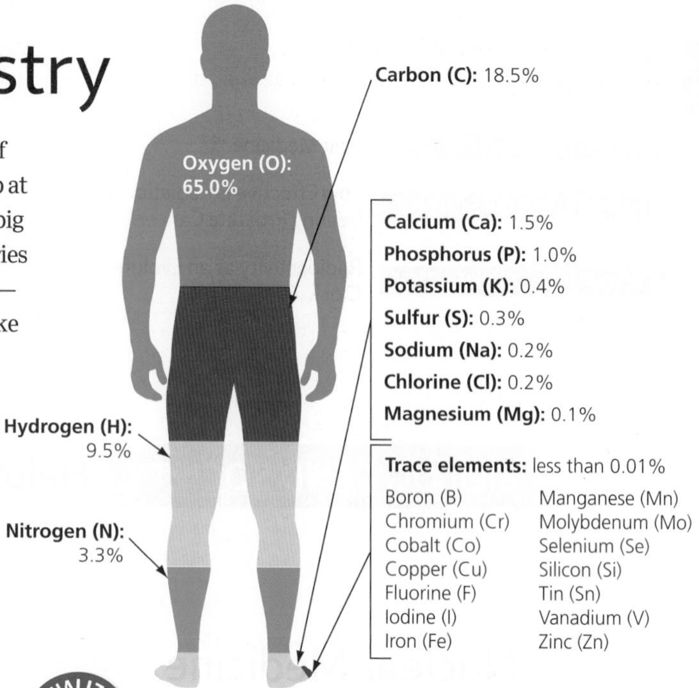

Matter: Elements and Compounds

You and everything that surrounds you are made of matter, the physical "stuff" of the universe. Matter is found on Earth in three physical states: solid, liquid, and gas. Defined more formally, **matter** is anything that occupies space and has mass. **Mass** is a measure of the amount of material in an object. All matter is composed of chemical elements. An **element** is a substance that cannot be broken down into other substances by chemical reactions. Think of it this way: When you burn wood, you are left with ashes. But when you burn ashes, you only get more ashes. That is because wood is a complex mixture of elements, while ashes consist only of a pure element (carbon) that cannot be broken down. There are 92 naturally occurring elements; examples are carbon, oxygen, and gold. Each element has a symbol derived from its English, Latin, or German name. For instance, the symbol for gold, Au, is from *aurum*, the Latin word for this element. All the elements—the 92 that occur naturally and others that are human-made—are listed in the **periodic table of the elements** (Figure 2.1), a fixture in any science lab (see Appendix B for a full version).

TOO LITTLE COPPER IN YOUR DIET CAUSES BRAIN DAMAGE, BUT TOO MUCH CAN HARM ORGANS.

Carbon (C): 18.5%
Oxygen (O): 65.0%
Calcium (Ca): 1.5%
Phosphorus (P): 1.0%
Potassium (K): 0.4%
Sulfur (S): 0.3%
Sodium (Na): 0.2%
Chlorine (Cl): 0.2%
Magnesium (Mg): 0.1%
Hydrogen (H): 9.5%
Nitrogen (N): 3.3%

Trace elements: less than 0.01%
Boron (B)   Manganese (Mn)
Chromium (Cr)   Molybdenum (Mo)
Cobalt (Co)   Selenium (Se)
Copper (Cu)   Silicon (Si)
Fluorine (F)   Tin (Sn)
Iodine (I)   Vanadium (V)
Iron (Fe)   Zinc (Zn)

VISUALIZING THE DATA

▲ **Figure 2.2 Chemical composition of the human body.** Just four elements make up 96% of your weight.

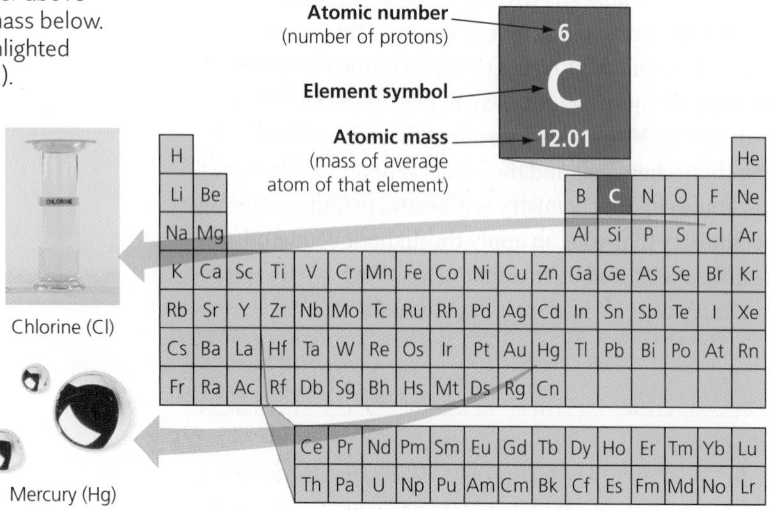

Of the naturally occurring elements, 25 are essential to humans. (Other organisms need fewer; plants, for example, typically need 17.) Four of these elements—oxygen (O), carbon (C), hydrogen (H), and nitrogen (N)—make up about 96% of the weight of the human body (Figure 2.2). Much of the remaining 4% is accounted for by seven elements, most of which are probably familiar to you, such as calcium (Ca). Calcium, important for building strong bones and teeth, is found abundantly in milk and dairy products as well as sardines and green, leafy vegetables (kale and broccoli, for example).

Less than 0.01% of your weight is made up of 14 trace elements. **Trace elements**, such as copper and iodine, are required in only very small amounts, but you cannot live without them. The average person needs a tiny speck of iodine each day. Iodine is an essential ingredient of hormones produced by the thyroid gland, located in the neck. An iodine deficiency causes the thyroid gland to enlarge, a condition called goiter. Therefore, consuming foods that are naturally rich in iodine—such as green vegetables, eggs, seafood, and dairy products—prevents goiter. The addition of iodine to table salt ("iodized salt") has nearly eliminated goiter in industrialized nations, but many thousands of people in developing nations are still affected (Figure 2.3). Another trace element is fluorine, which (in the form of fluoride) is added to dental products and drinking water and helps maintain healthy bones and teeth. Many food staples (such as flour) and prepared foods are fortified with trace mineral elements. Look at the side of a cereal box and

▼ **Figure 2.1 Abbreviated periodic table of the elements.** In the full periodic table (see Appendix B), each entry contains the element symbol in the center, with the atomic number above and the atomic mass below. The element highlighted here is carbon (C).

Atomic number
(number of protons)

Element symbol

Atomic mass
(mass of average atom of that element)

6
C
12.01

H																	He
Li	Be											B	C	N	O	F	Ne
Na	Mg											Al	Si	P	S	Cl	Ar
K	Ca	Sc	Ti	V	Cr	Mn	Fe	Co	Ni	Cu	Zn	Ga	Ge	As	Se	Br	Kr
Rb	Sr	Y	Zr	Nb	Mo	Tc	Ru	Rh	Pd	Ag	Cd	In	Sn	Sb	Te	I	Xe
Cs	Ba	La	Hf	Ta	W	Re	Os	Ir	Pt	Au	Hg	Tl	Pb	Bi	Po	At	Rn
Fr	Ra	Ac	Rf	Db	Sg	Bh	Hs	Mt	Ds	Rg	Cn						

Ce	Pr	Nd	Pm	Sm	Eu	Gd	Tb	Dy	Ho	Er	Tm	Yb	Lu
Th	Pa	U	Np	Pu	Am	Cm	Bk	Cf	Es	Fm	Md	No	Lr

Chlorine (Cl)

Mercury (Hg)

▼ **Figure 2.3 Diet and goiter.** Eating foods rich in iodine, a trace element, can prevent goiter, an enlargement of the thyroid gland.

you'll probably see iron listed; you can actually see the iron yourself if you crush the cereal and stir a magnet through it. Be grateful for this additive: It helps prevent anemia due to iron deficiency, one of the most common nutritional deficiencies among Americans. But too much of a trace element can also be a problem. For example, infants and older people can suffer from iron overload if too much is consumed.

Elements can combine to form **compounds**, substances that contain two or more different elements in a fixed ratio. In everyday life, compounds are much more common than pure elements. Table salt and water are both familiar examples of compounds. Table salt is sodium chloride, NaCl, consisting of equal parts of the elements sodium (Na) and chlorine (Cl). Table salt is a good example of an emergent property arising from interactions within a biological system: A compound (such as tasty salt) can have characteristics different from those of its elements (an explosive solid and a toxic gas). A molecule of water, H_2O, is another common compound; it has two atoms of hydrogen and one atom of oxygen. Most of the compounds in living organisms contain several different elements. DNA, for example, contains carbon, nitrogen, oxygen, hydrogen, and phosphorus. ☑

TABLE SALT COMBINES AN EXPLOSIVE SOLID AND A TOXIC GAS.

Atoms

Each element is made up of one kind of atom, and the atoms in one element are different from the atoms of other elements. An **atom** is the smallest unit of matter that still retains the properties of an element. In other words, the smallest amount of the element carbon is one carbon atom. Just how small is this "piece" of carbon? It would take about a million carbon atoms to stretch across the period at the end of this sentence.

The Structure of Atoms

Atoms are composed of subatomic particles, of which the three most important are protons, electrons, and neutrons. A **proton** is a subatomic particle with a single unit of positive electrical charge (+). An **electron** is a subatomic particle

with a single negative charge (−). A **neutron** is electrically neutral (has no charge).

Figure 2.4 shows a simplified model of an atom of the element helium (He), the lighter-than-air gas used to inflate party balloons. Each atom of helium has 2 neutrons (◉) and 2 protons (⊕) tightly packed into the **nucleus**, the atom's central core. Two electrons (⊖) move around the nucleus in a spherical cloud at nearly the speed of light. The negatively charged electron cloud is much bigger than the nucleus. If an atom were the size of a baseball stadium, the nucleus would be the size of a pea on the pitcher's mound and the electrons would be two gnats buzzing around the bleachers. When an atom has an equal number of protons and electrons, its net electrical charge is zero and so the atom is neutral.

All atoms of a particular element have the same unique number of protons. This number is the element's **atomic number**. An atom of helium, with 2 protons, has an atomic number of 2, and no other element has 2 protons. The periodic table of elements (Appendix B) lists elements in order of atomic number. Note that in these atoms the atomic number is also the number of electrons. A standard atom of any element has an equal number of protons and electrons, and thus its net electrical charge is 0 (zero). An atom's **mass number** is the sum of the number of protons and neutrons. For helium, the mass number is 4. The mass of a proton and the mass of a neutron are almost identical and are expressed in a unit of measurement called the dalton. Protons and neutrons each have masses close to 1 dalton. An electron has only about 1/2,000 the mass of a proton, so its mass is approximated as zero. An atom's **atomic mass**, which is listed in the periodic table as the bottom number (under the element symbol), is close to its mass number—the sum of its protons and neutrons—but may differ slightly because it represents an average of all the naturally occurring forms of that element.

Isotopes

All atoms of an element have the same atomic number (number of protons), but some atoms of that element may differ in mass number (protons + neutrons). The different **isotopes** of an element have the same number of protons and behave identically in chemical reactions, but they have different numbers of neutrons. The isotope carbon-12 (named for its mass number), which has 6 neutrons and 6 protons, makes up about 99% of all naturally occurring carbon. Most of the other 1% of carbon on Earth is the isotope carbon-13, which has 7 neutrons and 6 protons. A third isotope, carbon-14, which has 8 neutrons and 6 protons, occurs in minute quantities. All three isotopes have 6 protons—otherwise, they would not be carbon. Both carbon-12 and carbon-13 are stable

▼ **Figure 2.4 A simplified model of a helium atom.** This model shows the sub-atomic particles in an atom of helium. The electrons move very fast, creating a spherical cloud of negative charge surrounding the positively charged nucleus.

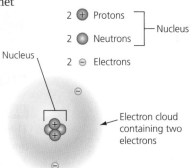

2 ⊕ Protons
2 ◉ Neutrons — Nucleus
2 ⊖ Electrons

Nucleus

Electron cloud containing two electrons

☑ **CHECKPOINT**

How many of the naturally occurring elements are used by your body? Which four are the most abundant in living cells?

■ *Answer: 25; oxygen, carbon, hydrogen, and nitrogen*

☑ **CHECKPOINT**

By definition, all atoms of carbon have exactly 6 _____, but the number of _____ varies from one isotope to another.

■ *Answer: protons; neutrons*

isotopes, meaning that their nuclei remain intact. The isotope carbon-14, on the other hand, is radioactive. A **radioactive isotope** is one in which the nucleus decays spontaneously, shedding particles and energy (radiation).

Uncontrolled exposure to high levels of radiation can be lethal because the particles and energy thrown off by radioactive atoms can damage molecules, especially DNA. Therefore, radiation from decaying isotopes can pose serious health risks. In 1986, the explosion of a nuclear reactor at Chernobyl, Ukraine, released large amounts of radioactive isotopes, killing 30 people within a few weeks. Millions of people in the surrounding areas were exposed, causing

an estimated 6,000 cases of thyroid cancer. The 2011 post-tsunami Fukushima nuclear disaster in Japan did not result in any immediate deaths due to radiation exposure, but scientists are carefully monitoring the people who live in the area for any long-term health consequences.

Although radioactive isotopes can cause harm when uncontrolled, they have many uses in biological research and medicine. The Biology and Society section discussed in general how radiation can be used to diagnose and treat diseases. The Process of Science section will explore one specific medical application of radiation: the treatment of prostate cancer. ☑

THE PROCESS OF SCIENCE | Helpful Radiation

How Effective Is Radiation in Treating Prostate Cancer?

BACKGROUND

The prostate is a gland within the male reproductive system that adds fluid to sperm during ejaculation. Among American men, prostate cancer is the most frequently diagnosed cancer (with over 160,000 new cases in 2017) and the second leading cause of cancer deaths (27,000 in 2017), after lung cancer.

METHOD

The most effective treatment for prostate cancer is surgical removal of the prostate. However, removal causes serious side effects. Another treatment option is radioactive seed implantation **(Figure 2.5a)**. In this treatment, several dozen small bits of radioactive metal (most commonly made from iodine-125), each the size of a grain of rice, are placed within the prostate using a needle **(Figure 2.5b)**. The isotopes used will decay into safe, nonradioactive metals in a few months or years. By carefully regulating the dosage, doctors attempt to abolish or slow the tumor without harming surrounding healthy tissues.

RESULTS

Is radioactive seed implantation therapy effective? A study published in 2014 involved over 1,000 British men with prostate cancer who were treated with iodine-125 seed

implants. Seeds were placed inside and around each tumor site. Five years later, 94% of the patients were free of cancer symptoms. While this number is impressive, how do we know if a similar number of men would have recovered even without the implants? Medical research trials include control groups whenever possible. However, there has never been a study involving patients who received radioactive seed implantation and those who received a placebo consisting of nonradioactive seeds because it would be unethical to withhold treatment from cancer patients. Nevertheless, comparisons with other studies that used different methods, such as surgery and external radiation, can be made **(Figure 2.5c)**. Such comparisons are problematic, though, because the groups of patients are not controlled for all characteristics. Therefore, studying the effects of radioactive seed implantation shows how researchers are sometimes unable to run properly controlled experiments when humans are involved.

Thinking Like a Scientist

How might a study on radioactive seed implantation in rats differ from a study on humans?

For the answer, see Appendix D.

▼ Figure 2.5 **Treatment of prostate cancer with radiation.**

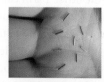

(a) **Radioactive seeds**

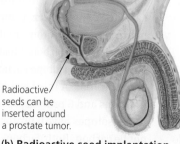

Radioactive seeds can be inserted around a prostate tumor.

(b) **Radioactive seed implantation**

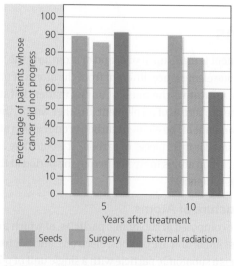

(c) **Comparing outcomes after three types of prostate treatment**

Seeds   Surgery   External radiation

Chemical Bonding and Molecules

Of the three subatomic particles we've discussed—protons, neutrons, and electrons—only electrons are directly involved in chemical reactions. The number of electrons in an atom therefore determines the chemical properties of that atom. Chemical reactions enable atoms to transfer or share electrons. These interactions usually result in atoms staying close together, held by attractions called **chemical bonds**. In this section, we will explore three types of chemical bonds: ionic, covalent, and hydrogen bonds.

Ionic Bonds

Table salt is an example of how the transfer of electrons can bond atoms together. As discussed earlier, the two ingredients of table salt are the elements sodium (Na) and chlorine (Cl). When a chlorine atom and a sodium atom are near each other, the chlorine atom strips an electron from the sodium atom (**Figure 2.6**). Before this electron transfer, both the sodium and chlorine atoms are electrically neutral. Because electrons are negatively charged, the electron transfer moves one unit of negative charge from sodium to chlorine. This action makes both atoms **ions**, atoms (or molecules) that are electrically charged as a result of gaining or losing electrons. In this case, the loss of an electron results in the sodium ion having a charge of +1, whereas chlorine's gain of an electron results in it having a charge of −1. Negatively charged ions often have names ending in "-ide," like "chloride" or "fluoride." The sodium ion (Na^+) and chloride ion (Cl^-) are held together by an

▼ **Figure 2.6 Electron transfer and ionic bonding.** When a sodium atom and a chlorine atom meet, the electron transfer between the two atoms results in two ions with opposite charges.

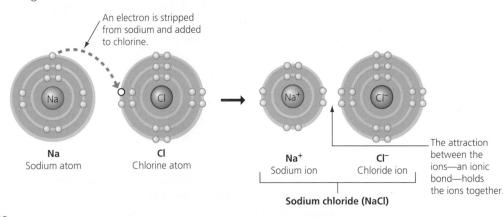

An electron is stripped from sodium and added to chlorine.

| Na | Cl | Na^+ | Cl^- |
| Sodium atom | Chlorine atom | Sodium ion | Chloride ion |

Sodium chloride (NaCl)

The attraction between the ions—an ionic bond—holds the ions together.

ionic bond, the attraction between oppositely charged ions. Compounds that are held together by ionic bonds are called ionic compounds, also known as salts. ☑

Covalent Bonds

In contrast to the complete *transfer* of electrons in ionic bonds, a **covalent bond** forms when two atoms *share* one or more pairs of electrons. Of the bonds we've discussed, covalent bonds are the strongest; these are the bonds that hold atoms together in a **molecule**. For example, in **Figure 2.7**, you can see that each of the two hydrogen atoms in a molecule of formaldehyde (CH_2O, a common disinfectant and preservative) shares one pair of electrons with the carbon atom. The oxygen atom shares two pairs of electrons with the carbon, forming a double bond. Notice that each atom of hydrogen (H) can form one covalent bond; oxygen (O) can form two; and carbon (C) can form four.

☑ **CHECKPOINT**

When a lithium ion (Li^+) joins a bromide ion (Br^-) to form lithium bromide, the resulting bond is a(n) _____ bond.

■ *Answer: ionic*

▼ **Figure 2.7 Alternative ways to represent a molecule.** A molecular formula, such as CH_2O, tells you the number of each kind of atom in a molecule but not how they are attached together. This figure shows four common ways of representing the arrangement of atoms in molecules.

Name (molecular formula)	Electron configuration	Structural formula	Space-filling model	Ball-and-stick model
	Shows how outermost electrons participate in chemical bonding.	Represents each covalent bond (a pair of shared electrons) with a line	Shows the shape of a molecule by symbolizing atoms with color-coded balls	Represents atoms with "balls" and bonds with "sticks"
Formaldehyde (CH_2O)	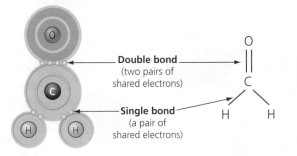			

Double bond (two pairs of shared electrons)

Single bond (a pair of shared electrons)

Hydrogen Bonds

A molecule of water (H_2O) consists of two hydrogen atoms joined to one oxygen atom by single covalent bonds. The covalent bonds between the atoms are represented as the "sticks" here in a ball-and-stick illustration, and the atoms are shown as the "balls":

However, the electrons are not shared equally between the oxygen and hydrogen atoms. The two yellow arrows shown in the space-filling model here indicate the stronger pull on the shared electrons that oxygen has compared with its hydrogen partners:

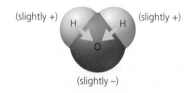

The unequal sharing of negatively charged electrons in a water molecule, combined with the molecule's V shape, makes the molecule polar. A **polar molecule** is one with an uneven distribution of charge that creates two poles, one positive pole and one negative pole. In the case of water, the oxygen end of the molecule has a slight negative charge, and the region around the two hydrogen atoms is slightly positive.

The polarity of water results in weak electrical attractions between neighboring water molecules. Because opposite charges attract, water molecules tend to orient such that a hydrogen atom from one water molecule is near the oxygen atom of an adjacent water molecule. These weak attractions are called **hydrogen bonds** (Figure 2.8). As you will see later in this chapter, the ability of water to form hydrogen bonds has many crucial implications for life

on Earth. And, as you'll learn in later chapters, molecules of DNA are held together by hydrogen bonds, which play a key role in the flow of information within living systems.

Chemical Reactions

The chemistry of life is dynamic. Your cells are constantly rearranging molecules by breaking existing chemical bonds and forming new ones in a "chemical square dance." Such changes in the chemical composition of matter are called **chemical reactions**. The transformation of matter, one of the key themes that unite all aspects of biology, can be seen in the countless chemical reactions that keep you alive.

An example of a chemical reaction is the breakdown of hydrogen peroxide, a common disinfectant that you may have poured on a cut:

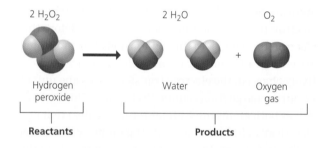

Let's translate the chemical shorthand: Two molecules of hydrogen peroxide ($2\ H_2O_2$) react to produce two molecules of water ($2\ H_2O$) and one molecule of oxygen (O_2, which is responsible for the fizzing that happens when hydrogen peroxide interacts with blood). The arrow in this equation indicates the conversion of the starting materials, the **reactants** ($2\ H_2O_2$), to the **products** ($2\ H_2O$ and O_2).

Notice that the same total numbers of hydrogen and oxygen atoms are present in reactants (to the left of the arrow) and products (to the right), although they are grouped differently. Chemical reactions cannot create or destroy matter; they can only rearrange it. These rearrangements usually involve the breaking of chemical bonds in reactants and the forming of new bonds in products.

This discussion of water molecules as the product of a chemical reaction is a good conclusion to this section on basic chemistry. Water is a substance so important in biology that we'll take a closer look at its life-supporting properties in the next section. ☑️

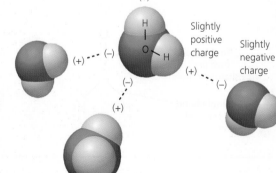

▶ **Figure 2.8 Hydrogen bonding in water.** The charged regions of the polar water molecules are attracted to oppositely charged areas of neighboring molecules. Each molecule can hydrogen-bond to a maximum of four partners.

Water and Life

Life on Earth began in water and evolved there for 3 billion years before spreading onto land. Modern life, even land-dwelling life, is still tied to water. Your cells are surrounded by a fluid that's composed mostly of water, and your cells range from 70% to 95% in water content.

The abundance of water is a major reason that Earth is habitable (Figure 2.9). Indeed, astrobiologists (scientists who search for life elsewhere in the universe) consider water vital to establishing and maintaining life on other planets. One of the primary missions of the Mars rovers is to search for liquid water; ice is abundant, and evidence suggests that liquid water has flowed on the surface in the past.

Water

The unique properties of water on which all life on Earth depends are a prime example of one of biology's overarching themes: the relationship of structure and function. The structure of water molecules—the polarity and the hydrogen bonding that results (see Figure 2.8)—explains most of water's life-supporting functions. We'll explore four of those properties here: the cohesive nature of water, the ability of water to moderate temperature, the biological significance of ice floating, and the versatility of water as a solvent.

The Cohesion of Water

Water molecules stick together as a result of hydrogen bonding. In a drop of water, any particular set of hydrogen bonds lasts for only a few trillionths of a second, yet at any instant, huge numbers of hydrogen bonds exist between molecules of liquid water. This tendency of molecules of the same kind to stick together, called **cohesion**, is much stronger for water than for most other liquids. The cohesion of water is important in the living world. Trees, for example, depend on cohesion to help transport water from their roots to their leaves (Figure 2.10).

Related to cohesion is surface tension, a measure of how difficult it is to stretch or break the surface of a liquid. Hydrogen bonds give water unusually high surface tension, making it behave as though it were coated with an invisible film (Figure 2.11).

▲ Figure 2.9 A watery world. In this photograph, you can see water as liquid (which covers three-quarters of Earth's surface), ice (as snow), and vapor (as clouds).

Evaporation from the leaves

Flow of water

▲ Figure 2.10 Cohesion and water transport in plants. The evaporation of water from leaves pulls water upward from the roots through microscopic tubes in the trunk of the tree. Because of cohesion, the pulling force is relayed through the tubes all the way down to the roots. As a result, water rises against the force of gravity.

Microscopic water-conducting tubes

Cohesion by hydrogen bonding between water molecules

Colorized SEM 150×

▼ Figure 2.11 A raft spider walking on water. The cumulative strength of hydrogen bonds between water molecules allows this spider to walk on pond water without breaking the surface.

Other liquids have much weaker surface tension; an insect, for example, could not walk on the surface of a cup of gasoline (which is why gardeners sometimes use gasoline to drown bugs removed from bushes).

▼ **Figure 2.12 Sweating as a mechanism of evaporative cooling.**

How Water Moderates Temperature

If you've ever burned your finger on a metal pot while waiting for the water in it to boil, you know that water heats up much more slowly than metal. In fact, because of hydrogen bonding, water has a stronger resistance to temperature change than most other substances.

When water is heated, the heat energy first disrupts hydrogen bonds and then makes water molecules jostle around faster. The temperature of the water doesn't go up until the water molecules start to speed up. Because heat is first used to break hydrogen bonds rather than raise the temperature, water absorbs and stores a large amount of heat while warming up only a few degrees. Conversely, when water cools, hydrogen bonds form in a process that releases heat. Thus, water can release a relatively large amount of heat to the surroundings while the water temperature drops only slightly.

Earth's giant water supply—the oceans, seas, lakes, rivers, and subsurface water—enables temperatures on the planet to stay within limits that permit life by storing a huge amount of heat from the sun during warm periods and giving off heat that warms the air during cold periods. That's why coastal areas generally have milder climates than inland regions. Water's resistance to temperature change also stabilizes ocean temperatures, creating a favorable environment for marine life. You may have noticed that the temperature of the ocean fluctuates much less than the air temperature.

Another way that water moderates temperature is by **evaporative cooling**. When a substance evaporates (changes from a liquid to a gas), the surface of the liquid that remains cools down. This cooling occurs because the molecules with the greatest energy (the "hottest" ones) tend to vaporize first. Think of it like this: If the five fastest runners on your track team quit school, it would lower the average speed of the remaining team. Evaporative cooling helps prevent some land-dwelling creatures from overheating; it's why sweating helps you dissipate excess body heat **(Figure 2.12)**. And the expression "It's not the heat; it's the humidity" has its basis in the fact that sweat evaporates more slowly when the air is already saturated with water vapor, so the cooling effect is inhibited.

The Biological Significance of Ice Floating

When most liquids get cold, their molecules move closer together. If the temperature is cold enough, the liquid freezes and becomes a solid. Water, however, behaves differently. When water molecules get cold enough, they move apart, with each molecule staying at "arm's length" from its neighbors, forming ice. A chunk of ice floats because it is less dense than the liquid water in which it is floating. Floating ice is a consequence of hydrogen bonding. In contrast to the short-lived and constantly changing hydrogen bonds in liquid water, the hydrogen bonds that form in solid ice last longer, with each molecule bonded to four neighbors. As a result, ice is a spacious crystal **(Figure 2.13)**.

▶ **Figure 2.13 Why ice floats.** Compare the tightly packed molecules in liquid water with the spaciously arranged molecules in the ice crystal. The less dense ice floats atop the denser water.

Liquid water
Hydrogen bonds constantly break and re-form.

Hydrogen bond

Ice
Stable hydrogen bonds hold molecules apart, making ice less dense than water.

How does floating ice help support life on Earth? When a deep body of water cools and a layer of ice forms on top, the floating ice acts as an insulating "blanket" over the liquid water, allowing life to persist under the frozen surface. But imagine what would happen if ice were denser than water: Ice would sink during winter. All ponds, lakes, and even the oceans would eventually freeze solid without the insulating protection of the top layer of ice. Then, during summer, only the upper few inches of the oceans would thaw. It's hard to imagine life surviving under such conditions.

Water as the Solvent of Life

If you've ever stirred sugar into coffee or added salt to soup, you know that you can dissolve sugar or salt in water. This results in a mixture known as a **solution**, a liquid consisting of a homogeneous mixture of two or more substances. The dissolving agent is called the **solvent**, and any substance that is dissolved is called a **solute**. When water is the solvent, the solution is called an **aqueous solution**. The fluids of organisms are aqueous solutions. For example, tree sap is an aqueous solution consisting of sugar and mineral solutes dissolved in water solvent.

Water can dissolve an enormous variety of solutes necessary for life, providing a medium for chemical reactions.

For example, water can dissolve salt ions, as shown in **Figure 2.14**. Each ion becomes surrounded by oppositely charged regions of water molecules. Solutes that are polar molecules, such as sugars, dissolve by orienting locally charged regions of their molecules toward water molecules in a similar way.

We have discussed four special properties of water, each a consequence of water's unique chemical structure. Next, we'll look at aqueous solutions in more detail. ✓

Acids, Bases, and pH

In aqueous solutions, most of the water molecules are intact. However, a very small percentage of the water molecules break apart into hydrogen ions (H^+) and hydroxide ions (OH^-). A balance of these two highly reactive ions is critical for the proper functioning of chemical processes within organisms.

A chemical compound that releases H^+ to a solution is called an **acid**. One example of a strong acid is hydrochloric acid (HCl), the acid in your stomach that aids in digestion of food. In solution, HCl breaks apart into the ions H^+ and Cl^-. A **base** (or alkali) is a compound that accepts H^+ ions and removes them from a solution. Some bases, such as sodium hydroxide (NaOH), do this by releasing OH^-, which combines with H^+ to form H_2O.

To describe the acidity of a solution, chemists use the **pH scale**, a measure of the hydrogen ion H^+ concentration in a solution. The scale ranges from 0 (most acidic) to 14 (most basic). Each pH unit represents a tenfold change in the concentration of H^+. For example, lemon juice at pH 2 has 100 times more H^+ than an equal amount of tomato juice at pH 4. Aqueous solutions that are neither acidic nor basic (such as pure water) are said to be neutral; they have a pH of 7. They do contain some H^+ and OH^-, but the concentrations of the two ions are

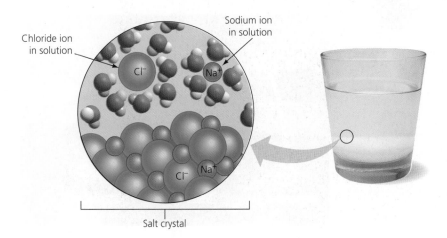

◀ **Figure 2.14 A crystal of table salt (NaCl) dissolving in water.** As a result of electrical charge attractions, H_2O molecules surround the sodium and chloride ions, dissolving the crystal in the process.

equal. The pH of the solution inside most living cells is close to 7. **Figure 2.15** provides an overview of the pH scale, with examples of acidic and basic solutions.

Even a slight change in pH can be harmful to an organism because the molecules in cells are extremely sensitive to H^+ and OH^- concentrations. Biological fluids contain **buffers**, substances that minimize changes in pH by accepting H^+ when that ion is in excess and donating H^+ when it is depleted. For example, buffer in contact lens solution helps protect the surface of the eye from potentially painful changes in pH. This buffering process, however, is not foolproof, and changes in environmental pH can profoundly affect ecosystems. For instance, about 25% of the carbon dioxide (CO_2) generated by people (primarily by burning fossil fuels) is absorbed by the oceans. When CO_2 dissolves in seawater, it reacts with water to form carbonic acid **(Figure 2.16)**, which lowers ocean pH. The resulting ocean acidification can alter marine environments. Oceanographers have calculated that the pH of the ocean is lower now than at any time in the past 420,000 years, and it is continuing to drop.

The effects of ocean acidification—including coral bleaching and changes in metabolism among a wide variety of sea creatures—are daunting reminders that the chemistry of life is linked to the chemistry of the environment. It reminds us, too, of the importance of interactions within biological systems. Industrial processes that affect the atmosphere in one region of the world can cause ecosystem changes underwater in another part of the world. ☑

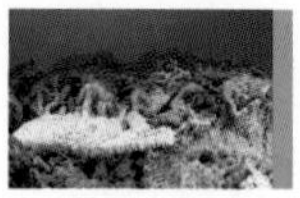

CAR EXHAUST IN BOSTON CONTRIBUTES TO DEATH OF CORAL IN AUSTRALIA.

▼ **Figure 2.15 The pH scale.** A solution having a pH of 7 is neutral, meaning that its H^+ and OH^- concentrations are equal. The lower the pH below 7, the more acidic the solution, or the greater its excess of H^+ compared with OH^-. The higher the pH above 7, the more basic the solution, or the greater the deficiency of H^+ relative to OH^-.

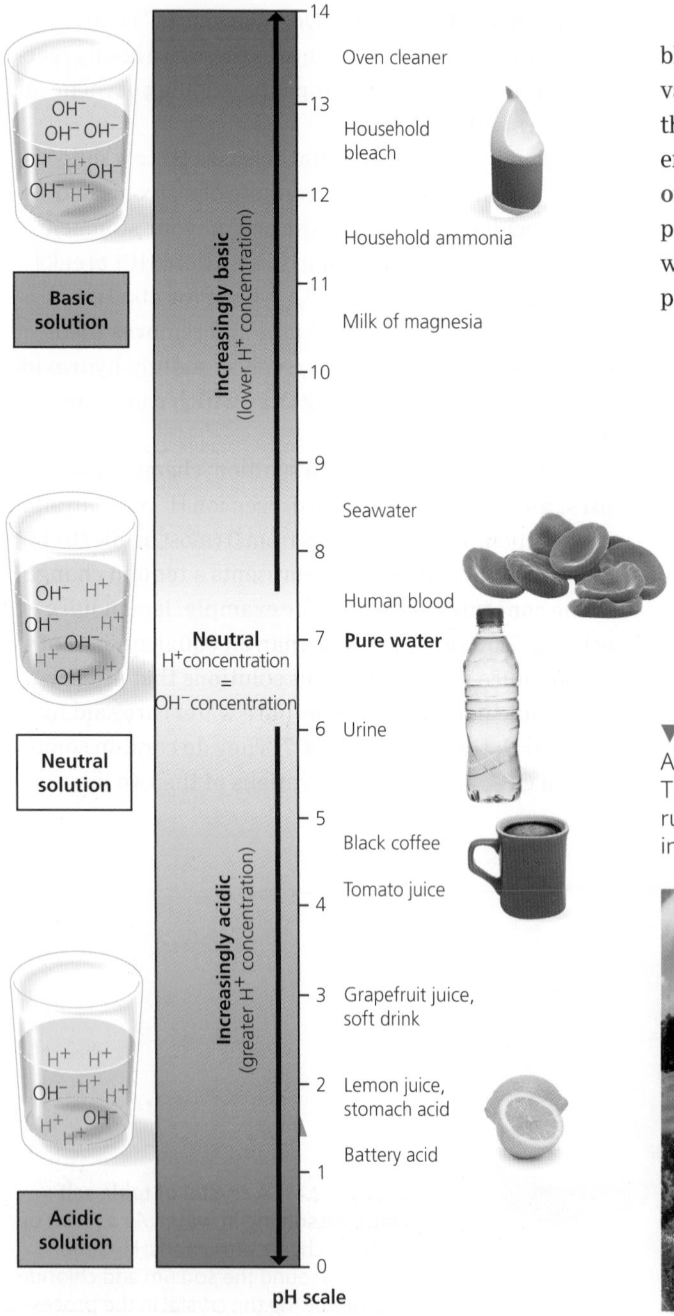

Basic solution

Increasingly basic (lower H^+ concentration)

- 14
- 13 — Oven cleaner
- Household bleach
- 12
- Household ammonia
- 11
- Milk of magnesia
- 10
- 9
- Seawater
- 8
- Human blood

Neutral
H^+ concentration
=
OH^- concentration

Neutral solution

- 7 — **Pure water**
- 6 — Urine
- 5
- Black coffee
- Tomato juice
- 4

Increasingly acidic (greater H^+ concentration)

- 3 — Grapefruit juice, soft drink
- 2 — Lemon juice, stomach acid
- Battery acid
- 1
- 0

Acidic solution

pH scale

▼ **Figure 2.16 Ocean acidification by atmospheric CO_2.** After dissolving in seawater, CO_2 reacts to form carbonic acid. This acid then undergoes further chemical reactions that disrupt coral growth. Such acidification can cause drastic changes in important marine ecosystems.

CO_2 Carbon dioxide

CO_2 Carbon dioxide + H_2O Water → H_2CO_3 Carbonic acid

Radioactivity as an Evolutionary Clock

Throughout this chapter, you have learned that radiation can be harmful when uncontrolled but is also helpful in food safety and in medicine. Another way that radiation can serve as a helpful tool involves the natural process of radioactive decay, which can be used to obtain important data about the evolutionary history of life on Earth.

Fossils—both preserved imprints and remains of dead organisms—are reliable chronological records of life because we can determine their ages through radiometric dating **(Figure 2.17)**, which is based on the decay of radioactive isotopes. The time it takes for 50% of a radioactive isotope to decay is called the **half-life** of the isotope. For example, carbon-14 is a radioactive isotope with a half-life of 5,700 years. It is present in trace amounts in the environment. **1** A living organism assimilates the different isotopes of an element in proportions that reflect their relative abundances in the environment. In this example, carbon-14 is taken up in trace quantities, along with much larger quantities of the more common carbon-12. **2** At the time of death, the organism ceases to take in carbon from the environment. From this moment forward, the amount of carbon-14 relative to carbon-12 in the fossil declines: The carbon-14 in the body decays into carbon-12, but no new carbon-14 is added. Because the half-life of carbon-14 is known, the ratio of the two isotopes (carbon-14 to carbon-12) is a reliable indicator of the age of the fossil. In this case, it takes 5,700 years for half of the radioactive carbon-14 to decay, half of the remainder is present after another 5,700 years, and so on. **3** A fossil's age can be estimated by measuring the ratio of the two isotopes to learn how many half-life reductions have occurred since it died. For example, if the ratio of carbon-14 to carbon-12 in this fossil was found to be one-eighth that of the environment, this fossil would be about 17,100 (5,700 × 3 half-lives) years old.

Using such techniques, scientists can estimate the ages of fossils and place them in an ordered sequence called the fossil record. As you'll see in Chapter 14, the fossil record is one of the most important and convincing sets of evidence that led Charles Darwin to formulate the theory of evolution by natural selection.

▼ **Figure 2.17 Radiometric dating.** Living organisms incorporate the carbon-14 isotope (represented here with blue dots). But no new carbon-14 is introduced once an organism dies, and the carbon-14 that remains slowly decays into carbon-12. By measuring the amount of carbon-14 in a fossil, scientists can estimate its age.

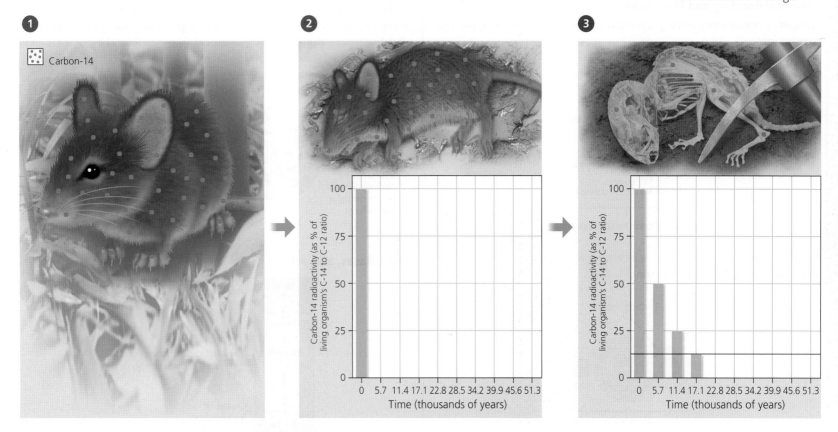

Chapter Review

SUMMARY OF KEY CONCEPTS

Some Basic Chemistry

Matter: Elements and Compounds

Matter consists of elements, which cannot be broken down, and compounds, which are combinations of two or more elements. Of the 25 elements essential for life, oxygen, carbon, hydrogen, and nitrogen are the most abundant in living matter.

Atoms

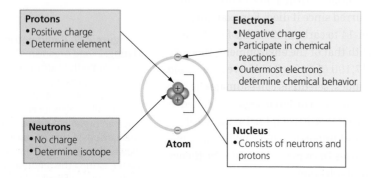

Protons
• Positive charge
• Determine element

Electrons
• Negative charge
• Participate in chemical reactions
• Outermost electrons determine chemical behavior

Neutrons
• No charge
• Determine isotope

Atom

Nucleus
• Consists of neutrons and protons

Chemical Bonding and Molecules

Transfer of one or more electrons produces attractions between oppositely charged ions:

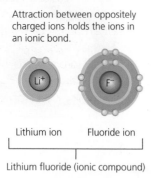

Attraction between oppositely charged ions holds the ions in an ionic bond.

Lithium ion   Fluoride ion

Lithium fluoride (ionic compound)

A molecule consists of two or more atoms connected by covalent bonds, which are formed by electron sharing:

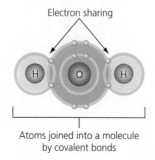

Electron sharing

H   O   H

Atoms joined into a molecule by covalent bonds

Water is a polar molecule; the slightly positively charged H atoms in one water molecule may be attracted to the partial negative charge of O atoms in neighboring water molecules, forming weak but important hydrogen bonds:

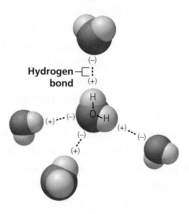

Hydrogen—⌐ ⁝
bond   (−)
           (+)

H
|
O—H
(+)···(−)   (+)···(−)
(−)
(+)

Chemical Reactions

By breaking bonds in reactants and forming new bonds in products, chemical reactions rearrange matter.

Water and Life

Water

The cohesion (sticking together) of water molecules is essential to life. Water moderates temperature by absorbing heat in warm environments and releasing heat in cold environments. Evaporative cooling also helps stabilize the temperatures of oceans and organisms. Ice floats because it is less dense than liquid water, and the insulating properties of floating ice prevent the oceans from freezing solid. Water is a very good solvent, dissolving a great variety of solutes to produce aqueous solutions.

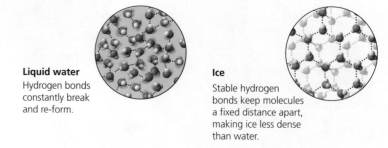

Liquid water
Hydrogen bonds constantly break and re-form.

Ice
Stable hydrogen bonds keep molecules a fixed distance apart, making ice less dense than water.

Acids, Bases, and pH

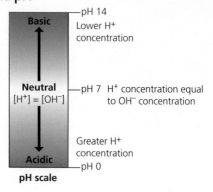

Basic — pH 14
Lower H⁺ concentration

Neutral — pH 7 H⁺ concentration equal
$[H^+] = [OH^-]$ to OH⁻ concentration

Greater H⁺ concentration

Acidic — pH 0

pH scale

Mastering Biology

For practice quizzes, BioFlix animations, MP3 tutorials, video tutors, and more study tools designed for this textbook, go to MasteringBiology™

SELF-QUIZ

1. An atom can be changed into an ion by adding or removing _____. An atom can be changed into a different isotope by adding or removing _____.

2. If you change the number of _____, the atom becomes a different element.

3. A sodium atom has 11 protons, and the most common isotope of sodium has 12 neutrons. A radioactive isotope of sodium has 11 neutrons. What are the atomic numbers and mass numbers of the stable and radioactive forms of sodium?

4. Why are radioactive isotopes useful as tracers in research on the chemistry of life?

5. What is chemically nonsensical about this structure?

$$H - C = C - H$$

6. Why is it unlikely that two neighboring water molecules would be arranged like this?

7. Which of the following is not a chemical reaction?
   a. Sugar ($C_6H_{12}O_6$) and oxygen gas (O_2) combine to form carbon dioxide (CO_2) and water (H_2O).
   b. Sodium metal and chlorine gas unite to form sodium chloride.
   c. Hydrogen gas combines with oxygen gas to form water.
   d. Ice melts to form liquid water.

8. Some people in your study group say they don't understand why water is considered a polar molecule. You explain that water is a polar molecule because
   a. the oxygen atom is found between the two hydrogen atoms.
   b. the oxygen atom attracts the hydrogen atoms.
   c. the oxygen end of the molecule has a slight negative charge, and the hydrogen end has a slight positive charge.
   d. both hydrogen atoms are at one end of the molecule, and the oxygen atom is at the other end.

9. Explain how the unique properties of water result from the fact that water is a polar molecule.

10. Explain why it is unlikely that life could exist on a planet where the main liquid is methane, a nonpolar molecule that does not form hydrogen bonds.

11. A can of cola consists mostly of sugar dissolved in water, with some carbon dioxide gas that makes it fizzy and makes the pH less than 7. Describe the cola using the following terms: solute, solvent, acidic, aqueous solution.

For answers to the Self Quiz, see Appendix D.

IDENTIFYING MAJOR THEMES

For each statement, identify which major theme is evident (the relationship of structure to function, information flow, pathways that transform energy and matter, interactions within biological systems, or evolution) and explain how the statement relates to the theme. If necessary, review the theme descriptions (see Chapter 1) and review the examples highlighted in blue in this chapter.

12. Cells are constantly rearranging molecules by breaking existing chemical bonds and forming new ones.

13. The polarity of water molecules helps explain water's ability to support life.

14. The release of CO_2 from a factory in one region can cause ocean acidification that affects coral reefs in another region.

For answers to Identifying Major Themes, see Appendix D.

THE PROCESS OF SCIENCE

15. *Helicobacter pylori*, a bacterium that causes gastric inflammation, produces urease, an enzyme that splits urea into carbon dioxide (CO_2) and ammonia. Humans do not produce urease. The "urea breath test" uses carbon isotope labeling to rapidly detect the presence of *Helicobacter pylori* in the intestinal tract of a patient. How might this test work?

16. **Interpreting Data** As shown in Figure 2.17, radiometric dating can be used to determine the age of biological materials. Carbon-14 makes up about 1 part per trillion of naturally occurring carbon. When an organism dies, its body contains 1 carbon-14 atom per trillion total carbon atoms. After 5,700 years (the half-life of carbon-14), its body contains one-half as much carbon-14; the other half has decayed and is no longer present. French scientists determined that prehistoric cave paintings were made using natural dyes about 13,000 years ago. How much carbon-14 must the scientists have found in the cave paintings to support this result? Express your answer in both percentage of carbon-14 remaining and in parts of carbon-14 per trillion.

BIOLOGY AND SOCIETY

17. Critically evaluate this statement: "It's paranoid to worry about contaminating the environment with chemical wastes; this stuff is just made of the same atoms that were already present in our environment."

18. A major source of the CO_2 that causes ocean acidification is emissions from coal-burning power plants. One way to reduce these emissions is to use nuclear power to produce electricity. The proponents of nuclear power contend that it is the only way that the U.S. can increase its energy production while reducing air pollution because nuclear power plants emit few airborne pollutants. What are some of the benefits of nuclear power? What are the possible costs and dangers? Should we increase our use of nuclear power to generate electricity? If a new power plant were to be built near your home, would you prefer it to be a coal-burning plant or a nuclear plant? Explain.

3 The Molecules of Life

WHAT MAKES CHOCOLATE SO GREAT? IT MELTS IN YOUR MOUTH BECAUSE COCOA BUTTER IS HIGH IN SATURATED FATS.

Why Macromolecules Matter

Every biological system, including your own body, can be viewed as a large collection of chemicals. Although living cells contain countless types of molecules, many of the most important ones can be grouped into a few broad categories. Understanding biological molecules sheds light on many important issues, such as the link between diet and health.

WHAT DOES YOUR HAIR HAVE IN COMMON WITH A STEAK? THEY ARE BOTH MADE OF PROTEINS.

THE STRUCTURE OF YOUR DNA IS NEARLY INDISTINGUISHABLE FROM THE DNA OF A MOSQUITO OR AN ELEPHANT. THE DIFFERENCES IN SPECIES RESULT FROM THE WAY NUCLEOTIDES ARE ARRANGED.

BIOLOGY AND SOCIETY   Lactose Intolerance

Got Lactose?

You've probably seen ads that associate a milk mustache with good health. Indeed, milk is a very healthy food: It's rich in protein, minerals, and vitamins and can be low in fat. But most people begin to lose the ability to digest milk at around age 2. In fact, for most adults, a glass of milk delivers a heavy dose of digestive discomfort that can include bloating, gas, and abdominal pain. These are symptoms of lactose intolerance, the inability to digest lactose, the main sugar found in milk.

For people with lactose intolerance, the problem starts when lactose enters the small intestine. To absorb this sugar, digestive cells must produce a molecule called lactase. Lactase is an enzyme, a protein that helps speed up chemical reactions—in this case, the breakdown of lactose into smaller sugars. During childhood, lactase levels in most people decline significantly. Lactose that is not broken down in the small intestine passes into the large intestine, where bacteria feed on it and belch out gaseous by-products. An accumulation of gas produces uncomfortable symptoms. Sufficient lactase can thus mean the difference between delight and discomfort when you eat an ice cream cone.

There is no treatment for the underlying cause of lactose intolerance: the decreased production of lactase. What, then, are the options for people who do not produce enough of this enzyme? The first option is avoiding lactose-containing foods; abstinence is always the best form of protection! Alternatively, substitutes are available, such as milk made from soy, almonds, or coconuts, or cow's milk that has been pretreated with lactase. If you are lactose intolerant and you crave ice cream, you can buy lactose-free ice cream or you can make your own from lactose-free cream. Also, lactase in pill form can be taken along with food to ease digestion by artificially providing the enzyme that the body naturally lacks.

Lactose intolerance illustrates one way the interplay of biological molecules can affect your health. Such molecular interactions, repeated in countless variations, drive all biological processes. In this chapter, we'll explore the structure and function of large molecules that are essential to life. We'll start with an overview of carbon and then examine four classes of molecules: carbohydrates, lipids, proteins, and nucleic acids. Along the way, we'll look at where these molecules occur in your diet and the important roles they play in your body.

Lactose-rich ice cream. Lactose digestion arises from the interactions between several classes of the body's molecules.

Organic Compounds

A cell is mostly made up of water, but the rest is mainly carbon-based molecules. When it comes to the chemistry of life, carbon plays the leading role. This is because carbon is unparalleled in its ability to form the skeletons of large, complex, diverse molecules that are necessary for life's functions. The study of carbon-based molecules, which are called **organic compounds**, lies at the heart of any study of life.

Carbon Chemistry

Carbon is a versatile molecule ingredient because an atom of carbon can share electrons with other atoms in four covalent bonds that can branch off in four directions. Because carbon can use one or more of its bonds to attach to other carbon atoms, it is possible to construct an endless diversity of carbon skeletons varying in size and branching pattern (**Figure 3.1**). Thus, molecules with multiple carbon "intersections" can form very elaborate shapes. The carbon atoms of organic compounds can also bond with other elements, most commonly hydrogen, oxygen, and nitrogen.

One of the simplest organic compounds is methane, CH_4, with a single carbon atom bonded to four hydrogen atoms (**Figure 3.2**). Methane is abundant in natural gas and is also produced by prokaryotes that live in swamps (in the form of swamp gas) and in the digestive tracts of grazing animals, such as cows. Larger organic compounds

▼ **Figure 3.2 Three representations of methane, a simple organic compound.**

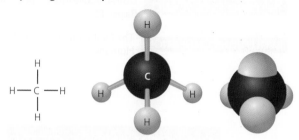

Structural formula   Ball-and-stick model   Space-filling model

(such as octane, with eight carbons) are the main molecules in the gasoline we burn in combustion engines. Organic compounds are also important fuels in your body; the energy-rich parts of fat molecules have a structure similar to gasoline (**Figure 3.3**).

The unique properties of an organic compound depend not only on its carbon skeleton but also on the atoms attached to the skeleton. In an organic compound, the groups of atoms directly involved in chemical reactions are called **functional groups**. Each functional group plays a particular role during chemical reactions. Two examples of functional groups are the hydroxyl group (—OH, found in alcohols such as isopropyl rubbing alcohol) and the carboxyl group (—COOH, found in all proteins). Many biological molecules have two or more functional groups. Keeping in mind this basic scheme—carbon skeletons with attached functional groups—we are now ready to explore how our cells make large molecules out of smaller ones.

▼ **Figure 3.1 Variations in carbon skeletons.** Notice that each carbon atom forms four bonds and each hydrogen atom forms one bond. One line represents a single bond (sharing of one pair of electrons), and two lines represent a double bond (sharing of two pairs of electrons).

Carbon skeletons vary in length

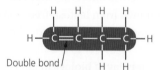

Double bond

Carbon skeletons may have double bonds, which can vary in location

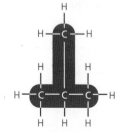

Carbon skeletons may be unbranched or branched

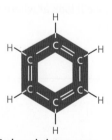

Carbon skeletons may be arranged in rings

▼ **Figure 3.3 Hydrocarbons as fuel.** Energy-rich organic compounds in gasoline provide fuel for engines, and energy-rich molecules in fats provide fuel for cells.

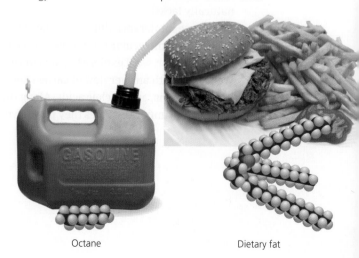

Octane   Dietary fat

Giant Molecules from Smaller Building Blocks

The important molecules of all living things—from bacteria to whales—fall into just four main classes: carbohydrates, such as those found in starchy foods like French fries and bagels; lipids, such as fats and oils; proteins, such as enzymes and the components of your hair; and nucleic acids: DNA and RNA. On a molecular scale, the members of three of these classes—carbohydrates, proteins, and nucleic acids—are gigantic; in fact, biologists call them **macromolecules** (*macro* means "big"). Despite the size of macromolecules, their structures can be easily understood because they are **polymers**, large molecules made by stringing together many smaller molecules called **monomers**. A polymer is like a lengthy train made from many box car monomers. Although at first the structure of a long train may seem very complex, it can be easily understood by asking two questions: What kind of cars make up the train? How are they joined together? Even a very long train may only have a few components (engine at one end, caboose at the other, passenger cars in the middle). Similarly, even large and complex biological macromolecules can be understood by considering which monomers make them up and how those monomers are attached together.

Cells link monomers together to form a polymer through a **dehydration reaction**; this is analogous to assembling a long train from the individual cars. As the name implies, a dehydration reaction involves removing a molecule of water **(Figure 3.4a)**. For each monomer added to a chain, a water molecule (H_2O) is formed by the release of two hydrogen atoms and one oxygen atom from the reactants—imagine a splash of water being released every time two train cars are coupled. This same type of dehydration reaction occurs regardless of the specific monomers involved and the type of polymer the cell is producing. Therefore, you'll see this type of reaction repeated many times throughout this chapter.

Organisms not only make macromolecules but also break them down. For example, you must digest macromolecules in food to make their monomers available to your cells, which can then rebuild the monomers into the macromolecules that make up your body. The food macromolecules are like trains being separated into individual box cars (monomers) that are then rearranged into new trains (your own body's molecules). The breakdown of polymers occurs by a process called **hydrolysis** **(Figure 3.4b)**. The term *hydrolysis* means water breakage (from the Greek *hydro*, water, and *lysis*, break). Cells break bonds between monomers by adding water to them, a process that is essentially the reverse of a dehydration reaction. A macromolecule "train" can be disassembled by adding water to break the connections between the monomer box cars. You learned in the Biology and Society section about one real-world example of a hydrolysis reaction: the breakdown of lactose into its monomers by the enzyme lactase. ☑

☑ CHECKPOINT

1. Chemical reactions that build polymers from _____ are called _____ reactions because they release a molecule of water.
2. The reverse type of reaction, one that breaks larger molecules down into smaller molecules, is called _____.

Answers: 1. monomers; dehydration 2. hydrolysis

▼ **Figure 3.4** **Synthesis and breakdown of polymers.** For simplicity, the only atoms shown in these diagrams are hydrogens and hydroxyl groups (—OH) in strategic locations.

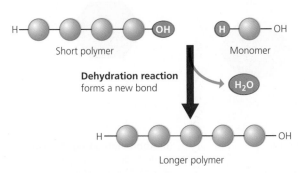

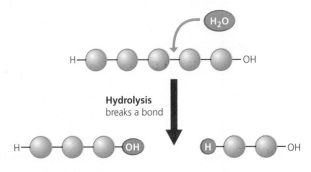

(a) **Building a polymer chain.** A polymer grows in length when an incoming monomer and the monomer at the end of the polymer each contribute atoms to form a water molecule. The monomers replace the lost covalent bonds with a bond between each other.

(b) **Breaking a polymer chain.** Hydrolysis reverses the process by adding a water molecule, which breaks the bond between two monomers, creating two smaller molecules from one bigger one.

Figure Walkthrough

Mastering **Biology**
goo.gl/rC8XZT

Large Biological Molecules

As previously noted, four classes of large biological molecules are found in all living organisms: carbohydrates, lipids, proteins, and nucleic acids. **For each class, we'll explore the structure and function of these large molecules by first learning about the smaller molecules used to build them.**

Carbohydrates

Carbohydrates, often referred to as "carbs," are a class of molecules that includes sugars and polymers of sugars. Some examples are the small sugar molecules dissolved in soft drinks and the long starch molecules in spaghetti and bread. In animals, carbohydrates are a primary source of dietary energy and raw material for manufacturing other kinds of organic compounds; in plants, carbohydrates serve as a building material for much of the plant body.

Monosaccharides

Simple sugars, or **monosaccharides** (from the Greek *mono*, single, and *sacchar*, sugar), are the monomers of carbohydrates; they cannot be broken down into smaller sugars. Common examples are glucose, found in energy drinks, and fructose, found in fruit **(Figure 3.5)**. Both of these simple sugars are also in honey. The molecular formula for glucose is $C_6H_{12}O_6$. Fructose has the same formula, but its atoms are arranged differently. Glucose and fructose are examples of **isomers**, molecules that have the same molecular formula but different structures. Isomers are like anagrams—words that contain the same letters in a different order, such as *heart* and *earth*. Because molecular shape is so important, seemingly minor differences in the arrangement of atoms give isomers different properties, such as how they react with other molecules. In this case, the rearrangement of functional groups makes fructose taste much sweeter than glucose.

It is convenient to draw monosaccharides as if their carbon skeletons were linear. However, many simple sugars form rings when dissolved in water **(Figure 3.6)**. You'll notice this ring shape in the depiction of the structure of many carbohydrates in this chapter.

Monosaccharides, particularly glucose, are the main fuel molecules for cellular work. Like an automobile engine consuming gasoline, your cells break down glucose molecules and extract their stored energy, giving off carbon dioxide as "exhaust." The rapid conversion of glucose to cellular energy is why a solution of glucose dissolved in water (often called dextrose) is given as an IV to sick or injured patients; the glucose provides an immediate energy source to tissues in need of repair. Cells also use the carbon skeletons of glucose as raw material for making other kinds of organic molecules. The use of simple sugars as both energy resources and organic building blocks clearly illustrates the theme of transformations of energy and matter that constantly occur in all biological systems. ☑

☑ **CHECKPOINT**

1. All carbohydrates consist of one or more _____, also called simple sugars. Name two examples of simple sugars.

2. Explain how glucose and fructose can have the same formula ($C_6H_{12}O_6$) but different properties.

■ *Answers: 1. monosaccharides; glucose and fructose 2. Different arrangements of atoms affect molecular shapes and properties.*

▼ **Figure 3.5 Monosaccharides (simple sugars).** Glucose and fructose are isomers, molecules with the same atoms arranged differently.

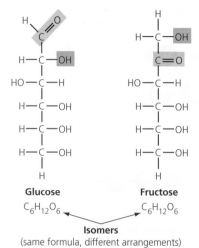

Glucose
$C_6H_{12}O_6$

Fructose
$C_6H_{12}O_6$

Isomers
(same formula, different arrangements)

▼ **Figure 3.6 The ring structure of glucose.**

(a) Linear and ring structures. The carbon atoms are numbered so you can relate the linear and ring versions of the molecule. As the double arrows indicate, ring formation is a reversible process, but at any instant in an aqueous solution, most glucose molecules are rings.

(b) Abbreviated ring structure. In this book, we'll use this abbreviated ring symbol for glucose. Each unmarked corner represents a carbon and its attached atoms.

Disaccharides

A **disaccharide**, or double sugar, is constructed from two monosaccharides by a dehydration reaction. The disaccharide lactose, sometimes called "milk sugar," is made from the monosaccharides glucose and galactose **(Figure 3.7)**. Another common disaccharide is maltose, naturally found in germinating seeds. It is used in making beer, liquor, malted milk shakes, and malted milk ball candy. A molecule of maltose consists of two glucose monomers joined together.

The most common disaccharide is sucrose (table sugar), which consists of a glucose monomer linked to a fructose monomer. Sucrose is the main carbohydrate in plant sap, and it nourishes all the parts of the plant. Sucrose can be extracted from the stems of sugarcane, but the vast majority of table sugar consumed in the United States is derived from the roots of sugar beets. Another common sweetener is high-fructose corn syrup (HFCS), made through a commercial process that uses an enzyme to convert natural glucose in corn syrup to the much sweeter fructose. HFCS is a clear, goopy liquid containing about 55% fructose; it is much cheaper than sucrose and easier to mix into drinks and processed foods.

The average American consumes a whopping 130 pounds of sugar each year. This national "sweet tooth" persists despite our growing awareness about how sugar can negatively affect our health. Sugar is a major cause of tooth decay, and overconsumption increases the risk of developing diabetes and heart disease. Moreover, high sugar consumption tends to replace eating more varied and nutritious foods. The description of sugars as "empty calories" is accurate in the sense that most sweeteners contain only negligible amounts of nutrients other than carbohydrates. Accordingly, food labels will be revised to list both total sugars and sugars added during food processing **(Figure 3.8)**. The government's dietary guidelines suggest that we all keep added sugars to less than 10% of our daily calories. For good health, we also require proteins, fats, vitamins, and minerals. And we need to include substantial amounts of complex carbohydrates—that is, polysaccharides—in our diet. Let's examine these macromolecules next. ✓

✓ **CHECKPOINT**

1. Table sugar, formally called _____, is an example of a _____, or double sugar.

2. How and why do manufacturers produce HFCS?

■ Answers: **1.** sucrose; disaccharide **2.** They convert glucose to the sweeter fructose. HFCS is cheaper and more easily blended with processed foods.

▼ **Figure 3.7 Disaccharide (double sugar) formation.**
To form a disaccharide, two simple sugars are joined by a dehydration reaction, in this case forming a bond between monomers of glucose and galactose to make the disaccharide lactose.

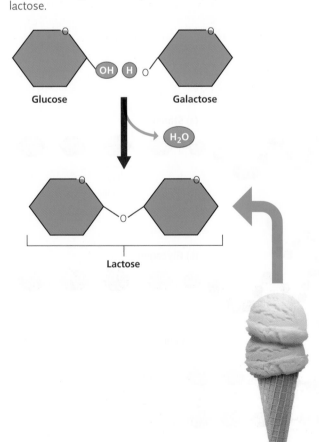

▼ **Figure 3.8 A revised food label displaying the amount of added sugars.**

Sugars added during manufacture should account for less than 10% of total daily calories.

Polysaccharides

Complex carbohydrates, or **polysaccharides**, are long chains of sugars—*poly*mers of mono*saccharides*. One familiar example is starch, a storage polysaccharide found in plants. **Starch** consists of long strings of glucose monomers **(Figure 3.9a)**. Starch granules serve as carbohydrate "storage tanks" from which plant cells can withdraw glucose for energy or building materials. Potatoes and grains, such as wheat, corn, and rice, are the major sources of starch in our diet. Animals digest starch when enzymes within the digestive system break the bonds between glucose monomers through hydrolysis reactions.

Animals store excess glucose in the form of a polysaccharide called **glycogen**. Like starch, glycogen is a polymer of glucose monomers, but glycogen is more extensively branched **(Figure 3.9b)**. Most of your glycogen is stored in liver and muscle cells, which break down the glycogen to release glucose when you need energy. This is why some athletes "carbo-load," consuming large amounts of starchy foods the night before an athletic event. The starch is converted to glycogen, which is then available for rapid use during physical activity the next day. After a day or so, unused glycogen is broken down and its energy is then used to make fats for long-term storage—so don't overdo the carbo-loading!

Cellulose, the most abundant organic compound on Earth, forms cable-like fibrils in the tough walls that enclose plant cells and is a major component of wood and other structural components of plants **(Figure 3.9c)**. We take advantage of that structural strength when we use timber as a building material. Cellulose is also a polymer of glucose, but its glucose monomers are linked together in a unique way. Unlike the glucose linkages in starch and glycogen, those in cellulose cannot be broken by any enzyme produced by animals. Grazing animals are able to derive nutrition from cellulose because microorganisms inhabiting their digestive tracts break it down. The cellulose in plant foods that you eat, commonly known as dietary fiber (your grandma might call it "roughage") passes through your digestive tract unchanged and therefore provides no nutritional benefits (or calories). But fiber does help keep your digestive system healthy. The passage of cellulose stimulates cells lining the digestive tract to secrete mucus, which allows food to pass smoothly. The health benefits of dietary fiber include lowering the risk of heart disease, diabetes, and gastrointestinal disease. However, most Americans do not get the recommended levels of fiber in their diet. Foods rich in fiber include fruits, vegetables, whole grains, bran, and beans.

▼ Figure 3.9 **Three common polysaccharides.**

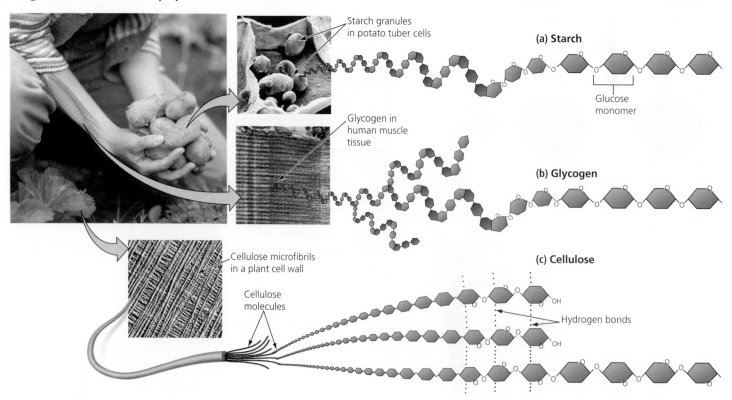

Starch granules in potato tuber cells

(a) Starch

Glucose monomer

Glycogen in human muscle tissue

(b) Glycogen

Cellulose microfibrils in a plant cell wall

Cellulose molecules

(c) Cellulose

OH

Hydrogen bonds

OH

Lipids

Almost all carbohydrates are **hydrophilic** ("water-loving") molecules that dissolve readily in water. In contrast, **lipids** are **hydrophobic** ("water-fearing"); they do not mix with water. You've probably seen this chemical behavior when you combine oil and vinegar: The oil, which is a type of lipid, separates from the vinegar, which is mostly water (**Figure 3.10**). If you shake vigorously, you can force a temporary mixture long enough to douse your salad, but what remains in the bottle will quickly separate. Lipids also differ from carbohydrates, proteins, and nucleic acids in that they are neither huge macromolecules nor are they necessarily polymers built from repeating monomers. Lipids are a diverse group of molecules made from different molecular building blocks. In this section, we'll look at two types of lipids: fats and steroids.

Fats

A typical **fat** consists of a glycerol molecule joined with three fatty acid molecules by dehydration reactions (**Figure 3.11a**). The resulting fat molecule is called a **triglyceride** (**Figure 3.11b**), a term you may hear in the results of a blood test. A fatty acid is a long molecule (typically a chain of 16–18 carbons) that stores a lot of energy. A pound of fat packs more than twice as much energy as a pound of carbohydrate. The downside to this energy efficiency is that it is very difficult for a person trying to lose weight to "burn off" excess fat. We stock these long-term food stores in specialized reservoirs called adipose cells, which swell and shrink when we deposit and withdraw fat from them. This adipose tissue, or body fat (also called blubber), not only stores energy but also cushions vital organs and insulates us, helping maintain a constant, warm body temperature.

Notice in Figure 3.11b that the bottom fatty acid bends where there is a double bond in the carbon skeleton. That fatty acid is **unsaturated** because it has fewer than the maximum number of hydrogens at the double bond. The other two fatty acids in the fat molecule lack double bonds in their tails. Those fatty acids are **saturated**, meaning that they contain the maximum number of hydrogen atoms, giving them a straight shape. A saturated fat is one with all three of its fatty acid tails saturated. If one or more of the fatty acids is unsaturated, then it's an unsaturated fat, like the one in Figure 3.11b. A polyunsaturated fat has several double bonds within its fatty acids.

▼ Figure 3.10 The separation of hydrophobic (oil) and hydrophilic (vinegar) components in salad dressing.

Oil (hydrophobic)

Vinegar (hydrophilic)

▼ Figure 3.11 The synthesis and structure of a triglyceride molecule.

Glycerol

Fatty acid

(a) A dehydration reaction linking a fatty acid to glycerol

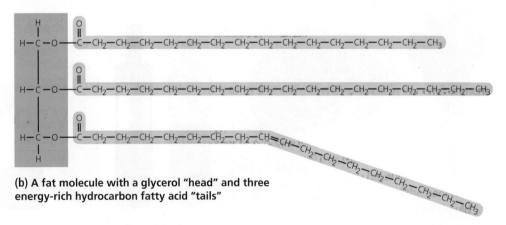

(b) A fat molecule with a glycerol "head" and three energy-rich hydrocarbon fatty acid "tails"

▶ **Figure 3.12 Relative amounts of different fats.** The total fat in the same serving size from three different sources is broken down into saturated fat, monounsaturated fat (unsaturated fats with just one double bond), and polyunsaturated fat (unsaturated fats with two or more double bonds).

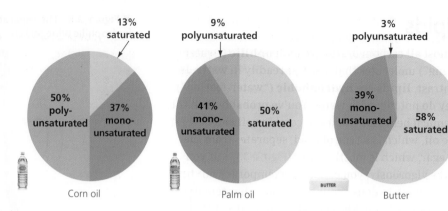

Most animal fats, such as butter and bacon grease, have a relatively high proportion of saturated fatty acids, as shown in the pie chart for butter in **Figure 3.12**. The linear shape of saturated fatty acids allows these molecules to stack easily (like bricks in a wall), so saturated fats tend to be solid at room temperature **(Figure 3.13)**. Diets rich in saturated fats may contribute to cardiovascular disease by promoting atherosclerosis. In this condition, lipid-containing deposits called plaque build up along the inside walls of blood vessels, reducing blood flow and increasing risk of heart attacks and strokes.

Most plant oils are relatively high in unsaturated fatty acids (see the pie chart for corn oil in Figure 3.12).

CHOCOLATE MELTS IN YOUR MOUTH BECAUSE COCOA BUTTER IS HIGH IN SATURATED FATS.

The bent shape of unsaturated fatty acids makes them less likely to form solids (imagine trying to build a wall using bent bricks!). Most unsaturated fats are liquid at room temperature. Fats that are primarily unsaturated include vegetable oils (such as corn and canola oil) and fish oils (such as cod liver oil).

Although plant oils tend to be low in saturated fat, tropical plant fats are an exception (see the pie chart for palm oil in Figure 3.12). Cocoa butter, a main ingredient in chocolate, contains a mix of saturated and unsaturated fat that gives chocolate a melting point near body temperature. Thus, chocolate stays solid at room temperature but melts in your mouth, creating a pleasing "mouth feel" that is one of the reasons chocolate is so appealing.

▼ **Figure 3.13 Types of fats.**

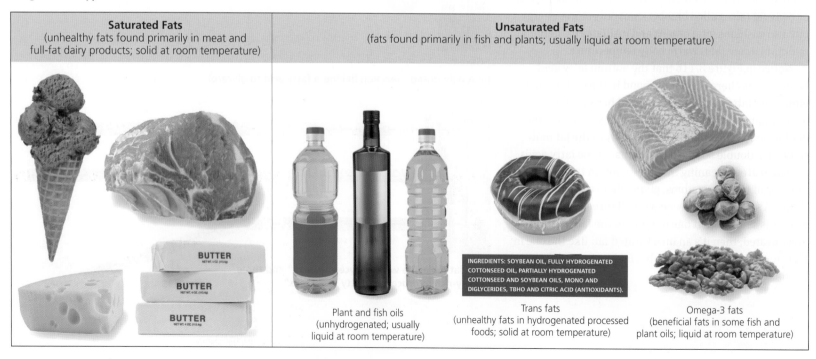

Saturated Fats (unhealthy fats found primarily in meat and full-fat dairy products; solid at room temperature)	**Unsaturated Fats** (fats found primarily in fish and plants; usually liquid at room temperature)

Plant and fish oils (unhydrogenated; usually liquid at room temperature)

INGREDIENTS: SOYBEAN OIL, FULLY HYDROGENATED COTTONSEED OIL, PARTIALLY HYDROGENATED COTTONSEED AND SOYBEAN OILS, MONO AND DIGLYCERIDES, TBHO AND CITRIC ACID (ANTIOXIDANTS).

Trans fats (unhealthy fats in hydrogenated processed foods; solid at room temperature)

Omega-3 fats (beneficial fats in some fish and plant oils; liquid at room temperature)

Food manufacturers can convert unsaturated fats to saturated fats by adding hydrogen, a process called **hydrogenation**. Hydrogenated oils don't spoil as quickly and work better in deep fryers. Unfortunately, hydrogenation can also create **trans fats**, a type of unsaturated fat that is particularly bad for your health. In 2015, the U.S. Food and Drug Administration determined that partially hydrogenated oils are not generally recognized as safe, a designation that will likely result in a near total phasing out of trans fats from the American food supply by 2018.

Although trans fats should generally be avoided and saturated fats limited, it is not true that *all* fats are unhealthy. In fact, some fats perform important functions within the body and are beneficial and even essential to a healthy diet. For example, fats containing omega-3 fatty acids have been shown to reduce the risk of heart disease and relieve the symptoms of arthritis and inflammatory bowel disease. Some natural sources of these beneficial fats are nuts and oily fish such as salmon. Other foods, such as eggs and pasta, can be supplemented with omega-3 fatty acids.

Steroids

Steroids are lipids that are very different from fats in structure and function. All steroids have a carbon skeleton with four fused rings. Different steroids vary in the functional groups attached to this set of rings, and these chemical variations affect their function. One common steroid is cholesterol, which has a bad reputation because of its association with cardiovascular disease. However, cholesterol is a key component of the membranes that surround your cells. It is also the "base steroid" from which your body produces other steroids, such as the hormones testosterone and estrogen (Figure 3.14), which are responsible for the development of male and female sex characteristics, respectively.

In human males, the steroid hormone testosterone causes buildup of muscle and bone mass during puberty and maintains masculine traits throughout life. Anabolic steroids are synthetic variants of testosterone that mimic some of its effects. Anabolic steroids are prescribed to treat diseases that cause muscle wasting, such as cancer and AIDS. However, some individuals abuse anabolic steroids to build up their muscles quickly. In recent years, many famous athletes have admitted using chemically modified ("designer") performance-enhancing anabolic steroids (Figure 3.15). Such revelations have raised questions about the validity of home run records and other athletic accomplishments.

Using anabolic steroids is indeed a fast way to increase body size beyond what hard work can produce. But at what cost? Steroid abuse may cause violent mood swings ("roid rage"), depression, liver damage, high cholesterol, shrunken testicles, reduced sex drive, and infertility. Symptoms related to sexuality occur because artificial anabolic steroids often cause the body to reduce its output of natural sex hormones. Most athletic organizations ban the use of anabolic steroids because of their many potential health hazards coupled with the unfairness of an artificial advantage. ✓

▲ **Figure 3.15 Steroids and the modern athlete.** Baseball player Mark McGwire is one of many famous athletes who have admitted to steroid use.

☑ **CHECKPOINT**

What steroid acts as the molecular building block of the human steroid hormones?

Answer: cholesterol

▼ **Figure 3.14 Examples of steroids.** The molecular structures are abbreviated by omitting the atoms that make up the rings. The subtle difference between testosterone and estrogen results in anatomical and physiological differences between male and female mammals. This example illustrates the importance of molecular structure to function.

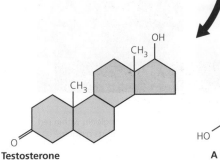

Testosterone

can be converted by the body to

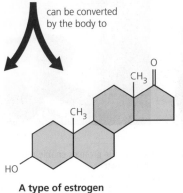

A type of estrogen

Cholesterol

79

Proteins

A **protein** is a polymer of amino acid monomers. Proteins account for more than 50% of the dry weight of most cells, and they are instrumental in almost everything cells do (Figure 3.16). Proteins are the "worker bees" of your body: Chances are, if something is getting done, there is a protein doing it. Your body has tens of thousands of different kinds of proteins, each with a unique three-dimensional shape corresponding to a specific function. In fact, proteins are the most structurally sophisticated molecules in your body.

YOUR HAIR AND STEAK ARE BOTH MADE OF PROTEINS.

The Monomers of Proteins: Amino Acids

All proteins are made by stringing together a common set of 20 kinds of amino acids. Every **amino acid** consists of a central carbon atom bonded to four covalent partners. Three of those attachments are common to all 20 amino acids: a carboxyl group ($—COOH$), an amino group ($—NH_2$), and a hydrogen atom. The variable component of amino acids is called the side chain (or R group, for radical group); it is attached to the fourth bond of the central carbon (Figure 3.17a). Each type of amino acid has a unique side chain which gives that amino acid its special chemical properties (Figure 3.17b). Some amino acids have very simple side chains; the amino acid glycine, for example, has a single hydrogen as its side chain. Other amino acids have more complex side chains, some with branches or rings within them. ☑

✅ CHECKPOINT

1. Which of the following is *not* made of protein: hair, muscle, cellulose, or enzymes?

2. What are the monomers of all proteins? What is the one part of an amino acid that varies?

■ *Answers: 1. Cellulose is a carbohydrate. 2. amino acids; the side chain*

▼ **Figure 3.17 Amino acids.** All amino acids share common functional groups but vary in their side chains.

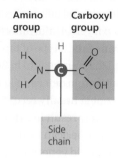

(a) The general structure of an amino acid.

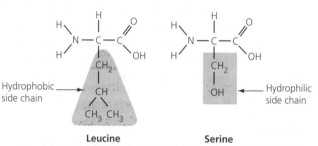

Leucine   **Serine**

(b) Examples of amino acids with hydrophobic and hydrophilic side chains. The side chain of the amino acid leucine is hydrophobic. In contrast, the side chain of the amino acid serine has a hydroxyl group (–OH), which is hydrophilic.

▼ **Figure 3.16 Some of the varied roles played by proteins.**

MAJOR TYPES OF PROTEINS				
Structural Proteins (provide support)	**Storage Proteins** (provide amino acids for growth)	**Contractile Proteins** (help movement)	**Transport Proteins** (help transport substances)	**Enzymes** (help chemical reactions)
Structural proteins form hair, ligaments, and horns.	Seeds and eggs are rich in storage proteins.	Contractile proteins enable muscles to contract.	The protein hemoglobin within red blood cells transports oxygen.	Some cleaning products use enzymes to help break down molecules.

Protein Shape

Cells link amino acid monomers together by—can you guess?—dehydration reactions. The bond that joins adjacent amino acids is called a **peptide bond** (**Figure 3.18**). The resulting long chain of amino acids is called a **polypeptide**. A functional protein is one or more polypeptide chains precisely twisted, folded, and coiled into a molecule of unique shape. The difference between a polypeptide and a protein is like the relationship between a long strand of yarn and a sweater. To be functional, the long fiber (the yarn) must be precisely knit into a specific shape (the sweater).

How can the huge variety of proteins in your body be made from just 20 kinds of amino acids? The answer is arrangement. Though the protein "alphabet" is slightly smaller than the English alphabet (just 20 "letters"), the "words" are much longer, with a typical polypeptide being hundreds or thousands of amino acids in length. Just as a word is a unique sequence of letters, each protein has a unique linear sequence of amino acids.

The amino acid sequence determines the protein's three-dimensional structure, which enables the protein to carry out its specific function. Nearly all proteins work by binding to another molecule. For example, the shape of lactase enables it to attach to lactose. **For all proteins, structure and function are interrelated.** What a protein does is a consequence of its shape. The twists and turns in the ribbon model in **Figure 3.19** represent the protein's shape that allows it to do its job.

▼ **Figure 3.18 Joining amino acids.** A dehydration reaction links adjacent amino acids by a peptide bond.

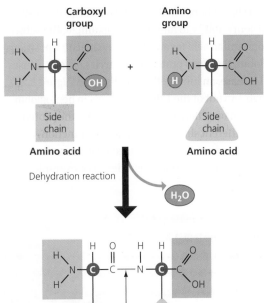

Dehydration reaction

Peptide bond

▼ **Figure 3.19 The structure of a protein.** The chain below, drawn in serpentine fashion so that it fits on the page, shows the amino acid sequence of a polypeptide found in lysozyme, an enzyme in your tears and sweat that helps prevent bacterial infections. The names of the amino acids are given as their three-letter abbreviations; for example, the amino acid alanine is abbreviated "Ala." This amino acid sequence folds into a protein of specific shape, shown in the two computer-generated representations. Without this specific shape, the protein could not perform its function.

One amino acid (alanine)

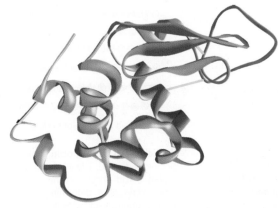

Here you can see how the polypeptide folds into a compact shape.

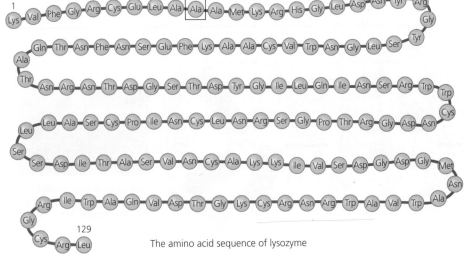

The amino acid sequence of lysozyme

This model allows you to see the details of the protein's structure.

Changing a single letter can drastically affect the meaning of a word—"tasty" versus "nasty," for instance. Similarly, even a slight change in the amino acid sequence can affect a protein's ability to function. For example, the substitution of one amino acid for another at a particular position in hemoglobin, the blood protein that carries oxygen, causes sickle-cell disease, an inherited blood disorder (Figure 3.20). Even though 145 out of the 146 amino acids in hemoglobin are correct, that one change is enough to cause the protein to fold into a different shape, which alters its function, which in turn causes disease. Misfolded proteins are associated with several severe brain disorders. For example, the diseases shown in **Figure 3.21** are all caused by prions, misfolded versions of normal brain proteins. Prions can infiltrate the brain, converting normally folded proteins into the abnormal shape. Clustering of the misfolded proteins eventually causes a fatal disruption of brain function.

In addition to dependence on the amino acid sequence, a protein's shape is sensitive to the environment. An unfavorable change in temperature, pH, or some other factor can cause a protein to unravel. If you cook an egg, the transformation of the egg white from clear to opaque is caused by proteins in the egg white coming apart. One of the reasons why extremely high fevers are so dangerous is that some proteins in the body lose their shape above 104°F.

What determines a protein's amino acid sequence? The amino acid sequence of each polypeptide chain is specified by a gene. And this relationship between genes and proteins brings us to this chapter's last category of large biological molecules: nucleic acids. ✓

CHECKPOINT

How can changing an amino acid alter the function of a protein?

■ *Answer: Changing an amino acid may alter the shape of the protein, which changes its function.*

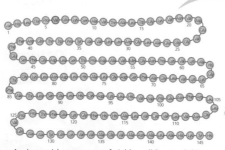
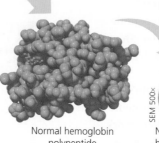

Amino acid sequence of normal hemoglobin

Normal hemoglobin polypeptide

Normal red blood cell

SEM 500×

(a) Normal hemoglobin. Red blood cells of humans are normally disk-shaped. Each cell contains millions of molecules of the protein hemoglobin, which transports oxygen from the lungs to other organs of the body.

Amino acid sequence of sickle-cell hemoglobin

Sickle-cell hemoglobin polypeptide

Sickled red blood cell

SEM 500×

▶ **Figure 3.20 A single amino acid substitution in a protein causes sickle-cell disease.**

(b) Sickle-cell hemoglobin. A slight change in the amino acid sequence of hemoglobin causes sickle-cell disease. The inherited substitution of one amino acid—valine in place of the amino acid glutamic acid—occurs at the sixth position of the polymer. Such abnormal hemoglobin molecules tend to crystallize, deforming some cells into a sickle shape. Someone with the disease suffers from dangerous episodes when the angular cells clog tiny blood vessels, impeding blood flow.

▶ **Figure 3.21 How an improperly folded protein can lead to brain disease.** Here, you can see how prion proteins bring about the destruction of brain tissue.

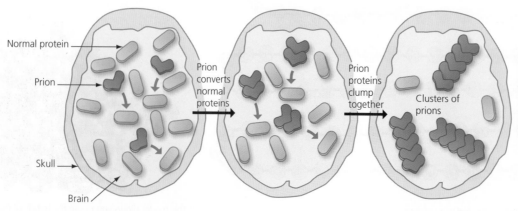

Normal protein

Prion

Skull

Brain

Prion converts normal proteins

Prion proteins clump together

Clusters of prions

Prions cause fatal weight loss in several animals, including deer, elk, and moose. They also cause mad cow disease.

Nucleic Acids

Nucleic acids are macromolecules that store information and provide the instructions for building proteins. The name *nucleic* comes from the fact that DNA is found in the nuclei of eukaryotic cells. There are two types of nucleic acids: **DNA** (which stands for <u>d</u>eoxyribo<u>n</u>ucleic <u>a</u>cid) and **RNA** (for <u>ribon</u>ucleic <u>a</u>cid). The genetic material that humans and all other organisms inherit from their parents consists of giant molecules of DNA. The DNA resides in the cell as one or more very long fibers called chromosomes. A **gene** is a unit of inheritance encoded in a specific stretch of DNA that programs the amino acid sequence of a polypeptide. Those programmed instructions, however, are written in a chemical code that must be translated from "nucleic acid language" to "protein language" **(Figure 3.22)**. A cell's RNA molecules help make this translation (discussed in detail in Chapter 10). This interaction of three types of molecules—DNA, RNA, and proteins—enables the transmission of hereditary information from one generation to the next, ensuring the continuity of life.

Nucleic acids are polymers made from monomers called **nucleotides (Figure 3.23)**. Each nucleotide contains three

▼ **Figure 3.22** **Building a protein.** Within the cell, a gene (a segment of DNA) provides the directions to build a molecule of RNA, which can then be translated into a protein.

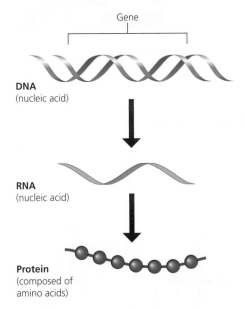

Gene

DNA
(nucleic acid)

RNA
(nucleic acid)

Protein
(composed of amino acids)

▼ **Figure 3.23** **A DNA nucleotide.** A DNA nucleotide monomer consists of three parts: a sugar (deoxyribose), a phosphate, and a nitrogenous (nitrogen-containing) base.

Nitrogenous base
(can be A, G, C, or T)

Connection to the next nucleotide in the chain

Phosphate group

Thymine (T)

Phosphate

Base

Sugar

Connection to the next nucleotide in the chain

Sugar (deoxyribose)

(a) Atomic structure

(b) Symbol used in this book

parts. At the center of each nucleotide is a five-carbon sugar (blue in the figure), deoxyribose in DNA and ribose in RNA. Attached to the sugar is a negatively charged phosphate group (yellow) containing a phosphorus atom bonded to oxygen atoms (PO_4^-). Also attached to the sugar is a nitrogen-containing base (green) made of one or two rings. The sugar and phosphate are the same in all nucleotides; only the base varies. Each DNA nucleotide has one of four possible nitrogenous bases: adenine (abbreviated A), guanine (G), cytosine (C), or thymine (T) **(Figure 3.24)**. Thus, all genetic information is written in a four-letter alphabet.

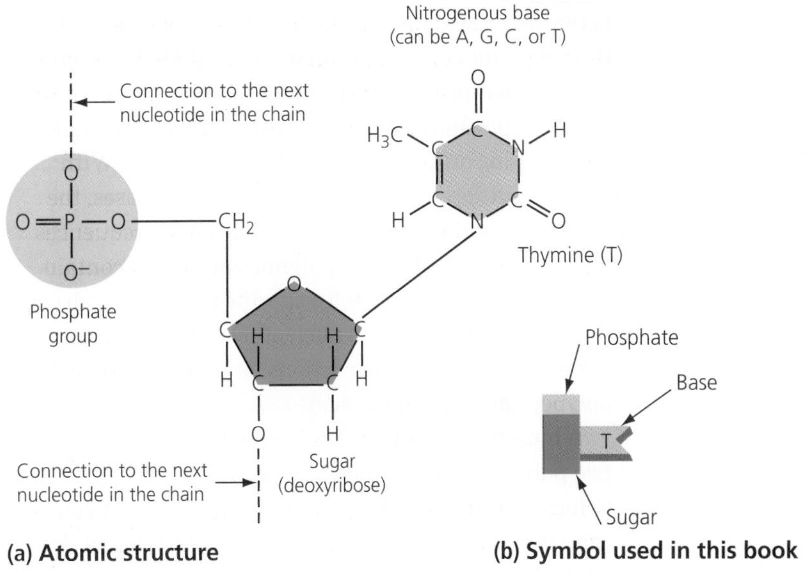

▼ **Figure 3.24** **The nitrogenous bases of DNA.** Notice that adenine and guanine have double-ring structures. Thymine and cytosine have single-ring structures.

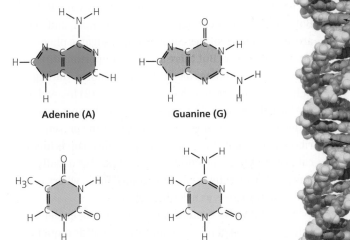

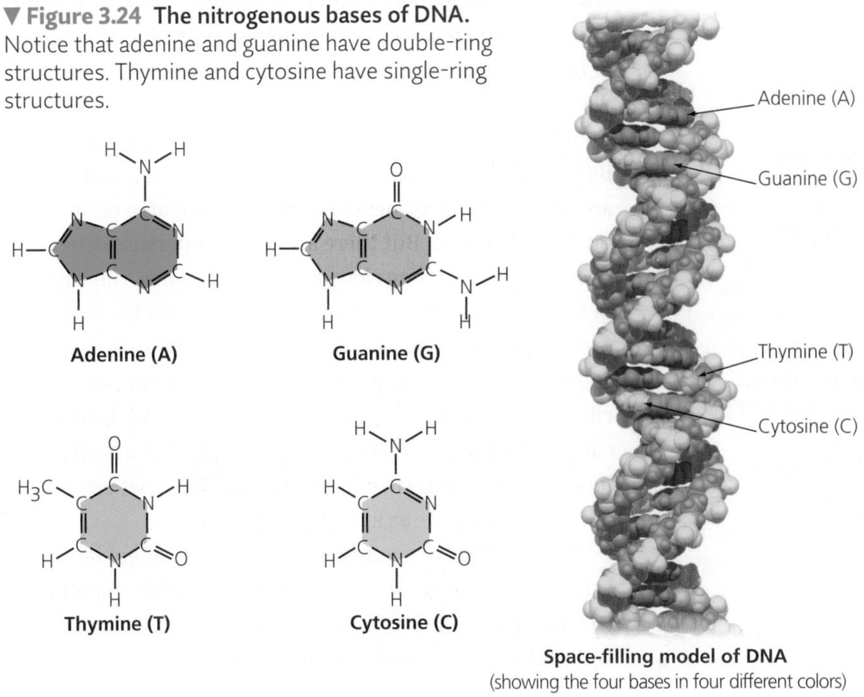

Adenine (A)

Guanine (G)

Thymine (T)

Cytosine (C)

Adenine (A)

Guanine (G)

Thymine (T)

Cytosine (C)

Space-filling model of DNA
(showing the four bases in four different colors)

THE STRUCTURE OF YOUR DNA IS NEARLY THE SAME AS THAT OF A MOSQUITO OR ELEPHANT.

Dehydration reactions link nucleotide monomers into long chains called polynucleotides (**Figure 3.25a**). In a polynucleotide, nucleotides are joined by covalent bonds between the sugar of one nucleotide and the phosphate of the next. This results in a **sugar-phosphate backbone**, a repeating pattern of sugar-phosphate-sugar-phosphate, with the bases (A, T, C, or G) hanging off the backbone like appendages. With different combinations of the four bases, the number of possible polynucleotide sequences is vast. One long polynucleotide may contain many genes, each a specific series of hundreds or thousands of nucleotides. This sequence is a code that provides instructions for building a specific polypeptide from amino acids.

A molecule of cellular DNA is double-stranded, with two polynucleotide strands coiled around each other to form a **double helix (Figure 3.25b)**. Think of a candy cane that has two intertwined spirals, one red and one white. In the central core of the helix (corresponding to the interior of the candy cane), the bases along one DNA strand hydrogen-bond to bases along the other strand. The bonds are individually weak, but collectively they zip the two strands together into a very stable double helix formation. To understand how DNA strands are bonded, think of Velcro, in which two strips are held together by hook-and-loop bonds, each of which is weak but which collectively form a tight grip. Because of the way the functional groups hang off the bases, the base pairing in a DNA double helix is specific: The base A can pair only with T, and G can pair only with C. Thus, if you know the sequence of bases along one DNA strand, you also know the sequence along the complementary strand in the double helix. This unique base pairing is the basis of DNA's ability to act as the molecule of inheritance (as discussed in Chapter 10).

There are many similarities between DNA and RNA. Both are polymers of nucleotides, for example, and both are made of nucleotides consisting of a sugar, a phosphate, and a base. But there are three important differences. (1) As its name *ribonucleic acid* denotes, the sugar in RNA is ribose rather than deoxyribose. (2) Instead of the base thymine, RNA has a similar but distinct base called uracil (U) (**Figure 3.26**). Except for the presence of ribose and uracil, an RNA polynucleotide chain is identical to a DNA polynucleotide chain. (3) RNA is usually found in living cells in single-stranded form, whereas DNA usually exists as a double helix.

Now that we've examined the structure of nucleic acids, we'll look at how a change in nucleotide sequence can affect protein production. To illustrate this point, we'll return to a familiar condition. ✅

✅ **CHECKPOINT**

1. DNA contains _____ polynucleotide strands, each composed of _____ kinds of nucleotides. (Provide two numbers.)

2. If one DNA strand has the sequence GAATGC, what is the sequence of the other strand?

3. Complete this biologist pickup line: "Of all the nucleic acids, RNA is my favorite, because it's got _____ in it!"

■ *Answers: 1. two; four 2. CTTACG 3. U*

▼ **Figure 3.25 The structure of DNA.** The base pairing in a DNA molecule is specific: A always pairs with T; G always pairs with C.

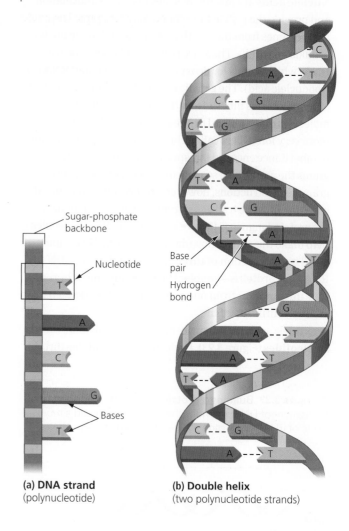

Sugar-phosphate backbone

Nucleotide

Bases

Base pair

Hydrogen bond

(a) **DNA strand**
(polynucleotide)

(b) **Double helix**
(two polynucleotide strands)

▼ **Figure 3.26 An RNA nucleotide.** Notice that this RNA nucleotide differs from the DNA nucleotide in Figure 3.23 in two ways: The RNA sugar is ribose rather than deoxyribose, and the base is uracil (U) instead of thymine (T).

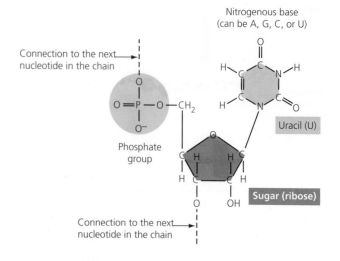

Connection to the next nucleotide in the chain

Nitrogenous base (can be A, G, C, or U)

Uracil (U)

Phosphate group

Sugar (ribose)

Connection to the next nucleotide in the chain

Does Lactose Intolerance Have a Genetic Basis?

BACKGROUND

Like all proteins, the enzyme lactase **(Figure 3.27a)** is encoded by a DNA gene. A reasonable hypothesis is that lactose-intolerant people have a defect in their lactase gene. However, this hypothesis is not supported by observation. Even though lactose intolerance runs in families, most lactose-intolerant people have a normal version of the lactase gene. What, then, is the basis for lactose intolerance?

METHOD

A group of scientists proposed that lactose intolerance is correlated with one nucleotide at a site on chromosome 2 that falls outside the lactase gene itself **(Figure 3.27b)**. To test this hypothesis, they examined the chromosomes of 196 lactose-intolerant people from nine families.

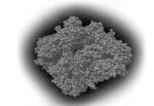

◄ **Figure 3.27 A genetic cause of lactose intolerance.** A research study showed a correlation between lactose intolerance and a nucleotide located outside of the lactase gene.

(a) Lactase enzyme

RESULTS

Their results showed a 100% correlation between lactose intolerance and having cytosine (C) rather than thymine (T) at the site **(Figure 3.27c)**. Depending on the nucleotide sequence within the distant region of the DNA molecule, the action of the lactase gene is ramped up or down (in a way that likely involves producing a regulatory protein that interacts with the nucleotides near the lactase gene). This study shows how a small change in a DNA nucleotide sequence outside of a gene itself can affect production of a protein and the well-being of an organism.

> **Thinking Like a Scientist**
>
> Explain how a change in DNA outside of a gene can still affect that gene.
>
> *For the answer, see Appendix D.*

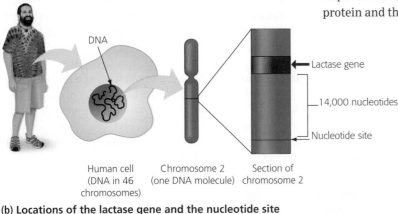

DNA

Human cell (DNA in 46 chromosomes) — Chromosome 2 (one DNA molecule) — Section of chromosome 2

Lactase gene
14,000 nucleotides
Nucleotide site

(b) Locations of the lactase gene and the nucleotide site

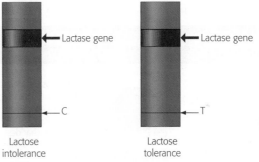

Lactase gene — C — Lactose intolerance

Lactase gene — T — Lactose tolerance

(c) The nucleotide difference determining lactose intolerance or lactose tolerance

The Evolution of Lactose Intolerance in Humans

Lactose intolerance is found in 80% of African Americans and Native Americans and 90% of Asian Americans, but only about 10% of Americans of northern European descent. From an evolutionary perspective, it is reasonable to infer that it is rare among northern Europeans because the ability to tolerate lactose offered a survival advantage to their ancestors. In northern Europe's relatively cold climate, herd animals were a main source of food. Domesticated cattle first appeared in northern Europe about 8,000 years ago. With dairy products available year-round, natural selection would have favored people with a mutation that kept the lactase gene switched on beyond infancy. In cultures where dairy products were not a dietary staple, natural selection would not favor such a mutation.

Researchers wondered whether the genetic basis for lactose tolerance might be present in other cultures that kept dairy herds. They compared the genetic makeup and lactose tolerance of 43 ethnic groups in East Africa, all of whom use dairy products **(Figure 3.28)**. They identified three other genetic changes that keep the lactase gene active. These mutations occurred around 7,000 years ago, when archaeological evidence shows domestication of cattle in these African regions.

This research shows that lactose intolerance is the human norm and that "lactose tolerance" is a relatively recent mutation. Genetic mutations that confer a selective advantage, such as surviving cold winters or drought by drinking milk, spread rapidly where the climate favored such changes. Whether you can digest milk is therefore an evolutionary record of the cultural history of your ancestors.

▼ **Figure 3.28** Harvested milk as a staple crop in northern Europe and Africa.

Chapter Review

SUMMARY OF KEY CONCEPTS

Organic Compounds

Carbon Chemistry

Carbon atoms can form large, complex, diverse molecules by bonding to four potential partners, including other carbon atoms. In addition to variations in the size and shape of carbon skeletons, organic compounds vary in the presence and locations of different functional groups.

Giant Molecules from Smaller Building Blocks

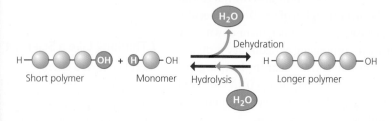

Large Biological Molecules

Large Biological Molecules	Functions	Components	Examples
Carbohydrates	Dietary energy; storage; plant structure	Monosaccharide	Monosaccharides: glucose, fructose; disaccharides: lactose, sucrose; polysaccharides: starch, cellulose
Lipids	Long-term energy storage (fats); hormones (steroids)	Components of a triglyceride	Fats (triglycerides); steroids (testosterone, estrogen)
Proteins	Enzymes, structure, storage, contraction, transport, etc.	Amino acid	Lactase (an enzyme); hemoglobin (a transport protein)
Nucleic acids	Information storage	Nucleotide	DNA, RNA

Carbohydrates

Simple sugars (monosaccharides) provide cells with energy and building materials. Double sugars (disaccharides), such as sucrose, consist of two monosaccharides joined by a dehydration reaction. Polysaccharides are long polymers of sugar monomers. Starch in plants and glycogen in animals are storage polysaccharides. The cellulose of plant cell walls, which is indigestible by animals, is an example of a structural polysaccharide.

Lipids

Lipids are hydrophobic. Fats, a type of lipid, are the major form of long-term energy storage in animals. A molecule of fat, or triglyceride, consists of three fatty acids joined by dehydration reactions to a molecule of glycerol. Most animal fats are saturated, meaning that their fatty acids have the maximum number of hydrogens. Plant oils contain mostly unsaturated fats, having fewer hydrogens in the fatty acids because of double bonding in the carbon skeletons. Steroids, including cholesterol and the sex hormones, are also lipids.

Proteins

There are 20 types of amino acids, the monomers of proteins. They are linked by dehydration reactions to form polymers called polypeptides. A protein consists of one or more polypeptides folded into a specific three-dimensional shape. The shape of a protein determines its function. Changing the amino acid sequence of a polypeptide may alter the shape and therefore the function of the protein. Shape is sensitive to environment, and if a protein loses its shape because of an unfavorable environment, its function may also be lost.

Nucleic Acids

Nucleic acids include RNA and DNA. DNA takes the form of a double helix, two DNA strands (polymers of nucleotides) held together by hydrogen bonds between nucleotide components called bases. There are four kinds of DNA bases: adenine (A), guanine (G), thymine (T), and cytosine (C). A always pairs with T, and G always pairs with C. These base-pairing rules enable DNA to act as the molecule of inheritance. RNA has U (uracil) instead of T.

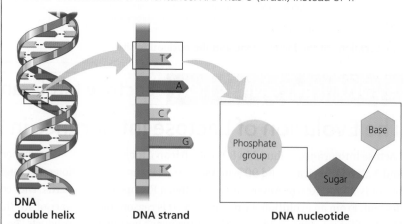

DNA double helix   DNA strand   DNA nucleotide

Mastering Biology

For practice quizzes, BioFlix animations, MP3 tutorials, video tutors, and more study tools designed for this textbook, go to Mastering Biology™

SELF-QUIZ

1. One isomer of methamphetamine is the addictive illegal drug known as "crank." Another isomer is a medicine for sinus congestion. How can you explain the differing effects of the two isomers?

2. Monomers are joined together to form larger polymers through _____. Such a reaction releases a molecule of _____.

3. Polymers are broken down into monomers through the chemical reaction called _____. Such a reaction consumes a molecule of _____.

4. Table sugar is _____.
   a. glucose, a monosaccharide
   b. glucose, a disaccharide
   c. sucrose, a monosaccharide
   d. sucrose, a disaccharide

5. When two molecules of glucose ($C_6H_{12}O_6$) are joined together by a dehydration reaction, what are the formulas of the two products? (*Hint*: No atoms are gained or lost.)

6. One molecule of dietary fat is made by joining three molecules of _____ to one molecule of _____. What is the formal name of the resulting molecule?

7. By definition, what type of fatty acid has double bonds?
   a. steroid   c. unsaturated
   b. triglyceride   d. saturated

8. Humans and other animals cannot digest wood because they
   a. cannot digest any carbohydrates.
   b. cannot chew it fine enough.
   c. lack the enzyme needed to break down cellulose.
   d. get no nutrients from it.

9. Explain how it could be possible to change an amino acid within a protein but not affect that protein's function.

10. Most proteins can easily dissolve in water. Where in the shape of a protein would you find hydrophobic amino acids?

11. A shortage of phosphorus in the soil would make it especially difficult for a plant to manufacture
    a. DNA.   c. cellulose.
    b. proteins.   d. fatty acids.

12. Nucleic acids are polymers of _____ monomers.

13. Name three similarities between DNA and RNA. Name three differences.

14. What is the structure of a gene? What is the function of a gene?

For answers to the Self Quiz, see Appendix D.

IDENTIFYING MAJOR THEMES

For each statement, identify which major theme is evident (the relationship of structure to function, information flow, pathways that transform energy and matter, interactions within biological systems, or evolution) and explain how the statement relates to the theme. If necessary, review the theme descriptions (see Chapter 1) and review the examples highlighted in blue in this chapter.

15. Plants use glucose as both a source of energy for cells and as a building block of their bodies.

16. Both starch and cellulose consist of joined monosaccharide monomers, but the way that they are joined makes starch digestible by humans and cellulose not.

17. Your genetic legacy is contained within the sequence of nucleotides on your chromosomes.

For answers to Identifying Major Themes, see Appendix D.

THE PROCESS OF SCIENCE

18. Scientists at the U.S. Food and Drug Administration (FDA) are testing a cake mix to see if it contains fat. Hydrolysis of the mix yields glucose, fructose, glycerol, amino acids, and molecules with long chains with a carboxyl group at one end. Is there fat in the mix?

19. Based on your knowledge of the different types of chemical bonds and of the molecular structures involved, predict which of the four families of macromolecules discussed in this chapter are soluble in water and which are not, and explain why.

20. **Interpreting Data** Read this food label for one cookie. One gram of fat packs 9 Calories, and 1 gram of carbohydrates or protein packs 4 Calories. The top of the label shows that each cookie contains 140 Calories in total. What percentage of the Calories in this cookie are from fat, carbohydrates, and protein?

Nutrition Facts		
8 servings per container		
Serving size 1 cookie 28 g/1 oz)		
Amount per serving		
Calories		**140**
		% Daily Value*
Total Fat 7g		11%
Saturated Fat 3g		15%
Trans Fat 0g		
Cholesterol 10mg		3%
Sodium 80mg		3%
Total Carbohydrate 18g		6%
Dietary Fiber 1g		4%
Total Sugars 10g		
Includes 10g Added Sugars		100%
Protein 2g		

BIOLOGY AND SOCIETY

21. Some athletes take anabolic steroids to build strength ("bulk up"). The health risks are extensively documented. What is your opinion about the ethics of using chemicals to enhance performance? Is it cheating? Explain.

22. Heart disease is the leading cause of death among people in the United States and other industrialized nations. Fast food is a major source of unhealthy fats that contribute significantly to heart disease. Imagine you're a juror sitting on a trial where a fast-food manufacturer is being sued for making a harmful product. To what extent do you think manufacturers of unhealthy foods should be held responsible for the health consequences of their products? As a jury member, how would you vote and why?

23. Each year, industrial chemists develop and test organic compounds for use as insecticides, fungicides, and weed killers. Is your general opinion of such chemicals positive or negative? What influences have shaped your feelings?

4 A Tour of the Cell

WHAT DO YOU HAVE IN COMMON WITH A MUSHROOM? AT THE CELLULAR LEVEL, QUITE A LOT!

CHAPTER CONTENTS

Why Cells Matter

Life begins at the cellular level. Although microscopic, a cell contains complex machinery capable of supporting all of life's processes. Any study of biology can therefore benefit from an examination of cellular structures.

WITHOUT THE CYTOSKELETON, YOUR CELLS WOULD COLLAPSE IN ON THEMSELVES, MUCH LIKE A BUILDING COLLAPSES WHEN THE INFRASTRUCTURE FAILS.

ENJOY THAT BUZZ? THE CAFFEINE THAT GIVES COFFEE A KICK ALSO PROTECTS COFFEE PLANTS FROM HERBIVORES.

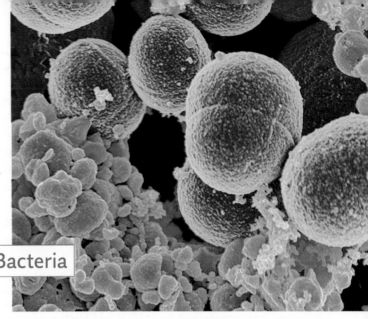

Colorized SEM 20,000×

Staphylococcus aureus. These bacteria (yellow) are evading destruction by human white blood cells (blue).

BIOLOGY AND SOCIETY Humans Versus Bacteria

Antibiotics: Drugs That Target Bacterial Cells

Antibiotics—drugs that disable or kill infectious bacteria—are marvels of modern medicine. The first antibiotic to be discovered was penicillin in 1928. Before this time, even relatively minor infections could be deadly. After the development of penicillin, a revolution in human health rapidly followed. Fatality rates of many diseases (such as bacterial pneumonia and surgical infections) plummeted, saving millions of lives. In fact, human health care improved so quickly and so profoundly that some doctors in the early 1900s predicted the end of infectious diseases altogether. Unfortunately, this did not come to pass—see the Evolution Connection section in Chapter 13 for a discussion of why infectious diseases were not so easily defeated.

The goal of antibiotic treatment is to knock out invading bacteria while doing no damage to the human host. How does an antibiotic zero in on its bacterial target among trillions of human cells? Most antibiotics are so precise because they bind to structures found only in bacterial cells. For example, the common antibiotics erythromycin and streptomycin bind to the bacterial ribosome, a vital cellular structure responsible for producing proteins. The ribosomes of humans are different enough from those of bacteria that the antibiotics bind only to bacterial ribosomes, leaving human ribosomes unaffected. Ciprofloxacin (commonly referred to as Cipro) is an antibiotic that targets an enzyme bacteria need to maintain their chromosome structure. Your cells can survive just fine in the presence of Cipro because human chromosomes have a sufficiently different makeup from bacterial chromosomes. Other drugs, such as penicillin, ampicillin, and bacitracin, disrupt the synthesis of cell walls, a structure found in most bacteria that is absent from the cells of humans and other animals. As you'll learn in this chapter, researchers continue to exploit the unique structures of bacterial cells to design and discover new antibiotics.

This discussion of how various antibiotics target bacteria underscores the main point of this chapter: To understand how life works—whether in bacteria or in your own body—you first need to learn about cells. On the scale of biological organization, cells occupy a special place: They are the simplest objects that can be alive. Nothing smaller than a cell is capable of displaying all of life's properties. In this chapter, we'll explore the microscopic structure and function of cells. Along the way, we'll further consider how the ongoing battle between humans and infectious bacteria is affected by the cellular structures present on both sides.

The Microscopic World of Cells

If you've ever gazed through a microscope, you've seen that every cell is a miniature marvel (Figure 4.1). If the world's most sophisticated jumbo jet were reduced to microscopic size, its complexity would pale next to a living cell.

Organisms are either single-celled, such as most prokaryotes and protists, or multicellular, such as plants, animals, and most fungi. Your own body is a cooperative society of trillions of cells of many specialized types. As you read this page, muscle cells allow you to scan your eye across the words, while sensory cells in your eye gather information and send it to brain cells, which interpret the words. Everything you do—every action and every thought—is possible because of processes that occur at the cellular level.

Figure 4.2 shows the size range of cells compared with objects both larger and smaller. Most cells are between

▼ **Figure 4.1 Types of micrographs.** Photographs taken with microscopes are called micrographs. Light microscopes can magnify up to about 1,000-fold. Electron microscopes use beams of electrons rather than light and can show objects about 100 times smaller than light microscopes.

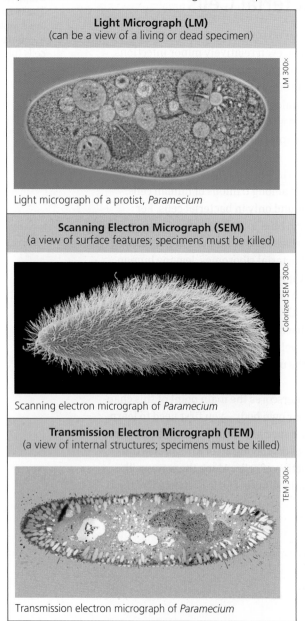

Light Micrograph (LM)
(can be a view of a living or dead specimen)

LM 300x

Light micrograph of a protist, *Paramecium*

Scanning Electron Micrograph (SEM)
(a view of surface features; specimens must be killed)

Colorized SEM 300x

Scanning electron micrograph of *Paramecium*

Transmission Electron Micrograph (TEM)
(a view of internal structures; specimens must be killed)

TEM 300x

Transmission electron micrograph of *Paramecium*

▼ **Figure 4.2 Size ranges.** Starting at the top of this scale with 10 m (10 meters) and going down, each measurement along the left side marks a tenfold decrease in size. Micrographs in this book have size notations (see Figure 4.1). For example, 300x means 300 times the original size.

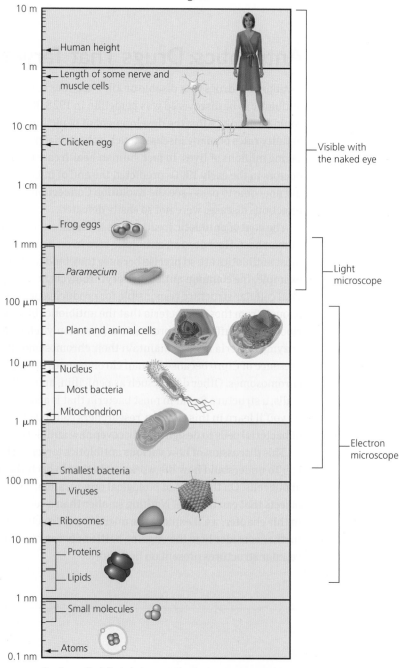

10 m
— Human height
1 m
— Length of some nerve and muscle cells
10 cm
— Chicken egg
1 cm
— Frog eggs
1 mm
— *Paramecium*
100 μm
— Plant and animal cells
10 μm
— Nucleus
— Most bacteria
— Mitochondrion
1 μm
— Smallest bacteria
100 nm
— Viruses
— Ribosomes
10 nm
— Proteins
— Lipids
1 nm
— Small molecules
— Atoms
0.1 nm

Visible with the naked eye

Light microscope

Electron microscope

See Appendix A for help in converting between measurements.

1 and 100 μm in diameter (yellow region of the figure) and are therefore visible only with a microscope. There are some interesting exceptions: An ostrich egg is a single cell about 6 inches across and weighing about 3 pounds, and nerve cells in giant squid can be more than 30 feet long!

How do new living cells arise? The **cell theory** states that all living things are composed of cells and that all cells come from earlier cells. So every cell in your body (and in every other living organism on Earth) was formed by division of a previously living cell. (That raises an obvious question: How did the first cell evolve? This fascinating topic is addressed in Chapter 15.) With that introduction, let's begin to explore the variety of cells found among life on Earth.

The Two Major Categories of Cells

The countless cells that exist on Earth can be divided into two basic types: prokaryotic cells and eukaryotic cells (**Table 4.1**). Biologists classify all life into three major groups called **domains**. Organisms of the domains Bacteria and Archaea are composed of **prokaryotic cells** and are called prokaryotes. Organisms of the domain Eukarya—including protists, plants, fungi, and animals—are composed of **eukaryotic**

cells and are called eukaryotes. Some (such as yeast) are microscopic, while others can be massive. Every organism you can see with your own eyes is a eukaryote.

All cells, whether prokaryotic or eukaryotic, have several features in common. They are all bounded by a barrier called a **plasma membrane**, which regulates the traffic of molecules between the cell and its surroundings. Inside all cells is a thick, jellylike fluid called the **cytosol**, in which cellular components are suspended. All cells have one or more **chromosomes** carrying genes made of DNA. And all cells have **ribosomes** that build proteins according to instructions from the genes. Because some structures are unique to bacteria, as mentioned in the Biology and Society section at the beginning of the chapter, some antibiotics—such as streptomycin—target prokaryotic ribosomes, crippling protein synthesis in the bacterial invaders but not in the eukaryotic host (you).

Although they have many similarities, prokaryotic and eukaryotic cells differ in several important ways. Fossil evidence shows that prokaryotes were the first life on Earth, appearing more than 3.5 billion years ago, and were Earth's sole inhabitants for over a billion years (see Figure 15.1). In contrast, the first eukaryotes did not appear until around 1.8 billion years ago. Prokaryotic cells are usually much smaller—about one-tenth the length of a typical eukaryotic cell— and are simpler in structure. Think of a prokaryotic cell as being like a

WHAT DO YOU HAVE IN COMMON WITH A MUSHROOM? AT THE CELLULAR LEVEL, QUITE A LOT!

Table 4.1	Comparing Prokaryotic and Eukaryotic Cells
Prokaryotic cells	Eukaryotic cells
First evolved approximately 3.5 billion years ago	First evolved approximately 2.1 billion years ago
Found in bacteria and archaea	Found in protists, plants, fungi, and animals
Smaller, simpler	Larger, more complex
Most have cell walls; some have capsules, fimbriae, and/or flagella.	Plant cells have cell walls; animal cells are surrounded by an extracellular matrix.
Have a plasma membrane	Have a plasma membrane
No membrane-bound organelles	Membrane-bound organelles (for example, nucleus, ER)
Have a nucleoid region containing a single circular chromosome	Have a nucleus containing one or more linear chromosomes
Have ribosomes	Have ribosomes

bicycle, whereas a eukaryotic cell is like a sports utility vehicle. Both a bike and an SUV get you from place to place, but one is much smaller and contains many fewer parts than the other. Similarly, prokaryotic cells and eukaryotic cells perform similar functions, but prokaryotic cells are much smaller and less complex. The most significant structural difference between the two types of cells is that eukaryotic cells have **organelles** ("little organs"), membrane-enclosed structures that perform specific functions; prokaryotic cells do not have organelles. The most important organelle is the **nucleus**, which houses most of a eukaryotic cell's DNA. The nucleus is surrounded by a double membrane. A prokaryotic cell lacks a nucleus; its DNA is coiled into a "nucleus-like" region called the **nucleoid**, which is not partitioned from the rest of the cell by membranes.

Consider this analogy: A eukaryotic cell is like an office building that is separated into cubicles. Within each cubicle, a specific function is performed, thus dividing the labor among many internal compartments. One cubicle may hold the accounting department, for example, while another is home to the sales force. The cubicle walls within eukaryotic cells are made from membranes that help maintain a unique chemical environment inside each cubicle. In contrast, the interior of a prokaryotic cell is like an open warehouse. The spaces for specific tasks within a prokaryotic warehouse are distinct, but they are not separated by physical barriers: Imagine desks for the sales and accounting departments arranged across an open floor.

Figure 4.3 depicts an idealized prokaryotic cell and a micrograph of an actual bacterium. Surrounding the plasma membrane of most prokaryotic cells is a rigid cell wall, which protects the cell and helps maintain its shape. Recall from the Biology and Society section that bacterial cell walls are the targets of some antibiotics. In some prokaryotes, a sticky outer coat called a capsule surrounds the cell wall. Capsules provide protection and help prokaryotes stick to surfaces and to other cells in a colony. For example, capsules help bacteria in your mouth stick together to form harmful dental plaque. Prokaryotes can have short projections called fimbriae, which can also attach to surfaces. Many prokaryotic cells have flagella, long projections that propel them through their liquid environment. ✓

An Overview of Eukaryotic Cells

All eukaryotic cells—whether from animals, plants, protists, or fungi—are fundamentally similar to one another and quite different from prokaryotic cells. The key difference is that eukaryotic cells are partitioned by membranes into organelles. The membranes allow each organelle to maintain specific chemical conditions that favor the metabolic tasks performed there. **Figure 4.4** provides overviews of an idealized animal cell and plant cell. No real cell looks quite like these idealized cells because living cells have many more copies of most of the structures

▼ **Figure 4.3 A prokaryotic cell.** A micrograph of *Helicobacter pylori* (left), a bacterium that causes stomach ulcers, is shown alongside a drawing of an idealized prokaryotic cell (right).

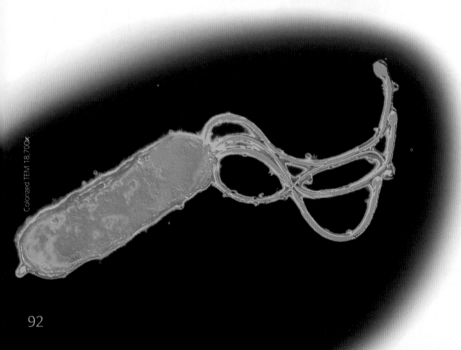

Colorized TEM 18,700x

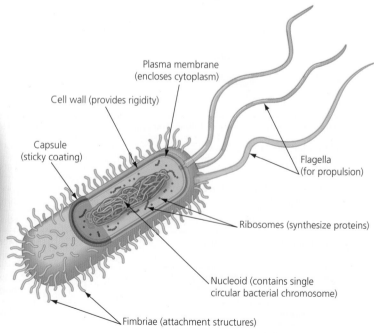

Plasma membrane (encloses cytoplasm)

Cell wall (provides rigidity)

Capsule (sticky coating)

Flagella (for propulsion)

Ribosomes (synthesize proteins)

Nucleoid (contains single circular bacterial chromosome)

Fimbriae (attachment structures)

shown; each of your cells has hundreds of mitochondria and millions of ribosomes, for example.

The parts of a cell form an integrated team, with the properties of life emerging as a result of the team members working together. **A cell beautifully illustrates the theme of interactions: It is a living unit that is greater than the sum of its parts.** Throughout this chapter we'll use miniature versions of Figure 4.4 as road maps, highlighting the structure we're discussing. Notice that the structures are color-coded; we'll use this color scheme throughout the book.

The region of the cell outside the nucleus and within the plasma membrane is called the **cytoplasm**. (This term is also used to refer to the interior of a prokaryotic cell.) The cytoplasm of a eukaryotic cell consists of various organelles suspended in the liquid cytosol. As you can see in Figure 4.4, most organelles are found in both animal and plant cells. But you'll notice some important differences between these two types of cells. For example, plant cells have chloroplasts (where photosynthesis occurs), a cell wall (which provides stiffness to plant structures), and a central vacuole. In the rest of this section, we'll take a closer look at the architecture of eukaryotic cells, beginning with the plasma membrane. ☑

▼ **Figure 4.4 An idealized animal cell and plant cell.** For now, the labels on the drawings are just words, but these cellular components will come to life as we take a closer look at how each part of the cell functions.

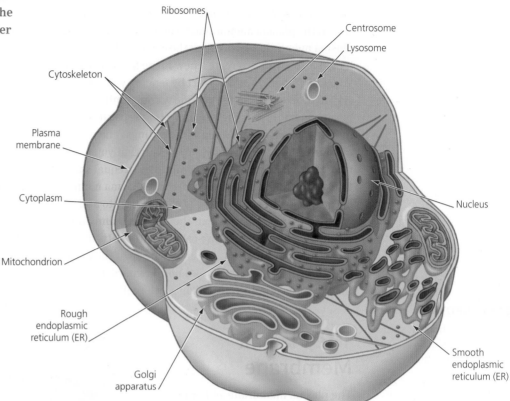

Idealized animal cell

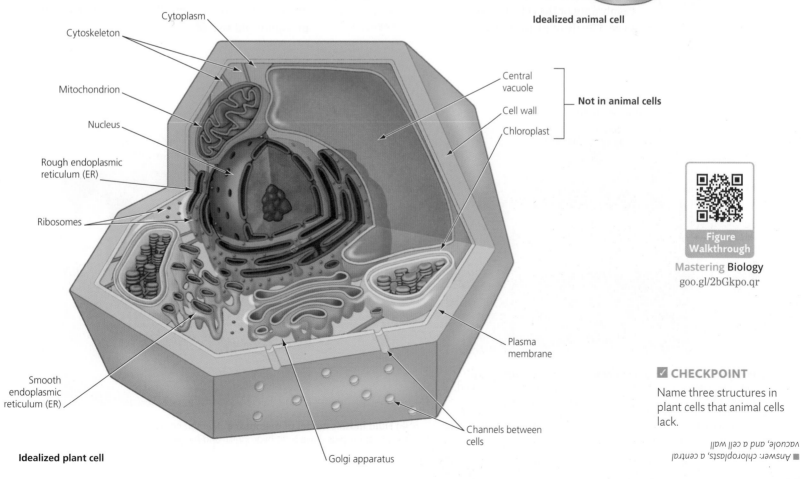

Idealized plant cell

Figure Walkthrough

Mastering Biology
goo.gl/2bGkpo.qr

☑ **CHECKPOINT**

Name three structures in plant cells that animal cells lack.

■ *Answer: chloroplasts, a central vacuole, and a cell wall*

93

Cell Surfaces

Surrounding the plasma membrane of a plant cell is a cell wall made from cellulose fibers, which are long chains of polysaccharides (see Figure 3.9c). The walls protect the cells, maintain cell shape, and keep cells from absorbing so much water that they burst. Plant cells are connected to each other by channels that pass through the cell walls, joining the cytoplasm of each cell to that of its neighbors. These channels allow water and other small molecules to move between cells, integrating the activities of a tissue.

Animal cells lack a cell wall, but most animal cells secrete a sticky coat called the **extracellular matrix**. Fibers made of the protein collagen (also found in your skin, cartilage, bones, and tendons) hold cells together in tissues and can also have protective and supportive functions. In addition, the surfaces of most animal cells contain cell junctions, structures that connect cells together into tissues, allowing the cells to function in a coordinated way. Next, in the Process of Science section, we'll see how the outer surfaces of bacterial cells are exploited by a new antibiotic. ☑

☑ **CHECKPOINT**

What polysaccharide is the primary component of plant cell walls?

■ Answer: cellulose

THE PROCESS OF SCIENCE | Humans Versus Bacteria

How Was the First 21st-Century Antibiotic Discovered?

BACKGROUND

As discussed in the Biology and Society section, antibiotics are drugs that can treat infections by killing or slowing the growth of bacteria. Many antibiotics disrupt cellular structures that are found in bacteria but not in human cells, such as cell walls and bacterial chromosomes. Because human cells lack these structures, the antibiotics will not harm the human host. Like many scientific breakthroughs, antibiotics were not created by scientists in the lab, but rather were discovered to already exist in nature.

Most of the antibiotics prescribed today were developed decades ago **(Figure 4.6a)**. Penicillin, for example, was first discovered in 1928, amoxicillin in 1972 and ciprofloxacin in 1987. As populations of bacteria are exposed to antibiotics, natural selection will favor those that can survive and reproduce. Thus, many early antibiotics have been rendered ineffective by the evolution of antibiotic resistance.

To combat the growing problem of antibiotic resistance, medical researchers are constantly trying to produce new antibiotics. Although some (such as ciprofloxacin) are human-made, most classes of antibiotics were discovered in the natural world. Penicillin and ampicillin, for example, are produced by fungi and bacteria (respectively) that grow naturally in soil.

METHOD

One of the newest natural antibiotics to be discovered is a molecule called teixobactin. What may be most interesting is how it was discovered. Soil samples from a grass field in Maine were placed in plastic devices with many tiny compartments. The soil was diluted so each compartment contained a single bacterium, thereby allowing different species of bacteria to be grown simultaneously. The devices were then buried in the original soil **(Figure 4.6b)**. Membranes allowed nutrients into each compartment while keeping other bacteria species out. After being grown into colonies, the bacteria in each compartment were tested for

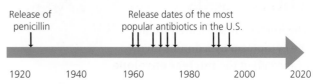

Release of penicillin

Release dates of the most popular antibiotics in the U.S.

1920 1940 1960 1980 2000 2020

(a) Release dates of the most widely used antibiotics in the United States

Compartments

(b) A device used to grow potentially helpful species of bacteria in soil

Chromosome
Cell wall
Ribosomes

(c) Bacterial structures commonly targeted by antibiotics

◀ **Figure 4.6 The search for new antibiotics.** The quest for drugs that target harmful bacteria is an ongoing effort.

the ability to kill the bacteria *Staphylococcus aureus,* which causes deadly MRSA (methicillin-resistant *S. aureus*) infections and tuberculosis.

RESULTS

When the researchers analyzed the soil samples, they observed that the bacterium most effective in killing *S. aureus* was a previously undiscovered species named *Eleftheria terrae*. They found that this bacterium produced teixobactin, which inhibits production of bacterial cell walls, leading to the destruction of the bacterial cell **(Figure 4.6c)**.

After the discovery of teixobactin in 2015, researchers expected that it will take several years until this new antibiotic can be tested against *S. aureus* in human clinical trials. If proven successful, it will be a good illustration of how a knowledge of cellular structures, combined with new technologies, can benefit human health.

Thinking Like a Scientist

In what way does the history of antibiotic discovery promote efforts to conserve natural habitats?

For the answer, see Appendix D.

The Nucleus and Ribosomes: Genetic Control of the Cell

If you think of the cell as a factory, then the nucleus is its control center. Here, the master plans are stored, orders are given, changes are made in response to external signals, and the process of making new factories is initiated. The factory supervisors are the genes, the inherited DNA molecules that direct almost all the business of the cell. Each gene is a stretch of DNA that stores the information necessary to produce a particular protein. Proteins can be likened to workers on the factory floor because they do most of the actual work of the cell.

The Nucleus

The nucleus is separated from the cytoplasm by a double membrane called the **nuclear envelope**

(Figure 4.7). Each membrane of the nuclear envelope is similar in structure to the plasma membrane: a phospholipid bilayer with associated proteins. Pores in the envelope allow certain materials to pass between the nucleus and the surrounding cytoplasm. Within the nucleus, long DNA molecules and associated proteins form fibers called **chromatin**. Each long chromatin fiber constitutes one chromosome (Figure 4.8). The number of chromosomes in a cell depends on the species. For example, each human body cell has 46 chromosomes, whereas rice cells have 24 and dog cells have 78 (see Figure 8.2 for more examples). The **nucleolus** (shown in Figure 4.7), a prominent structure within the nucleus, is the site where the components of ribosomes are made. We'll examine ribosomes next. ☑

☑ CHECKPOINT

What is the relationship between chromosomes, chromatin, and DNA?

■ *Answer: Chromosomes are made of chromatin, which is a combination of DNA and proteins.*

▼ Figure 4.7 **The nucleus.**

Chromatin fiber · Nuclear envelope · Nucleolus · Nuclear pore

TEM 8,800×

TEM 12,500×

Surface of nuclear envelope

Nuclear pores

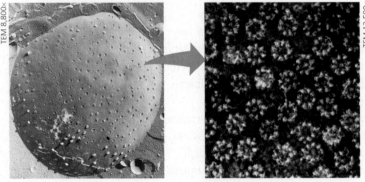

▼ Figure 4.8 **The relationship between DNA, chromatin, and a chromosome.**

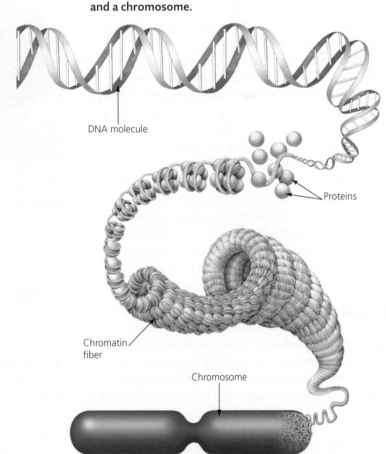

DNA molecule

Proteins

Chromatin fiber

Chromosome

Ribosomes

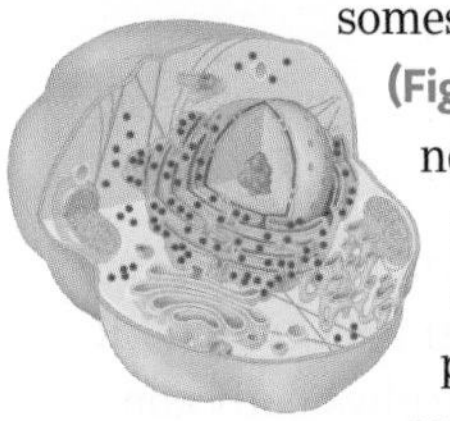

The small blue dots in the cells in Figure 4.4 and outside the nucleus in Figure 4.7 represent the ribosomes. Ribosomes are responsible for protein synthesis (Figure 4.9). In eukaryotic cells, the components of ribosomes are made in the nucleus and then transported through the pores of the nuclear envelope into the cytoplasm. It is in the cytoplasm that the ribosomes assemble and begin their work. Some ribosomes are suspended in the cytosol, making proteins that remain within the fluid of the cell. Other ribosomes are attached to the outside of the nucleus or an organelle called the endoplasmic reticulum (or ER) (Figure 4.10), making proteins that are incorporated into membranes or secreted by the cell. Free and bound ribosomes are structurally identical, and ribosomes can switch locations, moving between the endoplasmic reticulum and the cytosol. Cells that make a lot of proteins have many ribosomes. For example, each cell in your pancreas that produces digestive enzymes may contain a few million ribosomes.

▶ **Figure 4.9** **A computer model of a ribosome in the process of synthesizing a protein.**

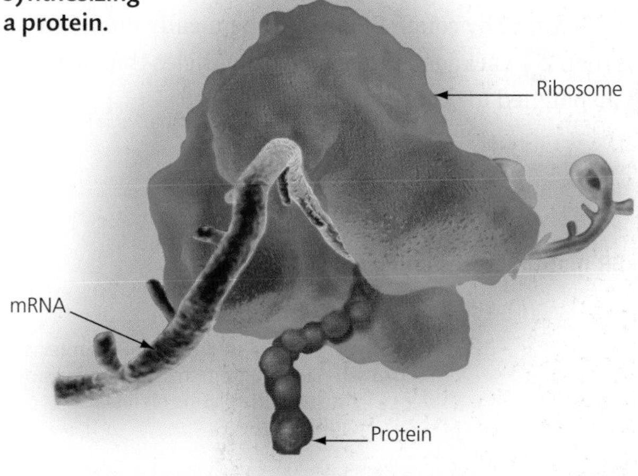

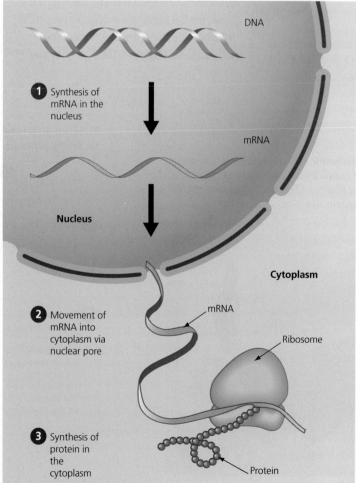

▼ **Figure 4.10** **ER-bound ribosomes.**

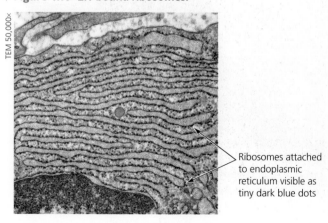

Ribosomes attached to endoplasmic reticulum visible as tiny dark blue dots

How DNA Directs Protein Production

Like a company executive, the DNA doesn't actually do any of the work of the cell. Instead, the DNA executive issues orders that result in work being done by the protein workers. Figure 4.11 shows the sequence of events during protein production in a eukaryotic cell (with the DNA and other structures being shown disproportionately large in relation to the nucleus). ❶ DNA transfers its coded information to a molecule called messenger RNA (mRNA). Like a middle manager, the mRNA molecule carries the order to "build this type of protein." ❷ The mRNA exits the nucleus through pores in the nuclear envelope and travels to the cytoplasm, where it binds to a ribosome. ❸ The ribosome moves along the mRNA, translating the genetic message into a protein with a specific amino acid sequence. (You'll learn how the message is translated in Chapter 10.) In this way, information carried by the DNA can direct the work of the entire cell without the DNA ever leaving the protective confines of the nucleus. The pathway from DNA to RNA to protein, involving several organelles, illustrates the importance of the flow of information within living cells. ☑

☑ **CHECKPOINT**

1. What is the function of ribosomes?
2. What is the role of mRNA in making a protein?

■ Answers: 1. protein synthesis 2. A molecule of mRNA carries the genetic message from a gene (DNA) to ribosomes that translate it into protein.

◀ **Figure 4.11** **DNA → RNA → protein.** Inherited genes in the nucleus control protein production and hence the activities of the cell.

DNA

❶ Synthesis of mRNA in the nucleus

mRNA

Nucleus

Cytoplasm

❷ Movement of mRNA into cytoplasm via nuclear pore

mRNA

Ribosome

❸ Synthesis of protein in the cytoplasm

Protein

The Endomembrane System: Manufacturing and Distributing Cellular Products

Like an office partitioned into cubicles, the cytoplasm of a eukaryotic cell is partitioned by organelle membranes (see Figure 4.4). Some organelles are physically connected, like two cubicles that share a door. Other organelles are linked by **vesicles**, sacs made of membrane that transfer membrane segments between organelles, like a mail cart delivering packages from one department to another. Together, these organelles form the **endomembrane system**. This system includes the nuclear envelope, the endoplasmic reticulum, the Golgi apparatus, lysosomes, and vacuoles.

The Endoplasmic Reticulum

The **endoplasmic reticulum (ER)** is one of the main manufacturing facilities within a cell. It produces an enormous variety of molecules. Connected to the nuclear envelope, the ER forms an extensive labyrinth of tubes and sacs running throughout the cytoplasm **(Figure 4.12)**. A membrane separates the internal ER compartment from the cytosol. There are two components that make up the ER: rough ER and smooth ER. These two types of ER are physically connected but differ in structure and function.

Rough ER

The "rough" in **rough ER** refers to ribosomes that stud the outside of its membrane. One of the functions of rough ER is to make phospholipids that are inserted into the ER membrane. In this way, the ER membrane grows, and portions of it can bubble off and be transferred to other parts of the cell. The ribosomes attached to the rough ER produce proteins that will be inserted into the growing ER membrane, transported to other organelles, and eventually exported. Cells that secrete a lot of protein—such as the cells of your salivary glands, which secrete enzymes into your mouth—are especially rich in rough ER. As shown in **Figure 4.13**, ❶ some products manufactured by rough ER are ❷ chemically modified and then ❸ packaged into transport vesicles that are ❹ dispatched to other locations in the cell.

▼ **Figure 4.12 The endoplasmic reticulum (ER).** In this drawing, the flattened sacs of rough ER and the tubes of smooth ER are connected. Notice that the ER is also connected to the nuclear envelope (the nucleus has been omitted from the illustration for clarity).

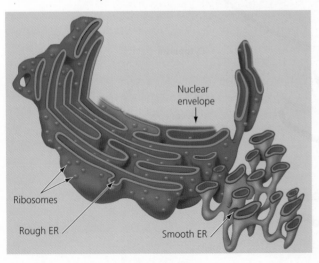

Nuclear envelope

Ribosomes

Rough ER

Smooth ER

▼ **Figure 4.13 How rough ER manufactures and packages secretory proteins.**

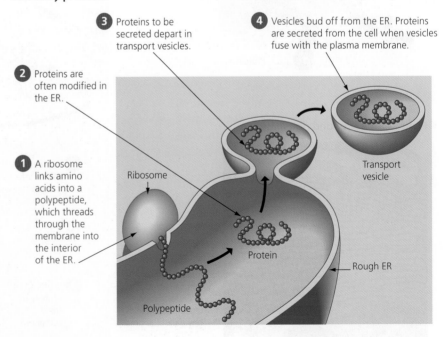

❸ Proteins to be secreted depart in transport vesicles.

❹ Vesicles bud off from the ER. Proteins are secreted from the cell when vesicles fuse with the plasma membrane.

❷ Proteins are often modified in the ER.

❶ A ribosome links amino acids into a polypeptide, which threads through the membrane into the interior of the ER.

Ribosome

Protein

Transport vesicle

Rough ER

Polypeptide

Smooth ER

The "smooth" in **smooth ER** refers to the fact that this organelle lacks the ribosomes that populate the surface of rough ER (see Figure 4.13). A diversity of enzymes built into the smooth ER membrane enables this organelle to perform many functions. One is the synthesis of lipids, including steroids (see Figure 3.14). For example, the cells in ovaries or testes that produce the steroid sex hormones are enriched with smooth ER. In liver cells, enzymes of the smooth ER detoxify circulating drugs such as barbiturates, amphetamines, and some antibiotics (which is why antibiotics don't remain in the bloodstream for long after you stop taking them). As liver cells are exposed to a drug, the amounts of smooth ER and its detoxifying enzymes increase. This can strengthen the body's tolerance of the drug, meaning that higher doses will be required in the future to achieve the desired effect. The growth of smooth ER in response to one drug can also increase tolerance of other drugs. For example, abusing sleeping pills may make other useful drugs less effective by accelerating their breakdown in the liver.

The Golgi Apparatus

Working in close partnership with the ER, the **Golgi apparatus**, an  organelle named for its discoverer (Italian scientist Camillo Golgi), receives, refines, stores, and distributes chemical products of the cell **(Figure 4.14)**. You can think of the Golgi apparatus as a detailing facility that receives shipments of newly manufactured goods (proteins), puts on finishing touches, stores the completed goods, and then ships them out when needed.

Products made in the ER reach the Golgi apparatus in transport vesicles. The Golgi apparatus consists of a stack of membrane plates, looking much like a pile of pita bread. ❶ One side of a Golgi stack serves as a receiving dock for vesicles from the ER. ❷ Proteins within a vesicle are usually modified by enzymes during their transit from the receiving to the shipping side of the Golgi apparatus. For example, molecular identification tags may be added that serve to mark and sort protein molecules into different batches for different destinations. ❸ The shipping side of a Golgi stack is a depot from which finished products can be carried in transport vesicles to other organelles or to the plasma membrane. Vesicles that bind with the plasma membrane transfer proteins to it or secrete finished products to the outside of the cell. ☑

☑ **CHECKPOINT**

1. What makes rough ER rough?
2. What is the relationship between the Golgi apparatus and the ER in a protein-secreting cell?

■ *Answers: 1. ribosomes attached to the membrane 2. The Golgi apparatus receives proteins from the ER through vesicles, processes them proteins, and then dispatches them in vesicles.*

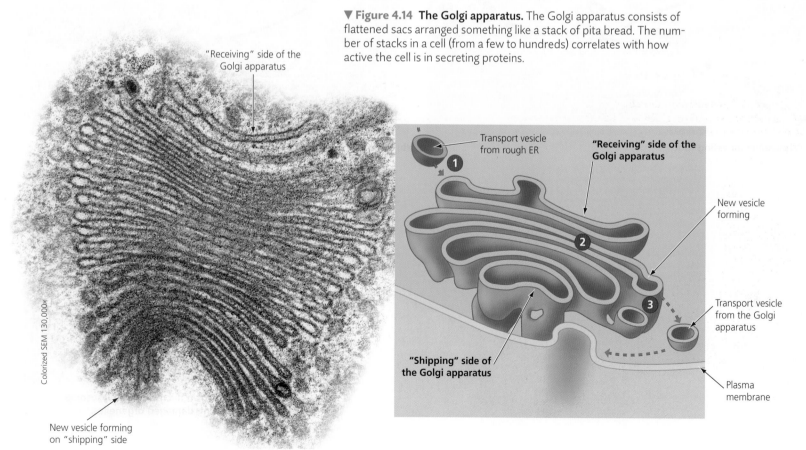

▼ **Figure 4.14 The Golgi apparatus.** The Golgi apparatus consists of flattened sacs arranged something like a stack of pita bread. The number of stacks in a cell (from a few to hundreds) correlates with how active the cell is in secreting proteins.

"Receiving" side of the Golgi apparatus

Colorized SEM 130,000×

New vesicle forming on "shipping" side

Transport vesicle from rough ER

"Receiving" side of the Golgi apparatus

New vesicle forming

"Shipping" side of the Golgi apparatus

Transport vesicle from the Golgi apparatus

Plasma membrane

Lysosomes

A **lysosome** is a membrane-enclosed sac of digestive enzymes. The enzymes and membranes of lysosomes are made by rough ER and processed in the Golgi apparatus. The lysosome provides a compartment where digestive enzymes can safely break down large molecules without unleashing the digestive enzymes on the cell itself.

Lysosomes perform several digestive functions. Many single-celled protists engulf nutrients into tiny cytoplasmic sacs called food vacuoles. Lysosomes fuse with the food vacuoles, exposing the food to digestive enzymes **(Figure 4.15a)**. Small molecules that result from this digestion, such as amino acids, leave the lysosome and nourish the cell. Lysosomes also help destroy harmful bacteria. For example, your white blood cells ingest bacteria into vacuoles, and lysosomal enzymes that are emptied into these vacuoles rupture the bacterial cell walls. In addition, without harming the cell, a lysosome can engulf and digest parts of another organelle, recycling it by making its molecules available for the construction of new organelles **(Figure 4.15b and c)**. With the help of lysosomes, a cell can thereby continually renew itself. Lysosomes also have sculpting functions in embryonic development. For example, lysosomes release enzymes that digest webbing between developing fingers in an early human embryo.

The importance of lysosomes to cell function and human health is made clear by hereditary disorders called lysosomal storage diseases. A person with such a disease is missing one or more of the digestive enzymes normally found within lysosomes. The abnormal lysosomes become filled with indigestible substances, and this eventually interferes with other cellular functions. Most of these diseases are fatal in early childhood. In Tay-Sachs disease, for example, lysosomes lack a lipid-digesting enzyme. As a result, nerve cells die as they accumulate excess lipids, ravaging the nervous system. Fortunately, lysosomal storage diseases are rare. ✓

☑ CHECKPOINT

How can defective lysosomes result in excess accumulation of a particular chemical compound in a cell?

■ *Answer: If the lysosomes lack an enzyme needed to break down the compound, the cell will accumulate an excess of that compound.*

▼ **Figure 4.15 Two functions of lysosomes.**

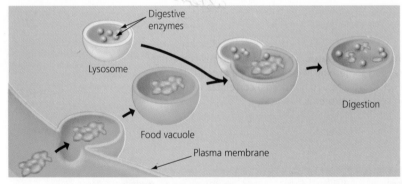

(a) **A lysosome digesting food**

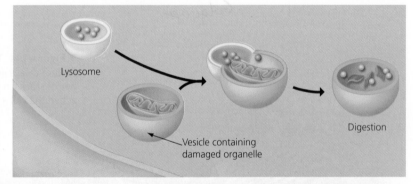

(b) **A lysosome breaking down the molecules of damaged organelles**

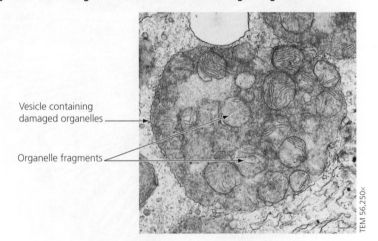

(c) **Electron micrograph of lysosome recycling damaged organelles.**

Vacuoles

Vacuoles are large vesicles with a variety of functions. For example, Figure 4.15a shows a food vacuole budding 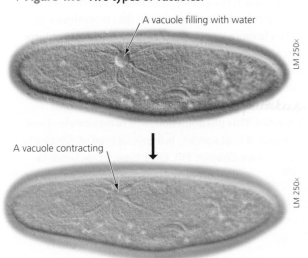 from the plasma membrane. Certain freshwater protists have contractile vacuoles that pump out excess water that flows into the cell (**Figure 4.16a**).

Another type of vacuole is a **central vacuole**, a versatile compartment that can account for more than half the volume of a mature plant cell (**Figure 4.16b**). A central vacuole stores organic nutrients, such as proteins stockpiled in the vacuoles of seed cells. It also contributes to plant growth by absorbing water and causing cells to expand. In the cells of flower petals, central vacuoles

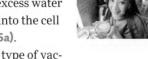

may contain pigments that attract pollinating insects. Central vacuoles may also contain poisons that protect against plant-eating animals. Some important crop plants produce and store large amounts of toxic chemicals—harmful to animals that might graze on the plant but useful to us—such as tobacco plants (which store nicotine) and coffee and tea plants (which store caffeine).

THE CAFFEINE THAT GIVES COFFEE A KICK PROTECTS COFFEE PLANTS FROM HERBIVORES.

Now that we've explored the organelles of the endomembrane system, **Figure 4.17** reviews how they are related. A product made in one part of the endomembrane system may exit the cell or become part of another organelle without crossing a membrane. Also, membrane made by the ER can become part of the plasma membrane through the fusion of a transport vesicle. So even the plasma membrane is related to the endomembrane system. ☑

☑ **CHECKPOINT**

Place the following cellular structures in the order they would be used in the production and secretion of a protein: Golgi apparatus, nucleus, plasma membrane, ribosome, transport vesicle.

■ *Answer: nucleus, ribosome, transport vesicle, Golgi apparatus, plasma membrane*

▼ **Figure 4.16 Two types of vacuoles.**

A vacuole filling with water

LM 250×

A vacuole contracting

LM 250×

(a) Contractile vacuole in *Paramecium*. A contractile vacuole fills with water and then contracts to pump the water out of the cell.

Central vacuole

Colorized TEM 1,700×

(b) Central vacuole in a plant cell. The central vacuole (colorized blue in this micrograph) is often the largest organelle in a mature plant cell.

▼ **Figure 4.17 Review of the endomembrane system.** The dashed arrows show some of the pathways of cell product distribution and membrane migration through transport vesicles.

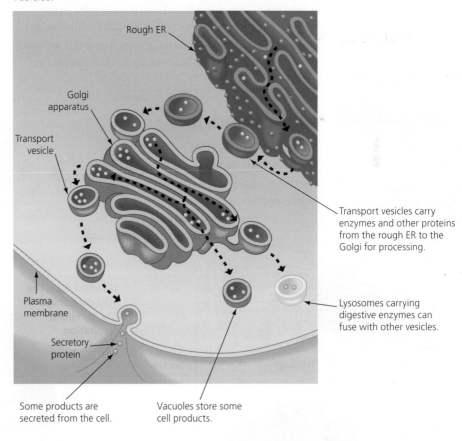

Rough ER

Golgi apparatus

Transport vesicle

Transport vesicles carry enzymes and other proteins from the rough ER to the Golgi for processing.

Plasma membrane

Lysosomes carrying digestive enzymes can fuse with other vesicles.

Secretory protein

Some products are secreted from the cell.

Vacuoles store some cell products.

Chloroplasts and Mitochondria: Providing Cellular Energy

One of the central themes of biology is the transformation of energy: how it enters living systems, is converted from one form to another, and is eventually given off as heat. To follow this flow of energy, we must consider the cellular power stations: chloroplasts and mitochondria.

Chloroplasts

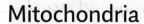

Most of the living world runs on energy from photosynthesis, the conversion of light energy from the sun to the chemical energy of sugar molecules. **Chloroplasts** are the photosynthetic organelles of plants and algae.

A chloroplast is divided into compartments by membranes, **(Figure 4.18)**. The innermost compartment holds a thick fluid called stroma, which contains DNA, ribosomes, and enzymes. A network of sacs called thylakoids is suspended in the stroma. The sacs are often stacked like poker chips; each stack is called a granum (plural, *grana*). The grana are solar power packs, converting light energy to chemical energy (as detailed in Chapter 7).

Mitochondria

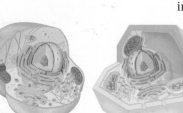

Mitochondria (singular, *mitochondrion*) are the organelles in which cellular respiration takes place; during cellular respiration, energy is harvested from sugars and

transformed into another form of chemical energy called ATP (adenosine triphosphate). Cells use molecules of ATP as a direct energy source. Mitochondria are found in almost all eukaryotic cells, including those of plants and animals.

A mitochondrion is enclosed by two membranes, separated by a narrow space. The inner membrane encloses a thick fluid called the mitochondrial matrix (**Figure 4.19**). The inner membrane has many infoldings called cristae. The cristae create a large surface area in which many molecules that function in cellular respiration are embedded, maximizing ATP output. (You'll learn more about how mitochondria work in Chapter 6.)

Besides providing cellular energy, mitochondria and chloroplasts share another feature: They contain their own DNA that encodes some proteins made by their own ribosomes. Each chloroplast and mitochondrion contains a single circular DNA chromosome that resembles a prokaryotic chromosome. In fact, mitochondria and chloroplasts can grow and pinch in two, reproducing themselves as many prokaryotes do. This and other evidence indicate that they evolved from ancient free-living prokaryotes that established residence within other, larger host prokaryotes. This phenomenon, where one species lives inside a host species, is a special type of symbiosis (see Chapter 16). Over time, mitochondria and chloroplasts became increasingly interdependent with the host prokaryote, eventually evolving into a single organism with inseparable parts. **The DNA found within mitochondria and chloroplasts likely includes remnants of this ancient evolutionary event.** ☑

▼ **Figure 4.18** The chloroplast: site of photosynthesis.

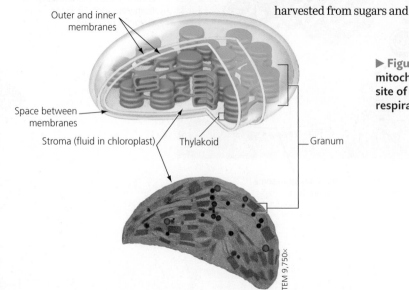

Outer and inner membranes

Space between membranes

Stroma (fluid in chloroplast)

Thylakoid

Granum

TEM 9,750x

▶ **Figure 4.19** The mitochondrion: site of cellular respiration.

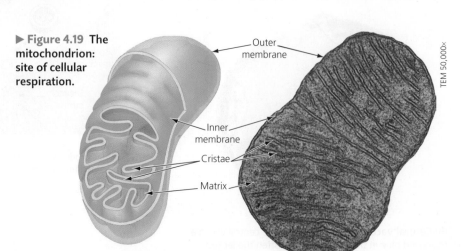

Outer membrane

Inner membrane

Cristae

Matrix

TEM 50,000x

The Cytoskeleton:
Cell Shape and Movement

If someone asked you to describe a house, you would most likely mention the various rooms and their locations. You probably would not think to mention the foundation and beams that support the house. Yet these structures perform an extremely important function. Similarly, cells have an infrastructure called the **cytoskeleton**, a network of protein fibers extending throughout the cytoplasm. The cytoskeleton serves as both skeleton and "muscles" for the cell, functioning in support and movement.

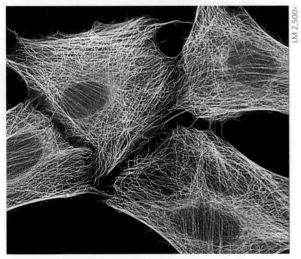

WITHOUT THE CYTOSKELETON, YOUR CELLS WOULD COLLAPSE IN ON THEMSELVES.

Maintaining Cell Shape

One function of the cytoskeleton is to give mechanical support to the cell and maintain its shape. This is especially important for animal cells, which lack rigid cell walls. The cytoskeleton contains several types of fibers made from different types of protein. One important type of fiber forms **microtubules**, hollow tubes of protein (**Figure 4.20a**). The other kinds of cytoskeletal fibers, called intermediate filaments and microfilaments, are thinner and solid.

Just as the bony skeleton of your body helps fix the positions of your organs, the cytoskeleton provides anchorage and reinforcement for many organelles in a cell. For instance, the nucleus is held in place by a "cage" of cytoskeletal filaments. Other organelles use the cytoskeleton for movement. For example, a lysosome might reach a food vacuole by gliding along a microtubule track. Microtubules growing from a region called a centrosome guide the movement of chromosomes when cells divide (by means of the mitotic spindle—see Chapter 8). Microfilaments are also involved in cell movements. Different filaments composed of different kinds of proteins contract, resulting in the crawling movement of the protist *Amoeba* (**Figure 4.20b**) and movement of some of our white blood cells. ☑

☑ **CHECKPOINT**

From which important class of biological molecules are the microtubules of the cytoskeleton made?

■ *Answer: protein*

▼ **Figure 4.20 The cytoskeleton.**

(b) Microtubules and movement. The crawling movement of an *Amoeba* is due to the rapid degradation and rebuilding of microtubules.

LM 250x

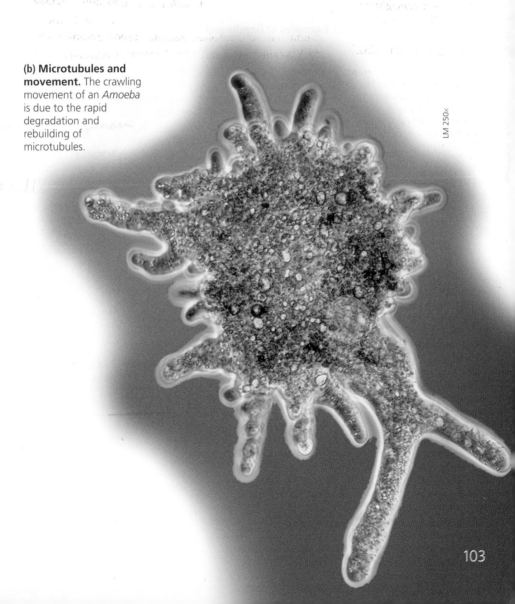

LM 2,500x

(a) Microtubules in the cytoskeleton. In this micrograph of animal cells, the cytoskeleton microtubules are labeled with a fluorescent yellow dye.

Flagella and Cilia

In some eukaryotic cells, microtubules are arranged into structures called flagella and cilia, extensions from a cell that aid in movement. Eukaryotic **flagella** (singular, *flagellum*) propel cells with an undulating, whiplike motion. They often occur singly, such as in human sperm cells (**Figure 4.21a**), but may also appear in groups on the outer surface of protists. **Cilia** (singular, *cilium*) are generally shorter and more numerous than flagella and move in a coordinated back-and-forth motion, like the rhythmic oars of a rowing team. Both cilia and flagella propel various protists through water. For example, *Paramecium* cilia are visible in the SEM in Figure 4.1. Though different in length, number per cell, and beating pattern, cilia and flagella have the same basic architecture.

Some cilia extend from nonmoving cells that are part of a tissue layer, moving fluid over the tissue's surface. For example, cilia lining your windpipe clean your respiratory system by sweeping mucus with trapped debris out of your lungs (**Figure 4.21b**). Tobacco smoke can paralyze these cilia, interfering with the normal cleansing mechanisms and allowing more toxin-laden smoke particles to reach the lungs. Frequent coughing—common in heavy smokers—then becomes the body's attempt to cleanse the respiratory system.

Because human sperm rely on flagella for movement, it's easy to understand why problems with flagella can lead to male infertility. Interestingly, some men with a type of hereditary sterility also suffer from respiratory problems. Because of a defect in the structure of their flagella and cilia, the sperm of men afflicted with this disorder cannot swim normally within the female reproductive tract to fertilize an egg (causing sterility), and their cilia do not sweep mucus out of their lungs (causing recurrent respiratory infections). ☑

☑ **CHECKPOINT**

Compare and contrast cilia and flagella.

■ *Answer: Cilia and flagella have the same basic structure, are made from microtubules, and aid in movement. Cilia are short and numerous and move back and forth. Flagella are longer, often occurring singly, and they undulate.*

▼ **Figure 4.21 Examples of flagella and cilia.**

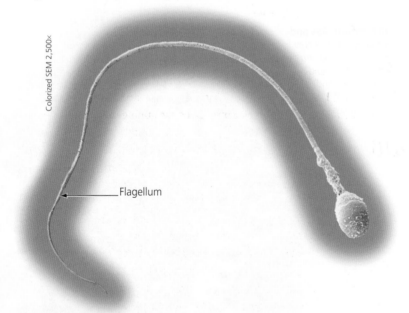

Colorized SEM 2,500×

Flagellum

(a) Flagellum of a human sperm cell. A eukaryotic flagellum undulates in a whiplike motion, driving a cell such as this sperm cell through its fluid environment.

Colorized SEM 3,000×

Cilia

(b) Cilia lining the respiratory tract. The cilia lining your respiratory tract sweep mucus with trapped debris out of your lungs. This helps keep your airway clear and prevents infections.

The Evolution of Bacterial Resistance in Humans

Individuals with variations that make them better suited for the local environment will survive and reproduce more often (on average) than those who lack such variations. When the advantageous variations have a genetic basis, the offspring of individuals with the variations will more often also have the favorable adaptions, giving them a survival and reproductive advantage. In this way, repeated over many generations, natural selection promotes evolution of the population.

Within a human population, the presence of a disease can provide a new basis for measuring those people who are best suited for survival in the local environment. For example, a recent evolutionary study examined people living in Bangladesh. This population has been exposed to the disease cholera—caused by an infectious bacterium—for millennia **(Figure 4.22)**. After cholera bacteria enter a victim's digestive tract (usually through contaminated drinking water), the bacteria produce a toxin that binds to intestinal cells. There, the toxin alters proteins in the plasma membrane, causing the cells to excrete fluid. The resulting diarrhea, which spreads the bacteria by shedding it back into the environment, can cause severe dehydration and death if untreated.

Because Bangladeshis have lived for so long in an environment that teems with cholera bacteria, one might expect that natural selection would favor those individuals who have some resistance to the bacteria. Indeed, recent studies of people from Bangladesh revealed mutations in several genes that appear to confer an increased resistance to cholera. Researchers discovered one mutation in a gene that encodes for the plasma membrane proteins that are the targets of the cholera bacteria. Although the mechanism is not yet understood, these genes appear to offer a survival advantage by making the proteins more resistant to attack by the cholera toxin. Because such genes offer a survival advantage within this population, they have slowly spread through the Bangladeshi population over the past 30,000 years. In other words, the Bangladeshi population is evolving increased resistance to cholera.

In addition to providing insight into the recent evolutionary past, data from this study reveal potential ways that humans might thwart the cholera bacterium. Perhaps pharmaceutical companies can exploit the proteins produced by the identified mutations to create a new generation of antibiotics. If so, this will represent another way that biologists have applied lessons learned from our understanding of evolution to improve human health. It also reminds us that we humans, like all life on Earth, are shaped by evolution due to changes in our environment, including the presence of infectious microorganisms that live around us.

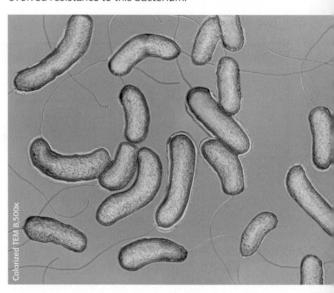

▼ **Figure 4.22** *Vibrio cholerae*, **the cause of the deadly disease cholera.** Some people living in Bangladesh have evolved resistance to this bacterium.

Colorized TEM 8,500x

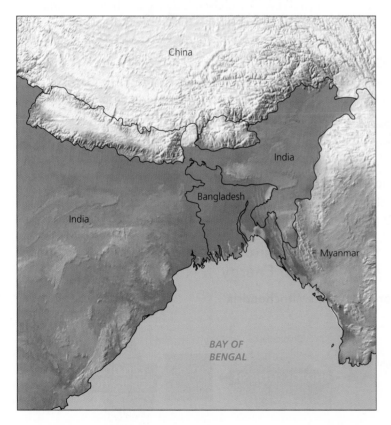

Chapter Review

SUMMARY OF KEY CONCEPTS

The Microscopic World of Cells

The Two Major Categories of Cells

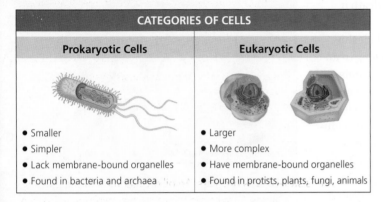

CATEGORIES OF CELLS	
Prokaryotic Cells	**Eukaryotic Cells**
• Smaller • Simpler • Lack membrane-bound organelles • Found in bacteria and archaea	• Larger • More complex • Have membrane-bound organelles • Found in protists, plants, fungi, animals

An Overview of Eukaryotic Cells

Membranes partition eukaryotic cells into functional compartments. The largest organelle is usually the nucleus. Other organelles are located in the cytoplasm, the region outside the nucleus and within the plasma membrane.

Membrane Structure

The Plasma Membrane

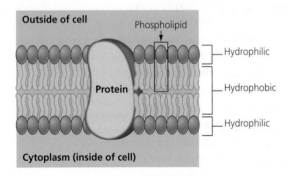

Cell Surfaces

The walls that encase plant cells support plants against the pull of gravity and also prevent cells from absorbing too much water. Animal cells are coated by a sticky extracellular matrix.

The Nucleus and Ribosomes: Genetic Control of the Cell

The Nucleus

An envelope consisting of two membranes encloses the nucleus. Within the nucleus, DNA and proteins make up chromatin fibers; each very long fiber is a single chromosome. The nucleus also contains the nucleolus, which produces components of ribosomes.

Ribosomes

Ribosomes produce proteins in the cytoplasm using messages produced by the DNA.

How DNA Directs Protein Production

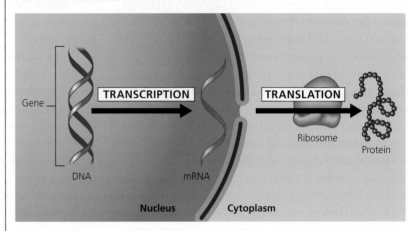

The Endomembrane System: Manufacturing and Distributing Cellular Products

The Endoplasmic Reticulum

The ER consists of membrane-enclosed tubes and sacs within the cytoplasm. Rough ER, named because of the ribosomes attached to its surface, makes membrane and secretory proteins. The functions of smooth ER include lipid synthesis and detoxification.

The Golgi Apparatus

The Golgi apparatus refines certain ER products and packages them in transport vesicles targeted for other organelles or export from the cell.

Lysosomes

Lysosomes, sacs containing digestive enzymes, aid digestion and recycling within the cell.

Vacuoles

Vacuoles include the contractile vacuoles that expel water from certain freshwater protists and the large, multifunctional central vacuoles of plant cells.

Chloroplasts and Mitochondria: Providing Cellular Energy

Chloroplasts and Mitochondria

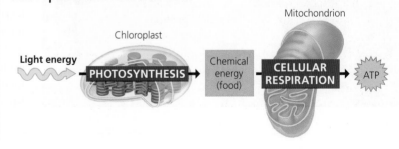

The Cytoskeleton: Cell Shape and Movement

Maintaining Cell Shape

Microtubules are an important component of the cytoskeleton, an organelle that gives support to, and maintains the shape of, cells.

Cilia and Flagella

Cilia and eukaryotic flagella are appendages that aid in movement, and they are made primarily of microtubules. Cilia are short and numerous and move the cell by coordinated beating. Flagella are long, often occur singly, and propel a cell with whiplike movements.

Mastering Biology

For practice quizzes, BioFlix animations, MP3 tutorials, video tutors, and more study tools designed for this textbook, go to Mastering Biology™

SELF-QUIZ

1. You look into a microscope and view an unknown cell. What might you see that would tell you that the cell is eukaryotic?
   a. DNA
   b. a nucleoid region
   c. a plasma membrane
   d. membrane-bound organelles

2. Explain how each word in the term *fluid mosaic* describes the structure of a membrane.

3. Identify which of the following structures includes all the others in the list: rough ER, smooth ER, endomembrane system, the Golgi apparatus.

4. A large amount of _____ ER can be found in liver cells to detoxify _____.

5. Why do cell walls make good targets for antibiotic drugs?

6. Name two similarities in the structure or function of chloroplasts and mitochondria. Name two differences.

7. Match the following organelles with their functions:
   a. nucleus   1. locomotion
   b. flagella   2. protein export
   c. mitochondria   3. gene control
   d. Golgi apparatus   4. digestion
   e. lysosomes   5. cellular respiration

8. DNA controls the cell by transmitting genetic messages that result in protein production. Place the following organelles in the order that represents the flow of genetic information from the DNA through the cell: nuclear pores, ribosomes, nucleus, rough ER, Golgi apparatus.

9. Compare and contrast cilia and flagella.

For answers to the Self Quiz, see Appendix D.

IDENTIFYING MAJOR THEMES

For each statement, identify which major theme is evident (the relationship of structure to function, information flow, pathways that transform energy and matter, interactions within biological systems, or evolution) and explain how the statement relates to the theme. If necessary, review the themes (see Chapter 1) and review the examples highlighted in blue in this chapter.

10. The genetic message contained in DNA is used to build proteins.

11. Several different organelles work together to carry out instructions in DNA.

12. Sunlight can be used to drive the photosynthesis of sugars.

For answers to Identifying Major Themes, see Appendix D.

THE PROCESS OF SCIENCE

13. Plant seeds store oils as droplets. An oil droplet membrane is a single layer of phospholipids. Draw a model for such a membrane. Explain why it is more stable than a bilayer.

14. Both chloroplasts and mitochondria contain rudimentary genomes, are able to divide independently from their host cells, and are bounded by double membranes. These similarities led to the idea that these organelles could have been derived from ancient prokaryotes that took up residence in eukaryotic ancestor cells. Which of the two organelles is more likely to have invaded the eukaryotic ancestor first and why?

15. **Interpreting Data** A population of bacteria may evolve resistance to a drug over time. Draw a graph that represents such a change. Label the x-axis as "time" and the y-axis as "population size." Draw a line on the graph that represents how the size of the population might change over time after the introduction of a new antibiotic. Label the point on your line where the new drug was introduced, and then indicate how the population size might change over time after that introduction.

BIOLOGY AND SOCIETY

16. Doctors at a university medical center removed John Moore's spleen, which is a standard treatment for his type of leukemia. The disease did not recur. Researchers kept the spleen cells alive in a nutrient medium. They found that some cells produced a blood protein that showed promise as a treatment for cancer and AIDS. The researchers patented the cells. Moore sued, claiming a share in profits from any products derived from his cells. The U.S. Supreme Court ruled against Moore, stating that his lawsuit "threatens to destroy the economic incentive to conduct important medical research." Moore argued that the ruling left patients "vulnerable to exploitation at the hands of the state." Do you think Moore was treated fairly? What else would you like to know about this case that might help you decide?

17. Scientists can manipulate living cells, changing their genetic composition and the way they function. Some companies have sought to patent engineered cell lines. Should society allow cells to be patented? Should the same patent rules apply to human and bacterial cells?

5 The Working Cell

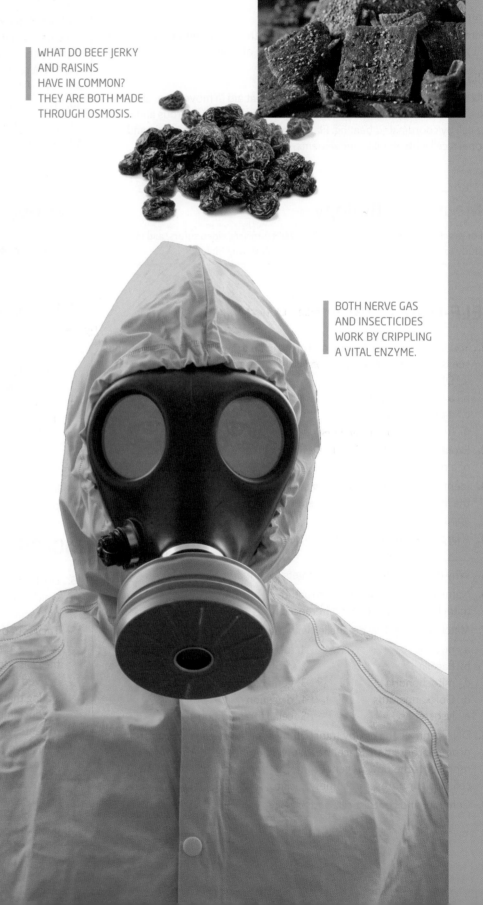

WHAT DO BEEF JERKY AND RAISINS HAVE IN COMMON? THEY ARE BOTH MADE THROUGH OSMOSIS.

Why Cellular Functions Matter

Life begins at the level of the cell: Nothing smaller is alive, and a single cell may contain all of the machinery necessary to carry out life's functions. Living cells are dynamic, performing work constantly. It is therefore important for any student of biology to understand how a working cell goes about the business of life.

BOTH NERVE GAS AND INSECTICIDES WORK BY CRIPPLING A VITAL ENZYME.

GET MOVING! YOU HAVE TO WALK MORE THAN 2 HOURS TO BURN THE CALORIES IN HALF A PEPPERONI PIZZA.

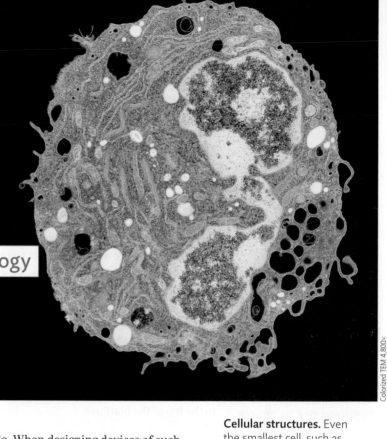

Colorized TEM 4,800×

BIOLOGY AND SOCIETY Nanotechnology

Harnessing Cellular Structures

Imagine a tiny robot with balls of carbon atoms for wheels, or a microchip carved onto an object 1,000 times smaller than a grain of sand. These are real-world examples of nanotechnology, the manipulation of materials at the molecular scale. When designing devices of such small size, researchers often turn to living cells for inspiration. After all, you can think of a cell as a machine that continuously and efficiently performs a variety of functions, such as movement, energy processing, and manufacturing a wide variety of products. Let's consider one example of cell-based nanotechnology and see how it relates to working cells.

Researchers at Cornell University are attempting to harvest the energy-producing capability of human sperm cells. Like other cells, a sperm cell generates energy by breaking down sugars and other molecules that pass through its plasma membrane. Enzymes within the cell release energy by breaking down glucose. The released energy is used to produce molecules of ATP. Within a living sperm, such ATP provides the energy that propels the cell through the female reproductive tract. In an attempt to harness this energy-producing system, the Cornell researchers attached three enzymes to a computer chip. The enzymes continued to function in this artificial system, producing energy from sugar. The hope is that a larger set of enzymes can eventually be used to power microscopic robots. Such nanorobots could use glucose from the bloodstream to power the delivery of drugs to body tissues, perhaps homing in on cancerous or otherwise abnormal cells. This example is only a glimpse into the incredible potential of new technologies inspired by working cells.

In this chapter, we'll explore three processes common to all living cells: energy metabolism, the use of enzymes to speed chemical reactions, and transport regulation by the plasma membrane. Along the way, we'll further consider nanotechnologies that mimic the natural activities of living cells.

Cellular structures. Even the smallest cell, such as this human white blood cell, is a miniature machine of startling complexity.

Some Basic Energy Concepts

Energy makes the world go round—both on a planetary scale and on a cellular scale. But what exactly is energy? Our first step in understanding the working cell is to learn a few basic concepts about energy.

Conservation of Energy

Energy is defined as the capacity to cause change. Some forms of energy are used to perform work, such as moving an object against an opposing force—for example, lifting a barbell against the force of gravity. Imagine a diver climbing to the top of a platform and then diving off **(Figure 5.1)**. To get to the top of the platform, the diver must perform work to overcome the opposing force of gravity. In this specific case, chemical energy from food is converted to **kinetic energy**, the energy of motion. The kinetic energy takes the form of muscle movement propelling the diver to the top of the platform.

What happens to the kinetic energy when the diver reaches the top of the platform? Does it disappear at that point? In fact, it does not. A physical principle known as **conservation of energy** explains that it is not possible to destroy or create energy. Energy can only be converted from one form to another. A power plant, for example, does not make energy; it merely converts it from one form (such as energy stored in coal) to a more convenient form (such as electricity). That's what happens in the diver's climb up the steps. The kinetic energy of muscle movement is stored as **potential energy**, the energy an object has because of its location or structure. The energy contained by water behind a dam and energy stored by a compressed spring are examples of potential energy. In our example, the diver at the top of the platform has potential energy because of his elevated location. Then the act of diving off the platform into the water converts the potential energy back to kinetic energy. **Life depends on countless similar transformations of energy from one form to another.** ✓

☑ **CHECKPOINT**

Can an object at rest have energy?

■ *Answer: Yes, it can have potential energy because of its location or structure.*

▶ **Figure 5.1 Energy transformations during a dive.**

On the platform, the diver has more potential energy.

Climbing the steps converts kinetic energy of muscle movement to potential energy.

Diving converts potential energy to kinetic energy.

In the water, the diver has less potential energy.

Heat

If energy cannot be destroyed, where has the energy gone in our example when the diver hits the water? The energy has been converted to **heat**, a type of kinetic energy contained in the random motion of atoms and molecules. The friction between the body and its surroundings generated heat in the air and then in the water.

All energy transformations generate some heat. Although releasing heat does not destroy energy, it does make it more difficult to harness for useful work. Heat is energy in its most disordered, chaotic form, the energy of aimless molecular movement.

Entropy is a measure of the amount of disorder, or randomness, in a system. Consider an analogy from your own room. It's easy to increase the chaos—in fact, it seems to happen spontaneously! But it requires the expenditure of significant energy to restore order once again.

Every time energy is converted from one form to another, entropy increases. The energy transformations during the climb up the ladder and the dive from the platform increased entropy as the diver emitted heat to the surroundings. To climb up the steps again for another dive, the diver must use additional stored food energy. This transformation will also create heat and therefore increase entropy. ☑

Chemical Energy

How can molecules derived from the food we eat provide energy for our working cells? The molecules of food, gasoline, and other fuels have a form of potential energy called **chemical energy**, which arises from the arrangement of atoms and can be released by a chemical reaction. Carbohydrates, fats, and gasoline have structures that make them especially rich in chemical energy.

Living cells and automobile engines use the same basic process to make the chemical energy stored in their fuels available for work **(Figure 5.2)**. In both cases, this process breaks organic fuel into smaller waste molecules that have much less chemical energy than the fuel molecules did, thereby releasing energy that can be used to perform work.

For example, the engine of an automobile mixes oxygen with gasoline (which is why all cars require an air intake system) in an explosive chemical reaction that breaks down the fuel molecules and pushes the pistons that eventually move the wheels. The waste products emitted from the car's exhaust pipe are mostly carbon dioxide and water. Only about 25% of the energy that an automobile engine extracts from its fuel is converted to the kinetic energy of the car's movement. Most of the rest is converted to heat—so much that the engine would

☑ **CHECKPOINT**

Which form of energy is most randomized and difficult to put to work?

■ *Answer: heat*

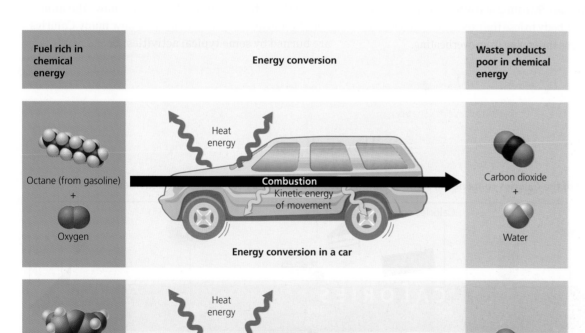

◀ **Figure 5.2 Energy transformations in a car and a cell.** In both a car and a cell, the chemical energy of organic fuel molecules is harvested using oxygen. This chemical breakdown releases energy stored in the fuel molecules and produces carbon dioxide and water. The released energy can be used to perform work.

Fuel rich in chemical energy — Energy conversion — **Waste products poor in chemical energy**

Octane (from gasoline) + Oxygen

Heat energy

Combustion
Kinetic energy of movement

Carbon dioxide + Water

Energy conversion in a car

Glucose (from food) + Oxygen

Heat energy

Cellular respiration
ATP
Energy for cellular work

Carbon dioxide + Water

Energy conversion in a cell

Figure Walkthrough

Mastering **Biology**
goo.gl/UWp9f9

melt if the car's radiator did not disperse heat into the atmosphere. That is why high-end performance cars need sophisticated air flow systems to avoid overheating.

Cells also use oxygen in reactions that release energy from fuel molecules. As in a car engine, the "exhaust" from such reactions in cells is mostly carbon dioxide and water. The combustion of fuel in cells is called cellular respiration, which is a more gradual and efficient "burning" of fuel compared with the explosive combustion in an automobile engine. Cellular respiration is the energy-releasing chemical breakdown of fuel molecules and the storage of that energy in a form the cell can use to perform work. (We will discuss the details of cellular respiration in Chapter 6.) You convert about 34% of your food energy to useful work, such as movement of your muscles. The rest of the energy released by the breakdown of fuel molecules generates body heat. Humans and many other animals can use this heat to keep the body at an almost constant temperature (37°C, or 98.6°F, in the case of humans), even when the surrounding air is much colder. You've probably noticed how quickly a crowded room warms up—it's all that released metabolic heat energy! The liberation of heat energy also explains why you feel hot after exercise. Sweating and other cooling mechanisms enable your body to lose the excess heat, much as a car's radiator keeps the engine from overheating.

YOU HAVE TO WALK MORE THAN 2 HOURS TO BURN THE CALORIES IN HALF A PEPPERONI PIZZA.

☑ CHECKPOINT

According to Figure 5.3, how long would you have to swim to burn off the energy contained in a hamburger?

■ *Answer: One hour (swimming burns 200 calories each half hour, and a hamburger is 400 calories)*

Food Calories

Read any packaged food label and you'll find the number of calories in each serving of that food. Calories are units of energy. A **calorie** (cal) is the amount of energy that can raise the temperature of 1 gram (g) of water by 1°C. You could actually measure the caloric content of a peanut by burning it under a container of water to convert all of the stored chemical energy to heat and then measuring the temperature increase of the water.

Calories are tiny units of energy, so using them to describe the fuel content of foods is not practical. Instead, it's conventional to use kilocalories (kcal), units of 1,000 calories. In fact, the Calories (capital C) on a food package are actually kilocalories. For example, one peanut has about 5 Calories. That's a lot of energy, enough to increase the temperature of 1 kg (a little more than a quart) of water by 5°C. And just a handful of peanuts contains enough Calories, if converted to heat, to boil 1 kg of water. In living organisms, of course, food isn't used to boil water but instead is used to fuel the activities of life. **Figure 5.3** shows the number of Calories in several foods and how many Calories are burned by some typical activities. ☑

▼ Figure 5.3 **A comparison of the Calories contained in some common foods and burned by some common activities.**

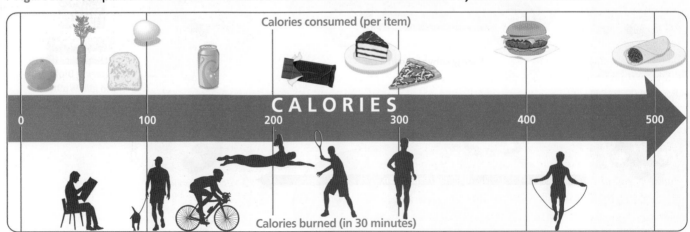

ATP and Cellular Work

The carbohydrates, fats, and other fuel molecules we obtain from food cannot be used directly as fuel for our cells. Instead, the chemical energy released by the breakdown of organic molecules during cellular respiration is used to generate molecules of ATP. These molecules of ATP then power cellular work. ATP acts like an energy shuttle, storing energy obtained from food and then releasing it as needed at a later time. **Such energy transformations are essential for all life on Earth.**

The Structure of ATP

The abbreviation ATP stands for adenosine triphosphate. **ATP** consists of an organic molecule called adenosine plus a tail of three phosphate groups **(Figure 5.4)**. The triphosphate tail is the "business" end of ATP, the part that provides energy for cellular work. Each phosphate group is negatively charged. Negative charges repel each other. The crowding of negative charges in the triphosphate tail contributes to the potential energy of ATP. It's analogous to storing energy by compressing a spring; if you release the spring, it will relax, and you can use that springiness to do some useful work. For ATP power, it is release of the phosphate at the tip of the triphosphate tail that makes energy available to working cells. What remains is **ADP**, adenosine diphosphate (two phosphate groups instead of three, shown on the right side of Figure 5.4).

Phosphate Transfer

When ATP drives work in cells by being converted to ADP, the released phosphate groups don't just fly off into space. ATP energizes other molecules in cells by transferring phosphate groups to those molecules. When a target molecule accepts the third phosphate group, it becomes energized and can then perform work in the cell. Imagine a bicyclist pedaling up a hill. In the muscle cells of the rider's legs, ATP transfers phosphate groups to motor proteins. The proteins then change shape, causing the muscle cells to contract **(Figure 5.5a)**. This contraction provides the mechanical energy needed to propel the rider. ATP also enables the transport of ions and other dissolved substances across the membranes of the rider's nerve cells **(Figure 5.5b)**, helping them send signals to her legs. And ATP drives the production of a cell's large molecules from smaller molecular building blocks **(Figure 5.5c)**.

▼ **Figure 5.4 ATP power.** Each ⓟ in the triphosphate tail of ATP represents a phosphate group, a phosphorus atom bonded to oxygen atoms. The transfer of a phosphate from the triphosphate tail to other molecules provides energy for cellular work.

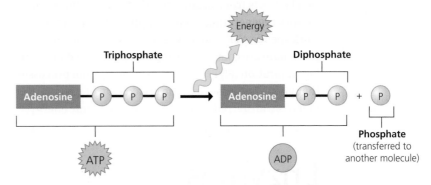

▼ **Figure 5.5 How ATP drives cellular work.** Each type of work shown here is powered when an enzyme transfers phosphate from ATP to a recipient molecule.

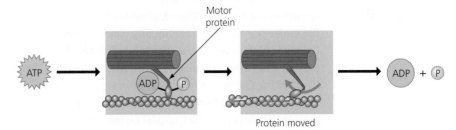

(a) **Motor protein performing mechanical work (moving a muscle fiber)**

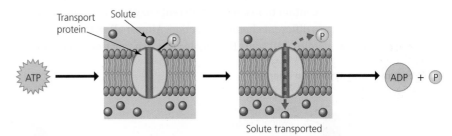

(b) **Transport protein performing transport work (importing a solute)**

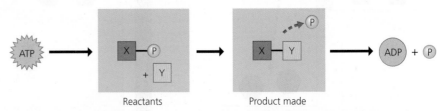

(c) **Chemical reactants performing chemical work (promoting a chemical reaction)**

☑ **CHECKPOINT**

1. Explain how ATP powers cellular work.
2. What is the source of energy for regenerating ATP from ADP?

■ *Answers:* **1.** *ATP transfers a phosphate to another molecule, increasing that molecule's energy.* **2.** *chemical energy harvested from sugars and other organic fuels via cellular respiration*

The ATP Cycle

Your cells spend ATP continuously. Fortunately, it is a renewable resource. ATP can be restored by adding a phosphate group back to ADP. That takes energy, like recompressing a spring. And that's where food enters the picture. The chemical energy that cellular respiration harvests from sugars and other organic fuels is put to work regenerating a cell's supply of ATP. Cellular work spends ATP, which is recycled when ADP and phosphate are combined using energy released by cellular respiration (**Figure 5.6**). Thus, energy from processes that yield energy, such as the breakdown of organic fuels, is transferred to processes that consume energy,

▼ **Figure 5.6** The ATP cycle.

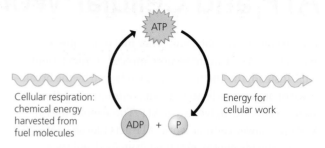

such as muscle contraction and other cellular work. The ATP cycle can run at an astonishing pace: Up to 10 million ATPs are consumed and recycled each second in a working muscle cell. ☑

Enzymes

A living organism contains a vast collection of chemicals, and countless chemical reactions constantly change the organism's molecular makeup. In a sense, a living organism is a complex "chemical square dance," with the molecular "dancers" continually changing partners through chemical reactions. The total of all the chemical reactions in an organism is called **metabolism**. **Illustrating the theme of system interactions, a cell's metabolism depends on the coordination of many molecular players.** In fact, almost no metabolic reactions occur without help. Most require the assistance of **enzymes**, molecules that speed up chemical reactions without being consumed by those reactions. All living cells contain thousands of different enzymes, each promoting a different chemical reaction. Almost all enzymes are proteins, but some RNA molecules also function as enzymes.

☑ **CHECKPOINT**

How does an enzyme affect the activation energy of a chemical reaction?

■ *Answer: An enzyme lowers the activation energy.*

Activation Energy

For a chemical reaction to begin, chemical bonds in the reactant molecules must be broken. (The first step in swapping partners during a square dance is to let go of your current partner's hand.) This process requires that the molecules absorb energy from their surroundings. In other words, for most chemical reactions, a cell has to spend a little energy first. You can easily relate this concept to your own life: It takes effort to clean your room, but this will save you more energy in the long run because you won't have to hunt for your belongings. The energy that must be invested to start a reaction is called **activation energy** because it activates the reactants and triggers the chemical reaction.

Enzymes enable metabolism to occur by reducing the amount of activation energy required to break the bonds of reactant molecules. Without an enzyme, the activation energy barrier might never be breached. For example, lactose (milk sugar) can sit for years without breaking down into its components. But if you add a small amount of the enzyme lactase, all the lactose will be broken down in seconds. If you think of the activation energy as a barrier to a chemical reaction, an enzyme's function is to lower that barrier (**Figure 5.7**). It does so by binding to reactant molecules and putting them under physical or chemical stress, making it easier to break their bonds and start a reaction. In our analogy of cleaning your room, this is like a friend offering to help you. You start and end in the same place whether solo or assisted, but your friend's help lowers your activation energy, making it more likely that you'll proceed. Next, we'll return to our theme of nanotechnology to see how enzymes can be engineered to be even more efficient. ☑

▼ **Figure 5.7** Enzymes and activation energy.

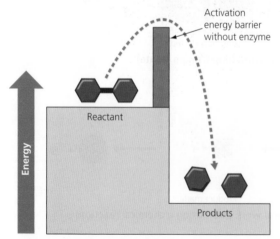

(a) Without enzyme. A reactant molecule must overcome the activation energy barrier before a chemical reaction can break the molecule into products.

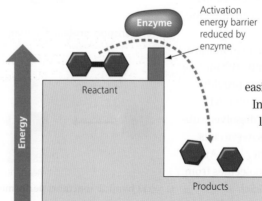

(b) With enzyme. An enzyme speeds the chemical reaction by lowering the activation energy barrier.

Can Enzymes Be Engineered?

BACKGROUND

Like all other proteins, enzymes are encoded by genes. Observations of genetic sequences suggest that many of our genes were formed through a type of molecular evolution: One ancestral gene duplicated, and the two copies diverged over time through random genetic changes, eventually becoming distinct genes.

METHOD

Scientists often use the natural world as inspiration for new experiments. For example, in 2015, a group of researchers from the University of British Columbia used a technique called directed evolution to produce a better version of a particular enzyme. This enzyme cuts the molecular tags that appear on the outside of red blood cells; these tags are responsible for the different blood types (A, B, AB, and O). If a patient receives a transfusion of blood that contains tags that are incompatible with the patient's blood, a potentially deadly rejection could result. The researchers hoped to alleviate such problems by producing an enzyme that efficiently removes the problematic tags from donated blood, thereby making all blood donations safe.

In their attempt to mimic the natural process of evolution, the researchers randomly mutated many copies of the gene that codes for the enzyme (Figure 5.8). Each mutated gene was screened for its ability to code for an enzyme that removes the molecular tags from red blood cells. Any genes that were not efficient were removed from the experimental pool. The genes for the enzymes that were most effective were kept, then subjected to several more rounds of duplication, mutation, and screening.

RESULTS

After many rounds of directed evolution, the researchers isolated a gene that codes for an enzyme that is 170 times more efficient at promoting the reaction than the original enzyme. Testing of this new enzyme is now underway. These results are another example of how scientists can mimic natural processes to modify cellular components into useful medical tools.

Thinking Like a Scientist

In what ways does directed evolution mimic natural selection? In what ways does it differ?

For the answer, see Appendix D.

▶ **Figure 5.8 Directed evolution of an enzyme.** By mutating the gene for glycoside hydrolase, which can remove compatibility proteins from the outside of red blood cells, researchers isolated a gene that codes for a much more effective enzyme.

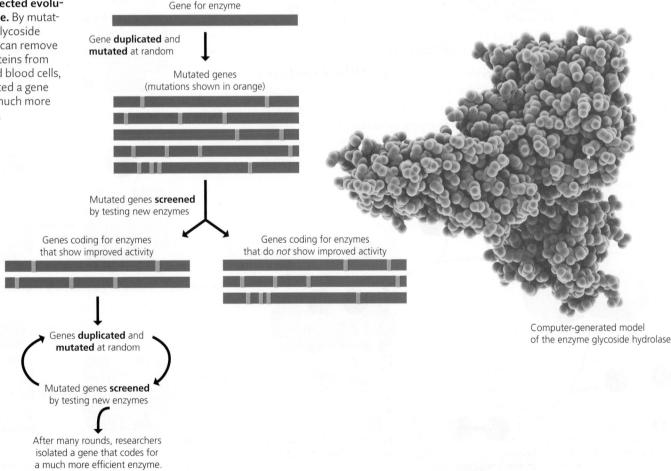

Gene for enzyme

Gene **duplicated** and **mutated** at random

Mutated genes (mutations shown in orange)

Mutated genes **screened** by testing new enzymes

Genes coding for enzymes that show improved activity

Genes coding for enzymes that do *not* show improved activity

Genes **duplicated** and **mutated** at random

Mutated genes **screened** by testing new enzymes

After many rounds, researchers isolated a gene that codes for a much more efficient enzyme.

Computer-generated model of the enzyme glycoside hydrolase

Enzyme Activity

An enzyme is very selective in the reaction it catalyzes. The specific molecule that an enzyme acts on is called the enzyme's **substrate**. Enzymes illustrate the close relationship between structure and function. Each kind of enzyme has a unique three-dimensional shape that determines what specific chemical reaction the enzyme promotes. A region of the enzyme called the **active site** has a shape and chemistry that fit the substrate molecule. The active site is typically a pocket or groove on the surface of the enzyme. When a substrate slips into this docking station, the active site changes shape slightly to embrace the substrate and catalyze the reaction. This interaction is called **induced fit** because the entry of the substrate induces the enzyme to change shape slightly, making the fit between substrate and active site snugger. Think of a handshake: As your hand makes contact with another hand, it changes shape slightly to make a better fit.

After the products are released from the active site, the enzyme can accept another molecule of substrate. In fact, the ability to function repeatedly is a key characteristic of enzymes. **Figure 5.9** follows the action of the enzyme lactase, which breaks down the disaccharide lactose (the substrate). This enzyme is underproduced or defective in lactose-intolerant people. Like lactase, many enzymes are named for their substrates, with an *-ase* ending. ☑

Enzyme Inhibitors

Certain molecules called **enzyme inhibitors** can inhibit a metabolic reaction by binding to an enzyme and disrupting its function **(Figure 5.10)**. Some of these inhibitors are substrate imposters that plug up the active site. (You can't shake a person's hand if someone else puts a banana in it first!) Other inhibitors bind to the enzyme at a site remote from the active site, but the binding changes the enzyme's shape. (Imagine trying to shake hands when someone is tickling your ribs, causing you to clench your hand.) In each case, an inhibitor disrupts the enzyme by altering its shape.

In some cases, the binding of an inhibitor is reversible. For example, if a cell is producing more of a certain product than it needs, that product may reversibly inhibit an enzyme required for its production. This feedback regulation keeps the cell from wasting resources by building an unneeded product.

Many beneficial drugs work by inhibiting enzymes. Penicillin blocks the active site of an enzyme that bacteria use in making cell walls. Ibuprofen inhibits an enzyme involved in sending pain signals. Many cancer drugs inhibit enzymes that promote cell division. Many toxins and poisons also work as inhibitors. Nerve gases (a form of chemical warfare) irreversibly bind to the active site of BOTH NERVE GAS AND INSECTICIDES WORK BY CRIPPLING A VITAL ENZYME.
an enzyme vital to transmitting nerve impulses, leading to rapid paralysis and death. Many pesticides are toxic to insects because they inhibit this same enzyme.

☑ **CHECKPOINT**

How does an enzyme recognize its substrate?

■ *Answer: The substrate and the enzyme's active site are complementary in shape and chemistry.*

▼ **Figure 5.9 How an enzyme works.** Our example is the enzyme lactase, named for its substrate, lactose.

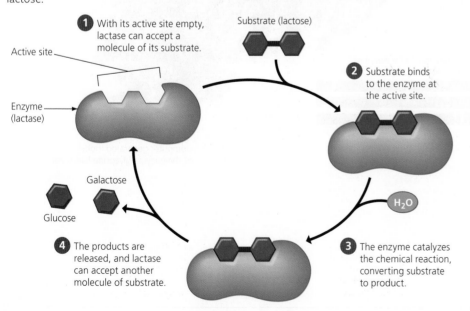

❶ With its active site empty, lactase can accept a molecule of its substrate.

Active site

Enzyme (lactase)

Substrate (lactose)

❷ Substrate binds to the enzyme at the active site.

H_2O

❸ The enzyme catalyzes the chemical reaction, converting substrate to product.

❹ The products are released, and lactase can accept another molecule of substrate.

Galactose

Glucose

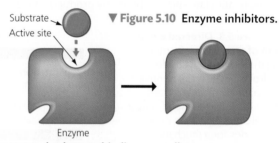

Substrate
Active site

▼ **Figure 5.10 Enzyme inhibitors.**

Enzyme

(a) **Enzyme and substrate binding normally**

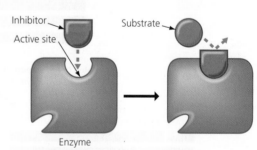

Inhibitor
Active site

Substrate

Enzyme

(b) **Enzyme inhibition by a substrate imposter**

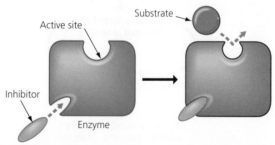

Active site

Substrate

Inhibitor

Enzyme

(c) **Inhibition of an enzyme by a molecule that causes the active site to change shape**

Membrane Function

So far, we have discussed how cells control the flow of energy and how enzymes affect the pace of chemical reactions. In addition to these vital processes, cells must also regulate the flow of materials to and from the environment. The plasma membrane consists of a double layer of fat with embedded proteins—a phospholipid bilayer (see Figure 4.5). **Figure 5.11** describes the major functions of these membrane proteins. Of all the functions shown in the figure, one of the most important is the regulation of transport in and out of the cell. A steady traffic of small molecules moves across a cell's plasma membrane in both directions. But this traffic flow is never willy-nilly. Instead, all biological membranes are selectively permeable—that is, they only allow certain molecules to pass. Let's explore this in more detail.

Passive Transport: Diffusion across Membranes

Molecules are restless. They constantly vibrate and wander randomly. One result of this motion is **diffusion**, the movement of molecules spreading out evenly into the available space. Each molecule moves randomly, but the overall diffusion of a population of molecules is usually directional, from a region where the molecules are more concentrated to a region where they are less concentrated. For example, imagine many molecules of perfume inside a bottle. If you remove the bottle top, every molecule of perfume will move randomly about, but the overall movement will be out of the bottle, and the whole room will eventually smell of the perfume. You could, with great effort, return the perfume molecules to its bottle, but the molecules would never all return spontaneously.

▼ **Figure 5.11 Primary functions of membrane proteins.** An actual cell may have just a few of the types of proteins shown here, and many copies of each particular protein may be present.

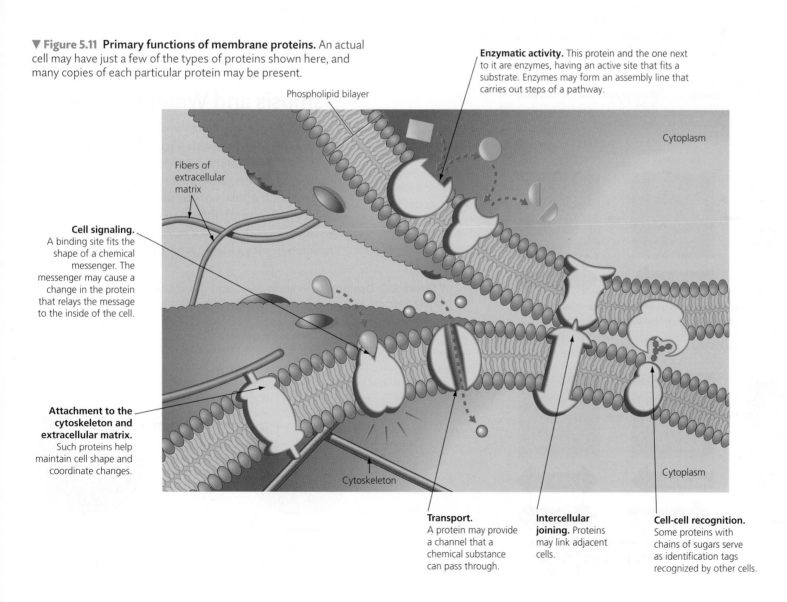

Phospholipid bilayer

Enzymatic activity. This protein and the one next to it are enzymes, having an active site that fits a substrate. Enzymes may form an assembly line that carries out steps of a pathway.

Cytoplasm

Fibers of extracellular matrix

Cell signaling. A binding site fits the shape of a chemical messenger. The messenger may cause a change in the protein that relays the message to the inside of the cell.

Attachment to the cytoskeleton and extracellular matrix. Such proteins help maintain cell shape and coordinate changes.

Cytoskeleton

Transport. A protein may provide a channel that a chemical substance can pass through.

Intercellular joining. Proteins may link adjacent cells.

Cell-cell recognition. Some proteins with chains of sugars serve as identification tags recognized by other cells.

Cytoplasm

For an example closer to a living cell, imagine a membrane separating pure water from a mixture of dye dissolved in water **(Figure 5.12)**. Assume that this membrane has tiny holes that allow dye molecules to pass. Although each dye molecule moves randomly, there will be a net migration across the membrane to the side that began as pure water. Movement of the dye will continue until both solutions have equal concentrations. After that, there will be a dynamic equilibrium: Molecules will still be moving, but at that point as many dye molecules move in one direction as in the other, so the concentration of dye molecules in water remains steady.

Diffusion of dye across a membrane is an example of **passive transport**—*passive* because no energy is needed for the diffusion to happen. In passive transport, a substance diffuses down its **concentration gradient**, from where the substance is more concentrated to where it is less concentrated. In our lungs, for example, there is more oxygen gas (O_2) in the air than in the blood. Therefore, oxygen moves by passive transport from the air into the bloodstream. But remember that the cell membrane is selectively permeable. For example, small molecules such as oxygen (O_2) generally pass through more readily than larger molecules such as amino acids. But the membrane

is relatively impermeable to even some very small substances, such as most ions, which are too hydrophilic to pass through the fatty phospholipid bilayer.

Substances that do not cross membranes spontaneously—or otherwise cross very slowly—can be transported by proteins that act as corridors for specific molecules (see Figure 5.11). This assisted transport is called **facilitated diffusion**. For example, water molecules can move through the plasma membrane of some cells by way of transport proteins—each of which can help 3 billion water molecules per second pass through! People with a rare mutation in the gene that encodes these water-transport proteins have defective kidneys that cannot reabsorb water; such people must drink 20 liters of water every day to prevent dehydration. On the flip side, a common complication of pregnancy is fluid retention, the culprit responsible for swollen ankles and feet, often caused by increased synthesis of water channel proteins. Other specific transport proteins move glucose across cell membranes 50,000 times faster than diffusion. Even at this rate, facilitated diffusion is a type of passive transport because it does not require the cell to expend energy. As in all passive transport, the driving force is the concentration gradient. ☑

CHECKPOINT

1. What does it mean to say that molecules move "down the concentration gradient"?

2. Why is facilitated diffusion a form of passive transport?

■ Answer: 1. The molecule moves from where it is more concentrated to where it is less concentrated. 2. It uses proteins to transport materials down a concentration gradient without expending energy.

▼ **Figure 5.12 Passive transport: diffusion across a membrane.** A substance will diffuse from where it is more concentrated to where it is less concentrated. Put another way, a substance tends to diffuse down its concentration gradient.

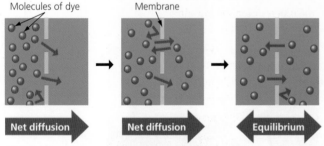

(a) Passive transport of one type of molecule. The membrane is permeable to these dye molecules, which diffuse down the concentration gradient. At equilibrium, the molecules are still restless, but the rate of transport is equal in both directions.

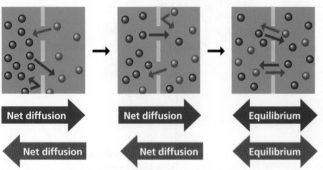

(b) Passive transport of two types of molecules. If solutions have two or more solutes, each will diffuse down its own concentration gradient.

Osmosis and Water Balance

The diffusion of water across a selectively permeable membrane is called **osmosis (Figure 5.13)**. A **solute** is a substance that is dissolved in a liquid solvent, and the resulting mixture is called a solution. For example, a solution of salt water contains salt (the solute) dissolved in water (the solvent). Imagine a membrane separating two solutions with different concentrations of a solute. The solution with a higher concentration of solute is

▼ **Figure 5.13 Osmosis.** A membrane separates two solutions with different sugar concentrations. Water molecules can pass through the membrane, but the sugar molecules cannot.

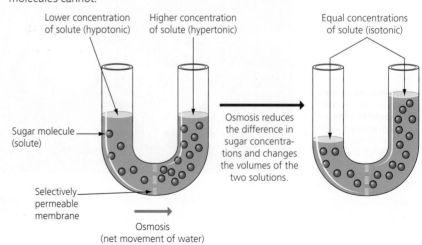

said to be **hypertonic** to the other solution. The solution with the lower solute concentration is said to be **hypotonic** to the other. Note that the hypotonic solution, by having the lower solute concentration, has the higher water concentration (less solute = more water). Therefore, water will diffuse across the membrane along its concentration gradient from an area of higher water concentration (hypotonic solution) to one of lower water concentration (hypertonic solution). This reduces the difference in solute concentrations and changes the volumes of the two solutions.

People can take advantage of osmosis to preserve foods. Salt is often applied to meats—like beef and fish—to cure them into jerky; the salt causes water to move out of food-spoiling bacteria and fungi. Food can also be preserved in honey because a high sugar concentration draws water out of food.

When the solute concentrations are the same on both sides of a membrane, water molecules will move at the same rate in both directions, so there will be no net change in solute concentration. Solutions of equal solute concentration are said to be **isotonic**. For example, many marine animals, such as sea stars and crabs, are isotonic to seawater, so that overall they neither gain nor lose water from the environment. In your body, red blood cells are isotonic to the bloodstream in which they flow.

Water Balance in Animal Cells

The survival of a cell depends on its ability to balance water uptake and loss. When an animal cell is immersed in an isotonic solution, the cell's volume remains constant because the cell gains water at the same rate that it loses water (**Figure 5.14a**, top). But what happens if an animal cell is in contact with a hypotonic solution, which has a lower solute concentration than the cell? Due to osmosis, the cell would gain water, swell, and possibly burst (lyse) like an overfilled water balloon (**Figure 5.14b**, top). A hypertonic environment is also harsh on an animal cell; the cell shrivels from water loss (**Figure 5.14c**, top).

For an animal to survive a hypotonic or hypertonic environment, the animal must have a way to balance the uptake and loss of water. The control of water balance is called **osmoregulation**. For example, a freshwater fish has kidneys and gills that work constantly to prevent an excessive buildup of water in the body. Humans can suffer consequences of osmoregulation failure. Dehydration (consumption of too little water) can cause fatigue and even death. Drinking too much water—called hyponatremia, or "water intoxication"—can also cause death by overdiluting necessary ions.

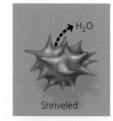
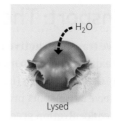
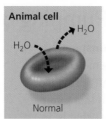

Animal cell
Normal | Lysed | Shriveled

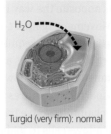
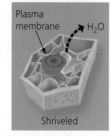
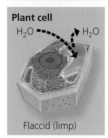

Plant cell
Plasma membrane
Flaccid (limp) | Turgid (very firm): normal | Shriveled

(a) Isotonic solution | **(b) Hypotonic solution** | **(c) Hypertonic solution**

◀ **Figure 5.14 Osmotic environments.** Animal cells (such as a red blood cell) and plant cells behave differently in different osmotic environments.

Water Balance in Plant Cells

Problems of water balance are somewhat different for cells that have rigid cell walls, such as those from plants, fungi, many prokaryotes, and some protists. A plant cell immersed in an isotonic solution is flaccid (floppy), and the plant wilts (Figure 5.14a, bottom). In contrast, a plant cell is turgid (very firm) and healthiest in a hypotonic environment, with a net inflow of water (Figure 5.14b, bottom). Although the elastic cell wall expands a bit, the back pressure it exerts prevents the cell from taking in too much water and bursting. Turgor is necessary for plants to retain their upright posture and the extended state of their leaves (**Figure 5.15**). However, in a hypertonic environment, a plant cell is no better off than an animal cell. As a plant cell loses water, it shrivels, and its plasma membrane pulls away from the cell wall (Figure 5.14c, bottom). This usually kills the cell. Thus, plant cells thrive in a hypotonic environment, whereas animal cells thrive in an isotonic one. ☑

BEEF JERKY AND RAISINS ARE BOTH MADE THROUGH OSMOSIS.

▼ **Figure 5.15 Plant turgor.** Watering a wilted plant will make it regain its turgor.

Wilted | Turgid

☑ CHECKPOINT

1. A hypotonic solution will cause a plant cell to become _____ .
2. What would happen if you placed one of your cells in pure water?

■ *Answers:* **1.** *turgid* **2.** *Water would rush in (because it is more concentrated outside the cell and less concentrated inside) and would burst the cell.*

Active Transport: The Pumping of Molecules across Membranes

In contrast to passive transport, **active transport** requires that a cell expend energy to move molecules across a membrane. Cellular energy (usually provided by ATP) is used to drive a transport protein that pumps a solute *against* the concentration gradient—that is, in the direction that is opposite the way it would naturally flow **(Figure 5.16)**. Movement against a force, like rolling a boulder uphill against gravity, requires a considerable expenditure of energy. Consider this analogy: During a storm, water spontaneously flows downhill into a basement (passive transport) but requires a powered sump pump to move it back uphill (active transport).

Active transport allows cells to maintain internal concentrations of small solutes that differ from environmental concentrations. For example, compared with its surroundings, an animal nerve cell has a much higher concentration of potassium ions and a much lower concentration of sodium ions. The plasma membrane helps maintain these differences by pumping sodium out of the

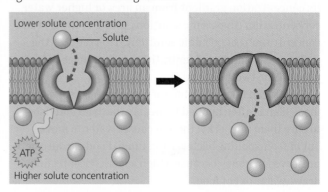

▼ **Figure 5.16 Active transport.** Transport proteins are specific in their recognition of atoms or molecules. This transport protein (purple) has a binding site that accepts only a certain solute. Using energy from ATP, the protein pumps the solute against its concentration gradient.

Lower solute concentration

Solute

ATP

Higher solute concentration

cell and potassium into the cell. This particular case of active transport (called the sodium-potassium pump) is vital to the nervous system of most animals. ☑

Exocytosis and Endocytosis: Traffic of Large Molecules

So far, we've focused on how water and small solutes enter and leave cells by moving through the plasma membrane. The story is different for large molecules such as proteins, which are much too big to fit through the membrane. Their traffic into and out of the cell depends on the ability of the cell to package large molecules inside sacs called vesicles. You have already seen an example of this: During protein production by the cell, secretory proteins exit the cell from transport vesicles that fuse with the plasma membrane, spilling the contents outside the cell (see Figures 4.13 and 4.17). That process is called **exocytosis (Figure 5.17)**. When you cry, for example, cells in your tear glands use exocytosis to export the salty tears. In your brain, the exocytosis of neurotransmitter chemicals such as dopamine helps neurons communicate.

In **endocytosis**, a cell takes in material by vesicles that bud inward **(Figure 5.18)**. For example, in a process called

phagocytosis ("cellular eating"), a cell engulfs a particle and packages it within a food vacuole. Other times, a cell "gulps" droplets of fluid into vesicles. Endocytosis can also be triggered by the binding of certain external molecules to specific receptor proteins built into the plasma membrane. This binding causes the local region of the membrane to form a vesicle that transports the specific substance into the cell. In human liver cells, this process is used to take up cholesterol from the blood. An inherited defect in the receptors on liver cells can lead to an inability to process cholesterol, which can lead to heart attacks at ages as young as 5. Cells of your immune system use endocytosis to engulf and destroy invading bacteria and viruses.

Because all cells have a plasma membrane, it is logical to infer that membranes first formed early in the evolution of life on Earth. In the final section of this chapter, we'll consider the evolution of membranes.

▼ **Figure 5.17 Exocytosis.**

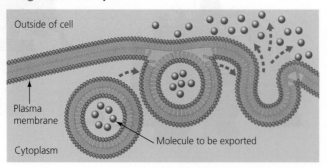

Outside of cell

Plasma membrane

Molecule to be exported

Cytoplasm

▼ **Figure 5.18 Endocytosis.**

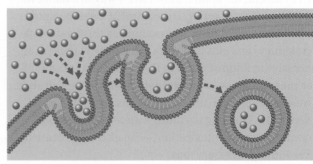

EVOLUTION CONNECTION | Nanotechnology

The Origin of Membranes

By simulating conditions found on the early Earth, scientists have been able to demonstrate that many of the molecules important to life can form spontaneously. (See Figure 15.3 and the accompanying text for a description of one such experiment.) Such results suggest that phospholipids, the key ingredients in all membranes, were probably among the first organic compounds that formed from chemical reactions on the early Earth. Once formed, they could self-assemble into simple membranes. When a mixture of phospholipids and water is shaken, for example, the phospholipids organize into bilayers, forming water-filled bubbles of membrane **(Figure 5.19)**. This assembly requires neither genes nor other information beyond the properties of the phospholipids themselves.

The tendency of lipids in water to spontaneously form membranes has led biomedical engineers to produce liposomes (a type of artificial vesicle) that can encase specific chemicals. In the future, these engineered liposomes may be used to deliver nutrients or medications to specific sites within the body. In fact, over a dozen drugs have been approved for delivery by liposomes, including ones that target fungal infections, influenza, and hepatitis. Thus, membranes—like the other cellular components discussed in the Biology and Society and the Process of Science sections—have inspired novel nanotechnologies.

The formation of membrane-enclosed collections of molecules would have been a critical step in the evolution of the first cells. A membrane can enclose a solution that is different in composition from its surroundings. A plasma membrane that allows cells to regulate their chemical exchanges with the environment is a basic requirement for life. Indeed, all cells are enclosed by a plasma membrane that is similar in structure and function—illustrating the evolutionary unity of life.

▼ **Figure 5.19** The spontaneous formation of membranes: a key step in the origin of life.

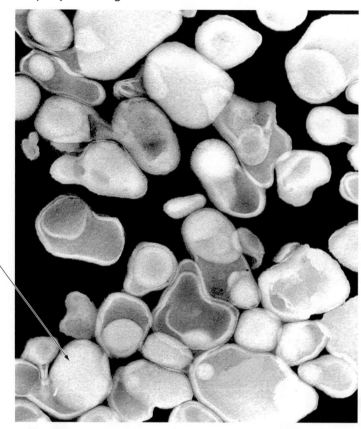

Water-filled bubble made of phospholipids

Colorized TEM 600×

Chapter Review

SUMMARY OF KEY CONCEPTS

Some Basic Energy Concepts

Conservation of Energy
Machines and organisms can transform kinetic energy (energy of motion) to potential energy (stored energy) and vice versa. In all such energy transformations, total energy is conserved. Energy cannot be created or destroyed.

Heat
Every energy transformation releases some randomized energy in the form of heat. Entropy is a measure of disorder, or randomness.

Chemical Energy
Molecules store varying amounts of potential energy in the arrangement of their atoms. Organic compounds are relatively rich in such chemical energy. The combustion of gasoline within a car's engine and the breakdown of glucose by cellular respiration within living cells are both examples of how the chemical energy stored in molecules can be converted to useful work.

Food Calories
Food Calories, actually kilocalories, are units used to measure the amount of energy in our foods and the amount of energy we expend in various activities.

ATP and Cellular Work

Your cells recycle ATP: As ATP is broken down to ADP to drive cellular work, new molecules of ATP are built from ADP using energy obtained from food.

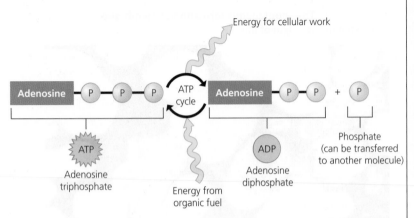

Enzymes

Activation Energy

Enzymes are biological catalysts that speed up metabolic reactions by lowering the activation energy required to break the bonds of reactant molecules.

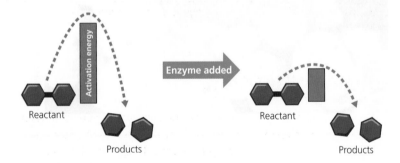

Enzyme Activity

The entry of a substrate into the active site of an enzyme causes the enzyme to change shape slightly, allowing for a better fit and thereby promoting the interaction of enzyme with substrate.

Enzyme Inhibitors

Enzyme inhibitors are molecules that can disrupt metabolic reactions by binding to enzymes, either at the active site or elsewhere.

Membrane Function

Proteins embedded in the plasma membrane perform a wide variety of functions, including regulating transport, anchoring to other cells or substances, promoting enzymatic reactions, and recognizing other cells.

Passive Transport, Osmosis, and Active Transport

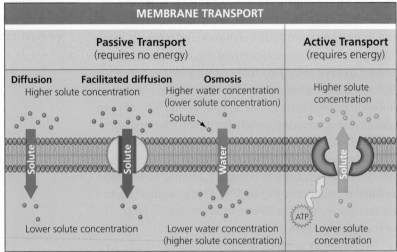

Most animal cells require an isotonic environment, with equal concentrations of water within and outside the cell. Plant cells need a hypotonic environment, which causes water to flow inward, keeping walled cells turgid.

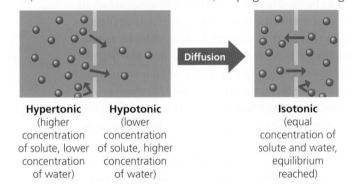

Exocytosis and Endocytosis: Traffic of Large Molecules

Exocytosis is the secretion of large molecules within vesicles. Endocytosis is the import of large substances by vesicles into the cell.

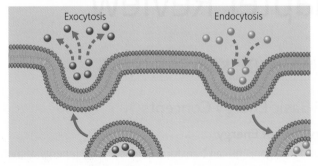

Mastering **Biology**

For practice quizzes, BioFlix animations, MP3 tutorials, video tutors, and more study tools designed for this textbook, go to Mastering Biology™

SELF-QUIZ

1. Describe the energy transformations that occur when a dog jumps onto a truck.

2. _____ is the capacity to perform work, while _____ is a measure of randomness.

3. The label on a tin of tomato soup says that it contains 56 Calories. If you could convert all of that energy to heat, you could raise the temperature of how much water by 15°C?

4. Why does removing a phosphate group from the triphosphate tail in a molecule of ATP release energy?

5. Your digestive system uses a variety of enzymes to break down large food molecules into smaller ones that your cells can assimilate. A generic name for a digestive enzyme is hydrolase. What is the chemical basis for that name? (*Hint:* Review Figure 3.4.)

6. Explain how an inhibitor can disrupt an enzyme's action without binding to the active site.

7. If someone at the other end of a room smokes a cigarette, you may breathe in some smoke. The movement of smoke is similar to what type of transport?
   a. osmosis
   b. diffusion
   c. facilitated diffusion
   d. active transport

8. Explain why it is not enough to say that a solution is "hypertonic."

9. What is the primary difference between passive and active transport in terms of concentration gradients?

10. Pumping sodium out of an animal nerve cell requires energy. What type of transport is this?
    a. active transport
    b. osmosis
    c. facilitated diffusion
    d. diffusion

For answers to the Self Quiz, see Appendix D.

IDENTIFYING MAJOR THEMES

For each statement, identify which major theme is evident (the relationship of structure to function, information flow, pathways that transform energy and matter, interactions within biological systems, or evolution) and explain how the statement relates to the theme. If necessary, review the themes (see Chapter 1) and review the examples highlighted in blue in this chapter.

11. You turn the crank of a well, raising a bucket of water from below ground to the surface.

12. Your ability to walk depends upon the coordination of many different enzymes and other cellular structures.

13. Enzymes unravel and stop functioning if the environment gets too hot.

For answers to Identifying Major Themes, see Appendix D.

THE PROCESS OF SCIENCE

14. HIV, the virus that causes AIDS, depends on an enzyme called reverse transcriptase to multiply. Reverse transcriptase reads a molecule of RNA and creates a molecule of DNA from it. A molecule of AZT, an anti-AIDS drug, has a shape very similar to (but a bit different than) that of the DNA base thymine. Propose a model for how AZT inhibits HIV.

15. Cyclic guanosine monophosphate (cGMP) is a signaling molecule that relaxes smooth muscle tissues, thereby allowing blood flow into the corpus cavernosum and triggering penile erection. It is degraded by the enzyme cGMP-specific phosphodiesterase type 5 (PDE5). Sildenafil (Viagra), used to treat erectile dysfunction, has a molecular structure similar to that of cGMP. Propose a model for how sildenafil prevents erectile dysfunction.

16. **Interpreting Data** The graph illustrates two chemical reactions. Which curve represents the reaction in the presence of the enzyme? What energy changes are represented by lines a, b, and c?

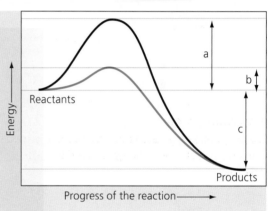

BIOLOGY AND SOCIETY

17. Many detox and diet plans recommend drinking large amounts of water—sometimes even up to 5 liters per day. Which properties of water may have led to this recommendation? Is a massively high intake of water always recommendable? How can drinking large amounts of water have a negative physiological effect? Compare recommendations for water intake found on different detoxing/dieting websites, and determine whether they are scientifically sound.

18. Nanotechnology devices can improve human health. But can they do harm or be abused? What regulations would allow for beneficial uses of nanorobots without allowing for harm?

19. Lead inhibits enzymes and can interfere with development of the nervous system. One battery manufacturer banned female employees of childbearing age from working in areas where they might be exposed to high levels of lead. The Supreme Court ruled the policy illegal. But many people are uncomfortable about the "right" to work in an unsafe environment. What rights and responsibilities of employers, employees, and government agencies are in conflict? What criteria should be used to decide who can work in a particular environment?

6 Cellular Respiration: Obtaining Energy from Food

CHAPTER CONTENTS

Why Cellular Respiration Matters

You can survive for weeks without eating and days without drinking, but you can only live for minutes without breathing. Why? Because every cell in your body relies on the energy created by using oxygen to break down glucose during the process of cellular respiration.

THE METABOLIC PROCESSES THAT PRODUCE ACID IN YOUR MUSCLES AFTER A HARD WORKOUT ARE SIMILAR TO THE PROCESSES THAT PRODUCE PEPPERONI, SOY SAUCE, YOGURT, AND BREAD.

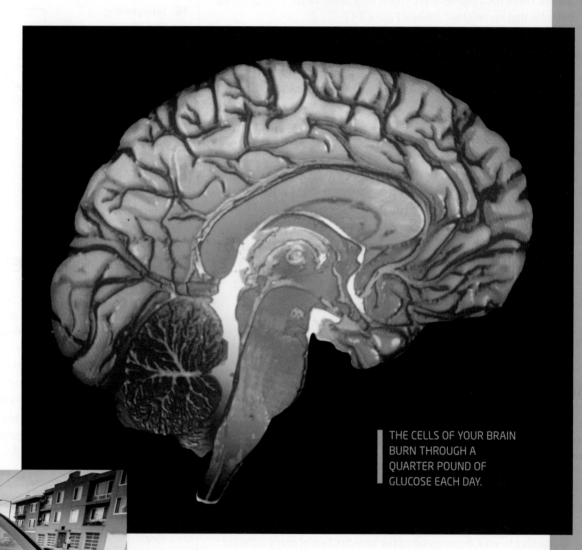

YOU HAVE SOMETHING IN COMMON WITH A SPORTS CAR: YOU BOTH REQUIRE AN AIR INTAKE SYSTEM TO BURN FUEL EFFICIENTLY.

THE CELLS OF YOUR BRAIN BURN THROUGH A QUARTER POUND OF GLUCOSE EACH DAY.

BIOLOGY AND SOCIETY Exercise Science

Getting the Most Out of Your Muscles

Serious athletes train extensively to reach the peak of their physical potential. A key aspect of athletic conditioning involves increasing aerobic capacity, the ability of the heart and lungs to deliver oxygen to body cells. For many endurance athletes, such as long-distance runners or cyclists, the rate at which oxygen is provided to working muscles is the limiting factor in their performance.

Why is oxygen so important? Whether you are exercising or just going about your daily tasks, your muscles need a continuous supply of energy to perform work. Muscle cells obtain this energy from the sugar glucose through a series of chemical reactions that depend upon a constant input of oxygen (O_2). Therefore, to keep moving, your body needs a steady supply of O_2. When there is enough oxygen reaching your cells to support their energy needs, metabolism is said to be aerobic. As your muscles work harder, you breathe faster and more deeply to inhale O_2. If you continue to pick up the pace, you will approach your aerobic capacity, the maximum rate at which O_2 can be taken in and used by your muscle cells and therefore the most strenuous exercise that your body can maintain aerobically. Exercise scientists can use oxygen-monitoring equipment to precisely determine the maximum possible aerobic output for any given person. Such data allow a well-trained athlete to stay within aerobic limits, ensuring the maximum possible output—in other words, his or her best effort.

If you work even harder and exceed your aerobic capacity, the demand for oxygen in your muscles will outstrip your body's ability to deliver it; metabolism then becomes anaerobic. With insufficient O_2, your muscle cells switch to an "emergency mode" in which they break down glucose very inefficiently and produce lactic acid as a by-product. As lactic acid and other wastes accumulate, muscle activity is impaired. Your muscles can work under these conditions for only a few minutes before they give out. When this happens (sometimes called "hitting the wall"), your muscles cannot function, and you will likely collapse, unable to even stand.

Every living organism depends on processes that provide energy. In fact, we need energy to walk, talk, and think—in short, to stay alive. The human body has trillions of cells, all hard at work, all demanding fuel continuously. In this chapter, you'll learn how cells harvest food energy and put it to work with the help of oxygen. Along the way, we'll consider the implications of how the body responds to exercise.

The science of exercise.
Endurance athletes must carefully monitor their efforts so that they maintain an aerobic pace over the long term.

Energy Flow and Chemical Cycling in the Biosphere

All life requires energy. In almost all ecosystems on Earth, this energy originates with the sun. During **photosynthesis**, the energy of sunlight is converted to the chemical energy of sugars and other organic molecules (as we'll discuss in Chapter 7). Photosynthesis takes place in the chloroplasts of plants and algae, as well as in some prokaryotes. All animals depend on this conversion for food and more. You're probably wearing clothing made of a product of photosynthesis—cotton. Most of our homes are framed with timber, which is wood produced by photosynthetic trees. Even textbooks are printed on a material (paper) that can be traced to photosynthesis in plants. But from an animal's point of view, photosynthesis is primarily about providing food.

Producers and Consumers

Plants and other **autotrophs** ("self-feeders") are organisms that make all their own organic matter—including carbohydrates, lipids, proteins, and nucleic acids—from nutrients that are entirely inorganic: carbon dioxide from the air and water and minerals from the soil. In other words, autotrophs make their own food; they don't need to eat to gain energy to power their cellular processes. In contrast, humans and other animals are **heterotrophs** ("other-feeders"), organisms that cannot make organic molecules from inorganic ones. Therefore, we must eat organic material to get our nutrients and provide energy for life's processes.

Most ecosystems depend entirely on photosynthesis for food. For this reason, biologists refer to plants and other autotrophs as **producers**. Heterotrophs, in contrast, are **consumers** because they obtain their food by eating plants or by eating animals that have eaten plants **(Figure 6.1)**. We animals and other heterotrophs depend on autotrophs for organic fuel and for the raw organic materials we need to build our cells and tissues. ☑

Chemical Cycling between Photosynthesis and Cellular Respiration

Within a plant, the chemical ingredients for photosynthesis are carbon dioxide (CO_2), a gas that passes from the air into a plant through tiny pores, and water (H_2O), absorbed from the soil by the plant's roots. Inside leaf cells, organelles called chloroplasts use light energy to rearrange the atoms of these ingredients to produce sugars—most importantly glucose ($C_6H_{12}O_6$)—and other organic molecules **(Figure 6.2)**. You can think of chloroplasts as tiny solar-powered sugar factories. A by-product of photosynthesis is oxygen gas (O_2) that is released through pores into the atmosphere.

▶ **Figure 6.1 Producer and consumer.** A giraffe (consumer) eating leaves produced by a photosynthetic plant (producer).

A chemical process called cellular respiration uses O_2 to convert the energy stored in the chemical bonds of sugars to another source of chemical energy called ATP. Cells expend ATP for almost all their work. In both plants and animals, the production of ATP during cellular respiration occurs mainly in the organelles called mitochondria (see Figure 4.19).

You might notice in Figure 6.2 that energy takes a one-way trip through an ecosystem, entering as sunlight and exiting as heat. Chemicals, in contrast, are recycled. Notice also in Figure 6.2 that the waste products of cellular respiration are CO_2 and H_2O—the very same ingredients used as inputs for photosynthesis. Plants store chemical energy through photosynthesis and then harvest this energy through cellular respiration. (Note that plants perform *both* photosynthesis to produce fuel molecules

and cellular respiration to burn them, while animals perform *only* cellular respiration.) Plants usually make more organic molecules than they need for fuel. This photosynthetic surplus provides material for the plant to grow or can be stored (as starch in potatoes, for example). Thus, when you consume a carrot, potato, or turnip, you are eating the energy reservoir that plants (if unharvested) would have used to grow the following spring.

People have always taken advantage of plants' photosynthetic abilities by eating them. More recently, engineers have managed to tap into this energy reserve to produce liquid biofuels, primarily ethanol (see Chapter 7 for a discussion of biofuels). But no matter the end product, you can trace the energy and raw materials for growth back to solar-powered photosynthesis. ☑

☑ **CHECKPOINT**

What is misleading about the following statement? "Plants perform photosynthesis, whereas animals perform cellular respiration."

■ *Answer: It implies that only animals perform cellular respiration, when in fact all life does.*

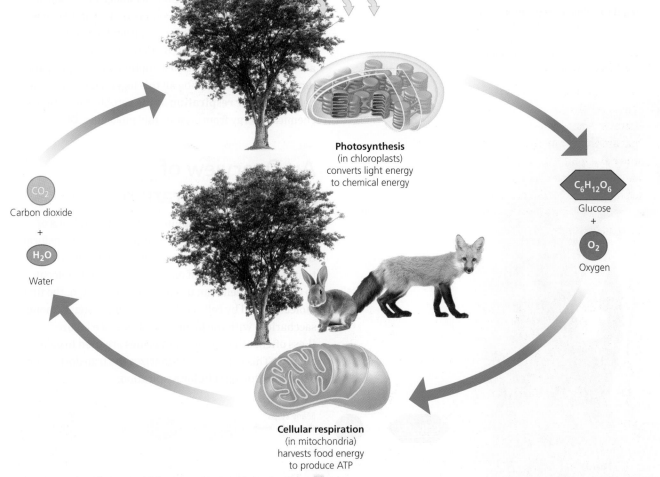

Sunlight energy enters ecosystem

Photosynthesis (in chloroplasts) converts light energy to chemical energy

CO_2

Carbon dioxide

+

H_2O

Water

$C_6H_{12}O_6$

Glucose

+

O_2

Oxygen

Cellular respiration (in mitochondria) harvests food energy to produce ATP

ATP drives cellular work

Heat energy exits ecosystem

◀ **Figure 6.2 Energy flow and chemical cycling in ecosystems.** Energy flows through an ecosystem, entering as sunlight and exiting as heat. In contrast, chemical elements are recycled within an ecosystem.

Figure Walkthrough

Mastering **Biology** goo.gl/SvLydh

Cellular Respiration: Aerobic Harvest of Food Energy

We usually use the word *respiration* to mean breathing. Although respiration on the organismal level should not be confused with cellular respiration, the two processes are closely related (Figure 6.3). Cellular respiration requires a cell to exchange two gases with its surroundings. The cell takes in oxygen in the form of the gas O_2. It gets rid of waste in the form of the gas carbon dioxide, or CO_2. Respiration, or breathing, results in the exchange of these same gases between your blood and the outside air. Oxygen present in the air you inhale diffuses across the lining of your lungs and into your bloodstream. And the CO_2 in your bloodstream diffuses into your lungs and exits your body when you exhale. Every molecule of CO_2 that you exhale was originally formed in one of the mitochondria of your body's cells. Internal combustion engines, like the ones found in cars, use O_2 (through the air intakes) to break down gasoline. A cell also requires O_2 to break down its fuel (see Figure 5.2). Cellular respiration—a biological version of internal combustion—is the main way that chemical energy is harvested from food and converted to ATP energy (see Figure 5.6). Cellular respiration is an **aerobic** process, which is just another way of saying that it requires oxygen. Putting all this together, we can now define **cellular respiration** as the aerobic harvesting of chemical energy from organic fuel molecules. ☑

YOU HAVE SOMETHING IN COMMON WITH A SPORTS CAR: YOU BOTH REQUIRE AN AIR INTAKE SYSTEM TO BURN FUEL EFFICIENTLY.

✅ CHECKPOINT

At both the organismal and cellular levels, respiration involves taking in the gas _____ and expelling the gas _____.

■ *Answer: O_2; CO_2*

▼ **Figure 6.3 How breathing is related to cellular respiration.** When you inhale, you breathe in O_2. The O_2 is delivered to your cells, where it is used in cellular respiration. Carbon dioxide, a waste product of cellular respiration, diffuses from your cells to your blood and travels to your lungs, where it is exhaled.

O_2

CO_2

Lungs

O_2 CO_2

O_2 CO_2

Cellular respiration **Muscle cells**

An Overview of Cellular Respiration

All living organisms depend on transformations of energy and matter. We see examples of such transformations throughout the study of life, but few are as important as the conversion of energy in fuel (food molecules) to a form that cells can use directly. Most often, the fuel molecule used by cells is glucose, a simple sugar (monosaccharide) with the formula $C_6H_{12}O_6$ (see Figure 3.6). (Less often, other organic molecules are used to gain energy.) This equation summarizes the transformation of glucose during cellular respiration:

$$C_6H_{12}O_6 + 6\ O_2 \xrightarrow{\text{Many steps}} 6\ CO_2 + 6\ H_2O + \text{approx. } 32\ ATP + \text{heat}$$

The series of arrows in this formula represents the fact that cellular respiration consists of many chemical steps. A specific enzyme catalyzes each reaction in the pathway, more than two dozen reactions in all. In fact, these reactions constitute one of the most important metabolic pathways for nearly every eukaryotic cell: those found in plants, fungi, protists, and animals. This pathway provides the energy these cells need to maintain the functions of life.

The many chemical reactions that make up cellular respiration can be grouped into three main stages: glycolysis, the citric acid cycle, and electron transport. **Figure 6.4** is a road map that will help you follow the three stages of respiration and see where each stage occurs in your cells. During **glycolysis**, a molecule of glucose is split into two molecules of a compound called pyruvic acid. The enzymes for glycolysis are located in the cytoplasm. The **citric acid cycle** (also called the Krebs cycle) completes the breakdown of glucose all the way to CO_2, which is then released as a waste product. The enzymes for the citric acid cycle are dissolved in the fluid within mitochondria. Glycolysis and the citric acid cycle generate a small amount of ATP directly. They generate much more ATP indirectly, by reactions that transfer electrons from fuel molecules to a molecule called NAD^+ (<u>n</u>icotinamide <u>a</u>denine <u>d</u>inucleotide) that cells make from niacin, a B vitamin. The electron transfer forms a molecule called **NADH** (the H represents the transfer of

hydrogen along with the electrons) that acts as a shuttle carrying high-energy electrons from one area of the cell to another. The third stage of cellular respiration is **electron transport**. Electrons captured from food by the NADH formed in the first two stages are stripped of their energy, a little bit at a time, until they are finally combined with oxygen to form water. The proteins and other molecules that make up electron transport chains are embedded within the inner membrane of the mitochondria. The transport of electrons from NADH to oxygen releases the energy your cells use to make most of their ATP.

The overall equation for cellular respiration shows that the atoms of the reactant molecules glucose and oxygen are rearranged to form the products carbon dioxide and water. But don't lose track of why this process occurs: The main function of cellular respiration is to generate ATP for cellular work. In fact, the process can produce around 32 ATP molecules for each glucose molecule consumed. ☑

CELLULAR RESPIRATION: AEROBIC HARVEST OF FOOD ENERGY

☑ **CHECKPOINT**

Which stages of cellular respiration take place in the mitochondria? Which stage takes place outside the mitochondria?

■ *Answer: the citric acid cycle and electron transport; glycolysis*

▶ **Figure 6.4** **A road map for cellular respiration.**

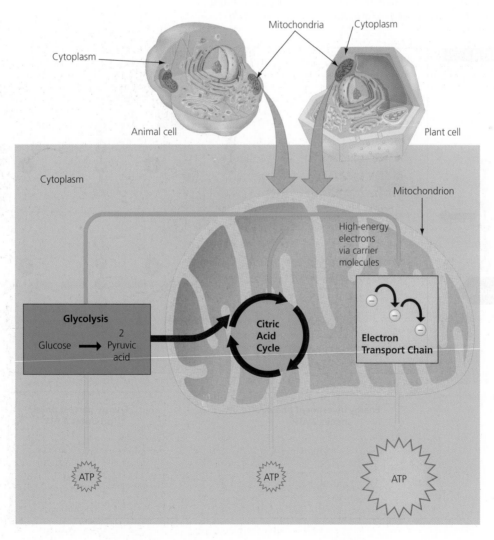

The Three Stages of Cellular Respiration

Now that you have a big-picture view of cellular respiration, let's examine the process in more detail. A small version of Figure 6.4 will help you keep the overall process of cellular respiration in plain view as we take a closer look at its three stages.

Stage 1: Glycolysis

The word *glycolysis* means "splitting of sugar" (**Figure 6.5**), and that's just what happens. **1** During glycolysis, a six-carbon glucose molecule is broken in

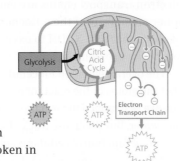

half, forming two three-carbon molecules. Notice in Figure 6.5 that the initial split requires an energy "investment" of two ATP molecules per glucose. **2** The three-carbon molecules then donate high-energy electrons to NAD⁺, forming NADH. **3** In addition to NADH, glycolysis also "banks" four ATP molecules directly when enzymes transfer phosphate groups from fuel molecules to ADP. Glycolysis thus produces a "profit" of two molecules of ATP per molecule of glucose (two invested, but four banked; this fact will become important during our discussion of fermentation later). What remains of the fractured glucose at the end of glycolysis are two molecules of pyruvic acid. The pyruvic acid still holds most of the energy of glucose, and that energy is harvested in the second stage of cellular respiration, the citric acid cycle.

▼ **Figure 6.5 Glycolysis.** In glycolysis, a team of enzymes splits glucose, eventually forming two molecules of pyruvic acid. After investing 2 ATP at the start, glycolysis generates 4 ATP directly. More energy will be harvested later from high-energy electrons used to form NADH and from the two molecules of pyruvic acid.

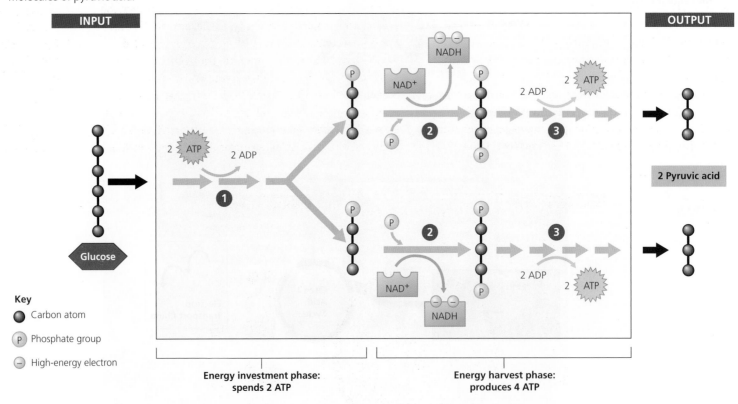

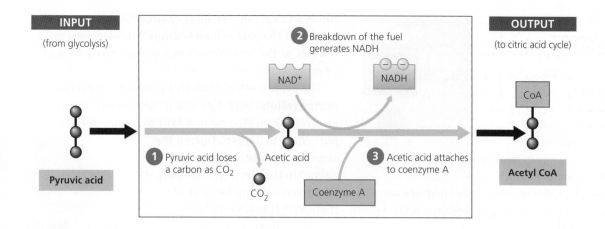

INPUT
(from glycolysis)

OUTPUT
(to citric acid cycle)

2 Breakdown of the fuel generates NADH

NAD⁺   NADH

CoA

1 Pyruvic acid loses a carbon as CO₂   Acetic acid   3 Acetic acid attaches to coenzyme A

Pyruvic acid

CO₂

Coenzyme A

Acetyl CoA

◄ **Figure 6.6 The link between glycolysis and the citric acid cycle: the conversion of pyruvic acid to acetyl CoA.** Remember that one molecule of glucose is split into two molecules of pyruvic acid. Therefore, the process shown here occurs twice for each starting glucose molecule.

Stage 2:
The Citric Acid Cycle

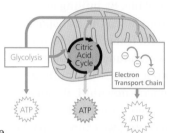

During glycolysis, one molecule of glucose is split into two molecules of pyruvic acid. But before pyruvic acid can be used by the citric acid cycle, it must be "groomed"—converted to a form the citric acid cycle can use **(Figure 6.6)**. 1 First, each pyruvic acid loses a carbon as CO₂. This is the first of this waste product we've seen so far in the breakdown of glucose. The remaining fuel molecules, each with only two carbons left, are called acetic acid (the acid that's in vinegar). 2 Electrons are stripped from these molecules and transferred to another molecule of NAD⁺, forming more NADH. 3 Finally, each acetic acid is attached to a molecule called coenzyme A (CoA), an enzyme derived from the B vitamin pantothenic acid, to form acetyl CoA. The

CoA escorts the acetic acid into the first reaction of the citric acid cycle. The CoA is then stripped and recycled.

The citric acid cycle finishes extracting the energy of sugar by dismantling the acetic acid molecules all the way down to CO₂ **(Figure 6.7)**. 1 Acetic acid joins a four-carbon acceptor molecule to form a six-carbon product called citric acid (for which the cycle is named). For every acetic acid molecule that enters the cycle as fuel, 2 two CO₂ molecules eventually exit as a waste product. Along the way, the citric acid cycle harvests energy from the fuel. 3 Some of the energy is used to produce ATP directly. However, the cycle captures much more energy in the form of 4 NADH and 5 a second, closely related electron carrier called FADH₂. 6 All the carbon atoms that entered the cycle as fuel are accounted for as CO₂ exhaust, and the four-carbon acceptor molecule is recycled. We have tracked only one acetic acid molecule through the citric acid cycle here. But because glycolysis splits glucose in two, the citric acid cycle occurs twice for each glucose molecule that fuels a cell. ☑

☑ **CHECKPOINT**

1. Two molecules of what compound are produced by glycolysis? Does this molecule enter the citric acid cycle?

2. Pyruvic acid must travel from the _____, where glycolysis takes place, to the _____, where the citric acid cycle takes place.

■ *Answers:* **1.** *Pyruvic acid; No; it is first converted to acetic acid.* **2.** *cytoplasm; mitochondria*

▶ **Figure 6.7
The citric acid cycle.**

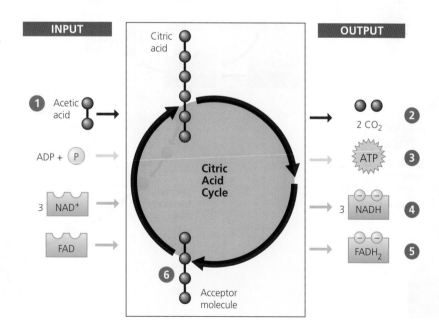

INPUT

Citric acid

OUTPUT

1 Acetic acid

ADP + P

3 NAD⁺

FAD

Citric Acid Cycle

2 CO₂ 2

ATP 3

3 NADH 4

FADH₂ 5

6

Acceptor molecule

Stage 3: Electron Transport

During cellular respiration, the electrons gathered from food molecules gradually "fall," losing energy at each step. In this way, cellular respiration

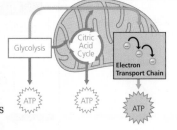

unlocks chemical energy in small amounts, bit by bit, that cells can put to productive use.

Electrons are transferred from glucose in food molecules to NAD⁺. This electron transfer converts NAD⁺ to NADH. Then NADH releases two electrons that enter an **electron transport chain**, a series of electron carrier molecules. This chain is like a bucket brigade, with each molecule passing an electron to the next molecule. With each exchange, the electron gives up a bit of energy. This downward cascade releases energy from the electron and uses it to make ATP (Figure 6.8).

Let's take a closer look at the path that electrons take (Figure 6.9). Each link in an electron transport chain is a molecule, usually a protein (shown as purple circles in Figure 6.9). In a series of reactions, each member of the chain transfers electrons. With each transfer, the electrons give up a small amount of energy that can then be used indirectly to generate ATP. The first molecule of the chain accepts electrons from NADH. Thus, NADH carries electrons from glucose and other fuel molecules and deposits them at the top of an electron transport chain. The molecule at the bottom of the chain finally "drops" the electrons to oxygen. At the same time, oxygen picks up hydrogen, forming water.

The overall effect of all this transfer of electrons during cellular respiration is a "downward" trip for electrons from glucose to NADH to an electron transport chain to oxygen. During the stepwise release of chemical energy in the electron transport chain, our cells make most of their ATP. It is actually oxygen, the "electron grabber," at the end, that makes it all possible. By pulling electrons down the transport chain from fuel molecules, oxygen functions somewhat like gravity pulling objects downhill. Because oxygen is the final electron acceptor, we cannot survive more than a few minutes without breathing. Viewed this way, drowning is deadly because it deprives cells of the final "electron grabbers" (oxygen) needed to drive cellular respiration.

▶ **Figure 6.8 Cellular respiration illustrated using a hard-hat analogy.** Splitting the bonds of a molecule of glucose provides the energy that boosts an electron from a low-energy state to a high-energy state.

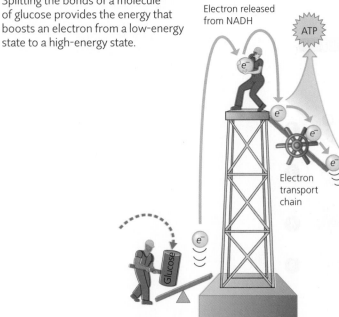

▼ **Figure 6.9 The role of oxygen in harvesting food energy.** In cellular respiration, electrons "fall" in small steps from food to oxygen, producing water. NADH transfers electrons from food to an electron transport chain. The attraction of oxygen to electrons "pulls" the electrons down the chain.

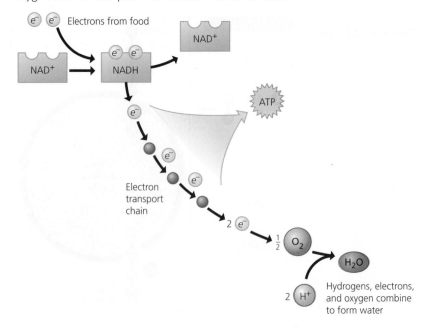

The molecules of electron transport chains are built into the inner membranes of mitochondria (see Figure 4.19). **Because these membranes are highly folded, their large surface area can accommodate thousands of copies of the electron transport chain—a good example of how biological structure fits function.** Each chain acts as a chemical pump that uses the energy released by the "fall" of electrons to move hydrogen ions (H⁺) across the inner mitochondrial membrane. This pumping causes ions to become more concen-  trated on one side of the membrane than on the other. Such a difference in concentration stores potential energy, similar to the way water can be stored behind a dam. There is a tendency for hydrogen ions to gush back to where they are less concentrated, just as there is a tendency for water to flow downhill. The inner membrane temporarily "dams" hydrogen ions.

The energy of dammed water can be harnessed to perform work. Gates in a dam allow the water to rush downhill, turning giant turbines, and this work can be used to generate electricity. Your mitochondria have structures that act like turbines. Each of these miniature machines, called an **ATP synthase**, is constructed from proteins built into the inner mitochondrial membrane, adjacent to the proteins of the electron transport chains. **Figure 6.10** shows a simplified view of how the energy previously stored in NADH and FADH₂ can now be used to generate ATP. ❶ NADH and ❷ FADH₂ transfer electrons to an electron transport chain. ❸ The electron transport chain uses this energy supply to pump H⁺ across the inner mitochondrial membrane. ❹ Oxygen pulls electrons down the transport chain. ❺ The H⁺ concentrated on one side of the membrane rushes back "downhill" through an ATP synthase. This action spins a component of the ATP synthase, just as water turns the turbines in a dam. ❻ The rotation activates parts of the synthase molecule that attach phosphate groups to ADP molecules to generate ATP.

The poison cyanide produces its deadly effect by binding to one of the protein complexes in the electron transport chain (marked with a skull-and-crossbones symbol in Figure 6.10). When bound there, cyanide blocks the passage of electrons to oxygen. This blockage is like clogging the outflow channel of a dam. As a result, no H⁺ gradient is generated, and no ATP is made. Cells stop working, and the organism dies. ✓

▼ **Figure 6.10   How electron transport drives ATP synthase machines.**

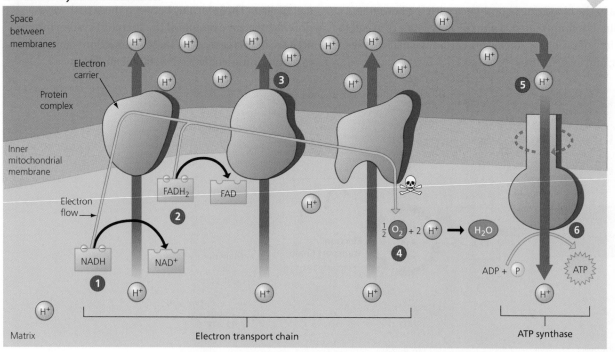

The Results of Cellular Respiration

When taking cellular respiration apart to see how all the molecular nuts and bolts of its metabolic machinery work, it's easy to lose sight of its overall function: to generate about 32 molecules of ATP per molecule of glucose (the actual number can vary by a few, depending on the organism and molecules involved). **Figure 6.11** will help you keep track of the ATP molecules generated. As we discussed, glycolysis and the citric acid cycle each contribute 2 ATP by directly making it. The rest of the ATP molecules are produced by ATP synthase, powered by the "fall" of electrons from food to oxygen. The electrons are carried from the organic fuel to electron transport chains by NADH and $FADH_2$. Each electron pair "dropped" down a transport chain from NADH or $FADH_2$ can power the synthesis of a few ATP. You can visualize the process like this: Energy flows from glucose to carrier molecules and ultimately to ATP.

We have seen that glucose can provide the energy to make the ATP our cells use for all their work. All of the energy-consuming activities of your body—moving your muscles, maintaining your heartbeat and temperature, and even the thinking that goes on within your brain—can be traced back to ATP and, before that, the glucose that was used to make it. The importance of glucose is underscored by the severity of diseases in which glucose balance is disturbed. Diabetes, which affects more than 422 million people worldwide, is caused by an inability to properly regulate glucose levels in the blood due to problems with the hormone insulin. If left untreated, a glucose imbalance can lead to a variety of problems, including cardiovascular disease, coma, and even death.

THE CELLS OF YOUR BRAIN BURN THROUGH A QUARTER POUND OF GLUCOSE EACH DAY.

☑ CHECKPOINT

Which stage of cellular respiration produces the majority of ATP?

■ Answer: electron transport

▶ **Figure 6.11 A summary of ATP yield during cellular respiration.**

But even though we have concentrated on glucose as the fuel that is broken down during cellular respiration, respiration is a versatile metabolic furnace that can "burn" many other kinds of food molecules. **Figure 6.12** diagrams some metabolic routes for the use of carbohydrates, fats, and proteins as fuel for cellular respiration. Taken together, all of these food molecules make up your calorie-burning metabolism. **The interplay between these pathways provides a clear example of the theme of system interactions; in this case, all of these interactions contribute to maintaining a balanced metabolism. ☑**

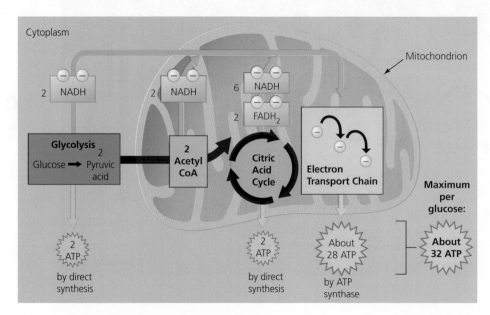

▼ **Figure 6.12 Energy from food.** The monomers from carbohydrates (polysaccharides and sugars), fats, and proteins can all serve as fuel for cellular respiration.

Fermentation:
Anaerobic Harvest of Food Energy

Although you must breathe to stay alive, some of your cells can work for short periods without oxygen. This **anaerobic** ("without oxygen") harvest of food energy is called fermentation.

Fermentation in Human Muscle Cells

You know by now that as your muscles work, they require a constant supply of ATP, which is generated by cellular respiration. As long as your blood provides your muscle cells with enough O_2 to keep electrons "falling" down transport chains in mitochondria, your muscles will work aerobically.

But under strenuous conditions, your muscles can spend ATP faster than your bloodstream can deliver O_2; when this happens, your muscle cells begin to work anaerobically. After functioning anaerobically for about 15 seconds, muscle cells will begin to generate ATP by the process of fermentation. **Fermentation** relies on glycolysis, the first stage of cellular respiration. Glycolysis does not require O_2 but does produce two ATP molecules for each glucose molecule broken down to pyruvic acid. That isn't very efficient compared with the 32 or so ATP molecules each glucose molecule generates during cellular respiration, but it can energize muscles for a short burst of activity. However, in such situations your cells will have to consume more glucose fuel per second because so much less ATP per glucose molecule is generated under anaerobic conditions. **Fermentation in muscle cells may be a vestige of evolution (see the Evolution Connection section) and provides a survival advantage even today.**

To harvest food energy during glycolysis, NAD^+ must be present to receive electrons (see Figure 6.9). This is no problem under aerobic conditions because the cell regenerates NAD^+ when NADH drops its electron cargo down electron transport chains to O_2. However, this recycling of NAD^+ cannot occur under anaerobic conditions because there is no O_2 to accept the electrons. Instead, NADH disposes of electrons by adding them to the pyruvic acid produced by glycolysis **(Figure 6.13)**. This restores NAD^+ and keeps glycolysis working.

The addition of electrons to pyruvic acid produces a waste product called lactic acid. The lactic acid by-product is eventually transported to the liver, where liver cells convert it back to pyruvic acid. Exercise scientists have long speculated about the role that lactic acid plays in muscle fatigue, as you'll see next. ☑

☑ **CHECKPOINT**

How many molecules of ATP can be produced from one molecule of glucose during fermentation?

■ *Answer: two*

▼ Figure 6.13 **Fermentation: producing lactic acid.**
Glycolysis produces ATP even in the absence of O_2. This process requires a continuous supply of NAD^+ to accept electrons from glucose. The NAD^+ is regenerated when NADH transfers the electrons it removed from food to pyruvic acid, thereby producing lactic acid (or other waste products, depending on the species of organism).

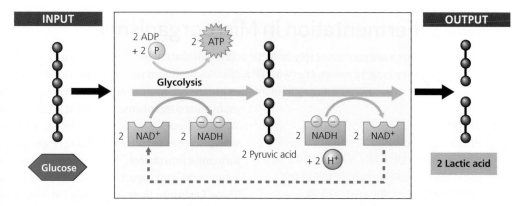

THE PROCESS OF SCIENCE   Exercise Science

What Causes Muscle Burn?

BACKGROUND

You probably know that your muscles burn after hard exercise ("Feel the burn!"). But what causes the burn? This question was investigated by one of the founders of the field of exercise science, a British biologist named A.V. Hill. In fact, Hill won a 1922 Nobel Prize for his investigations of muscle contraction.

Hill knew that muscles produce lactic acid under anaerobic conditions. In 1929, Hill developed a technique for electrically stimulating dissected frog muscles in a laboratory solution. He wondered if a buildup of lactic acid would cause muscle activity to stop.

METHOD

Hill's experiment tested frog muscles under two different sets of conditions (Figure 6.14). First, he showed that muscle performance declined when lactic acid could not diffuse away from the muscle tissue. Next, he showed that when lactic acid was allowed to diffuse away, performance improved significantly. These results led Hill to the conclusion that the buildup of lactic acid is the primary cause of muscle failure under anaerobic conditions.

RESULTS

Given his scientific stature—he was considered the world's leading authority on muscle activity—Hill's conclusion went unchallenged for many decades. Gradually, however, evidence that contradicted Hill's results began to accumulate. For example, the effect that Hill demonstrated did not appear to occur at human body temperature. And certain people who are unable to accumulate lactic acid have muscles that

fatigue *more* rapidly, which is the opposite of what you would expect. Recent experiments have directly refuted Hill's conclusions. Some research indicates that increased levels of other ions may be to blame. Therefore, the cause of muscle fatigue remains hotly debated.

The changing view of lactic acid's role in muscle fatigue illustrates an important point about the process of science: Scientific belief is dynamic and subject to constant adjustment as new evidence is uncovered. This would not have surprised Hill, who himself observed that all scientific hypotheses may become obsolete, and that changing conclusions in light of new evidence is necessary for the advancement of science.

▼ **Figure 6.14** A. V. Hill's 1929 apparatus for measuring muscle fatigue.

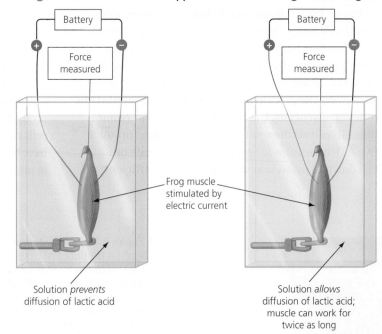

Battery

Force measured

Battery

Force measured

Frog muscle stimulated by electric current

Solution *prevents* diffusion of lactic acid

Solution *allows* diffusion of lactic acid; muscle can work for twice as long

Thinking Like a Scientist

The explanation for the cause of muscle burn has changed over time as new evidence has accumulated. Does this represent a failure of the scientific method?

For the answer, see Appendix D.

Fermentation in Microorganisms

Our muscles cannot rely on lactic acid fermentation for very long. However, the two ATP molecules produced per glucose molecule during fermentation are enough to sustain many microorganisms. We have domesticated such microbes to transform milk into cheese, sour cream, and yogurt. These foods owe their

THE METABOLIC PROCESSES THAT PRODUCE ACID IN YOUR MUSCLES AFTER A HARD WORKOUT ARE SIMILAR TO THE PROCESSES THAT PRODUCE PEPPERONI, SOY SAUCE, YOGURT, AND BREAD.

sharp or sour flavor mainly to lactic acid. The food industry also uses fermentation to produce soy sauce from soybeans, to pickle cucumbers, olives, and cabbage, and to produce meat products like sausage, pepperoni, and salami.

Yeast, a microscopic fungus, is capable of both cellular respiration and fermentation. When kept in an anaerobic environment, yeast cells ferment sugars and other foods to stay alive. As they do, the yeast produce ethyl alcohol as a waste product instead of lactic acid

(Figure 6.15). This alcoholic fermentation also releases CO_2. For thousands of years, people have put yeast to work producing alcoholic beverages such as beer and wine using airtight barrels and vats, where the lack of oxygen forces the yeast to ferment glucose into ethanol. And as every baker knows, the CO_2 bubbles from fermenting yeast also cause bread dough to rise. (The alcohol produced in fermenting bread is burned off during baking.) ✓

✓ CHECKPOINT

What kind of acid builds up in human muscle during strenuous activity?

■ Answer: lactic acid

▼ **Figure 6.15 Fermentation: producing ethyl alcohol.** The alcohol produced by yeast as bread rises is burned off during baking.

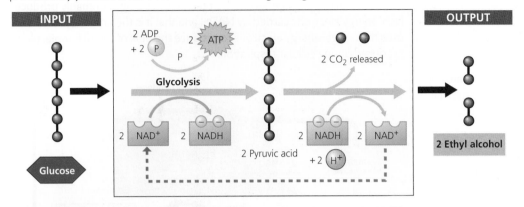

EVOLUTION CONNECTION | Exercise Science

The Importance of Oxygen

In the Biology and Society and the Process of Science sections, we were reminded of the important role that oxygen plays during aerobic exercise. But in the last section on fermentation, we learned that exercise can continue on a limited basis even under anaerobic (oxygen-free) conditions. Both aerobic and anaerobic respiration start with glycolysis, the splitting of glucose to form pyruvic acid. Glycolysis is thus the universal energy-harvesting process of life. If you looked inside a bacterial cell, one of your body cells, or any other living cell, you would find the metabolic machinery of glycolysis.

The universality of glycolysis in life on Earth has an evolutionary basis. Ancient prokaryotes probably used glycolysis to make ATP long before oxygen was present in Earth's atmosphere. The oldest known fossils of bacteria date back more than 3.5 billion years and resemble some living photosynthetic bacteria. Evidence indicates, however, that significant levels of O_2, formed as a by-product of bacterial photosynthesis, did not accumulate in the atmosphere until about 2.7 billion years ago (Figure 6.16). For almost 1 billion years, prokaryotes must have generated ATP exclusively from glycolysis, a process that does not require oxygen.

The fact that glycolysis occurs in almost all organisms suggests that it evolved very early in ancestors common to all the domains of life. The location of glycolysis within the cell also implies great antiquity; the

pathway does not require any of the membrane-enclosed organelles of the eukaryotic cell, which evolved about a billion years after the prokaryotic cell. Glycolysis is a metabolic heirloom from early cells that continues to function in fermentation and as the first stage in the breakdown of organic molecules by cellular respiration. The ability of our muscles to function anaerobically can therefore be viewed as a vestige of our ancient ancestors, who relied exclusively on this metabolic pathway.

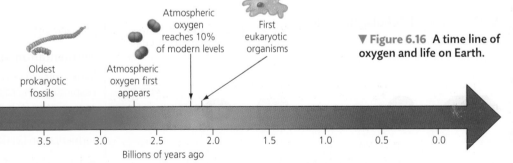

▼ **Figure 6.16** A time line of oxygen and life on Earth.

Chapter Review

SUMMARY OF KEY CONCEPTS

Energy Flow and Chemical Cycling in the Biosphere

Producers and Consumers

Autotrophs (producers) make organic molecules from inorganic nutrients by photosynthesis. Heterotrophs (consumers) must consume organic material and obtain energy by cellular respiration.

Chemical Cycling between Photosynthesis and Cellular Respiration

The molecular outputs of cellular respiration—CO_2 and H_2O—are the molecular inputs of photosynthesis, and vice versa. While these chemicals cycle through an ecosystem, energy flows through, entering as sunlight and exiting as heat.

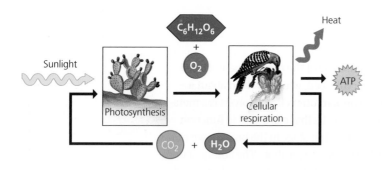

Cellular Respiration: Aerobic Harvest of Food Energy

An Overview of Cellular Respiration

The overall equation of cellular respiration simplifies a great many chemical steps into one formula:

The Three Stages of Cellular Respiration

Cellular respiration occurs in three stages. During glycolysis, a molecule of glucose is split into two molecules of pyruvic acid, producing two molecules of ATP and two high-energy electrons stored in NADH. During the citric acid cycle, what remains of glucose is completely broken down to CO_2, producing a bit of ATP and a lot of high-energy electrons stored in NADH and $FADH_2$. The electron transport chain uses the high-energy electrons to pump H^+ across the inner mitochondrial membrane, eventually handing them off to O_2, producing H_2O. Backflow of H^+ across the membrane powers the ATP synthases, which produce ATP from ADP.

The Results of Cellular Respiration

You can follow the flow of molecules through the process of cellular respiration in the diagram. Notice that the first two stages primarily produce high-energy electrons carried by NADH and that it is the final stage that uses these high-energy electrons to produce the bulk of the ATP molecules produced during cellular respiration.

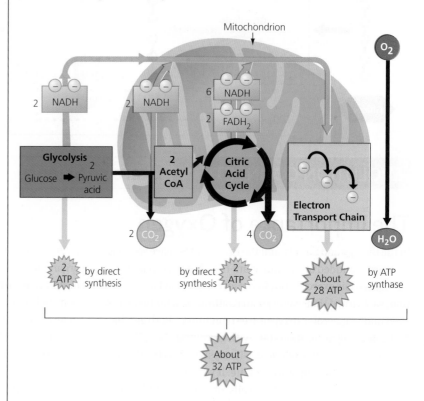

Fermentation: Anaerobic Harvest of Food Energy

Fermentation in Human Muscle Cells

When muscle cells consume ATP faster than O_2 can be supplied for cellular respiration, the conditions become anaerobic, and muscle cells will begin to regenerate ATP by fermentation. The waste product under these anaerobic conditions is lactic acid. The ATP yield per glucose is much lower during fermentation (2 ATP) than during cellular respiration (about 32 ATP).

Fermentation in Microorganisms

Yeast and some other organisms can survive with or without O_2. Wastes from fermentation can be ethyl alcohol, lactic acid, or other compounds, depending on the species.

Mastering Biology

For practice quizzes, BioFlix animations, MP3 tutorials, video tutors, and more study tools designed for this textbook, go to Mastering Biology™

SELF-QUIZ

1. Which of the following statements is a correct distinction between autotrophs and heterotrophs?
   a. Only heterotrophs require chemical compounds from the environment.
   b. Cellular respiration is unique to heterotrophs.
   c. Only heterotrophs have mitochondria.
   d. Only autotrophs can live on nutrients that are entirely inorganic.

2. Why are plants called producers? Why are animals called consumers?

3. How is your breathing related to your cellular respiration?

4. Of the three stages of cellular respiration, which produces the most ATP molecules per glucose?

5. _____ is regenerated during fermentation.

6. The poison cyanide acts by blocking a key step in the electron transport chain. Knowing this, explain why cyanide kills so quickly.

7. Cells can harvest the most chemical energy from which of the following?
   a. an NADH molecule
   b. a glucose molecule
   c. six CO_2 molecules
   d. two pyruvic acid molecules

8. _____ is a metabolic pathway common to both fermentation and cellular respiration.

9. Physicians find that a child is born with a rare disease in which mitochondria are missing from certain skeletal muscle cells but her muscle cells function. Not surprisingly, they also find that
   a. the muscles contain large amounts of lactic acid following even mild physical exercise.
   b. the muscles contain large amounts of carbon dioxide following even mild physical exercise.
   c. the muscles require extremely high levels of oxygen to function.
   d. the muscle cells cannot split glucose to pyruvic acid.

10. A glucose-fed yeast cell is moved from an aerobic environment to an anaerobic one. For the cell to continue to generate ATP at the same rate, approximately how much glucose must it consume in the anaerobic environment compared with the aerobic environment?

For answers to the Self-Quiz, see Appendix D.

IDENTIFYING MAJOR THEMES

For each statement, identify which major theme is evident (the relationship of structure to function, information flow, pathways that transform energy and matter, interactions within biological systems, or evolution) and explain how the statement relates to the theme. If necessary, review the themes (see Chapter 1) and review the examples highlighted in blue in this chapter.

11. The highly folded membranes of the mitochondria make these organelles well suited to carry out the huge number of chemical reactions required for cellular respiration to proceed.

12. Cellular respiration and photosynthesis are linked, with each process using inputs created by the other.

13. Your body uses many different intersecting chemical pathways that, all together, constitute your metabolism.

For answers to Identifying Major Themes, see Appendix D.

THE PROCESS OF SCIENCE

14. The process of cellular respiration, which breaks down glucose into CO_2 and H_2O and generates approximately 32 molecules of ATP, takes place in different compartments of the cell—glycolysis occurs in the cytoplasm, the citric acid cycle in the mitochondrial matrix, and the electron transport chain in the inner mitochondrial membrane. In order for this process to get completed, several molecules need to cross the mitochondrial membranes. Which are these molecules, and in which directions do they move? What is the need for this elaborate arrangement?

15. **Interpreting Data** Basal metabolic rate (BMR) is the amount of energy that must be consumed by a person at rest to maintain his or her body weight. BMR depends on several factors, including sex, age, height, and weight. The following graph shows the BMR for a 6'0" 45-year-old male. For this person, how does BMR correlate with weight? How many more calories must a 250-pound man consume to maintain his weight than a 200-pound man? Why does BMR depend on weight?

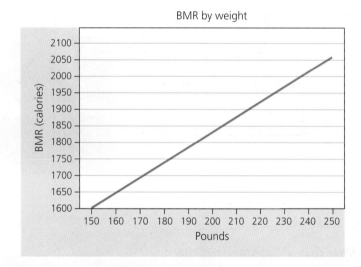

BMR by weight

BIOLOGY AND SOCIETY

16. Delivery of oxygen to muscles is the limiting factor for many athletes. Some try blood doping (injecting oxygenated blood). Others train at high altitude. Both approaches increase the number of red blood cells. Why is doping considered cheating? How would you enforce antidoping rules?

17. Various cultures have been using fermentation for a long time to produce and conserve foods. Compile a list of the different types of fermentation and to which end they are used.

18. Alcohol consumption during pregnancy can cause a series of birth defects called fetal alcohol syndrome (FAS). Symptoms include head and facial irregularities, heart defects, intellectual disability, and behavioral problems. The U.S. Surgeon General's Office recommends that pregnant women abstain from alcohol and liquor bottles have a warning label. If you were a server in a restaurant and a pregnant woman ordered a daiquiri, how would you respond? Is it her right to make those decisions about her unborn child's health? Do you bear any responsibility? Is a restaurant responsible for monitoring dietary habits of customers?

7 Photosynthesis: Using Light to Make Food

Why Photosynthesis Matters

Do you like to eat? We humans can trace every morsel of our food back to plants. By capturing the energy of sunlight and using it to create organic materials, plants performing photosynthesis feed the world.

COVER UP! PROTECTING YOURSELF FROM SHORT WAVELENGTHS OF LIGHT CAN BE LIFESAVING.

WANT TO DO SOMETHING SIMPLE TO COMBAT GLOBAL CLIMATE CHANGE? PLANT A TREE—YOU'LL BE GLAD YOU DID!

NEARLY ALL LIFE ON EARTH—INCLUDING YOU—CAN TRACE ITS SOURCE OF ENERGY BACK TO THE SUN.

BIOLOGY AND SOCIETY Solar Energy

Tapping into the sun.
Rooftop solar panels are becoming a common sight.

A Solar Revolution

When you hear the term *solar power*, you may think of rooftop rows of solar panels. Indeed, the electricity that photovoltaic (PV) solar panels produce from sunlight can be used directly, stored in batteries, or transferred to the national electric grid. The advantages of solar panels are many: Sunlight is free, plentiful, and abundant across the whole planet; the technology is quite reliable because it uses no moving parts; and no pollution is created. It's not surprising, then, that residential photovoltaic systems are becoming more popular.

Photovoltaic panels are just one of many ways that humans can harness the sun to produce energy. Living organisms that capture sunlight through the process of photosynthesis can also serve as energy sources. For example, plant matter can be burned directly, as with stoves or boilers that burn wood or wood pellets. Plant material can also be used to produce biofuels such as bioethanol and biodiesel. You may have noticed a sticker on a gas pump declaring the percentage of ethanol in that gasoline; most cars today run on a blend of 85% gasoline and 15% ethanol. Many car manufacturers are producing "flexible-fuel" vehicles that can run on any combination of gasoline and bioethanol.

All of these methods of capturing the energy in sunlight share an important trait: They are all renewable—that is, they are derived from sources that are naturally replenished on a human time scale. In addition to solar energy, renewable energy (also called green energy) can be derived from wind through turbines, water through hydroelectric dams, and geothermal sources through heat pumps. In fact, the majority of new electricity-generating capacity installed in 2015 was renewable, accounting for nearly $300 billion in investment. About one-quarter of all electricity is now generated from renewable sources, and this percentage is rising steadily. One nation that is leading this effort is Iceland, which generates 99.8% of its electricity from renewable sources.

How do you benefit from the solar revolution? Society's ever-increasing demand for fossil fuels (oil, gasoline, diesel, and other products derived from petroleum) has serious consequences. Burning fossil fuels releases pollution, which can cause respiratory diseases and acid precipitation, and carbon dioxide (CO_2), which contributes to global climate change. Additionally, a barrel of oil removed from the ground can never be replaced. Therefore, your health and the health of the planet benefit from using green energy instead of fossil fuels.

Nearly all life on Earth taps into the energy of the sun. Photosynthesis is the process by which plants, algae, and other organisms use light to make sugars from carbon dioxide—sugars that are food for these organisms and the starting point for most of our own food. In this chapter, we'll first examine some basic concepts of photosynthesis; then we'll look at the specific mechanisms involved in this process.

The Basics of Photosynthesis

The process of photosynthesis is the ultimate source of energy for nearly every ecosystem on Earth. **Photosynthesis** is a process whereby plants, algae (which are protists), and certain bacteria transform light energy into chemical energy, using carbon dioxide and water as starting materials and releasing oxygen gas as a by-product. The chemical energy produced through photosynthesis is stored in the bonds of sugar molecules. Organisms that generate their own food are called autotrophs (see Chapter 6). Plants and other organisms that do this by photosynthesis—photoautotrophs—are the producers for most ecosystems (Figure 7.1). Photoautotrophs not only feed us but also clothe us (as the source of cotton fibers), house us (wood), and provide energy for warmth, light, and transportation (biofuels). The fact that nearly all organisms—including you—can trace their source of energy back to the sun clearly illustrates that the ability to transform energy and matter is vital to the existence of life on Earth.

NEARLY ALL LIFE ON EARTH CAN TRACE ITS SOURCE OF ENERGY BACK TO THE SUN.

Chloroplasts: Sites of Photosynthesis

Photosynthesis in plants and algae occurs within light-absorbing organelles called **chloroplasts** (see Figure 4.18). All green parts of a plant have chloroplasts and thus can carry out photosynthesis. In most plants, however, the leaves have the most chloroplasts (about 300 million chloroplasts in a piece of leaf the size of a nickel). Their green color is from **chlorophyll**, a pigment (light-absorbing molecule) in the chloroplasts that plays a central role in converting solar energy to chemical energy.

Chloroplasts are concentrated in the interior cells of leaves (Figure 7.2), with a typical cell containing 30–40 chloroplasts. Carbon dioxide (CO_2) enters a leaf, and oxygen (O_2) exits, by way of tiny pores called **stomata** (singular, *stoma*, meaning "mouth"). The carbon dioxide that enters the leaf is the source of carbon for much of the body of the plant, including the sugars and starches that we eat. So the bulk of the body of a plant derives from the air, not the soil. As proof of this idea, consider hydroponics, a means of growing plants using only air and water; no soil whatsoever is involved. In addition to carbon dioxide, photosynthesis requires water, which is absorbed by the plant's roots and transported to the leaves, where veins carry it to the photosynthetic cells.

Membranes within the chloroplast form the framework where many of the reactions of photosynthesis occur. Like a mitochondrion, a chloroplast has a double-membrane envelope. The chloroplast's inner membrane encloses a compartment filled with **stroma**, a thick fluid. (It's easy to confuse two terms associated with photosynthesis: *Stomata* are pores through which gases are exchanged,

▼ Figure 7.1 A diversity of photoautotrophs.

PHOTOSYNTHETIC AUTOTROPHS		
Plants (mostly on land)	**Photosynthetic Protists** (aquatic)	**Photosynthetic Bacteria** (aquatic)
Forest plants	Kelp, a large, multicellular alga	Micrograph of cyanobacteria

LM 375×

and *stroma* is the fluid within the chloroplast.) Suspended in the stroma are interconnected membranous sacs called **thylakoids**. The thylakoids are concentrated in stacks called **grana** (singular, *granum*). The chlorophyll molecules that capture light energy are built into the thylakoid membranes. **The structure of a chloroplast—with its stacks of disks—aids its function by providing a large surface area for the reactions of photosynthesis.** ☑

▼ **Figure 7.2 Journey into a leaf.** This series of blowups takes you into a leaf's interior, then into a plant cell, and finally into a chloroplast, the site of photosynthesis.

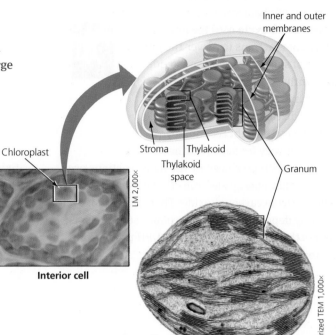
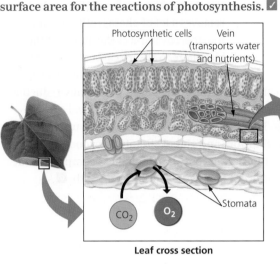

Leaf cross section

Interior cell

Chloroplast

☑ **CHECKPOINT**

Photosynthesis takes place within organelles called _____ using gases that are exchanged through pores called _____.

■ *Answer: chloroplasts; stomata*

An Overview of Photosynthesis

The following chemical equation, simplified to highlight the relationship between photosynthesis and cellular respiration, provides a summary of the reactants and products of photosynthesis:

$$6\ CO_2 + 6\ H_2O \xrightarrow{\text{Light energy}} C_6H_{12}O_6 + 6\ O_2$$

Notice that the reactants of photosynthesis—carbon dioxide (CO_2) and water (H_2O)—are the same as the waste products of cellular respiration (see Figure 6.2). Also notice that photosynthesis produces what respiration uses—glucose ($C_6H_{12}O_6$) and oxygen (O_2). In other words, photosynthesis recycles the "exhaust" of cellular respiration and rearranges its atoms to produce food and oxygen. Photosynthesis is a chemical transformation that requires a lot of energy, and sunlight absorbed by chlorophyll provides that energy.

Recall that cellular respiration is a process of electron transfer (see Chapter 6). A "fall" of electrons from food molecules to oxygen to form water releases the energy that mitochondria can use to make ATP (see Figure 6.9). The opposite occurs in photosynthesis: Electrons are boosted "uphill" and added to carbon dioxide to produce sugar. Hydrogen is moved along with the electrons being

transferred from water to carbon dioxide. This transfer of hydrogen requires the chloroplast to split water molecules into hydrogen and oxygen. The hydrogen is transferred along with electrons to carbon dioxide to form sugar. The oxygen escapes through stomata in leaves into the atmosphere as O_2, a waste product of photosynthesis.

The overall equation for photosynthesis is a simple summary of a complex process. Like many energy-producing processes within cells, photosynthesis is a multistep chemical pathway, with each step in the path producing products that are used as reactants in the next step. For a better overview, let's look at the two stages of photosynthesis: the light reactions and the Calvin cycle **(Figure 7.3)**.

In the **light reactions**, chlorophyll in the thylakoid membranes absorbs solar energy (the "photo" part of photosynthesis), which is then converted to the chemical energy of ATP (the molecule

Figure
Walkthrough

Mastering **Biology**
goo.gl/3rETcw

▼ **Figure 7.3
A road map for photosynthesis.**

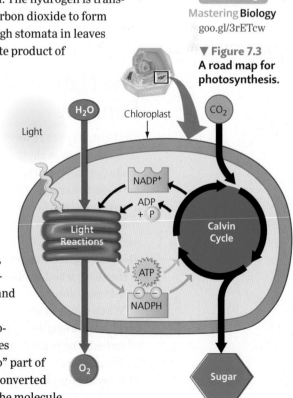

143

that drives most cellular work) and **NADPH** (an electron carrier). During the light reactions, water is split, providing a source of electrons and giving off O_2 gas as a by-product.

The **Calvin cycle** uses the products of the light reactions to power the production of sugar from carbon dioxide (the "synthesis" part of photosynthesis). The enzymes that drive the Calvin cycle are dissolved in the stroma. ATP generated by the light reactions provides the energy for sugar synthesis. And the NADPH produced by the light reactions provides the high-energy electrons that drive the synthesis of glucose from carbon dioxide. Thus, the Calvin cycle indirectly depends on light to produce sugar because it requires the supply of ATP and NADPH produced by the light reactions.

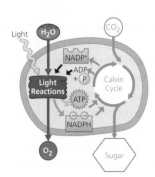

WANT TO DO SOMETHING SIMPLE TO COMBAT GLOBAL CLIMATE CHANGE? PLANT A TREE— YOU'LL BE GLAD YOU DID!

The initial incorporation of carbon from CO_2 into organic compounds is called **carbon fixation**. This process has important implications for global climate, because the removal of carbon from the air and its incorporation into plant material can help reduce the concentration of carbon dioxide in the atmosphere. Deforestation, which removes a lot of photosynthetic plant life, thereby reduces the ability of the biosphere to absorb carbon. Planting new forests can have the opposite effect of fixing carbon from the atmosphere, potentially reducing the effect of the gases that contribute to global climate change. **The relationship between photosynthesis, carbon fixation, and global climate is a good example of how interactions between biological components at many different levels affect all life on Earth.** ☑

The Light Reactions: Converting Solar Energy to Chemical Energy

Chloroplasts are solar-powered sugar factories. Let's look at how they use the energy in sunlight to store chemical energy.

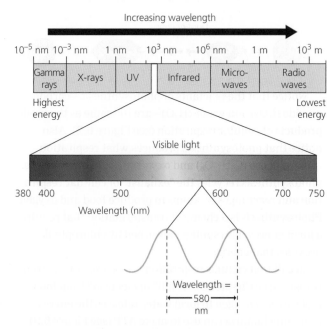

The Nature of Sunlight

Sunlight, like other forms of radiation, travels through space as rhythmic waves, like the ripples made by a pebble dropped into a pond. The distance between the crests of two adjacent waves is called a **wavelength**. The full range of radiation, from the very short wavelengths of gamma rays to the very long wavelengths of radio signals, is called the **electromagnetic spectrum** (Figure 7.4). Visible light is the small fraction of the spectrum that our eyes see as different colors.

When sunlight shines on a pigmented material, certain wavelengths (colors) of the visible light are absorbed. The colors that are not absorbed are reflected by the material. For example, we see a pair of jeans as blue because pigments in the fabric absorb the other colors, leaving only blue light to be reflected to our

▼ **Figure 7.4 The electromagnetic spectrum.** The middle of the figure expands the thin slice of the spectrum that is visible to us as different colors of light, from about 380 nanometers (nm) to about 750 nm in wavelength. The bottom of the figure shows waves of one particular wavelength of visible light.

eyes. In the 1800s, plant biologists discovered that only certain wavelengths of light are used by plants, as we'll see next.

What Colors of Light Drive Photosynthesis?

BACKGROUND

Scientific breakthroughs often occur after careful observations of natural phenomena. In 1883, German biologist Theodor Engelmann noticed that certain aquatic bacteria tend to cluster in areas with higher oxygen concentrations. He formed a hypothesis that oxygen-seeking bacteria would congregate near algae that were performing the most photosynthesis and therefore producing the most oxygen. He conducted a simple but elegant experiment to determine which wavelengths (colors) of light are used during photosynthesis.

METHOD

Engelmann placed a string of freshwater algal cells within a drop of water on a microscope slide and used a prism to expose the cells to the color spectrum. He then added oxygen-seeking bacteria to the water **(Figure 7.5)**.

RESULTS

The experiment showed that most bacteria congregated around algae exposed to blue-violet and red-orange light. Other experiments have since verified that those wavelengths are mainly responsible for photosynthesis. Variations of this classic experiment are still performed. For example, biofuel researchers test different species of algae to determine which wavelengths result in optimal fuel production. Biofuel facilities of the future may use a variety of species that take advantage of the full spectrum of light that shines down on them.

Engelmann used a prism to shine the color spectrum of visible light onto a slide containing algal cells in water.

He added oxygen-seeking bacteria to the water. They congregated wherever the most O_2 gas was produced.

◀ **Figure 7.5** Investigating how light wavelength affects photosynthesis.

Most bacteria congregated around algal cells exposed to blue-violet and red-orange light.

Thinking Like a Scientist

Why was Engelmann's use of a prism with natural sunlight a key to his discovery?

For the answer, see Appendix D.

Chloroplast Pigments

The selective absorption of light by leaves explains why they appear green to us; light of that color is poorly absorbed by chloroplasts and is thus reflected or transmitted toward the observer **(Figure 7.6)**. Energy cannot be destroyed, so the absorbed energy must be converted to other forms. Chloroplasts contain several different pigments that absorb light of different wavelengths. Chlorophyll *a*,

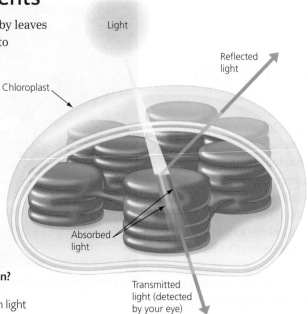

▶ **Figure 7.6 Why are leaves green?**
Chlorophyll and other pigments in chloroplasts reflect or transmit green light while absorbing other colors.

145

the pigment that participates directly in the light reactions, absorbs mainly blue-violet and red light. A very similar molecule, chlorophyll *b*, absorbs mainly blue and orange light. Chlorophyll *b* does not participate directly in the light reactions, but it conveys absorbed energy to chlorophyll *a*, which then puts the energy to work in the light reactions.

Chloroplasts also contain a family of yellow-orange pigments called carotenoids, which absorb mainly blue-green light. Some carotenoids have a protective function: They dissipate excess light energy that would otherwise damage chlorophyll. Some carotenoids are human nutrients: beta-carotene (a bright orange/red pigment found in pumpkins, sweet potatoes, and carrots) is converted to vitamin A in the body, and lycopene (a bright red pigment found in tomatoes, watermelon, and red peppers) is an antioxidant that is being studied for potential anti-cancer properties. Additionally, the spectacular colors of fall foliage in some parts of the world are due partly to the yellow-orange light reflected from carotenoids **(Figure 7.7)**. The decreasing temperatures in autumn cause a decrease in the levels of chlorophyll, allowing the colors of the longer-lasting carotenoids to be seen in all their fall glory.

All of these chloroplast pigments are built into the thylakoid membranes (see Figure 7.2). There the pigments are organized into light-harvesting complexes called photosystems, our next topic. ☑

▲ **Figure 7.7 Photosynthetic pigments.** Falling autumn temperatures cause a decrease in the levels of green chlorophyll within the foliage of leaf-bearing trees. This decrease allows the colors of the carotenoids to be seen.

☑ **CHECKPOINT**

What is the specific name of the pigment that absorbs energy during the light reactions?

■ *Answer: chlorophyll a*

How Photosystems Harvest Light Energy

Thinking about light as waves explains most of light's properties. However, light also behaves as discrete packets of energy called photons. A **photon** is a fixed quantity of light energy. The shorter the wavelength of light, the greater the energy of a photon. A photon of violet light, for example, packs nearly twice as much energy as a photon of red light. This is why short-wavelength light—such as ultraviolet light and X-rays—can be damaging; photons at these wavelengths carry enough energy to damage proteins and DNA, potentially leading to cancerous mutations.

PROTECTING YOURSELF FROM SHORT WAVELENGTHS OF LIGHT CAN BE LIFESAVING.

When a pigment molecule absorbs a photon, one of the pigment's electrons gains energy. This electron (e⁻) is now said to be "excited"; that is, the electron has been raised from its starting state (called the ground state) to an excited state. The excited state is highly unstable, so an excited electron usually gives off its excess energy and falls back to its ground state almost immediately **(Figure 7.8a)**. Most pigments release heat energy as their light-excited electrons fall back to their ground state. (That's why a surface with a lot of pigment, such as a black driveway, gets so hot on a sunny day.) But some pigments emit light as well as heat after absorbing photons.

▼ **Figure 7.8 Excited electrons in pigments.**

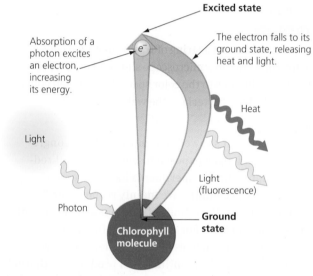

(a) Absorption of a photon

(b) Fluorescence of a glow stick. Breaking a vial within a glow stick starts a chemical reaction that excites electrons within a fluorescent dye. As the electrons fall from their excited state to the ground state, the excess energy is emitted as light.

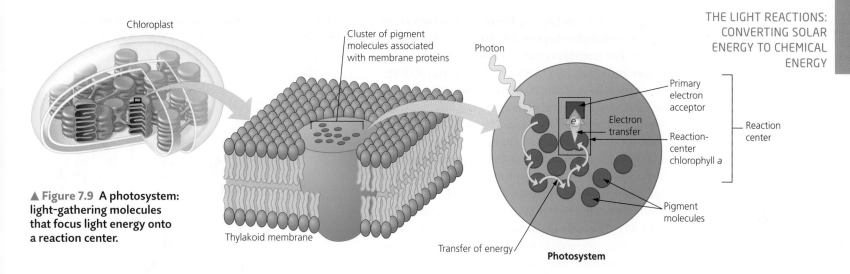

▲ Figure 7.9 A photosystem: light-gathering molecules that focus light energy onto a reaction center.

The fluorescent light emitted by a glow stick is caused by a chemical reaction that excites electrons of a fluorescent dye (Figure 7.8b). The excited electrons quickly fall back down to their ground state, releasing energy in the form of fluorescent light.

In the thylakoid membrane, chlorophyll molecules are organized with other molecules into photosystems. Each **photosystem** has a cluster of a few hundred pigment molecules (Figure 7.9). This cluster of pigment molecules functions as a light-gathering antenna. When a photon strikes one of the pigment molecules, the energy jumps from molecule to molecule until it arrives at the reaction center of the photosystem. The reaction center consists of chlorophyll a molecules that sit next to another molecule called a primary electron acceptor. This primary electron acceptor traps the light-excited electron from the chlorophyll a in the reaction center. As you'll see next, another team of molecules built into the thylakoid membrane then uses that trapped energy to make ATP and NADPH. ✓

☑ **CHECKPOINT**

What is the role of a reaction center during photosynthesis?

■ Answer: A reaction center transfers a light-excited photon from pigment molecules to molecules that can use this trapped energy to drive chemical reactions.

How the Light Reactions Generate ATP and NADPH

Two photosystems cooperate in the light reactions (Figure 7.10). ❶ Photons excite electrons in the chlorophyll of the first photosystem. These photons are then trapped by the primary electron acceptor. This photosystem then replaces the lost electrons by extracting new ones from water. This is the step that releases O_2 during photosynthesis. ❷ Energized electrons from the first photosystem pass down an electron transport chain to the second photosystem. The chloroplast uses the energy released by this electron "fall" to make ATP. ❸ The second photosystem transfers its light-excited electrons to $NADP^+$, converting it to NADPH.

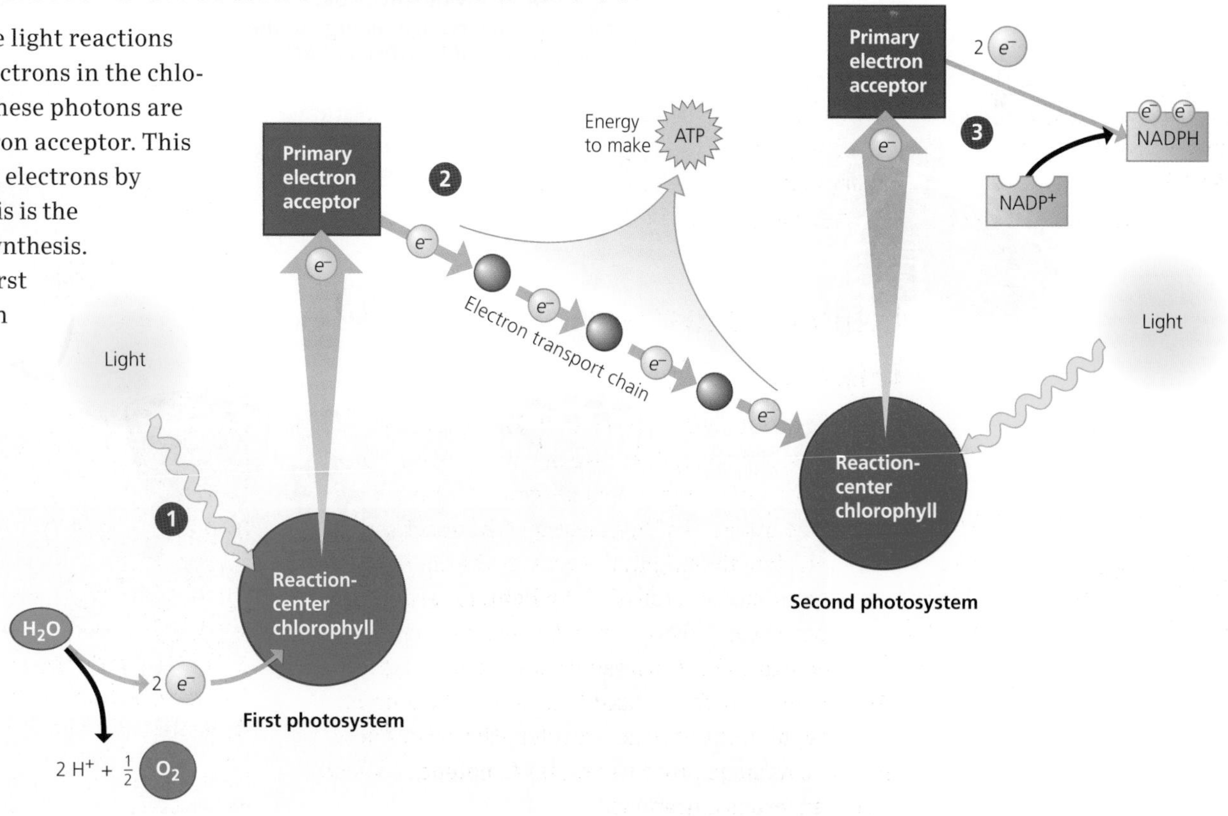

▶ Figure 7.10 The light reactions of photosynthesis. The orange arrows trace a light-driven flow of electrons from H_2O to NADPH. These electrons also produce ATP.

☑ CHECKPOINT

1. Why is water required as a reactant in photosynthesis? (*Hint*: Review Figures 7.10 and 7.11.)

2. In addition to conveying electrons between photosystems, the electron transport chains of chloroplasts provide the energy for the synthesis of _____.

■ *Answers:* **1.** *The splitting of water provides electrons for converting* NAD⁺ *to NADPH.* **2.** *ATP*

Figure 7.11 shows the location of the light reactions in the thylakoid membrane. The two photosystems and the electron transport chain that connects them transfer electrons from H_2O to $NADP^+$, producing NADPH. Notice that the mechanism of ATP production during the light reactions is very similar to the mechanism we saw in cellular respiration (see Figure 6.10). In both cases, an electron transport chain pumps hydrogen ions (H^+) across a membrane—the inner mitochondrial membrane in the case of respiration and the thylakoid membrane in photosynthesis. And in both cases, ATP synthases use the energy stored by the H^+ gradient to make ATP. The main difference is that food provides the high-energy electrons in cellular respiration, whereas light-excited electrons flow down the transport chain during photosynthesis. The traffic of electrons shown in Figures 7.10 and 7.11 is analogous to the cartoon in **Figure 7.12**.

We have seen how the light reactions absorb solar energy and convert it to the chemical energy of ATP and NADPH. Notice again, however, that the light reactions did not produce any sugar. That's the job of the Calvin cycle, as we'll see next. ☑

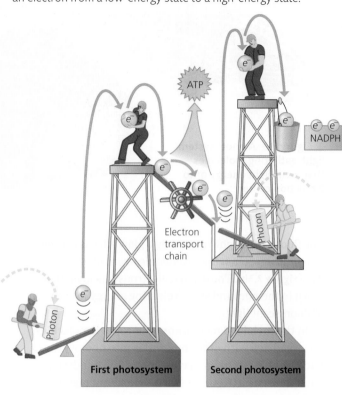

▼ **Figure 7.12 The light reactions illustrated using a hard-hat analogy.** The energy of sunlight is used to boost an electron from a low-energy state to a high-energy state.

First photosystem

Second photosystem

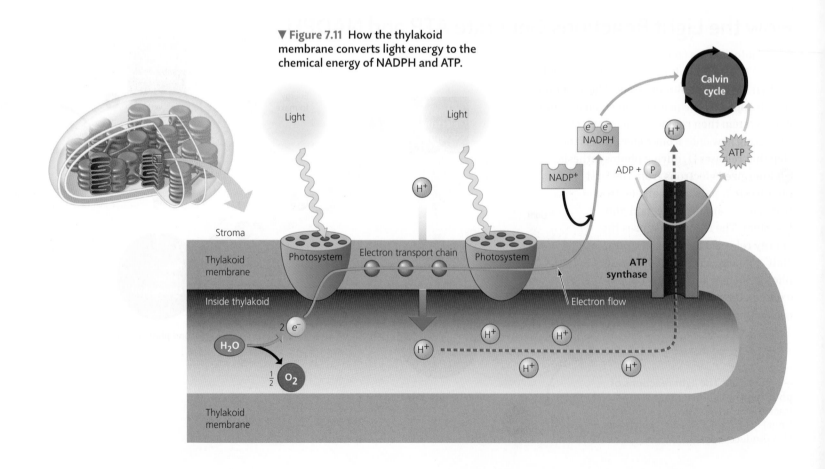

▼ **Figure 7.11 How the thylakoid membrane converts light energy to the chemical energy of NADPH and ATP.**

The Calvin Cycle: Making Sugar from Carbon Dioxide

If chloroplasts are solar-powered sugar factories, then the Calvin cycle is the actual sugar-manufacturing machinery. This process is called a cycle because its starting material is regenerated. With each turn of the cycle, there are chemical inputs and outputs. The inputs are CO_2 from the air as well as two products of the light reactions: ATP (which provides energy) and NADPH (which provides electrons). Using carbon from CO_2, energy from ATP, and high-energy electrons from NADPH, the Calvin cycle constructs an energy-rich sugar molecule called glyceraldehyde 3-phosphate (G3P). The plant cell can then use G3P as the raw material to make the glucose and other organic compounds (such as cellulose and starch) that it needs. **Figure 7.13** presents the basics of the Calvin cycle, emphasizing inputs and outputs. Each ● symbol represents a carbon atom, and each Ⓟ symbol represents a phosphate group. ☑

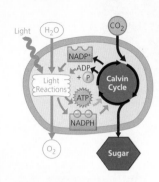

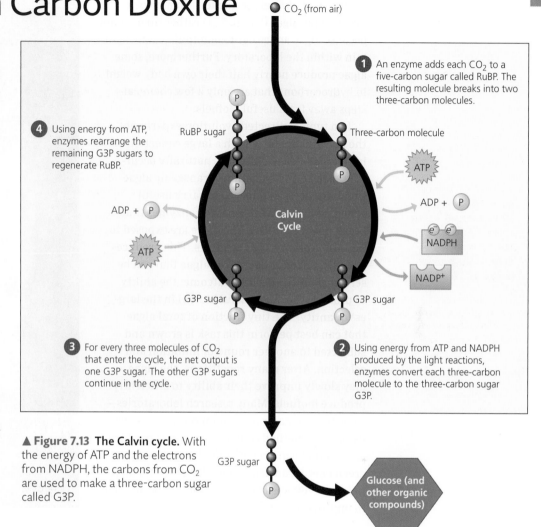

1 An enzyme adds each CO_2 to a five-carbon sugar called RuBP. The resulting molecule breaks into two three-carbon molecules.

4 Using energy from ATP, enzymes rearrange the remaining G3P sugars to regenerate RuBP.

3 For every three molecules of CO_2 that enter the cycle, the net output is one G3P sugar. The other G3P sugars continue in the cycle.

2 Using energy from ATP and NADPH produced by the light reactions, enzymes convert each three-carbon molecule to the three-carbon sugar G3P.

▲ **Figure 7.13 The Calvin cycle.** With the energy of ATP and the electrons from NADPH, the carbons from CO_2 are used to make a three-carbon sugar called G3P.

Creating a Better Biofuel Factory

Throughout this chapter, we've studied how plants convert solar energy to chemical energy by photosynthesis. Such transformations are vital to our welfare and to Earth's ecosystems. As discussed in the Biology and Society section, scientists are attempting to tap into the energy of photosynthesis to produce biofuels. But the production of biofuels is highly inefficient. In fact, it is usually far more costly to produce biofuels than to extract the equivalent amount of fossil fuels.

Biomechanical engineers are working to solve this dilemma by turning to an obvious example: evolution by natural selection. In nature, organisms with genes that make them better suited to their local environment will, on average, more often survive and pass those genes on to the next generation. Repeated over many generations, genes that enhance survival within that environment will become more common, and the species evolves.

When trying to solve an engineering problem, scientists can impose their own desired outcomes using a process called directed evolution (see the Process of Science section in Chapter 5 for another example).

☑ **CHECKPOINT**

What is the function of NADPH in the Calvin cycle?

■ *Answer: It provides the high-energy electrons that are added to CO_2 to form G3P (a sugar).*

During this process, scientists in the laboratory (instead of the natural environment) determine which organisms are the fittest. Directed evolution of biofuel production often involves microscopic algae **(Figure 7.14)** rather than plants because algae are easier to manipulate and maintain within the laboratory. Furthermore, some algae produce nearly half their own body weight in hydrocarbons that are only a few chemical steps away from useful biofuels.

In a typical directed evolution experiment, the researcher starts with a large collection of individual alga—sometimes naturally occurring species and sometimes transgenic algae that have been engineered to carry useful genes, such as fungal genes for enzymes that break down cellulose. The algae are exposed to mutation-promoting chemicals. This produces a highly varied collection of algae that can be screened for the desired outcome: the ability to produce the most useful biofuel in the largest quantity. The tiny fraction of total algae that can best perform this task is grown and subjected to another round of mutation and selection. After many repetitions, the algae may slowly improve their ability to efficiently produce biofuels. Many research laboratories—some within major petroleum companies—are using such methods and may someday produce an alga that can provide the ultimate source of green energy, an achievement that would highlight how lessons from evolution can be applied to improve our lives.

▼ **Figure 7.14 Microscopic biofuel factories.** This researcher is monitoring a reaction chamber in which microscopic algae are using light to produce biofuels.

Chapter Review

SUMMARY OF KEY CONCEPTS

The Basics of Photosynthesis

Photosynthesis is a process whereby light energy is transformed into chemical energy. During the process, carbon dioxide and water are used to make sugar molecules.

Chloroplasts: Sites of Photosynthesis

Chloroplasts contain a thick fluid called stroma surrounding a network of membranes called thylakoids.

An Overview of Photosynthesis

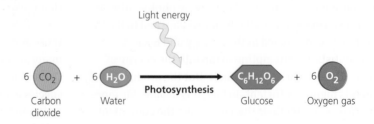

Light energy

$6\ CO_2$ + $6\ H_2O$ → $C_6H_{12}O_6$ + $6\ O_2$

Photosynthesis

Carbon dioxide   Water   Glucose   Oxygen gas

The overall process of photosynthesis can be divided into two stages connected by energy-carrying and electron-carrying molecules:

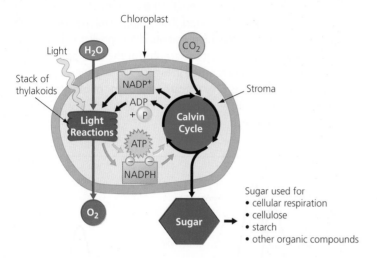

The Light Reactions: Converting Solar Energy to Chemical Energy

The Nature of Sunlight

Visible light is part of the electromagnetic spectrum. It travels through space as waves. Different wavelengths of light appear as different colors. Shorter wavelengths carry more energy.

Chloroplast Pigments

Pigment molecules absorb light energy of certain wavelengths and reflect other wavelengths. We see the reflected wavelengths as the color of the pigment. Several chloroplast pigments absorb light of various wavelengths and convey it to other pigments, but it is the green pigment chlorophyll *a* that participates directly in the light reactions.

How Photosystems Harvest Light Energy and How the Light Reactions Generate ATP and NADPH

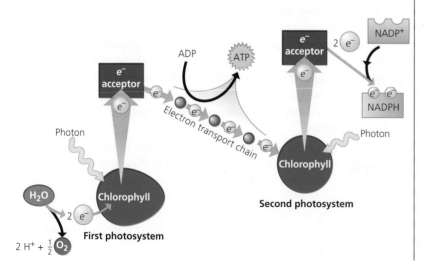

The Calvin Cycle: Making Sugar from Carbon Dioxide

Within the stroma (fluid) of the chloroplast, carbon dioxide from the air and ATP and NADPH produced during the light reactions are used to produce G3P, an energy-rich sugar molecule that can be used to make glucose and other organic molecules.

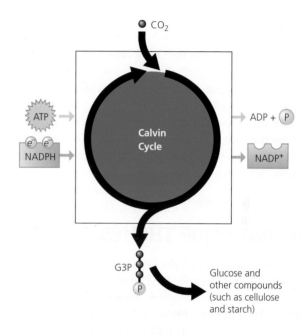

Mastering Biology

For practice quizzes, BioFlix animations, MP3 tutorials, video tutors, and more study tools designed for this textbook, go to Mastering Biology™

SELF-QUIZ

1. The light reactions take place in the structures of the chloroplast called the _____, while the Calvin cycle takes place in the _____.

2. In terms of the spatial organization of photosynthesis within the chloroplast, what is the advantage of the light reactions producing NADPH and ATP on the stroma side of the thylakoid membrane?

3. Which of the following equations best summarizes photosynthesis?
   a. $6\,CO_2 + 6\,H_2O + 6\,O_2 \rightarrow C_6H_{12}O_6$
   b. $6\,CO_2 + 6\,H_2O \rightarrow C_6H_{12}O_6 + 6\,O_2$
   c. $6\,CO_2 + 6\,O_2 \rightarrow C_6H_{12}O_6 + 6\,H_2O$
   d. $C_6H_{12}O_6 + 6\,O_2 \rightarrow 6\,CO_2 + 6\,H_2O$

4. Explain how the name "photosynthesis" describes what this process accomplishes.

5. What color of light is the least effective in driving photosynthesis? Why?

6. When light strikes chlorophyll molecules, they lose electrons, which are ultimately replaced by splitting molecules of _____.

7. Which of the following are produced by reactions that take place in the thylakoids and are consumed by reactions in the stroma?
   a. CO_2 and H_2O
   b. $NADP^+$ and ADP
   c. ATP and NADPH
   d. glucose and O_2

8. The carbon atoms that enter the Calvin cycle as CO_2 are used to make _____.

9. Of the following metabolic processes, which is common to photosynthesis and cellular respiration?
   a. reactions that convert light energy to chemical energy
   b. reactions that split H_2O molecules and release O_2
   c. reactions that store energy by pumping H^+ across membranes
   d. reactions that convert CO_2 to sugar

For answers to the Self Quiz, see Appendix D.

IDENTIFYING MAJOR THEMES

For each statement, identify which major theme is evident (the relationship of structure to function, information flow, pathways that transform energy and matter, interactions within biological systems, or evolution) and explain how the statement relates to the theme. If necessary, review the themes (see Chapter 1) and review the examples highlighted in blue in this chapter.

10. Most ecosystems depend on the process of photosynthesis to convert the energy in sunlight into high-energy organic molecules, which then provide fuel for other organisms.

11. The reactions of photosynthesis are driven by membrane-bound enzymes, and the folded shape of the thylakoid greatly increases the area on which these enzymes can operate.

12. Deforestation in Asia can alter the climate in North America.

For answers to Identifying Major Themes, see Appendix D.

THE PROCESS OF SCIENCE

13. Oxygen is a highly reactive element that easily combines with (oxidizes) other elements and compounds. For example, it forms five different oxides with iron, which is by mass the most abundant element on Earth. Given its highly reactive nature, how is oxygen maintained at a level of 16% in the Earth's atmosphere? Based on the information in this chapter, draw a cycle of reactions that keeps the concentration of oxygen steady and describes its flow between different interconnected systems of the biosphere. Was the atmospheric concentration of oxygen higher or lower before the emergence of life?

14. Suppose you wanted to discover whether the oxygen atoms in the glucose produced by photosynthesis come from H_2O or from CO_2. Explain how you could use a radioactive isotope to find out.

15. **Interpreting Data** In the following graph, called an absorption spectrum, each line is made by shining light of varying wavelengths through a sample. For each wavelength, the amount of that light absorbed by the sample is recorded. This graph combines three such measurements, one each for the pigments chlorophyll *a*, chlorophyll *b*, and the carotenoids. Notice that the graphs for the chlorophyll pigments match the data presented in Figure 7.5. Imagine a plant that lacks chlorophyll and relies only on carotenoids for photosynthesis. What colors of light would work best for this plant? How would this plant appear to your eye?

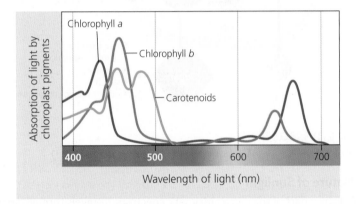

BIOLOGY AND SOCIETY

16. There is general agreement within the scientific community that Earth is getting warmer because of an intensified greenhouse effect resulting from increased CO_2 emissions from industry, vehicles, and the burning of forests. Global climate change is influencing agriculture, causing polar ice to melt, and altering weather patterns. In response to these threats, 196 parties negotiated the Paris Agreement in 2015. Soon after, 195 members signed an agreement to follow its provisions. However, in June 2017, U.S. President Donald Trump announced that the U.S. would withdraw from the agreement, stating that this withdrawal was meant to help U.S. businesses and workers, especially those in fossil fuel industries. Do you agree with this decision? In what ways might efforts to reduce greenhouse gases hurt the economy? How can those costs be weighed against the costs of global climate change? Would you vote to stay in the Paris Agreement or withdraw from it? Explain.

17. In this age of ecological awareness, generation of energy from sunlight is gaining popularity. Drawbacks of this technology are the use of toxic heavy metals in some photovoltaic cells and its relatively low-cost efficiency. Given that plants are experts in harvesting sunlight, would it not make sense to use their photosystems to convert sunlight into energy? Can you think of a device that performs this feat?

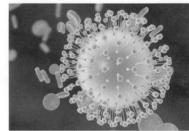
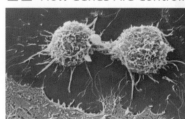

8 Cellular Reproduction: Cells from Cells

Why Cellular Reproduction Matters

About 9 months before you were born, you were just a single cell. From that microscopic beginning, every cell in your body was created through cell reproduction. Even today, millions of cells are reproducing themselves in your body every second—and errors in the process can be deadly.

KEEPING TRACK OF CHROMOSOMES DURING CELL DIVISION IS VITAL: DUPLICATING THE WRONG NUMBER OF CHROMOSOMES IS ALMOST ALWAYS FATAL.

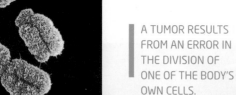

IF STRETCHED OUT, THE DNA IN ANY ONE OF YOUR CELLS WOULD BE TALLER THAN YOU!

A TUMOR RESULTS FROM AN ERROR IN THE DIVISION OF ONE OF THE BODY'S OWN CELLS.

Nucleus
DNA
Cell

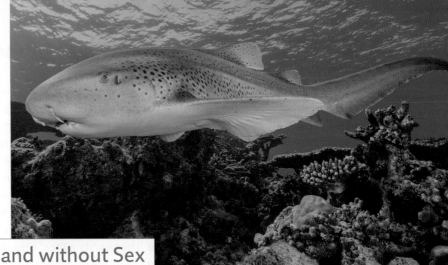

BIOLOGY AND SOCIETY   Life with and without Sex

Zebra shark. The zebra shark can be found living around coral reefs and sandy bottoms in the Indian and Pacific Oceans.

Virgin Birth of a Shark

Biologists at the Reef HQ Aquarium in Queensland, Australia helped care for Leonie, a female zebra shark (*Stegostoma fasciatum*) captured from the wild. To promote a breeding program, a male shark was introduced to Leonie's tank, resulting in several litters of offspring. When the aquarium decided to end the breeding program, the male shark was removed, and Leonie lived in the tank with a daughter.

No one was surprised that Leonie continued to lay eggs. After all, sharks, like chickens (or, for that matter, humans), produce infertile eggs even in the absence of sex. But Leonie's keepers were shocked when, in 2015, she laid eggs with viable embryos in them. One of those eggs hatched into Cleo, a healthy daughter. This is surprising because the vast majority of animal species create offspring only through sexual reproduction, involving the union of a male's sperm with a female's egg. But, despite the fact that Leonie was last in the company of a male three years prior, she produced a normal daughter. Cleo remains on display at the aquarium today.

DNA analysis confirmed that Cleo's genes came solely from her mother. The birth had resulted from parthenogenesis, the production of offspring by a female without involvement of a male. Parthenogenesis is rare among vertebrates (animals with backbones), although it has been documented in about 70 species as diverse as lizards (including the Komodo dragon, the world's largest lizard), domesticated birds (such as chickens and turkeys), and frogs. (In case you were wondering, there are no documented cases of parthenogenesis among humans or other mammals.) It is extremely rare, however, for a single female to produce offspring via both sexual reproduction and parthenogenesis. Leonie's case proved that zebra sharks can switch between two reproductive modes—the first documented case of this ability among sharks. Biologists are investigating the evolutionary basis of this phenomenon and considering what implications it may have on efforts to repopulate endangered species such as the zebra sharks and Komodo dragons.

The ability of organisms to procreate best distinguishes living things from nonliving matter. All organisms—from bacteria to lizards to you—are the result of repeated cell divisions. Therefore, the perpetuation of life depends on cell division, the production of new cells. In this chapter, we'll look at how individual cells are copied and then see how cell reproduction underlies the process of sexual reproduction. Throughout our discussion, we'll consider examples of asexual and sexual reproduction among both plants and animals.

What Cell Reproduction Accomplishes

When you hear the word *reproduction*, you probably think of the birth of new organisms. But reproduction actually occurs much more often at the cellular level. Consider the skin on your arm. Skin cells are constantly reproducing themselves and moving outward toward the surface, replacing dead cells that have rubbed off. This renewal of your skin goes on throughout your life. And when your skin is injured, reproduction of cells helps heal the wound.

When a cell undergoes reproduction, the process is called **cell division**. The two "daughter" cells that result from cell division are genetically identical to each other and to the original "parent" cell. (Biologists traditionally use the word *daughter* in this context to refer to offspring cells even though cells lack gender.) Before the parent cell splits into two, it duplicates its **chromosomes**, the structures that contain most of the cell's DNA. Then, during cell division, each daughter cell receives one identical set of chromosomes from the original parent cell.

As summarized in **Figure 8.1**, cell division plays several important roles in the lives of organisms. For example, within your body, millions of cells must divide every second to replace damaged or lost cells. Another function of cell division is growth. All of the trillions of cells in your body are the result of repeated cell divisions that began in your mother's body with a single fertilized egg cell.

Another vital function of cell division is reproduction. Many single-celled organisms, such as amoebas, reproduce by dividing in half, and the offspring are genetic replicas of the parent. This is an example of

asexual reproduction, the creation of genetically identical offspring by a single parent, without the participation of sperm and egg. Offspring produced by asexual reproduction inherit all their chromosomes from a single parent and are thus genetic duplicates. An individual that reproduces asexually gives rise to a **clone**, a group of genetically identical individuals.

Many multicellular organisms can reproduce asexually as well. For example, some sea star species can grow new individuals from fragmented pieces. If you've ever grown a houseplant from a clipping, you've observed asexual reproduction in plants. In asexual reproduction, there is one simple principle of inheritance: The lone parent and each of its offspring have identical genes. The type of cell division responsible for asexual reproduction and for the growth and maintenance of multicellular organisms is called mitosis.

Sexual reproduction is different; it requires fertilization of an egg by a sperm. The production of **gametes**—egg and sperm—involves a special type of cell division called meiosis, which occurs only in reproductive organs. As we'll discuss later, a gamete has only half as many chromosomes as the parent cell that gave rise to it.

In summary, two kinds of cell division are involved in sexually reproducing organisms: mitosis for growth and maintenance and meiosis for reproduction. To define and distinguish these processes, the remainder of the chapter is divided into two main sections: one on mitosis and one on meiosis. ✓

☑ **CHECKPOINT**

Ordinary cell division produces two daughter cells that are genetically identical. Name three functions of this type of cell division. Which of these functions occur in your body?

■ *Answer: Cell replacement, growth of an organism, and asexual reproduction of an organism. Only the first two occur in your body.*

▶ **Figure 8.1 Three functions of cell division by mitosis.**

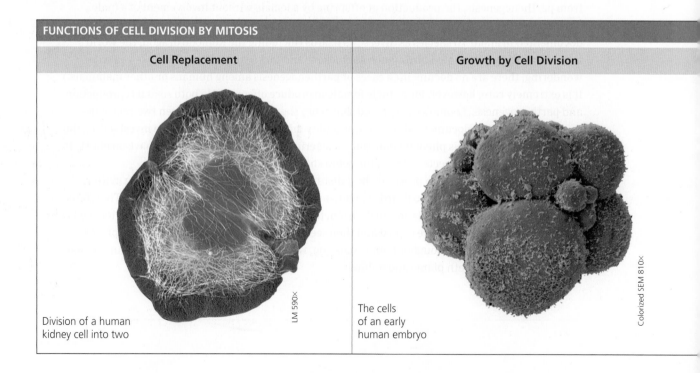

FUNCTIONS OF CELL DIVISION BY MITOSIS

Cell Replacement	Growth by Cell Division
Division of a human kidney cell into two	The cells of an early human embryo

LM 590×

Colorized SEM 810×

The Cell Cycle and Mitosis

Almost all the genes of a eukaryotic cell—around 21,000 in humans—are located on chromosomes in the cell nucleus. (The main exceptions are genes on small DNA molecules found in mitochondria and chloroplasts.) Because chromosomes are the lead players in cell division, we'll focus on them before turning our attention to the cell as a whole.

IF STRETCHED OUT, THE DNA IN ANY ONE OF YOUR CELLS WOULD BE TALLER THAN YOU!

Eukaryotic Chromosomes

Each eukaryotic chromosome contains one very long DNA molecule, typically bearing thousands of genes. The number of chromosomes in a eukaryotic cell depends on the species **(Figure 8.2)**. For example, human body cells have 46 chromosomes, while the body cells of a dog have 78 and those of a koala have 16. Chromosomes are made up of a material called **chromatin**, fibers composed of roughly equal amounts of DNA and protein molecules. The protein molecules help organize the chromatin and help control the activity of its genes.

Most of the time, the chromosomes exist as thin fibers that are much longer than the nucleus they are stored in. In fact, if fully extended, the DNA in just one of your cells would be more than 6 feet long! Chromatin in this state is too thin to be seen using a light

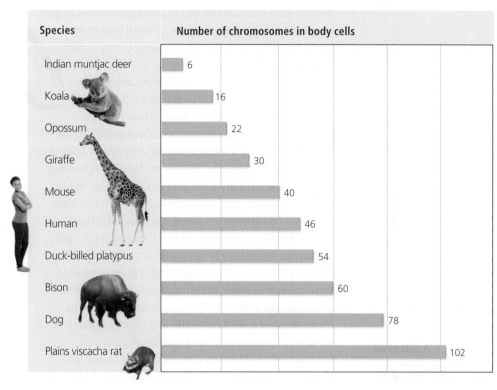

▼ **Figure 8.2 The number of chromosomes in the cells of selected mammals.** Notice that humans have 46 chromosomes and that the number of chromosomes does not correspond to the size or complexity of an organism.

Species	Number of chromosomes in body cells
Indian muntjac deer	6
Koala	16
Opossum	22
Giraffe	30
Mouse	40
Human	46
Duck-billed platypus	54
Bison	60
Dog	78
Plains viscacha rat	102

Asexual Reproduction

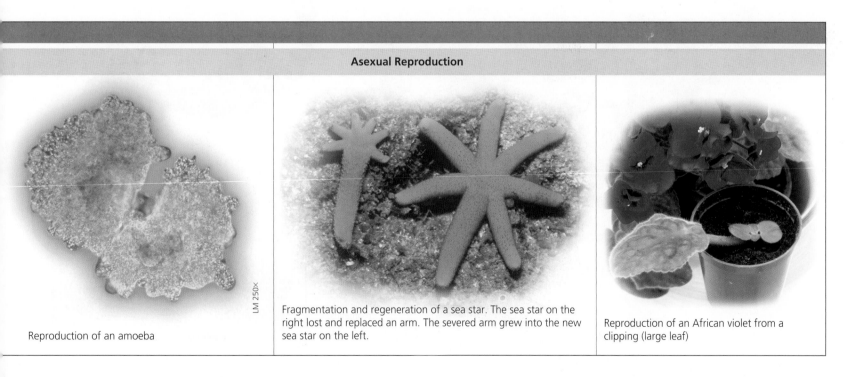

Reproduction of an amoeba

LM 250×

Fragmentation and regeneration of a sea star. The sea star on the right lost and replaced an arm. The severed arm grew into the new sea star on the left.

Reproduction of an African violet from a clipping (large leaf)

microscope. As a cell prepares to divide, its chromatin fibers coil up, forming compact chromosomes that become visible under a light microscope (**Figure 8.3**).

Such long molecules of DNA can fit into the tiny nucleus because within each chromosome the DNA is packed into an elaborate, multilevel system of coiling and folding. A crucial aspect of DNA packing is the association of the DNA with small proteins called **histones**. Why is it necessary for a cell's chromosomes to be compacted in this way? Imagine that your belongings are spread out around your room. If you had to move, you would gather up all your things and put them in small containers. Similarly, a cell must compact its DNA before it can move it to a new cell.

Figure 8.4 presents a simplified model of DNA packing. First, histones attach to the DNA. In electron micrographs, the combination of DNA and histones has the appearance of beads on a string. Each "bead," called a **nucleosome**, consists of DNA that is wound around several histone molecules. When not dividing, the DNA of active genes takes on this lightly packed arrangement. When preparing to divide, chromosomes condense even more: The beaded string itself wraps, loops, and folds into a tight, compact structure, as you can see in the chromosome at the bottom of the figure. Viewed as a whole, Figure 8.4 gives a sense of how successive levels of coiling and folding enable a huge amount of DNA to fit into a cell's tiny nucleus. If you think of a DNA molecule as a very long piece of yarn, a chromosome is like a bundle of yarn: one very long piece folded into a tight package for easier handling. **The structure of chromosomes therefore varies,** depending on the function the chromosomes are performing: loose and unwound when the cell is undergoing its normal functions, but compact and tightly wound when the cell is dividing.

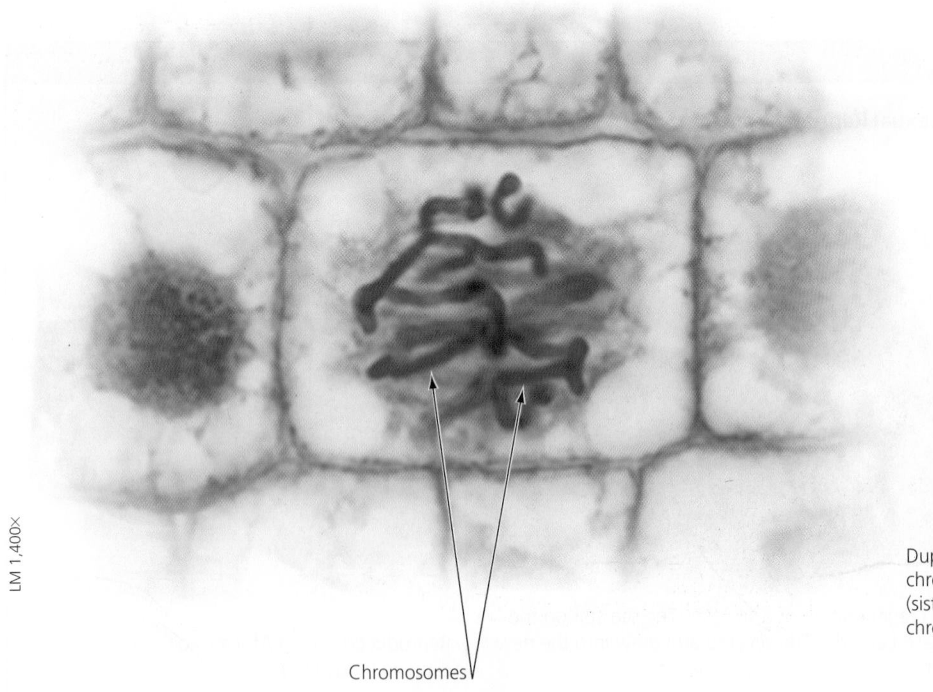

▼ **Figure 8.4** **DNA packing in a eukaryotic chromosome.** Successive levels of coiling of DNA and associated proteins ultimately result in highly compacted chromosomes. The fuzzy appearance of the final chromosome at the bottom arises from the intricate twists and folds of the chromatin fibers.

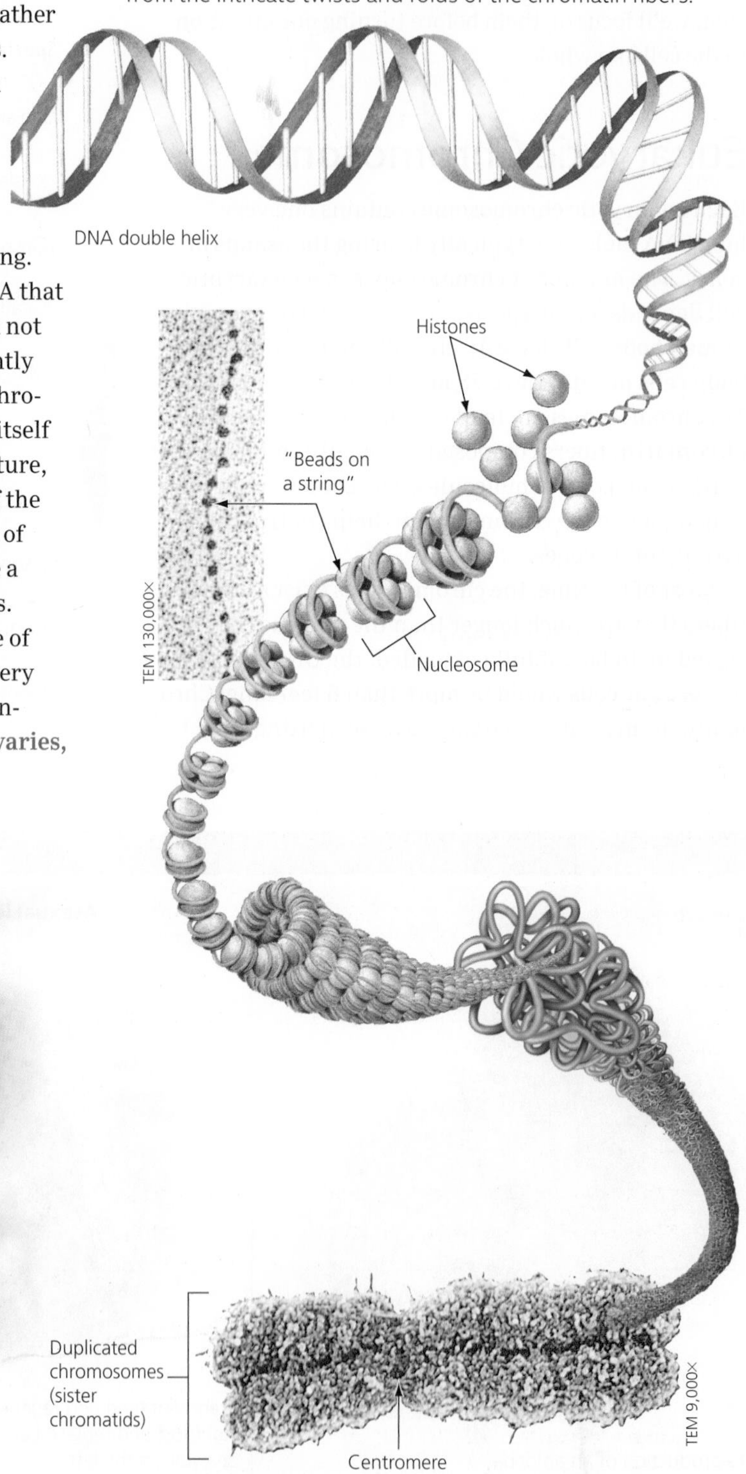

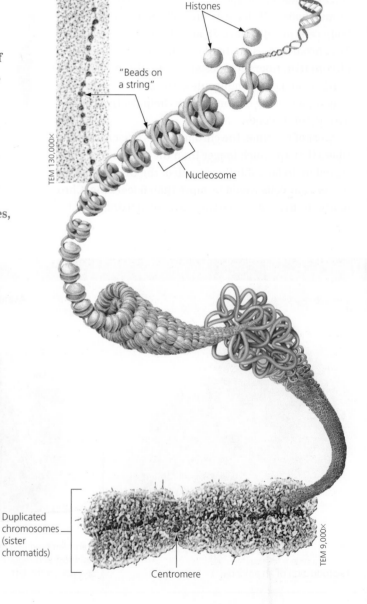

DNA double helix

Histones

TEM 130,000×

"Beads on a string"

Nucleosome

Duplicated chromosomes (sister chromatids)

Centromere

TEM 9,000×

▼ **Figure 8.3** **A plant cell just before division, with chromosomes colored by stains.**

LM 1,400×

Chromosomes

Duplicating Chromosomes

Think of the chromosomes as being like a detailed instruction manual on how to run a cell. During division, the original cell must pass on a copy of the manual to the new cell while also retaining a copy for itself. Duplicating chromosomes is therefore a vital step in the flow of genetic information from one generation to the next.

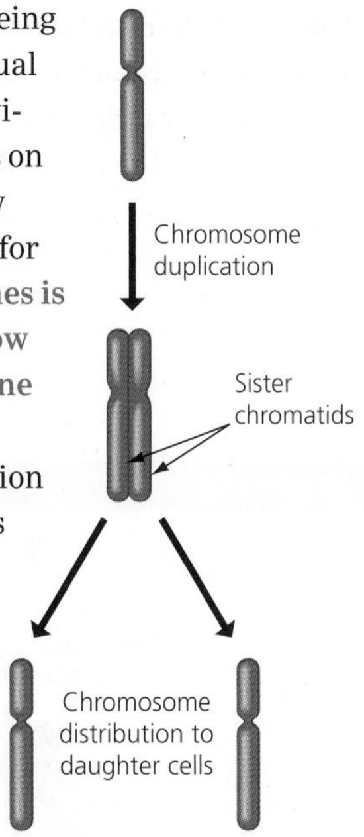

Before a cell begins the division process, it must duplicate all its chromosomes. The DNA molecule of each chromosome is copied through the process of DNA replication (discussed in detail in Chapter 10), and new histone protein molecules attach as needed. The result is that—at this point—each chromosome consists of identical copies called **sister chromatids**, which contain the same genes. At the bottom of Figure 8.4, the two sister chromatids are joined together most tightly at a narrow "waist" called the **centromere**.

When the cell divides, the sister chromatids of a duplicated chromosome separate from each other (**Figure 8.5**). When a chromatid is separated from its sister, it is considered a full-fledged chromosome, and it is identical to the original chromosome. One of the new chromosomes goes to one daughter cell, and the other goes to the other daughter cell. In this way, each daughter cell receives a complete and identical set of chromosomes. A dividing human skin cell, for example, has 46 duplicated chromosomes, and each of the two daughter cells that result from it has 46 single chromosomes.

The Cell Cycle

The rate at which a cell divides depends on its role within the organism's body. Some cells divide once a day, others divide less often, and some highly specialized cells, such as mature muscle cells, do not divide at all.

The **cell cycle** is the ordered sequence of events that extends from the time a cell is first formed from a dividing parent cell until its own division into two cells. Think of the cell cycle as the "lifetime" of a cell, from its "birth" to its own reproduction. As **Figure 8.6** shows, most of the

cell cycle is spent in **interphase**. Interphase is a time when a cell goes about its usual business, performing its normal functions within the organism. For example, during interphase a cell in your stomach lining might make and release enzyme molecules that aid in digestion. While in interphase, a cell roughly doubles everything in its cytoplasm. It increases its supply of proteins, increases the number of many of its organelles (such as mitochondria and ribosomes), and grows in size. Typically, interphase lasts for at least 90% of the cell cycle.

Interphase (see Figure 8.6) can be divided into three subphases: the G_1 phase ("first gap"), the S phase ("synthesis" of DNA—also known as DNA replication), and the G_2 phase ("second gap"). Although the G phases are called "gaps," cells are actually quite active and grow throughout all three subphases of interphase. The chromosomes are duplicated during the S phase, which typically lasts about half of interphase. At the beginning of the S phase, each chromosome is single. At the end of this subphase, after DNA replication, the chromosomes are doubled, each consisting of two sister chromatids joined along their lengths. During the G_2 phase, the cell completes preparations for cell division.

The part of the cell cycle when the cell is actually dividing is called the **mitotic (M) phase**. It includes two overlapping stages, mitosis and cytokinesis. In **mitosis**, the nucleus and its contents, most importantly the duplicated chromosomes, divide and are evenly distributed, forming two daughter nuclei. During **cytokinesis**, the cytoplasm (along with all the organelles) is divided in two. The combination of mitosis and cytokinesis produces two genetically identical daughter cells, each fully equipped with a nucleus, cytoplasm, organelles, and plasma membrane. ☑

☑ CHECKPOINT

1. A duplicated chromosome consists of two sister _____ joined together at the _____.

2. What are the two broadest divisions of the cell cycle? What two processes are involved in the actual duplication of the cell?

■ Answers: 1. chromatids; centromere. 2. interphase and the mitotic phase; mitosis and cytokinesis

▲ **Figure 8.5 Duplication and distribution of a single chromosome.** During cell reproduction, the cell duplicates each chromosome and distributes the two copies to the daughter cells.

(Figure 8.5 labels: Chromosome duplication; Sister chromatids; Chromosome distribution to daughter cells)

▼ **Figure 8.6 The eukaryotic cell cycle.** The cell cycle extends from the "birth" of a cell (just after the point indicated by the double-headed arrow at the bottom of the cycle), resulting from cell reproduction, to the time the cell itself divides in two. (During interphase, the chromosomes are diffuse masses of thin fibers; they do not actually appear in the rodlike form you see here.)

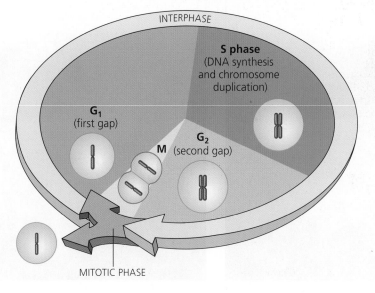

INTERPHASE

S phase (DNA synthesis and chromosome duplication)

G_1 (first gap)

G_2 (second gap)

M

MITOTIC PHASE

Mitosis and Cytokinesis

Figure 8.7 illustrates the cell cycle for an animal cell using drawings, descriptions, and fluorescent light micrographs. The micrographs running along the bottom row of the page show dividing cells from a salamander, with chromosomes depicted in blue. The drawings in the top row include details that are not visible in the micrographs. In these cells, we illustrate just four chromosomes to keep the process a bit simpler to follow; remember that one of your cells actually contains 46 chromosomes. The text within the figure describes the events occurring

▼ **Figure 8.7 Cell reproduction: a dance of the chromosomes.** After the chromosomes duplicate during interphase, the elaborately choreographed stages of mitosis—prophase, metaphase, anaphase, and telophase—distribute the duplicate sets of chromosomes to two separate nuclei. Cytokinesis then divides the cytoplasm, yielding two genetically identical daughter cells.

INTERPHASE	PROPHASE

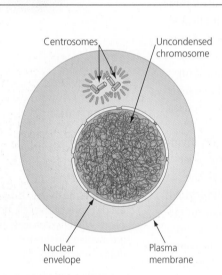

Centrosomes
Uncondensed chromosome
Nuclear envelope
Plasma membrane

Interphase is the period of cell growth when the cell makes new molecules and organelles. At the point shown here, late interphase (G₂), the cytoplasm contains two centrosomes. Within the nucleus, the chromosomes are duplicated, but they cannot be distinguished individually because they are still in the form of loosely packed chromatin fibers.

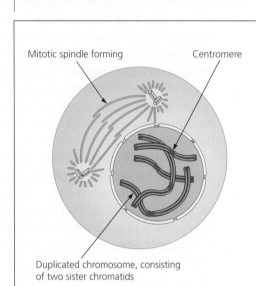

Mitotic spindle forming
Centromere
Duplicated chromosome, consisting of two sister chromatids

During prophase, changes occur in both the nucleus and cytoplasm. In the nucleus, the chromatin fibers coil, so that the chromosomes become thick enough to be seen individually with a light microscope. Each chromosome exists as two identical sister chromatids that are joined together at the narrow "waist" of the

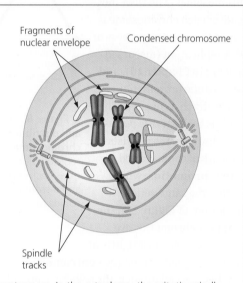

Fragments of nuclear envelope
Condensed chromosome
Spindle tracks

centromere. In the cytoplasm, the mitotic spindle begins to form. Late in prophase, the nuclear envelope breaks into pieces. The spindle tracks attach to the centromeres of the chromosomes and move the chromosomes toward the center of the cell.

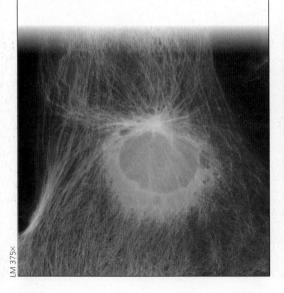

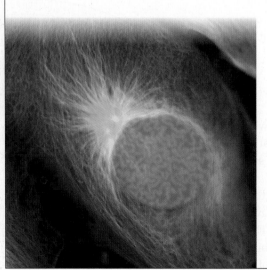

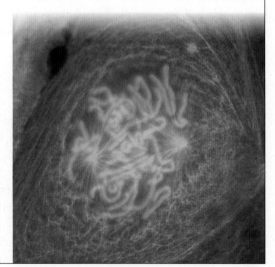

at each stage. Study this figure carefully (it has a lot of information and it's important!) and notice the striking changes in the nucleus and other cellular structures.

Biologists distinguish four main stages of mitosis: **prophase**, **metaphase**, **anaphase**, and **telophase**. The timing of these stages is not precise, and they overlap a bit. Think of stages in your own life—infancy, childhood, adulthood, old age—and you'll realize that the division between stages isn't always clear, and the timings of the stages vary from person to person; so it is with the stages of mitosis.

The chromosomes are the stars of the mitotic drama, and their movements depend on the **mitotic spindle**, a football-shaped structure of microtubule tracks (colored green in the figure) that guides the separation of the two sets of daughter chromosomes. The tracks of spindle microtubules grow from structures within the cytoplasm called centrosomes.

METAPHASE	ANAPHASE	TELOPHASE

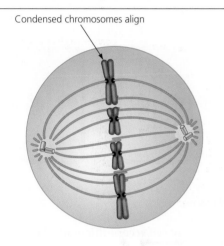

Condensed chromosomes align

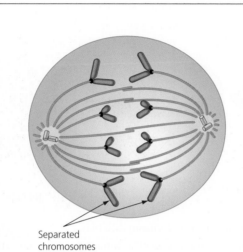

Separated chromosomes

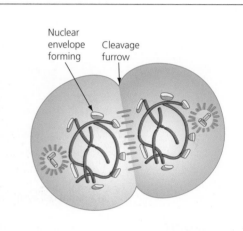

Nuclear envelope forming

Cleavage furrow

The mitotic spindle is now fully formed. The centromeres of all the chromosomes line up between the two poles of the spindle. For each chromosome, the tracks of the mitotic spindle attached to the two sister chromatids pull toward opposite poles. This tug of war keeps the chromosomes in the middle of the cell.

Anaphase begins suddenly when the sister chromatids of each chromosome separate. Each is now considered a full-fledged (daughter) chromosome. The chromosomes move toward opposite poles of the cell as the spindle tracks shorten. Simultaneously, the tracks not attached to chromosomes lengthen, pushing the poles farther apart and elongating the cell.

Telophase begins when the two groups of chromosomes have reached opposite ends of the cell. Telophase is the reverse of prophase: Nuclear envelopes form, the chromosomes uncoil, and the spindle disappears. Mitosis, the division of one nucleus into two genetically identical daughter nuclei, is now finished. Cytokinesis, the division of the cytoplasm, usually occurs with telophase. In animals, a cleavage furrow pinches the cell in two, producing two daughter cells.

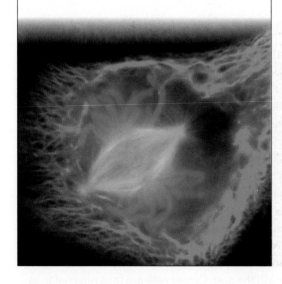

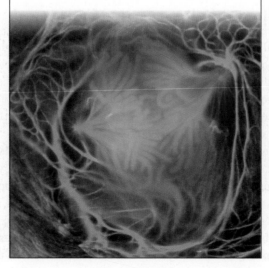

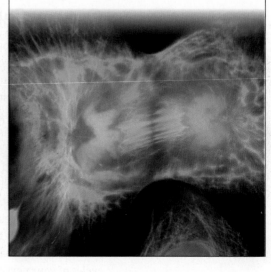

▼ **Figure 8.8** Cytokinesis in animal and plant cells.

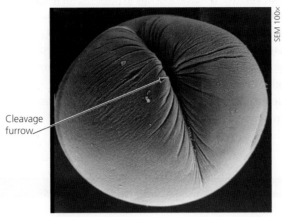

SEM 100×

Cleavage
furrow

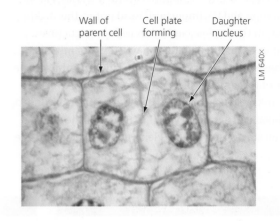

Wall of
parent cell

Cell plate
forming

Daughter
nucleus

LM 640×

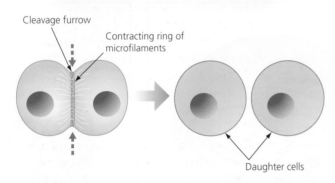

Cleavage furrow

Contracting ring of
microfilaments

Daughter cells

(a) Animal cell cytokinesis

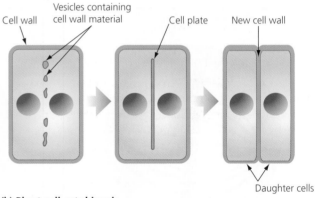

Cell wall

Vesicles containing
cell wall material

Cell plate

New cell wall

Daughter cells

(b) Plant cell cytokinesis

☑ **CHECKPOINT**

An organism called a
plasmodial slime mold is
one huge cytoplasmic mass
with many nuclei. Explain
how a variation in the cell
cycle could cause this
"monster cell" to arise.

■ *Answer: Mitosis occurs repeatedly*
without cytokinesis.

Cytokinesis, the division of the cytoplasm into two cells, usually begins during telophase, overlapping the end of mitosis. In animal cells, the cytokinesis process is known as **cleavage**. The first sign of cleavage is the appearance of an indentation called a cleavage furrow. A ring of microfilaments in the cytoplasm just under the plasma membrane contracts, like the pulling of a drawstring on a hooded sweatshirt, deepening the furrow and pinching the parent cell in two **(Figure 8.8a)**.

Cytokinesis in a plant cell occurs differently. Vesicles containing cell wall material collect at the middle of the cell. The vesicles fuse, forming a membranous disk called the **cell plate**. The cell plate grows outward, accumulating more cell wall material as more vesicles join it. Eventually, the membrane of the cell plate fuses with the plasma membrane, and the cell plate's contents join the parental cell wall. The result is two daughter cells **(Figure 8.8b)**. ☑

Cancer Cells: Dividing Out of Control

For a plant or animal to grow and maintain its tissues normally, it must control the timing of cell division—speeding up, slowing down, or turning the process off or on as needed. The sequential events of the cell cycle are directed by a **cell cycle control system** that consists of specialized proteins within the cell. These proteins integrate information from the environment and from other body cells and send "stop" and "go-ahead" signals at certain key points during the cell cycle. For example, the cell cycle normally halts within the G₁ phase of interphase unless the cell receives a go-ahead signal through certain control proteins. If that signal never arrives, the cell will switch into a permanently nondividing state. The cell cycles of your nerve and muscle cells, for example, are

arrested this way. If the go-ahead signal is received and the G_1 checkpoint is passed, the cell will usually complete the rest of the cycle. **Therefore, the reproductive behavior of cells—whether they will divide or not—results from interactions of many different molecules.**

What Is Cancer?

Cancer, which currently claims the lives of one out of every five people in the United States and other industrialized nations, is a disease of the cell cycle. Cancer cells do not respond normally to the cell cycle control system; they divide excessively and may invade other tissues of the body. If unchecked, cancer cells may continue to divide until they kill the host. Cancer cells are thus referred to as "immortal" because, unlike other human cells, they will never cease dividing. In fact, thousands of laboratories around the world today use a laboratory strain of human cells that were originally obtained from a woman named Henrietta Lacks, who died of cervical cancer in 1951.

The abnormal behavior of cancer cells begins when a single cell undergoes genetic changes (mutations) in one or more genes that encode for proteins in the cell cycle control system. These changes cause the cell to grow abnormally. The immune system generally recognizes and destroys such cells. However, if the cell evades destruction, it may proliferate to form a **tumor**, an abnormally growing mass of body cells. If the abnormal cells remain at the original site, the lump is called a **benign tumor**. Benign tumors can cause problems if they grow large and disrupt certain organs, such as the brain, but often they can be completely removed by surgery and are rarely deadly.

A TUMOR RESULTS FROM AN ERROR IN THE DIVISION OF ONE OF THE BODY'S OWN CELLS.

In contrast, a **malignant tumor** is one that has the potential to spread into neighboring tissues and other parts of the body, forming new tumors **(Figure 8.9)**. A malignant tumor may or may not have actually begun to spread, but if it does, it will soon displace normal tissue and interrupt organ function. A person with a malignant tumor is said to have **cancer**. The spread of cancer cells beyond their original site is called **metastasis**, and such cells are said to metastasize. Cancers are named according to where they originate. Liver cancer, for example, always begins in liver tissue and may metastasize from there.

Cancer Treatment

Once a tumor starts growing in the body, how can it be treated? There are three main types of cancer treatment. Surgery to remove a tumor is usually the first step. For many benign tumors, surgery alone may be sufficient. If not, the next step is usually **radiation therapy**. Cancerous tumors are exposed to high-energy radiation, which harms cancer cells more than normal cells. Radiation therapy is often effective against malignant tumors that have not yet spread. However, radiation can damage normal body cells enough to produce side effects, such as nausea and hair loss. **Chemotherapy**, the use of drugs to disrupt cell division, is used to treat tumors that have spread throughout the body. Some chemotherapy drugs prevent cell division by interfering with the mitotic spindle. Other drugs prevent the mitotic spindle from forming in the first place. Chemotherapy often has significant side effects because it contacts and may damage many different body tissues.

Frontiers of Cancer Treatment

Medical researchers are constantly searching for new ways to combat cancerous cells. One particularly promising area is immunotherapy, treatments that use the body's immune system to attack tumors. Immunotherapy can be accomplished by boosting the body's natural immunity in general or by boosting specific immune components that attack cancer cells. Alternatively, immune components (such as proteins specifically designed to recognize and help destroy cancer cells) may be created in the lab and injected into a patient. This type of therapy has been effective against cancers of the breast, stomach, and some forms of leukemia and lymphoma. Many cancer researchers believe that immunotherapy represents the next major breakthrough in cancer treatment that could save many lives in the coming decades.

Cancer Prevention and Survival

Although cancer can strike anyone, there are certain lifestyle changes you can make to reduce your chances of developing cancer or increase your chances of surviving it. Not smoking, exercising adequately (usually defined as at least 150 minutes of moderate exercise each week), avoiding overexposure to the sun, and eating a high-fiber, low-fat diet can all help reduce the likelihood of getting cancer. Seven types of cancer can be easily detected: skin and oral (by physical exam), breast (by self-exams or mammograms for higher-risk women and women 50 and older), prostate (by rectal exam), cervical (by Pap smear), testicular (by self-exam), and colon (by colonoscopy). Regular visits to the doctor can help identify tumors early, which is the best way to increase the chance of successful treatment. ✓

✓ **CHECKPOINT**

What differentiates a benign tumor from a malignant tumor?

■ Answer: A benign tumor remains at its point of origin, whereas a malignant tumor can spread.

▼ Figure 8.9 **Growth and metastasis of a malignant tumor of the breast.**

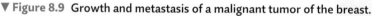

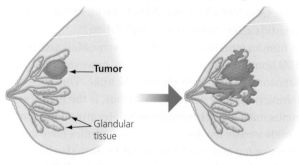

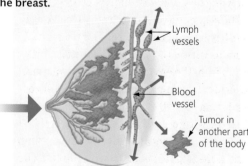

A tumor grows from a single cancer cell.

Cancer cells invade neighboring tissue.

Metastasis: Cancer cells spread through lymph and blood vessels to other parts of the body.

Meiosis, the Basis of Sexual Reproduction

Only maple trees produce more maple trees; only goldfish make more goldfish; and only people make more people. These simple facts of life have been recognized for thousands of years and are reflected in the age-old saying, "Like begets like." But in a strict sense, "Like begets like" applies only to asexual reproduction, where offspring inherit all their DNA from a single parent. Asexual offspring are exact genetic replicas of that one parent and of each other, and their appearances are very similar.

The family photo in **Figure 8.10** makes the point that in a sexually reproducing species like does not exactly beget like. You probably resemble your biological parents more closely than you resemble strangers, but you do not look exactly like your parents or a sibling—unless you are an identical twin. Each offspring of sexual reproduction inherits a unique combination of genes from its two parents, and this combined set of genes programs a unique combination of traits. As a result, sexual reproduction can produce tremendous variety among offspring.

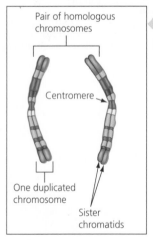

▲ **Figure 8.10 The varied products of sexual reproduction.** Every child inherits a unique combination of genes from his or her parents and displays a unique combination of traits.

Sexual reproduction depends on the cellular processes of meiosis and fertilization. But before exploring these processes, we need to return to chromosomes and the role they play in the life cycle of sexually reproducing organisms.

Homologous Chromosomes

If we examine cells from different individuals of a single species—sticking to one sex, for now—we find that they have the same number and types of chromosomes. Viewed with a microscope, your chromosomes would look exactly like those of Beyoncé (if you're a woman) or Jay Z (if you're a man).

A typical body cell, called a **somatic cell**, has 46 chromosomes in humans. A technician can break open a human cell in metaphase of mitosis, stain the chromosomes with dyes, take a picture with the aid of a microscope, and arrange the chromosomes in matching pairs by size. The resulting display is called a **karyotype (Figure 8.11)**. Notice in the figure that each chromosome is duplicated, with two sister chromatids joined along their length. Within the white box, for example, each "stick" is actually a pair of sister chromatids stuck together (as shown in the drawing to the left). Notice also that almost every chromosome has a twin that resembles it in length and centromere position; in the figure, the white box surrounds one set of twin chromosomes. The two chromosomes of such a matching pair, called **homologous chromosomes**, carry genes controlling the same inherited characteristics. For example, if a gene influencing freckles is located at a particular place on one chromosome—within the yellow band in the drawing in Figure 8.11, for instance—then the homologous chromosome has that same gene in the same location. However, the two homologous chromosomes may have different versions of the same gene. Let's restate this concept because it often confuses students: A pair of homologous chromosomes has two nearly identical chromosomes, each of which consists of two identical sister chromatids after chromosome duplication.

In human females, the 46 chromosomes fall neatly into 23 homologous pairs. For a male, however, the chromosomes in one pair do not look alike. This non-matching pair, only partly homologous, is the male's sex chromosomes. **Sex chromosomes** determine a person's sex (male versus female). In mammals, males have one X chromosome and one Y chromosome. Females have two X chromosomes. (Other organisms have different

▼ **Figure 8.11 Pairs of homologous chromosomes in a human male karyotype.** This karyotype shows 22 completely homologous pairs (autosomes) and a 23rd pair that consists of an X chromosome and a Y chromosome (sex chromosomes). With the exception of X and Y, the homologous chromosomes of each pair match in size, centromere position, and staining pattern.

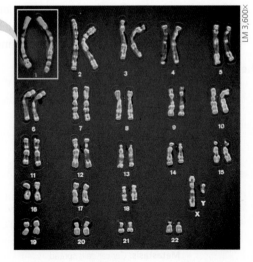

systems; in this chapter, we focus on humans.) The remaining chromosomes (44 in humans), found in both males and females, are called **autosomes**. For both autosomes and sex chromosomes, you inherited one chromosome of each pair from your mother and the other from your father.

Gametes and the Life Cycle of a Sexual Organism

The **life cycle** of a multicellular organism is the sequence of generation-to-generation stages from fertilization to the production of its own offspring. Having two sets of chromosomes, one inherited from each parent, is a key factor in the life cycle of humans and all other species that reproduce sexually. **Figure 8.12** shows the human life cycle, emphasizing the number of chromosomes.

Humans (as well as most other animals and many plants) are **diploid** organisms because all typical body cells (somatic cells) contain pairs of homologous chromosomes. In other words, all your chromosomes come in matching sets. This is similar to shoes in your closet: You may have 46 shoes, but they are organized as 23 pairs, with the members of each pair being nearly identical to each other. The total number of chromosomes, 46 in humans, is the diploid number (abbreviated 2n). The gametes, egg and sperm cells, are not diploid. Made by meiosis in an ovary or testis, each gamete has a single set of chromosomes: 22 autosomes plus a sex chromosome, either X or Y. A cell with a single chromosome set is called a **haploid** cell; it has only one member of each pair of homologous chromosomes. To visualize the haploid state, imagine your closet containing only one shoe from each pair. For humans, the haploid number, n, is 23.

In the human life cycle, a haploid sperm fuses with a haploid egg in a process called **fertilization**. The resulting fertilized egg, called a **zygote**, is diploid. It has two sets of chromosomes, one set from each parent. The life cycle is completed as a sexually mature adult develops from the zygote. Mitotic cell division ensures that all somatic cells of the human body receive a copy of all of the zygote's 46 chromosomes. Thus, every one of the trillions of cells in your body can trace its ancestry back through mitotic divisions to the single zygote produced when your father's sperm and your mother's egg fused about nine months before you were born (although you probably don't want to dwell on those details!).

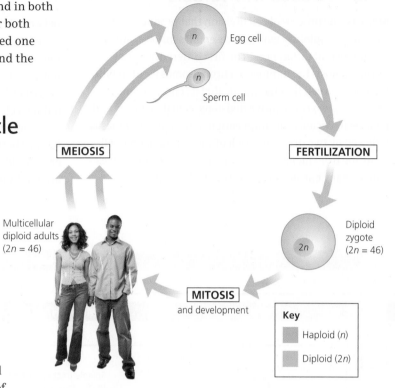

Figure Walkthrough

Mastering **Biology**
goo.gl/AJGKBd

◄ **Figure 8.12 The human life cycle.** In each generation, the doubling of chromosome number that results from fertilization is offset by the halving of chromosome number during meiosis.

Producing haploid gametes by meiosis keeps the chromosome number from doubling in every generation. To illustrate, **Figure 8.13** tracks one pair of homologous chromosomes. **1** Each of the chromosomes is duplicated during interphase (before mitosis). **2** The first division, meiosis I, segregates the two chromosomes of the homologous pair, packaging them in separate (haploid) daughter cells. But each chromosome is still doubled. **3** Meiosis II separates the sister chromatids. Each of the four daughter cells is haploid and contains only a single chromosome from the pair of homologous chromosomes.

▼ **Figure 8.13 How meiosis halves chromosome number.**

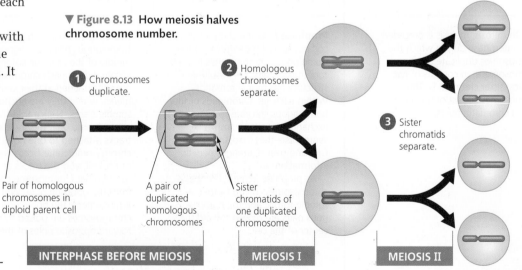

The Process of Meiosis

Meiosis, the process of cell division that produces haploid gametes in diploid organisms, resembles mitosis, but with two important differences. The first difference is that during meiosis the number of chromosomes is cut in half. In meiosis, a cell that has duplicated its chromosomes undergoes two consecutive divisions, called meiosis I and meiosis II. Because one duplication of the chromosomes is followed by two divisions, each of the four daughter cells resulting from meiosis has a haploid set of chromosomes—half as many chromosomes as the starting cell.

The second difference between meiosis and mitosis is an exchange of genetic material—pieces of chromosomes—between homologous chromosomes. This exchange, called crossing over, occurs during the first prophase of meiosis. We'll look more closely at crossing over later. For now, study **Figure 8.14**, including the text below it, which describes the stages of meiosis in detail for a hypothetical animal cell containing four chromosomes.

As you go through Figure 8.14, keep in mind the difference between homologous chromosomes and sister chromatids: The two chromosomes of a homologous pair are individual chromosomes that were inherited from

▼ **Figure 8.14**
The stages of meiosis.

MEIOSIS I: HOMOLOGOUS CHROMOSOMES SEPARATE

INTERPHASE	PROPHASE I	METAPHASE I	ANAPHASE I

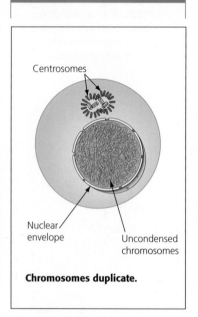

Chromosomes duplicate.

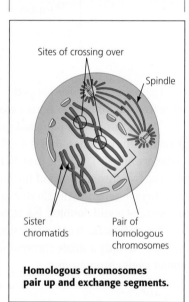

Homologous chromosomes pair up and exchange segments.

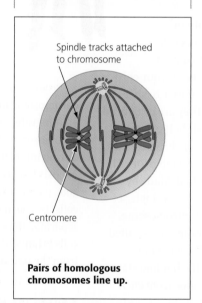

Pairs of homologous chromosomes line up.

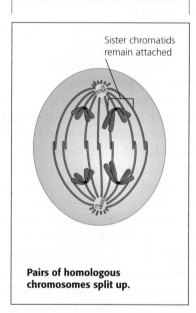

Pairs of homologous chromosomes split up.

As with mitosis, meiosis is preceded by an interphase during which the chromosomes duplicate. Each chromosome then consists of two identical sister chromatids. The chromosomes consist of uncondensed chromatin fibers.

Prophase I As the chromosomes coil up, special proteins cause the homologous chromosomes to stick together in pairs. The resulting structure has four chromatids. Within each set, chromatids of the homologous chromosomes exchange corresponding segments—they "cross over." Crossing over rearranges genetic information.

As prophase I continues, the chromosomes coil up further, a spindle forms, and the homologous pairs are moved toward the center of the cell.

Metaphase I At metaphase I, the homologous pairs are aligned in the middle of the cell. The sister chromatids of each chromosome are still attached at their centromeres, where they are anchored to spindle tracks. Notice that for each chromosome pair, the spindle tracks attached to one homologous chromosome come from one pole of the cell, and the tracks attached to the other chromosome come from the opposite pole. With this arrangement, the homologous chromosomes are poised to move toward opposite poles of the cell.

Anaphase I The attachment between the homologous chromosomes of each pair breaks, and the chromosomes now migrate toward the poles of the cell. *In contrast to mitosis, the sister chromatids migrate as a pair instead of splitting up.* They are separated not from each other but from their homologous partners.

different parents, one from the mother and one from the father. The members of a pair of homologous chromosomes in Figure 8.14 (and later figures) are identical in size and shape but colored in the illustrations differently (red versus blue) to remind you that they differ in this way. In the interphase just before meiosis, each chromosome duplicates to form sister chromatids that remain together until anaphase of meiosis II. Before crossing over occurs, sister chromatids are identical and carry the same versions of all their genes. ☑

Meiosis II in a lily cell

☑ **CHECKPOINT**

If a single diploid somatic cell with 18 chromosomes undergoes meiosis and produces sperm, the result will be _____ sperm, each with _____ chromosomes. (Provide two numbers.)

■ *Answer: four, nine*

MEIOSIS II: SISTER CHROMATIDS SEPARATE

TELOPHASE I AND CYTOKINESIS

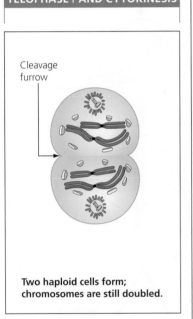

Cleavage furrow

Two haploid cells form; chromosomes are still doubled.

PROPHASE II

METAPHASE II

ANAPHASE II

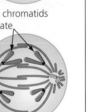

Sister chromatids separate

TELOPHASE II AND CYTOKINESIS

Haploid daughter cells forming

During another round of cell division, the sister chromatids finally separate; four haploid daughter cells result, containing single chromosomes.

Telophase I and Cytokinesis In telophase I, the chromosomes arrive at the poles of the cell. When they finish their journey, each pole has a haploid chromosome set, although each chromosome is still in duplicate form. Usually, cytokinesis occurs along with telophase I, and two haploid daughter cells are formed.

The Process of Meiosis II Meiosis II is essentially the same as mitosis. The important difference is that meiosis II starts with a haploid cell that has **not** undergone chromosome duplication during the preceding interphase.

During prophase II, a spindle forms and moves the chromosomes toward the middle of the cell. During metaphase II, the chromosomes are aligned as they are in mitosis, with the tracks attached to the sister chromatids of each chromosome coming from opposite poles.

In anaphase II, the centromeres of sister chromatids separate, and the sister chromatids of each pair move toward opposite poles of the cell. In telophase II, nuclei form at the cell poles, and cytokinesis occurs at the same time. There are now four haploid daughter cells, each with single chromosomes.

Review: Comparing Mitosis and Meiosis

You have now learned the two ways that cells of eukaryotic organisms divide (Figure 8.15). Mitosis—which provides for growth, tissue repair, and asexual reproduction—produces daughter cells that are genetically identical to the parent cell. Meiosis, needed for sexual reproduction, yields genetically unique haploid daughter cells—cells with only one member of each homologous chromosome pair.

For both mitosis and meiosis, the chromosomes duplicate only once, in the preceding interphase. Mitosis involves one division of the nucleus and cytoplasm (duplication, then division in half), producing two diploid cells. Meiosis involves two nuclear and cytoplasmic divisions

▶ **Figure 8.15 Comparing mitosis and meiosis.** The events unique to meiosis occur during meiosis I: In prophase I, duplicated homologous chromosomes pair along their lengths, and crossing over occurs between homologous (nonsister) chromatids. In metaphase I, pairs of homologous chromosomes (rather than individual chromosomes) are aligned at the center of the cell. During anaphase I, sister chromatids of each chromosome stay together and go to the same pole of the cell as homologous chromosomes separate. At the end of meiosis I, there are two haploid cells, but each chromosome still has two sister chromatids.

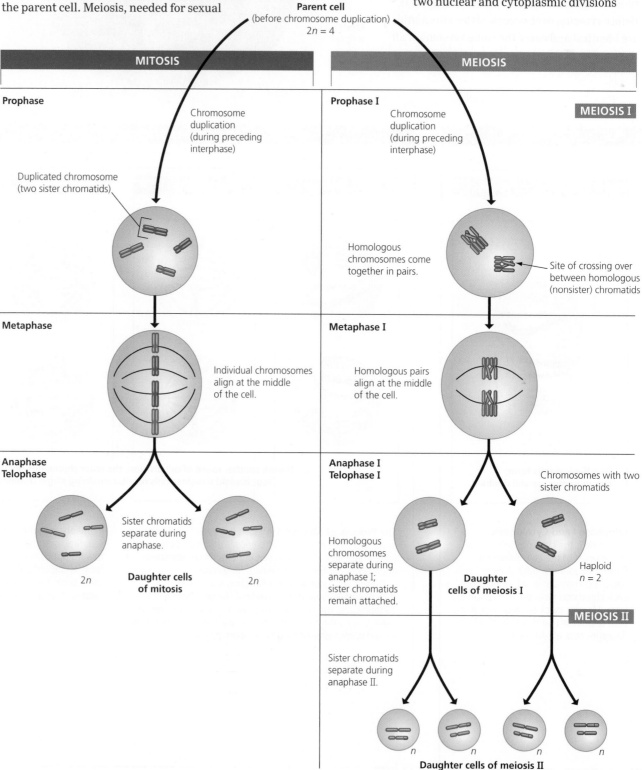

Parent cell
(before chromosome duplication)
$2n = 4$

MITOSIS

Prophase

Chromosome duplication (during preceding interphase)

Duplicated chromosome (two sister chromatids)

Metaphase

Individual chromosomes align at the middle of the cell.

Anaphase Telophase

Sister chromatids separate during anaphase.

$2n$ **Daughter cells of mitosis** $2n$

MEIOSIS

MEIOSIS I

Prophase I

Chromosome duplication (during preceding interphase)

Homologous chromosomes come together in pairs.

Site of crossing over between homologous (nonsister) chromatids

Metaphase I

Homologous pairs align at the middle of the cell.

Anaphase I Telophase I

Chromosomes with two sister chromatids

Homologous chromosomes separate during anaphase I; sister chromatids remain attached.

Daughter cells of meiosis I

Haploid $n = 2$

MEIOSIS II

Sister chromatids separate during anaphase II.

n n n n

Daughter cells of meiosis II

(duplication, division in half, then division in half again), yielding four haploid cells.

In comparing mitosis and meiosis, Figure 8.15 traces these two processes for a diploid parent cell with four chromosomes. As before, homologous chromosomes are those matching in size. Imagine that the red chromosomes were inherited from the mother and the blue chromosomes from the father. Notice that all the events unique to meiosis occur during meiosis I. Meiosis II is virtually identical to mitosis in that it separates sister chromatids. But unlike mitosis, meiosis II yields daughter cells with a haploid set of chromosomes. ✓

The Origins of Genetic Variation

As discussed earlier, offspring that result from sexual reproduction are genetically different from their parents and from one another. How does meiosis produce such genetic variation?

Independent Assortment of Chromosomes

Figure 8.16 illustrates one way in which meiosis contributes to genetic variety. The figure shows how the arrangement of homologous chromosomes at metaphase of meiosis I affects the resulting gametes. Once again, our example is from a hypothetical diploid organism with four chromosomes (two pairs of homologous chromosomes), with colors used to differentiate homologous chromosomes (red for chromosomes inherited from the mother and blue for chromosomes from the father).

When aligned during metaphase I, the side-by-side orientation of each homologous pair of chromosomes is a matter of chance. Either the red or blue chromosome may be on the left or right. Thus, in this example, there are two possible ways that the chromosome pairs can align during metaphase I. In possibility 1, the chromosome pairs are oriented with both red chromosomes on the same side (blue/red and blue/red). In this case, each of the gametes produced at the end of meiosis II has only red or only blue chromosomes (combinations a and b). In possibility 2, the chromosome pairs are oriented differently (blue/red and red/blue). This arrangement produces gametes with one red and one blue chromosome (combinations c and d). Thus, with the two possible arrangements shown in this example, the organism will produce gametes with four different combinations of chromosomes. For a species with more than two pairs of chromosomes, such as humans, every chromosome pair orients independently of all the others at metaphase I. (Chromosomes X and Y behave as a homologous pair in meiosis.)

☑ **CHECKPOINT**
True or false: Both mitosis and meiosis are preceded by chromosome duplication.

■ *Answer: true*

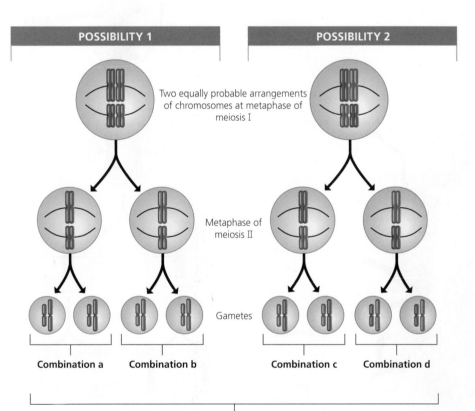

| POSSIBILITY 1 | POSSIBILITY 2 |

Two equally probable arrangements of chromosomes at metaphase of meiosis I

Metaphase of meiosis II

Gametes

Combination a   Combination b   Combination c   Combination d

Because possibilities 1 and 2 are equally likely, the four possible types of gametes will be made in approximately equal numbers.

◀ **Figure 8.16 Results of alternative arrangements of chromosomes at metaphase of meiosis I.** The arrangement of chromosomes at metaphase I determines which chromosomes will be packaged together in the haploid gametes.

For any species, the total number of chromosome combinations that can appear in gametes is 2^n, where n represents the haploid number. For the hypothetical organism in Figure 8.16, $n = 2$, so the number of chromosome combinations is 2^2, or 4. For a human ($n = 23$), there are 2^{23}, or about 8 million, possible chromosome combinations! This means that every gamete a person produces contains one of about 8 million possible combinations of maternal and paternal chromosomes. When you consider that a human egg cell with about 8 million possibilities is fertilized at random by a human sperm cell with about 8 million possibilities (Figure 8.17), you can see that a single man and a single woman can produce zygotes with 64 trillion combinations of chromosomes!

Crossing Over

So far, we have focused on genetic variety in gametes and zygotes at the whole-chromosome level. We'll now take a closer look at **crossing over**, the exchange of corresponding segments between nonsister chromatids of homologous chromosomes, which occurs during prophase I of meiosis. **Figure 8.18** shows crossing over between two homologous chromosomes and the resulting gametes. At the time that crossing over begins, very early in prophase I, homologous chromosomes are closely paired all along their lengths, with a precise gene-by-gene alignment.

The exchange of segments between nonsister chromatids—one maternal chromatid and one paternal chromatid of a homologous pair—adds to the genetic variety resulting from sexual reproduction. In Figure 8.18, if there were no crossing over, meiosis could produce only two types of gametes: the ones ending up with chromosomes that exactly match the parents' chromosomes, either all blue or all red (as in Possibility 1 in Figure 8.16). With crossing over, gametes arise with chromosomes that are partly from the mother and partly from the father. These chromosomes are called "recombinant" because they result from genetic recombination, the production

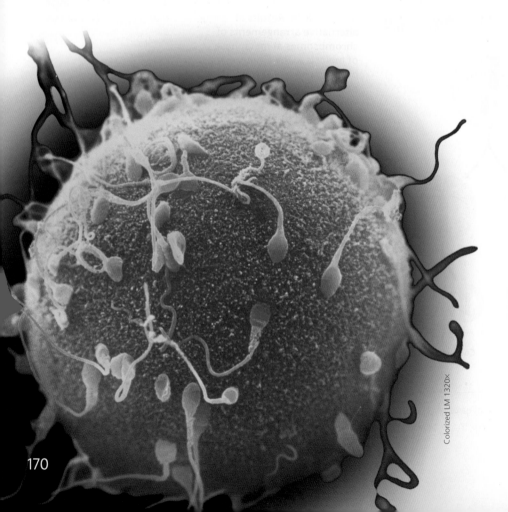

▼ **Figure 8.17 The process of fertilization: a close-up view.** Here you see many human sperm contacting an egg. Only one sperm can add its chromosomes to produce a zygote.

Colorized LM 1320x

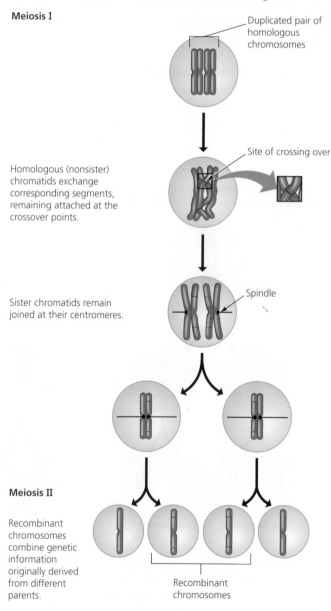

▼ **Figure 8.18 The results of crossing over during meiosis for a single pair of homologous chromosomes.** A real cell has multiple pairs of homologous chromosomes that produce a huge variety of recombinant chromosomes in the gametes.

Meiosis I

Duplicated pair of homologous chromosomes

Site of crossing over

Homologous (nonsister) chromatids exchange corresponding segments, remaining attached at the crossover points.

Sister chromatids remain joined at their centromeres.

Spindle

Meiosis II

Recombinant chromosomes combine genetic information originally derived from different parents.

Recombinant chromosomes

of gene combinations different from those carried by the parental chromosomes.

Because most chromosomes contain thousands of genes, a single crossover event can affect many genes.

When we also consider that multiple crossovers can occur in each pair of homologous chromosomes, it's not surprising that gametes and the offspring that result from them are so incredibly varied. ☑

☑ **CHECKPOINT**

Name two events during meiosis that contribute to genetic variety among gametes. During what stages of meiosis does each occur?

■ *Answer: crossing over between homologous chromosomes during prophase I and independent orientation/assortment of the pairs of homologous chromosomes at metaphase I*

THE PROCESS OF SCIENCE | Life with and without Sex

Do All Animals Have Sex?

BACKGROUND

As discussed in the Biology and Society section, some species such as zebra sharks can reproduce through both sexual and asexual methods. Although some animal species can reproduce asexually, very few animals reproduce *only* asexually. In fact, biologists have traditionally considered asexual reproduction an evolutionary dead end (for reasons we'll discuss in the Evolution Connection section at the end of the chapter).

To investigate a case in which asexual reproduction seemed to be the norm, researchers from Harvard University studied a group of animals called bdelloid rotifers (Figure 8.19a). This class of nearly microscopic freshwater invertebrates includes about 460 species. Despite hundreds of years of observations, no one has ever found bdelloid rotifer males or evidence of sexual reproduction. Has this entire class of animals reproduced solely by asexual means for tens of millions of years?

METHOD

In most species, the two versions of a gene in a pair of homologous chromosomes are very similar due to the constant trading of genes during sexual reproduction. If a species has survived without sex for millions of years, the researchers reasoned, then changes in the DNA sequences of homologous genes should accumulate independently,

and the two versions of the genes should have significantly diverged from each other over time. The researchers therefore compared the sequences of a series of genes in bdelloid rotifers and in other closely related rotifers who were known to reproduce sexually.

RESULTS

Their results were striking (Figure 8.19b). Among non-bdelloid rotifers that reproduce sexually, homologous versions of the genes were 99.5% identical, differing by only 0.5% on average. In contrast, the versions of the same gene in bdelloid rotifers were only 46 to 96% identical, differing by between 4 and 54%.

These data provided strong evidence that bdelloid rotifers have evolved for millions of years without sex. Recent research has suggested that these fascinating creatures may indeed rarely undergo some type of previously undiscovered sexual reproduction, but precisely what it is and how it works remain unknown. The study of bdelloid rotifers illustrates a general principle: Exploration of unusual cases can provide general insights into nature. In this case, studies of a tiny, obscure creature are shedding light on one of the biggest questions in all of animal biology: Why have sex?

Thinking Like a Scientist

A botanist noticed that a tree growing in a monastery in Brazil produced seedless oranges. Every navel orange tree alive today is derived from this single mutant. What common principle do navel oranges and bdelloid rotifers illustrate?

For the answer, see Appendix D.

▼ Figure 8.19 **A study of bdelloid rotifers.**

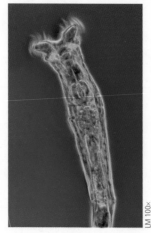

(a) **One of over 460 species of bdelloid rotifers**

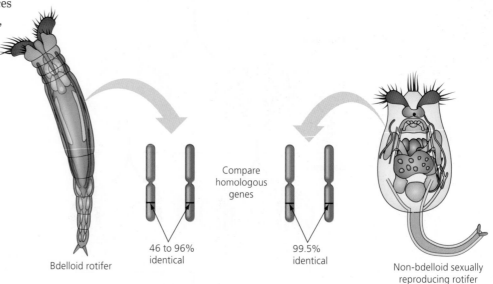

Compare homologous genes

Bdelloid rotifer

46 to 96% identical

99.5% identical

Non-bdelloid sexually reproducing rotifer

(b) **An experiment to determine whether bdelloid rotifer chromosomes behave like those of sexually reproducing rotifers**

When Meiosis Goes Wrong

So far, our discussion of meiosis has focused on the process as it normally and correctly occurs. But what happens when there is an error in the process? Such a mistake can result in genetic abnormalities that range from mild to severe to fatal.

How Accidents During Meiosis Can Alter Chromosome Number

Within the human body, meiosis occurs repeatedly as the testes or ovaries produce gametes. Almost always, chromosomes are distributed to daughter cells without any errors. But occasionally there is a mishap, called a **nondisjunction**, in which the members of a chromosome pair fail to separate at anaphase. Nondisjunction can occur during meiosis I or II **(Figure 8.20)**. In either case, gametes with abnormal numbers of chromosomes are the result.

Figure 8.21 shows what can happen when an abnormal gamete produced by nondisjunction unites with a normal gamete during fertilization. When a normal sperm fuses with an egg cell with an extra chromosome, the result is a zygote with a total of $2n + 1$ chromosomes. Because mitosis duplicates the

chromosomes as they are, the abnormality will be passed to all embryonic cells. If the organism survives, it will have an abnormal karyotype and probably a medical disorder caused by the abnormal number of genes. Nondisjunction is estimated to be involved in 10–30% of human conceptions and is the main reason for pregnancy loss. Next, we'll examine one particular case of survival nondisjunction. ☑

CHECKPOINT

Explain how nondisjunction in meiosis could result in a diploid gamete.

■ *Answer: A diploid gamete would result if there were nondisjunction of all the chromosomes during meiosis I or II.*

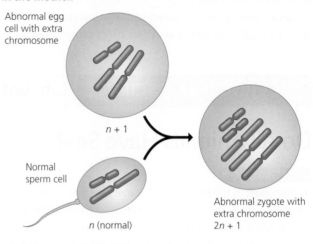

▼ **Figure 8.21 Fertilization after nondisjunction in the mother.**

Abnormal egg cell with extra chromosome

$n + 1$

Normal sperm cell

n (normal)

Abnormal zygote with extra chromosome
$2n + 1$

▶ **Figure 8.20 Two types of nondisjunction.** In both examples in the figure, the cell at the top is diploid (2*n*), with two pairs of homologous chromosomes.

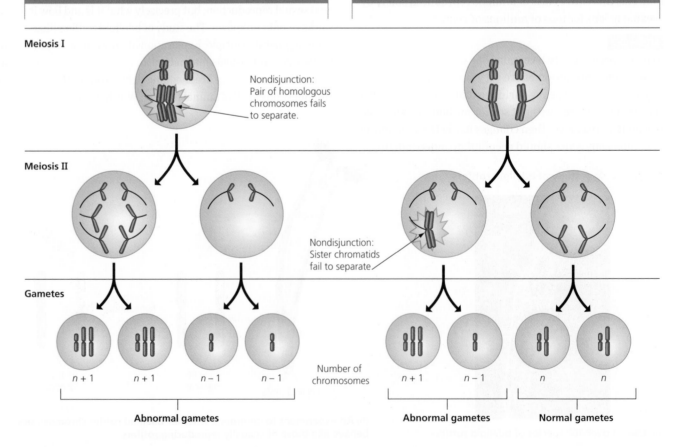

NONDISJUNCTION IN MEIOSIS I	NONDISJUNCTION IN MEIOSIS II

Meiosis I

Nondisjunction: Pair of homologous chromosomes fails to separate.

Meiosis II

Nondisjunction: Sister chromatids fail to separate.

Gametes

Number of chromosomes

$n + 1$ $n + 1$ $n - 1$ $n - 1$ $n + 1$ $n - 1$ n n

Abnormal gametes   Abnormal gametes   Normal gametes

Down Syndrome: An Extra Chromosome 21

Figure 8.11 showed a normal human complement of 23 pairs of chromosomes. Compare it with the karyotype in **Figure 8.22**. Besides having two X chromosomes (because it's from a female), the karyotype in Figure 8.21 has three number 21 chromosomes. This person is triploid for chromosome 21 (instead of having the usual diploid condition) and therefore has 47 chromosomes. This condition is called **trisomy 21**.

▼ **Figure 8.22   Trisomy 21 and Down syndrome.** This child displays the characteristic facial features of Down syndrome. The karyotype (bottom) shows trisomy 21; notice the three copies of chromosome 21.

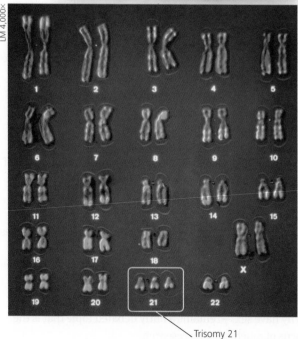

LM 4,000X

Trisomy 21

In most cases, a human embryo with an atypical number of chromosomes is spontaneously aborted (miscarried) long before birth, often before the woman is even aware that she is pregnant. In fact, some doctors speculate that miscarriages due to genetic defects occur in nearly one-quarter of all pregnancies, although this number is difficult to verify. However, some aberrations in chromosome number seem to upset the genetic balance less drastically, and people with such abnormalities can survive. These people usually have a characteristic set of symptoms, called a syndrome. A person with trisomy 21 has a condition called **Down syndrome** (named after John Langdon Down, an English physician who first described this condition in 1866).

Trisomy 21 affects about 1 out of every 850 children and is the most common chromosome number abnormality and most common serious birth defect in the United States. Down syndrome includes characteristic facial features—frequently a fold of skin at the inner corner of the eye, a round face, and a flattened nose—as well as short stature, heart defects, and susceptibility to leukemia and Alzheimer's disease. People with Down syndrome usually have a life span shorter than normal. They also exhibit varying degrees of developmental delays. However, with proper care, many people with Down syndrome live to middle age or beyond, and many are socially adept, live independently, and hold jobs. Although no one is sure why, the risk of Down syndrome increases with the age of the mother, climbing to about 1% risk for mothers at age 40. Therefore, the fetuses of pregnant women age 35 and older are candidates for chromosomal prenatal screenings (see Chapter 9). ☑

Abnormal Numbers of Sex Chromosomes

Trisomy 21 is an autosomal nondisjunction. Nondisjunction in meiosis can also lead to abnormal numbers of sex chromosomes, X and Y. Unusual numbers of sex chromosomes seem to upset the genetic balance less than unusual numbers of autosomes. This may be because the Y chromosome is very small and carries relatively few genes. Furthermore, mammalian cells normally operate with only one functioning X chromosome because other copies of the chromosome become inactivated in each cell (see Chapter 11).

☑ **CHECKPOINT**

What does it mean to refer to a disease as a "syndrome," as with AIDS?

■ *Answer: A syndrome displays a set of multiple symptoms rather than just a single symptom.*

DUPLICATING THE WRONG NUMBER OF CHROMOSOMES IS ALMOST ALWAYS FATAL.

Table 8.1	Abnormalities of Sex Chromosome Number in Humans	
Sex Chromosomes	Syndrome	Symptoms
XXY	Klinefelter syndrome (male)	Sterile; underdeveloped testes; secondary female characteristics
XYY	None (normal male)	Slightly taller than average
XXX	None (normal female)	Slightly taller than average; slight risk of learning disabilities
XO	Turner syndrome (female)	Sterile; immature sex organs

Table 8.1 lists the most common human sex chromosome abnormalities. An extra X chromosome in a male, making him XXY, produces a condition called Klinefelter syndrome. If untreated, men with this disorder have abnormally small testes, are sterile, and often have breast enlargement and other feminine body contours. These symptoms can be reduced through administration of the sex hormone testosterone. Klinefelter syndrome is also found in men with more than three sex chromosomes, such as XXYY, XXXY, or XXXXY. These abnormal numbers of sex chromosomes result from multiple nondisjunctions; such men are more likely to have developmental disabilities than XY or XXY men.

Human males with a single extra Y chromosome (XYY) do not have any well-defined syndrome, although they tend to be taller than average. Females with an extra X chromosome (XXX) cannot be distinguished from XX females except by karyotype. Such women tend to be slightly taller than average and have a higher risk of learning disabilities.

Females who are lacking an X chromosome are designated XO; the O indicates the absence of a second sex chromosome. These women have Turner syndrome. They have a characteristic appearance, including short stature and often a web of skin extending between the neck and shoulders. Women with Turner syndrome are of normal intelligence but are sterile. If left untreated, they have poor development of breasts and other secondary sex characteristics. Administration of estrogen can alleviate those symptoms. The XO condition is the sole known case where having only 45 chromosomes is not fatal in humans.

Notice the crucial role of the Y chromosome in determining a person's sex. In general, having at least one Y chromosome produces biological maleness, regardless of the number of X chromosomes. The absence of a Y chromosome results in biological femaleness. ☑

☑ **CHECKPOINT**

Why is a person more likely to survive with an abnormal number of sex chromosomes than an abnormal number of autosomes?

■ *Answer: because the Y chromosome is very small and extra X chromosomes are inactivated*

EVOLUTION CONNECTION
Life with and without Sex

The Advantages of Sex

Throughout this chapter, we've examined cell division within the context of reproduction. Like the zebra shark discussed in the Biology and Society section, many species (including a few dozen animal species, but many more within the plant kingdom) can reproduce both sexually and asexually (**Figure 8.23**). An important advantage of asexual reproduction is that there is no need for a partner. Asexual reproduction may thus confer an evolutionary advantage when organisms are sparsely distributed (on an isolated island, for example) and unlikely to meet a mate. Furthermore, if an organism is superbly suited to a stable environment, asexual reproduction has the advantage of passing on its entire genetic legacy intact. Asexual reproduction also eliminates the need to expend energy forming gametes and copulating with a partner.

In contrast to plants, the vast majority of animals reproduce by sexual means. There are exceptions, such as the few species that can reproduce by parthenogenesis and the bdelloid rotifers discussed in the Process of Science section. But most animals reproduce only through sex. In fact, asexual reproduction seems to be an evolutionary "dead end" in the animal kingdom. Therefore, sex must enhance evolutionary fitness. But how? The answer remains elusive. Most hypotheses focus on the unique combinations of genes formed during meiosis and fertilization. By producing offspring of varied genetic makeup, sexual reproduction may enhance survival by speeding adaptation to a changing environment. Another idea is that shuffling genes during sexual reproduction might reduce the incidence of harmful genes more rapidly and allow for disadvantageous mutations to be more quickly eliminated from the gene pool. But for now, one of biology's most basic questions—Why have sex?—remains a hotly debated topic that is the focus of much ongoing research.

▼ **Figure 8.23 Sexual and asexual reproduction.** Many plants, such as this strawberry, have the ability to reproduce both sexually (through flowers that produce fruit) and asexually (through runners).

Runner

Chapter Review

SUMMARY OF KEY CONCEPTS

What Cell Reproduction Accomplishes

Cell reproduction, also called cell division, produces genetically identical cells:

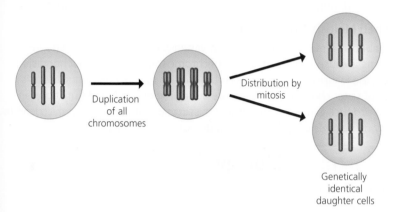

Duplication of all chromosomes → Distribution by mitosis → Genetically identical daughter cells

Some organisms use mitosis (ordinary cell division) to reproduce. This is called asexual reproduction, and it results in offspring that are genetically identical to the lone parent and to each other. Mitosis also enables multicellular organisms to grow and develop and to replace damaged or lost cells. Organisms that reproduce sexually, by the union of a sperm with an egg cell, carry out meiosis, a type of cell division that yields gametes with only half as many chromosomes as body (somatic) cells.

The Cell Cycle and Mitosis

Eukaryotic Chromosomes

The genes of a eukaryotic genome are grouped into multiple chromosomes in the nucleus. Each chromosome contains one very long DNA molecule, with many genes, that is wrapped around histone proteins. Individual chromosomes are coiled up and therefore visible with a light microscope only when the cell is in the process of dividing; otherwise, they are in the form of thin, loosely packed chromatin fibers.

Duplicating Chromosomes

Because chromosomes contain the information needed to control cellular processes, they must be copied and distributed to daughter cells. Before a cell starts dividing, the chromosomes duplicate, producing sister chromatids (containing identical DNA) joined together at the centromere.

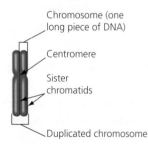

Chromosome (one long piece of DNA)

Centromere

Sister chromatids

Duplicated chromosome

The Cell Cycle

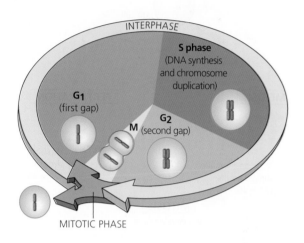

INTERPHASE

S phase (DNA synthesis and chromosome duplication)

G₁ (first gap)

G₂ (second gap)

M

MITOTIC PHASE

Mitosis and Cytokinesis

Mitosis is divided into four phases: prophase, metaphase, anaphase, and telophase. At the start of mitosis, the chromosomes coil up and the nuclear envelope breaks down (prophase). Next, a mitotic spindle made of microtubule tracks moves the chromosomes to the middle of the cell (metaphase). The sister chromatids then separate and are moved to opposite poles of the cell (anaphase), where two new nuclei form (telophase). Cytokinesis overlaps the end of mitosis. In animals, cytokinesis occurs by cleavage, which pinches the cell in two. In plants, a membranous cell plate divides the cell in two. Mitosis and cytokinesis produce genetically identical cells.

Cancer Cells: Dividing Out of Control

When the cell cycle control system malfunctions, a cell may divide excessively and form a tumor. Cancer cells may grow to form malignant tumors, invade other tissues (metastasize), and even kill the host. Surgery can remove tumors, and radiation and chemotherapy are effective as treatments because they interfere with cell division. You can increase the likelihood of surviving some forms of cancer through lifestyle changes and regular screenings.

Meiosis, the Basis of Sexual Reproduction

Homologous Chromosomes

The somatic cells (body cells) of each species contain a specific number of chromosomes; human cells have 46, made up of 23 pairs of homologous chromosomes. The chromosomes of a homologous pair carry genes for the same characteristics at the same places. Mammalian males have X and Y sex chromosomes (only partly homologous), and females have two X chromosomes.

Gametes and the Life Cycle of a Sexual Organism

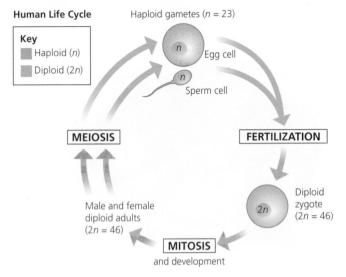

Human Life Cycle

Key
- Haploid (*n*)
- Diploid (*2n*)

Haploid gametes (*n* = 23)

n Egg cell

n Sperm cell

MEIOSIS

FERTILIZATION

Male and female diploid adults (*2n* = 46)

MITOSIS
and development

2n Diploid zygote (*2n* = 46)

The Process of Meiosis

Meiosis, like mitosis, is preceded by chromosome duplication. But in meiosis, the cell divides twice to form four daughter cells. The first division, meiosis I, starts with the pairing of homologous chromosomes. In crossing over, homologous chromosomes exchange corresponding segments. Meiosis I separates the members of the homologous pairs and produces two daughter cells, each with one set of (duplicated) chromosomes. Meiosis II is essentially the same as mitosis; in each of the cells, the sister chromatids of each chromosome separate.

Review: Comparing Mitosis and Meiosis

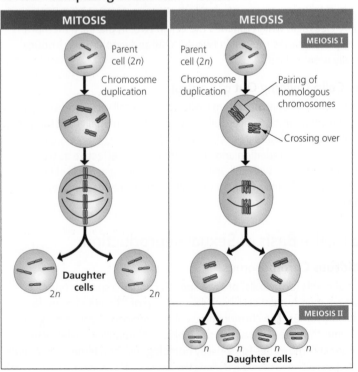

MITOSIS

Parent cell (*2n*)

Chromosome duplication

Daughter cells
2n *2n*

MEIOSIS

MEIOSIS I

Parent cell (*2n*)

Chromosome duplication

Pairing of homologous chromosomes

Crossing over

MEIOSIS II

Daughter cells
n *n* *n* *n*

The Origins of Genetic Variation

Because the chromosomes of a homologous pair come from different parents, they carry different versions of many of their genes. The large number of possible arrangements of chromosome pairs at metaphase of meiosis I leads to many different combinations of chromosomes in eggs and sperm. Random fertilization of eggs by sperm greatly increases the variation. Crossing over during prophase of meiosis I increases variation still further.

When Meiosis Goes Wrong

An abnormal number of chromosomes can cause problems. Down syndrome is caused by an extra copy of chromosome 21, a result of nondisjunction, the failure of homologous chromosomes to separate during meiosis I or of sister chromatids to separate during meiosis II. Nondisjunction can also produce gametes with extra or missing sex chromosomes, which lead to varying degrees of malfunction but do not usually affect survival.

Mastering Biology

For practice quizzes, BioFlix animations, MP3 tutorials, video tutors, and more study tools designed for this textbook, go to Mastering Biology™

SELF-QUIZ

1. Which of the following is not a function of mitosis in humans?
   a. repair of wounds
   c. production of gametes from diploid cells
   b. growth
   d. replacement of lost or damaged cells

2. In what sense are the two daughter cells produced by mitosis identical?

3. Why is it hard to observe individual chromosomes in interphase?

4. A biochemist observes cells growing in the laboratory. A cell that completes the cell cycle without undergoing cytokinesis will
   a. have less genetic material than it started with.
   b. not have completed anaphase.
   c. have its chromosomes lined up in the middle of the cell.
   d. have two nuclei.

5. What phases of mitosis are opposites in terms of changes in the nucleus?

6. Complete the following table to compare mitosis and meiosis.

	Mitosis	Meiosis
a. Number of chromosomal duplications		
b. Number of cell divisions		
c. Number of daughter cells produced		
d. Number of chromosomes in daughter cells		
e. How chromosomes line up during metaphase		
f. Genetic relationship of daughter cells to parent cells		
g. Functions performed in the human body		

7. If an intestinal cell in a dog contains 78 chromosomes, a dog sperm cell would contain _____ chromosomes.

8. A micrograph of a dividing cell from a mouse shows 19 chromosomes, each consisting of two sister chromatids. During which stage of meiosis could this micrograph have been taken? (Explain your answer.)

9. Tumors that remain at their site of origin are called _____, and tumors that can migrate to other body tissues are called _____.

10. A diploid body (somatic) cell from a fruit fly contains eight chromosomes. This means that _____ different combinations of chromosomes are possible in its gametes.

11. Although nondisjunction is a random event, there are many more people with an extra chromosome 21, which causes Down syndrome, than people with an extra chromosome 3 or chromosome 16. Explain.

For answers to the Self Quiz, see Appendix D.

IDENTIFYING MAJOR THEMES

For each statement, identify which major theme is evident (the relationship of structure to function, information flow, pathways that transform energy and matter, interactions within biological systems, or evolution) and explain how the statement relates to the theme. If necessary, review the themes (see Chapter 1) and review the examples highlighted in blue in this chapter.

12. Passing along genetic instructions from one generation to the next requires a precise duplication of the chromosomes.

13. The combined actions of many different proteins determine whether a cell will divide or not.

14. By examining the compactness of a chromosome region (loose versus tightly wound), you can gain insight into whether the genes in that region are actively being used.

For answers to Identifying Major Themes, see Appendix D.

THE PROCESS OF SCIENCE

15. A mule is the offspring of a horse and a donkey. A donkey sperm contains 31 chromosomes and a horse egg has 32, so the zygote has 63. The zygote develops normally. However, a mule is sterile; meiosis cannot occur normally in its testes or ovaries. Explain why mitosis is normal in cells containing both horse and donkey chromosomes, but the mixed set interferes with meiosis.

16. You prepare a slide with a thin slice of an onion root tip. You see the following view in a light microscope. Identify the stage of mitosis for each of the outlined cells, a–d.

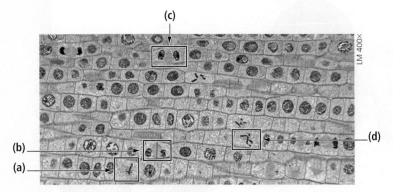

17. **Interpreting Data** The graph shows the incidence of Down syndrome in the offspring of normal parents as the age of the mother increases. For women under the age of 30, how many infants with Down syndrome are born per 1,000 births? At age 40? At age 50? How many times more likely is a 50-year-old woman to give birth to a baby with Down syndrome than a 30-year-old woman?

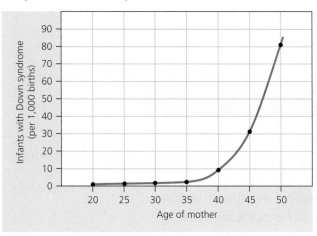

BIOLOGY AND SOCIETY

18. If an endangered species can reproduce by parthenogenesis, what implications might this have on efforts to repopulate that species? How might a parthenogenesis program harm such a species?

19. Every year, about a million Americans are diagnosed with cancer. This means that about 75 million Americans now living will eventually have cancer, and one in five will die of the disease. There are many kinds of cancers and many causes, such as smoking, overexposure to ultraviolet rays, a high-fat and low-fiber diet, and some workplace chemicals. Hundreds of millions of dollars are spent each year on searching for treatments, yet far less money is spent on prevention. Should we devote more resources to treating cancer or to preventing it? Explain.

20. The risk of trisomy 21 and other chromosomal abnormalities increases if a woman is conceiving late. Are there other considerations that make pregnancy at an older age risky? Should modern medical strategies to achieve pregnancy, such as hormone fertility treatment and artificial insemination, be limited to women below a certain age?

21. The anticancer drug Taxol freezes the mitotic spindle after it forms, which may stop a tumor from growing. It is made from a chemical in the bark of the Pacific yew, a tree found mainly in the northwestern United States. It has fewer side effects than many other anticancer drugs and seems effective against some hard-to-treat cancers of the ovary and breast. Another drug, vinblastine, prevents the mitotic spindle from forming. It was first obtained from the periwinkle plant, native to the tropical rain forests of Madagascar. Given these examples, preserving biodiversity may be the key to discovering lifesaving anticancer drugs. Such drugs are often discovered in developing areas, but because they are expensive, they primarily benefit people in developed regions. Do you see a conflict in this? Explain.

177

9 Patterns of Inheritance

Why Genetics Matters

Every trait that an individual is born with results from inherited genes. Understanding genetics is therefore vital to such diverse fields as forensics, agriculture, and evolution. It is also the basis for all inborn diseases that affect humans, dogs, and all other life on Earth.

YOUR SEX WAS ENTIRELY DETERMINED BY YOUR FATHER.

SOME FEATURES ARE DETERMINED SOLELY BY GENES, SOME BY ENVIRONMENT, AND SOME BY BOTH.

YOUR CHANCES OF INHERITING AND PASSING ALONG A GENETIC DISEASE CAN OFTEN BE DETERMINED FROM YOUR FAMILY HISTORY.

Breeding a best friend.
Dogs, such as this Cavalier King Charles spaniel, are one of humankind's longest running genetic experiments.

BIOLOGY AND SOCIETY   Dog Breeding

Darwin's Dogs

Purebred pooches are living proof that dogs are more than our best friends: They are also one of our longest-running genetic experiments. Evidence (which we'll explore at the end of this chapter) suggests that people have selected and mated dogs with preferred traits for more than 15,000 years. For example, nearly every modern Cavalier King Charles spaniel can trace its ancestry back to a single pair that were brought to the United States by a breeder in 1952. These original dogs were chosen due to their desired traits, both physical (silky coat, long elegant ears, and gentle eyes) and behavioral (obedient, sweet, and gentle). Similar choices were made for every modern dog breed. Over thousands of years, such genetic tinkering has led to the incredible variety of body types and behaviors we have today, from huge, docile Great Danes to tiny, spunky Chihuahuas.

Given their long and well-documented history (thanks to the American Kennel Club's system of pedigree registration), dogs make an ideal case study in genetics. Several groups of researchers are working to uncover how genes have influenced canines as they evolved from ancestral wolves. One of the largest efforts, called Darwin's Dogs, is led by researchers at the University of Massachusetts Medical School. Their project is a public effort; anyone can submit a DNA sample from their pet and fill out a questionnaire about that dog's characteristics. The data they collect are already providing valuable insights into neurological behavior—both in normal dogs and in those who suffer from diseases.

Although people have been applying genetics for thousands of years—by breeding food crops (such as wheat, rice, and corn) as well as domesticated animals (such as cows, sheep, and goats)—the biological principles underlying genetics have only recently been understood. In this chapter, you will learn the basic rules of how genetic traits are passed from generation to generation and how the behavior of chromosomes (the topic of Chapter 8) accounts for these rules. In the process, you will learn how to predict the ratios of offspring with particular traits. Along the way, we'll consider many examples of how genetic principles can help us understand the biology of humans, dogs, plants, and many other familiar creatures.

Genetics and Heredity

Heredity is the transmission of traits from one generation to the next. **Genetics**, the scientific study of heredity, began in the 1860s, when a monk named Gregor Mendel (**Figure 9.1**) deduced its fundamental principles by breeding garden peas. Mendel lived and worked in an abbey in Brunn, Austria (now Brno, in the Czech Republic). Strongly influenced by his study of physics, mathematics, and chemistry at the University of Vienna, his research was both experimentally and mathematically rigorous, and these qualities were largely responsible for his success.

In a paper published in 1866, Mendel correctly argued that parents pass on to their offspring discrete genes (which he called "heritable factors") that are responsible for inherited traits, such as purple flowers or round seeds in pea plants. (It is interesting to note that Mendel's publication came just seven years after Darwin's 1859 publication of *On the Origin of Species*, making the 1860s a banner decade in the advent of modern biology.) In his paper, one of the most influential in the history of biology, Mendel stressed that genes retain their individual identities generation after generation. That is, genes are like playing cards: A deck may be shuffled, but all the cards retain their identities, and no card is ever blended with another. Similarly, genes may be sorted, but each gene retains its intact identity.

▲ **Figure 9.1 Gregor Mendel.**

wanted to fertilize one plant with pollen from a different plant, he pollinated the plants by hand, as shown in **Figure 9.3**. Thus, Mendel was always sure of the parentage of his new plants.

Each of the characters Mendel chose to study, such as flower color, occurred in two distinct traits. Mendel worked with his plants until he was sure he had purebred varieties—that is, varieties for which self-fertilization produced offspring all identical to the parent. For instance, he identified a purple-flowered variety that, when self-fertilized, always produced offspring plants that had all purple flowers.

Next Mendel was ready to ask what would happen when he crossed different purebred varieties with each other. For example, what offspring would result if plants

In an Abbey Garden

Mendel chose to study garden peas because they were easy to grow and they came in many readily distinguishable varieties. For example, one variety has purple flowers and another variety has white flowers. A heritable feature that varies among individuals, such as flower color, is called a **character**. Each variant of a character, such as purple or white flowers, is called a **trait**.

Perhaps the most important advantage of pea plants as an experimental model was that Mendel could strictly control their reproduction. The petals of a pea flower (**Figure 9.2**) almost completely enclose the egg-producing organ (the carpel) and the sperm-producing organs (the stamens). Consequently, in nature, pea plants usually self-fertilize because sperm-carrying pollen grains released from the stamens land on the tip of the egg-containing carpel of the same flower. Mendel could ensure self-fertilization by covering a flower with a small bag so that no pollen from another plant could reach the carpel. When he

▼ **Figure 9.2 The structure of a pea flower.** To reveal the reproductive organs—the stamens and carpel—one of the petals is not shown in this drawing.

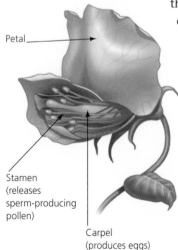

Petal

Stamen (releases sperm-producing pollen)

Carpel (produces eggs)

▼ **Figure 9.3 Mendel's technique for cross-fertilizing pea plants.**

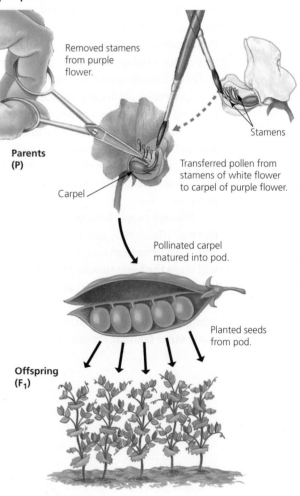

Removed stamens from purple flower.

Stamens

Parents (P)

Carpel

Transferred pollen from stamens of white flower to carpel of purple flower.

Pollinated carpel matured into pod.

Planted seeds from pod.

Offspring (F₁)

with purple flowers and plants with white flowers were cross-fertilized as shown in Figure 9.3? The offspring of two different purebred varieties are called **hybrids**, and the cross-fertilization itself is referred to as a genetic **cross**. The purebred parents are called the **P generation**, and their hybrid offspring are the F_1 **generation** (F for *filial*, from the Latin for "son" or "daughter"). When F_1 plants self-fertilize or fertilize each other, their offspring are the F_2 **generation**. ☑

Mendel's Law of Segregation

Mendel performed many experiments in which he tracked the inheritance of characters, such as flower color, that occur as two alternative traits **(Figure 9.4)**. The results led him to formulate several hypotheses about inheritance. Let's look at some of his experiments and follow the reasoning that led to his hypotheses.

▼ **Figure 9.4 The seven characters of pea plants studied by Mendel.** Each character comes in the two alternative traits shown here.

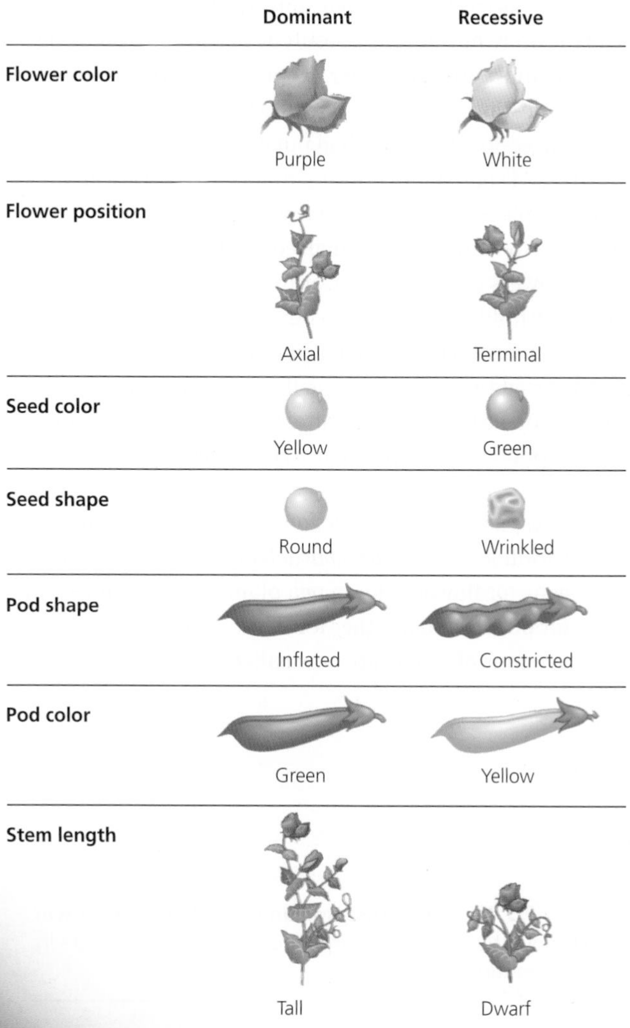

	Dominant	Recessive
Flower color	Purple	White
Flower position	Axial	Terminal
Seed color	Yellow	Green
Seed shape	Round	Wrinkled
Pod shape	Inflated	Constricted
Pod color	Green	Yellow
Stem length	Tall	Dwarf

☑ **CHECKPOINT**

Imagine you have two purebred dogs, one male and one female. Use the correct genetic names to describe the next two possible generations.

■ *Answer: The purebred parents are the P generation. Their puppies are the F_1 generation. If those puppies are bred with each other, they would produce the F_2 generation.*

Inheritance of a Single Character

Figure 9.5 shows a cross between a purebred pea plant with purple flowers and a purebred pea plant with white flowers. Mendel saw that the F_1 plants all had purple flowers. Was the factor responsible for inheritance of white flowers now lost as a result of the cross? By mating the F_1 plants with each other, Mendel found the answer to this question to be no. Of the 929 F_2 plants he bred, about three-fourths (705) had purple flowers and one-fourth (224) had white flowers; that is, there were about three purple F_2 plants for every white plant, or a 3:1 ratio of purple to white. Mendel figured out that the gene for white flowers did not disappear in the F_1 plants but was somehow hidden or masked when the purple-flower factor was present. He also deduced that the F_1 plants must have carried two factors for the flower-color character, one for purple and one for white. From these results and others, Mendel developed four hypotheses:

1. *There are alternative versions of genes that account for variations in inherited characters.* For example, the gene for flower color in pea plants exists in one form for purple and another for white. The alternative versions of a gene are now called **alleles**.

▼ **Figure 9.5 Mendel's cross tracking one character (flower color).** Note the 3:1 ratio of purple flowers to white flowers in the F_2 generation.

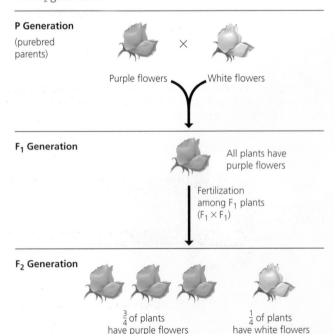

P Generation
(purebred parents)

Purple flowers × White flowers

F_1 Generation

All plants have purple flowers

Fertilization among F_1 plants ($F_1 \times F_1$)

F_2 Generation

$\frac{3}{4}$ of plants have purple flowers $\frac{1}{4}$ of plants have white flowers

2. *For each character, an organism inherits two alleles of a gene, one from each parent.* These alleles may be the same or different. An organism that has two identical alleles for a gene is said to be **homozygous** for that gene (and is called a homozygote for that trait). An organism that has two different alleles for a gene is said to be **heterozygous** for that gene (and is a heterozygote).

3. *If an organism has two different alleles for a gene, one allele determines the organism's appearance and is called the* **dominant allele**; *the other allele has no noticeable effect on the organism's appearance and is called the* **recessive allele**. Geneticists use uppercase italic letters (such as P) to represent dominant alleles and lowercase italic letters (such as p) to represent recessive alleles.

4. *A sperm or egg carries only one allele for each inherited character because the two alleles for a character segregate (separate) from each other during the production of gametes.* This statement is called the **law of segregation**. When sperm and egg unite at fertilization, each contributes its alleles, restoring the paired condition in the offspring.

Figure 9.6 illustrates Mendel's law of segregation, which explains the inheritance pattern shown in Figure 9.5. Mendel's hypotheses predict that when alleles segregate during gamete formation in the F_1 plants, half the gametes will receive a purple-flower allele (P) and the other half a white-flower allele (p). During pollination among the F_1 plants, the gametes unite randomly. An egg with a purple-flower allele has an equal chance of being fertilized by a sperm with a purple-flower allele or one with a white-flower allele (that is, a P egg may fuse with a P sperm or a p sperm). Because the same is true for an egg with a white-flower allele (a p egg with a P sperm or p sperm), there are a total of four equally likely combinations of sperm and egg.

The diagram at the bottom of Figure 9.6, called a **Punnett square**, repeats the cross shown in Figure 9.5 in a way that highlights the four possible combinations of gametes and the resulting four possible offspring in the F_2 generation. Each square represents an equally probable product of fertilization. For example, the box in the upper right corner of the Punnett square shows the genetic combination resulting from a p sperm fertilizing a P egg.

According to the Punnett square, what will be the physical appearance of these F_2 offspring? One-fourth of the plants have two alleles specifying purple flowers (PP); clearly, these plants will have purple flowers. One-half (two-fourths) of the F_2 offspring have inherited one allele for purple flowers and one allele for white flowers (Pp); like

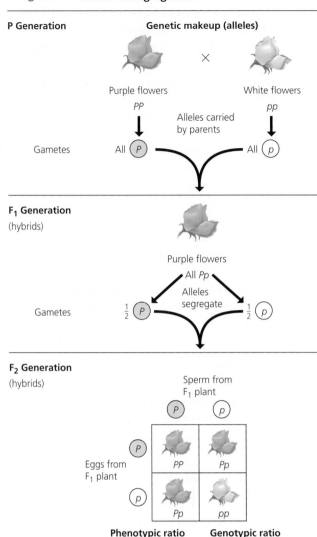

P Generation — Genetic makeup (alleles)

Purple flowers
PP

White flowers
pp

Alleles carried by parents

Gametes — All ⓟ — All ⓟ

F₁ Generation
(hybrids)

Purple flowers
All *Pp*

Alleles segregate

Gametes — ½ ⓟ — ½ ⓟ

F₂ Generation
(hybrids)

Sperm from F₁ plant
ⓟ   ⓟ

Eggs from F₁ plant
ⓟ
ⓟ

PP	*Pp*
Pp	*pp*

Phenotypic ratio
3 purple:1 white

Genotypic ratio
1 *PP*:2 *Pp*:1 *pp*

the F₁ plants, these plants will also have purple flowers, the dominant trait. (Note that *Pp* and *pP* are equivalent and usually written as *Pp*.) Finally, one-fourth of the F₂ plants have inherited two alleles specifying white flowers (*pp*) and will express this recessive trait. Thus, Mendel's model accounts for the 3:1 ratio that he observed in the F₂ generation.

Geneticists distinguish between an organism's observable traits, called its **phenotype** (such as purple or white flowers), and its genetic makeup, called its **genotype** (such as *PP*, *Pp*, or *pp*). Now we can see that Figure 9.5 shows only the phenotypes and Figure 9.6 both the genotypes and phenotypes in our sample cross. For the F₂ plants, the ratio of plants with purple flowers to those with white flowers (3:1) is called the phenotypic ratio. The genotypic ratio is 1(*PP*):2(*Pp*):1(*pp*).

Mendel found that each of the seven characters he studied had the same inheritance pattern: One of two

parental traits disappeared in the F₁ generation, only to reappear in one-fourth of the F₂ offspring. The underlying mechanism is explained by Mendel's law of segregation: *Pairs of alleles segregate (separate) during gamete formation; the fusion of gametes at fertilization creates allele pairs again.* Research since Mendel's day has established that the law of segregation applies to all sexually reproducing organisms, including dogs and people.

The Relationship between Alleles and Homologous Chromosomes

Before continuing with Mendel's experiments, let's consider how an understanding of chromosomes (see Chapter 8) fits with what we've said about genetics so far. The diagram in **Figure 9.7** shows a pair of homologous chromosomes—chromosomes that carry alleles of the same genes. Recall that every diploid cell, whether from a pea plant or a person, has pairs of homologous chromosomes. One member of each pair comes from the organism's female parent, and the other member of each pair comes from the male parent. **Therefore, transmission of genetic information from one generation to the next can be accounted for by the genes found on homologous chromosomes.**

Each labeled band on the chromosomes in the figure represents a gene **locus** (plural, *loci*), a specific location of a gene along the chromosome. You can see the connection between Mendel's law of segregation and homologous chromosomes: Alleles (alternative versions) of a gene reside at the same locus on homologous chromosomes. However, the two chromosomes may bear either identical alleles (such as the *P*/*P* and *a*/*a* loci) or different alleles (as in the *B*/*b* locus) at any one locus. In other words, the organisms may be homozygous or heterozygous for the gene at that locus. We will return to the chromosomal basis of Mendel's law later in the chapter. ☑

Figure Walkthrough

Mastering Biology
goo.gl/HwihMJ

☑ CHECKPOINT

1. Genes come in different versions called _____. What term describes the condition where the two copies are identical? What term describes the condition where the two copies are different?

2. If two plants have the same genotype, must they have the same phenotype? If two plants have the same phenotype, must they have the same genotype?

3. You carry two alleles for every trait. Where did these alleles come from?

Answers: 1. alleles; homozygous; heterozygous. 2. Yes; No: One could be homozygous for the dominant allele, whereas the other is heterozygous. 3. One is from your father through his sperm and one is from your mother through her egg.

▼ Figure 9.7 **The relationship between alleles and homologous chromosomes.** The matching colors of corresponding loci highlight the fact that homologous chromosomes carry alleles for the same genes at the same positions along their lengths.

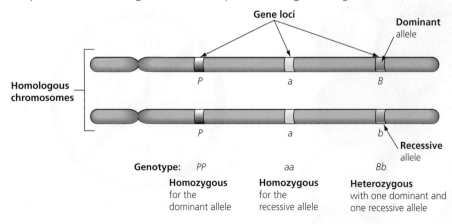

Gene loci

Dominant allele

Homologous chromosomes

P   *a*   *B*

P   *a*   *b*

Recessive allele

Genotype: *PP*   *aa*   *Bb*

Homozygous for the dominant allele

Homozygous for the recessive allele

Heterozygous with one dominant and one recessive allele

Mendel's Law of Independent Assortment

Mendel deduced his law of segregation by following one character (flower color in our example) through two generations. A cross between two individuals that are heterozygous for one character (*Pp* × *Pp* in this case) is called a **monohybrid cross**. By tracking monohybrid crosses, Mendel knew that the allele for round seed shape (designated *R*) was dominant to the allele for wrinkled seed shape (*r*) and that the allele for yellow seed color (*Y*) was dominant to the allele for green seed color (*y*). What would result from a **dihybrid cross**, a cross between two organisms that are each heterozygous for two characters being followed?

Mendel crossed homozygous plants having round-yellow seeds (genotype *RRYY*) with plants having wrinkled-green seeds (*rryy*). As shown in **Figure 9.8**,

the union of *RY* and *ry* gametes from the P generation yielded hybrids heterozygous for both characters (*RrYy*)—that is, dihybrids. As we would expect, all of these offspring, the F_1 generation, had round-yellow seeds (the two dominant traits). But were the two characters transmitted from parents to offspring as a package, or was each character inherited independently of the other?

The question was answered when Mendel crossed the F_1 plants with each other. If the genes for the two characters were inherited together (Figure 9.8a), then the F_1 hybrids would produce only the same two kinds of gametes that they received from their parents. In that case, the F_2 generation would show a 3:1 phenotypic ratio (three plants with round-yellow seeds for every one with wrinkled-green seeds), as in the Punnett square in Figure 9.8a. If, however, the two seed characters sorted independently, then the F_1 generation would produce four gamete genotypes—*RY*, *rY*, *Ry*, and *ry*—in equal quantities. The Punnett square in Figure 9.8b shows all

▼ **Figure 9.8** **Testing alternative hypotheses for gene assortment in a dihybrid cross.**
Only the hypothesis of independent assortment was supported by Mendel's data.

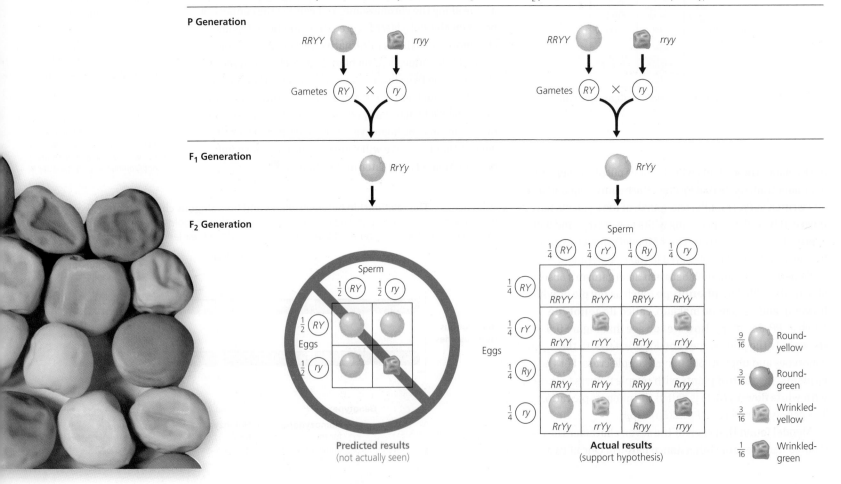

(a) Hypothesis: Dependent assortment
This hypothesis leads to the prediction that F_2 plants will have seeds that match the parents, either round-yellow or wrinkled-green.

(b) Hypothesis: Independent assortment
This hypothesis leads to the prediction that F_2 plants will have four different seed phenotypes.

possible combinations of alleles that can result in the F_2 generation from the union of four kinds of sperm with four kinds of eggs. If you study the Punnett square, you'll see that it predicts nine different genotypes in the F_2 generation. These nine genotypes will produce four different phenotypes in a ratio of 9:3:3:1.

The Punnett square in Figure 9.8b also reveals that a dihybrid cross is equivalent to two monohybrid crosses occurring simultaneously. From the 9:3:3:1 ratio, we can see that there are 12 plants in the F_2 generation with round seeds compared to 4 with wrinkled seeds, and 12 yellow-seeded plants compared to four green-seeded ones. These 12:4 ratios each reduce to 3:1, which is the F_2 ratio for a monohybrid cross. Mendel tried his seven pea characters in various dihybrid combinations and always observed a 9:3:3:1 ratio (or two simultaneous 3:1 ratios) of phenotypes in the F_2 generation. These results supported the hypothesis that *each pair of alleles segregates independently of the other pairs of alleles during gamete formation*. In other words, the inheritance of one character has no effect on the inheritance of another. This is called Mendel's **law of independent assortment**.

For another application of the law of independent assortment, examine the dog breeding experiment described in **Figure 9.9**. The inheritance of two characters in Labrador retrievers is controlled by separate genes: black versus chocolate coat color, and normal vision versus the eye disorder progressive retinal atrophy (PRA). Black Labs have at least one copy of an allele called *B*. The *B* allele is dominant to *b*, so only the coats of dogs with genotype *bb* are chocolate in color. The allele that causes PRA, called *n*, is recessive to allele *N*, which is necessary for normal vision. Thus, only dogs of genotype *nn* become blind from PRA. If you mate two doubly heterozygous (*BbNn*) Labs (bottom of Figure 9.9), the phenotypic ratio of the offspring (F_2) is 9:3:3:1. These results resemble the F_2 results in Figure 9.8, demonstrating that the coat color and PRA genes are inherited independently. ☑

☑ **CHECKPOINT**

In Figure 9.9, what is the ratio of black coat to chocolate coat in Labs? Of normal vision to blindness?

■ Answer: 3:1; 3:1

▼ **Figure 9.9** Independent assortment of genes in Labrador retrievers.
Blanks in the genotypes indicate alleles that can be either dominant or recessive.

(a) Possible phenotypes of Labrador retrievers

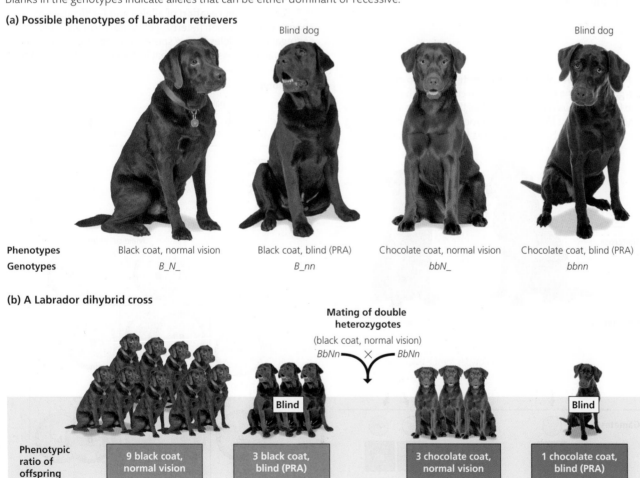

Phenotypes	Black coat, normal vision	Black coat, blind (PRA)	Chocolate coat, normal vision	Chocolate coat, blind (PRA)
Genotypes	B_N_	B_nn	bbN_	bbnn

(b) A Labrador dihybrid cross

Mating of double heterozygotes
(black coat, normal vision)
BbNn ✕ BbNn

Phenotypic ratio of offspring	9 black coat, normal vision	3 black coat, blind (PRA)	3 chocolate coat, normal vision	1 chocolate coat, blind (PRA)

Using a Testcross to Determine an Unknown Genotype

Suppose you have a chocolate Lab. Consulting Figure 9.9, you can tell that its genotype must be *bb*, the only combination of alleles that produces the chocolate-coat phenotype. But what if you have a black Lab? It could have one of two possible genotypes—*BB* or *Bb*—and there is no way to tell which genotype is the correct one by looking at the dog. To determine your dog's genotype, you could perform a **testcross**, a mating between an individual of dominant phenotype but unknown genotype (your black Lab) and a homozygous recessive individual—in this case, a *bb* chocolate Lab.

Figure 9.10 shows the offspring that could result from such a mating. If, as shown on the left, the black Lab parent's genotype is *BB*, we would expect all the offspring to be black, because a cross between genotypes *BB* and *bb* can produce only *Bb* offspring. On the other hand, if the black Lab parent is *Bb*, we would expect both black (*Bb*) and chocolate (*bb*) offspring. Thus, the appearance of the offspring may reveal the original black dog's genotype. ☑

▼ **Figure 9.10 A Labrador retriever testcross.** To determine the genotype of a black Lab, it can be crossed with a chocolate Lab (homozygous recessive, *bb*). If all the offspring are black, the black parent most likely had genotype *BB*. If any of the offspring are chocolate, the black parent must be heterozygous (*Bb*).

The Rules of Probability

Mendel's strong background in mathematics served him well in his studies of inheritance. For instance, he understood that genetic crosses obey the rules of probability— the same rules that apply when tossing coins, rolling dice, or drawing cards. An important lesson we can learn from tossing coins is that for each and every toss, the probability of heads is $\frac{1}{2}$. Even if heads has landed five times in a row, the probability of the next toss coming up heads is still $\frac{1}{2}$. In other words, the outcome of any particular toss is unaffected by what has happened on previous attempts. Each toss is an independent event.

If two coins are tossed simultaneously, the outcome for each coin is an independent event, unaffected by the other coin. What is the chance that both coins will land heads-up? The probability of such a dual event is the product of the separate probabilities of the independent events—for the coins, $\frac{1}{2} \times \frac{1}{2} = \frac{1}{4}$. This idea is called the **rule of multiplication**, and it holds true for independent events that occur in genetics as well as coin tosses (**Figure 9.11**). In our dihybrid cross of Labradors (see Figure 9.9), the genotype of the F_1 dogs for coat color was *Bb*. What is the probability that a particular F_2 dog will

▼ **Figure 9.11 Segregation of alleles and fertilization as chance events.** When heterozygotes (*Bb*) form gametes, segregation of alleles during sperm and egg formation is like two separately tossed coins (that is, two independent events).

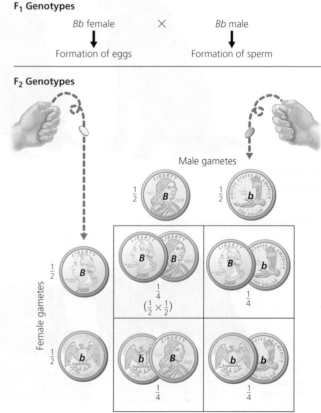

have the *bb* genotype? To produce a *bb* offspring, both egg and sperm must carry the *b* allele. The probability that an egg from a *Bb* dog will have the *b* allele is $\frac{1}{2}$, and the probability that a sperm will have the *b* allele is also $\frac{1}{2}$. By the rule of multiplication, the probability that two *b* alleles will come together at fertilization is $\frac{1}{2} \times \frac{1}{2} = \frac{1}{4}$. This is exactly the answer given by the Punnett square in Figure 9.11. If we know the genotypes of the parents, we can predict the probability for any genotype among the offspring. By applying the rules of probability to segregation and independent assortment, we can solve some rather complex genetics problems. ☑

Family Pedigrees

Researchers working with pea plants or Labrador retrievers can perform testcrosses. But geneticists who study people obviously cannot control the mating of their research participants. Instead, they must analyze the results of matings that have already occurred. First, a geneticist collects as much information as possible about a family's history for a trait. Then the researcher assembles this information into a family tree, called a **pedigree**. You may associate pedigrees with purebred animals such as racehorses and champion dogs, but they can be used to represent human matings just as well. To analyze a pedigree, the geneticist applies logic and Mendel's laws.

Let's apply this approach to the example in **Figure 9.12**, a pedigree tracing the incidence of freckles versus no freckles, a trait believed to be controlled by a single gene. The letter *F* stands for the dominant allele

for freckles, and *f* symbolizes the recessive allele for no freckles. What is the relationship between phenotypes and genotypes? The dominant phenotype (freckled) results from having either one or two dominant alleles (*F*) in the genotype: *Ff* or *FF*. If both alleles are recessive (*ff* in this case), the phenotype is recessive (no freckles, in this case).

In the pedigree, ☐ represents a male, ◯ represents a female, colored symbols (◼ and ⬤) indicate that the person does not have the trait being investigated (in this case, does not have freckles), and an unshaded symbol represents a person who does have the trait (has freckles). The earliest (oldest) generation is at the top of the pedigree, and the most recent generation is at the bottom.

By applying Mendel's laws, we can deduce that the allele for no freckles is recessive because that is the only way that Karen does not have freckles even though both of her parents (Hal and Ina) have freckles. We can therefore label all the individuals without freckles in the pedigree (that is, all those with colored circles or squares) as homozygous recessive (*ff*).

Mendel's laws enable us to deduce the genotypes for most of the people in the pedigree. For example, Hal and Ina must have carried the *f* allele (which they passed on to Karen) along with the *F* allele that gave them freckles. The same must be true of Aaron and Betty because they both have freckles but Fred and Gabe do not. For each of these cases, Mendel's laws and simple logic allow us to definitively assign a genotype.

Notice that we cannot deduce the genotype of every member of the pedigree. For example, Lisa must have

☑ **CHECKPOINT**

Using a standard 52-card deck, what is the probability of being dealt an ace? What about being dealt an ace or a king? What about being dealt an ace and then another ace?

■ Answer: $\frac{1}{13}$ (4 aces/52 cards); $\frac{2}{13}$ ($\frac{8}{52}$); $\frac{4}{52} \times \frac{3}{51}$ (because there are 3 aces left in a deck with 51 cards remaining) = 0.0045, or $\frac{1}{221}$

▼ **Figure 9.12** A family pedigree showing inheritance of freckles versus no freckles.

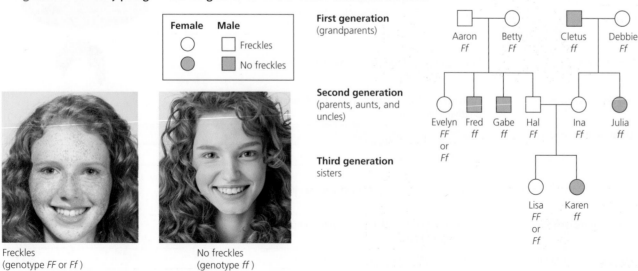

Female	Male	
◯	☐	Freckles
⬤	◼	No freckles

Freckles
(genotype *FF* or *Ff*)

No freckles
(genotype *ff*)

First generation
(grandparents)

Aaron *Ff* Betty *Ff* Cletus *ff* Debbie *Ff*

Second generation
(parents, aunts, and uncles)

Evelyn *FF* or *Ff* Fred *ff* Gabe *ff* Hal *Ff* Ina *Ff* Julia *ff*

Third generation
sisters

Lisa *FF* or *Ff* Karen *ff*

at least one *F* allele, but she could be *FF* or *Ff*. We cannot distinguish between these two possibilities using the available data. Perhaps future generations will provide the data that solve this mystery.

Keep in mind that the word *dominant* in genetics does not mean that a dominant phenotype is either normal or more common than a recessive phenotype. Instead, dominance means that a heterozygous genotype (in this pedigree, *Ff*) results in the dominant phenotype. Traits that are most common in nature are called **wild type**. Wild type has nothing to do with dominance; a recessive trait may in fact be more common in the population than a dominant one. For example, the absence of freckles (*ff*) is more common than their presence (*FF* or *Ff*). ☑

Human Traits Controlled by a Single Gene

The genetic bases of many human characters—such as eye and hair color—are complex and poorly understood. In fact, some traits that were long believed to be controlled by a single gene—such as free versus attached earlobes or the ability to curl the tongue—are now thought to have a more complex genetic basis. However, Mendel's laws can be applied to the inheritance of human traits that are generally agreed to be determined by a single gene, such as freckles or a widow's peak hairline. The presence of a widow's peak is dominant to a straight hairline **(Figure 9.13)**.

The human genetic disorders listed in **Table 9.1** are known to be inherited as dominant or recessive traits

▼ **Figure 9.13** Example of an inherited human trait thought to be controlled primarily by a single gene.

DOMINANT TRAIT	RECESSIVE TRAIT
Widow's peak	Straight hairline

Table 9.1	Some Autosomal Disorders in People	
Disorder	**Major Symptoms**	
Recessive Disorders		
Albinism	Lack of pigment in skin, hair, and eyes	
Cystic fibrosis	Excess mucus in lungs, digestive tract, liver; increased susceptibility to infections; death in early childhood unless treated	
Phenylketonuria (PKU)	Accumulation of phenylalanine in blood; lack of normal skin pigment; developmental disabilities unless treated	
Sickle-cell disease	Misshapen red blood cells; damage to many tissues	
Tay-Sachs disease	Lipid accumulation in brain cells; developmental deficiencies; death in childhood	
Dominant Disorders		
Achondroplasia	Dwarfism	
Alzheimer's disease (one type)	Mental deterioration; usually strikes late in life	
Huntington's disease	Uncontrollable movements; cognitive impairments; occurs in middle age	
Hypercholesterolemia	Excess cholesterol in blood; heart disease	

controlled by a single gene. These disorders therefore show simple inheritance patterns like the ones Mendel studied in pea plants. The genes involved are all located on autosomes, chromosomes other than the sex chromosomes X and Y.

Recessive Disorders

Most genetic mutations that cause medical disorders are recessive. The genetic conditions caused by recessive alleles range in severity from harmless to life-threatening. Most people who have recessive disorders are born to normal parents who are both heterozygotes—that is, parents who are **carriers** of the recessive allele for the disorder but appear normal themselves.

Using Mendel's laws, we can predict the fraction of affected offspring that is likely to result from a mating between two carriers. Suppose two heterozygous carriers of albinism (*Aa*) have a child. What is the probability that the child will display albinism? As the Punnett square in **Figure 9.14** shows, each child of two carriers has a $\frac{1}{4}$ chance of inheriting two recessive alleles. Thus, we can say that about one-fourth of the children of this couple are likely to display albinism. We can also say that a child with normal pigmentation from such a family has a $\frac{2}{3}$ chance of being a carrier (that is, on average, two out of three of the offspring with the normal pigmentation phenotype will be *Aa*). We can apply this same method of pedigree analysis and prediction to any genetic trait controlled by a single gene.

The most common lethal genetic disease in the United States is cystic fibrosis (CF). Affecting about 30,000 Americans, the recessive CF allele is carried by about 1 in 31 Americans—which means that about

10.5 million Americans are carriers. A person with two copies of this allele has cystic fibrosis, which is characterized by an excessive secretion of very thick mucus from the lungs, pancreas, and other organs. This mucus can interfere with breathing, digestion, and liver function and makes the person vulnerable to recurrent bacterial infections. Although there is no cure for CF, a special diet, antibiotics to prevent infection, frequent pounding of the chest and back to clear the lungs, and other treatments can have a profound impact on the health of the affected person. Once invariably fatal in childhood, advances in CF treatment have raised the median survival age of Americans with CF to nearly 40. ☑

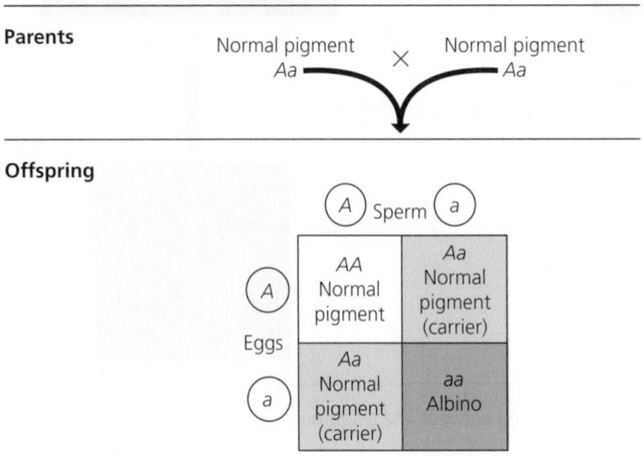

Dominant Disorders

A number of human disorders are caused by dominant alleles. Some are harmless, such as extra fingers and toes or webbed fingers and toes. A more serious disorder caused by a dominant allele is achondroplasia, a form of dwarfism in which the head and torso develop normally but the arms and legs are short. The homozygous dominant genotype (*AA*) causes death of the embryo, and therefore only heterozygotes (*Aa*), individuals with a single copy of the defective allele, have this disorder. This also means that a person with achondroplasia has a 50% chance of passing the condition on to any children **(Figure 9.15)**. Therefore, all those who do not have achondroplasia, more than 99.99% of the population, are homozygous for the recessive allele (*aa*). This example makes it clear that a dominant allele is not necessarily more common in a population than the corresponding recessive allele.

Dominant alleles that cause lethal disorders are much less common than recessive alleles that cause lethal diseases. One example is the allele that causes Huntington's disease, a degeneration of the nervous system that usually does not begin until middle age. Once the deterioration of the nervous system begins, it is irreversible and inevitably fatal. Because the allele for Huntington's disease is dominant, any child born to a parent with the allele has a 50% chance of inheriting the allele and the disorder. This example makes it clear that a dominant allele is not necessarily "better" than the corresponding recessive allele.

One way that geneticists study human traits is to find similar genes in animals, which can then be studied in greater detail through controlled matings and other experiments. Let's return to the chapter thread—dog breeding—to investigate the legs of man's best friend.

☑ **CHECKPOINT**

A man and a woman who are both carriers of cystic fibrosis have three children without cystic fibrosis. If the couple has a fourth child, what is the probability that the child will have the disorder?

■ *Answer: $\frac{1}{4}$ (The genotypes and phenotypes of their other children are irrelevant.)*

▼ **Figure 9.14** Predicted offspring when both parents are carriers for albinism, a recessive disorder.

Parents

Normal pigment
Aa × Normal pigment
Aa

Offspring

	Sperm *A*	Sperm *a*
Eggs *A*	*AA* Normal pigment	*Aa* Normal pigment (carrier)
Eggs *a*	*Aa* Normal pigment (carrier)	*aa* Albino

▼ **Figure 9.15** A Punnett square illustrating a family with and without achondroplasia.

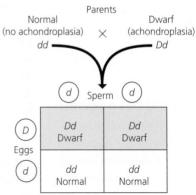

Parents

Normal (no achondroplasia) *dd* × Dwarf (achondroplasia) *Dd*

	Sperm *d*	Sperm *d*
Eggs *D*	*Dd* Dwarf	*Dd* Dwarf
Eggs *d*	*dd* Normal	*dd* Normal

189

THE PROCESS OF SCIENCE | Dog Breeding

What Is the Genetic Basis of Short Legs in Dogs?

BACKGROUND

It's obvious that dogs come in a wide variety of physical types. In fact, domesticated dogs display the greatest range of phenotypes of any mammal. One of the most striking features that distinguishes some breeds is chondrodysplasia, a condition that affects the growth of bones in the leg. The resulting shortened, curved bones are a defining characteristic of a few dog breeds (**Figure 9.16a**). Through test crosses, breeders have long known that the short-legged trait is dominant, but nothing was known about the cause of the phenotype.

METHOD

A group of researchers set out to discover the genetic basis of the short-legged phenotype. They used an automated gene chip (see Figure 11.10) to examine the DNA of 95 dogs from 7 short-legged breeds (the experimental group) and 702 dogs from 64 breeds with long legs (the control group). They compared the results to identify any differences between the two groups at thousands of sites across the dog genome (**Figure 9.16b**).

RESULTS

One location on chromosome 18 stood out for being strongly associated with short legs. Closer examination of the region surrounding that location revealed a gene that codes for a protein called fibroblast growth factor 4. The protein produced by this gene is known to be associated with the growth of legs during embryonic development. The researchers identified a specific change in the chromosome that corresponded to short legs. Interestingly, they were able to link the effect of this gene in dogs to a related protein associated with the most common form of human dwarfism. This experiment shows how animal models may provide insight into genetic conditions in humans.

▼ **Figure 9.16** The genetic basis of chondrodysplasia in dogs.

(a) Some examples of short-legged and long-legged breeds

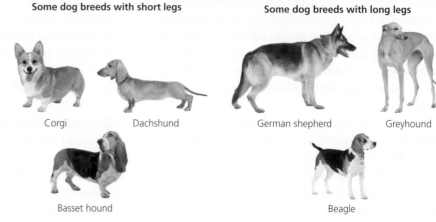

Some dog breeds with short legs

Corgi   Dachshund

Basset hound

Some dog breeds with long legs

German shepherd   Greyhound

Beagle

(b) Comparing DNA from different dog breeds

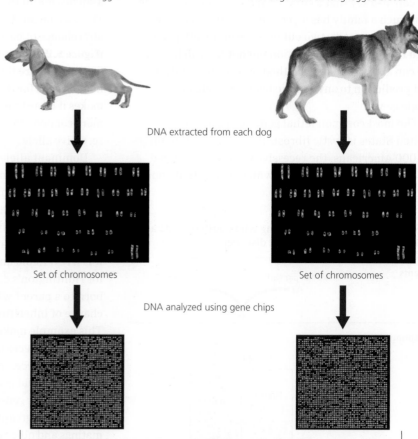

Experimental group:
95 dogs from 7 short-legged breeds

Control group:
702 dogs from 64 long-legged breeds

DNA extracted from each dog

Set of chromosomes   Set of chromosomes

DNA analyzed using gene chips

Results compared between the two groups

Thinking Like a Scientist

Why might it be easier to find the genetic basis for a physical condition in dogs than to do so in humans?

For the answer, see Appendix D.

Genetic Testing

In the past, the onset of symptoms was the only way to know if a person had inherited a genetic disease. Modern technologies offer several ways to identify about 30 treatable genetic diseases before they appear. For example, tests can distinguish between people who have no disease-causing alleles and those who are heterozygous carriers of such diseases as Tay-Sachs and sickle-cell. Other parents may know that a dominant but late-appearing disease, such as Huntington's disease, runs in their family. Such people may benefit from genetic tests for dominant alleles. Information from genetic testing can inform decisions about family planning.

Several technologies exist for detecting genetic disorders during pregnancy. Genetic testing before birth usually requires the collection of fetal cells. In amniocentesis, a physician uses a needle to extract about 2 teaspoons of the fluid that bathes the developing fetus (Figure 9.17). In chorionic villus sampling, a physician inserts a narrow, flexible tube through the mother's vagina and into her cervix, removing some placental tissue. After cells are obtained, they can be tested for genetic diseases.

Because amniocentesis and chorionic villus sampling have risks, these techniques are usually reserved for situations in which the possibility of a genetic disease is high—for example, if the mother is older than 35 or has had a child with a genetic condition in the past. Alternatively, blood tests on the mother at 15 to 20 weeks of pregnancy can help identify fetuses at risk for certain birth defects. The most widely used blood test measures levels of a protein called alpha-fetoprotein (AFP) in the mother's blood; abnormally high levels may indicate developmental defects in the fetus, while abnormally low levels may indicate Down syndrome. For a more complete risk profile, a woman's doctor may order a "quad screen test," which measures AFP as well as other hormones produced by the placenta. Abnormal levels of these substances in the maternal blood may also point to a risk of Down syndrome. Newer genetic screening procedures involve isolating tiny amounts of fetal cells or DNA released into the mother's bloodstream. These newer technologies, which are noninvasive, more accurate, and can be performed earlier and more safely than other tests, are gradually replacing more invasive screening methods.

As genetic testing becomes more routine, geneticists are working to make sure that the tests do not cause more problems than they solve. Current practice is to only screen for genetic diseases for which early intervention will improve outcomes. Geneticists stress that patients seeking genetic testing should receive counseling both before and after to explain the test and to help them cope with the results. Identifying a genetic disease early can give families choices, allow for earlier treatments, and provide time to prepare—emotionally, medically, and financially. Advances in biotechnology offer possibilities for reducing human suffering, but not before key ethical issues are resolved. The dilemmas posed by human genetics reinforce one of this book's themes: the immense social implications of biology. ☑

☑ CHECKPOINT

Peter is a 28-year-old man whose father died of Huntington's disease. Peter's mother shows no signs of the disease. What is the probability that Peter has inherited Huntington's disease?

■ Answers: $\frac{1}{2}$

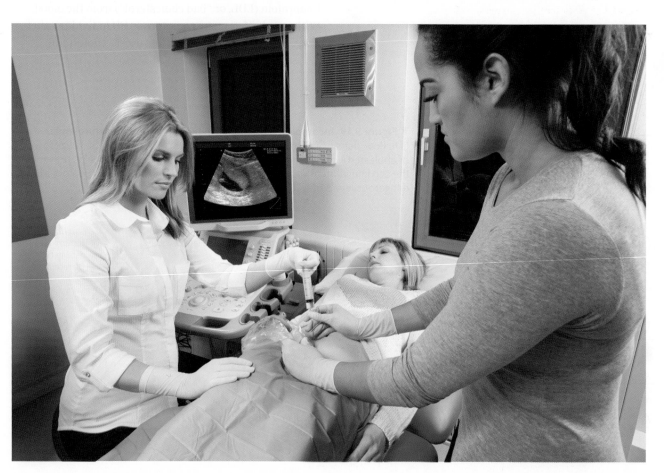

◀ **Figure 9.17**
Amniocentesis. A doctor uses a sonogram to help guide the removal of fetal cells for genetic testing.

Variations on Mendel's Laws

Mendel's two laws explain inheritance in terms of genes that are passed along from generation to generation according to simple rules of probability. These laws are valid for all sexually reproducing organisms, including garden peas, Labrador retrievers, and people. But just as the basic rules of musical harmony cannot account for all the rich sounds of a symphony, Mendel's laws stop short of explaining some patterns of genetic inheritance. In fact, for most sexually reproducing organisms, cases in which Mendel's rules can strictly account for the patterns of inheritance are relatively rare. More often, the observed inheritance patterns are more complex. Next, we'll look at several extensions to Mendel's laws that help account for this complexity.

Incomplete Dominance in Plants and People

The F₁ offspring of Mendel's pea crosses always looked like one of the two parent plants. In such situations, the dominant allele has the same effect on the phenotype whether present in one or two copies. But for some characters, the appearance of F₁ hybrids falls between the phenotypes of the two parents, an effect called **incomplete dominance**. For instance, when red snapdragons are crossed with white snapdragons, all the F₁ hybrids have pink flowers (**Figure 9.18**). And in the F₂ generation, the genotypic ratio and the phenotypic ratio are the same: 1:2:1.

We also see examples of incomplete dominance in people. One case involves a recessive allele (*h*) that causes hypercholesterolemia, dangerously high levels of cholesterol in the blood. Normal individuals are homozygous dominant, *HH*. Heterozygotes (*Hh*) have blood cholesterol levels about twice what is normal. Such heterozygotes are very prone to cholesterol buildup in artery walls and may have heart attacks from blocked heart arteries by their mid-30s. Hypercholesterolemia is even more serious in homozygous individuals (*hh*). Homozygotes have about five times the normal amount of blood cholesterol and may have heart attacks as early as age 2. If we look at the molecular basis for hypercholesterolemia, we can understand the intermediate phenotype of heterozygotes (**Figure 9.19**). The *H* allele specifies a cell-surface receptor protein that liver cells use to mop up excess low-density lipoprotein (LDL, or "bad cholesterol") from the blood. With only half as many receptors as *HH* individuals, heterozygotes can remove much less excess cholesterol.

▼ **Figure 9.18 Incomplete dominance in snapdragons.**
Compare this diagram with Figure 9.6, where one of the alleles displays complete dominance.

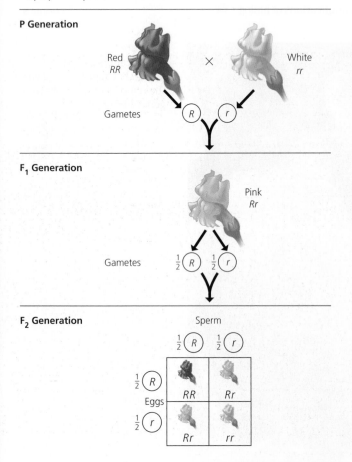

P Generation

Red *RR* × White *rr*

Gametes *R* *r*

F₁ Generation

Pink *Rr*

Gametes ½ *R* ½ *r*

F₂ Generation

Sperm

½ *R* ½ *r*

Eggs ½ *R* *RR* *Rr*

½ *r* *Rr* *rr*

▼ **Figure 9.19 Incomplete dominance in human hypercholesterolemia.**
LDL (low-density lipoprotein) receptors on liver cells promote the breakdown of cholesterol carried in the bloodstream by LDL. This process helps prevent the accumulation of cholesterol in the arteries. Having too few receptors allows dangerous levels of LDL to build up in the blood.

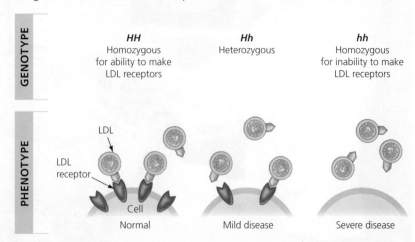

GENOTYPE	*HH* Homozygous for ability to make LDL receptors	*Hh* Heterozygous	*hh* Homozygous for inability to make LDL receptors

PHENOTYPE

LDL

LDL receptor

Cell

Normal | Mild disease | Severe disease

ABO Blood Groups: An Example of Multiple Alleles and Codominance

So far, we have discussed inheritance patterns involving only two alleles per gene (*H* versus *h*, for example). But most genes can be found in populations in more than two forms, known as multiple alleles. Although each individual carries, at most, two different alleles for a particular gene, in cases of multiple alleles, more than two possible alleles exist in the population.

The **ABO blood groups** in humans involve three alleles of a single gene. Various combinations of these three alleles produce four phenotypes: A person's blood type may be A, B, AB, or O. These letters refer to two carbohydrates, designated A and B, that may be found on the surface of red blood cells (**Figure 9.20**). A person's red blood cells may be coated with carbohydrate A (giving them type A blood), carbohydrate B (type B), both (type AB), or neither (type O). (In case you are wondering, the "positive" and "negative" notations associated with blood types—referred to as the Rh blood group system—are due to inheritance of a separate, unrelated gene.)

Matching compatible blood groups is critical for safe blood transfusions. If a donor's blood cells have a carbohydrate (A or B) that is foreign to the recipient, then the recipient's immune system produces blood proteins called antibodies that bind to the foreign carbohydrates and cause the donor blood cells to clump together, potentially killing the recipient.

The four blood groups result from various combinations of the three different alleles: I^A (for the ability to make substance A), I^B (for B), and i (for neither A nor B). Each person inherits one of these alleles from each parent. Because there are three alleles, there are six possible genotypes, as listed in Figure 9.21. Both the I^A and I^B alleles are dominant to the i allele. Thus, I^AI^A and I^Ai people have type A blood, and I^BI^B and I^Bi people have type B. Recessive homozygotes (*ii*) have type O blood with neither carbohydrate. Finally, people of genotype I^AI^B make *both* carbohydrates. In other words, the I^A and I^B alleles are **codominant**, meaning that both alleles are expressed in heterozygous individuals (I^AI^B) who have type AB blood. Notice that type O blood reacts with no others, making such a person a universal donor. A person with type AB blood is a universal receiver. Be careful to distinguish codominance (the expression of both alleles) from incomplete dominance (the expression of one intermediate trait). ☑

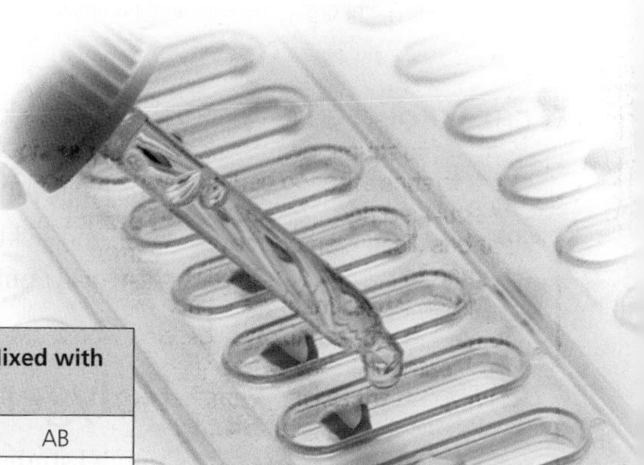

▼ **Figure 9.20 Multiple alleles for the ABO blood groups.** The three versions of the gene responsible for blood type may produce carbohydrate A (allele I^A), carbohydrate B (allele I^B), or neither carbohydrate (allele i). Because each person carries two alleles, six genotypes are possible that result in four different phenotypes. The clumping reaction that occurs between antibodies and foreign blood cells is the basis of blood-typing (shown in the photograph at right) and of the adverse reaction that occurs when someone receives a transfusion of incompatible blood.

Blood Group (Phenotype)	Genotypes	Red Blood Cells	Antibodies Present in Blood	Reactions When Blood from Groups Below Is Mixed with Antibodies from Groups at Left			
				O	A	B	AB
A	I^AI^A or I^Ai	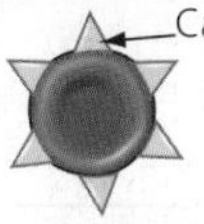 Carbohydrate A	Anti-B	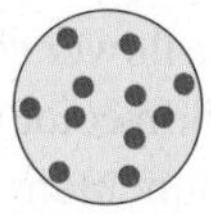	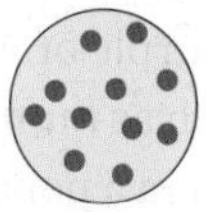	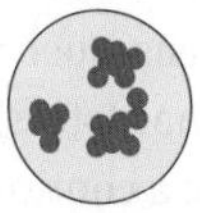	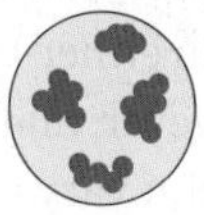
B	I^BI^B or I^Bi	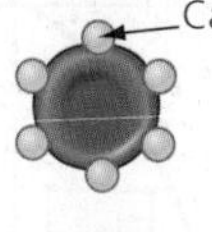 Carbohydrate B	Anti-A	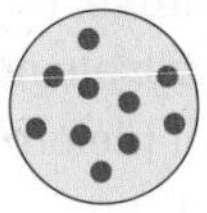	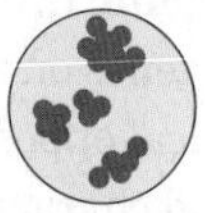	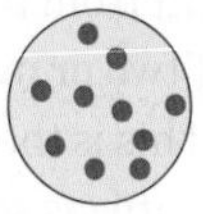	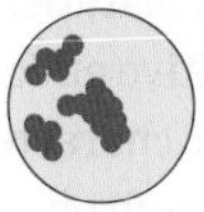
AB	I^AI^B	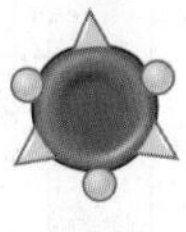	—	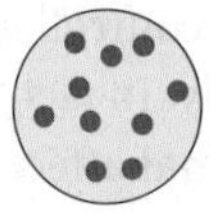	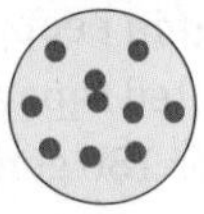	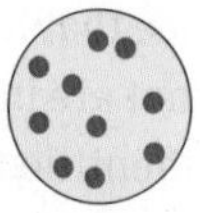	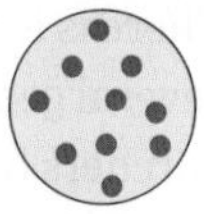
O	*ii*	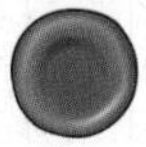	Anti-A Anti-B	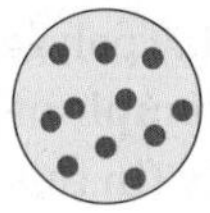	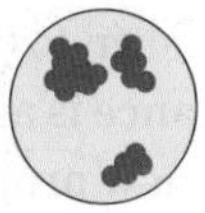	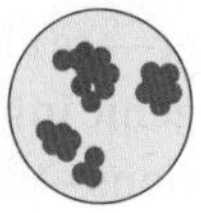	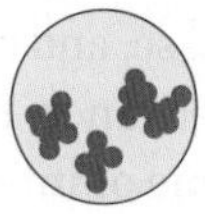

Pleiotropy and Sickle-Cell Disease

Our genetic examples to this point have been cases in which each gene specifies only one hereditary character. But in many cases, one gene influences several characters, a property called **pleiotropy**.

An example of pleiotropy in humans is sickle-cell disease (sometimes called sickle-cell anemia), a disorder characterized by a diverse set of symptoms.

The direct effect of the sickle-cell allele is to make red blood cells produce abnormal hemoglobin proteins (see Figure 3.20). These abnormal molecules tend to link together and crystallize, especially when the oxygen content of the blood is lower than usual because of high altitude, overexertion, or respiratory ailments. As the hemoglobin crystallizes, the normally disk-shaped red blood cells deform to a sickle shape with jagged edges **(Figure 9.21)**. As is so often the case in biology, **altering structure affects function.** Because of their shape, sickled cells do not flow smoothly in the blood and tend to accumulate and clog tiny blood vessels. This clogging of blood vessels causes a cascade of symptoms: Blood flow to body parts is reduced, resulting in periodic fever, severe pain, and damage to the heart, brain, and kidneys. The abnormal sickled cells are destroyed by the body, causing anemia and general weakness. Blood transfusions and drugs may relieve some of the symptoms, but there is no cure, and sickle-cell disease kills about 100,000 people in the world annually. ☑

▼ **Figure 9.21 Sickle-cell disease: multiple effects of a single human gene.**

Individual homozygous for sickle-cell allele

Sickle-cell (abnormal) hemoglobin

Abnormal hemoglobin crystallizes into long, flexible chains, causing red blood cells to become sickle-shaped.

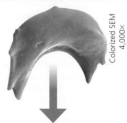

Colorized SEM 4,000×

Sickled cells can lead to a cascade of symptoms, such as weakness, pain, organ damage, and paralysis.

☑ **CHECKPOINT**

How does sickle-cell disease demonstrate the concept of pleiotropy?

■ *Answer: Homozygotes for the sickle-cell allele have abnormal hemoglobin, and its effect on the shape of red blood cells leads to a cascade of traits affecting many organs of the body.*

Polygenic Inheritance

Mendel studied genetic characters that could be classified on an either-or basis, such as purple or white flower color. However, many characters, such as human skin color and height, vary along a continuum in a population. Many such features result from **polygenic inheritance**, the additive effects of two or more genes on a single phenotypic character. (This is the logical opposite of pleiotropy, in which one gene affects several characters.)

There is evidence that height in people is controlled by several genes that are inherited separately. In fact, researchers have identified over 180 genes that affect human height. Many diseases—including diabetes, heart disease, and cancer—are also known to display polygenic inheritance. **Polygenic inheritance is an example of the theme of interactions—that a new property (in this case, a polygenic trait) can result from the interactions of many smaller parts (genes).**

To illustrate this principle, let's consider three hypothetical genes, with a tall allele for each (*A*, *B*, and *C*) contributing one "unit" of tallness to the phenotype and being incompletely dominant to the other alleles (*a*, *b*, and *c*). A person who is *AABBCC* would be very tall, while an *aabbcc* individual would be very short. An *AaBbCc* person would be of intermediate height. Because the alleles have an additive effect, the genotype *AaBbCc* would produce the same height as any other genotype with just three tallness alleles, such as *AABbcc*. The Punnett square in **Figure 9.22** shows all possible genotypes from a mating of two triple

▼ **Figure 9.22 A model for polygenic inheritance of height.**

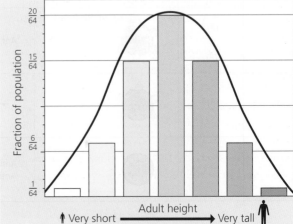

heterozygotes (*AaBbCd*). The row of figures below the Punnett square shows the seven height phenotypes that would theoretically result. This hypothetical example shows how inheritance of three genes could lead to seven different versions of a trait at the frequencies indicated by the bars in the graph at the bottom of the figure.

Epigenetics and the Role of Environment

If we examine a real human population for height, we would see more height phenotypes than just seven. The true range might be similar to the entire distribution of height under the bell-shaped curve in Figure 9.22. In fact, no matter how carefully we characterize the genes for height, a purely genetic description will always be incomplete. This is because height is also influenced by environmental factors, such as nutrition and exercise.

SOME FEATURES ARE DETERMINED BY GENES, SOME BY ENVIRONMENT, AND SOME BY BOTH.

Many phenotypic characters result from a combination of heredity and environment. For example, the leaves of a tree all have the same genotype, but they vary in size, shape, and color, depending on exposure to wind and sun and the tree's nutritional state. For people, exercise alters build; experience improves performance on intelligence tests; and social and cultural factors can greatly affect appearance. As geneticists learn more and more about our genes, it's becoming clear that many human characters— such as risk of heart disease, cancer, alcoholism, and schizophrenia—are influenced by both genes *and* environment.

Whether human characters are more influenced by genes or by the environment—nature or nurture—is a very hotly contested issue. For some characters, such as the ABO blood group, a given genotype mandates an exact phenotype, and the environment plays no role whatsoever. In contrast, the number of blood cells in a drop of your blood varies quite a bit, depending on such factors as the altitude, your physical activity, and whether or not you have a cold.

Spending time with identical twins will convince anyone that environment, which can include effects of personal choices, can influence a person's traits

(Figure 9.23). In general, only genetic influences are inherited, and any effects of the environment are not usually passed to the next generation. In recent years, however, biologists have begun to recognize the importance of **epigenetic inheritance**, the transmission of traits by mechanisms not directly involving DNA sequence. For example, DNA and/or protein components of chromosomes can be chemically modified by adding or removing chemical groups. Over a lifetime, the environment plays a role in these changes, which may explain how one identical twin can suffer from a genetically based disease whereas the other twin does not, despite their identical genomes. Recent research has shown that young identical twins are essentially indistinguishable in terms of their epigenetic markers, but epigenetic differences accumulate as they age, resulting in substantial differences in gene expression between older twins. Epigenetic modifications—and the changes in gene activity that result—may even be carried on to the next generation. For example, some research suggests that epigenetic changes may underlie certain instincts in animals, allowing behaviors learned by one generation (such as avoiding certain stimuli) to be passed on to the next generation through chromosomal modifications. Unlike alterations to the DNA sequence, chemical changes to the chromosomes can be reversed (by processes that are not yet fully understood). Research into the importance of epigenetic inheritance is a very active area of biology, and new discoveries that alter our understanding of genetics are sure to follow. ☑

✓ **CHECKPOINT**

If you produced cloned mice, how would you predict the number of epigenetic differences among the clones to change as they age?

■ *Answer: You would expect the number of epigenetic differences to be low at first and then gradually increase as the clones age.*

◄ **Figure 9.23** As a result of environmental influences, including choices such as hair style and color, even identical twins can look different.

The Chromosomal Basis of Inheritance

▶ **Figure 9.24** The chromosomal basis of Mendel's laws.

It was not until many years after Mendel's death that biologists understood the significance of his work. Cell biologists worked out the processes of mitosis and meiosis in the late 1800s (see Chapter 8). Then, around 1900, researchers began to notice parallels between the behavior of chromosomes and the behavior of genes. One of biology's most important concepts began to emerge.

The **chromosome theory of inheritance** states that genes are located at specific positions (loci) on chromosomes and that the behavior of chromosomes during meiosis and fertilization accounts for inheritance patterns. Indeed, it is chromosomes that undergo segregation and independent assortment during meiosis and thus account for Mendel's

laws. **Figure 9.24** correlates the results of the dihybrid cross in Figure 9.8b with the movement of chromosomes through meiosis. Starting with two purebred parental plants, the diagram follows two genes on different chromosomes—one for seed shape (alleles R and r) and one for seed color (alleles Y and y)—through the F_1 and F_2 generations. ✔

Linked Genes

Realizing that genes are on chromosomes when they segregate leads to some important conclusions about the ways genes can be inherited. The number of genes in a cell is far greater than the number of chromosomes; each

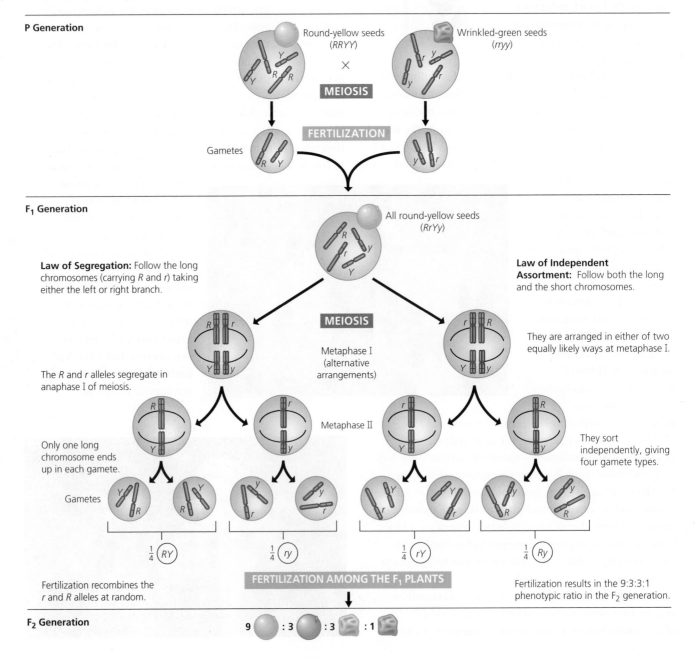

P Generation

Round-yellow seeds ($RRYY$) × Wrinkled-green seeds ($rryy$)

MEIOSIS

FERTILIZATION

Gametes

F₁ Generation

All round-yellow seeds ($RrYy$)

Law of Segregation: Follow the long chromosomes (carrying R and r) taking either the left or right branch.

The R and r alleles segregate in anaphase I of meiosis.

Law of Independent Assortment: Follow both the long and the short chromosomes.

They are arranged in either of two equally likely ways at metaphase I.

MEIOSIS

Metaphase I (alternative arrangements)

Metaphase II

Only one long chromosome ends up in each gamete.

They sort independently, giving four gamete types.

Gametes

$\frac{1}{4}$ (RY) $\frac{1}{4}$ (ry) $\frac{1}{4}$ (rY) $\frac{1}{4}$ (Ry)

Fertilization recombines the r and R alleles at random.

FERTILIZATION AMONG THE F₁ PLANTS

Fertilization results in the 9:3:3:1 phenotypic ratio in the F_2 generation.

F₂ Generation

9 : 3 : 3 : 1

chromosome therefore carries hundreds or thousands of genes. Genes located near each other on the same chromosome, called **linked genes**, tend to travel together during meiosis and fertilization. Such genes are often inherited as a set and therefore often do not follow Mendel's law of independent assortment. For closely linked genes, the inheritance of one *does* in fact correlate with the inheritance of the other, producing results that do not follow the standard ratios predicted by Punnett squares. In contrast, genes located far apart on the same chromosome usually do sort independently because of crossing over (see Figure 8.18).

The patterns of genetic inheritance we've discussed so far have always involved genes located on autosomes, not on the sex chromosomes. We're now ready to look at the role of sex chromosomes and the inheritance patterns exhibited by the characters they control. As you'll see, genes located on sex chromosomes produce some unusual patterns of inheritance. ☑

Sex Determination in Humans

Many animals, including all mammals, have a pair of sex chromosomes—designated X and Y— that determine an individual's sex **(Figure 9.25)**. Individuals with one X chromosome and one Y chromosome are males; XX individuals are females. Human males and females both have 44 autosomes (chromosomes other than sex chromosomes). As a result of chromosome segregation during meiosis, each gamete contains one

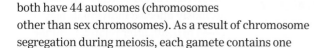

YOUR SEX WAS ENTIRELY DETERMINED BY YOUR FATHER.

sex chromosome and a haploid set of autosomes (22 in humans). All eggs contain a single X chromosome. Of the sperm cells, half contain an X chromosome and half contain a Y chromosome. An offspring's sex depends on whether the sperm cell that fertilizes the egg bears an X or a Y.

Sex-Linked Genes

Besides bearing genes that determine sex, the sex chromosomes also contain genes for characters unrelated to maleness or femaleness. A gene located on a sex chromosome is called a **sex-linked gene**. The human X chromosome contains approximately 1,100 genes, whereas the Y chromosome contains only 78, about half of which are expressed only in the testes. Therefore, most sex-linked genes are found on the X chromosome.

A number of human conditions, including red-green colorblindness, hemophilia, and a type of muscular dystrophy, result from sex-linked recessive alleles. Red-green colorblindness is a common sex-linked disorder that is caused by a malfunction of light-sensitive cells in the eyes. (Colorblindness is actually a large class of disorders involving several sex-linked genes, but we will focus on just one specific type of colorblindness here.) A person with normal color vision can see more than 150 colors. In contrast, someone with red-green colorblindness can see fewer than 25. **Figure 9.26** shows a simple test for red-green colorblindness. Mostly males are affected, but heterozygous females have some defects, too.

THE CHROMOSOMAL BASIS OF INHERITANCE

☑ **CHECKPOINT**

What are linked genes? Why do they often disobey Mendel's law of independent assortment?

■ *Answer: Genes located near each other on the same chromosome. Because they often travel together during meiosis and fertilization, linked genes tend to be inherited together rather than independently.*

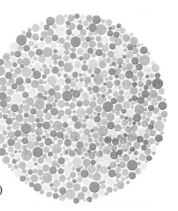

▲ **Figure 9.26 A test for red-green colorblind-ness.** Can you see a green numeral 7 against the reddish background? If not, you probably have some form of red-green colorblindness, a sex-linked trait.

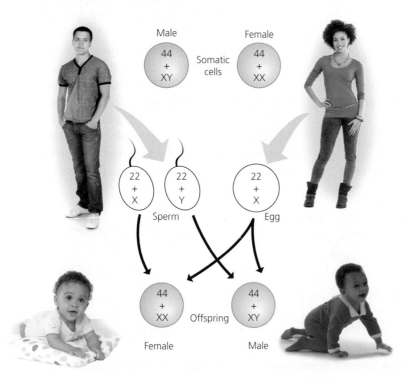

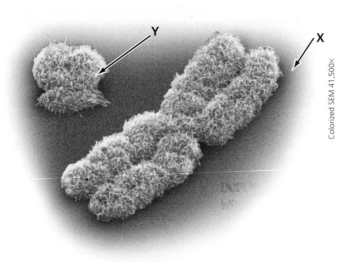

▲ **Figure 9.25 The chromosomal basis of sex determination in humans.** The micrograph above shows human X and Y chromosomes in a duplicated state.

Because they are located on the sex chromosomes, sex-linked genes exhibit unusual inheritance patterns. **Figure 9.27a** illustrates what happens when a colorblind male has offspring with a homozygous female with normal color vision. All the children have normal color vision, suggesting that the allele for wild-type (normal) color vision is dominant. If a female carrier mates with a male who has normal color vision, the classic 3:1 phenotypic ratio of normal color vision to colorblindness appears among the children (**Figure 9.27b**). However, there is a surprising twist: The colorblind trait shows up only in males. All the females have normal vision, while half the males are colorblind and half are normal. This is because the gene involved in this inheritance pattern is located exclusively on the X chromosome; there is no corresponding locus on the Y. Thus, females (XX) carry two copies of the gene for this character, and males (XY) carry only one. Because the colorblindness allele is recessive, a female will be colorblind only if she receives that allele on both X chromosomes (**Figure 9.27c**). For a male, however, a single copy of the recessive allele confers colorblindness. For this reason, recessive sex-linked traits are expressed much more frequently in men than in women. For example, colorblindness is about 20-fold more common among males than among females. This inheritance pattern is why people often say that certain genes "skip a generation"—because they are passed from a male (in generation 1) to a female carrier (who does not express it in generation 2) back to a male (in generation 3).

Hemophilia is a sex-linked recessive trait with a long, well-documented history. Hemophiliacs bleed excessively

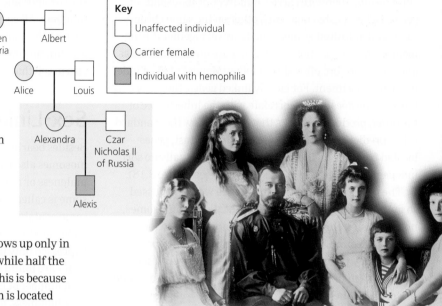

▲ **Figure 9.28** **Hemophilia in the royal family of Russia.** The photograph shows Queen Victoria's granddaughter Alexandra; Alexandra's husband, Nicholas, who was the last czar of Russia; their son Alexis; and their daughters. Notice that Alexis inherited the allele for hemophilia that had been passed down for three generations.

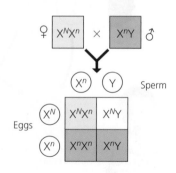

YOUR CHANCES OF INHERITING AND PASSING ALONG A GENETIC DISEASE CAN OFTEN BE DETERMINED FROM YOUR FAMILY HISTORY.

when injured because they have inherited an abnormal allele for a factor involved in blood clotting. The most seriously affected individuals may bleed to death after relatively minor bruises or cuts. A high incidence of hemophilia plagued the royal families of Europe. Queen Victoria (1819–1901) of England was a carrier of the hemophilia allele. She passed it on to one of her sons and two of her daughters. Through marriage, her daughters then introduced the disease into the royal families of Prussia, Russia, and Spain. In this way, the former practice of strengthening international alliances by marriage effectively spread hemophilia through the royal families of several nations (**Figure 9.28**). ☑

☑ CHECKPOINT

1. What is meant by a sex-linked gene?

2. White eye color is a recessive sex-linked trait in fruit flies. If a white-eyed *Drosophila* female is mated with a red-eyed (wild-type) male, what do you predict for the numerous offspring?

■ *Answers:* **1.** *a gene that is located on a sex chromosome, usually the X chromosome* **2.** *All female offspring will be heterozygous ($X^R X^r$), with red eyes; all male offspring will be white-eyed ($X^r Y$).*

▼ **Figure 9.27** **Inheritance of colorblindness, a sex-linked recessive trait.** We use an uppercase N for the dominant, normal color-vision allele and n for the recessive, colorblind allele. To indicate that these alleles are on the X chromosome, we show them as superscripts to the letter X. The Y chromosome does not have a gene locus for vision; therefore, the male's phenotype results entirely from the sex-linked gene on his single X chromosome.

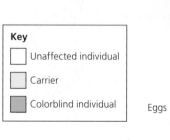

Key

☐ Unaffected individual

☐ Carrier

▨ Colorblind individual

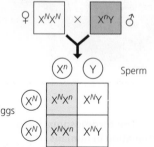

(a) Normal female × colorblind male

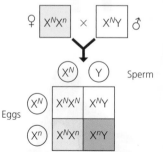

(b) Carrier female × normal male

(c) Carrier female × colorblind male

Barking Up the Evolutionary Tree

As we've seen throughout this chapter, dogs are more than man's best friend: They are also one of our longest-running genetic experiments. About 15,000 years ago, in East Asia, people began to cohabit with ancestral canines that were predecessors of both modern wolves and dogs. As people moved into permanent, geographically isolated settlements, populations of canines were separated from one another and eventually became inbred.

Different groups of people chose dogs with different traits. A 2010 study indicated that small dogs were first bred within early agricultural settlements of the Middle East around 12,000 years ago. Elsewhere, herders selected dogs for controlling flocks, while hunters chose dogs for retrieving prey. Continued over millennia, such genetic tinkering has resulted in a diverse array of dog body types and behaviors. In each breed, a distinct genetic makeup results in a distinct set of physical and behavioral traits.

As discussed in the Process of Science section, our understanding of canine evolution took a big leap forward when researchers sequenced the complete genome of a dog. Using the genome sequence and other data, canine geneticists produced an evolutionary tree based on a genetic analysis of 85 breeds (Figure 9.29). The analysis shows that the canine family tree includes a series of well-defined branch points. Each fork represents a purposeful selection by people that produced a genetically distinct subpopulation with specific desired traits.

The genetic tree shows that the most ancient breeds, those most closely related to the wolf, are Asian species such as the shar-pei and Akita. Subsequent genetic splits created distinct breeds in Africa (basenji), the Arctic (Alaskan malamute and Siberian husky), and the Middle East (Afghan hound and saluki). The remaining breeds, primarily of European ancestry, were developed most recently and can be grouped by genetic makeup into those bred for guarding (for example, the rottweiler), herding (such as sheepdogs), and hunting (including the Labrador retriever and beagle). The formulation of an evolutionary tree for the domestic dog shows that new technologies can provide important insights into genetic and evolutionary questions about life on Earth.

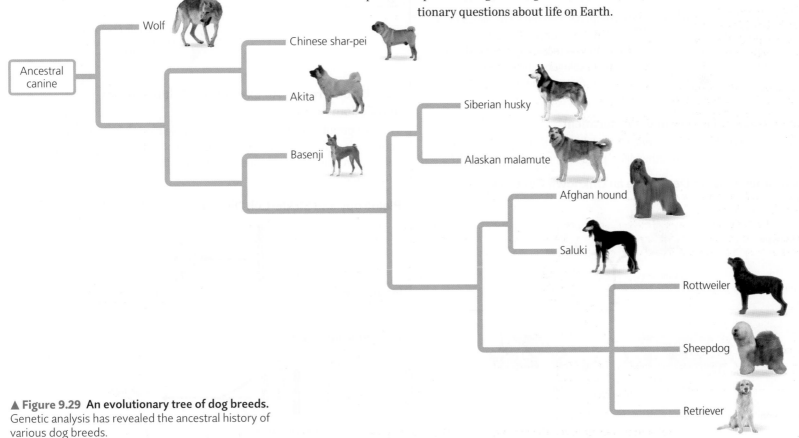

▲ **Figure 9.29 An evolutionary tree of dog breeds.**
Genetic analysis has revealed the ancestral history of various dog breeds.

Chapter Review

SUMMARY OF KEY CONCEPTS

Genetics and Heredity

Gregor Mendel, the first to study the science of heredity by analyzing patterns of inheritance, emphasized that genes retain permanent identities.

In an Abbey Garden

Mendel started with purebred varieties of pea plants representing two alternative variants of a hereditary character. He then crossed the different varieties and traced the inheritance of traits from generation to generation.

Mendel's Law of Segregation

Pairs of alleles separate during gamete formation; fertilization restores the pairs.

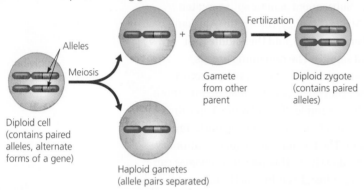

If an individual's genotype (genetic makeup) has two different alleles for a gene and only one influences the organism's phenotype (appearance), that allele is said to be dominant and the other allele recessive. Alleles of a gene reside at the same locus, or position, on homologous chromosomes. When the allele pair matches, the organism is homozygous; when the alleles are different, the organism is heterozygous.

Mendel's Law of Independent Assortment

By following two characters at once, Mendel found that the alleles of a pair segregate independently of other allele pairs during gamete formation.

Using a Testcross to Determine an Unknown Genotype

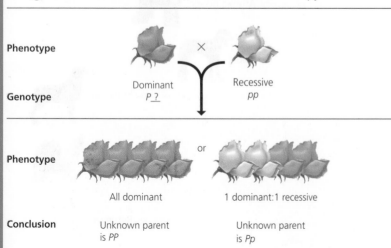

The Rules of Probability

Inheritance follows the rules of probability. The chance of inheriting a recessive allele from a heterozygous parent is $\frac{1}{2}$. The chance of inheriting it from both of two heterozygous parents is $\frac{1}{2} \times \frac{1}{2} = \frac{1}{4}$, illustrating the rule of multiplication for calculating the probability of two independent events.

Family Pedigrees

Geneticists can use family pedigrees to determine patterns of inheritance and individual genotypes.

Human Traits Controlled by a Single Gene

For traits that vary within a population, the one most commonly found in nature is called the wild type. Many inherited disorders in humans are controlled by a single gene with two alleles. Most of these disorders, such as cystic fibrosis, are caused by autosomal recessive alleles. A few, such as Huntington's disease, are caused by dominant alleles.

Variations on Mendel's Laws

Incomplete Dominance in Plants and People

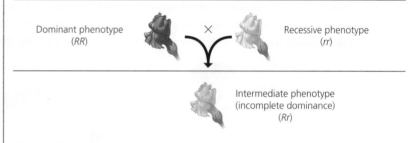

ABO Blood Groups: An Example of Multiple Alleles and Codominance

Within a population, there are often multiple kinds of alleles for a character, such as the three alleles for the ABO blood groups. The alleles determining blood groups are codominant; that is, both are expressed in a heterozygote.

Pleiotropy and Sickle-Cell Disease

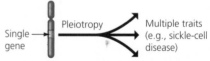

In pleiotropy, one gene (such as the sickle-cell disease gene) can affect many characters (such as the multiple symptoms of the disease).

Polygenic Inheritance

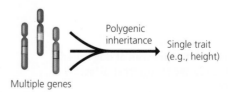

Epigenetics and the Role of Environment

Many phenotypic characters result from a combination of genetic and environmental effects, but only genetic influences are biologically heritable. Epigenetic inheritance, the transmission of traits from one generation to the next by means of chemical modifications to DNA and proteins, may explain how environmental factors can affect genetic traits.

The Chromosomal Basis of Inheritance

Genes are located on chromosomes. The behavior of chromosomes during meiosis and fertilization accounts for inheritance patterns.

Linked Genes

Certain genes are linked: They tend to be inherited as a set because they lie close together on the same chromosome.

Sex Determination in Humans

In humans, sex is determined by whether a Y chromosome is present. A person who inherits two X chromosomes develops as a female. A person who inherits one X and one Y chromosome develops as a male.

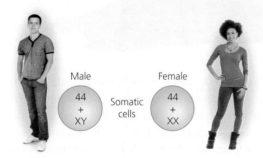

Male

44
+
XY

Somatic cells

Female

44
+
XX

Sex-Linked Genes

The inheritance of genes on the X chromosome reflects the fact that females have two homologous X chromosomes, but males have only one. Most sex-linked human disorders, such as red-green colorblindness and hemophilia, are due to recessive alleles and are seen mostly in males. A male receiving a single sex-linked recessive allele from his mother will have the disorder; a female has to receive the allele from both parents to be affected.

Sex-Linked Traits			
Female: Two alleles	Genotype	$X^N X^N$	$X^N X^n$	$X^n X^n$
	Phenotype	Normal female	Carrier female	Affected female (rare)

Male: One allele	Genotype	$X^N Y$	$X^n Y$
	Phenotype	Normal male	Affected male

Mastering Biology

For practice quizzes, BioFlix animations, MP3 tutorials, video tutors, and more study tools designed for this textbook, go to MasteringBiology™

SELF-QUIZ

1. The genetic makeup of an organism is called its _____, and the physical traits of an organism are called its _____.

2. Which of Mendel's laws is represented by each statement?
   a. Alleles of each pair of homologous chromosomes separate independently during gamete formation.
   b. Alleles segregate during gamete formation; fertilization creates pairs of alleles once again.

3. Edward was found to be heterozygous (Ss) for the sickle-cell trait. The alleles represented by the letters S and s are
   a. on the X and Y chromosomes.
   b. linked.
   c. on homologous chromosomes.
   d. both present in each of Edward's sperm cells.

4. Whether an allele is dominant or recessive depends on
   a. how common the allele is, relative to other alleles.
   b. whether it is inherited from the mother or the father.
   c. whether it or another allele determines the phenotype when both are present.
   d. whether or not it is linked to other genes.

5. Two fruit flies with eyes of the usual red color are crossed, and their offspring are as follows: 77 red-eyed males, 71 ruby-eyed males, 152 red-eyed females. The gene that controls whether eyes are red or ruby is _____, and the allele for ruby eyes is _____.
   a. autosomal (carried on an autosome); dominant
   b. autosomal; recessive
   c. sex-linked; dominant
   d. sex-linked; recessive

6. All the offspring of a white hen and a black rooster are gray. The simplest explanation for this pattern of inheritance is
   a. pleiotropy.   c. codominance.
   b. sex linkage.   d. incomplete dominance.

7. A man who has type B blood and a woman who has type A blood could have children of which of the following phenotypes? (*Hint*: Review Figure 9.20.)
   a. A, B, or O   c. AB or O
   b. AB only   d. A, B, AB, or O

8. Duchenne muscular dystrophy is a sex-linked recessive disorder characterized by a progressive loss of muscle tissue. Neither Rudy nor Carla has Duchenne muscular dystrophy, but their first son does have it. If the couple has a second child, what is the probability that he or she will also have the disease?

9. Adult height in people is at least partially hereditary; tall parents tend to have tall children. But people come in a range of sizes, not just tall or short. What extension of Mendel's model could produce this variation in height?

10. A purebred brown mouse is repeatedly mated with a purebred white mouse, and all their offspring are brown. If two of these brown offspring are mated, what fraction of the F_2 mice will be brown?

11. How could you determine the genotype of one of the brown F$_2$ mice in problem 10? How would you know whether a brown mouse is homozygous? Heterozygous?

12. Tim and Jan both have freckles (a dominant trait), but their son Michael does not. Show with a Punnett square how this is possible. If Tim and Jan have two more children, what is the probability that *both* of them will have freckles?

13. Incomplete dominance is seen in the inheritance of hypercholesterolemia. Mack and Toni are both heterozygous for this character, and both have elevated levels of cholesterol. Their daughter Katerina has a cholesterol level six times normal; she is apparently homozygous, *hh*. What fraction of Mack and Toni's children are likely to have elevated but not extreme levels of cholesterol, like their parents? If Mack and Toni have one more child, what is the probability that the child will suffer from the more serious form of hypercholesterolemia seen in Katerina?

14. Why was Henry VIII wrong to blame his wives for always producing daughters?

15. Both parents of a boy are phenotypically normal, but their son suffers from hemophilia, a sex-linked recessive disorder. Draw a pedigree that shows the genotypes of the three individuals. What fraction of the couple's children are likely to suffer from hemophilia? What fraction are likely to be carriers?

16. Heather was surprised to discover that she suffered from red-green colorblindness. She told her biology professor, who said, "Your father is colorblind, too, right?" How did her professor know this? Why did her professor not say the same thing to the colorblind males in the class?

17. In rabbits, black hair depends on a dominant allele, *B*, and brown hair on a recessive allele, *b*. Short hair is due to a dominant allele, *S*, and long hair to a recessive allele, *s*. If a true-breeding black short-haired male is mated with a brown long-haired female, describe their offspring. What will be the genotypes of the offspring? If two of these F$_1$ rabbits are mated, what phenotypes would you expect among their offspring? In what proportions?

For answers to the Self Quiz, see Appendix D.

IDENTIFYING MAJOR THEMES

For each statement, identify which major theme is evident (the relationship of structure to function, information flow, pathways that transform energy and matter, interactions within biological systems, or evolution) and explain how the statement relates to the theme. If necessary, review the themes (see Chapter 1) and review the examples highlighted in blue in this chapter.

18. The sickle-cell mutation shows how changing the shape of a protein (in this case, hemoglobin) affects how that protein works.

19. Your skin color is determined by many different proteins acting together in combination with environmental factors.

20. For each pair of your homologous chromosomes, one member derives from your mother, and the other from your father, ensuring the transmission of genes from one generation to the next.

For answers to Identifying Major Themes, see Appendix D.

THE PROCESS OF SCIENCE

21. In 1981, a stray cat with unusual curled-back ears was adopted by a family in Lakewood, California. Hundreds of descendants of this cat have since been born, and cat fanciers have developed the "curl" cat into a show breed. The curl allele is apparently dominant and carried on an autosome. Suppose you owned the first curl cat and wanted to develop a purebred variety. Describe tests that would determine whether the curl gene is dominant or recessive and whether it is autosomal or sex-linked.

22. **Interpreting Data** As shown in the Punnett square below, albinism is caused by an autosomal recessive allele. Two parents who do not show any signs of albinism but are carriers could therefore have a child who displays albinism because that child could inherit one recessive albinism-causing gene from each parent. Imagine that a male with albinism mates with a female who does not display albinism. We know the male with albinism must have the genotype *aa*, but the female could be either *Aa* or *AA*. Such a mating is essentially a testcross, like the one shown in Figure 9.10. If the parents' first child does not display albinism, can you say with certainty what the mother's genotype must be? What if the couple has four children (none twins), none displaying albinism—can you say with certainty the mother's genotype? What would it take for a definitive genotype to be assigned?

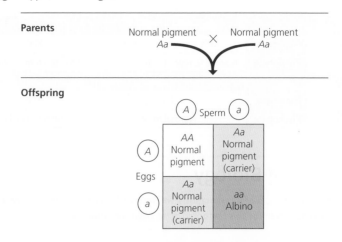

BIOLOGY AND SOCIETY

23. There are now nearly 200 recognized breeds of dog, from the affenpinscher to the Yorkshire terrier. But several of these suffer from medical problems due to the inbreeding required to establish the breed. For example, nearly every Cavalier King Charles spaniel (discussed in the Biology and Society section) suffers from heart murmurs caused by a genetically defective heart valve. Such problems are likely to remain as long as the organizations that oversee dog breeding maintain strict pedigree requirements. Some people are suggesting that every breed be allowed to mix with others to help introduce new gene lines free of the congenital defects. Why do you think the governing societies are resistant to such crossbreed mixing? What would you do if you were in charge of addressing the genetic defects that currently plague some breeds?

24. In the section on prenatal genetic testing, we have briefly assessed the possible impact of this diagnostic technology on the parents of the unborn baby and the need to provide counseling before and after the test. However, genetic testing is also likely to have an impact on the society, for example, by changing the way we perceive and deal with congenital disorders and genetic defects. In your opinion, what are the social risks associated with this technology?

25. Many infertile couples turn to in vitro fertilization to try to have a baby. In this technique, sperm and ova are collected and used to create eight-cell embryos for implantation into a woman's uterus. At the eight-cell stage, one of the fetal cells can be removed without causing harm to the developing fetus. Once removed, the cell can be genetically tested. Some couples may know that a particular genetic disease runs in their family. They might wish to avoid implanting any embryos with the disease-causing genes. Do you think this is an acceptable use of genetic testing? What if a couple wanted to use genetic testing to select embryos for traits unrelated to disease, such as freckles? Do you think that couples undergoing in vitro fertilization should be allowed to perform whatever genetic tests they wish? Or do you think that there should be limits on what tests can be performed? How do you draw the line between genetic tests that are acceptable and those that are not?

10 The Structure and Function of DNA

CHAPTER CONTENTS

MAD COW DISEASE AND OTHER DEADLY AILMENTS ARE CAUSED BY ODDLY SHAPED PROTEINS.

Why Molecular Biology Matters

DNA is the molecule of heredity and the chemical basis for evolution. Its structure and function are therefore among the key topics in the study of life. While learning about DNA, you'll explore some of the most important processes that underlie every inborn trait.

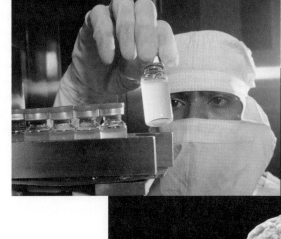

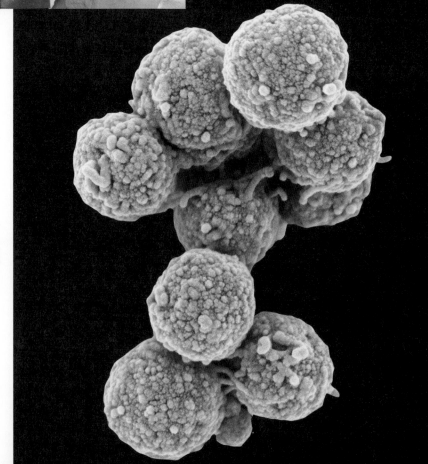

ALL LIFE ON EARTH SHARES A UNIVERSAL GENETIC CODE, SO YOUR DNA COULD BE USED TO GENETICALLY MODIFY BACTERIA AND VICE VERSA.

ONE "TYPO" IN DNA CAN RESULT IN SICKLE-CELL DISEASE.

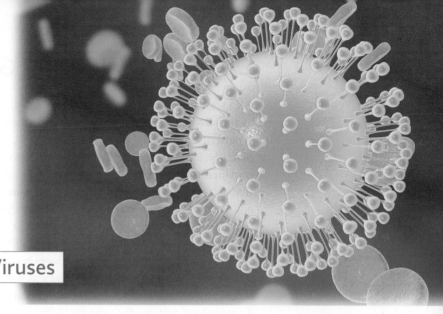

BIOLOGY AND SOCIETY Deadly Viruses

A computer illustration of the Zika virus. Spikes made of protein enable the virus to recognize a host cell.

The Global Threat of Zika Virus

In 2015, an alarming number of babies were born in Brazil with severe damage to their central nervous systems and sensory organs. The affected babies had neurological problems (such as underdeveloped brains and seizures), slow growth, difficulty feeding, and joint and muscle problems. After a frantic search, health officials discovered a link between these abnormalities and exposure to a little-known pathogen: the Zika virus. By 2016, when the United Nations World Health Organization (WHO) issued a worldwide health emergency, Zika virus and Zika-related health problems in newborns began appearing in warm, humid regions of the United States and many other countries.

The Zika virus was first discovered to infect humans in 1952 and had been identified in African monkeys a few years earlier. Zika virus can be transmitted to humans by one species of mosquito. It can also be spread between sexual partners. But Zika virus is not dangerous to most healthy adults. In fact, some people feel just fine after being infected, while others have mild symptoms like aches or a fever. However, Zika virus can be spread from mother to fetus. Unfortunately, developing babies are particularly vulnerable to the virus's effects.

Health agencies have few weapons against Zika virus. There is no vaccine, and medicines can only treat symptoms. Nighttime mosquito netting and staying indoors after dusk can offer protection against many mosquito-borne diseases, but the mosquitoes that carry Zika virus bite both night and day. Public awareness campaigns aimed at avoiding mosquito bites and eliminating mosquito breeding grounds (such as stagnant water) have been implemented in Zika-prone areas. In November of 2016, WHO declared that the Zika global health emergency was over, not because Zika is gone, but because it is expected to be a long-term problem, the "new normal" rather than an emergency.

The Zika virus, like all viruses, consists of a relatively simple structure of nucleic acid (RNA in this case) and protein. Viruses operate by hijacking our own cells and turning them into virus factories. Combating any virus therefore requires a detailed understanding of life at the molecular level. In this chapter, we will explore the structure of life's most important molecule—DNA—to learn how it replicates, mutates, and controls the cell by directing the synthesis of RNA and protein.

DNA: Structure and Replication

DNA was known to be a chemical component of cells by the late 1800s, but Gregor Mendel and other early geneticists had no idea what role (if any) DNA played in heredity. By the early 1950s, a series of discoveries had convinced the scientific world that DNA was the molecule that acts as the hereditary material. What came next was one of the most celebrated quests in the history of science: the effort to figure out the structure of DNA. A good deal was already known about DNA. Scientists had identified all its atoms and knew how they were bonded to one another. What was not understood was the specific three-dimensional arrangement of atoms that gives DNA its unique properties—the capacity to store genetic information, copy it, and pass it from generation to generation. The race was on to discover the link between the structure and function of this important molecule. We will describe that momentous discovery shortly. First, let's review the underlying chemical structure of DNA and its chemical cousin RNA.

DNA and RNA Structure

Both DNA and RNA are nucleic acids, which consist of long chains (polymers) of chemical units (monomers) called **nucleotides**. (For an in-depth refresher, see Figures 3.22–3.26.) A diagram of a nucleotide polymer, or **polynucleotide**, is shown in **Figure 10.1**. Polynucleotides can be very long and may have any sequence of the four different types of nucleotides (abbreviated A, C, T, and G), so a tremendous variety of polynucleotide chains is possible.

Nucleotides are joined together by covalent bonds between the sugar of one nucleotide and the phosphate of the next. This results in a repeating pattern of sugar-phosphate-sugar-phosphate, which is known as a **sugar-phosphate backbone**. The nitrogenous bases are arranged like ribs that project from this backbone. You can think of a polynucleotide as a long ladder split in half lengthwise, with rungs that come in four colors.

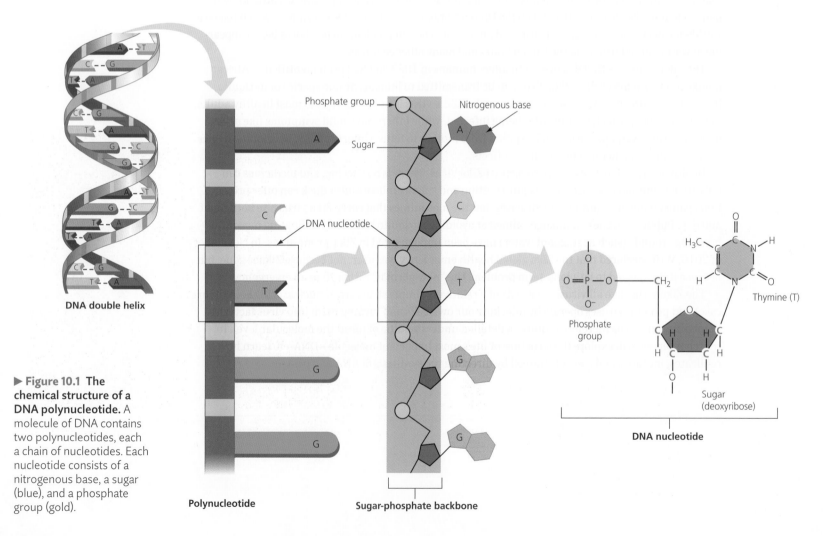

▶ **Figure 10.1 The chemical structure of a DNA polynucleotide.** A molecule of DNA contains two polynucleotides, each a chain of nucleotides. Each nucleotide consists of a nitrogenous base, a sugar (blue), and a phosphate group (gold).

DNA double helix

Polynucleotide

Sugar-phosphate backbone

Phosphate group

Nitrogenous base

Sugar

DNA nucleotide

Phosphate group

Thymine (T)

Sugar (deoxyribose)

DNA nucleotide

The sugars and phosphates make up the side of the ladder, with the sugars acting as the half-rungs.

Moving from left to right across Figure 10.1, we can zoom in to see that each nucleotide consists of three components: a nitrogenous base, a sugar (blue), and a phosphate group (gold). Examining a single nucleotide even more closely, we see the chemical structure of its three components. The phosphate group, with a phosphorus atom (P) at its center, is the source of the *acid* in nucleic acid. Each phosphate has a negative charge on one of its oxygen atoms. The sugar has five carbon atoms (shown in red), four in its ring and one extending above the ring. The ring also includes an oxygen atom. The sugar is called *deoxyribose* because, compared with the sugar ribose, it is missing an oxygen atom. The full name for **DNA** is *deoxyribonucleic acid*, with *nucleic* referring to DNA's location in the nuclei of eukaryotic cells. The nitrogenous base (thymine, in our example) has a ring of nitrogen and carbon atoms with various chemical groups attached. Nitrogenous bases are basic (having a high pH, the opposite of acidic), hence their name.

The four nucleotides found in DNA differ only in their nitrogenous bases (see Figure 3.24 for a review). The bases can be divided into two types. **Thymine (T)** and **cytosine (C)** are single-ring structures. **Adenine (A)** and **guanine (G)** are larger, double-ring structures. Instead of thymine, RNA has a similar base called **uracil (U)**. And RNA contains a slightly different sugar than DNA (ribose instead of deoxyribose, accounting for the names RNA versus DNA). Other than that, RNA and DNA polynucleotides have the same chemical structure. **Figure 10.2** is a computer graphic of a piece of RNA polynucleotide about 20 nucleotides long. ☑

Cytosine
Uracil
Adenine
Guanine
Phosphate
Sugar (ribose)

▶ **Figure 10.2 An RNA polynucleotide.** The yellow used for the phosphorus atoms and the blue of the sugar atoms make it easy to spot the sugar-phosphate backbone.

☑ **CHECKPOINT**

Compare and contrast the chemical components of DNA and RNA.

■ *Answer: Both are polymers of nucleotides (a sugar + a nitrogenous base + a phosphate group). In RNA, the sugar is ribose; in DNA, it is deoxyribose. Both RNA and DNA have the bases A, G, and C, but DNA has T and RNA has U.*

Watson and Crick's Discovery of the Double Helix

The partnership that solved the puzzle of DNA structure began soon after a 23-year-old recently graduated American Ph.D. named James D. Watson journeyed to Cambridge University in England. There, a more senior scientist, Francis Crick, was studying protein structure. While visiting the laboratory of Maurice Wilkins at King's College in London, Watson saw an X-ray image of DNA produced by Wilkins's colleague Rosalind Franklin. The data produced by Franklin turned out to be the key to the puzzle. A careful study of the image enabled Watson to figure out that the basic shape of DNA is a helix (spiral) with a uniform diameter. The thickness of the helix suggested that it was made up of two polynucleotide strands—in other words, a **double helix**. But how were the nucleotides arranged in the double helix?

Using wire models, Watson and Crick began trying to construct a double helix that conformed to all known data about DNA (**Figure 10.3**). Watson placed the backbones on the outside of the model, forcing the nitrogenous bases to swivel to the interior of the molecule. As he did this, it occurred to him that the four kinds of bases must pair in a specific way. This idea of specific base pairing was a flash of inspiration that enabled Watson and Crick to solve the DNA puzzle.

▼ **Figure 10.3 Discoverers of the double helix.**

James Watson (left) and **Francis Crick.** The discoverers of the structure of DNA are shown in 1953 with their model of the double helix.

Rosalind Franklin Using X-rays, Franklin generated some of the key data that provided insight into the structure of DNA.

At first, Watson imagined that the bases paired like with like—for example, A with A, C with C. But that kind of pairing did not fit with the fact that the DNA molecule has a uniform diameter. An AA pair (made of two double-ringed bases) would be almost twice as wide as a CC pair (made of two single-ringed bases), resulting in an uneven molecule. It soon became apparent that a double-ringed base on one strand must always be paired with a single-ringed base on the opposite strand. Moreover, Watson and Crick realized that the chemical structure of each kind of base dictated the pairings even more specifically.

Each base has protruding chemical groups that can best form hydrogen bonds with only one appropriate partner. This would be like having four colors of snap-together puzzle pieces and realizing that only certain colors can snap together (e.g., red can only snap together with blue). Similarly, adenine can best form hydrogen bonds with thymine, and guanine with cytosine. In the biologist's shorthand: A pairs with T, and G pairs with C. A is also said to be "complementary" to T, and G to C.

If you imagine a polynucleotide strand as a half ladder, then you can picture the model of the DNA double helix proposed by Watson and Crick as a full ladder twisted into a spiral (**Figure 10.4**). Figure 10.5 shows three more detailed representations of the double helix. The ribbon-like diagram in **Figure 10.5a** symbolizes the bases with shapes that emphasize their complementarity. **Figure 10.5b** is a more chemically precise version showing only four base pairs, with the helix untwisted and the individual hydrogen bonds specified by dashed lines. **Figure 10.5c** is a computer model showing part of a double helix in detail.

Although the base-pairing rules dictate the side-by-side combinations of bases that form the rungs of the double helix, they place no restrictions on the sequence of nucleotides along the length of a DNA strand. In fact, the sequence of bases can vary in countless ways, and each gene has a unique order of nucleotides.

In 1953, Watson and Crick rocked the scientific world with a brief paper proposing their molecular model for

▶ **Figure 10.4 A rope-ladder model of a double helix.** The ropes at the sides represent the sugar-phosphate backbones. Each rung stands for a pair of bases connected by hydrogen bonds.

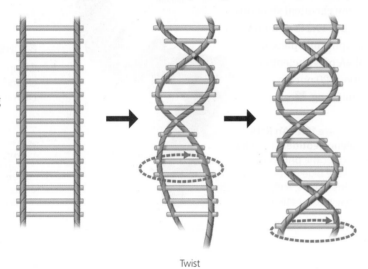

Twist

▼ **Figure 10.5 Three representations of DNA.**

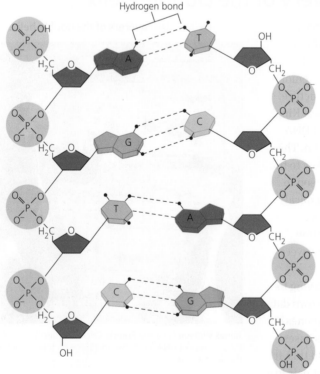

Hydrogen bond

(a) Ribbon model. The sugar-phosphate backbones are blue ribbons, and the bases are complementary shapes in shades of green and orange.

(b) Atomic model. In this more chemically detailed structure, you can see the individual hydrogen bonds (dashed lines). You can also see that the strands run in opposite directions: Notice that the sugars on the two strands are upside down with respect to each other.

(c) Computer model. Each atom is shown as a sphere, creating a space-filling model.

DNA. Few milestones in the history of biology have had as broad an impact as their double helix, with its AT and CG base pairing. In 1962, Watson, Crick, and Wilkins received the Nobel Prize for their work. (Sadly, Rosalind Franklin died of multiple cancers in 1958 at the age of 38 and was thus ineligible for the prize, which is only given to living scientists. Some suspect that her work with X-ray radiation may have caused her illness.)

In their 1953 paper, Watson and Crick wrote that the structure they proposed "immediately suggests a possible copying mechanism for the genetic material." In other words, the structure of DNA points toward a molecular explanation for life's unique properties of reproduction and inheritance. **The arrangement of parts within DNA affects its actions within a cell, an example of the relationship of structure to function, as we'll see next.**

DNA Replication

Every cell contains a DNA "cookbook" that provides complete information on how to make and maintain that cell. When a cell reproduces, it must duplicate this information, providing one copy to the new offspring cell while keeping one copy for itself. Thus, each cell must have a means of copying the DNA instructions. In a clear demonstration of how the structure of a biological system can provide insight into its function, Watson and Crick's model of DNA suggests that each DNA strand serves as a mold, or template, to guide reproduction of the other strand. If you know the sequence of bases in one strand of the double helix, you can very easily determine the sequence of bases in the other strand by applying the base-pairing rules: A pairs with T (and T with A), and G pairs with C (and C with G). For example, if one polynucleotide has the sequence AGTC, then the complementary polynucleotide in that DNA molecule must have the sequence TCAG.

Figure 10.6 shows how this model can account for the direct copying of a piece of DNA. The two strands of parental DNA separate, and each becomes a template for the assembly of a complementary strand from a supply of free nucleotides. The nucleotides are lined up one at a time along the template strand in accordance with the base-pairing rules. Enzymes link the nucleotides to form new DNA strands. The completed new molecules, identical to the parental molecule, are known as daughter DNA molecules (no gender should be inferred from this name).

The process of DNA replication requires the cooperation of more than a dozen enzymes and other proteins. **DNA polymerases** are the enzymes that make the covalent bonds between the nucleotides of a new DNA strand. As an incoming nucleotide base-pairs

with its complement on the template strand, a DNA polymerase adds it to the end of the growing daughter strand. The process is both fast (a rate of 50 nucleotides per second is typical) and amazingly accurate, with fewer than one in a billion bases incorrectly paired. In addition to their roles in DNA replication, DNA polymerases and some of the associated proteins can repair DNA that has been damaged by toxic chemicals or high-energy radiation, such as X-rays and ultraviolet light.

DNA replication begins on a double helix at specific sites, called origins of replication. Replication then proceeds in both directions, creating what are called replication "bubbles" **(Figure 10.7)**. The parental DNA strands open up as daughter strands elongate on both sides of each bubble. The DNA molecule of a typical eukaryotic chromosome has many origins where replication can start simultaneously, shortening the total time needed for the process. Eventually, all the bubbles merge, yielding two completed double-stranded daughter DNA molecules.

DNA replication ensures that all the body cells in a multicellular organism carry the same genetic information. It is also the means by which genetic information is passed along to offspring. ☑

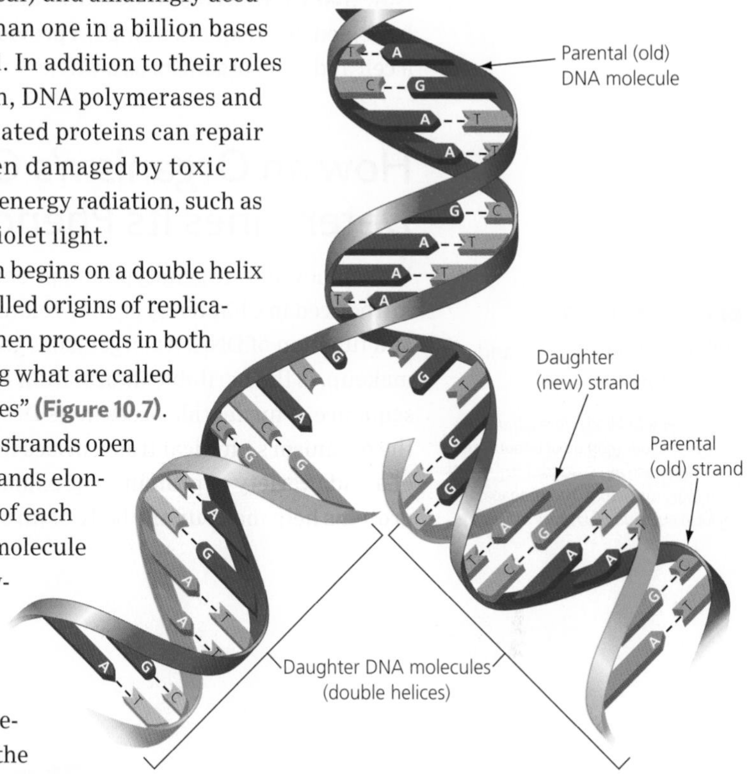

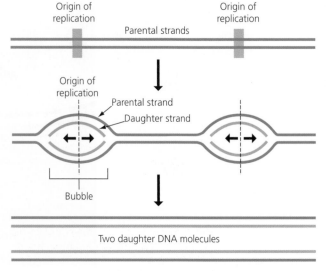

Parental (old) DNA molecule

Daughter (new) strand

Parental (old) strand

Daughter DNA molecules (double helices)

▲ **Figure 10.6 DNA replication.** Replication results in two daughter DNA molecules, each consisting of one old strand and one new strand. The parental DNA untwists as its strands separate, and the daughter DNA rewinds as it forms.

▼ **Figure 10.7 Multiple "bubbles" in replicating DNA.**

Origin of replication

Origin of replication

Parental strands

Origin of replication

Parental strand

Daughter strand

Bubble

Two daughter DNA molecules

☑ **CHECKPOINT**

1. How does complementary base pairing make DNA replication possible?

2. What enzymes connect nucleotides together during DNA replication?

From DNA to RNA to Protein

Now that we've seen how the structure of DNA allows it to be copied, let's explore how DNA provides instructions to a cell and to an organism as a whole.

How an Organism's Genotype Determines Its Phenotype

We can now define genotype and phenotype (terms first introduced in Chapter 9) as they relate to the structure and function of DNA. An organism's *genotype*, its genetic makeup, is the heritable information contained in the sequence of nucleotide bases in its DNA. The *phenotype*, the organism's physical traits, arises from the actions of a wide variety of proteins. For example, structural proteins help make up the body of an organism, and

enzymes catalyze the chemical reactions that are necessary for life.

DNA specifies the synthesis of proteins. However, a gene does not build a protein directly. Instead, DNA dispatches instructions in the form of RNA, which in turn programs protein synthesis. This fundamental concept in biology is summarized in **Figure 10.8**. The molecular "chain of command" is from DNA in the nucleus (purple area in the figure) to RNA to protein synthesis in the cytoplasm (blue area). The two stages are **transcription**, the transfer of genetic information from DNA into an RNA molecule, and **translation**, the transfer of the information from RNA into a polypeptide (protein strand). **Therefore, the relationship between genes and proteins is one of information flow. The function of a DNA gene is to dictate the production of a polypeptide.** ☑

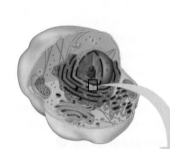

▼ Figure 10.8 **The flow of genetic information in a eukaryotic cell.** A sequence of nucleotides in the DNA is transcribed into a molecule of RNA in the cell's nucleus. The RNA travels to the cytoplasm, where it is translated into the specific amino acid sequence of a protein.

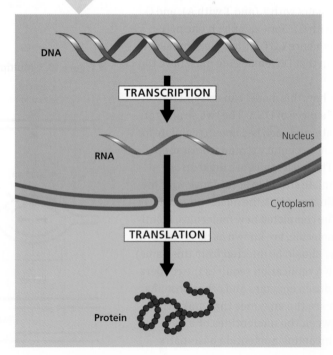

DNA

TRANSCRIPTION

RNA

Nucleus

Cytoplasm

TRANSLATION

Protein

From Nucleotides to Amino Acids: An Overview

Genetic information in DNA is transcribed into RNA and then translated into polypeptides, which then fold into proteins. But how do these processes occur? Transcription and translation are linguistic terms, and it is useful to think of nucleic acids and proteins as having languages. To understand how genetic information passes from genotype to phenotype, we need to see how the chemical language of DNA is translated into the different chemical language of proteins.

What exactly is the language of nucleic acids? Both DNA and RNA are polymers made of nucleotide monomers strung together in specific sequences that convey information, much as specific sequences of letters convey information in English. In DNA, the monomers are the four types of nucleotides, which differ in their nitrogenous bases (A, T, C, and G). The same is true for RNA, although it has the base U instead of T.

The language of DNA is written as a linear sequence of nucleotide bases, such as the blue sequence you see on the enlarged DNA strand in **Figure 10.9**. Every gene is made up of a specific sequence of bases, with special sequences marking the beginning and the end. A typical gene is a few thousand nucleotides in length.

When a segment of DNA is transcribed, the result is an RNA molecule. The process is called transcription because the nucleic acid language of DNA has simply been rewritten (transcribed) as a sequence of bases of RNA; the language is still that of nucleic acids. The nucleotide bases of the RNA molecule are complementary to those on the DNA strand. As you will soon see, this is because the RNA was synthesized using the DNA as a template.

Translation is the conversion of the nucleic acid language to the polypeptide language. Like nucleic acids, polypeptides are straight polymers, but the monomers that make them up—the letters of the polypeptide alphabet—are the 20 amino acids common to all organisms (represented as purple shapes in Figure 10.9). The sequence of nucleotides of the RNA molecule dictates the sequence of amino acids of the polypeptide. But remember, RNA is only a messenger; the genetic information that dictates the amino acid sequence originates in DNA.

What are the rules for translating the RNA message into a polypeptide? In other words, what is the correspondence between the nucleotides of an RNA molecule and the amino acids of a polypeptide? Keep in mind that there are only four different kinds of nucleotides in DNA (A, G, C, T) and RNA (A, G, C, U). During translation, these four must somehow specify 20 amino acids. If each nucleotide base coded for one amino acid, only 4 of the 20 amino acids could be accounted for. In fact, triplets of bases are the smallest "words" of uniform length that can specify all the amino acids. There can be 64 (that is, 4^3) possible code words of this type—more than enough to specify the 20 amino acids. Indeed, there are enough triplets to allow more than one coding for each amino acid. For example, the base triplets AAA and AAG both code for the same amino acid.

Experiments have verified that the flow of information from gene to protein is based on a triplet code. The genetic instructions for the amino acid sequence of a polypeptide chain are written in DNA and RNA as a series of three-base words called **codons**. Three-base codons in the DNA are transcribed into complementary three-base codons in the RNA, and then the RNA codons are translated into amino acids that form a polypeptide. As summarized in Figure 10.9, one DNA codon (three nucleotides) → one RNA codon (three nucleotides) → one amino acid. Next we turn to the codons themselves. ✔

☑ CHECKPOINT
How many nucleotides are necessary to code for a polypeptide that is 100 amino acids long?

■ *Answer: 300*

▼ **Figure 10.9 Transcription of DNA and translation of codons.** This figure focuses on a small region of a DNA molecule. The enlarged segment from one strand of the molecule shows a specific sequence of bases. The red strand and purple chain represent the products of transcription and translation, respectively.

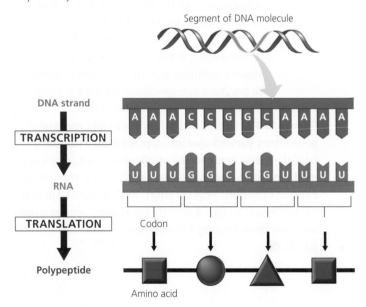

Segment of DNA molecule

DNA strand

A A A C C G G C A A A A

TRANSCRIPTION

RNA

U U U G G C C G U U U U

TRANSLATION

Codon

Polypeptide

Amino acid

The Genetic Code

The **genetic code** is the set of rules that convert a nucleotide sequence in RNA to an amino acid sequence. As **Figure 10.10** shows, 61 of the 64 triplets code for amino acids. The triplet AUG has a dual function: It codes for the amino acid methionine (abbreviated Met, green in the figure) and can also provide a signal for the start of a polypeptide chain. Three codons (UAA, UAG, and UGA—red in the figure) do not designate amino acids. They are the stop codons that instruct the ribosomes to end the polypeptide.

Notice in Figure 10.10 that a given RNA triplet always specifies a given amino acid. For example, although codons UUU and UUC both specify phenylalanine (Phe), neither of them ever represents any other amino acid. The codons in the figure are the triplets found in RNA. They have a straightforward, complementary relationship to the codons in DNA. The nucleotides making up the codons occur in a linear order along the DNA and RNA, with no gaps separating the codons.

The genetic code is nearly universal, shared by organisms from the simplest bacteria to the most complex plants and animals. **The universality of the genetic vocabulary suggests that it arose very early in evolution and was passed on over the eons to all the organisms living on Earth today.** In fact, such universality is the key to modern DNA technologies. Because diverse organisms share a common genetic code, it is possible to program one species to produce a protein from another species by transplanting DNA **(Figure 10.11)**. This allows scientists to mix and match genes from various species—a procedure with many useful genetic engineering applications in agriculture, medicine, and research (see Chapter 12 for further discussion of genetic engineering). Besides having practical purposes, a shared genetic vocabulary also reminds us of the evolutionary kinship that connects all life on Earth. ✓

ALL LIFE SHARES A GENETIC CODE, SO YOUR DNA COULD BE USED TO MODIFY BACTERIA.

☑ **CHECKPOINT**

An RNA molecule contains the nucleotide sequence CCAUUUACG. Using Figure 10.10, translate this sequence into the corresponding amino acid sequence.

■ Answer: Pro-Phe-Thr

▼ **Figure 10.10** **The dictionary of the genetic code, listed by RNA codons.** Practice using this dictionary by finding the codon UGG. It is the only codon for the amino acid tryptophan, Trp. Notice that the codon AUG not only stands for the amino acid methionine (Met, highlighted in green) but also functions as a signal to "start" translating the RNA at that place. Three of the 64 codons (highlighted in red) function as "stop" signals that mark the end of a genetic message, but do not encode any amino acids.

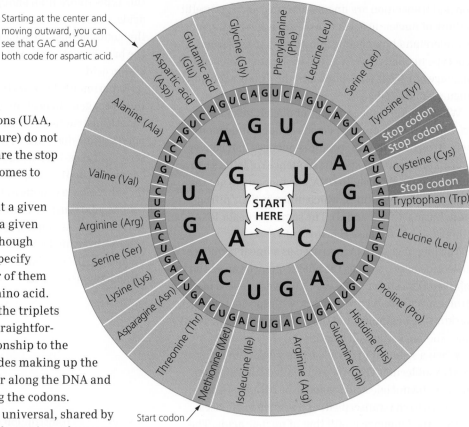

Starting at the center and moving outward, you can see that GAC and GAU both code for aspartic acid.

Start codon

▼ **Figure 10.11** **A mouse expressing a foreign gene.** This glowing rodent is the result of researchers incorporating a jelly (jellyfish) gene for a protein called green fluorescent protein (GFP) into the DNA of a laboratory mouse.

Transcription: From DNA to RNA

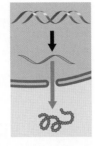

Let's look more closely at transcription, the transfer of genetic information from DNA to RNA. If you think of your DNA as a cookbook, then transcription is the process of copying one recipe onto an index card (a molecule of RNA) for immediate use. **Figure 10.12a** is a close-up view of this process. As with DNA replication, the two DNA strands must first separate at the place where the process will start. In transcription, however, only one of the DNA strands serves as a template for the newly forming RNA molecule; the other strand is unused. The nucleotides that make up the new RNA molecule take their place one at a time along the DNA template strand by forming hydrogen bonds with the nucleotide bases there. Notice that the RNA nucleotides follow the usual base-pairing rules, except that U, rather than T, pairs with A. The RNA nucleotides are linked by the transcription enzyme **RNA polymerase**.

Figure 10.12b is an overview of the transcription of an entire gene. Special sequences of DNA nucleotides tell the RNA polymerase where to start and where to stop the transcribing process.

① Initiation of Transcription

The "start transcribing" signal is a nucleotide sequence called a **promoter**, which is located in the DNA at the beginning of the gene. A promoter is a specific place where RNA polymerase attaches. The first phase of transcription, called initiation, is the attachment of RNA polymerase to the promoter and the start of RNA synthesis.

For any gene, the promoter dictates which of the two DNA strands is to be transcribed (the particular strand varies from gene to gene).

② RNA Elongation

During the second phase of transcription, elongation, the RNA grows longer. As RNA synthesis continues, the RNA strand peels away from its DNA template, allowing the two separated DNA strands to come back together in the region already transcribed.

③ Termination of Transcription

In the third phase, termination, the RNA polymerase reaches a special sequence of bases in the DNA template called a **terminator**. This sequence signals the end of the gene. At this point, the polymerase molecule detaches from the RNA molecule and the gene, and the DNA strands rejoin.

In addition to producing RNA that encodes amino acid sequences, transcription makes two other kinds of RNA that are involved in building polypeptides. We discuss these kinds of RNA a little later. ☑

☑ CHECKPOINT

How does RNA polymerase "know" where to start transcribing a gene?

■ *Answer: It recognizes the gene's promoter, a specific nucleotide sequence.*

▼ **Figure 10.12 Transcription.**

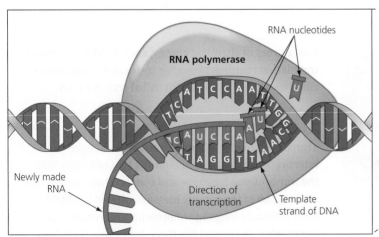

(a) A close-up view of transcription. As RNA nucleotides base-pair one by one with DNA bases on one DNA strand (called the template strand), the enzyme RNA polymerase (orange) links the RNA nucleotides into an RNA chain.

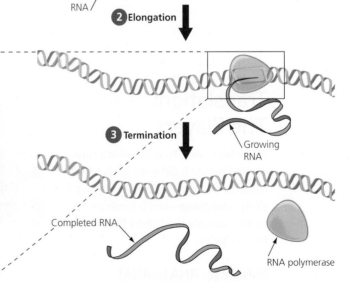

(b) Transcription of a gene. The transcription of an entire gene occurs in three phases: initiation, elongation, and termination of the RNA. The section of DNA where the RNA polymerase starts is called the promoter; the place where it stops is called the terminator.

The Processing of Eukaryotic RNA

In the cells of prokaryotes, which lack nuclei, the RNA transcribed from a gene immediately functions as **messenger RNA (mRNA)**, the molecule that is translated into protein. But this is not the case in eukaryotic cells. The eukaryotic cell not only localizes transcription in the nucleus but also modifies, or processes, the RNA transcripts there before they move to the cytoplasm for translation by the ribosomes.

One kind of RNA processing is the addition of extra nucleotides to the ends of the RNA transcript. These additions, called the cap and tail, mark the mRNA for export from the nucleus and help ribosomes recognize the RNA as mRNA.

Another type of RNA processing is made necessary in eukaryotes by noncoding stretches of nucleotides that interrupt the nucleotides that actually code for amino acids. It is as if nonsense words were randomly interspersed within a recipe that you copied. Most genes of plants and animals, it turns out, include such internal noncoding regions, which are called **introns**. The coding regions—the parts of a gene that are expressed—are called **exons**. As **Figure 10.13** illustrates, both exons and introns are transcribed from DNA into RNA. However, before the RNA leaves the nucleus, the introns are removed, and the exons are joined to produce an mRNA molecule with a continuous coding sequence. This cutting-and-pasting process is called **RNA splicing**. RNA splicing is believed to play a significant role in humans in allowing our approximately 21,000 genes to produce several times this

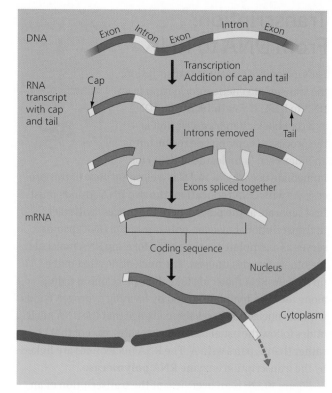

▲ **Figure 10.13 The production of messenger RNA (mRNA) in a eukaryotic cell.** Note that the molecule of mRNA that leaves the nucleus is substantially different from the molecule of RNA that was first transcribed from the gene. In the cytoplasm, the coding sequence of the final mRNA will be translated.

number of polypeptides. This is accomplished by varying the exons that are included in the final mRNA.

With capping, tailing, and splicing completed, the "final draft" of eukaryotic mRNA is ready for translation. ☑

CHECKPOINT

Why is a final mRNA often shorter than the DNA gene that coded for it?

■ *Answer: because introns are removed from the RNA*

Translation: The Players

As we have seen, translation is a conversion between different languages—from the nucleic acid language to the protein language—and it involves more elaborate machinery than transcription.

Messenger RNA (mRNA)

The first important ingredient required for translation is the mRNA produced by transcription. Once it is present, the machinery used to translate mRNA requires enzymes

and sources of chemical energy, such as ATP. In addition, translation requires two other important components: ribosomes and a kind of RNA called transfer RNA.

Transfer RNA (tRNA)

Translation of any language into another requires an interpreter, someone or something that can recognize the words of one language and convert them to the other. Translation of the genetic message carried in mRNA into the amino acid language of proteins also requires an interpreter. To convert the three-letter words (codons) of nucleic acids to the amino acid words of proteins, a cell uses a molecular interpreter, a type of RNA called **transfer RNA (tRNA)**, depicted in **Figure 10.14**.

▼ **Figure 10.14 The structure of tRNA.** At one end of the tRNA is the site where an amino acid will attach (purple), and at the other end is the three-nucleotide anticodon where the mRNA will attach (light green).

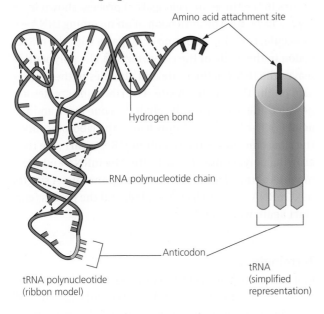

Amino acid attachment site

Hydrogen bond

RNA polynucleotide chain

Anticodon

tRNA polynucleotide
(ribbon model)

tRNA
(simplified
representation)

Ribosomes

Ribosomes are the organelles in the cytoplasm that coordinate the functioning of mRNA and tRNA and actually make polypeptides. As you can see in **Figure 10.15a**, a ribosome consists of two subunits. Each subunit is made up of proteins and a considerable amount of yet another kind of RNA, **ribosomal RNA (rRNA)**. A fully assembled ribosome has a binding site for mRNA on its small subunit and binding sites for tRNA on its large subunit. **Figure 10.15b** shows how two tRNA molecules get together with an mRNA molecule on a ribosome. One of the tRNA binding sites, the P site, holds the tRNA carrying the growing polypeptide chain, while another, the A site, holds a tRNA carrying the next amino acid to be added to the chain. The anticodon on each tRNA base-pairs with a codon on the mRNA. The subunits of the ribosome act like a vise, holding the tRNA and mRNA molecules close together. The ribosome can then connect the amino acid from the tRNA in the A site to the growing polypeptide. ☑

☑ CHECKPOINT
What is an anticodon?

■ *Answer: An anticodon is the base triplet of a tRNA molecule that couples the tRNA to a complementary codon in the mRNA. The base pairing of anticodon to codon is a key step in translating mRNA to a polypeptide.*

A cell that is producing proteins has in its cytoplasm a supply of amino acids. But amino acids themselves cannot recognize the codons arranged in sequence along messenger RNA. It is up to the cell's molecular interpreters, tRNA molecules, to match amino acids to the appropriate codons to form the new polypeptide. To perform this task, tRNA molecules must carry out two distinct functions: (1) pick up the appropriate amino acids and (2) recognize the appropriate codons in the mRNA. **The unique structure of tRNA molecules enables them to perform both functions.**

As shown on the left in Figure 10.14, a tRNA molecule is made of a single strand of RNA—one polynucleotide chain—consisting of about 80 nucleotides. The chain twists and folds upon itself, forming several double-stranded regions in which short stretches of RNA base-pair with other stretches. At one end of the folded molecule is a special triplet of bases called an **anticodon**. The anticodon triplet is complementary to a codon triplet on mRNA. During translation, the anticodon on the tRNA recognizes a particular codon on the mRNA by using base-pairing rules. At the other end of the tRNA molecule is a site where one specific kind of amino acid attaches. Although all tRNA molecules are similar, there are slightly different versions of tRNA for each amino acid.

▼ **Figure 10.15 The ribosome.**

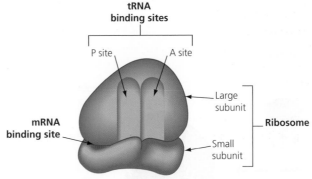

tRNA
binding sites

P site   A site

Large
subunit

Ribosome

mRNA
binding site

Small
subunit

(a) A simplified diagram of a ribosome. Notice the two subunits and sites where mRNA and tRNA molecules bind.

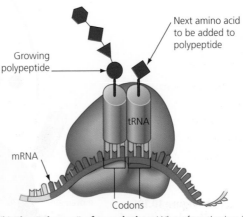

Next amino acid
to be added to
polypeptide

Growing
polypeptide

tRNA

mRNA

Codons

(b) The "players" of translation. When functioning in polypeptide synthesis, a ribosome holds one molecule of mRNA and two molecules of tRNA. The growing polypeptide is attached to one of the tRNAs.

Translation: The Process

Translation is divided into the same three phases as transcription: initiation, elongation, and termination.

Initiation

This first phase brings together the mRNA, the first amino acid with its attached tRNA, and the two subunits of a ribosome. An mRNA molecule, even after splicing, is longer than the genetic message it carries (**Figure 10.16**). Nucleotide sequences at either end of the molecule (pink) are not part of the message, but along with the cap and tail in eukaryotes, they help the mRNA bind to the ribosome. The initiation process determines exactly where translation will begin so that the mRNA codons will be translated into the correct sequence of amino acids. Initiation occurs in two steps, as shown in **Figure 10.17 ❶**. An mRNA molecule binds to a small ribosomal subunit. A special initiator tRNA then binds to the **start codon**, where translation is to begin on the mRNA. The initiator tRNA carries the amino acid methionine (Met); its anticodon, UAC, binds to the start codon, AUG ❷. A large ribosomal subunit binds to the small one, creating a functional ribosome. The initiator tRNA fits into the P site on the ribosome.

Elongation

Once initiation is complete, amino acids are added one by one to the first amino acid. Each addition occurs in the three-step elongation process shown in **Figure 10.18**. ❶ The anticodon of an incoming tRNA molecule, carrying its amino acid, pairs with the mRNA codon in the A site of the ribosome. ❷ The polypeptide leaves the tRNA in the P site and attaches to the amino acid on the tRNA in the A site. The ribosome creates a new peptide bond. Now the chain has one more amino acid. ❸ The P site tRNA now leaves the ribosome, and the ribosome moves the remaining tRNA, carrying the growing polypeptide, to the P site. The mRNA and tRNA move as a unit. This movement brings into the A site the next mRNA codon to be translated, and the process can start again with step 1.

Termination

Elongation continues until a **stop codon** reaches the ribosome's A site. Stop codons—UAA, UAG, and UGA—do not code for amino acids but instead tell translation to stop. The completed polypeptide, typically several hundred amino acids long, is freed, and the ribosome splits back into its subunits. ☑

▼ **Figure 10.16** A molecule of mRNA.

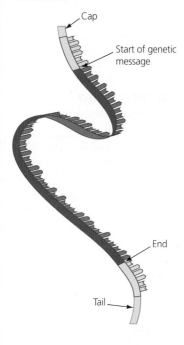

▼ **Figure 10.17 The initiation of translation.**

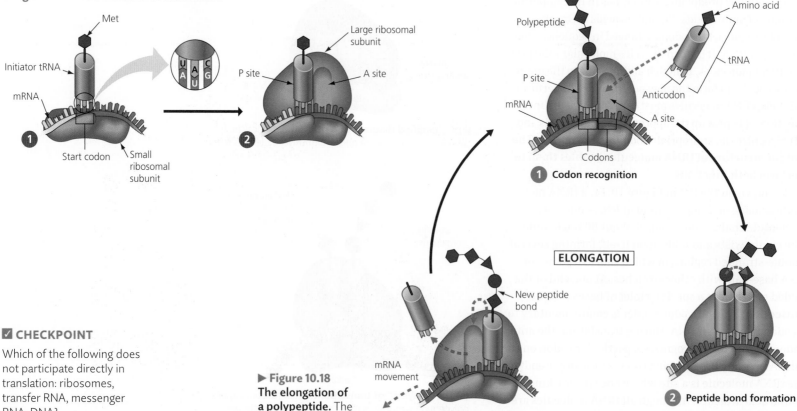

▶ **Figure 10.18**
The elongation of a polypeptide. The dashed red arrows indicate movement.

☑ **CHECKPOINT**

Which of the following does not participate directly in translation: ribosomes, transfer RNA, messenger RNA, DNA?

■ Answer: DNA

Review:
DNA → RNA → Protein

Figure 10.19 reviews the flow of genetic information in the cell, from DNA to RNA to protein. In eukaryotic cells, transcription (DNA → RNA) occurs in the nucleus, and the RNA is processed before it enters the cytoplasm. Translation (RNA → protein) is rapid; a single ribosome can make an average-sized polypeptide in less than a minute. As it is made, a polypeptide coils and folds, assuming its final three-dimensional shape.

What is the overall significance of transcription and translation? These are the processes whereby genes control the structures and activities of cells—or, more broadly,

the way the genotype produces the phenotype. The flow of information originates with the specific sequence of nucleotides in a DNA gene. The gene dictates the transcription of a complementary sequence of nucleotides in mRNA. In turn, the information within the mRNA specifies the sequence of amino acids in a polypeptide. Finally, the proteins that form from the polypeptides determine the appearance and capabilities of the cell and organism.

For decades, the DNA → RNA → protein pathway was believed to be the sole means by which genetic information controls traits. In recent years, however, this notion has been challenged by discoveries that point to more complex roles for RNA. (We will explore some of these special properties of RNA in Chapter 11.) ☑

☑ **CHECKPOINT**

1. Transcription is the synthesis of _____, using _____ as a template.
2. Translation is the synthesis of _____, with one _____ determining each amino acid in the sequence.
3. Which organelle coordinates translation?

■ *Answers: **1.** mRNA; DNA **2.** protein (polypeptides); codon **3.** ribosomes*

▼ **Figure 10.19 A summary of transcription and translation.** This figure summarizes the main stages in the flow of genetic information from DNA to protein in a eukaryotic cell.

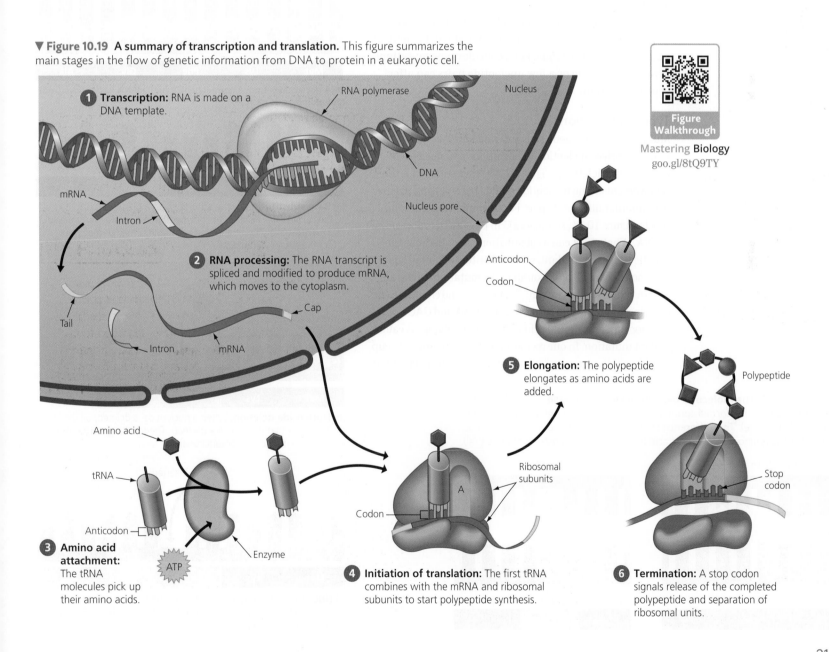

Figure Walkthrough

Mastering **Biology**
goo.gl/8tQ9TY

1 **Transcription:** RNA is made on a DNA template.

RNA polymerase

Nucleus

DNA

mRNA

Intron

Nucleus pore

Tail

Intron

mRNA

Cap

2 **RNA processing:** The RNA transcript is spliced and modified to produce mRNA, which moves to the cytoplasm.

Anticodon

Codon

5 **Elongation:** The polypeptide elongates as amino acids are added.

Polypeptide

Amino acid

tRNA

Ribosomal subunits

Stop codon

Anticodon

A

Codon

Enzyme

ATP

3 **Amino acid attachment:** The tRNA molecules pick up their amino acids.

4 **Initiation of translation:** The first tRNA combines with the mRNA and ribosomal subunits to start polypeptide synthesis.

6 **Termination:** A stop codon signals release of the completed polypeptide and separation of ribosomal units.

Mutations

Many inherited traits can be understood in molecular terms. For instance, sickle-cell disease can be traced to a change in a single amino acid in one of the polypeptides in the hemoglobin protein (see Figure 3.20). This difference is caused by a single nucleotide modification in the DNA coding for that polypeptide **(Figure 10.20)**.

Any change to the genetic information of a cell or virus is called a **mutation**. Mutations can involve large regions of a chromosome or just a single nucleotide pair, as in sickle-cell disease. Occasionally, a base substitution leads to an improved protein or one with new capabilities that enhance the success of the mutant organism and its descendants. Much more often, though, mutations are harmful. Think of a mutation as a typo in a recipe; occasionally, such a typo might lead to an improved recipe, but much more often it will be neutral, mildly bad, or disastrous. Let's consider how mutations involving only one or a few nucleotide pairs can affect gene translation.

Types of Mutations

Mutations within a gene can be divided into two general categories: nucleotide substitutions and nucleotide insertions or deletions **(Figure 10.21)**. A substitution is the replacement of one nucleotide and its base-pairing partner with another nucleotide pair. For example, in the second row in Figure 10.21, A replaces G in the fourth codon of the mRNA. What effect can a substitution have? Because the genetic code is redundant, some substitution mutations have no effect at all. For example, if a mutation causes an mRNA codon to change from GAA to GAG, no change in the protein product would result because GAA and GAG both code for the same amino acid (Glu). Such a change is called a silent mutation. In our recipe example, changing "1¼ cup sugar" to "1¼ cup sugor" would probably be translated the

ONE "TYPO" IN DNA CAN RESULT IN SICKLE-CELL DISEASE.

same way, just like the translation of a silent mutation does not change the meaning of the message.

Other substitutions involving a single nucleotide do change one amino acid to another. Such mutations are called missense mutations. For example, if a mutation causes an mRNA codon to change from GGC to AGC, the resulting protein will have a serine (Ser) instead of a glycine (Gly) at this position. Some missense mutations have little or no effect on the shape or function of

▼ **Figure 10.21 Three types of mutations and their effects.** Mutations are changes in DNA. In each case, you can see the effect of the mutation on the mRNA and polypeptide product.

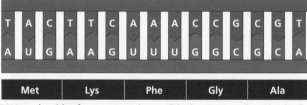

DNA nucleotides from a normal gene (blue), mRNA nucleotides (red), and amino acids (purple)

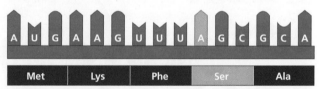

(a) Base substitution. Here, an A replaces a G in the fourth codon of the mRNA. The result in the polypeptide is a serine (Ser) instead of a glycine (Gly). This amino acid substitution may or may not affect the protein's function.

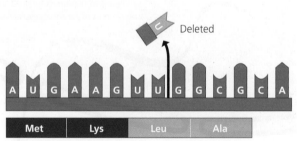

(b) Nucleotide deletion. When a nucleotide is deleted, all the codons from that point on are misread. The resulting polypeptide is likely to be completely nonfunctional.

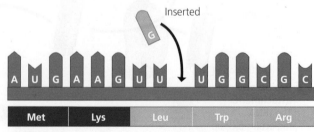

(c) Nucleotide insertion. As with a deletion, inserting one nucleotide disrupts all codons that follow, most likely producing a nonfunctional polypeptide.

▼ **Figure 10.20 The molecular basis of sickle-cell disease.** The sickle-cell allele differs from its normal counterpart, a gene for hemoglobin, by only one nucleotide (orange). This difference changes the mRNA codon from one that codes for the amino acid glutamic acid (Glu) to one that codes for valine (Val).

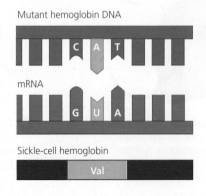

the resulting protein, just as changing a recipe from "1¼ cups sugar" to "1⅓ cups sugar" may have little effect on the final product. However, other substitutions, as we saw in the sickle-cell case, will cause changes in the protein that prevent it from performing normally. This would be like changing "1¼ cups sugar" to "6¼ cups sugar"— a single change big enough to ruin the whole recipe.

Some substitutions, called nonsense mutations, change an amino acid codon into a stop codon. For example, if an AGA (Arg) codon is mutated to a UGA (stop) codon, the result will be a prematurely terminated protein, which probably will not function properly. In our recipe analogy, this would be like stopping food preparation before the end of the recipe, which is almost certainly going to ruin the dish.

Because mRNA is read as a series of triplets during translation, adding or subtracting nucleotides may alter the triplet grouping of the genetic message. Such a mutation, called a frameshift mutation, occurs whenever the number of nucleotides inserted or deleted is not a multiple of three. All the nucleotides after the insertion or deletion will be regrouped into different codons. Consider this recipe example: Add one cup egg nog. Deleting the second letter produces an entirely nonsensical message—ado nec upe ggn og—which will not produce a useful product. Similarly, a frameshift mutation most often produces a nonfunctioning polypeptide.

Mutagens

Mutations can occur in a number of ways. Spontaneous mutations result from random errors during DNA replication or recombination. Other sources of mutation are physical and chemical agents called **mutagens**. The most common physical mutagen is high-energy radiation, such as X-rays and ultraviolet (UV) light. Chemical mutagens are of various types. One type, for example, consists of chemicals that are similar to normal DNA bases but that base-pair incorrectly when incorporated into DNA.

Because many mutagens can act as carcinogens, agents that cause cancer, you would do well to avoid them as much as possible. What can you do to avoid exposure to mutagens? Several lifestyle practices can help, including not smoking and wearing protective clothing and sunscreen to minimize direct exposure to the sun's UV rays. But such precautions are not foolproof, and it is not possible to avoid mutagens (such as UV radiation and secondhand smoke) entirely.

Although mutations are often harmful, they can also be beneficial, both in nature and in the laboratory. Mutations are one source of the rich diversity of genes in the living world, a diversity that makes evolution by natural selection possible **(Figure 10.22)**. Mutations are also essential tools for geneticists. Whether naturally occurring or created in the laboratory, mutations are responsible for the different alleles needed for genetic research. ✓

☑ CHECKPOINT

1. What would happen if a mutation changed a start codon to some other codon?

2. What happens when one nucleotide is lost from the middle of a gene?

Answers: 1. mRNA transcribed from the mutated gene would be nonfunctional because ribosomes would not initiate translation. 2. In the mRNA, the reading of the triplets downstream from the deletion is shifted, leading to a long string of incorrect amino acids in the polypeptide.

▼ **Figure 10.22 Mutations and diversity.** Mutations are one source of the diversity of life visible in this scene from a coral reef in Bali, Indonesia.

Viruses and Other Noncellular Infectious Agents

Viruses share some of the characteristics of living organisms, such as having genetic material in the form of nucleic acid packaged within a highly organized structure. A virus is generally not considered alive, however, because it is not cellular and cannot reproduce on its own. (See Figure 1.11 to review the properties of life.) A **virus** is an infectious particle consisting of little more than "genes in a box": a bit of nucleic acid wrapped in a protein coat and, in some cases, an envelope of membrane **(Figure 10.23)**. Unlike the genomes of all living cells, a viral genome may consist of DNA or RNA, and may be single- or double-stranded. A virus cannot reproduce on its own; it can multiply only by infecting a living cell and directing the cell's molecular machinery to make more viruses. In this section, we'll look at viruses that infect different types of host organisms, starting with bacteria.

Bacteriophages

Viruses that attack bacteria are called **bacteriophages** ("bacteria-eaters"), or **phages** for short. **Figure 10.24** shows a micrograph of a bacteriophage called T4 infecting an *Escherichia coli* bacterium. The phage consists of a molecule of DNA enclosed within an elaborate structure made of proteins. The "legs" of the phage bend when they touch the cell surface. The tail is a hollow rod enclosed in a springlike sheath. As the legs bend, the spring compresses, the bottom of the rod punctures the cell membrane, and the viral DNA passes from inside the head of the virus into the cell.

Once they infect a bacterium, most phages enter a reproductive cycle called the **lytic cycle**. The lytic cycle gets its name from the fact that after many copies of the phage are produced within the bacterial cell, the bacterium lyses (breaks open). Some viruses can also reproduce by an alternative route—the **lysogenic cycle**. During a lysogenic cycle, viral DNA replication occurs without phage production or the death of the cell.

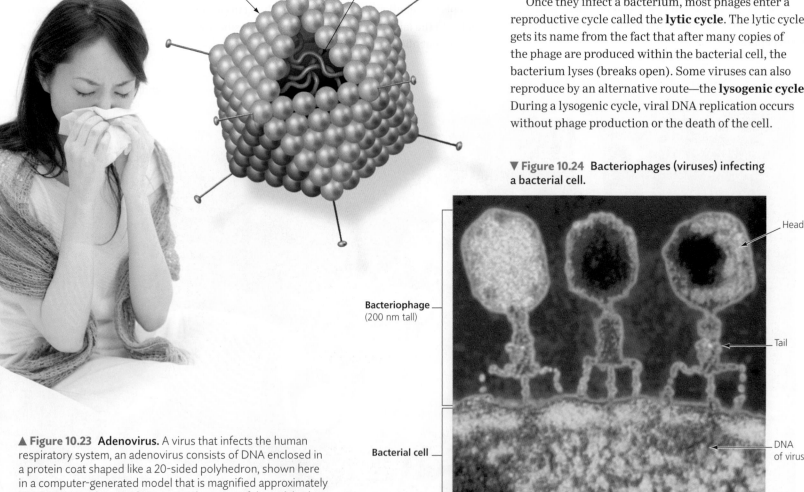

Protein coat

DNA

▲ **Figure 10.23 Adenovirus.** A virus that infects the human respiratory system, an adenovirus consists of DNA enclosed in a protein coat shaped like a 20-sided polyhedron, shown here in a computer-generated model that is magnified approximately 500,000 times the actual size. At each corner of the polyhedron is a protein spike, which helps the virus attach to a susceptible cell.

▼ **Figure 10.24 Bacteriophages (viruses) infecting a bacterial cell.**

Head

Bacteriophage (200 nm tall)

Tail

Bacterial cell

DNA of virus

Colorized TEM 225,000×

Figure 10.25 illustrates the two kinds of cycles for a phage named lambda that can infect *E. coli* bacteria. At the start of infection, **①** lambda binds to the outside of a bacterium and injects its DNA inside. **②** The injected lambda DNA forms a circle. In the lytic cycle, this DNA immediately turns the cell into a virus-producing factory. **③** The cell's own machinery for DNA replication, transcription, and translation is hijacked by the virus and used to produce copies of the virus. **④** The cell lyses, releasing the new phages.

In the lysogenic cycle, **⑤** the viral DNA is inserted into the bacterial chromosome. Once there, the phage DNA is referred to as a **prophage**, and most of its genes are inactive. Survival of the prophage depends on the reproduction of the cell where it resides. **⑥** The host cell replicates the prophage DNA along with its cellular DNA and then, upon dividing, passes on both the prophage and the cellular DNA to its two daughter cells. A single infected bacterium can quickly give rise to a large population of bacteria that all carry prophages. The prophages may remain in the bacterial cells indefinitely. **⑦** Occasionally, however, a prophage leaves its chromosome; this event may be triggered by environmental conditions such as exposure to a mutagen. Once separate, the lambda DNA usually switches to the lytic cycle, which results in the production of many copies of the virus and lysing of the host cell.

Sometimes the few prophage genes active in a lysogenic bacterial cell can cause medical problems. For example, the bacteria that cause diphtheria, botulism, and scarlet fever would be harmless to people if it were not for the prophage genes they carry. Certain of these genes direct the bacteria to produce toxins that make people ill. ☑

☑ **CHECKPOINT**

Describe one way that some viruses can perpetuate their genes without immediately destroying the cells they infect.

■ *Answer: Some viruses can insert their DNA into the DNA of the cell they infect (the lysogenic cycle). The viral DNA is replicated along with the cell's DNA every time the cell divides.*

Phage lambda

E. coli

▼ **Figure 10.25 Alternative phage reproductive cycles.** Certain phages can undergo alternative reproductive cycles. After entering the bacterial cell, the phage DNA can either integrate into the bacterial chromosome (lysogenic cycle) or immediately start the production of progeny phages (lytic cycle), destroying the cell. Once it enters a lysogenic cycle, the phage's DNA may be carried in the host cell's chromosome for many generations.

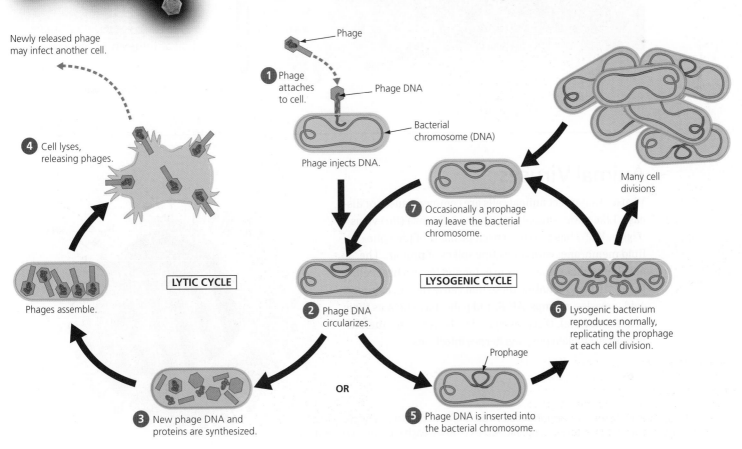

Newly released phage may infect another cell.

Phage

① Phage attaches to cell.

Phage DNA

Bacterial chromosome (DNA)

Phage injects DNA.

④ Cell lyses, releasing phages.

LYTIC CYCLE

⑦ Occasionally a prophage may leave the bacterial chromosome.

LYSOGENIC CYCLE

Many cell divisions

Phages assemble.

② Phage DNA circularizes.

⑥ Lysogenic bacterium reproduces normally, replicating the prophage at each cell division.

Prophage

③ New phage DNA and proteins are synthesized.

OR

⑤ Phage DNA is inserted into the bacterial chromosome.

Plant Viruses

Viruses that infect plant cells can stunt plant growth and diminish crop yields. Most known plant viruses have RNA rather than DNA as their genetic material. Many of them, like the tobacco mosaic virus (TMV) shown in **Figure 10.26**, are rod-shaped with a spiral arrangement of proteins surrounding the nucleic acid. TMV, which infects tobacco and related plants, causing discolored spots on the leaves, was the first virus ever discovered (in 1930).

To infect a plant, a virus must first get past the plant's epidermis, an outer protective layer of cells. For this reason, a plant damaged by wind, chilling, injury, or insects is more susceptible to infection than a healthy plant. Some insects carry and transmit plant viruses, and farmers and gardeners may unwittingly spread plant viruses through the use of pruning shears and other tools.

There is no cure for most viral plant diseases, and agricultural scientists focus on preventing infection and on breeding or genetically engineering varieties of crop plants that resist viral infection. In Hawaii, for example, the spread of papaya ringspot potyvirus (PRSV) by aphids wiped out the native papaya (Hawaii's second largest crop) in certain island regions. But since 1998, farmers have been able to plant a genetically engineered PRSV-resistant strain of papaya, and papayas have been reintroduced into their old habitats. ☑

▼ **Figure 10.26 Tobacco mosaic virus.** The photo shows the mottling of leaves in tobacco mosaic disease. The rod-shaped virus causing the disease has RNA as its genetic material.

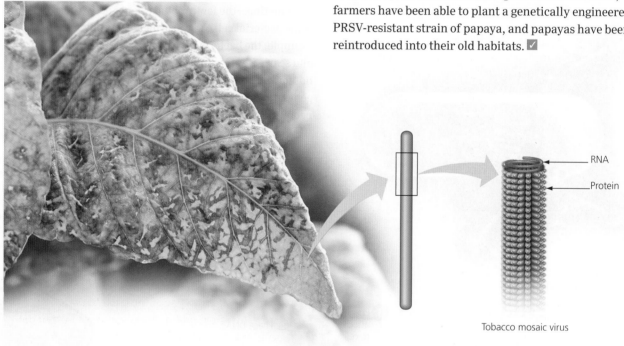

Tobacco mosaic virus

Animal Viruses

Viruses that infect animal cells are common causes of disease. Like many animal viruses, the influenza (flu) virus **(Figure 10.27)** has an outer envelope made of phospholipid membrane, with projecting spikes of protein. The envelope enables the virus to enter and leave a host cell. Many viruses, including those that cause the flu, common cold, measles, mumps, AIDS, and polio, have RNA as their genetic material. Diseases caused by DNA viruses include hepatitis, chicken pox, and herpes infections.

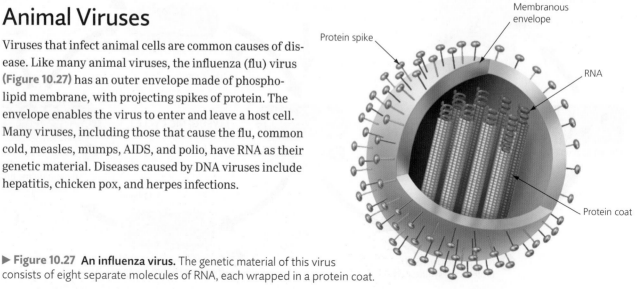

▶ **Figure 10.27 An influenza virus.** The genetic material of this virus consists of eight separate molecules of RNA, each wrapped in a protein coat.

Figure 10.28 shows the reproductive cycle of the mumps virus, a typical RNA virus. Once a common childhood disease characterized by fever and swelling of the salivary glands, mumps has become quite rare in industrialized nations due to widespread vaccination. When the virus contacts a susceptible cell, protein spikes on its outer surface attach to receptor proteins on the cell's plasma membrane. **1** The viral envelope fuses with the cell's membrane, allowing the protein-coated RNA to enter the cytoplasm. **2** Enzymes then remove the protein coat. **3** An enzyme that entered the cell as part of the virus uses the virus's RNA genome as a template for making complementary strands of RNA. The new strands have two functions: **4** They serve as mRNA for the synthesis of new viral proteins, and **5** they serve as templates for synthesizing new viral genome RNA. **6** The new coat proteins assemble around the new viral RNA. **7** Finally, the viruses leave the cell by cloaking themselves in plasma membrane. In other words, the virus obtains its envelope from the cell, budding off the cell without necessarily rupturing it.

Not all animal viruses reproduce in the cytoplasm. For example, herpesviruses—different strains of which cause chicken pox, shingles, cold sores, and genital herpes—are enveloped DNA viruses that reproduce in a host cell's nucleus, and they get their envelopes from the cell's nuclear membrane. Copies of the herpesvirus DNA usually remain behind in the nuclei of certain nerve cells. There they remain dormant until some sort of stress, such as a cold, sunburn, or emotional stress, triggers virus production, resulting in unpleasant symptoms. Once acquired, herpes infections may flare up repeatedly throughout a person's life. More than 75% of American adults carry herpes simplex 1 (which causes cold sores), and more than 20% carry herpes simplex 2 (which causes genital herpes).

The amount of damage a virus causes the body depends partly on how quickly the immune system responds to fight the infection and partly on the ability of the infected tissue to repair itself. We usually recover completely from colds because our respiratory tract tissue can efficiently replace damaged cells by mitosis. In contrast, the poliovirus attacks nerve cells, which are not usually replaceable. The damage to such cells by polio is permanent. In such cases, the only medical option is to prevent the disease with vaccines. Indeed, the development of vaccines was one of the great triumphs of medicine. This effort continues, as you'll see next in the Process of Science section. ☑

▼ **Figure 10.28 The reproductive cycle of an enveloped virus.** This virus is the one that causes mumps. Like the flu virus, it has a membranous envelope with protein spikes, but its genome is a single molecule of RNA.

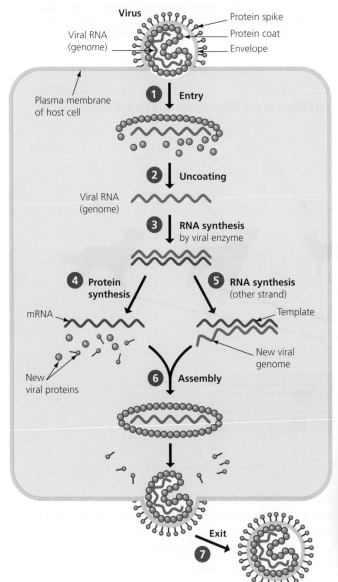

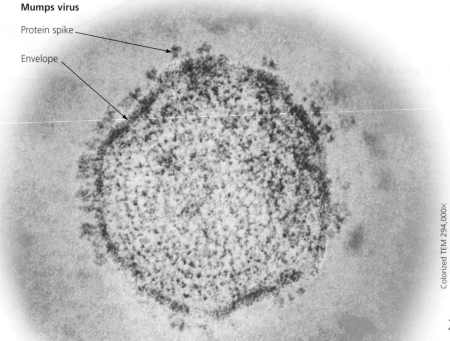

☑ CHECKPOINT

Why is infection by herpesvirus permanent?

■ *Answer: because herpesvirus leaves viral DNA in the nuclei of nerve cells*

Can DNA and RNA Vaccines Protect Against Viruses?

BACKGROUND

West Nile virus first appeared in the United States in 1999. Most people with the virus do not become ill. But in some cases, the virus causes a potentially fatal swelling of the central nervous system. The virus is spread by mosquitoes. There is no vaccine and no cure for those who become ill.

Medical researchers are hoping to improve our arsenal against West Nile virus by developing DNA and RNA vaccines. Vaccines trigger an immune response to a harmless molecule that mimics some part of the attacking pathogen. Once exposed, the immune system is primed to fight the disease if it detects the real pathogen in the future. A traditional vaccine contains the trigger protein. An RNA or DNA vaccine contains a copy of a viral gene and lets the patient's body make the specific trigger protein.

METHOD

Researchers tested an RNA vaccine containing a gene from West Nile virus using cats and dogs, which, like most mammals, are susceptible to the virus **(Figure 10.29a)**. The cats were given high doses of the vaccine, low doses, or placebos (ineffective treatments that serve as a control). The dogs received only low doses or placebos. All animals received booster vaccines 28 days after their first dose. After that time, the animals were exposed to mosquitoes carrying the virus and tested for the presence of infection.

RESULTS

The results were striking **(Figure 10.29b)**. No animals became infected with the virus after a high dose of vaccine. Clearly, the new vaccine was quite effective, at least among cats and dogs. Although no DNA or RNA vaccines have yet been approved for human use, this type of research suggests that they may became a standard tool in the near future.

TREATMENT	% OF CATS INFECTED	% OF DOGS INFECTED
Placebo	82	93
Low vaccine dose	12.5	0
High vaccine dose	0	(no dose given)

(b) Effects of the RNA vaccine on cats and dogs

▶ **Figure 10.29** Testing an RNA vaccine against West Nile virus.

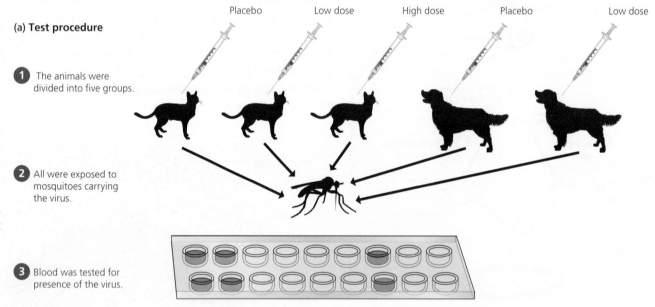

(a) Test procedure

Placebo   Low dose   High dose   Placebo   Low dose

1 The animals were divided into five groups.

2 All were exposed to mosquitoes carrying the virus.

3 Blood was tested for presence of the virus.

Thinking Like a Scientist

Why was it important for some dogs and cats to be given placebo injections?

For the answer, see Appendix D.

HIV, the AIDS Virus

The devastating disease **AIDS** (acquired immunodeficiency syndrome) is caused by **HIV** (human immunodeficiency virus), an RNA virus with some nasty twists. In outward appearance, HIV **(Figure 10.30)** resembles the mumps or flu virus. Its envelope enables HIV to enter and leave a cell much the way the mumps virus does. But HIV has a different mode of reproduction. It is a **retrovirus**, an RNA virus that reproduces by means of a DNA molecule, the reverse of the usual DNA → RNA flow of genetic information. These viruses carry molecules of an enzyme called **reverse transcriptase**, which catalyzes reverse transcription: the synthesis of DNA on an RNA template.

▼ **Figure 10.30** HIV, the AIDS virus.

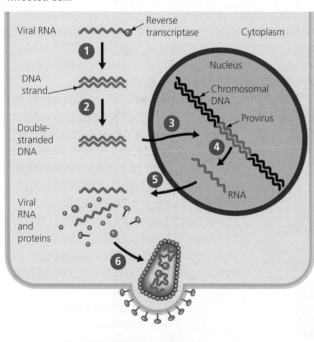

Envelope
Surface protein
Protein coat
RNA (two identical strands)
Reverse transcriptase

Figure 10.31 illustrates what happens after HIV RNA is uncoated in the cytoplasm of a cell. The reverse transcriptase (green) ❶ uses the RNA as a template to make a DNA strand and then ❷ adds a second, complementary DNA strand. ❸ The resulting double-stranded viral DNA then enters the cell nucleus and inserts itself into the chromosomal DNA,

▼ **Figure 10.31** The behavior of HIV nucleic acid in an infected cell.

Viral RNA
Reverse transcriptase
Cytoplasm
DNA strand
Nucleus
Chromosomal DNA
Provirus
Double-stranded DNA
RNA
Viral RNA and proteins

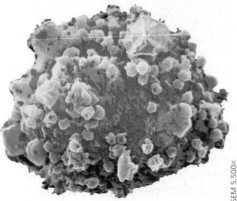

HIV (red dots) infecting a white blood cell

SEM 5,500×

becoming a **provirus**. Occasionally, the provirus is ❹ transcribed into RNA ❺ and translated into viral proteins. ❻ New viruses assembled from these components eventually leave the cell and can then infect other cells. This is the standard reproductive cycle for retroviruses.

HIV infects and eventually kills several kinds of white blood cells that are important in the body's immune system. The loss of such cells causes the body to become susceptible to other infections that it would normally be able to fight off. Such secondary infections cause the syndrome (a collection of symptoms) that eventually kills AIDS patients. Since it was first recognized in 1981, HIV has infected tens of millions of people worldwide, resulting in millions of deaths.

Although there is as yet no cure for AIDS, its progression can be slowed by two categories of anti-HIV drugs. Both types of medicine interfere with the reproduction of the virus. The first type inhibits the action of enzymes called proteases, which help produce the final versions of HIV proteins. The second type, which includes the drug AZT, inhibits the action of the HIV enzyme reverse transcriptase. The key to AZT's effectiveness is its shape. The shape of a molecule of AZT is very similar to the shape of part of the T (thymine) nucleotide **(Figure 10.32)**. In fact, AZT's shape is so similar to the T nucleotide that AZT can bind to reverse transcriptase, essentially taking the place of T. But unlike thymine, AZT cannot be incorporated into a growing DNA chain. Thus, AZT "gums up the works," interfering with the synthesis of HIV DNA. Because this synthesis is an essential step in the reproductive cycle of HIV, AZT may block the spread of the virus within the body.

Many HIV-infected people in the United States and other industrialized countries take a "drug cocktail" that contains both reverse transcriptase inhibitors and protease inhibitors, and the combination seems to be much more effective than the individual drugs in keeping the virus at bay and extending patients' lives. In fact, the death rate from HIV infection can be lowered by 80% with proper treatment. However, even in combination, the drugs do not completely rid the body of the virus. Typically, HIV reproduction and the symptoms of AIDS return if a patient discontinues the medications. Because AIDS has no cure yet, prevention (namely, avoiding unprotected sex and staying away from needle sharing) is the only healthy option. ☑

☑ **CHECKPOINT**

Why is HIV called a retrovirus?

■ *Answer: Because it synthesizes DNA from its RNA genome. This is the reverse ("retro") of the usual DNA → RNA information flow.*

▼ **Figure 10.32** AZT and the T nucleotide. The anti-HIV drug AZT (right) has a chemical shape very similar to part of the T (thymine) nucleotide of DNA.

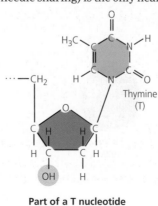

Part of a T nucleotide

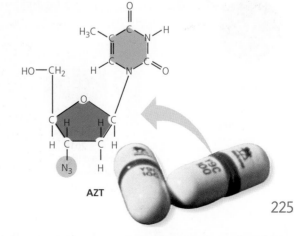

AZT

MAD COW
DISEASE IS
CAUSED BY
ODDLY SHAPED
PROTEINS.

☑ **CHECKPOINT**

What makes prions so
unusual as pathogens?

■ Answer: Prions, unlike any other
infectious agent, have no nucleic
acid (DNA or RNA).

Prions

Prions are infectious proteins that cause brain diseases in
several animal species. While a virus contains DNA
or RNA, a prion consists solely of a misfolded form
of a normal brain protein. When the prion gets into
a cell containing the normal form of the protein, the
prion somehow converts normal protein molecules
to misfolded versions. The misfolded proteins then
clump together, disrupting brain functions.

Diseases caused by prions include scrapie in
sheep; chronic wasting disease in deer and elk; mad cow
disease, which infected more than 2 million cattle in the
United Kingdom in the 1980s; and Creutzfeldt-Jakob

disease in humans, an incurable and inevitably fatal deterioration of the brain. An early 1900s New Guinea epidemic
of kuru, another human disease caused by prions, was
halted after anthropologists identified the cause—ritualistic
cannibalism of the brain—and convinced locals to stop that
practice.

Prions incubate at least 10 years before symptoms
develop. This can prevent timely identification of sources
of infection. Additionally, prions are not destroyed in
food by normal heating. The only hope for developing
effective treatments lies in understanding the process
of infection.

To close the chapter, let's revisit some other noncellular
threats to human health: emerging viruses. ☑

EVOLUTION CONNECTION | Deadly Viruses

Emerging Viruses

Viruses that suddenly come to the attention of medical
scientists are called **emerging viruses (Figure 10.33)**.
We've already explored Zika virus (first recognized in
Brazil in 2015) and West Nile virus (which first appeared
in North America in 1999). Although each virus had persisted at low levels for many years, each became a much
greater threat quite suddenly.

How do viruses give rise to new diseases? First, they can
evolve into more dangerous forms. Although viruses are not

alive, they are subject to natural selection, which is accelerated by high mutation rates. Unlike DNA, RNA has no mechanisms to repair copying errors, so RNA viruses can mutate
rapidly. Some mutations enable viruses to infect people who
had developed resistance to the ancestral strain. This is why
we need yearly flu vaccines: Mutations create new influenza
virus strains to which people have no immunity.

Second, viral diseases can spread from one host species to
another. Scientists estimate that about three-quarters of new
human diseases originated in other animals.
When humans hunt, live, or raise livestock in
new habitats, the risk increases. HIV (which
causes AIDS) may have started as a slightly
different virus in chimpanzees. Human
hunters were probably infected when they
butchered infected animals. As the virus
mutated in the human hosts, strains that
out-competed other varieties for human host
cells became increasingly common.

Third, viral diseases from a small, isolated population can spread, leading to an
epidemic. AIDS went unnamed and virtually
ignored for decades. Several factors, including international travel, intravenous drug
use, sexual activity, and delayed effective
action allowed it to become a global scourge.

Nobel Prize winner Joshua Lederberg
warned: "We live in evolutionary competition with microbes. There is no guarantee
that we will be the survivors." If we are to
be victorious in the fight against emerging
viruses, we must understand molecular
biology and evolutionary processes.

▼ **Figure 10.33**
**A sample of
major emerging
virus outbreaks
of the past
100 years.**

VISUALIZING THE DATA

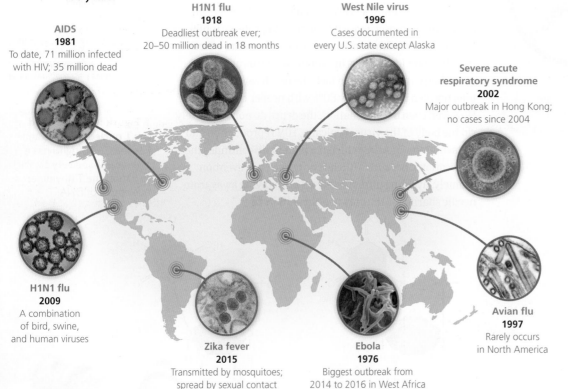

**AIDS
1981**
To date, 71 million infected
with HIV; 35 million dead

**H1N1 flu
1918**
Deadliest outbreak ever;
20–50 million dead in 18 months

**West Nile virus
1996**
Cases documented in
every U.S. state except Alaska

**Severe acute
respiratory syndrome
2002**
Major outbreak in Hong Kong;
no cases since 2004

**H1N1 flu
2009**
A combination
of bird, swine,
and human viruses

**Zika fever
2015**
Transmitted by mosquitoes;
spread by sexual contact

**Ebola
1976**
Biggest outbreak from
2014 to 2016 in West Africa

**Avian flu
1997**
Rarely occurs
in North America

Chapter Review

DNA: Structure and Replication

DNA and RNA Structure

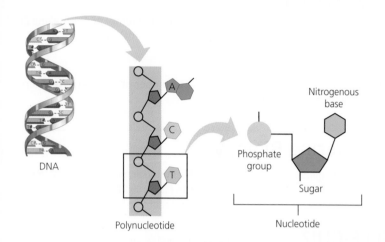

DNA

Polynucleotide

Nucleotide

Nitrogenous base

Phosphate group

Sugar

	DNA	RNA
Nitrogenous base	C G A T	C G A U
Sugar	Deoxy-ribose	Ribose
Number of strands	2	1

Watson and Crick's Discovery of the Double Helix

Watson and Crick worked out the three-dimensional structure of DNA: two polynucleotide strands wrapped around each other in a double helix. Hydrogen bonds between bases hold the strands together. Each base pairs with a complementary partner: A with T, and G with C.

DNA Replication

The structure of DNA, with its complementary base pairing, allows it to function as the molecule of heredity through DNA replication.

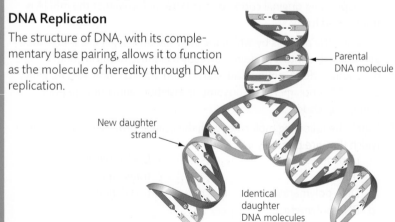

Parental DNA molecule

New daughter strand

Identical daughter DNA molecules

From DNA to RNA to Protein

How an Organism's Genotype Determines Its Phenotype

The information constituting an organism's genotype is carried in the sequence of its DNA bases. The genotype controls phenotype through the expression of proteins.

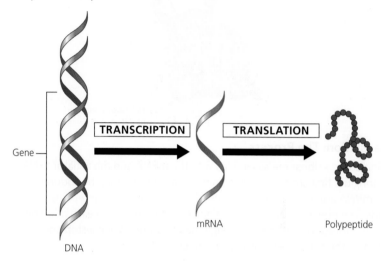

Gene

TRANSCRIPTION

TRANSLATION

DNA

mRNA

Polypeptide

From Nucleotides to Amino Acids: An Overview

The DNA of a gene is transcribed into RNA using the usual base-pairing rules, except that an A in DNA pairs with U in RNA. In the translation of a genetic message, each triplet of nucleotide bases in the RNA, called a codon, specifies one amino acid in the polypeptide.

The Genetic Code

In addition to codons that specify amino acids, the genetic code has one codon that is a start signal and three that are stop signals for translation.

Transcription: From DNA to RNA

In transcription, RNA polymerase binds to the promoter of a gene, opens the DNA double helix there, and catalyzes the synthesis of an RNA molecule using one DNA strand as a template. As the single-stranded RNA transcript peels away from the gene, the DNA strands rejoin.

The Processing of Eukaryotic RNA

The RNA transcribed from a eukaryotic gene is processed before leaving the nucleus to serve as messenger RNA (mRNA). Introns are spliced out, and a cap and tail are added.

Translation: The Players

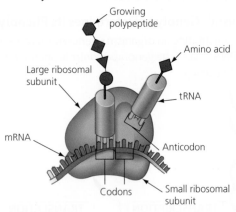

- Growing polypeptide
- Amino acid
- Large ribosomal subunit
- tRNA
- mRNA
- Anticodon
- Codons
- Small ribosomal subunit

Translation: The Process

In initiation, a ribosome assembles with the mRNA and the initiator tRNA bearing the first amino acid. Beginning at the start codon, the codons of the mRNA are recognized one by one by tRNAs bearing succeeding amino acids. The ribosome bonds the amino acids together. With each addition, the mRNA moves by one codon through the ribosome. When a stop codon is reached, the completed polypeptide is released.

Review: DNA → RNA → Protein

The sequence of codons in DNA, through the sequence of codons in mRNA, spells out the primary structure of a polypeptide.

Mutations

Mutations are changes in the DNA base sequence, caused by errors in DNA replication, recombination, or mutagens. Substituting, deleting, or inserting nucleotides in a gene has varying effects on the polypeptide and organism.

Type of Mutation	Effect
Substitution of one DNA base for another	**Silent** mutations result in no change to amino acids.
	Missense mutations swap one amino acid for another.
	Nonsense mutations change an amino acid codon to a stop codon.
Insertions or **deletions** of DNA nucleotides	**Frameshift** mutations can alter the triplet grouping of codons and greatly change the amino acid sequence.

Viruses and Other Noncellular Infectious Agents

Viruses are infectious particles consisting of genes packaged in protein.

Bacteriophages

When phage DNA enters a lytic cycle inside a bacterium, it is replicated, transcribed, and translated. The new viral DNA and protein molecules then assemble into new phages, which burst from the cell. In the lysogenic cycle, phage DNA inserts into the cell's chromosome and is passed on to genera-tions of daughter cells. Much later, it may initiate phage production.

Plant Viruses

Viruses that infect plants can be a serious agricultural problem. Most have RNA genomes. Viruses enter plants through breaks in the plant's outer layers.

Animal Viruses

Many animal viruses, such as flu viruses, have RNA genomes; others, such as hepatitis viruses, have DNA. Some animal viruses "steal" a bit of cell membrane as a protective envelope. Some, such as the herpesvirus, can remain latent inside cells for long periods.

HIV, the AIDS Virus

HIV is a retrovirus. Inside a cell it uses its RNA as a template for making DNA, which is then inserted into a chromosome.

Prions

Prions are infectious proteins that cause a number of degenerative brain diseases in humans and other animals.

Mastering Biology

For practice quizzes, BioFlix animations, MP3 tutorials, video tutors, and more study tools designed for this textbook, go to Mastering Biology™

SELF-QUIZ

1. A molecule of DNA contains two polymer strands called _____, made by bonding together many monomers called _____.

2. Name the three parts of every nucleotide.

3. List these terms in order of size from largest to smallest: chromosome, codon, gene, nucleotide.

4. A scientist inserts a radioactively labeled DNA molecule into a bacterium. The bacterium replicates this DNA molecule and distributes one daughter molecule (double helix) to each of two daughter cells. How much radioactivity will the DNA in each of the two daughter cells contain? Why?

5. Which mRNA nucleotide triplet encodes the amino acid tryptophan (see Figure 10.10)? During translation, an amino-acid-conjugated tRNA binds to an mRNA nucleotide triplet via its anticodon. What is the nucleotide sequence of the tryptophan's tRNA anticodon? What is the corresponding original codon on the DNA molecule that the mRNA is transcribed from?

6. Describe the process by which the information in a gene is transcribed and translated into a protein. Correctly use these terms in your description: tRNA, amino acid, start codon, transcription, mRNA, gene, codon, RNA polymerase, ribosome, translation, anticodon, peptide bond, stop codon.

7. Match the following molecules with the cellular process or processes in which they are primarily involved.
   - a. ribosomes
   - b. tRNA
   - c. DNA polymerases
   - d. RNA polymerase
   - e. mRNA
   1. DNA replication
   2. transcription
   3. translation

8. A geneticist finds that a particular mutation has no effect on the polypeptide encoded by the gene. This mutation probably involves
   a. deletion of one nucleotide.
   b. alteration of the start codon.
   c. insertion of one nucleotide.
   d. substitution of one nucleotide.

9. Scientists have discovered how to put together a bacteriophage with the protein coat of phage A and the DNA of phage B. If this composite phage were allowed to infect a bacterium, the phages produced in the cell would have
   a. the protein of A and the DNA of B.
   b. the protein of B and the DNA of A.
   c. the protein and DNA of A.
   d. the protein and DNA of B.

10. How do some viruses reproduce without ever having DNA?

11. HIV requires an enzyme called _____ to convert its RNA genome to a DNA version. Why is this enzyme a particularly good target for anti-AIDS drugs? (*Hint*: Would you expect such a drug to harm the human host?)

For answers to the Self Quiz, see Appendix D.

IDENTIFYING MAJOR THEMES

For each statement, identify which major theme is evident (the relationship of structure to function, information flow, pathways that transform energy and matter, interactions within biological systems, or evolution) and explain how the statement relates to the theme. If necessary, review the themes (see Chapter 1) and review the examples highlighted in blue in this chapter.

12. Nearly every organism on Earth shares the identical genetic code, indicating that this scheme arose very early in the history of life.

13. The shape of a tRNA molecule, with its anticodon on one end and amino acid attachment site at the other end, hints at how the molecule acts during translation.

14. Genes carry the instructions needed to build an RNA and then a protein.

For answers to Identifying Major Themes, see Appendix D.

THE PROCESS OF SCIENCE

15. *Acetabularia* are enormously large (2–4 cm long), single-celled green algae that look somewhat like mushrooms. They consist of a cap, a stalk, and a root-like structure called "rhizoid" that contains the large cell nucleus. In 1943, Joachim Hämmerling exchanged the nuclei of an *Acetabularia mediterranea* (which forms a flat cap) with that of an *Acetabularia crenulata* (which forms a castellated cap). What result would you expect from this experiment? Which basic concept does it confirm?

16. In 1958, Matthew Meselson and Franklin Stahl grew bacteria in a medium enriched with the rare, heavy nitrogen isotope ^{15}N. The DNA extracted from these bacteria can be separated from the DNA extracted from the bacteria grown in normal, mostly ^{14}N-containing medium by

density gradient ultracentrifugation—the two DNA fractions separate into distinct layers within the gradient. What did Meselson and Stahl observe when they transferred bacteria that were initially cultured in ^{15}N medium to ^{14}N medium just long enough for one more round of cell division? What does this experiment demonstrate?

17. **Interpreting Data** The graph shows the number of cases per week of Zika, Dengue, and Chikungunya virus in Puerto Rico during the period from November 1, 2015 to April 14, 2016. The same mosquitoes spread all three viruses. Did all three diseases show a similar pattern during this time period? Explain.

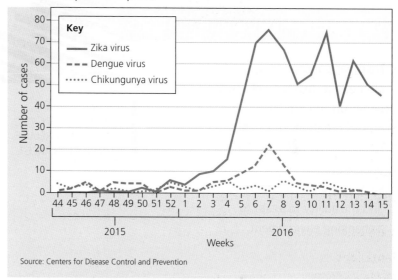

Source: Centers for Disease Control and Prevention

BIOLOGY AND SOCIETY

18. Scientists at the National Institutes of Health (NIH) have worked out thousands of sequences of genes and the proteins they encode, and similar analyses are being carried out at universities and private companies. Knowledge of the nucleotide sequences of genes might be used to treat genetic defects or produce lifesaving medicines. The NIH and some U.S. biotechnology companies have applied for patents on their discoveries. In Britain, the courts have ruled that a naturally occurring gene cannot be patented. Do you think individuals and companies should be able to patent genes and gene products? Before answering, consider the following: What are the purposes of a patent? How might the discoverer of a gene benefit from a patent? How might the public benefit? What negative effects might result from patenting genes?

19. Your college roommate seeks to improve her appearance by visiting a tanning salon. How would you explain the dangers of this to her?

20. Flu vaccines have been shown to be safe, are very reliable at reducing the risk of hospitalization or death from influenza, and are inexpensive. Should children be required to obtain a flu vaccine before going to school? What about hospital workers before reporting to work? Defend your answers to these questions.

11 How Genes Are Controlled

Why Gene Regulation Matters

No organism uses all its genes all the time. Instead, cells turn genes on and off depending on the situation. The ability to regulate genes therefore underlies all of life's processes, such as growth, metabolism, and the ability to adapt to a changing environment.

SIMPLE CHANGES IN LIFESTYLE CAN DRAMATICALLY REDUCE YOUR RISK OF CANCER.

CLONING MAY HELP SAVE THE GIANT PANDA FROM EXTINCTION.

A DNA CHIP MAY SOON BECOME A DIAGNOSTIC TOOL AS COMMON AS X-RAYS.

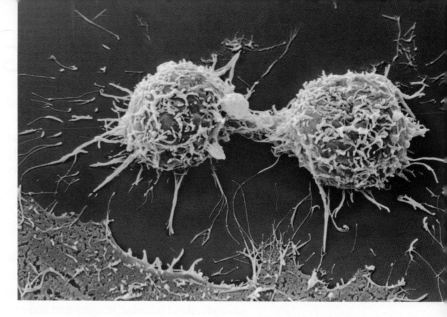

Human cancer cells.
These cells from a cancerous cervical tumor have lost the ability to control their growth.

BIOLOGY AND SOCIETY Cancer

Breast Cancer and Chemotherapy

First, some bad news: About one out of every eight women will develop breast cancer. But there is also some good news: If a woman's breast cancer is treated in its earliest stages, her chance of surviving five years or more is nearly 100%. On the other hand, if her breast cancer is not treated until it has spread throughout her body, the five-year survival rate is less than 25%. Treatment typically involves surgery, followed by hormone therapy and radiation to kill cancer cells. Chemotherapy may serve as a final step. Unfortunately, chemotherapy has many negative side effects, such as nausea and hair loss, and in a small percentage of women, long-term nerve, blood, or heart disorders.

Cancer specialists traditionally determine whether a woman should have chemotherapy by using clinical factors, such as the size of her tumor and how many lymph nodes had cancer cells in them. As an analogy, suppose that you knew a car with no brakes was headed toward a cliff. Should you shoot out the tires? Is the intervention worth the risk? The clinical factors are like looking at how far the car is from the edge of the cliff. The closer to the edge, the more it makes sense to use an extreme intervention. However, a woman may have a large tumor and several affected lymph nodes because she has had a slow-growing cancer for a long time. Her risk may actually be low. Giving her chemotherapy might be an unnecessary intervention, like shooting out the tires on a car that is never going to reach the cliff.

How can doctors tell which cases are risky enough to warrant chemotherapy? What determines how fast cancer cells spread and therefore the risk posed by a particular tumor? First, it's helpful to know how mutations lead to cancer. Many cancer-associated genes encode proteins that turn other genes on or off. When these genes are mutated, the proteins malfunction and the cell may become cancerous. Scientists can now tell which genes are mutated in a given tumor. This information is like knowing the size of the engine of the car headed for the cliff. It allows medical professionals to predict the potential growth rate of the cancer. Soon, the genes of all cancer patients may be evaluated in this way, allowing therapy to be optimized for each patient. Such research may someday provide a strong barrier between cars and the edge of the cliff.

The ability to properly control which genes are active at any given time is crucial to normal cell function. How genes are controlled and how the regulation of genes affects cells and organisms—including ways that gene regulation affects your own life—are the subjects of this chapter.

How and Why Genes Are Regulated

Every cell in your body—and, indeed, all the cells in the body of every sexually reproducing organism—was produced through successive rounds of mitosis starting from the zygote, the original cell that formed after fusion of sperm and egg. Mitosis exactly duplicates the chromosomes. Therefore, every cell in your body has the same DNA as the zygote. To put it another way: Every somatic (body) cell contains every gene. However, the cells in your body are specialized in structure and function; a neuron, for example, looks and acts nothing like a red blood cell. But if every cell contains identical genetic instructions, how do cells develop differently from one another? To help you understand this idea, imagine that every restaurant in your hometown uses the same cookbook. If that were the case, how could each restaurant develop a unique menu? The answer is obvious: Even though each restaurant has the same cookbook, different restaurants pick and choose different recipes from this book to prepare. Similarly, cells with the same genetic information can develop into different types of cells through **gene regulation**, mechanisms that turn on certain genes while other genes remain turned off. Regulating gene activity allows for specialization of cells within the body, just as regulating which recipes are used allows for varying menus in multiple restaurants.

As an example of gene regulation, consider the development of a single-celled zygote into a multicellular organism. During embryonic growth, groups of cells follow different paths, and each group becomes a particular kind of tissue. In the mature organism, each cell type—neuron or red blood cell, for instance—has a different pattern of turned-on genes.

What does it mean to say that genes are turned on or off? Genes determine the nucleotide sequence of specific mRNA molecules, and mRNA in turn determines the sequence of amino acids in proteins (in summary: DNA → RNA → protein; see Chapter 10). A gene that is turned on is being transcribed into mRNA, and that message is being translated into specific proteins. The overall process by which genetic information flows from genes to proteins is called **gene expression**. **The control of gene expression makes it possible for cells to produce specific kinds of proteins when and where they are needed, allowing cells to respond quickly and efficiently to information from the environment.**

As an illustration of this principle, **Figure 11.1** shows the patterns of gene expression for four genes in three different specialized cells of an adult human. Note that the genes for "housekeeping" enzymes, such as those that provide energy through glycolysis, are "on" in all the cells. In contrast, the genes for some proteins, such as insulin and hemoglobin, are expressed only by particular kinds of cells. One protein, hemoglobin, is not expressed in any of the cell types shown in the figure. ✓

Gene Regulation in Bacteria

To understand how a cell can regulate gene expression, consider the relatively simple case of bacteria. In the

☑ **CHECKPOINT**

If your blood cells and skin cells have the same genes, how can they be so different?

■ *Answer: Each cell type expresses different genes than the other cell type.*

▶ Figure 11.1 **Patterns of gene expression in three types of human cells.** Different types of cells express different combinations of genes. The specialized proteins whose genes are represented here are an enzyme involved in glucose digestion; an antibody, which aids in fighting infection; insulin, a hormone made in the pancreas; and the oxygen transport protein hemoglobin, which is expressed only in red blood cells.

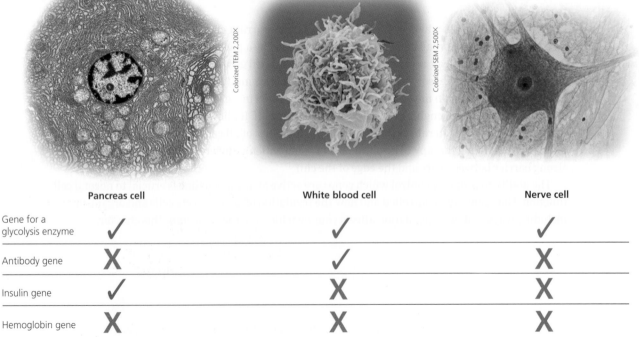

	Pancreas cell	White blood cell	Nerve cell
Gene for a glycolysis enzyme	✓	✓	✓
Antibody gene	X	✓	X
Insulin gene	✓	X	X
Hemoglobin gene	X	X	X

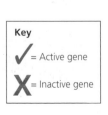

Key

✓ = Active gene

X = Inactive gene

course of their lives, bacteria must regulate their genes in response to environmental changes. For example, when a nutrient is plentiful, bacteria do not squander valuable resources to make the nutrient from scratch. Bacterial cells that can conserve resources and energy have a survival advantage over cells that are unable to do so. Thus, natural selection has favored bacteria that express only the genes whose products are needed by the cell.

Imagine an *Escherichia coli* bacterium living in your intestines. It will be bathed in various nutrients, depending on what you eat. If you drink a milk shake, for example, there will be a sudden rush of the sugar lactose. In response, *E. coli* will express three genes for enzymes that enable the bacterium to absorb and digest this sugar. After the lactose is gone, these genes are turned off; the bacterium does not waste its energy continuing to produce these enzymes when they are not needed. Thus, a bacterium can adjust its gene expression to changes in the environment. **Such regulation is at the heart of metabolism, the chemical reactions that transform energy and matter within all cells.**

How does a bacterium "know" if lactose is present or not? In other words, how does the presence or absence of lactose influence the activity of the genes that code for the lactose enzymes? The key is the way the three lactose-digesting genes are organized: They are adjacent in the DNA and turned on and off as a single unit. This regulation is achieved through short stretches of DNA that help turn all three genes on and off at once, coordinating their expression. Such a cluster of related genes and sequences that control them is called an **operon (Figure 11.2)**. The operon considered here, the *lac* (short for lactose) operon, illustrates principles of gene regulation that apply to a wide variety of prokaryotic genes.

How do DNA control sequences turn genes on or off? One control sequence, called a **promoter** (green in the figure), is the site where the enzyme RNA polymerase attaches and initiates transcription—in our example, transcription of the genes for lactose-digesting enzymes. Between the promoter and the enzyme genes, a DNA segment called an **operator** (yellow) acts as a switch that is turned on or off, depending on whether a specific protein is bound there. The operator and protein together determine whether RNA polymerase can attach to the promoter and start transcribing the genes (light blue). In the *lac* operon, when the operator switch is turned on, all the enzymes needed to metabolize lactose are made at once.

The top half of Figure 11.2 shows the *lac* operon in "off" mode, its status when there is no lactose available. Transcription is turned off because ❶ a protein called a **repressor** ( ) binds to the operator ( ) and ❷ physically blocks the attachment of RNA polymerase ( ) to the promoter ( ).

The bottom half of Figure 11.2 shows the operon in "on" mode, when lactose is present. The lactose ( )

interferes with attachment of the *lac* repressor to the operator by ❶ binding to the repressor and ❷ changing the repressor's shape. **As we so often see in biological systems, structure and function are related.** In this case, altering the repressor's shape changes how it acts. In its new shape ( ), the repressor cannot bind to the operator, and the operator switch remains on. ❸ RNA polymerase is no longer blocked, so it can now bind to the promoter and from there ❹ transcribe the genes for the lactose enzymes into mRNA. ❺ Translation produces all three lactose enzymes (purple).

Many operons have been identified in bacteria. Some are quite similar to the *lac* operon, whereas others have somewhat different mechanisms of control. For example, operons that control amino acid synthesis cause bacteria to stop making these molecules when they are already present in the environment, saving materials and energy for the cells. In these cases, the amino acid *activates* the repressor. Armed with a variety of operons, *E. coli* and other prokaryotes can thrive in frequently changing environments. ✓

☑ **CHECKPOINT**

A mutation in *E. coli* makes the *lac* operator unable to bind the active repressor. How would this mutation affect the cell? Why would this effect be a disadvantage?

■ *Answer: The cell would wastefully produce the enzymes for lactose metabolism continuously, even in the absence of lactose.*

▼ **Figure 11.2** The *lac* operon of *E. coli.*

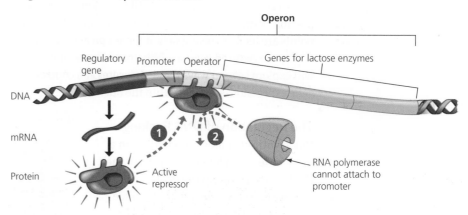

Operon turned off (lactose absent)

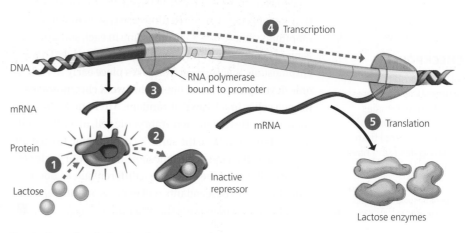

Operon turned on (lactose inactivates repressor)

233

Gene Regulation in Eukaryotic Cells

Eukaryotes, especially multicellular ones, have more sophisticated mechanisms than bacteria for regulating the expression of their genes. This is not surprising because a prokaryote, being a single cell, does not have different types of specialized cells, such as neurons and red blood cells. Therefore, it does not require the elaborate regulation of gene expression that leads to cell specialization in multicellular eukaryotic organisms.

The pathway from gene to protein in eukaryotic cells is a long one, providing a number of points where the process can be turned on or off, speeded up or slowed down. Picture the series of pipes that carry water from your local reservoir to a faucet in your home. At various points, valves control the flow of water. We use this analogy in **Figure 11.3** to illustrate the flow of genetic information from a eukaryotic chromosome—a reservoir of genetic information—to an active protein that has been made in the cell's cytoplasm. The multiple mechanisms that control gene expression are analogous to the control valves in your water pipes. In the figure, each control knob indicates a gene expression "valve." All these knobs represent possible control points, although only one or a few control points are likely to be important for a typical protein.

Using a reduced version of Figure 11.3 as a guide, we will explore several ways that eukaryotes can control gene expression, starting within the nucleus.

The Regulation of DNA Packing

Eukaryotic chromosomes may be in a more or less condensed state, with the DNA and accompanying proteins more or less tightly wrapped together (see Figure 8.4). DNA packing tends to prevent gene expression by preventing RNA polymerase and other transcription proteins from binding to the DNA.

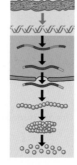

Cells may use DNA packing for the long-term inactivation of genes. One intriguing case is seen in female mammals, where one X chromosome in each somatic cell is highly compacted and almost entirely inactive. This X chromosome inactivation first takes place early in embryonic development, when one of the two X chromosomes in each cell is inactivated at random. After one X chromosome is inactivated in each embryonic cell, all of that cell's descendants will have the same X chromosome turned off. Consequently, if one X chromosome in the embryonic cell has one allele and the other has a different allele, about half of the cell's descendants will express one allele, while the other half will express the other allele **(Figure 11.4)**. ☑

☑ CHECKPOINT

Would a gene on the X chromosome be expressed more in human females (who have two copies of the X chromosome) than in human males (who have one copy)?

■ *Answer: No, because in females one of the X chromosomes in each cell is inactivated.*

▼ **Figure 11.3 The gene expression "pipeline" in a eukaryotic cell.** Each valve in the pipeline represents a stage at which the pathway from gene to functioning protein can be regulated, turned on or off, or speeded up or slowed down. Throughout this discussion we will use a miniature version of this figure to keep track of the stages as they are discussed.

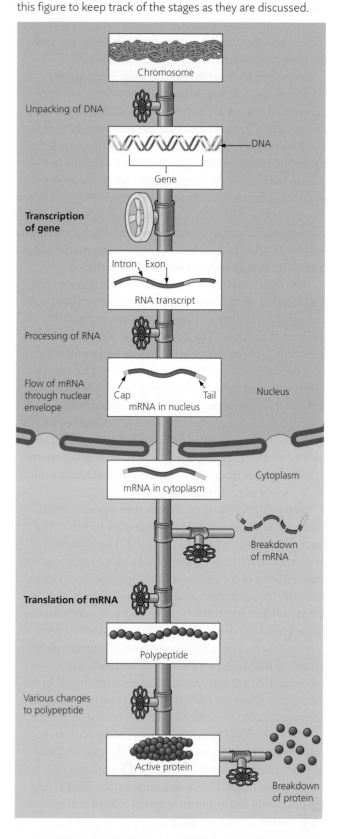

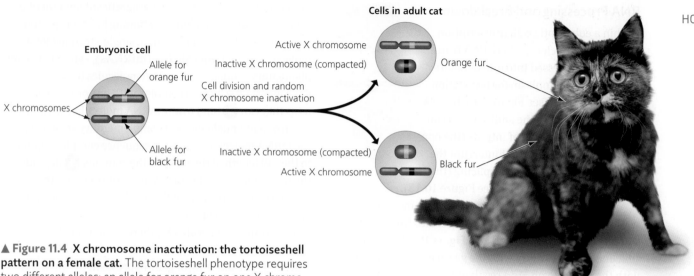

▲ **Figure 11.4 X chromosome inactivation: the tortoiseshell pattern on a female cat.** The tortoiseshell phenotype requires two different alleles: an allele for orange fur on one X chromosome and an allele for black fur on the other X chromosome. Orange patches are formed by populations of cells in which the X chromosome with the allele for orange fur is active; black patches have cells in which the X chromosome with the allele for black fur is active.

The Initiation of Transcription

The most important stage for regulating gene expression is the determination of whether transcription is initiated or not. That is why this control point is emphasized with the large yellow valve in Figure 11.3. In both prokaryotes and eukaryotes, regulatory proteins bind to DNA and turn the transcription of genes on and off. Unlike prokaryotic genes, however, most eukaryotic genes are not grouped into operons. Instead, each eukaryotic gene usually has its own promoter and other control sequences.

Transcriptional regulation in eukaryotes is complex, typically involving many proteins (**Figure 11.5**). To do its job, RNA polymerase requires the assistance of proteins called **transcription factors**. Some are essential for transcribing all genes, and others are specific to a few or just one gene. Transcription factors (purple in the figure) bind to noncoding DNA sequences called **enhancers** (yellow) and help RNA polymerase (orange) bind to the promoter (green). Genes coding for related enzymes, such as those in a metabolic pathway, may share a specific kind of enhancer (or collection of enhancers), allowing these genes to be activated at the same time. (Not shown in the figure are repressor proteins, which may bind to DNA sequences called **silencers**, inhibiting the start of transcription.)

In fact, repressor proteins that turn genes off are less common in eukaryotes than **activators**, proteins that turn genes on by binding to DNA. Activators act by making it easier for RNA polymerase to bind to the promoter. The use of activators is efficient because a typical animal or plant cell needs to turn on (transcribe) only a small percentage of its genes, those required for the cell's specialized structure and function. The default state for most genes in multicellular eukaryotes seems to be off; research indicates that a typical human cell expresses only about 20% of its protein-coding genes at any given time. ☑

☑ **CHECKPOINT**

Of all the control points of DNA expression shown in Figure 11.3, which is under the tightest regulation?

■ *Answer: the initiation of transcription*

▼ **Figure 11.5 A model for turning on a eukaryotic gene.** Transcription is regulated by noncoding DNA sequences called enhancers. DNA-bending proteins help bring the enhancers close to the transcription site. Once this is done, transcription factors help RNA polymerase bind to the promoter, where it begins transcription of the gene.

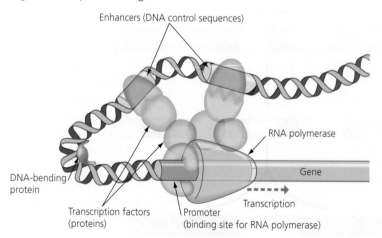

Enhancers (DNA control sequences)

RNA polymerase

Gene

DNA-bending protein

Transcription factors (proteins)

Promoter (binding site for RNA polymerase)

Transcription

RNA Processing and Breakdown

Within a eukaryotic cell, transcription occurs in the nucleus, where RNA transcripts are processed into mRNA before moving to the cytoplasm for translation by the ribosomes (see Figure 10.19). RNA processing includes the addition of a cap and a tail, the removal of introns (the non-coding DNA segments that interrupt the genetic message), and RNA splicing (the splicing together of exons) (see Figure 10.13).

Within a cell, exon splicing can occur in more than one way, generating different mRNA molecules from the same starting RNA molecule. Notice in **Figure 11.6**, for example, that one mRNA ends up with the green exon and the other with the brown exon. As a result of this process, called **alternative RNA splicing**, an organism can produce more than one type of polypeptide from a single gene. A typical human gene contains about ten exons; nearly all genes are spliced in at least two different ways, and some are spliced hundreds of different ways.

After an mRNA is produced in its final form, its "lifetime" can be highly variable, from hours to weeks to months. Controlling the timing of mRNA breakdown provides another opportunity for regulation. But all mRNAs are eventually broken down and their parts recycled. ☑

microRNAs

The vast majority of human DNA does not code for proteins. This DNA has long been thought to be lacking any genetic information. In fact, many biologists used to refer to these regions as "junk DNA" because they performed no discernible function. However, a significant amount of the genome is transcribed into functioning but non–protein-coding RNAs. For example, small, single-stranded RNA molecules, called **microRNAs (miRNAs)**, can bind to complementary sequences on mRNA molecules **(Figure 11.7)**. Each miRNA ❶ forms a complex with one or more proteins that can ❷ bind to any mRNA molecule with at least seven or eight nucleotides of complementary sequence. If the mRNA molecule contains a sequence complementary to the full length of the miRNA, the complex ❸ degrades the target mRNA. If the mRNA molecule matches the sequence along just part of the miRNA, the complex ❹ blocks its translation.

In addition to microRNAs, there is another class of small RNA molecules called small interfering RNAs (siRNAs). The blocking of gene expression by siRNAs is called **RNA interference (RNAi)**. Researchers can take advantage of siRNAs to artificially control gene expression. For example, injecting siRNA into a cell can turn off expression of a gene with a sequence that matches the siRNA. RNAi, therefore, allows researchers to disable specific genes in order to investigate their functions.

Biologists are excited about these recent discoveries, which hint at a large, diverse population of RNA molecules in the cell that play crucial roles in regulating gene expression but have gone largely unnoticed until recently. Our improved understanding may lead to important clinical applications. For example, in 2009, researchers discovered a particular miRNA that is essential for the proper functioning of the pancreas. Without it, insulin-producing beta cells die, which can lead to diabetes.

☑ CHECKPOINT

After a gene is transcribed in the nucleus, name three ways that the RNA may be processed.

■ *Answer: by the addition of cap and tail, the removal of introns, and the splicing of exons*

▼ **Figure 11.6 Alternative RNA splicing: producing multiple mRNAs from the same gene.** Two different cells can use the same DNA gene to synthesize different mRNAs and proteins. In this example, one mRNA has ended up with exon 3 (brown) and the other with exon 4 (green). These mRNAs, which are just two of many possible outcomes, can then be translated into different proteins.

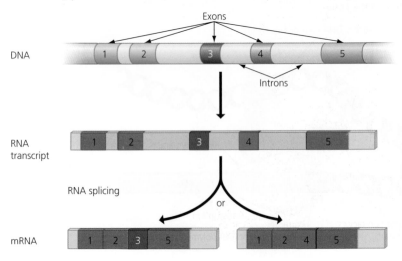

▼ **Figure 11.7 Regulation of gene expression by an miRNA.**

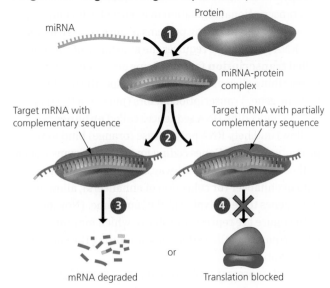

The Initiation of Translation

The process of translation—in which an mRNA is used to make a protein—offers additional opportunities for control by regulatory molecules. Red blood cells, for instance, have a protein that prevents the translation of hemoglobin mRNA unless the cell has a supply of heme, an iron-containing chemical group essential for hemoglobin function.

Protein Activation and Breakdown

The final opportunities for regulating gene expression occur after translation. For example, the hormone insulin is synthesized as one long, inactive polypeptide that must be chopped into pieces before it comes active. Other proteins require chemical modification before they become active.

Another control mechanism operating after translation is the selective breakdown of proteins. Some proteins that trigger metabolic changes in cells are broken down within a few minutes or hours. This regulation allows a cell to adjust the kinds and amounts of its proteins in response to changes in its environment.

Cell Signaling

So far, we have considered gene regulation only within a single cell. **Within a multicellular organism, information must be communicated between cells.** For example, a cell can produce and secrete chemicals, such as hormones, that affect gene regulation in another cell. This allows the organism as a whole to alter its activities in response to signals from the environment. Consider an analogy from your own experience: In grade school, did you ever station a classmate near the door to signal the teacher's return? Information from outside the room (the teacher's approach) was used to alter behavior within the classroom ("Stop messing around!"). In a similar way, cells use protein "lookouts" to convey information into the cell, resulting in changes to cellular functions.

A signal molecule can act by binding to a receptor protein and initiating a **signal transduction pathway**, a series of molecular changes that converts a signal received outside a cell to a specific response inside the target cell. **Figure 11.8** shows an example of cell-to-cell signaling in which the target cell's response is the transcription (turning on) of a gene. ❶ First, the signaling cell secretes the signal molecule ( ). ❷ This molecule binds to a specific receptor protein ( ) embedded in the target cell's plasma membrane. ❸ The binding activates a signal transduction pathway consisting of a series of relay proteins (green) within the target cell. Each relay molecule activates the next. ❹ The last relay molecule

in the series activates a transcription factor ( ) that ❺ triggers the transcription of a specific gene. ❻ Translation of the mRNA produces a protein that can then perform the function originally called for by the signal. ☑

▼ **Figure 11.8 A cell-signaling pathway that turns on a gene.** The coordination of cellular activities in a multicellular organism depends on cell-to-cell signaling that helps regulate genes.

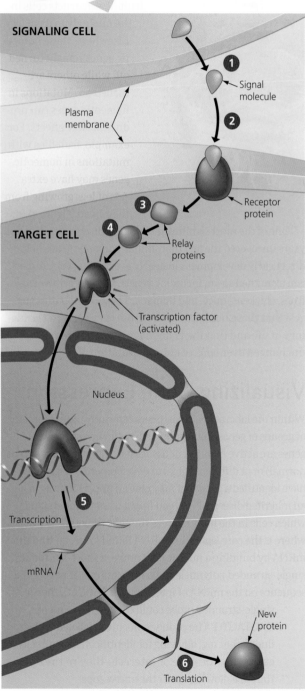

☑ **CHECKPOINT**

How can a signal molecule from one cell alter gene expression in a target cell without entering the target cell?

■ *Answer: by binding to a receptor protein in the membrane of the target cell and triggering a signal transduction pathway that activates transcription factors*

Figure Walkthrough

Mastering **Biology**
goo.gl/tztnY6

Homeotic Genes

Cell-to-cell signaling and the control of gene expression are particularly important during early embryonic development, when a single-celled zygote develops into a multicellular organism. **Interactions between the cells of an embryo through chemical signals coordinate the process of development. Master control genes called homeotic genes regulate groups of other genes that determine what body parts will develop in which locations.** For example, one set of homeotic genes in fruit flies instructs cells in the midbody to form legs. Elsewhere, these homeotic genes remain turned off, while others are turned on. Mutations in homeotic genes can produce bizarre effects. For example, fruit flies with mutations in homeotic genes may have extra sets of legs growing from their head **(Figure 11.9)**.

One of the most significant biological discoveries in recent years uncovered the fact that similar homeotic genes help direct early development in nearly every eukaryotic organism examined so far, including yeasts, plants, earthworms, frogs, chickens, mice, and humans. **These similarities suggest that these homeotic genes arose very early in the history of life and that the genes have remained remarkably unchanged over eons of animal evolution.** ☑

▶ **Figure 11.9 The effect of homeotic genes.** A mutant fruit fly has an extra pair of legs growing out of its head as a result of a mutation in a homeotic (master control) gene.

Eye

Antenna

SEM 50x

Extra pair of legs

SEM 50x

☑ CHECKPOINT

How can a mutation in just one homeotic gene drastically affect an organism's physical appearance?

■ *Answer: Because homeotic genes control many other genes, a single change can affect the expression of many of the proteins that control appearance.*

Visualizing Gene Expression

Within the laboratory, researchers can study the expression of groups of genes. For example, they can investigate which genes are active in different tissues (such as cancerous versus normal) or at different stages of development. Imagine you have identified a gene that may play an important role in an inherited disease. First, you might want to understand which cells in the body express the gene—in other words, where is the corresponding mRNA found? You can find the mRNA by building a nucleic acid probe, a short, synthetic, single-stranded polynucleotide. For example, if part of the sequence on the mRNA of interest is CUCAUCAC, then a single-stranded probe could contain the sequence GAGTAGTG. The probe molecule is labeled with a fluorescent tag, allowing for identification of all cells expressing the gene of interest—they're the ones that are glowing under the microscope!

Rather than studying just one gene, researchers can study many or even all genes at once using DNA microarrays. A **DNA microarray** (also called a DNA chip or gene chip) is a glass slide with many tiny wells, each containing a different fragment of single-stranded DNA that derives from a particular gene. The wells are arranged in a tightly spaced array, or grid.

During a DNA microarray study, a researcher collects all of the mRNA transcribed in a particular type of cell at a given moment. This collection of mRNA is mixed with reverse transcriptase, a viral enzyme that produces DNA that is complementary to each mRNA sequence. These fragments are called **complementary DNAs (cDNAs)** because each one is complementary to one of the mRNAs. The cDNAs are synthesized using nucleotides that have been modified to fluoresce (glow). The fluorescent cDNA collection represents all of the genes being actively transcribed in that particular cell at that particular time. The fluorescently labeled cDNA mixture is added to the DNA fragments of the microarray. If a molecule in the cDNA mixture is complementary to a DNA fragment at a particular location on the grid, the cDNA molecule binds to it, producing a detectable glow in the microarray. Often, the cDNAs from two samples (for example, two tissues) are labeled with molecules that emit different colors and tested on the same microarray **(Figure 11.10)**. The pattern of glowing spots enables the researcher to determine which genes are being transcribed in one tissue compared with another.

DNA microarrays hold great promise in medical research. One study showed that DNA microarray data can classify different types of leukemia into specific subtypes based on the activity of 17 genes. It may become standard practice for every cancer patient to have DNA microarray analysis to discover the specific mutations involved in their cancer.

A DNA CHIP MAY BECOME A TOOL AS COMMON AS X-RAYS.

▼ **Figure 11.10 Visualizing gene expression using a DNA microarray.**

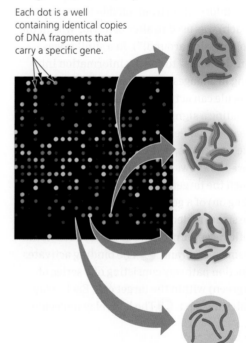

Each dot is a well containing identical copies of DNA fragments that carry a specific gene.

The genes in the red wells are expressed in one tissue and bind the red cDNAs.

The genes in the green wells are expressed in the other tissue and bind the green cDNAs.

The genes in the yellow wells are expressed in both tissues and bind both red and green cDNAs, appearing yellow.

The genes in the black wells are not expressed in either tissue and do not bind either cDNA.

Cloning Plants and Animals

Now that we have examined how gene expression is regulated, we will devote the rest of this chapter to discussing how gene regulation affects two important processes: cloning and cancer.

The Genetic Potential of Cells

One of the most important take-home lessons from this chapter is that all body cells contain a complete set of genes, even if they are not expressing all of them. If you've ever grown a plant from a small cutting, you've seen evidence of this yourself: A single differentiated plant cell can undergo cell division and give rise to a complete adult plant, a task that requires a complete set of genes. On a larger scale, the technique described in **Figure 11.11** can be used to produce hundreds or thousands of genetically identical organisms—clones—from the cells of a single plant.

Plant cloning is now used extensively in agriculture. For some plants, such as orchids, cloning is the only commercially practical means of reproducing plants. In other cases, cloning has been used to reproduce a plant with specific desirable traits, such as high fruit yield or resistance to disease. Seedless plants (such as seedless grapes, watermelons, and oranges) cannot reproduce sexually, leaving cloning as the sole means of mass-producing these common foods. In fact, every navel orange in supermarkets today is a clone of a single tree that grew on the grounds of a Brazilian monastery in the early 1800s.

Is this sort of cloning possible in animals? A good indication that some animal cells can also tap into their full genetic potential is **regeneration**, the regrowth of lost body parts. When a salamander loses a tail, for example, certain cells in the tail stump reverse their differentiated state, divide, and then differentiate again to give rise to a new tail. Many other animals, especially among the invertebrates (sea stars and sponges, for example), can regenerate lost parts, and isolated pieces of a few relatively simple animals can dedifferentiate and then develop into an entirely new organism (see Figure 8.1).

▼ **Figure 11.11 Test-tube cloning of an orchid.** Tissue removed from the stem of an orchid plant and placed in growth medium may begin dividing and eventually grow into an adult plant. The new plant is a genetic duplicate of the parent plant. This process proves that mature plant cells can reverse their differentiation and develop into all the specialized cells of an adult plant.

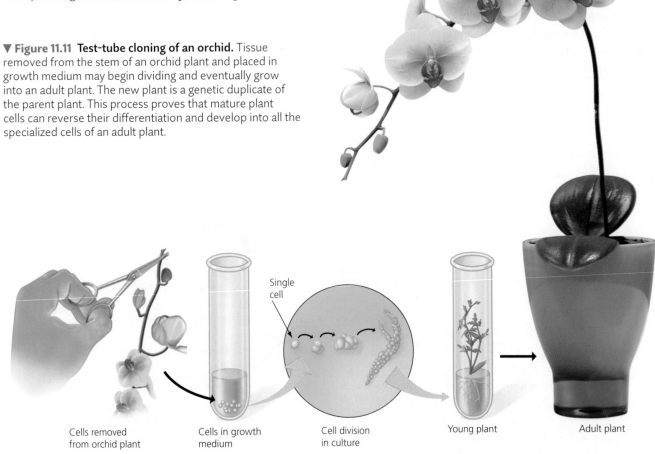

Cells removed from orchid plant | Cells in growth medium | Single cell | Cell division in culture | Young plant | Adult plant

Reproductive Cloning of Animals

Animal cloning is achieved through a procedure called **nuclear transplantation (Figure 11.12)**. First performed in the 1950s on frog embryos and in the 1990s on adult mammals, nuclear transplantation involves replacing the nucleus of an egg cell or a zygote with a nucleus from an adult body cell. If properly stimulated, the recipient cell may then begin to divide. Repeated cell divisions form a hollow ball of about 100 cells. At this point, the cells may be used for different purposes, as indicated by the two branches in Figure 11.12.

If the animal to be cloned is a mammal, further development requires implanting the early embryo into the uterus of a surrogate mother (Figure 11.12, upper branch). The resulting animal will be a clone (genetic copy) of the donor. This type of cloning is called **reproductive cloning** because it results in the birth of a new animal.

In 1996, researchers used reproductive cloning to produce the first mammal cloned from an adult cell, a sheep named Dolly. The researchers fused specially treated sheep cells with eggs from which they had removed the nuclei. After several days of growth, the resulting embryos were implanted in the uteruses of surrogate mothers. One of the embryos developed into Dolly—and, as expected, Dolly resembled the nucleus donor, not the egg donor or the surrogate mother.

CLONING MAY HELP SAVE THE GIANT PANDA FROM EXTINCTION.

Practical Applications of Reproductive Cloning

Since the first success in 1996, researchers have cloned many species of mammals, including mice, horses, dogs, mules, cows, pigs, rabbits, ferrets, camels, goats, and cats **(Figure 11.13a)**. Why would anyone want to do this? In agriculture, farm animals with specific sets of desirable traits might be cloned to produce identical herds. In research, genetically identical animals can provide perfect "control animals" for experiments. The pharmaceutical industry is experimenting with cloning animals for potential medical use **(Figure 11.13b)**. For example, researchers have produced pig clones that lack a gene for a protein that can cause immune system rejection in humans. Organs from such pigs may one day be used in human patients who require life-saving transplants. Other animals (such as dogs and cats) are cloned to serve as pets.

Perhaps the most intriguing practical application of the technique of reproductive cloning is to restock populations of endangered animals. Some of the rare animals that have been cloned are a banteng (an Asian cow) and a gaur (an Asian ox) **(Figure 11.13c)**. Meanwhile, the scientists who cloned Dolly are now trying to clone a giant panda in an effort to help maintain this unique species.

In 2003, a banteng was cloned using frozen cells from a zoo-raised banteng that had died 23 years prior. Scientists obtained banteng skin tissue from "The Frozen Zoo," a facility in San Diego, California, where samples from rare or endangered animals are stored for

▼ **Figure 11.12 Cloning by nuclear transplantation.** In nuclear transplantation, a nucleus from an adult body cell is injected into a nucleus-free egg cell. The resulting embryo may then be used to produce a new organism (reproductive cloning, shown in the upper branch) or to provide stem cells (therapeutic cloning, lower branch).

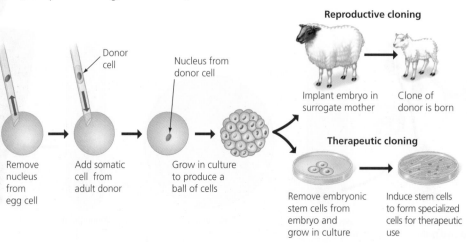

conservation. The scientists transplanted nuclei from the frozen cells into nucleus-free eggs from dairy cows (which are a separate species). The resulting embryos were implanted into surrogate cows, leading to the birth of a healthy baby banteng. This success shows that it is possible to produce a baby even when a female of the donor species is unavailable. Scientists may someday use similar cross-species methods to clone an animal from a recently extinct species.

The use of cloning to repopulate endangered species holds tremendous promise. However, cloning may also create new problems. Conservationists argue that cloning may detract from efforts to preserve natural habitats. They correctly point out that cloning does not increase genetic diversity and is therefore not as beneficial to endangered species as natural reproduction. In addition, an increasing body of evidence suggests that cloned animals are less healthy than animals produced by fertilization: Many cloned animals exhibit defects such as susceptibility to obesity, pneumonia, liver failure, and

premature death. Dolly, the cloned sheep, for example, was euthanized in 2003 after suffering complications from a lung disease usually seen only in much older sheep. She was 6 years old, while her breed has a life expectancy of 12 years. Some evidence suggests that chromosomal changes in the cloned animals are the cause, but the effects of cloning on animal health are still being investigated.

Human Cloning

The cloning of various mammals has heightened speculation that humans could be cloned. Critics point out the many practical and ethical objections to human cloning. Practically, cloning of mammals is extremely difficult and inefficient. Only a small percentage of cloned embryos (usually less than 10%) develop normally, and they appear less healthy than naturally born kin. Ethically, the discussion about whether or not people should be cloned—and if so, under what circumstances—is far from settled. Meanwhile, the research and the debate continue. ☑

☑ **CHECKPOINT**
Imagine that mouse coat color is always passed down from parent to offspring. Suppose a nucleus from an adult body cell of a black mouse is injected into an egg removed from a white mouse, and then the embryo is implanted into a brown mouse. What would be the color of the resulting cloned mice?

■ *Answer: black, the color of the nucleus donor*

▼ Figure 11.13 **Reproductive cloning of mammals.**

(a) The first clone. Dolly the sheep, shown in 1996 with her lone parent, was the first mammal cloned from an adult cell.

(b) Cloning for medical use. These piglets are clones of a pig that was genetically modified to lack a protein that causes transplant rejection in humans.

(c) Clones of endangered animals

Banteng

Gaur

241

Therapeutic Cloning and Stem Cells

The lower branch of Figure 11.12 shows the process of **therapeutic cloning**. The purpose of this procedure is not to create a living organism but rather to produce embryonic stem cells.

Embryonic Stem Cells

In mammals, **embryonic stem cells (ES cells)** are obtained by removing cells from a several-day-old embryo (which, at this stage, is a ball of cells). The removed cells are then grown in laboratory culture. Embryonic stem cells can divide indefinitely, and under the right conditions—such as the presence of certain growth-stimulating proteins—can (hypothetically) develop into a wide variety of different specialized cells **(Figure 11.14)**. If scientists can discover the right conditions, they may be able to grow cells for the repair of injured or diseased organs. Some people speculate, for example, that embryonic stem cells may one day be used to replace cells damaged by spinal cord injuries or heart attacks. The use of embryonic stem cells in therapeutic cloning is controversial, however, because removing them destroys the embryo.

Umbilical Cord Blood Banking

Another source of stem cells is blood collected from the umbilical cord and placenta at birth **(Figure 11.15)**. Such stem cells appear to be partially differentiated. In 2005,

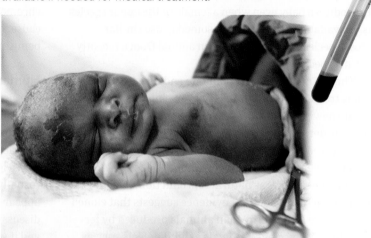

▼ **Figure 11.15 Umbilical cord blood banking.** Just after birth, a doctor inserts a needle into the umbilical cord and extracts ¼ to ½ cup of blood. The umbilical cord blood (inset), rich in stem cells, is frozen and kept in a blood bank, where it is available if needed for medical treatment.

doctors reported that an infusion of umbilical cord blood stem cells appeared to cure some babies of Krabbe disease, a fatal inherited disorder of the nervous system. Other people have received cord blood as a treatment for leukemia. To date, however, most attempts at umbilical cord blood therapy have not been successful. At present, the American Academy of Pediatrics recommends cord blood banking only for babies born into families with a known genetic risk.

Adult Stem Cells

Adult stem cells can also generate replacements for some of the body's cells. Adult stem cells are further along the road to differentiation than embryonic stem cells and can therefore give rise to only a few related types of specialized cells. For example, stem cells in bone marrow generate different kinds of blood cells. Adult stem cells from donor bone marrow have long been used as a source of immune system cells in patients whose own immune systems have been destroyed by disease or cancer treatments. Adult animals have only tiny numbers of stem cells, but scientists are learning to identify and isolate these cells from various tissues and, in some cases, to grow them in culture.

Because no embryonic tissue is involved in their harvest, adult stem cells are less ethically problematic than embryonic stem cells. However, many researchers hypothesize that only embryonic stem cells are likely to lead to groundbreaking advances in human health because they are more versatile. Recent research has shown that some adult cells, such as human skin cells, may be reprogrammed to act like embryonic stem cells. In the near future, such cells may prove to be both therapeutically useful and ethically clear. ✓

▼ **Figure 11.14 Differentiation of embryonic stem cells in culture.** Scientists hope to someday discover growth conditions that will stimulate cultured stem cells to differentiate into specialized cells.

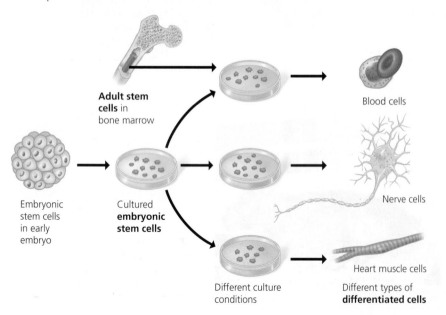

Adult stem cells in bone marrow

Blood cells

Embryonic stem cells in early embryo

Cultured **embryonic stem cells**

Nerve cells

Different culture conditions

Heart muscle cells

Different types of **differentiated cells**

The Genetic Basis of Cancer

Cancer occurs when cells escape from the control mechanisms that normally limit their growth and division (as introduced in Chapter 8). This escape involves changes in gene expression.

Genes That Cause Cancer

One of the earliest clues to the role of genes in cancer was the discovery in 1911 of a virus that causes cancer in chickens. Viruses that cause cancer can become permanent residents in host cells by inserting their nucleic acid into the DNA of host chromosomes. Over the last century, researchers have identified a number of viruses that harbor cancer-causing genes. One example is the human papillomavirus (HPV), which can be transmitted through sexual contact and is associated with several types of cancer, including cervical cancer.

Oncogenes and Tumor-Suppressor Genes

In 1976, American molecular biologists J. Michael Bishop, Harold Varmus, and their colleagues made a startling discovery. They found that a cancer-causing chicken virus contains a cancer-causing gene that is an altered version of a normal chicken gene. A gene that causes cancer is called an **oncogene** ("tumor gene"). Subsequent research has shown that the chromosomes of many animals, including humans, contain genes that can be converted to oncogenes. A normal gene with the potential to become an oncogene is called a **proto-oncogene**. (These terms can be confusing, so let's repeat them: A *proto-oncogene* is a normal, healthy gene that, if changed, can become a cancer-causing *oncogene*.) A cell can acquire an oncogene from a virus or from the mutation of one of its own proto-oncogenes.

How can a change in a gene cause cancer? Searching for the normal roles of proto-oncogenes in the cell, researchers found that many of these genes code for **growth factors**—proteins that stimulate cell division—or for other proteins that affect the cell cycle. When all these proteins are functioning normally, in the right amounts at the right times, they help keep the rate of cell division at an appropriate level. When they malfunction—if a growth factor becomes hyperactive, for example—cancer (uncontrolled cell growth) may result.

For a proto-oncogene to become an oncogene, a mutation must occur in the cell's DNA. **Figure 11.16** illustrates three kinds of changes in DNA that can produce active oncogenes. In all three cases, abnormal gene expression stimulates the cell to divide excessively.

Changes in genes whose products inhibit cell division are also involved in cancer. These genes are called **tumor-suppressor genes** because the proteins they encode normally help prevent uncontrolled cell growth (**Figure 11.17**). Any mutation that keeps a growth-inhibiting protein from being made or from functioning may contribute to the development of cancer. Researchers have identified many mutations in both tumor-suppressor and growth factor genes that are associated with cancer, as we'll discuss next.

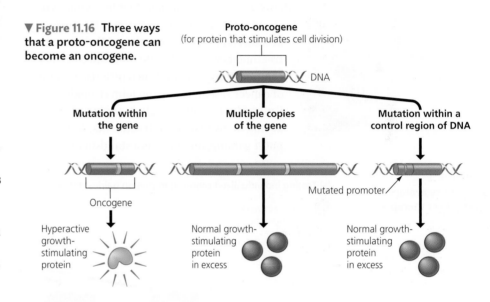

▼ **Figure 11.16** Three ways that a proto-oncogene can become an oncogene.

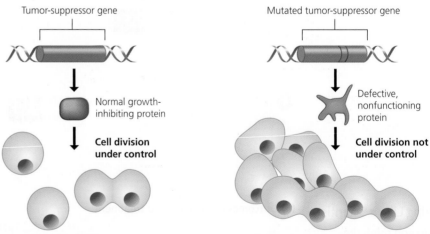

▼ **Figure 11.17** Tumor-suppressor genes.

(a) Normal cell growth. A tumor-suppressor gene normally codes for a protein that inhibits cell growth and division. Such genes help prevent cancerous tumors from arising or spreading.

(b) Uncontrolled cell growth (cancer). When a mutation in a tumor-suppressor gene makes its protein defective, cells that are usually under the control of the normal protein may divide excessively, forming a tumor.

Can Avatars Improve Cancer Treatment?

BACKGROUND

Imagine visiting your doctor because you have generalized stomach pain. You'd be quite surprised if your doctor suggested a treatment without first making a definitive diagnosis. And yet, cancer treatment usually proceeds in such a manner. Typically, a doctor knows the general type of cell involved in a tumor and will try treatments that work on that broad category of cancer. But taking an antacid won't cure appendicitis. Similarly, treatment that helps one type of cancer may not help patients with cancer arising from another mutation. Misdirected treatments waste valuable time and can cause complications. To avoid such pitfalls, a group of cancer researchers from Johns Hopkins University conducted a study that used mouse avatars to test potential treatments. As in computer gaming, an avatar is a stand-in for the real participant. In this case, live mice were standing in for the cancer patients.

METHOD

Researchers implanted cells from the tumors of 14 patients with high-risk cancers into mouse avatars **(Figure 11.18a)**. After treating each mouse with a different drug, they determined which treatments were effective at fighting the tumors. The therapies that showed the most promise were then given to the human patients.

RESULTS

In every case, treating a patient with a drug that was first shown to be effective in the avatar provided benefit to the human **(Figure 11.18b)**. Often, the treatment that was most effective would not have been identified without the use of avatars. In the future, such personalized therapies using an avatar may become standard treatment for otherwise deadly cancers.

▶ **Figure 11.18**
Using mouse avatars to test cancer therapies.

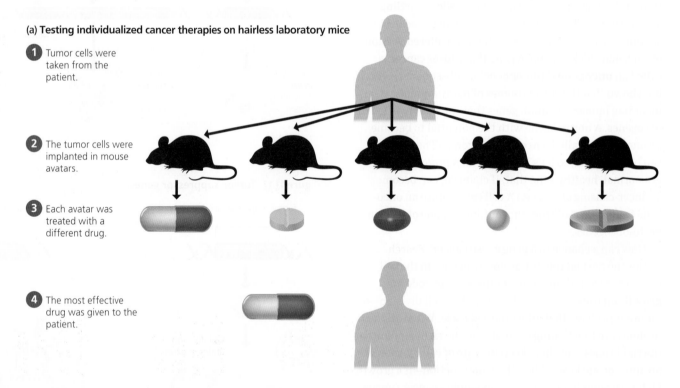

(a) Testing individualized cancer therapies on hairless laboratory mice

1 Tumor cells were taken from the patient.

2 The tumor cells were implanted in mouse avatars.

3 Each avatar was treated with a different drug.

4 The most effective drug was given to the patient.

Thinking Like a Scientist

Why do you think the researchers didn't just induce cancers in the mice and then test the drugs?

For the answer, see Appendix D.

(b) Outcomes of avatar-based treatment

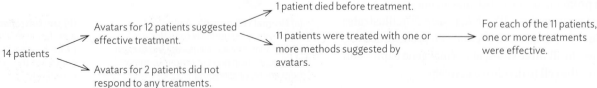

14 patients

Avatars for 12 patients suggested effective treatment.

Avatars for 2 patients did not respond to any treatments.

1 patient died before treatment.

11 patients were treated with one or more methods suggested by avatars.

For each of the 11 patients, one or more treatments were effective.

▼ **Figure 11.19** Stepwise development of colon cancer.

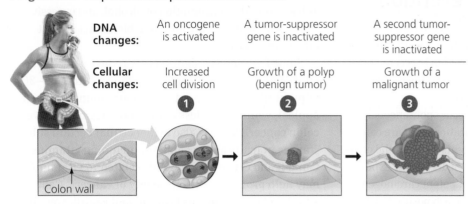

DNA changes:	An oncogene is activated	A tumor-suppressor gene is inactivated	A second tumor-suppressor gene is inactivated
Cellular changes:	Increased cell division	Growth of a polyp (benign tumor)	Growth of a malignant tumor

Colon wall

The Progression of a Cancer

Nearly 150,000 Americans will be stricken by cancer of the colon this year. One of the best-understood types, it illustrates a principle about how cancer develops: More than one mutation is needed to produce a cancer cell. As in many cancers, development of colon cancer is gradual.

As shown in **Figure 11.19**, ❶ colon cancer begins when an oncogene arises through mutation, causing unusually frequent division of normal-looking cells in the colon lining. ❷ Later, additional DNA mutations (such as the inactivation of a tumor-suppressor gene) cause the growth of a small benign tumor (called a polyp) in the colon wall. The cells of the polyp look normal, although they divide unusually frequently. If detected during a colonoscopy, suspicious polyps can usually be removed before they become a serious risk. ❸ Further mutations lead to formation of a malignant tumor—one that can metastasize (spread). It typically takes at least six mutations (usually creating at least one active oncogene and disabling at least one tumor-suppressor gene) before a cell becomes cancerous.

The development of a malignant tumor is accompanied by a gradual accumulation of mutations that convert proto-oncogenes to oncogenes and knock out tumor-suppressor genes **(Figure 11.20)**. The requirement for several DNA mutations explains why cancers can take a long time to develop. It may also help explain why the longer we live, the more likely we are to accumulate mutations.

Inherited Cancer

The fact that multiple genetic changes are required to produce a full-fledged cancer cell helps explain the observation that cancers can run in families. An individual inheriting an oncogene or a mutant version of a tumor-suppressor gene is one step closer to accumulating the necessary mutations for cancer to develop than is an individual without any such mutations. Geneticists are therefore devoting much effort to identifying inherited cancer mutations so that predisposition to certain cancers can be detected early in life.

About 15% of colorectal cancers, for example, involve inherited mutations. There is also evidence that inheritance plays a role in 5—10% of patients with breast cancer, a disease that strikes one out of every ten American women **(Figure 11.21)**. Mutations in either or both of two genes—called *BRCA1* (pronounced "braca-1") and *BRCA2*—are found in at least half of inherited breast cancers. Both *BRCA1* and *BRCA2* are considered tumor-suppressor genes because the normal versions protect against breast cancer. A woman who inherits one mutant *BRCA1* allele has a 60% probability of developing breast cancer before the age of 50, compared with only a 2% probability for an individual lacking the mutations. Tests using DNA sequencing can now detect these mutations. Surgical removal of the breasts and/or ovaries is the only preventive option currently available to women who carry the mutant genes. ☑

▼ **Figure 11.21**
Breast cancer. In 2013, at age 37, the actress Angelina Jolie underwent a preventive double mastectomy after learning she had a mutant *BRCA1* gene. Jolie's mother, grandmother, and aunt all died young from breast or ovarian cancer.

☑ CHECKPOINT

How can a mutation in a tumor-suppressor gene contribute to the development of cancer?

■ Answer: A mutated tumor-suppressor gene may produce a defective protein unable to function in a pathway that normally inhibits cell division and therefore normally suppresses tumors.

▶ **Figure 11.20 Accumulation of mutations in the development of a cancer cell.** Mutations leading to cancer accumulate in a lineage of cells. In this figure, colors distinguish the normal cells from cells with one or more mutations, leading to increased cell division and cancer. Once a cancer-promoting mutation occurs (orange band on chromosome), it is passed to all the descendants of the cell carrying it.

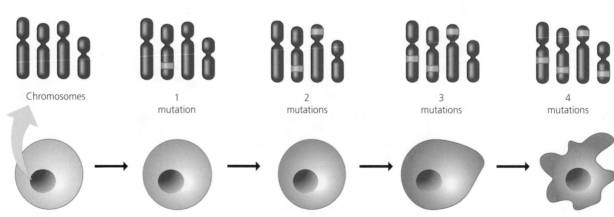

Chromosomes

	1 mutation	2 mutations	3 mutations	4 mutations

Normal cell

Malignant cell

Cancer Risk and Prevention

Cancer is the second-leading cause of death (after heart disease) in most industrialized countries. Death rates due to certain forms of cancer have decreased in recent years, but the overall cancer death rate is still on the rise, currently increasing at about 1% per decade.

Although some cases of cancer occur spontaneously, most often a cancer arises from mutations that are caused by **carcinogens**, cancer-causing agents found in the environment. Mutations often result from decades of exposure to carcinogens. One of the most potent carcinogens is ultraviolet (UV) radiation. Excessive exposure to UV radiation from the sun can cause skin cancer, including a deadly type called melanoma. You can decrease your risk by using sun protection (clothing, lotion, hats, etc.).

The one substance known to cause more cases and types of cancer than any other is tobacco. By a wide margin, more people die from lung cancer (nearly 156,000 Americans in 2017) than from any other form of cancer. Most tobacco-related cancers are due to smoking cigarettes, but smoking cigars, inhaling secondhand smoke, and smokeless tobacco also pose risks. As **Figure 11.22** indicates, tobacco use, sometimes in combination with alcohol consumption, causes several types of cancer. Exposure to some of

SIMPLE CHANGES
IN LIFESTYLE CAN
DRAMATICALLY
REDUCE YOUR RISK
OF CANCER.

the most lethal carcinogens is often a matter of choice: Tobacco use, the consumption of alcohol, and excessive time spent in the sun are all avoidable behaviors that affect cancer risk.

Some food choices significantly reduce a person's odds of developing cancer. For instance, eating 20–30 g of plant fiber daily (about the amount found in seven apples), while eating less animal fat, may help prevent colon cancer. There is also evidence that certain substances in fruits and vegetables, including vitamins C and E and certain compounds related to vitamin A, may help protect against a variety of cancers. Cabbage and its relatives, such as broccoli and cauliflower, are thought to be especially rich in substances that help prevent cancer, although some of the specific substances have not yet been identified. Determining how diet influences cancer has become an important focus of nutrition research. It is encouraging that we can help reduce our risk of acquiring some of the most common forms of cancer by the choices we make in our daily lives. The battle against cancer is being waged on many fronts, and there is reason for optimism in the progress being made. Next, in the Evolution Connection section, you'll see how knowledge of evolutionary processes may be applied to cancer treatment. ☑

☑ **CHECKPOINT**

Of all known behavioral factors, which one causes the most cancer cases and deaths?

■ *Answer: tobacco use*

▼ **Figure 11.22 Cancer in the United States in 2017.** Each data point (blue dot) corresponds to the number of deaths and number of new diagnoses for a particular type of cancer. Risk factors are anything that increases the likelihood of developing that cancer.

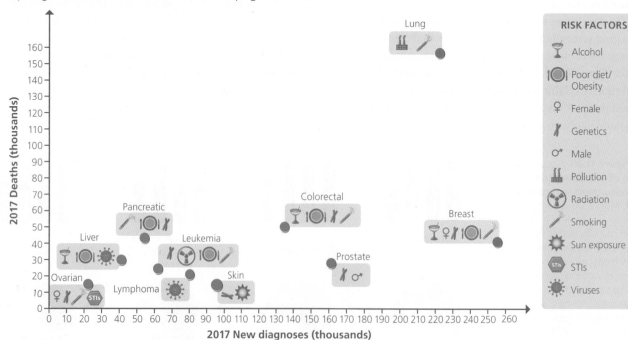

Data from American Cancer Society, *Cancer Facts and Figures 2017.*

The Evolution of Cancer in the Body

The theory of evolution describes natural selection acting on populations. Recently, medical researchers have been using an evolutionary perspective to gain insight into the development of tumors, such as the bone tumor shown in **Figure 11.23**. A tumor can be thought of as a population of cancer cells. Just as a population of organisms is driven to evolve by natural selection, a population of cancer cells may evolve in response to cancer treatments.

Recall that there are several assumptions behind Darwin's theory of natural selection (see Chapter 1). Let's consider how each one can be applied to cancer. First, all evolving populations have the potential to produce more offspring than can be supported by the environment. Cancer cells, with their uncontrolled growth, clearly demonstrate overproduction. Second, there must be variation among individuals of the population. Studies of tumor cell DNA show genetic variability within tumors. Third, variations in the population

must affect survival and reproductive success. Indeed, the accumulation of mutations in cancer cells renders them less susceptible to normal mechanisms of reproductive control. Mutations that enhance survival of malignant cancer cells are passed on to that cell's descendants. In short, a tumor evolves.

Viewing the progression of cancer through the lens of evolution helps explain why there is no easy "cure" for cancer but may also pave the way for novel therapies. For example, some researchers are attempting to "prime" tumors for treatment by increasing the reproductive success of only those cells that will be susceptible to a chemotherapy drug. Our understanding of cancer, like all other aspects of biology, benefits from an evolutionary perspective.

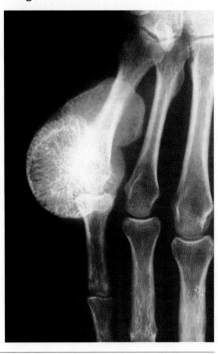

▼ **Figure 11.23** X-ray of a hand, revealing a large bone tumor.

Chapter Review

SUMMARY OF KEY CONCEPTS

How and Why Genes Are Regulated

The various types of cells in a multicellular organism owe their distinctiveness to different combinations of genes being turned on and off by gene regulation in each cell type.

Gene Regulation in Bacteria

An operon is a cluster of genes with related functions together with a promoter and other non–protein-coding DNA sequences that control their transcription. For example, the *lac* operon allows *E. coli* to produce enzymes for lactose use only when the sugar is present.

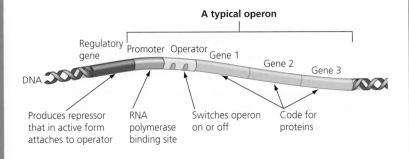

A typical operon

Regulatory gene
Promoter
Operator
Gene 1
Gene 2
Gene 3
DNA

Produces repressor that in active form attaches to operator
RNA polymerase binding site
Switches operon on or off
Code for proteins

Gene Regulation in Eukaryotic Cells

In the nucleus of eukaryotic cells, there are several possible control points in the pathway of gene expression.

- DNA packing tends to block gene expression by preventing access of transcription proteins to the DNA. An extreme example is X chromosome inactivation in the cells of female mammals.

- The most important control point in both eukaryotes and prokaryotes occurs at the start of gene transcription. Various regulatory proteins interact with DNA and with each other to turn the transcription of eukaryotic genes on or off.

- There are also opportunities for the control of eukaryotic gene expression after transcription, when introns are cut out of the RNA and a cap and tail are added to process RNA transcripts into mRNA.

- In the cytoplasm, presence of microRNAs may block the translation of an mRNA, and various proteins may regulate the start of translation.

- Finally, the cell may activate the finished protein in various ways (for instance, by cutting out portions or chemical modification). Eventually, the protein may be selectively broken down.

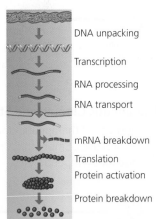

DNA unpacking
Transcription
RNA processing
RNA transport
mRNA breakdown
Translation
Protein activation
Protein breakdown

Cell Signaling

Cell-to-cell signaling is key to the development and functioning of multicellular organisms. Signal transduction pathways convert molecular messages to cell responses, such as the transcription of particular genes.

Homeotic Genes

Evidence for the evolutionary importance of gene regulation is apparent in homeotic genes, master genes that regulate other genes that in turn control embryonic development.

Visualizing Gene Expression

Researchers can visualize which genes are active in which tissues using a variety of techniques, including the use of probes and DNA microarrays.

Cloning Plants and Animals

The Genetic Potential of Cells

Most differentiated cells retain a complete set of genes, so an orchid plant, for example, can be made to grow from a single orchid cell. Under controlled conditions, animals can also be cloned.

Reproductive Cloning of Animals

Nuclear transplantation is a procedure whereby a donor cell nucleus is inserted into an egg from which the nucleus has been removed. First demonstrated in frogs in the 1950s, reproductive cloning was used in 1996 to clone a sheep from an adult cell and has since been used to create many other cloned animals.

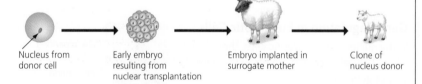

Nucleus from donor cell → Early embryo resulting from nuclear transplantation → Embryo implanted in surrogate mother → Clone of nucleus donor

Therapeutic Cloning and Stem Cells

The purpose of therapeutic cloning is to produce embryonic stem cells for medical uses. Embryonic, umbilical cord, and adult stem cells all show promise for therapeutic uses.

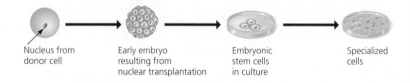

Nucleus from donor cell → Early embryo resulting from nuclear transplantation → Embryonic stem cells in culture → Specialized cells

The Genetic Basis of Cancer

Genes That Cause Cancer

Cancer cells, which divide uncontrollably, can result from mutations in genes whose protein products regulate the cell cycle.

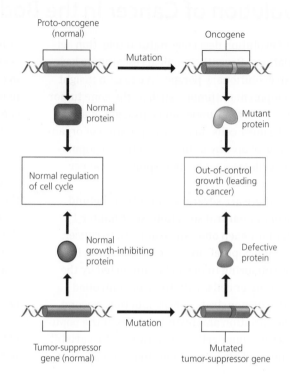

Proto-oncogene (normal) → Mutation → Oncogene

Normal protein / Mutant protein

Normal regulation of cell cycle / Out-of-control growth (leading to cancer)

Normal growth-inhibiting protein / Defective protein

Tumor-suppressor gene (normal) → Mutation → Mutated tumor-suppressor gene

Many proto-oncogenes and tumor-suppressor genes code for proteins active in signal transduction pathways regulating cell division. Mutations of these genes cause malfunction of the pathways. Cancer results from a series of genetic changes in a cell lineage. Researchers have identified many genes that, when mutated, promote the development of cancer.

Cancer Risk and Prevention

Reducing exposure to carcinogens (which induce cancer-causing mutations) and making other healthful lifestyle choices can help reduce cancer risk.

Mastering Biology

For practice quizzes, BioFlix animations, MP3 tutorials, video tutors, and more study tools designed for this textbook, go to Mastering Biology™

SELF-QUIZ

1. Your bone cells, muscle cells, and skin cells look different because
   a. different kinds of genes are present in each kind of cell.
   b. they are present in different organs.
   c. different genes are active in each kind of cell.
   d. different mutations have occurred in each kind of cell.

2. A group of prokaryotic genes with related functions that are regulated as a single unit, along with the control sequences that perform this regulation, is called a(n) _____.

3. The regulation of gene expression must be more complex in multicellular eukaryotes than in prokaryotes because
   a. eukaryotic cells are much larger.
   b. in a multicellular eukaryote, different cells are specialized.
   c. prokaryotes are restricted to stable environments.
   d. eukaryotes have fewer genes, so each gene must do several jobs.

4. A eukaryotic gene was inserted into the DNA of a bacterium. The bacterium then transcribed this gene into mRNA and translated the mRNA into protein. The protein produced was useless and contained many more amino acids than the protein made by the eukaryotic cell. Why?
   a. The mRNA was not spliced as it is in eukaryotes.
   b. Eukaryotes and prokaryotes use different genetic codes.
   c. Repressor proteins interfered with transcription and translation.
   d. Ribosomes were not able to bind to tRNA.

5. How does DNA packing in chromosomes prevent gene expression?

6. Is the genetic information in an animal cloned by nuclear transplantation identical to that of the nuclear donor?

7. The most common procedure for cloning an animal is _____.

8. What is learned from a DNA microarray?

9. Which of the following is a substantial difference between embryonic stem cells and the stem cells found in adult tissues?
   a. In laboratory culture, only adult stem cells are immortal.
   b. In nature, only embryonic stem cells give rise to all the different types of cells in the organism.
   c. Only adult stem cells can be made to differentiate in the laboratory.
   d. Only embryonic stem cells are in every tissue of the adult body.

10. Name three potential sources of stem cells.

11. What are the possible uses of reproductive cloning?

12. A mutation in one gene may cause a major change in the body of a fruit fly. Yet it takes many genes to produce a wing or leg. How can a change in one gene cause a big change? What are such genes called?

For answers to the Self Quiz, see Appendix D.

IDENTIFYING MAJOR THEMES

For each statement, identify which major theme is evident (the relationship of structure to function, information flow, pathways that transform energy and matter, interactions within biological systems, or evolution) and explain how the statement relates to the theme. If necessary, review the theme descriptions (see Chapter 1) and review the examples highlighted in blue in this chapter.

13. Changing the shape of the *lac* repressor affects how the repressor acts.

14. A cell can produce and secrete chemicals, such as hormones, that affect gene regulation in another cell.

15. Master control genes regulate other genes that determine what body parts will develop in which locations.

For answers to Identifying Major Themes, see Appendix D.

THE PROCESS OF SCIENCE

16. Study the depiction of the *lac* operon in Figure 11.2. Normally, the genes are turned off when lactose is not present. Lactose activates the genes, which code for enzymes that enable the cell to use lactose. Predict how the following mutations would affect the function of the operon in the presence and absence of lactose:
    a. mutation of regulatory gene; repressor will not bind to lactose
    b. mutation of operator; repressor will not bind to operator
    c. mutation of regulatory gene; repressor will not bind to operator

17. The human body has a far greater variety of proteins than genes, highlighting the importance of alternative RNA splicing. Suppose you have samples of two types of adult cells from one person. Design an experiment using microarrays to determine whether different gene expression is due to alternative RNA splicing.

18. Because a cat must have both orange and non-orange alleles to be tortoiseshell (see Figure 11.4), we would expect only female cats, which have two X chromosomes, to be tortoiseshell. Normal male cats (XY) can carry only one of the two alleles. Male tortoiseshell cats are rare and usually sterile. What might be their genotype?

19. Design two complementary experimental approaches that test whether the *Antennapedia* gene is a homeotic (master control) gene of leg development in the fruit fly.

20. **Interpreting Data** Review Figure 11.22. We can estimate the deadliness of each type of cancer by dividing the number of deaths by number of cases. (Although someone diagnosed may not die the same year, it's a useful approximation.) If nearly everyone diagnosed with a certain cancer dies, that ratio will be near 1 (100% deadly). If many more people receive diagnoses than die, the ratio will be near 0 (near 0% deadly). Which region of the graph represents the more deadly cancers? The least deadly? Calculate the diagnosis/death rate for different cancers.

BIOLOGY AND SOCIETY

21. The possibility of cloning humans is a highly controversial topic. Experts warn of the low success rates of cloning mammals, potential health problems, and a potentially shortened life span. However, there are also advocates of human cloning who pointed out the possibility of eliminating defective genes and of generating specific cells from a patient for therapeutic purposes. What are your opinions on human cloning?

22. There are genetic tests for several types of "inherited cancer." The results cannot usually predict that someone will get cancer. Rather, they indicate only an increased risk of developing cancer. For many cancers, lifestyle changes cannot decrease risk. Therefore, some people consider the tests useless. If your close family had a history of cancer and a test were available, would you get screened? What would you do with this information?

12 DNA Technology

CHAPTER CONTENTS

ARE MORE GENES BETTER? YOU HAVE ABOUT THE SAME NUMBER OF GENES AS A MICROSCOPIC WORM, AND ONLY HALF AS MANY AS A RICE PLANT.

SOMEDAY SOON, GENETICALLY MODIFIED POTATOES COULD SAVE TENS OF THOUSANDS OF CHILDREN FROM DEATH BY CHOLERA.

Why DNA Technology Matters

The products and services provided by genetic engineering have a significant impact on our lives and our society. DNA technologies have many practical applications, from the food we eat to the drugs we take to other products we use every day. Furthermore, such technologies are used to gather evidence for forensic, historical, and scientific investigations.

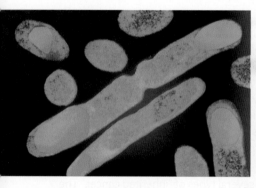

HUG A MICROBE TODAY: MILLIONS OF PEOPLE WITH DIABETES LIVE HEALTHIER LIVES THANKS TO INSULIN MADE BY BACTERIA.

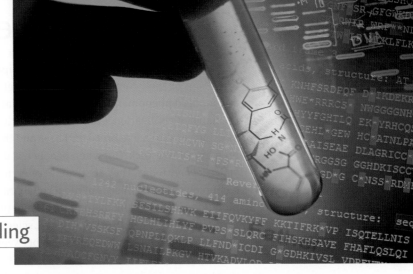

BIOLOGY AND SOCIETY   DNA Profiling

A DNA profile. Even minuscule bits of evidence can provide a DNA profile.

Using DNA to Establish Guilt and Innocence

In 1964, teenager Mary Sullivan was found dead in her new apartment, just days after she had moved to Boston. She had been raped and strangled. She was one of at least 11 victims of an unknown killer nicknamed the Boston Strangler, who terrorized the city for two years. Albert DeSalvo, who had a long criminal history, confessed to the crimes, but later denied guilt. DeSalvo was killed by a fellow inmate in 1973 while in prison on separate rape convictions. For decades, many wondered whether the real Boston Strangler was still free. The answer would have to wait for the development of modern DNA technologies.

Forensic DNA analysis hinges on a simple fact: The cells of every person (except identical twins) contain unique DNA. The DNA that codes for proteins is extremely similar in all people, but other DNA in our cells does not code for anything and is highly variable from person to person. This hyper-variable DNA is used for DNA profiling, which is an analysis of DNA samples to determine whether they come from the same person.

Almost 50 years after Mary Sullivan was murdered, the Boston Police Department reopened the investigation. Forensic scientists analyzed long-stored evidence from the crime, including a blanket with traces of semen. The evidence contained samples of the criminal's DNA and the victim's DNA. To isolate some of the criminal's DNA, scientists used technology that targets the Y chromosome, found only in males (and hence likely belonging to the perpetrator). The Y chromosome from the Mary Sullivan crime scene was compared with that of DeSalvo's nephew. After this comparison showed a partial match (as expected from a close relative), DeSalvo's body was exhumed for testing. The results were clear: DeSalvo was, in fact, the Boston Strangler. This conclusion brought closure to the case and provided a measure of solace to several grieving families.

DNA technology has revolutionized the field of forensics. Beyond the courtroom, DNA technology has led to some remarkable scientific advances: Crop plants have been genetically modified to produce their own insecticides; human genes are being compared with those of other animals to help shed light on what makes us distinctly human; and significant advances have been made toward detecting and curing fatal genetic diseases. This chapter will describe these and other uses of DNA technology and explain how various DNA techniques are performed. We'll also examine some of the social, legal, and ethical issues that lie at the intersection of biology and society.

Genetic Engineering

You may think of **biotechnology**, the manipulation of organisms or their components to make useful products, as a modern phenomenon, but it actually dates back to the dawn of civilization. Consider such ancient practices as using yeast to make bread and beer and the selective breeding of livestock. But when people use the term *biotechnology* today, they are usually referring to DNA technology, which consists of modern laboratory techniques for studying and manipulating genetic material. Using the methods of DNA technology, scientists can modify specific genes and move them between organisms as different as bacteria, plants, and animals. Organisms that have acquired one or more genes by artificial means are called **genetically modified (GM) organisms**. If the newly acquired gene is from another organism, typically of another species, the recombinant organism is called a **transgenic organism**.

In the 1970s, the field of biotechnology exploded with the invention of methods for making recombinant DNA in the laboratory. Scientists construct **recombinant DNA** by combining pieces of DNA from two different sources—often from different species—in order to form a single DNA molecule. Recombinant DNA technology is widely used in **genetic engineering**, the direct manipulation of genes for practical purposes. Scientists have genetically engineered bacteria to mass-produce a variety of useful chemicals, from cancer drugs to pesticides. Scientists have also transferred genes from bacteria to plants and from one animal species to another **(Figure 12.1)**. Such engineering can serve a variety of purposes, from basic research (What does this gene do?) to medical applications (Can we create animal models for a human disease?). ☑

Recombinant DNA Techniques

Although genetic engineering can be performed on a variety of organisms, bacteria (*Escherichia coli*, in particular) are the workhorses of modern biotechnology. To manipulate genes in the laboratory, biologists often use bacterial **plasmids**, which are small, circular DNA molecules that duplicate separately from the larger bacterial chromosome **(Figure 12.2)**. Because plasmids can carry virtually any gene and are passed from one generation of bacteria to the next, they are key tools for **DNA cloning**, the production of multiple identical copies of a piece of DNA. DNA cloning methods are central to most genetic engineering tasks.

How to Clone a Gene

Consider a typical genetic engineering challenge: A researcher at a pharmaceutical company identifies a gene of interest that codes for a valuable protein, such as a potential new drug. The biologist wants to manufacture the protein on a large scale. **Figure 12.3** illustrates a way to accomplish this by using recombinant DNA techniques.

☑ **CHECKPOINT**

What is biotechnology? What is recombinant DNA?

■ *Answer: the manipulation of organisms or their parts to produce a useful product; a molecule containing DNA from two different sources, often different species*

▼ **Figure 12.1 Researchers produced glowing fish by transferring a gene for a fluorescent protein originally obtained from jellies ("jellyfish").**

▼ **Figure 12.2 Bacterial plasmids.** The micrograph shows a bacterial cell that has been ruptured, revealing one long chromosome and several smaller plasmids. The inset is an enlarged view of a single plasmid.

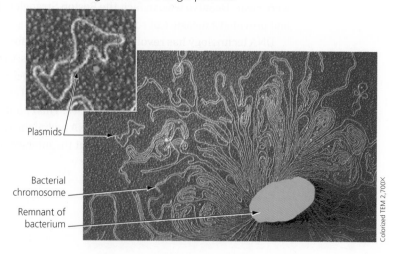

Plasmids

Bacterial chromosome

Remnant of bacterium

Colorized TEM 2,700X

To start, the biologist isolates two kinds of DNA: bacterial plasmids that will serve as **vectors** (gene carriers) and DNA from another organism that includes the gene of interest (shown in yellow in the figure). This foreign DNA may be from any type of organism, even a human. **1** The DNA from the two sources is joined, resulting in recombinant DNA plasmids. **2** The recombinant plasmids are then mixed with bacteria. Under the right conditions, the bacteria take up the recombinant plasmids. **3** Each bacterium, carrying its recombinant plasmid, is allowed to reproduce through cell division to form a **clone** of cells, a population of genetically identical cells descended from a single original cell. In this clone, each bacterium carries a copy of the gene of interest. When DNA cloning involves a gene-carrying segment of DNA (as it does here), it is known as **gene cloning**. As the

bacteria multiply, the foreign gene that is carried by the recombinant plasmid is also copied. **4** The transgenic bacteria with the gene of interest can then be grown in tanks, producing the protein in marketable quantities. The products of gene cloning may be copies of the gene itself, to be used in additional genetic engineering projects, or the protein product of the cloned gene, to be harvested and used. ☑

Cutting and Pasting DNA with Restriction Enzymes

As shown in Figure 12.3, recombinant DNA is created by combining two ingredients: a bacterial plasmid and the gene of interest. To understand how these DNA molecules are spliced together, you need to learn how enzymes cut and paste DNA.

☑ **CHECKPOINT**

Why are plasmids valuable tools for the production of recombinant DNA?

■ Answer: Plasmids can carry virtu-ally any foreign gene and are repli-cated by their bacterial host cells.

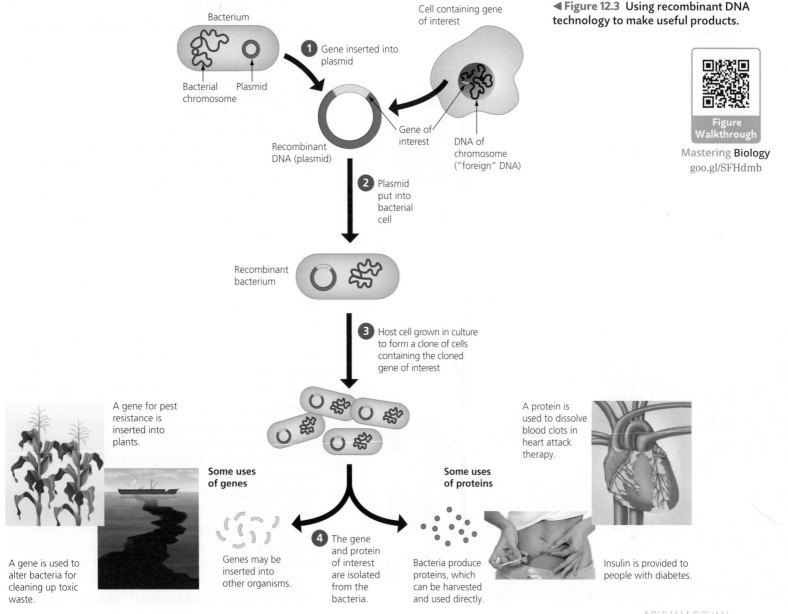

◀ **Figure 12.3 Using recombinant DNA technology to make useful products.**

Mastering Biology
goo.gl/SFHdmb

Bacterium
Bacterial chromosome   Plasmid

Cell containing gene of interest

1 Gene inserted into plasmid

Recombinant DNA (plasmid)

Gene of interest

DNA of chromosome ("foreign" DNA)

2 Plasmid put into bacterial cell

Recombinant bacterium

3 Host cell grown in culture to form a clone of cells containing the cloned gene of interest

A gene for pest resistance is inserted into plants.

A gene is used to alter bacteria for cleaning up toxic waste.

Some uses of genes

Genes may be inserted into other organisms.

4 The gene and protein of interest are isolated from the bacteria.

Some uses of proteins

Bacteria produce proteins, which can be harvested and used directly.

A protein is used to dissolve blood clots in heart attack therapy.

Insulin is provided to people with diabetes.

✓ **CHECKPOINT**

If you mix a restriction enzyme that cuts within the sequence AATC with DNA of the sequence CGAATCTAGCAATCGCGA, how many restriction fragments will result? (For simplicity, the sequence of only one of the two DNA strands is listed.)

■ *Answer: Cuts at two restriction sites will yield three restriction fragments.*

The cutting tools used for making recombinant DNA are bacterial enzymes called **restriction enzymes**. Biologists have identified hundreds of restriction enzymes, each recognizing a particular short DNA sequence, usually four to eight nucleotides long. For example, one restriction enzyme only recognizes the DNA sequence GAATTC, whereas another recognizes GGATCC. The DNA sequence recognized by a particular restriction enzyme is called a **restriction site**. After a restriction enzyme binds to its restriction site, it cuts the two strands of the DNA by breaking chemical bonds at specific points within the sequence, like a pair of highly specific molecular scissors.

The top of **Figure 12.4** shows a piece of DNA (blue) that contains one restriction site for a restriction enzyme. ❶ The restriction enzyme cuts the DNA strands between the bases A and G in the recognition sequence, producing pieces of DNA called **restriction fragments**. The staggered cuts yield two double-stranded DNA fragments with single-stranded ends, called "sticky ends." Sticky ends are the key to joining DNA restriction fragments originating from different sources. ❷ Next, a piece of DNA from another source (yellow) is added. Notice that the yellow DNA has single-stranded ends identical in base sequence to the sticky ends on the blue DNA because the same restriction enzyme was used to cut both types of DNA. ❸ The complementary ends on the blue and yellow fragments stick together by base pairing. ❹ The union between the blue and yellow fragments is then made permanent by the "pasting" enzyme **DNA ligase**. This enzyme connects the DNA pieces into continuous strands by forming bonds between adjacent nucleotides. The final outcome is a single molecule of recombinant DNA. The process just described explains what happens in step 1 of Figure 12.3. ✓

Gene Editing

A new DNA technology, called the **CRISPR-Cas9 system**, allows the nucleotide sequence of specific genes to be edited in living cells. Such editing can reveal the function of the gene or possibly even correct a genetic mutation. CRISPR-Cas9 is becoming one of the most important tools available for genetic engineering.

The CRISPR-Cas9 system, like restriction enzymes, was discovered as a natural component of prokaryotic cells. Scientists noticed that bacterial genomes have short repetitive DNA sequences with different stretches of "spacer DNA" between the repeats. Each spacer sequence corresponds to DNA from a virus that has infected the cell. A bacterial protein called Cas9 associated with the repeats can identify and cut viral DNA, thereby defending the bacterium against infection.

Cas9 cuts double-stranded DNA molecules (as restriction enzymes do). However, while a given restriction enzyme recognizes only one DNA sequence, the Cas9 protein will cut any DNA sequence that is complementary to an associated molecule of guide RNA. Cas9 is like a guided missile, with an RNA molecule as the guidance system.

To alter a cell, a scientist introduces a Cas9–guide RNA complex into it. The guide RNA is complementary to a target DNA sequence, such as a gene. After Cas9 cuts both strands of the target gene, DNA repair enzymes randomly insert nucleotides as they reconnect the target DNA. In this way, researchers can "knock out" (disable) a given gene. By observing the altered cell or organism, researchers may be able to determine the function of the knocked-out gene.

The CRISPR-Cas9 system can also be used to edit a gene **(Figure 12.5)**. Researchers can introduce a segment from the normal gene along with the Cas9–guide RNA complex. After Cas9 cuts the target DNA, repair enzymes use the normal DNA as a template to repair the target

▶ **Figure 12.4 Cutting and pasting DNA.** The production of recombinant DNA requires two enzymes: a restriction enzyme, which cuts the original DNA molecules into pieces, and DNA ligase, which pastes the pieces together.

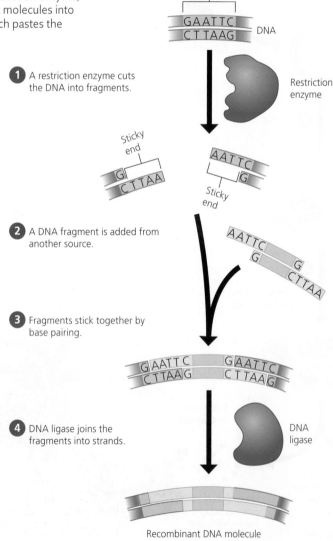

Recognition site (recognition sequence) for a restriction enzyme

GAATTC
CTTAAG   DNA

❶ A restriction enzyme cuts the DNA into fragments.

Restriction enzyme

Sticky end

G
CTTAA

AATTC
G

Sticky end

❷ A DNA fragment is added from another source.

AATTC
G   G
            CTTAA

❸ Fragments stick together by base pairing.

❹ DNA ligase joins the fragments into strands.

GAATTC   GAATTC
CTTAAG   CTTAAG

DNA ligase

Recombinant DNA molecule

▼ Figure 12.5 **The CRISPR-Cas9 system.** By combining an RNA "homing" sequence with an enzyme, specific genes can be inactivated or editing within living cells.

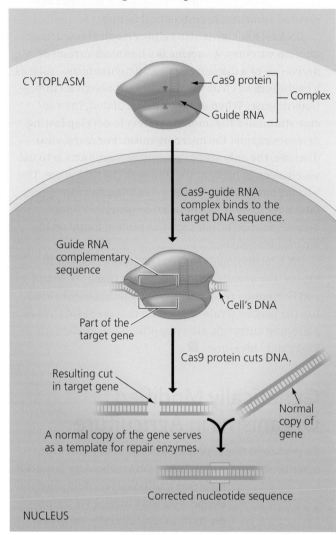

CYTOPLASM

Cas9 protein

Guide RNA

Complex

Cas9-guide RNA complex binds to the target DNA sequence.

Guide RNA complementary sequence

Cell's DNA

Part of the target gene

Cas9 protein cuts DNA.

Resulting cut in target gene

Normal copy of gene

A normal copy of the gene serves as a template for repair enzymes.

Corrected nucleotide sequence

NUCLEUS

DNA where it was cut. In this way, the CRISPR-Cas9 system acts like the "search and replace" function of a word processor, potentially fixing mutations in cells.

In 2014, researchers used the CRISPR-Cas9 system to fix a genetic defect in mice. They altered live mouse cells to correct a faulty gene that causes tyrosinemia, a disease affecting metabolism of the amino acid tyrosine, which can lead to organ dysfunction and developmental disabilities. A study in 2015 involved mice that carried a mutation in a gene that codes for dystrophin, a protein essential for muscle function. Researchers infected the mice with a virus carrying the Cas9–guide RNA complex. The virus infected the muscle cells and removed a region of the gene containing the dystrophin mutation. The gene, now lacking the mutation, produced normal dystrophin proteins, which allowed the muscles to function properly. Researchers hope to apply this technique to humans carrying a similar mutation that causes the disease Duchenne muscular dystrophy.

There are many hurdles to clear before the CRISPR-Cas9 system can be tried in humans, but the technique is sparking the interest of researchers and physicians all around the world. In the next section, we'll explore some ways that genetic modifications affect your life. ☑

Medical Applications

By transferring the gene for a desired protein into a bacterium, yeast, or other kind of cell that is easy to grow in culture, scientists can produce large quantities of useful proteins that are present naturally only in small amounts. In this section, you'll learn about some applications of recombinant DNA technology.

Humulin is human insulin produced by genetically modified bacteria (**Figure 12.6**). In humans, insulin is a protein normally made by the pancreas. Insulin functions as a hormone and helps regulate the level of glucose in the blood. If the body fails to produce enough insulin, the result is type 1 diabetes. There is no cure, so people with this disease must inject themselves daily with doses of insulin for the rest of their lives.

Because human insulin is not readily available, diabetes was historically treated using insulin from cows and pigs. This treatment was problematic, however. Pig and cow insulins can cause allergic reactions in people because their chemical structures differ slightly from that of human insulin. In addition, by the 1970s, the supply of beef and pork pancreas available for insulin extraction could not keep up with the demand.

In 1978, scientists working at a biotechnology company chemically synthesized DNA fragments and linked them to form the two genes that code for the two polypeptides that make up human insulin. They then inserted these artificial genes into *E. coli* host cells. Under proper growing conditions, the transgenic bacteria cranked out large quantities of the human protein. In 1982, Humulin hit the market as the world's first genetically engineered pharmaceutical product. Today, it is produced around the clock in gigantic fermentation vats filled with a liquid culture of bacteria. Each day, more than 4 million people with diabetes use the insulin collected, purified, and packaged at such facilities (**Figure 12.7**).

Insulin is just one of many human proteins produced by genetically modified bacteria. Another example is human growth hormone (HGH). Abnormally low levels of this hormone during childhood and adolescence can cause dwarfism.

☑ **CHECKPOINT**

What is the function of the guide RNA in the CRISPR-Cas9 system?

■ *Answer: It guides the complex to the proper location in the genome.*

▼ Figure 12.6 **Humulin, human insulin produced by genetically modified bacteria.**

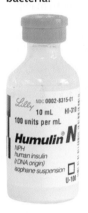

MILLIONS OF PEOPLE WITH DIABETES LIVE HEALTHIER LIVES THANKS TO INSULIN MADE BY BACTERIA.

▼ Figure 12.7 **A factory that produces genetically engineered insulin.**

Because growth hormones from other animals are not effective in people, HGH was an early target of genetic engineers. Before genetically engineered HGH became available in 1985, children with an HGH deficiency could only be treated with scarce and expensive supplies of HGH obtained from human cadavers. Another genetically engineered protein helps dissolve blood clots. If administered shortly after a stroke, it reduces the risk of additional strokes and heart attacks.

Although bacteria can produce many human proteins, some proteins can only be made by eukaryotic cells, such as cells from fungi, animals, and plants. Common baker's yeast is currently used to produce proteins used as medicines, including the hepatitis B vaccine, an antimalarial drug, and interferons used to treat cancer and viral infections. In 2015, scientists announced they had transferred 23 genes (from bacteria, plants, and animals) into yeast that allow the recombinant fungi to convert sugar into the painkiller drug hydrocodone. Genetically modified mammalian cells are used to produce erythropoietin (EPO), a hormone used to treat anemia by stimulating production of red blood cells. Researchers have also developed transgenic plant cells that can produce human drugs. The drug factories of the future may be carrots because they are easily grown in culture and are unlikely to be contaminated by human pathogens (such as viruses).

Genetically modified whole animals are also used to produce drugs. **Figure 12.8** shows a transgenic goat that carries a gene for an enzyme called lysozyme. This enzyme, found naturally in breast milk, has antibacterial properties. In another example, the gene for a human blood protein has been inserted into the genome of a goat so that the protein is secreted in the goat's milk. The protein is then purified from the milk. Because

transgenic animals are difficult to produce, researchers may create a single transgenic animal and then breed or clone it. The resulting herd of transgenic animals could serve as a grazing pharmaceutical factory.

DNA technology is also helping medical researchers develop vaccines. A vaccine is a harmless variant or derivative of a disease-causing microorganism—such as a bacterium or virus—that is used to prevent an infectious disease. When a person is inoculated, the vaccine stimulates the immune system to develop lasting defenses against the microorganism. For many viral diseases, the only way to prevent serious harm is to use vaccination to prevent the illness in the first place. The vaccine against hepatitis B, a disabling and sometimes fatal liver disease, is produced by genetically engineered yeast cells that secrete a protein found on the virus's outer surface.

DNA technologies can also identify causes of illnesses. For example, the Centers for Disease Control and Prevention regularly uses DNA technology to identify the precise strain of bacteria that is causing a food poisoning outbreak, allowing officials to implement food safety measures. ☑

Genetically Modified Organisms in Agriculture

Since ancient times, people have selectively bred crops to make them more useful. Today, DNA technology is quickly replacing traditional breeding programs as scientists work to improve the productivity of agriculturally important plants and animals.

In the United States today, nearly all of our corn, soybean, and cotton crops are genetically modified. **Figure 12.9** shows corn that has been genetically engineered to resist attack by an insect called the European corn borer. Growing insect-resistant plants reduces the need for chemical insecticides. In another example, modified strawberry plants produce bacterial proteins that act as a natural antifreeze, protecting the delicate plants from the damages of cold weather. Potatoes and rice have been engineered to produce harmless proteins derived from the cholera bacterium; researchers hope that these modified foods will one day serve as an edible vaccine against cholera, a disease that kills thousands of children in developing nations every year. In India, the insertion of a natural but rare saltwater resistance gene has enabled new varieties of rice to thrive in water three times as salty as seawater, allowing food to be grown in drought-stricken or flooded regions.

GENETICALLY MODIFIED POTATOES COULD SAVE MANY CHILDREN FROM DEATH BY CHOLERA.

☑ **CHECKPOINT**

Bacteria are easier to manipulate than yeast. So why are yeast used to produce some human proteins?

■ *Answer: Some genes require eukaryotic cells in order to be properly made.*

▶ **Figure 12.8 A genetically modified goat.**

▼ **Figure 12.9 Genetically modified corn.** The corn plants in this field carry a bacterial gene that helps prevent infestation by the European corn borer (inset).

Scientists are also using genetic engineering to improve the nutritional value of crop plants **(Figure 12.10)**. One example is "golden rice 2," a transgenic variety of rice that carries genes from daffodils and corn. This rice could help prevent vitamin A deficiency and resulting blindness, especially in developing nations that depend on rice as a staple crop. Cassava, a starchy root crop that is a staple for nearly 1 billion people in developing nations, has similarly been modified to produce increased levels of iron and beta-carotene (which is converted to vitamin A in the body). However, controversy surrounds the use of GM foods, as we'll discuss at the end of the chapter.

Genetic engineers are targeting agricultural animals as well as plant crops. Scientists might, for example, identify in one variety of cattle a gene that causes the development of larger muscles (which make up most of the meat we eat) and transfer it to other cattle or even to chickens. Researchers have genetically modified pigs to carry a roundworm gene whose protein converts less healthy fatty acids to omega-3 fatty acids. Meat from the modified pigs contains four to five times as much healthy omega-3 fat as regular pork. In 2015, researchers replaced a gene in dairy cows with one from Angus cattle to produce cattle that lack horns, saving the bulls from painful dehorning. Similar gene-editing techniques produce improved varieties of goats (for meat and cashmere wool), pigs (for agriculture and pets), and dogs. A type of Atlantic salmon has been genetically modified to reach market size in half the normal time (18 months versus 3 years) and to grow twice as large. In late 2015, the FDA approved the sale of this GMO salmon to U.S. consumers, declaring that it is as safe and nutritious as traditional salmon. Although it could be years before the GMO salmon reaches store shelves, this is the first time a transgenic animal product was allowed to be sold as food in the United States. ☑

☑ **CHECKPOINT**

What is a genetically modified organism?

■ Answer: one that carries DNA introduced through artificial means

▼ **Figure 12.10 Genetically modified staple crops.** "Golden rice 2," the yellow grains shown here (left) alongside ordinary rice, has been genetically modified to produce high levels of beta-carotene, a molecule that the body converts to vitamin A. Transgenic cassava (right), a starchy root crop that serves as the main food source for nearly a billion people, has been modified to produce extra nutrients.

Human Gene Therapy

We've seen that bacteria, fungi, plants, and nonhuman animals can be genetically modified—so what about humans? **Human gene therapy** is intended to treat disease by introducing new genes into an afflicted person. In some cases, a mutant version of a gene may be replaced or supplemented with the normal allele, potentially correcting a genetic disorder, perhaps permanently. In other cases, genes are inserted and expressed only long enough to treat a medical problem.

Figure 12.11 summarizes one approach to human gene therapy. The procedure closely resembles the gene cloning process shown in steps 1 through 3 of Figure 12.3, but in this instance human cells, rather than bacteria, are the targets. ❶ A gene from a normal person is cloned, converted to an RNA version, and then inserted into the RNA genome of a harmless virus. ❷ Bone marrow cells are taken from the patient and infected with the recombinant virus. ❸ The virus inserts a DNA copy of its genome, including the normal human gene, into the DNA of the patient's cells. ❹ The engineered cells are then injected back into the patient. The normal gene is transcribed and translated within the patient's body, producing the desired protein. Ideally, the nonmutant version of the gene would be inserted into cells that multiply throughout a person's life. Bone marrow cells, which include the stem cells that give rise to all the types of blood cells, are prime candidates. If the procedure succeeds, the cells will multiply permanently and produce a steady supply of the missing protein, curing the patient.

The promise of gene therapy exceeds actual results, but there have been some successes. In 2009, an international research team conducted a trial that focused on a form of progressive blindness linked to a defect in a gene responsible for producing light-detecting pigments in the eye. The researchers found that an injection of a virus carrying the normal gene into one eye of affected children improved vision in that eye, sometimes enough to allow normal function, without significant side effects. The other eye was left untreated as a control.

From 2000 to 2011, gene therapy cured 22 children with severe combined immunodeficiency (SCID), a fatal inherited disease caused by a defective gene that prevents development of the immune system, requiring patients to remain isolated within protective "bubbles." Unless treated with a bone marrow transplant, which is effective only 60% of the time, SCID patients quickly die from infections that most of us easily fend off. In these cases, researchers periodically removed immune system cells from the patients' blood, infected them with a

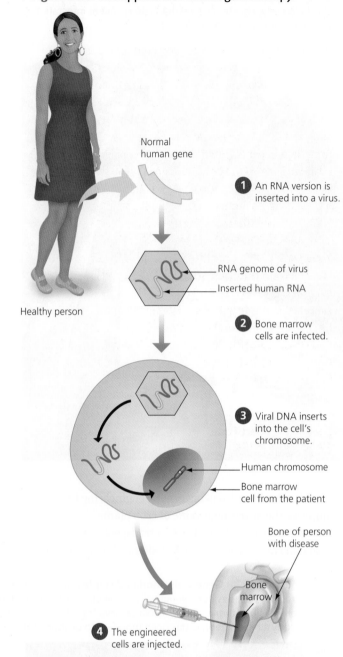

▼ **Figure 12.11 One approach to human gene therapy.**

Normal human gene

Healthy person

❶ An RNA version is inserted into a virus.

RNA genome of virus

Inserted human RNA

❷ Bone marrow cells are infected.

❸ Viral DNA inserts into the cell's chromosome.

Human chromosome

Bone marrow cell from the patient

Bone of person with disease

Bone marrow

❹ The engineered cells are injected.

virus engineered to carry the normal allele of the defective gene, then reinjected the blood into the patient. The treatment cured the patients of SCID, but there were serious side effects: Four of the treated patients developed leukemia, and one died after the inserted gene activated an oncogene (see Chapter 11), creating cancerous blood cells.

Two other illnesses that have been treated with gene therapy, if only in a few patients, are a degenerative disease of the nervous system and a blood disorder involving a hemoglobin gene. Research on gene therapy continues, with tougher guidelines for safe and effective application. ☑

☑ **CHECKPOINT**

Why are bone marrow stem cells ideally suited as targets for gene therapy?

■ *Answer: because bone marrow stem cells multiply throughout a person's life*

DNA Profiling and Forensic Science

When a crime is committed, body fluids (such as blood or semen) or small pieces of tissue (such as skin beneath a victim's fingernails) may be left at the scene or on the victim or assailant. As discussed in the Biology and Society section, such evidence can be examined by **DNA profiling**, the analysis of DNA samples to determine whether they come from the same individual. Indeed, DNA profiling has rapidly transformed the field of **forensics**, the scientific analysis of evidence for crime scene investigations and other legal proceedings. To produce a DNA profile, scientists compare DNA sequences that vary from person to person.

Figure 12.12 presents an overview of a typical investigation using DNA profiling. **1** First, DNA samples are isolated from the crime scene, suspects, victims, or other evidence. **2** Next, selected sequences from each DNA sample are amplified (copied many times) to produce a large sample of DNA fragments. **3** Finally, the amplified DNA fragments are compared. Together, these steps provide data about which samples are from the same individual and which samples are unique.

▼ **Figure 12.12 Overview of DNA profiling.** In this example, DNA from suspect 1 does not match DNA found at the crime scene, but DNA from suspect 2 does match.

DNA Profiling Techniques

In this section, we'll look at how a DNA profile is made. Researchers use three basic techniques: the polymerase chain reaction, short tandem repeat analysis, and gel electrophoresis.

The Polymerase Chain Reaction (PCR)

The **polymerase chain reaction (PCR)** is a technique by which a specific segment of DNA can be amplified—that is, targeted and copied quickly and precisely. Through PCR, a scientist can obtain enough DNA from even minute amounts of blood or other tissue to allow a DNA profile to be constructed. In fact, a microscopic sample with as few as 20 cells can be sufficient for PCR amplification.

In principle, PCR is simple. A DNA sample is mixed with nucleotides, the DNA replication enzyme DNA polymerase, and a few other ingredients. The solution is then exposed to cycles of heating (to separate the DNA strands) and cooling (to allow double-stranded DNA to re-form). During these cycles, specific regions of each molecule of DNA are replicated, doubling the amount of that DNA (Figure 12.13). The result of this chain reaction

▼ **Figure 12.13**
DNA amplification by PCR. The polymerase chain reaction (PCR) is a method for making many copies of a segment of DNA. Each round of PCR, performed on a thermal cycler (shown at top), doubles the total quantity of DNA.

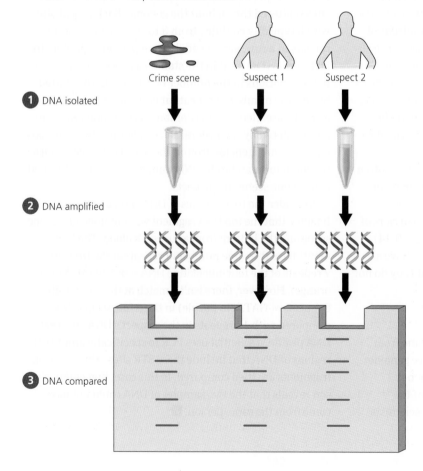

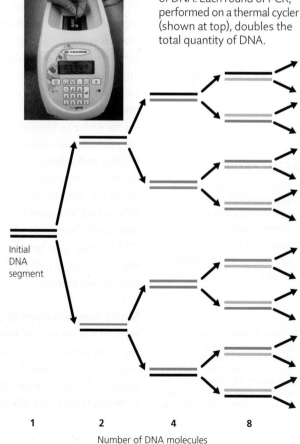

is an exponentially growing population of identical DNA molecules. The key to automated PCR is an unusually heat-stable DNA polymerase, first isolated from prokaryotes living in hot springs (see Figure 15.16). Unlike most proteins—which denature, or fall apart, at high temperatures—this enzyme can withstand the heat at the start of each cycle.

DNA molecules are typically very long, but usually only a very small target region of a large DNA molecule needs to be amplified. The key to amplifying one particular segment of DNA and no others is the use of **primers**, short (usually 15–20 nucleotides long), chemically synthesized single-stranded DNA molecules. For each experiment, specific primers are chosen that are complementary to sequences found only at each end of the target sequence. The primers thus bind to sequences that flank the target sequence, marking the start and end points for the segment of DNA to be amplified. Beginning with a single DNA molecule and the appropriate primers, automated PCR can generate hundreds of billions of copies of the desired sequence in a few hours.

In addition to forensic applications, PCR can be used in the treatment and diagnosis of disease. For example, because the sequence of the genome of HIV (the virus that causes AIDS) is known, PCR can be used to amplify, and thus detect, HIV in blood or tissue samples. In fact, PCR is often the best way to detect this otherwise elusive virus. Medical scientists can now diagnose hundreds of human genetic disorders by using PCR with primers that target the genes associated with these disorders. The amplified DNA product is then studied to reveal the presence or absence of the disease-causing mutation. Among the genes for human diseases that have been identified are those for sickle-cell disease, hemophilia, cystic fibrosis, Huntington's disease, and Duchenne muscular dystrophy. Individuals afflicted with such diseases can often be identified before the onset of symptoms, even before birth, allowing for preventative medical care to begin. PCR can also be used to identify symptomless carriers of potentially harmful recessive alleles (see Figure 9.14). Parents may thus be informed of whether they have a risk of bearing a child with a rare disease that they do not themselves display.

Short Tandem Repeat (STR) Analysis

How do you prove that two samples of DNA come from the same person? You could compare the entire genomes found in the two samples. But such an approach is impractical because the DNA of two humans of the same sex is 99.9% identical. Instead, forensic scientists typically compare about a dozen short segments of noncoding repetitive DNA that are known to vary from person to person. Have you ever seen a puzzle in a magazine that presents two nearly identical photos and asks you to find the few differences between them? In a similar way, scientists can focus on the few areas of difference in the human genome, ignoring the identical majority.

Repetitive DNA, which makes up much of the DNA that lies between genes in humans, consists of nucleotide sequences that are present in multiple copies in the genome. Some of this DNA consists of short sequences repeated many times tandemly (one after another); such a series of repeats in the genome is called a **short tandem repeat (STR)**. For example, one person might have the sequence AGAT repeated 12 times in a row at one place in the genome, the sequence GATA repeated 35 times at a second place, and so on; another person is likely to have the same sequences at the same places but with a different number of repeats. By focusing on STRs, forensic scientists are able to compare the tiny fraction of the genome that is most likely to vary from person to person.

STR analysis is a method of DNA profiling that compares the lengths of STR sequences at specific sites in the genome. The standard STR analysis procedure used by U.S. law enforcement agencies compares the number of repeats of specific four-nucleotide DNA sequences at 13 sites scattered throughout the genome. Each repeat site, which typically contains from 3 to 50 four-nucleotide repeats in a row, varies widely from person to person. In fact, some STRs used in the standard procedure have up to 80 variations in the number of repeats. In the United States, the number of repeats at each site is entered into a database called CODIS (Combined DNA Index System) administered by the Federal Bureau of Investigation. Law enforcement agencies around the world can access CODIS to search for matches to DNA samples they have obtained from crime scenes or suspects.

Consider the two samples of DNA shown in **Figure 12.14**. Imagine that the top DNA segment was obtained at a crime scene and the bottom from a suspect's blood. The two segments have the same number of repeats at the first site: 7 repeats of the four-nucleotide DNA sequence AGAT (in orange). However, there isn't a match at the second site: 8 repeats of GATA (in purple) in the crime scene DNA, compared with 12 repeats in the suspect's DNA. To create a DNA profile, a scientist uses PCR to specifically amplify the regions of DNA that include these STR sites. The resulting fragments are then compared. In this case, that comparison reveals that the two samples of DNA could not have come from the same person. ☑

☑ CHECKPOINT

1. Why is only the slightest trace of DNA at a crime scene often sufficient for forensic analysis?

2. What are STRs, and why are they useful for DNA profiling?

■ Answers: 1. because PCR can be used to produce enough molecules for analysis 2. STRs are nucleotide sequences repeated many times in a row. STRs are valuable for DNA profiling because different people have different numbers of repeats at the various STR sites.

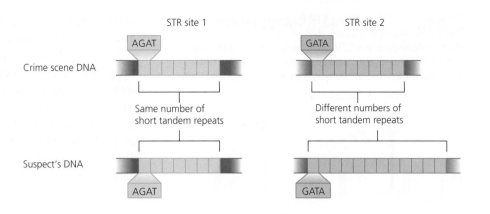

▶ **Figure 12.14 Short tandem repeat (STR) sites.** Scattered throughout the genome, STR sites contain tandem repeats of four-nucleotide sequences. The number of repetitions at each site can vary from individual to individual. In this figure, both DNA samples have the same number of repeats (7) at the first STR site, but different numbers (8 versus 12) at the second.

Gel Electrophoresis

DNA profiling by STR analysis depends upon comparing lengths of DNA fragments. This can be accomplished by **gel electrophoresis**, a method for sorting macromolecules—usually proteins or nucleic acids—primarily by electrical charge and size. **Figure 12.15** shows how gel electrophoresis separates DNA fragments obtained from different sources. A sample with many copies of the DNA from each source is placed in a separate well (hole) at one end of a gel, a thin slab of jellylike material that acts as a molecular sieve. A negatively charged electrode is then attached to this end of the gel and a positive electrode to the other end. Because the phosphate (PO_4^-) groups of nucleotides give DNA fragments a negative charge, the fragments move through the gel toward the positive pole. Imagine a small animal scampering through a thicket of vines, while a large animal plods much more slowly. Similarly, shorter DNA fragments move farther through a gel than do longer DNA fragments. Gel electrophoresis thus separates DNA fragments by length. When the current is turned off, a series of bands (blue smudges in the photo)is left in each column of the gel. Each band is a collection of DNA fragments of the same length. The DNA bands can be made visible by staining, by exposure onto photographic film (if radioactively labeled), or by measuring fluorescence (if labeled with a fluorescent dye). ✓

▼ **Figure 12.15 Gel electrophoresis of DNA molecules.** The photo shows DNA fragments of various sizes visibly stained on the gel.

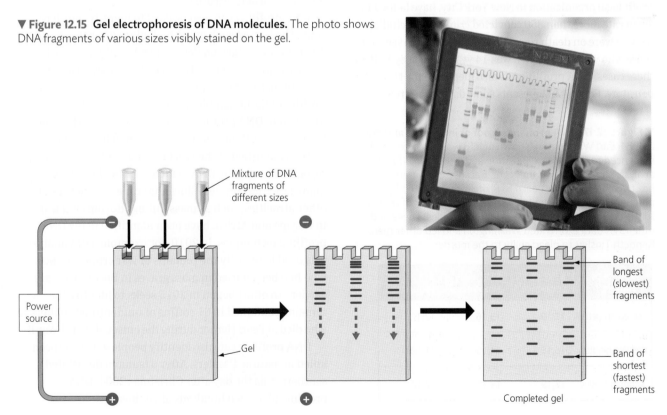

▼ **Figure 12.16 Visualizing STR fragment patterns.** This figure shows the bands that would result from gel electrophoresis of the STR sites illustrated in Figure 12.14. Notice that one of the bands from the crime scene DNA does not match one of the bands from the suspect's DNA.

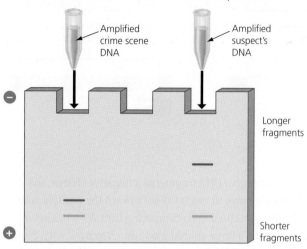

Figure 12.16 shows the gel that would result from using gel electrophoresis to separate the DNA fragments from the example in Figure 12.14. This figure simplifies the process; an actual STR analysis uses more than two sites and uses a different method to visualize the results. The differences in the locations of the bands reflect the different lengths of the DNA fragments. This gel would provide evidence that the crime scene DNA did not come from the suspect.

DNA profiling can provide evidence of guilt or innocence. As of 2017, lawyers at the Innocence Project, a nonprofit legal organization in New York City, have helped to exonerate more than 350 convicted criminals, including 20 who were on death row. The average sentence served by those who were exonerated was 14 years. In nearly half of these cases, DNA profiling has also identified the true perpetrators. **Figure 12.17** presents some data from a real case

▼ **Figure 12.17 DNA profiling: proof of innocence and guilt.** In 1984, Earl Washington (shown at left) was convicted and sentenced to death for a 1982 rape and murder. In 2000, STR analysis showed he was innocent. Because every person has two chromosomes, each STR site is represented by two numbers of repeats. The table shows the number of repeats for three STR DNA sequences in three samples. These and other STR data exonerated Washington and led other man, Kenneth Tinsley, to plead guilty to the murder.

Source of sample	STR sequence 1	STR sequence 2	STR sequence3
Semen on victim	17,19	13,16	12,12
Earl Washington	16,18	14,15	11,12
Kenneth Tinsley	17,19	13,16	12,12

in which STR analysis proved a convicted man innocent and helped identify the true perpetrator.

Just how reliable is a genetic profile? In forensic cases using STR analysis with the 13 standard sites, the probability of two people having identical DNA profiles is somewhere between one chance in 10 billion and one in several trillion. (The exact probability depends on the frequency of the particular DNA sequences in the general population.) Thus, despite problems that can still arise from insufficient data, human error, or flawed evidence, genetic profiles are now accepted as compelling evidence by legal experts and scientists alike.

Investigating Murder, Paternity, and Ancient DNA

Since DNA profiling was introduced in 1986, it has become a standard tool of forensics and has provided crucial evidence in many famous investigations. After the death of terrorist leader Osama bin Laden in 2011, U.S. Special Forces members obtained a sample of his DNA. Within hours, a military laboratory in Afghanistan had compared the tissue against samples obtained from several of bin Laden's relatives, including a sister who had died of brain cancer in a Boston hospital in 2010. Although facial recognition and an eyewitness identification provided preliminary evidence, it was DNA that provided a conclusive match, officially ending the long hunt for the notorious terrorist.

DNA profiling can also be used to identify murder victims. The largest such effort took place after the World Trade Center attack on September 11, 2001. Forensic scientists in New York City worked for years to identify more than 20,000 samples of victims' remains. DNA profiles of tissue samples from the disaster site were matched to DNA profiles from tissue known to be from the victims or their relatives. More than half of the victims identified at the World Trade Center site were recognized solely by DNA evidence, providing closure to many grieving families. Since that time, the victims of other atrocities, such as mass killings during civil wars in Europe and Africa, have been identified using DNA profiling techniques. In 2010, for example, DNA analysis was used to identify the remains of war crime victims who had been buried in mass graves in Bosnia 15 years earlier. An effort begun in 2015 seeks to identify the remains contained in 61 coffins of unidentified soldiers who died at Pearl Harbor during the outset of World War II.

DNA profiling can also identify people who have been killed in natural disasters. After a tsunami devastated southern Asia the day after Christmas 2004, DNA profiling identified hundreds of victims.

Comparing the DNA of a mother, her child, and a purported father can settle a question of paternity. Sometimes paternity is of historical interest: DNA profiling proved that Thomas Jefferson or a close male relative fathered a child with an enslaved woman, Sally Hemings. Similarly, tests conducted in 2015 confirmed a woman's story that Warren G. Harding had fathered her child while he was president. In another historical case, researchers investigated whether any descendants of Marie Antoinette (Figure 12.18), one-time queen of France, survived the French Revolution. DNA extracted from a preserved heart said to belong to her son was compared with DNA extracted from a lock of Marie's hair. A DNA match proved that her last known heir had, in fact, died in jail during the revolution. More recently, a former backup singer for James Brown, the "Godfather of Soul," sued his estate after his death, claiming that her child was Brown's son. A DNA paternity test proved her claim, and 25% of Brown's estate was awarded to the mother and child.

DNA profiling can also help protect endangered species by proving the origin of contraband animal products. For example, analysis of seized elephant tusks can pinpoint the location of the poaching, allowing enforcement officials to increase surveillance and prosecute those responsible. In 2014, three tiger poachers in India were sentenced to five years in jail after DNA profiling matched the dead tigers' flesh to tissue under the poachers' fingernails.

Modern methods of DNA profiling are so specific and powerful that the DNA samples can be in a partially degraded state. Such advances are revolutionizing the study of ancient remains. For example, a 2014 study of DNA extracted from five mummified Egyptian heads (dating from 800 B.C. to 100 A.D.) was used to deduce the geographic origins of the mummies, as well as to identify pathogens that had infected those people before they died. Another study determined that DNA extracted from a 27,000-year-old Siberian mammoth was 98.6% identical to DNA from living African elephants. Other studies involving a large collection of mammoth samples suggested that the last populations of the huge beasts migrated from North America to Siberia, where separate species interbred before dying out several thousand years ago.

▼ **Figure 12.18 Marie Antoinette.** DNA profiling proved that Louis (depicted with his mother in this 1785 painting), the son of the Queen of France, did not survive the French Revolution.

Bioinformatics

In the past decade, new experimental techniques have generated enormous volumes of data related to DNA sequences. The need to make sense of an ever-increasing flood of information has given rise to the field known as **bioinformatics**, the application of computational methods to the storage and analysis of biological data. In this section, we'll explore some of the methods by which sequence data are accumulated, as well as many of the practical ways such knowledge can be used.

DNA Sequencing

Researchers can exploit the principle of complementary base pairing to determine the complete nucleotide sequence of small DNA molecules. This process is known as **DNA sequencing**. In one standard procedure, called "next-generation sequencing," DNA is cut into fragments of around 300 nucleotides, and then thousands or hundreds of thousands of these fragments are sequenced simultaneously (Figure 12.19). This technology is rapid and inexpensive, making it possible to sequence more than 2 billion nucleotides in one day! This is an example of "high-throughput" DNA technology, which is currently the method of choice for studies where massive numbers of DNA

▼ **Figure 12.19 A DNA sequencer.** This high-throughput DNA-sequencing machine can process half a billion bases in a single 10-hour run.

samples—even representing an entire genome or a collection of genomes—are being sequenced.

More recently, scientists have improved or replaced next-generation sequencing. Several groups have been working on "third-generation sequencing," wherein a single, very long DNA molecule is sequenced on its own. The idea is to move a single strand of DNA through a very small pore in a membrane (a nanopore) while administering an electrical current that will detect the nitrogenous bases one by one. For each type of base, the electrical current is interrupted for a slightly different length of time, allowing the base sequence to be determined nucleotide by nucleotide.

In 2015, the first nanopore sequencer went on the market; this device is the size of a small candy bar and connects to a computer through a USB port. Software allows the immediate identification and analysis of the sequence. This is one of many approaches to increase the rate and cut the cost of sequencing, while allowing the methodology to move out of the laboratory and into the field.

Next-generation and third-generation sequencing techniques are ushering in a new era of faster, more affordable sequencing. Taken to their logical extreme, these techniques can be applied to whole genomes—our next topic.

Genomics

Improved DNA-sequencing techniques have transformed how we explore fundamental biological questions about evolution and how life works. A major leap forward occurred in 1995 when a team of scientists determined the nucleotide sequence of the entire genome of *Haemophilus influenzae*, a bacterium that causes several human diseases, including pneumonia and meningitis. **Genomics**, the study of complete sets of genes (genomes), was born.

The first targets of genomics research were bacteria, which have relatively little DNA. Researchers then studied more complex organisms with much larger genomes. Baker's yeast (*Saccharomyces cerevisiae*) was the first eukaryote to have its full sequence determined, and the roundworm *Caenorhabditis elegans* was the first multicellular organism. Other sequenced animals include the fruit fly (*Drosophila melanogaster*) and lab rat (*Rattus norvegicus*), both model organisms for genetics research. Among the sequenced plants are *Arabidopsis thaliana*, a type of mustard plant used as a model organism, and rice (*Oryza sativa*), one of the most economically important crops.

The genomes of thousands of species have been published, and tens of thousands more are in progress **(Figure 12.20)**. The majority of organisms sequenced are prokaryotes, including more than 4,000 bacterial species and nearly 200 archaea. Hundreds of eukaryotic genomes—including protists, fungi, plants, and animals both invertebrate and vertebrate—have been completed. Genome sequences have been determined for cells from several cancers, for ancient humans, and for the many bacteria that live in the human intestine. **As the ultimate repository of the genetic information from which all of life's inherited characteristics develop, genomes hold the key to our genetic identity.** ☑

☑ **CHECKPOINT**

Based on Figure 12.20, about how many nucleotides and genes are contained in the human genome?

■ *Answer: about 3 billion nucleotides and 21,000 genes*

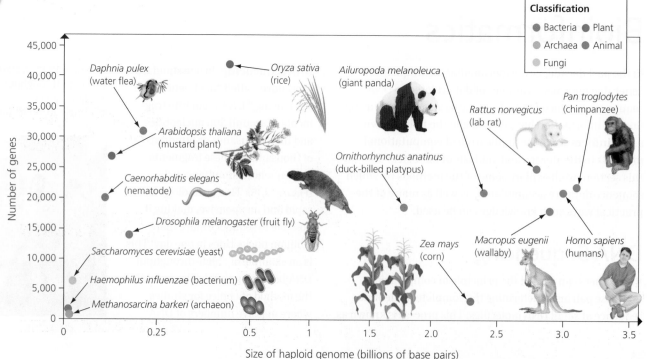

▼ **Figure 12.20 Comparing whole genomes.** Notice that the size or complexity of an organism does not always correspond to the size of the genome or the number of genes.

Genome-Mapping Techniques

Genomes are most often sequenced using a technique called the **whole-genome shotgun method (Figure 12.21)**. The first step is to chop the entire genome into fragments using restriction enzymes. Next, all the fragments are cloned and sequenced. Finally, computers running specialized mapping software reassemble the millions of overlapping short sequences into a single continuous sequence for every chromosome—an entire genome.

The DNA sequences determined by many research groups in the United States are deposited in GenBank, a database available to anyone through the Internet. You can browse it yourself at the National Center for Biotechnology Information: www.ncbi.nlm.nih.gov. As of 2017, GenBank includes the sequences of more than 200 billion base pairs of DNA! The database is constantly updated, and the amount of data it contains doubles every 18 months. Any sequence in the database can be retrieved and analyzed. For example, software can compare a collection of sequences from different species and diagram them as an evolutionary tree based on the sequence relationships. Bioinformatics has thereby revolutionized evolutionary biology by opening a vast new reservoir of data that can test evolutionary hypotheses. Next, we'll discuss a particularly notable example of a sequenced animal genome—our own.

The Human Genome

The **Human Genome Project** was a massive scientific endeavor to determine the nucleotide sequence of all the DNA in the human genome and to identify the location and sequence of every gene. The project began in 1990 as an effort by government-funded researchers from six nations. Several years into the project, private companies joined the effort. At the completion of the project, more than 99% of the genome had been determined to 99.999% accuracy. (There remain a few hundred gaps of unknown sequence that require special methods to figure out.) This ambitious project has provided a wealth of data that may illuminate the genetic basis of what it means to be human.

The chromosomes in the human genome (22 autosomes plus the X and Y sex chromosomes) contain approximately 3 billion nucleotide pairs of DNA. If you imagine this sequence printed in letters (A, T, C, and G) the same size you see on this page, the sequence would fill a stack of books 18 stories high! However, the biggest surprise from the Human Genome Project is the relatively small number of human genes—currently estimated to be about 21,000—very close to the number found in a roundworm!

Like the genomes of most complex eukaryotes, only a small amount of total human DNA consists of genes that code for proteins, tRNAs, or rRNAs. Most complex eukaryotes have a huge amount of noncoding DNA—about 98.5%

▼ **Figure 12.21 Genome sequencing.** In the photo at the bottom, a technician performs a step in the whole-genome shotgun method (depicted in the diagram).

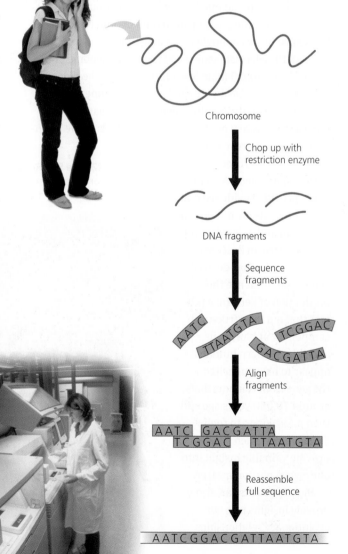

Chromosome

Chop up with restriction enzyme

DNA fragments

Sequence fragments

AATC TTAATGTA TCGGAC GACGATTA

Align fragments

AATC GACGATTA
TCGGAC   TTAATGTA

Reassemble full sequence

AATCGGACGATTAATGTA

YOU HAVE ABOUT
THE SAME NUMBER
OF GENES AS A
WORM, AND ONLY
HALF AS MANY AS
A RICE PLANT.

of human DNA is of this type. Some of this noncoding DNA is made up of gene control sequences such as promoters, enhancers, and microRNAs (see Chapter 11). Other noncoding regions include introns and repetitive DNA (some of which is used in DNA profiling). Some noncoding DNA is important to our health, with certain regions known to carry disease-causing mutations. But the function of most noncoding DNA remains unknown.

The human genome sequenced by government-funded scientists was actually a reference genome compiled from a group of individuals. As of today, the complete genomes of many individuals have been completed. Whereas sequencing the first human genome took 13 years and cost $100 million, we are rapidly approaching the day when an individual's genome can be sequenced in a matter of hours for less than $1,000.

Scientists have even begun to gather sequence data from our extinct relatives. Neanderthals (*Homo neanderthalensis*) appeared at least 300,000 years ago in Europe and Asia and survived until about 30,000 years ago. Modern humans (*Homo sapiens*) first appeared in Africa around 200,000 years ago and spread into Europe and Asia around 50,000 years ago—meaning that modern humans and Neanderthals most likely comingled for a long time.

In 2013, scientists sequenced the entire genome of a 130,000-year-old female Neanderthal (*Homo neanderthalensis*). Using DNA extracted from a toe bone found in a Siberian cave, the resulting genome was nearly as complete as that from a modern human. Analysis of the Neanderthal genome revealed evidence of interbreeding with *Homo sapiens*. A 2014 study provided evidence that many present-day humans of European and Asian descent (but not African descent) carry Neanderthal-derived genes that influence the production of keratin, a protein that is a key structural component of hair, nails, and skin. Modern humans appear to have inherited the gene from Neanderthals around 70,000 years ago and then passed it on to their descendants. Such studies provide valuable insight into our own evolutionary tree.

Bioinformatics can also provide insights into our evolutionary relationships

with nonhuman animals. In 2005, researchers completed the genome sequence for our closest living relative on the evolutionary tree of life, the chimpanzee (*Pan troglodytes*). Comparisons with human DNA revealed that we share 96% of our genome. Genomic scientists are currently finding and studying the important differences, shedding scientific light on the age-old question of what makes us human. **By comparing humans with related species both living (chimpanzees) and extinct (Neanderthals), researchers are shedding light on the recent evolutionary history of our own species.**

The potential benefits of knowing many human genomes are enormous. Thus far, more than 2,000 disease-associated genes have been identified. A recent example involved Behcet's disease, a painful and life-threatening illness that involves swelling of blood vessels throughout the body. Researchers have long known that this disease is found most commonly among people living along the ancient trade route in Asia called the Silk Road **(Figure 12.22)**. In 2013, researchers conducted a genome-wide search for genetic differences among Turkish people with and without the disease. They discovered four regions of the genome that are associated with the disease. The nearby genes are implicated in the immune system's ability to destroy invading microorganisms, to recognize infection sites, and to regulate autoimmune diseases. Interestingly, the function of the fourth gene has never been identified, but its close association with Behcet's disease may help researchers pinpoint its role. Next, in the Process of Science section, we'll examine how genomic analysis solved a medical mystery.

▼ **Figure 12.22 The Silk Road.** Behcet's disease is most commonly found along the Silk Road, part of which is shown in this map using modern place names.

Key	
——	= Silk Road

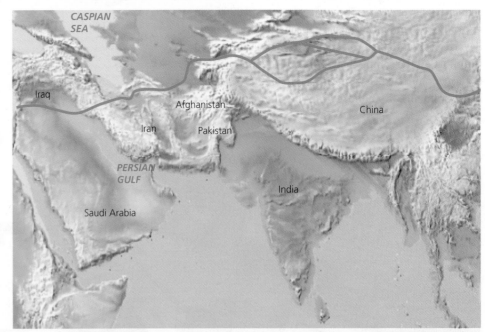

Did Nic Have a Deadly Gene?

BACKGROUND

When infant Nic Volker stopped breast feeding, he became a medical mystery. Although previously healthy, he now cried in agony after meals and began to waste away. He developed ulcers between his intestines and skin and had to be fed through a nasal tube. After two years of mistaken diagnoses and 100 surgeries, everyone was desperate. Doctors hypothesized that Nic had a rare mutation that made his immune system attack his digestive system. But how could they prove that idea?

METHOD

A team of doctors at Medical College of Wisconsin sequenced Nic's entire protein-coding genome, a radical and expensive procedure in 2009. When doctors compared Nic's genome with other sequenced human genomes, they found over 16,000 mutations, the vast majority of which were not medically relevant. The challenge was to eliminate all mutations except for the one that was causing the symptoms. They developed new software to filter the mutations, eliminating ones that did not lead to malfunctioning proteins or were not involved in the digestion or immune systems. Then they reduced the list to rare mutations.

RESULTS

After eliminating thousands of mutations, the *XIAP* gene remained. This gene consists of about 500 nucleotides that code for the XIAP protein. The doctors found that Nic had a single base substitution, which produced a tyrosine amino acid where cysteine should be in the protein. In this case, as in many others, researchers gained insights

through comparison with other animals. Data from several species confirmed that cysteine is always at that key location on the XIAP protein **(Figure 12.23)**. The diagnosis allowed the team to save Nic's life by replacing his bone marrow with donor cells that contained functional *XIAP* genes. Today, similar procedures are using personal genome sequencing to save the lives of many children with rare mutations.

Nic Volker with a DNA-sequencing machine

▼ **Figure 12.23 The sequence of amino acids in one region of the XIAP protein.** This table shows the sequence of amino acids (using one-letter abbreviations) in the XIAP protein in patient Nic and other organisms. At a key location (highlighted on the top line), Nic had a base substitution that resulted in tyrosine (symbol: Y) instead of cysteine.

Amino acids (one-letter abbreviations)

Nic	G	D	Q	V	Q	C	F	C	Y	G	G	K	L	K	N	W	E
Healthy human	G	D	Q	V	Q	C	F	C	C	G	G	K	L	K	N	W	E
Chicken	D	D	Q	V	Q	A	F	C	C	G	G	K	L	K	N	W	E
Zebra fish	D	D	N	V	Q	C	F	C	C	G	G	G	L	S	G	W	E
Frog	R	D	H	V	K	C	F	H	C	D	G	G	L	R	N	W	E
Housefly	L	D	H	V	K	C	V	W	C	N	G	V	I	A	K	W	E

Thinking Like a Scientist

Why would the scientists compare Nic's *XIAP* gene with genes in animals like flies and frogs?

For the answer, see Appendix D.

Applied Genomics

Solving the mystery of Nic Volker's rare mutation is a good example of applied genomics. In another medical application, a 2013 study used DNA sequencing to prove that cancerous skin cells that had spread to the brain had done so after fusing with red blood cells provided by a bone marrow donor. These results provided researchers with new insight into how cancer spreads throughout the body. Sequence data also provided strong evidence that a Florida dentist transmitted HIV to several patients and that a single natural strain of West Nile virus can infect both birds and people.

Applied genomics can also be used to investigate criminal cases. In 2001, a 63-year-old Florida man died from inhalation anthrax, a disease caused by breathing

spores of the bacterium *Bacillus anthracis*. Because he was the first victim of this disease in the United States since 1976 (and coming so soon after the 9/11 terrorist attacks the month before), his death was immediately considered suspicious. By the end of the year, four more people had died from inhaling anthrax. Law enforcement officials realized that someone was sending anthrax spores through the mail **(Figure 12.24)**. The United States was facing an unprecedented bioterrorist attack.

▶ **Figure 12.24 The 2001 anthrax attacks.** In 2001, some envelopes containing anthrax spores caused five deaths.

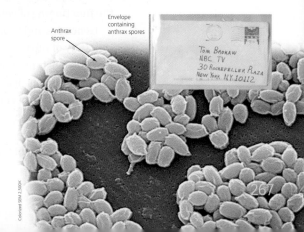

Envelope containing anthrax spores

Anthrax spore

In the investigation that followed, one of the most helpful clues turned out to be the anthrax spores themselves. Investigators sequenced the genomes of the mailed anthrax spores. They quickly established that all of the mailed spores were genetically identical to a laboratory subtype stored in a single flask at the U.S. Army Medical Research Institute of Infectious Diseases in Fort Detrick, Maryland. Based in part on this evidence, the FBI named an army research scientist as a suspect in the case. Although never charged, the suspect committed suicide in 2008, and the case officially remains unsolved.

Systems Biology

The computational power provided by the tools of bioinformatics allows the study of whole sets of genes and their interactions, as well as the comparison of genomes from different species. Genomics is a rich source of new insights into fundamental questions about genome organization, regulation of gene expression, embryonic development, and evolution.

Technological advances have also led to metagenomics, the study of DNA from an environmental sample. When obtained directly from the environment, a sample will likely contain genomes from many species. After the whole sample is sequenced, computer software sorts out the partial sequences from different species and assembles them into the individual specific genomes. So far, this approach has been applied to microbial communities found in environments as diverse as the Sargasso Sea and the human intestine. A 2012 study cataloged the astounding diversity of the human "microbiome"—the many species of bacteria that coexist within and upon our bodies and that contribute to our survival. The ability to sequence the DNA of mixed populations eliminates the need to culture each species separately in the lab, making it more efficient to study microbial species.

The successes in genomics have encouraged scientists to begin similar systematic studies of the full protein sets that genomes encode (proteomes), an approach called **proteomics (Figure 12.25)**. The number of different proteins in humans far exceeds the number of different genes (about 100,000 proteins versus about 21,000 genes). And because proteins, not genes, actually carry out the activities of the cell, scientists must study when and where proteins are produced and how they interact to understand the functioning of cells and organisms.

Genomics and proteomics enable biologists to approach the study of life from an increasingly global perspective. Biologists are now compiling catalogs of genes and proteins—that is, listings of all the "parts" that contribute to the operation of cells, tissues, and organisms. As such catalogs become complete, researchers are shifting their attention from the individual parts to how these parts work together in biological systems. This approach, called

systems biology, aims to model the dynamic behavior of whole biological systems based on the study of the interactions among the system's parts. Because of the vast amounts of data generated in these types of studies, advances in computer technology and bioinformatics have been crucial in making systems biology possible.

Such analyses may have many practical applications. For example, proteins associated with specific diseases may be used to aid diagnosis (by developing tests that search for a particular combination of proteins) and treatment (by designing drugs that interact with the proteins involved). As high-throughput techniques become more rapid and less expensive, they are increasingly being applied to the problem of cancer. The Cancer Genome Atlas project is a simultaneous investigation by multiple research teams of a large group of interacting genes and gene products. This project aims to determine how changes in biological systems lead to cancer. A three-year pilot project set out to find all the common mutations in three types of cancer—lung, ovarian, and brain—by comparing gene sequences and patterns of gene expression in cancer cells with those in normal cells. The results confirmed the role of several genes suspected to be linked to cancer and identified a few previously unknown ones, suggesting possible new targets for therapies. The research approach proved so fruitful for these three types of cancer that the project has been extended to ten other types of cancer, chosen because they are common and often lethal in humans.

Systems biology is a very efficient way to study emergent properties, novel properties that arise at each successive level of biological complexity as a result of the arrangement of building blocks at the underlying level. **The more we can learn about the arrangement and interactions of the components of genetic systems, the deeper will be our understanding of whole organisms.** ☑

☑ **CHECKPOINT**

What is the difference between genomics and proteomics?

■ *Answer: Genomics concerns the complete set of an organism's genes, whereas proteomics concerns the complete set of an organism's proteins.*

▼ **Figure 12.25 Proteomics.** Each peak on this three-dimensional graph represents one protein separated by gel electrophoresis. The height of the peak correlates with the amount of that protein. By identifying every protein in a sample, researchers can gain a fuller understanding of the complete biological system.

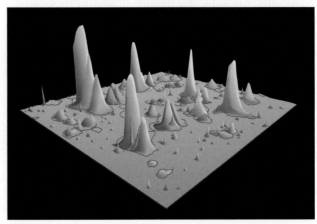

Safety and Ethical Issues

As soon as scientists realized the power of DNA technology, they began to worry about potential dangers. Early concerns focused on the possibility of creating hazardous new disease-causing organisms. What might happen, for instance, if cancer-causing genes were transferred into infectious bacteria or viruses? To address such concerns, scientists developed a set of guidelines that have become formal government regulations in the United States and some other nations.

One safety measure in place is a set of strict laboratory procedures to protect researchers from infection by engineered microorganisms and to prevent microorganisms from accidentally leaving the laboratory (**Figure 12.26**). In addition, strains of microorganisms to be used in recombinant DNA experiments are genetically crippled to ensure that they cannot survive outside the laboratory. As a further precaution, certain obviously dangerous experiments have been banned.

▼ **Figure 12.26 Maximum-security laboratory.** A scientist at the Centers for Disease Control wears a biohazard suit as he studies the virus that caused the 1918 flu pandemic, which killed over 50 million people worldwide.

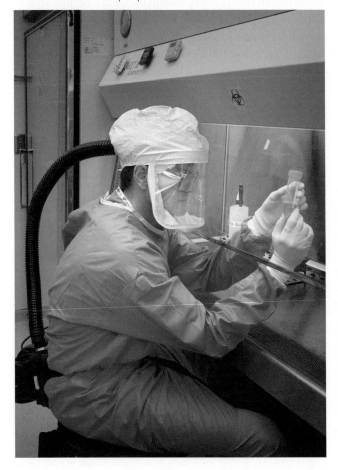

The Controversy over Genetically Modified Foods

Today, most public concern about possible hazards centers on genetically modified (GM) foods. GM strains account for a significant percentage of several staple crops in the United States, Argentina, and Brazil; together, these nations account for more than 80% of the world's supply of GM crops. The production of GM crops is often more profitable because the modifications improve yields and increase the size of harvest. Advocates for GM foods point out that transgenic crops will become increasingly necessary to combat starvation as more food must be grown on less space.

Controversy about the safety of these foods is a significant political issue (**Figure 12.27**). For example, the European Union has suspended the introduction into the market of new GM crops and considered banning the import of all GM foodstuffs. In the United States and other nations where the GM revolution initially proceeded relatively unnoticed, mandatory labeling of GM foods is now being debated.

Advocates of a cautious approach are concerned that crops carrying genes from other species might harm the environment or be hazardous to human health (by, for example, introducing new allergens, molecules that can cause allergic reactions, into foods). A major worry is that

▼ **Figure 12.27 Opposition to genetically modified organisms (GMOs).** Protesters in Oregon voice their displeasure about GMOs.

transgenic plants might pass their new genes to close relatives in nearby wild areas. We know that lawn and crop grasses, for example, commonly exchange genes with wild relatives through pollen transfer. If domestic plants carrying genes for resistance to herbicides, diseases, or insect pests pollinated wild plants, the offspring might become "superweeds" that would be very difficult to control. However, researchers may be able to prevent the escape of such plant genes in various ways—for example, by engineering plants so that they cannot breed. Concern has also been raised that the widespread use of GM seeds may reduce natural genetic diversity, leaving crops susceptible to catastrophic die-offs in the event of a sudden change to the environment or introduction of a new pest. Although the U.S. National Academy of Sciences released a study finding no scientific evidence that transgenic crops pose any special health or environmental risks, the authors of the study also recommended more stringent long-term monitoring to watch for unanticipated environmental impacts.

Negotiators from 130 nations (including the United States) agreed on a Biosafety Protocol that requires exporters to identify GM organisms present in bulk food shipments and allows importing nations to decide whether the shipments pose environmental or health risks. The United States declined to sign the agreement, but it went into effect anyway because the majority of nations were in favor of it. Since then, European nations have, on occasion, refused crops from the United States and other nations for fear that they contain GM crops, leading to trade disputes.

Governments and regulatory agencies throughout the world are grappling with how to facilitate the use of biotechnology in agriculture, industry, and medicine while ensuring that new products and procedures are safe. In the United States, all genetic engineering projects are evaluated for potential risks by a number of regulatory agencies, including the Food and Drug Administration, the Environmental Protection Agency, the National Institutes of Health, and the Department of Agriculture. ☑

CHECKPOINT

☑ **CHECKPOINT**

What is the main concern about adding genes for herbicide resistance to crop plants?

Answer: the possibility that the genes could escape, through cross-pollination, to wild plants that are closely related to the crop species

Ethical Questions Raised by Human DNA Technologies

Human DNA technologies raise legal and ethical questions—few of which have clear answers. Consider, for example, how the treatment of dwarfism with injections of human growth hormone (HGH) produced by genetically engineered cells might be extended beyond its current use. Should parents of short but hormonally normal children be able to seek HGH treatment to make their kids taller? If not, who decides which children are "tall

enough" to be excluded from treatment? In addition to technical challenges, human gene therapy also provokes ethical questions. Some critics believe that tampering with human genes in any way is immoral or unethical. Other observers see no fundamental difference between the transplantation of genes into somatic cells and the transplantation of organs.

Genetic engineering of gametes (sperm or ova) and zygotes has been accomplished in lab animals. It has not been attempted in humans because such a procedure raises very difficult ethical questions. Should we try to eliminate genetic defects in our children and their descendants? Should we interfere with evolution in this way? From a long-term perspective, the elimination of unwanted versions of genes from the gene pool could backfire. **Genetic variety is a necessary ingredient for the adaptation of a species as environmental conditions change with time.** Genes that are damaging under some conditions may be advantageous under others (one example is the sickle-cell allele—see the Evolution Connection in Chapter 17). Are we willing to risk making genetic changes that could be detrimental to our species in the future?

Similarly, advances in genetic profiling raise privacy issues **(Figure 12.28)**. If we were to create a DNA profile of every person at birth, then theoretically we could match nearly every violent crime to a perpetrator because it is virtually impossible for someone to commit a violent crime without leaving behind DNA evidence. But are we, as a society, prepared to sacrifice our genetic privacy, even for such worthwhile goals? In 2014, the U.S. Supreme Court, by a 5—4 vote, upheld the practice of collecting DNA samples from suspects at the time of their arrest (before they had been convicted). Ruling that obtaining DNA is "like fingerprinting and photographing, a legitimate police booking procedure that is reasonable under the Fourth Amendment," the Supreme Court

▼ **Figure 12.28 Genetic information and privacy.** Collecting genomic data raises privacy concerns that affect everyone.

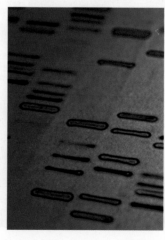

decision will likely usher in an era of expanded use of DNA profiling in many aspects of police work.

As more information becomes available about our personal genetic makeup, some people question whether greater access to this information is always beneficial. For example, mail-in kits (Figure 12.29) have become available that can tell healthy people their relative risk of developing various diseases (such as Parkinson's and Crohn's) later in life. Some argue that such information helps families to prepare. Others worry that the tests prey on our fears without offering any real benefit because certain diseases, such as Parkinson's, are not currently preventable or treatable. Other tests, however, such as for breast cancer risk, may help a person make changes that can prevent disease. How can we identify truly useful tests?

There is also a danger that information about disease-associated genes could be abused. One issue is the possibility of discrimination and stigmatization. In response, the U.S. Congress passed the Genetic Information Nondiscrimination Act of 2008. Title I of the act prohibits insurance companies from requesting or requiring genetic information during an application for health insurance. Title II provides similar protections in employment.

A much broader ethical question is how do we really feel about wielding one of nature's powers—the evolution of new organisms? Some might ask if we have any right

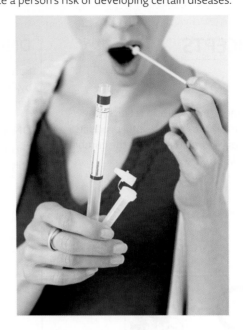

▼ Figure 12.29 **Personalized genetic testing.** This kit can be used to send saliva for genetic analysis. The results can indicate a person's risk of developing certain diseases.

to alter an organism's genes—or to create new organisms. DNA technologies raise many complex issues that have no easy answers. It is up to you, as a participating citizen, to make informed choices. ☑

☑ **CHECKPOINT**

Why does genetically modifying a human gamete raise different ethical questions than genetically modifying a human somatic (body) cell?

■ *Answer: A genetically modified somatic cell will affect only the patient. Modifying a gamete will affect an unborn individual as well as all of his or her descendants.*

EVOLUTION CONNECTION | DNA Profiling

The Y Chromosome as a Window on History

The human Y chromosome passes essentially intact from father to son. Therefore, by comparing Y DNA, researchers can learn about the ancestry of human males. DNA profiling can thus provide data about recent human evolution.

Geneticists have discovered that about 8% of males currently living in central Asia have Y chromosomes of striking genetic similarity. Further analysis traced their common genetic heritage to a single man living about 1,000 years ago. In combination with historical records, the data led to the speculation that the Mongolian ruler Genghis Khan (Figure 12.30) may have been responsible for the spread of the chromosome to nearly 16 million men living today. A similar study of Irish men suggested that nearly 10% of them were descendants of Niall of the Nine Hostages, a warlord who lived during the 1400s. Another study of Y DNA seemed to confirm the claim by the Lemba people of southern Africa that they are descended from ancient Jews (Figure 12.31). Sequences of Y DNA distinctive of the Jewish priestly

caste called Kohanim are found at high frequencies among the Lemba.

Comparison of Y chromosome DNA profiles is part of a larger effort to learn more about the human genome. Other research efforts are extending genomic studies to many more species. These studies will advance our understanding of all aspects of biology, including health and ecology, as well as evolution. In fact, comparisons of the completed genome sequences of bacteria, archaea, and eukaryotes first supported the theory that these are the three fundamental domains of life—a topic we discuss further in the next unit, Evolution and Diversity.

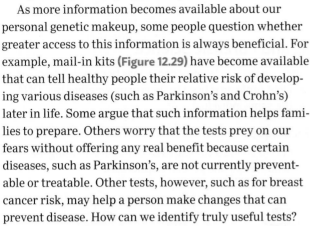

▲ Figure 12.30 **Genghis Khan.**

▶ Figure 12.31 **Lemba people of southern Africa.**

Chapter Review

SUMMARY OF KEY CONCEPTS

Genetic Engineering

DNA technology, the manipulation of genetic material, is a relatively new branch of biotechnology, the use of organisms to make helpful products. DNA technology often involves the use of recombinant DNA, the combination of nucleotide sequences from two different sources.

Recombinant DNA Techniques

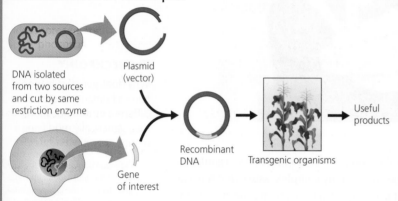

DNA isolated from two sources and cut by same restriction enzyme

Plasmid (vector)

Gene of interest

Recombinant DNA

Transgenic organisms

Useful products

Gene Editing

The CRISPR-Cas9 system can be used to deactivate or edit genes within living cells.

Medical Applications

By transferring a human gene into a bacterium or other easy-to-grow cell, scientists can mass-produce valuable human proteins to be used as drugs or vaccines.

Genetically Modified Organisms in Agriculture

Recombinant DNA techniques have been used to create genetically modified organisms, those that carry artificially introduced genes. Nonhuman cells have been engineered to produce human proteins, genetically modified food crops, and transgenic farm animals. A transgenic organism is one that carries artificially introduced genes, typically from a different species.

Human Gene Therapy

A virus can be modified to include a normal human gene. If this virus is injected into the bone marrow of a person suffering from a genetic disease, the normal human gene may be transcribed and translated, producing a normal human protein that may cure the genetic disease. This technique has been used in gene therapy trials involving a number of inherited diseases. There have been both successes and failures to date, and research continues.

DNA Profiling and Forensic Science

Forensics, the scientific analysis of legal evidence, has been revolutionized by DNA technology. DNA profiling is used to determine whether two DNA samples come from the same individual.

DNA Profiling Techniques

Short tandem repeat (STR) analysis compares DNA fragments using the polymerase chain reaction (PCR) and gel electrophoresis.

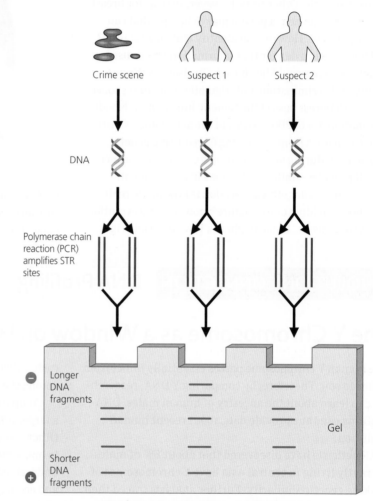

Crime scene

Suspect 1

Suspect 2

DNA

Polymerase chain reaction (PCR) amplifies STR sites

Longer DNA fragments

Shorter DNA fragments

Gel

DNA fragments compared by gel electrophoresis
(Bands of shorter fragments move faster toward the positive pole.)

Investigating Murder, Paternity, and Ancient DNA

DNA profiling can be used to establish innocence or guilt of a criminal suspect, identify victims, determine paternity, and contribute to basic research.

Bioinformatics

DNA Sequencing
Automated machines can now sequence many thousands of DNA nucleotides per hour.

Genomics
Advances in DNA sequencing have ushered in the era of genomics, the study of complete genome sets.

Genome-Mapping Techniques
The whole-genome shotgun method involves sequencing DNA fragments from an entire genome and then assembling the sequences.

The Human Genome
The nucleotide sequence of the human genome is providing a wealth of useful data. The 24 different chromosomes of the human genome contain about 3 billion nucleotide pairs and 21,000 genes. The majority of the genome consists of noncoding DNA.

Applied Genomics
Comparing genomes can aid criminal investigations and basic biological research.

Systems Biology
Success in genomics has given rise to proteomics, the systematic study of the full set of proteins found in organisms. Genomics and proteomics both contribute to systems biology, the study of how many parts work together within complex biological systems.

Safety and Ethical Issues

The Controversy over Genetically Modified Foods
The debate about genetically modified crops centers on whether they might harm humans or damage the environment by transferring genes through cross-pollination with other species.

Ethical Questions Raised by Human DNA Technologies
As members of society we must become educated about DNA technologies so that we can intelligently address the ethical questions raised by their use.

Mastering Biology

For practice quizzes, BioFlix animations, MP3 tutorials, video tutors, and more study tools designed for this textbook, go to Mastering Biology™

SELF-QUIZ

1. Which of the following best describes recombinant DNA?
   a. DNA that results from bacterial conjugation
   b. DNA that includes pieces from two different sources
   c. An alternate form of DNA that is the product of a mutation
   d. DNA that carries a translocation
2. The enzyme that is used to bind DNA fragments together is _____.
3. In making recombinant DNA, what is the benefit of using a restriction enzyme that cuts DNA in a staggered fashion?
4. A paleontologist has recovered a bit of organic material from the 400-year-old preserved skin of an extinct dodo. She would like to compare DNA from the sample with DNA from living birds. The most useful method for initially increasing the amount of dodo DNA available for testing is _____.
5. Why do DNA fragments containing STR sites from different people tend to migrate to different locations during gel electrophoresis?
6. Gel electrophoresis separates DNA fragments on the basis of differences in their _____.
   a. nucleotide sequences
   b. lengths
   c. hydrogen bonds between base pairs
   d. pH
7. After a gel electrophoresis procedure is run, the pattern of bars in the gel shows
   a. the order of bases in a particular gene.
   b. the presence of various-sized fragments of DNA.
   c. the order of genes along particular chromosomes.
   d. the exact location of a specific gene in the genome.
8. Name the steps of the whole-genome shotgun method.
9. Put the following steps of human gene therapy in the correct order.
   a. Virus is injected into patient.
   b. Human gene is inserted into a virus.
   c. Normal human gene is isolated and cloned.
   d. Normal human gene is transcribed and translated in the patient.

For answers to the Self Quiz, see Appendix D.

IDENTIFYING MAJOR THEMES

For each statement, identify which major theme is evident (the relationship of structure to function, information flow, pathways that transform energy and matter, interactions within biological systems, or evolution) and explain how the statement relates to the theme. If necessary, review the themes (see Chapter 1) and review the examples highlighted in blue in this chapter.

10. Comparisons of DNA sequences can reveal not just recent paternity, but patterns of ancient lineages, such as when humans diverged from other primates.

11. Bioinformatic tools allow for examinations of all the parts that constitute a living organism (such as proteins and genes), allowing their interrelationships to be studied.

12. Studying whole genomes reveals how genes from one generation can affect the appearance of the next.

For answers to Identifying Major Themes, see Appendix D.

THE PROCESS OF SCIENCE

13. Large-scale sequencing of the mitochondrial DNA of humans across the globe revealed significantly higher sequence diversity among Africans than among non-Africans. What does this tell us about the origin of humans?

14. **Interpreting Data** When comparing genomes from different species, biologists often calculate the genome density, the number of genes per number of nucleotides in the genome. Refer to Figure 12.20. You can estimate the gene density of each species by dividing the number of genes by the size of the genome (usually expressed in Mb, which is mega base pairs, or 1 million base pairs). Using a spreadsheet, estimate the gene density for every species in the figure. (Don't forget that 1 billion = 1,000 million; for example, humans have 3,000 Mb.) How does the gene density of bacteria compare to humans? Humans and roundworms have nearly the same number of genes, but how do the gene densities of these two species compare? Can you identify any general correlation between gene density and the size or complexity of an organism?

15. Listed below are 4 of the 13 genome sites used to create a standard DNA profile. Each site consists of a number of short tandem repeats: sets of 4 nucleotides repeated in a row within the genome. For each site, the number of repeats found at that site for this individual are listed.

Chromosome number	Genetic site	# of repeats
3	D3S1358	4
5	D5S818	10
7	D7S820	5
8	D8S1179	22

Imagine that you perform a PCR procedure to create a DNA profile for this individual. Which of the following four gels correctly represents the DNA profile of this person?

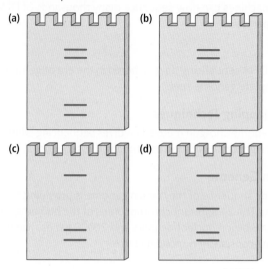

BIOLOGY AND SOCIETY

16. In the not-too-distant future, gene therapy may be used to treat many inherited disorders. What do you think are the most serious ethical issues to face before human gene therapy is used on a large scale? Explain.

17. Today, it is fairly easy to make transgenic plants and animals. What are some safety and ethical issues raised by this use of recombinant DNA technology? What are some dangers of introducing genetically engineered organisms into the environment? What are some reasons for and against leaving such decisions to scientists? Who should decide?

18. In October 2002, the government of the African nation of Zambia announced that it was refusing to distribute 15,000 tons of corn donated by the United States, enough corn to feed 2.5 million Zambians for three weeks. The government rejected the corn because it was likely to contain genetically modified kernels. The government made the decision after its scientific advisers concluded that the studies of the health risks posed by GM crops "are inconclusive." Do you agree with Zambia's decision? Why or why not? Consider that Zambia was facing food shortages, and 35,000 Zambians were expected to starve to death over the next six months. How do the risks posed by GM crops compare with the risk of starvation?

19. From 1977 to 2000, 12 convicts were executed in Illinois. During that same period, 13 death row inmates were exonerated based on DNA evidence. In 2000, the governor of Illinois declared a moratorium on all executions in his state because the death penalty system was "fraught with errors." Do you support the Illinois governor's decision? What rights should death penalty inmates have with regard to DNA testing of old evidence? Who should pay for this additional testing?

Unit 3

Evolution and Diversity

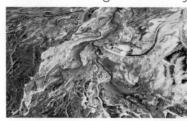

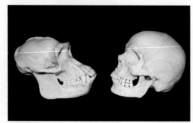

13 How Populations Evolve

DO YOU LIKE BIG, JUICY TOMATOES? IF IT WEREN'T FOR ARTIFICIAL SELECTION, YOU'D BE EATING TOMATOES THE SIZE OF BLUEBERRIES.

Why Evolution Matters

The abundant diversity of life on Earth, more than a million species, is the product of past evolution. Evolution is also happening right now, somewhere near you—perhaps even within your own body.

THE CHEETAH—THE FASTEST ANIMAL ON EARTH—MAY BE RACING TOWARD EXTINCTION. MANY ENDANGERED SPECIES ARE DOOMED BY THEIR LACK OF GENETIC DIVERSITY.

SCRATCHING YOUR HEAD OVER BUG INFESTATIONS? BECAUSE OF NATURAL SELECTION, HEAD LICE, BED BUGS, AND MOSQUITOES ARE INCREASINGLY HARD TO KILL.

A malaria ward in Kisii, Kenya.

BIOLOGY AND SOCIETY | Evolution in Action

Mosquitoes and Evolution

What does actor George Clooney have in common with George Washington, Ernest Hemingway, Christopher Columbus, and Mother Teresa? They all survived bouts with malaria, a disease caused by a microscopic parasite that is one of the worst killers in human history. In 1955, the World Health Organization (WHO) launched a campaign to eradicate malaria. Their strategy focused on killing the mosquitoes that carry the parasite from person to person. DDT, a recently developed pesticide in wide use at the time, was deployed in massive spraying operations. But in many locations, early success was followed by rebounding mosquito populations. Although the lethal chemical killed most of the mosquitoes immediately, survivors gave rise to new DDT-resistant populations—an example of evolution in action. Clearly, a single pesticide could not wipe out all the disease-carrying mosquitoes. WHO also learned that the parasite, too, was evolving. Drugs that once cured malaria became less and less effective as resistance evolved in parasite populations. Eradication of the disease is no longer viewed as imminent, but by using a judicious combination of mosquito-control strategies, public health agencies have made progress in the battle against malaria. WHO also monitors the evolution of drug resistance in the parasite's populations around the world.

As you'll learn in this chapter, malaria is not the only disease that has become more difficult to cure as a result of evolution. Dozens of species of bacteria and other microorganisms are increasingly resistant to antibiotics because of evolution. And malaria is not the only disease carried by mosquitoes—Zika virus, dengue (also called breakbone fever, for the intense pain it causes), Chikungunya virus, yellow fever, and West Nile virus are among the others. If you don't live in a region inhabited by the relatively few mosquito species that carry these diseases, you may not be familiar with them. That may soon change, however. As climate change brings rising temperatures, changing patterns of rainfall, and increased flooding to many regions, disease-carrying mosquitoes are expanding their range.

An understanding of evolution informs all of biology, from exploring life's molecules to analyzing ecosystems. And applications of evolutionary biology are transforming medicine, agriculture, biotechnology, and conservation biology. In this chapter, you'll learn how the process of evolution works and read about verifiable, measurable examples of evolution that affect our world.

The Diversity of Life

For all of human history, people have named, described, and classified the inhabitants of the natural world. As trade and exploration connected all regions of the planet, these tasks became increasingly complex. For example, a scholar who sought to describe all the types of plants known to the Greeks in 300 B.C. had about 500 species to distinguish. Today, scientists recognize roughly 400,000 plant species.

By the 1700s, it was clear that a unified system of naming and classifying was needed. The scientific community eventually agreed on a scheme introduced by Carolus Linnaeus, a Swedish scientist. His system, which is still in use today, is the basis for **taxonomy**, the branch of biology concerned with identifying, naming, and classifying species. The Linnaean system includes a method of

naming species and a hierarchical classification of species into broader groups of organisms.

Naming and Classifying the Diversity of Life

In the Linnaean system, each species is given a two-part Latinized name, or **binomial**. The first part of a binomial is the **genus** (plural, *genera*), a group of closely related species. For example, the genus of large cats is *Panthera*. The second part of a binomial is used to distinguish species within a genus. The two parts must be used together to name a species. In our example, the scientific name for the leopard is *Panthera pardus*. Notice that the first letter of the genus is capitalized and that the whole binomial is italicized. For instance, a fish in the genus *Tosanoides* that was discovered in Hawaii in 2016 was named *Tosanoides obama*, in honor of the former U.S. president, who was born in the state.

The Linnaean binomial solved the problem of the ambiguity of common names. Referring to an animal as a squirrel or to a plant as a daisy is not specific—there are many species of squirrels and daisies. In addition, people in different regions may use the same common name for different species. For example, the flowers called bluebells in Scotland, England, Texas, and the eastern United States are actually four unrelated species.

Linnaeus also introduced a system for grouping species into a hierarchy of categories. The first step of this classification is built into the binomial. For example, the genus *Panthera* contains three other species: the lion (*Panthera leo*), the tiger (*Panthera tigris*), and the jaguar (*Panthera onca*). Beyond the grouping of species within genera, taxonomy extends to progressively broader categories of classification. It places similar genera in the same **family**, puts families into **orders**, orders into **classes**, classes into **phyla** (singular, *phylum*), phyla into **kingdoms**, and kingdoms into **domains**. **Figure 13.1** places the leopard in this taxonomic scheme of groups within groups. The resulting classification of a particular organism is somewhat like a postal address identifying a person in a particular apartment, in a building with many apartments, on a street with many apartment buildings, in a city with many streets, and so on.

Grouping organisms into broader categories is a way to structure our understanding of the world. However, the criteria used to define more inclusive groups such as families, orders, and classes are ultimately arbitrary. After you learn about the processes by which the diversity of life evolved, we will introduce a classification system based on an understanding of evolutionary relationships (see Chapter 14). ✓

▼ **Figure 13.1 Hierarchical classification.** Taxonomy classifies species—the least inclusive groups—into increasingly broad categories. *Panthera pardus* is one of four species (indicated here with yellow boxes) in the genus *Panthera*; *Panthera* is a genus (orange box) in the family Felidae, and so on.

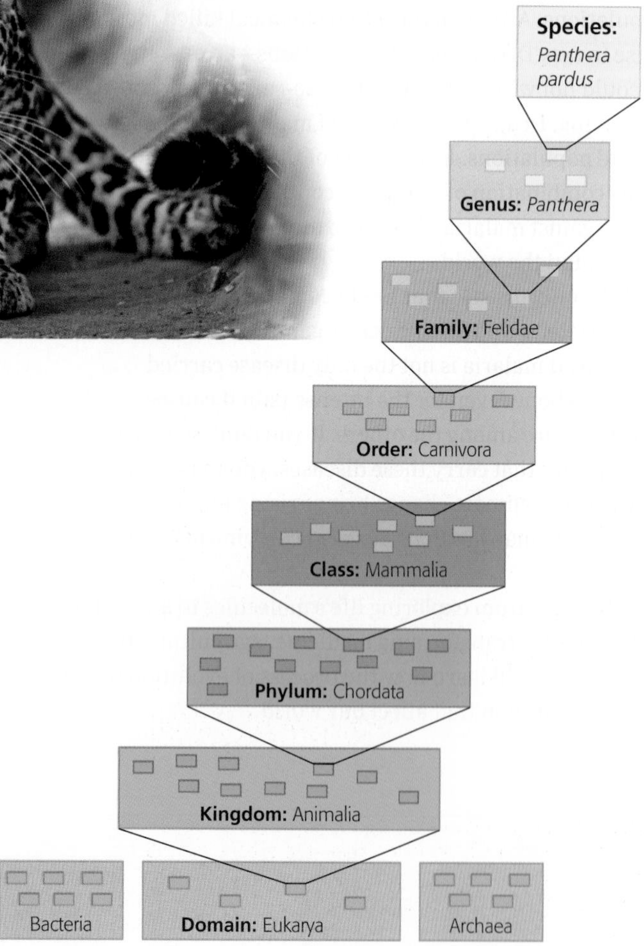

Species: *Panthera pardus*

Genus: *Panthera*

Family: Felidae

Order: Carnivora

Class: Mammalia

Phylum: Chordata

Kingdom: Animalia

Bacteria | **Domain:** Eukarya | Archaea

✓ **CHECKPOINT**

Which pair of animals is more closely related to each other, *Ursus americanus* and *Ursus maritimus* or *Ursus americanus* and *Bufo americanus*?

■ *Answer: Ursus americanus (black bear) and Ursus maritimus (polar bear), which belong to the same genus. The species name americanus distinguishes each animal from other members of its genus. (Bufo americanus is the American toad.)*

Explaining the Diversity of Life

Although early naturalists and philosophers sought to describe and organize the diversity of life, they also sought to explain its origin. The explanation accepted by present-day biologists is the evolutionary theory proposed by Charles Darwin in his best known book, *On the Origin of Species by Means of Natural Selection*, published in 1859. Before we introduce Darwin's theory, however, let's take a brief look at the scientific and cultural context that made the theory of evolution such a radical idea in Darwin's time.

The Idea of Fixed Species

The Greek philosopher Aristotle, whose ideas had an enormous impact on Western culture, generally held the view that species are fixed, permanent forms that do not change over time. Judeo-Christian culture reinforced this idea with a literal interpretation of the biblical book of Genesis, which tells the story of each form of life being individually created in its present-day form. In the 1600s, religious scholars used biblical accounts to estimate the age of Earth at 6,000 years. Thus, the idea that all living species came into being relatively recently and are unchanging in form dominated the intellectual climate of the Western world for centuries.

At the same time, however, naturalists were grappling with the interpretation of **fossils**—imprints or remains of organisms that lived in the past. Although fossils were thought to be the remains of living creatures, many were puzzling. For example, if "snakestones" **(Figure 13.2a)** were the coiled bodies of snakes, then why were none ever found with an intact head? And could some fossils represent species that had become extinct? Stunning discoveries in the early 1800s, including fossilized skeletons of a gigantic sea creature dubbed an ichthyosaur, or fish-lizard **(Figure 13.2b)**, convinced many naturalists that extinctions had indeed occurred.

Lamarck and Evolutionary Adaptations

Fossils told of other changes in the history of life, too. Naturalists compared fossil forms with living species and noted patterns of similarities and differences. In the early 1800s, French naturalist Jean-Baptiste Lamarck suggested that the best explanation for these observations is that life evolves. Lamarck explained evolution as the refinement of traits that equip organisms to perform successfully in their environments. He proposed that by using or not using its body parts, an individual may develop certain traits that it passes on to its offspring. For example, some birds have powerful beaks that enable them to crack tough seeds. Lamarck suggested that these strong beaks are the cumulative result of ancestors exercising their beaks during feeding and passing that acquired beak power on to offspring. However, simple observations provide evidence against the inheritance of acquired traits: An athlete who builds up strength through weight training will not pass enhanced biceps on to his or her children. Although Lamarck's idea of how species evolve was mistaken, his proposal that species evolve as a result of interactions between organisms and their environments helped set the stage for Darwin. ☑

☑ CHECKPOINT

How do fossils contradict the idea of fixed species?

■ *Answer: They don't match any living creature.*

▼ **Figure 13.2** **Fossils that perplexed naturalists in the 1800s.**

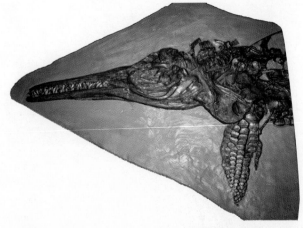

(a) **"Snakestone."** This fossil, once thought to be a snake, is actually a mollusc called an ammonite, an extinct relative of the present-day nautilus (see Figure 17.13). Ammonites of this type ranged in size from about several inches to more than 7 feet in diameter.

(b) **Icthyosaur skull and paddle-like forelimb.** These marine reptiles—some more than 50 feet long—ruled the oceans for 155 million years before becoming extinct about 90 million years ago. The enormous eye is thought to be an adaptation to the dim light of the deep sea.

Charles Darwin and *The Origin of Species*

Although Charles Darwin was born more than 200 years ago—on the very same day as Abraham Lincoln—his work had such an extraordinary impact that many scientists mark his birthday with a celebration of his contributions to biology. How did Darwin become a rock star of science?

As a boy, Darwin was fascinated with nature. He loved collecting insects and fossils, as well as reading books about nature. His father, an eminent physician, could see no future for his son as a naturalist and sent him to medical school. But young Darwin, finding medicine boring and surgery before the days of anesthesia horrifying, quit medical school. His father then enrolled him at Cambridge University with the intention that he should become a clergyman. After college, however, Darwin returned to his childhood interests rather than following the career path mapped out by his father. At the age of 22, he began a sea voyage on the HMS *Beagle* that helped him frame his theory of evolution.

Darwin's Journey

The *Beagle* was a survey ship. Although it stopped at many locations around the world, its main task was charting poorly known stretches of the South American coast (**Figure 13.3**). Darwin, a skilled naturalist, spent most of his time on shore doing what he enjoyed most—exploring the natural world. He collected thousands of specimens of fossils and living plants and animals. He also kept detailed journals of his observations. For a naturalist from a small, temperate country, seeing the glorious diversity of unfamiliar life-forms on other continents was a revelation. He carefully noted the characteristics of plants and animals that made them well suited to such diverse environments as the jungles of Brazil, the grasslands of Argentina, the towering peaks of the Andes, and the desolate and frigid lands at the southern tip of South America.

Observations

Many of Darwin's observations indicated that geographic proximity is a better predictor of relationships among organisms than similarity of environment. For example, the plants and animals living in temperate regions of South America more closely resembled species living in tropical regions of that continent than species living in similarly temperate regions of Europe. And the South American fossils Darwin found, although clearly

▼ **Figure 13.3 The voyage of the *Beagle*.**

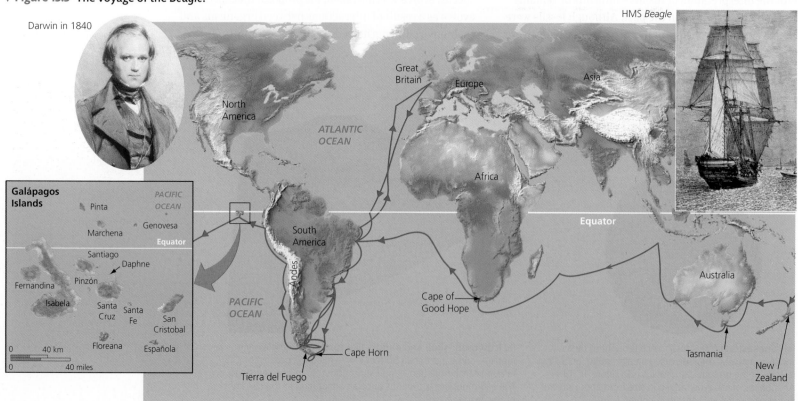

examples of species different from living ones, were distinctly South American in their resemblance to the contemporary plants and animals of that continent. For instance, he collected fossilized armor plates resembling those of living armadillo species. Paleontologists later reconstructed the creature to which the armor belonged—which turned out to be an extinct armadillo the size of a Volkswagen Beetle.

Darwin was particularly intrigued by the geographic distribution of organisms on the Galápagos Islands. The Galápagos are relatively young volcanic islands about 900 kilometers (540 miles) off the Pacific coast of South America. Most of the animals that inhabit these remote islands are found nowhere else in the world, but they resemble South American species.

Darwin noticed that Galápagos marine iguanas—with a flattened tail that aids in swimming—are similar to, but distinct from, land-dwelling iguanas on the islands and on the South American mainland. Furthermore, each island had its own distinct variety of giant tortoise (**Figure 13.4**), the strikingly unique inhabitants for which the islands were named (*galápago* means "tortoise" in Spanish).

New Insights

While on his voyage, Darwin was strongly influenced by the newly published *Principles of Geology* by Scottish geologist Charles Lyell. The book presented the case for an ancient Earth sculpted over millions of years by gradual geological processes that continue today. Darwin witnessed the power of natural forces to change Earth's surface firsthand when he experienced an earthquake that raised part of the coastline of Chile almost a meter.

By the time Darwin returned to Great Britain five years after the *Beagle* first set sail, he had begun to seriously doubt that Earth and all its living organisms had been specially created only a few thousand years earlier. As he reflected on his observations, analyzed his specimen collections, and discussed his work with colleagues, he concluded that the evidence was better explained by the hypothesis that present-day species are the descendants of ancient ancestors that they still resemble in some ways. **Over time, differences gradually accumulated by a process that Darwin called "descent with modification," his phrase to describe evolution.** Unlike others who had explored the idea that organisms had changed over time, however, Darwin also proposed a scientific mechanism for *how* life evolves, a process he called natural selection. In **natural selection**, individuals with certain inherited traits are more likely to survive and reproduce than are individuals with other traits. He hypothesized that as the descendants of a remote ancestor spread into various habitats over millions and millions of years, natural selection resulted in diverse modifications, or **evolutionary adaptations**, that fit them to specific ways of life in their environment.

▼ **Figure 13.4 Two varieties of Galápagos tortoise.**

(a) The thick, domed shell and short neck and legs are characteristic of tortoises found on wetter islands that have more abundant, dense vegetation.

(b) Saddleback shells have an arch at the front, which allows the long neck to emerge. This, along with longer legs, enables the tortoise to stretch higher to reach the scarce vegetation on dry islands.

☑ CHECKPOINT

What was the most sig-
nificant difference between
Darwin's theory of evolution
and the ideas about evolution that had been proposed
previously?

■ *Answer: Darwin also proposed a
mechanism (natural selection) for
how evolution occurs.*

Darwin's Theory

Darwin spent the next two decades compiling and writing about evidence for evolution. He realized that his ideas would cause an uproar, however, and he delayed publishing. Eventually, Darwin learned that Alfred Russel Wallace, a British naturalist doing fieldwork in Indonesia, had conceived a hypothesis almost identical to Darwin's. Not wanting to have his life's work eclipsed by Wallace's, Darwin finally published *The Origin of Species*, a book that supported his hypothesis with immaculate logic and hundreds of pages of evidence drawn from observations and experiments in biology, geology, and paleontology. The hypothesis of evolution set forth in *The Origin of Species* has since generated predictions that have been tested and verified by more than 150 years of research. Consequently, scientists regard Darwin's concept of evolution by means of natural selection as a **theory**—a widely accepted explanatory idea that is broader in scope than a hypothesis, generates new hypotheses, and is supported by a large body of evidence.

In the next several pages, we examine lines of evidence for Darwin's theory of **evolution**, the idea that living species are descendants of ancestral species that were different from present-day ones. Then we will return to natural selection as the mechanism for evolutionary change. With our current understanding of how this mechanism works, we extend Darwin's definition of evolution to include genetic changes in a population from generation to generation. ☑

Evidence of Evolution

Evolution leaves observable signs. Such clues to the past are essential to any historical science. Historians of human civilization can study written records from earlier times. But they can also piece together the evolution of societies by recognizing vestiges of the past in modern cultures. Even if we did not know from written documents that Spaniards colonized the Americas, we would deduce this from the Hispanic stamp on Latin American culture. Similarly, biological evolution has left evidence in fossils, as well as in today's organisms.

Evidence from Fossils

Fossils—imprints or remains of organisms that lived in the past—document differences between past and present organisms and show that many species have become extinct. The soft parts of a dead organism usually decay rapidly, but the hard parts of an animal that are rich in minerals, such as the bones and teeth of vertebrates and the shells of clams and snails, may remain as fossils. **Figure 13.5** (facing page) illustrates some of the ways that organisms can fossilize.

Not all fossils are the actual remnants of organisms. Some, such as the ammonite in Figure 13.2a, are casts. A cast forms when a dead organism that was buried in sediment decomposes and leaves an empty "mold" that is later filled by minerals dissolved in water. The minerals harden within the mold, making a replica of the organism. You may have seen crime scene investigators on TV shows use fast-acting plaster in the same way to make casts of footprints or tire tracks. Fossils may also be imprints, such as footprints or burrows, that remain after the organism decays. **Paleontologists** (scientists who study fossils) also eagerly examine coprolites—fossilized feces—for clues about the diets and digestive systems of extinct animals.

In rare instances, an entire organism, including its soft parts, is encased in a medium that prevents decomposition. Examples include insects trapped in amber (fossilized tree resin) and mammoths, bison, and even prehistoric humans frozen in ice or preserved in bogs.

Many fossils are found in fine-grained sedimentary rocks formed from the sand or mud that settles to the bottom of seas, lakes, swamps, and other aquatic habitats, covering dead organisms. Over millions of years, new layers of sediment are deposited atop older ones and compress them into layers of rock called strata (singular, *stratum*). Thus, the fossils in a particular stratum provide a glimpse of some of the organisms that lived in the area at the time the layer formed. Because younger strata are on top of older layers, the relative ages of fossils can be determined by the layer in which they are found. (Radiometric dating—see Figure 2.17—can be used to determine the approximate ages of fossils.) As a result, the sequence in which fossils appear within layers of sedimentary rocks is a historical record of life on Earth. The **fossil record** is this ordered sequence of fossils as they appear in the rock layers, marking the passage of geologic time (see Figure 14.13).

Of course, as Darwin acknowledged, the fossil record is incomplete. Many of Earth's organisms did not live in areas that favor fossilization. Many fossils that did form were in rocks later distorted or destroyed by geologic processes. Furthermore, not all fossils that have been preserved are accessible to paleontologists. Even with its limitations, however, the fossil record is remarkably detailed. And although the incompleteness of the fossil record may seem frustrating (wouldn't it be nice to have all our questions answered!), it makes paleontology an unexpectedly thrilling occupation. Like a mystery series in which new clues are uncovered in each episode, the thousands of fossils newly discovered each year give paleontologists new opportunities to test hypotheses about how the diversity of life evolved. ☑

☑ **CHECKPOINT**

Why are older fossils generally in deeper rock layers than younger fossils?

■ *Answer: Sedimentation places younger rock layers on top of older ones.*

▼ **Figure 13.5 A fossil gallery.**

Sedimentary fossils are formed when minerals seep into and replace organic matter. This petrified (stone) tree in Arizona's Petrified Forest National Park is about 190 million years old.

Sedimentary rocks are the richest hunting grounds for paleontologists, scientists who study fossils. This researcher is excavating a fossilized dinosaur skeleton from sandstone in Dinosaur National Monument, located in Utah and Colorado.

This 45-million-year-old insect is embedded in amber (hardened resin from a tree).

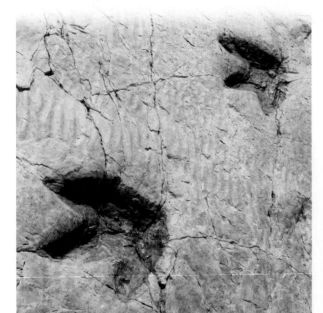

A dinosaur left these footprints 120 million years ago in what is now northern Spain. Biologists often study footprints to learn how extinct animals moved.

Scientists examine a 40,000-year-old baby mammoth found frozen in the ice of northern Russia.

Evidence from Homologies

A second type of evidence for evolution comes from analyzing similarities among different organisms. Evolution is a process of descent with modification—characteristics present in an ancestral organism are altered over time by natural selection as its descendants face different environmental conditions. In other words, evolution is a remodeling process. As a result, related species can have characteristics that have an underlying similarity yet function differently. Similarity resulting from common ancestry is known as **homology**.

Darwin cited the anatomical similarities among vertebrate forelimbs as evidence of common ancestry. As **Figure 13.6** shows, the same skeletal elements make up the forelimbs of humans, cats, whales, and bats. The functions of these forelimbs differ. A whale's flipper does not do the same job as a bat's wing, so if these structures had been uniquely engineered, then we would expect that their basic designs would be very different. The logical explanation instead is that the arms, forelegs, flippers, and wings of these different mammals are variations on an anatomical structure of an ancestral organism that over millions of years has become adapted to different functions. Biologists call such anatomical similarities in different organisms homologous structures—features that often have different functions but are structurally similar because of common ancestry.

As a result of advances in **molecular biology**, the study of the molecular basis of genes and gene expression, present-day scientists have a much deeper understanding of homologies than Darwin did. **Just as your genetic background is recorded in the DNA you inherit from your parents, the evolutionary history of each species is documented in the DNA inherited from its ancestral species in a continuous flow of genetic information across time.** If two species have homologous genes with sequences that match closely, biologists conclude that these sequences must have been inherited from a relatively recent common ancestor. Conversely, the greater the number of sequence differences between species, the more distant is their last common ancestor. Molecular comparisons between diverse organisms have allowed biologists to develop and test hypotheses about the evolutionary divergence of major branches on the tree of life.

Darwin's boldest hypothesis was that all life-forms are related. Molecular biology provides strong evidence for this claim: All forms of life use the same genetic language of DNA and RNA, and the genetic code—how RNA triplets are translated into amino acids—is essentially universal (see Figure 10.10). Thus, it is likely that all species descended from common ancestors that used this code. Because of these molecular homologies, bacteria engineered with human genes can produce human proteins such as insulin and human growth hormone. But molecular homologies go

▼ **Figure 13.6 Homologous structures: anatomical signs of descent with modification.** The forelimbs of all mammals are constructed from the same skeletal elements. (Homologous bones in each of these four mammals are colored the same.) The hypothesis that all mammals descended from a common ancestor predicts that their forelimbs, though diversely adapted, would be variations on a common anatomical theme.

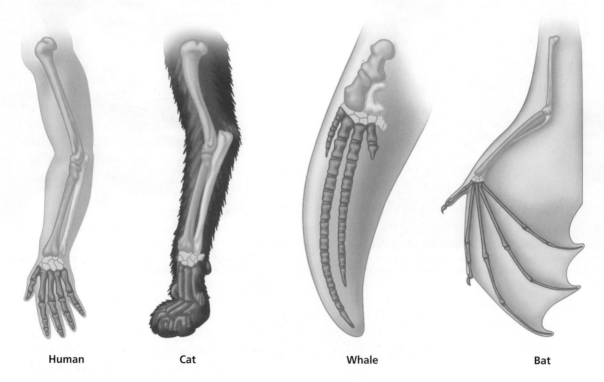

| Human | Cat | Whale | Bat |

beyond a shared genetic code. For example, organisms as dissimilar as humans and bacteria share homologous genes inherited from a very distant common ancestor.

Geneticists have also uncovered hidden molecular homologies. Organisms may retain genes that have lost their function through mutations, even though homologous genes in related species are fully functional. Many of these inactive genes have been identified in humans. For example, one such gene encodes an enzyme known as GLO that is used in making vitamin C. Almost all mammals have a metabolic pathway to make this essential vitamin from glucose. Although humans and other primates have functional genes for the first three steps in the pathway, the inactive GLO gene prevents us from making vitamin C—we must get sufficient amounts in our diet to maintain health.

Some of the most interesting homologies are "leftover" structures that are of marginal or perhaps no importance to the organism. These **vestigial structures** are remnants of features that served important functions in the organism's ancestors. For example, the small pelvis and hind-leg bones of ancient whales are vestiges (traces) of their walking ancestors. The eye remnants that are buried under scales in blind species of cave fishes—a vestige of their sighted ancestors—are another example. Humans have vestigial structures, too. When we're cold or agitated, we often get goose bumps caused by small muscles under the skin that make the body hair stand on end. The same response is more visible (and more functional) in a bird that fluffs up its feathery insulation (see Figure 18.9) or a cat that bristles when threatened.

An understanding of homology can also explain observations about embryonic development that are otherwise puzzling. For example, comparing early stages of development in different animal species reveals similarities not visible in adult organisms **(Figure 13.7)**. At some point in their development, all vertebrate embryos have a tail posterior to the anus, as well as structures called pharyngeal (throat) pouches. These pouches are homologous structures that ultimately develop to have very different functions, such as gills in fishes and parts of the ears and throat in humans.

Next, we'll see how homologies help us trace evolutionary descent. ☑

Evolutionary Trees

Darwin was the first to visualize the history of life as a tree in which patterns of descent branch off from a common trunk—the first organism—to the tips of millions of twigs representing the species living today. At each fork of the evolutionary tree is an ancestor common to all evolutionary branches extending from that fork. Closely related species share many traits because their lineage of common descent traces to a recent fork of the tree of life. Biologists illustrate these patterns of descent with an **evolutionary tree**, although today they often turn the trees sideways so they can be read from left to right (as we do in this book).

Homologous structures, both anatomical and molecular, can be used to determine the branching sequence of an evolutionary tree. Some homologous characters, such as the genetic code, are shared by all species because they date to the deep ancestral past. In contrast, traits that evolved more recently are shared by smaller groups of organisms. For example, all tetrapods (from the Greek *tetra*, four, and *pod*, foot) have the same basic limb bone structure illustrated in Figure 13.6, but their ancestors do not.

Figure 13.8 (on the next page) is an evolutionary tree of tetrapods (amphibians, mammals, and reptiles, including birds) and their closest living relatives, the lungfishes. In this diagram, each branch point represents the common ancestor of all species that descended from it. For example, lungfishes and all tetrapods are descended from ancestor ❶. Three homologies are marked by the blue dots on the tree—tetrapod limbs, the amnion (a protective embryonic membrane), and feathers. Tetrapod limbs were present in common ancestor ❷ and hence are found in its descendants (the tetrapods). The amnion was present in ancestor ❸ and thus is shared only by mammals and reptiles, which are known as amniotes. Feathers were present only in ancestor ❻ and hence are found only in birds.

Evolutionary trees are hypotheses reflecting our current understanding of patterns of evolutionary descent. Some trees, such as the one in Figure 13.8, are based on a convincing combination of fossil, anatomical, and molecular data. Others are more speculative because sufficient data are not yet available. ☑

☑ **CHECKPOINT**

How does the need for dietary vitamin C show that humans are more closely related to other primates than to other mammals?

■ *Answer: Humans and primates need vitamin C because a gene needed to make it in the body is nonfunctional. A homologous gene in most other mammals is functional. Thus, the gene's function must have been lost by mutation in an early primate or primate ancestor and inherited in that form by both humans and other primates.*

▼ **Figure 13.7 Evolutionary signs from comparative embryology.** At the early stage of development shown here, the kinship of vertebrates is unmistakable. Notice, for example, the pharyngeal pouches and tails in both the chicken embryo and the human embryo.

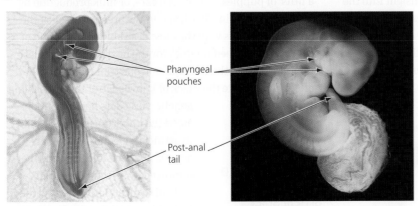

Pharyngeal pouches

Post-anal tail

Chicken embryo　　　　**Human embryo**

▼ Figure 13.8 An evolutionary tree of tetrapods (four-limbed animals).

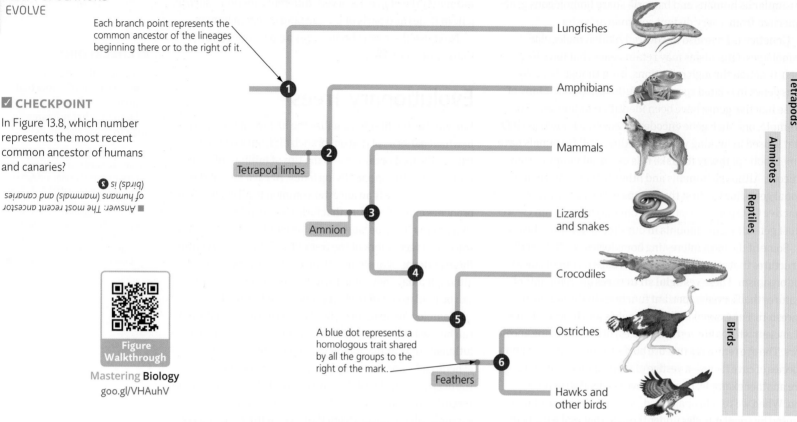

Each branch point represents the common ancestor of the lineages beginning there or to the right of it.

Tetrapod limbs

Amnion

A blue dot represents a homologous trait shared by all the groups to the right of the mark.

Feathers

Lungfishes

Amphibians

Mammals

Lizards and snakes

Crocodiles

Ostriches

Hawks and other birds

Tetrapods

Amniotes

Reptiles

Birds

☑ CHECKPOINT

In Figure 13.8, which number represents the most recent common ancestor of humans and canaries?

■ *Answer: The most recent ancestor of humans (mammals) and canaries (birds) is* **3**.

Figure Walkthrough

Mastering **Biology**

goo.gl/VHAuhV

Natural Selection as the Mechanism for Evolution

Now that you have learned about the lines of evidence supporting Darwin's theory of descent with modification, let's look at Darwin's explanation of *how* life evolves. Because he hypothesized that species formed gradually over long periods of time, Darwin knew that he would not be able to study the evolution of new species by direct observation. But he did have a way to gain insight into the process of incremental change: the practices used by plant and animal breeders.

All domesticated plants and animals are the products of selective breeding from wild ancestors. For example, the baseball-sized tomatoes grown today are very different from their Peruvian ancestors, which were not much larger than blueberries. Having conceived the notion that **artificial selection**—the selective breeding of domesticated plants and animals to promote the occurrence of desirable traits in the offspring—was

the key to understanding evolutionary change, Darwin bred pigeons to gain firsthand experience. He acquired further insight through conversations with farmers about livestock breeding. He learned that artificial selection has two essential components: variation and heritability. Variation among individuals, for example, differences in coat type in a litter of puppies, size of corn ears, or milk production by individual cows in a herd, allows the breeder to select the animals or plants with the most desirable combination of characters as breeding stock for the next generation. Heritability refers to the transmission of a trait from parent to offspring. Despite their lack of knowledge of the underlying genetics, breeders had long understood the importance of heritability in artificial selection—if a character is not heritable, it cannot be improved by selective breeding.

Unlike most naturalists, who looked for consistency of traits as a

IF IT WEREN'T FOR ARTIFICIAL SELECTION, TOMATOES WOULD BE THE SIZE OF BLUEBERRIES.

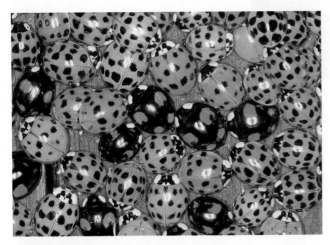

▲ **Figure 13.9** **Color variation within a population of Asian lady beetles.**

Ribbon of eggs

▲ **Figure 13.10** **Overproduction of offspring.** This sea slug, a mollusc related to snails (see Figure 17.13), is laying thousands of eggs embedded in the white ribbon around its body. Only a tiny fraction of the eggs will actually give rise to offspring that survive and reproduce.

way to classify organisms, Darwin was a careful observer of variations between individuals. He knew that individuals in natural populations have small but measurable differences, for example, variations in color and markings **(Figure 13.9)**. But what forces in nature determined which individuals became the breeding stock for the next generation?

Darwin found inspiration in an essay written by economist Thomas Malthus, who contended that much of human suffering—disease, famine, and war—was the consequence of human populations increasing faster than food supplies and other resources. Darwin applied Malthus's idea to populations of plants and animals, reasoning that the resources of any given environment are limited. The production of more individuals than the environment can support leads to a struggle for existence, with only some offspring surviving in each generation **(Figure 13.10)**. Of the many eggs laid, young born, and seeds spread, only a tiny fraction complete development and leave offspring themselves. The rest are eaten, starved, diseased, unmated, or unable to reproduce for other reasons. The essence of natural selection is this unequal reproduction. In the process of natural selection, individuals whose traits better enable them to obtain food, escape predators, or tolerate physical conditions will survive and reproduce more successfully, passing these adaptive traits to their offspring.

Darwin reasoned that if artificial selection can bring about significant change in a relatively short period of time, then natural selection could modify species considerably over hundreds or thousands of generations. Over vast spans of time, many traits that adapt a population to its environment will accumulate. If the environment changes, however, or if individuals move to a new environment, natural selection will select for adaptations to

these new conditions, sometimes producing changes that result in the origin of a completely new species.

Natural Selection in Action

Look at any natural environment and you will see the products of natural selection—adaptations that suit organisms to their environment. But can we see natural selection in action? Yes, indeed! Biologists have documented evolutionary change in thousands of scientific studies. The evolution of pesticide resistance in mosquitoes, which you learned about in the chapter introduction, is an unsettling example of natural selection in action. Pesticides control insects and prevent them from transmitting diseases or eating crops. But pesticide resistance has evolved in hundreds of insect species. Whenever a new type of pesticide is used to control pests, the outcome is similar. Let's take a closer look at how natural selection works by first examining pesticide-resistant insects.

In Figure 13.11, you'll see that mosquitoes are first sprayed with pesticide. A relatively small amount of poison initially kills most of the insects in the population, but a few individuals carry an allele (alternative form of a gene) that enables them to survive the chemical attack. These genetically resistant survivors reproduce and pass the allele for pesticide resistance to their offspring. Thus, subsequent applications of the same pesticide are less and less effective as the proportion of pesticide-resistant individuals increases in each generation.

BECAUSE OF NATURAL SELECTION, HEAD LICE, BED BUGS, AND MOSQUITOES ARE INCREASINGLY HARD TO KILL.

Key Points about Natural Selection

Before we move on, let's summarize how natural selection works in bringing about evolutionary change.

Natural selection affects individual organisms—in **Figure 13.11**, each mosquito either survived the pesticide or was killed by it. However, individuals do not evolve. Rather, it is the population—the group of organisms—that evolves over time as adaptive traits become more common in the group and other traits change or disappear. Evolution refers to generation-to-generation changes in populations.

Natural selection can amplify or diminish only heritable traits. Although an organism may, during its lifetime, acquire characters that help it survive, such acquired characters cannot be passed on to offspring. Natural selection is more an editing process than a creative mechanism. A pesticide does not create new alleles that allow insects to survive. Rather, the presence of the pesticide leads to natural selection for insects in the population that already have those alleles.

Natural selection is not goal-directed; it does not lead to perfectly adapted organisms. Whereas artificial selection is a deliberate attempt by humans to produce individuals with specific traits, natural selection is the result of environmental factors that vary from place to place and over time. A trait that is favorable in

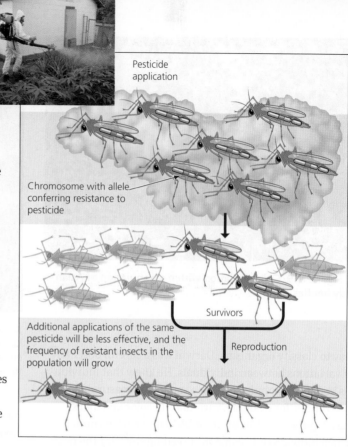

Pesticide application

Chromosome with allele conferring resistance to pesticide

Survivors

Reproduction

Additional applications of the same pesticide will be less effective, and the frequency of resistant insects in the population will grow

▲ **Figure 13.11 Evolution of pesticide resistance in insect populations.**

one situation may be useless—or even detrimental—in different circumstances. And some adaptations are compromises. ☑

The Evolution of Populations

In *The Origin of Species*, Darwin provided evidence that life on Earth has evolved over time, and he proposed that natural selection, in favoring some heritable traits over others, was the primary mechanism for that change. But how do the variations that are the raw material for natural selection arise in a population? And how are these variations passed along from parents to offspring? Darwin did not know that Gregor Mendel (see Figure 9.1) had already answered these questions. Although both men lived and worked at around the same time, Mendel's work was largely ignored by the scientific community in his life. Its rediscovery in 1900 set the stage for understanding the genetic differences on which evolution is based.

Sources of Genetic Variation

You have no trouble recognizing friends in a crowd. Each person has a unique genome, reflected in individual phenotypic variations such as appearance and other traits.

Indeed, individual variation occurs in all species, as illustrated by the snails in **Figure 13.12**. In addition to obvious physical differences, such as the snails' colors and patterns, most populations have a great deal of phenotypic variation that can be observed only at the molecular level, such as an enzyme that detoxifies a pesticide. Of course, not all variation in a population is heritable. Phenotype—the expressed traits of an organism—results from a combination of the genotype, which is inherited, and many environmental influences. For instance, if you have dental work to straighten and whiten your teeth, you will not pass your environmentally produced smile to your offspring. Only the genetic component of variation is relevant to natural selection. Many of the characters that vary in a population result from the combined effect of several genes. Other features, such as Mendel's purple and white pea flowers and human blood types, are determined by a single gene locus, with different alleles producing distinct phenotypes. But where do these alleles come from?

▼ **Figure 13.12 Variation in a snail population.** Phenotypic variation in brown-lipped snails includes diverse colors and banding patterns.

Mutation

New alleles originate by mutation, a change in the nucleotide sequence of DNA. Thus, mutation is the ultimate source of the genetic variation that serves as raw material for evolution. In multicellular organisms, however, only mutations in cells that produce gametes can be passed to offspring and affect a population's genetic variability.

A change as small as a single nucleotide in a protein-coding gene can have a significant effect on phenotype, as in sickle-cell disease (see Figure 9.21). An organism is a refined product of thousands of generations of past selection, and a random change in its DNA is not likely to improve its genome any more than randomly changing some words on a page is likely to improve a story. In fact, mutation that affects a protein's function will probably be harmful. On rare occasions, however, a mutated allele may actually improve the adaptation of an individual to its environment and enhance its reproductive success. This kind of effect is more likely when the environment is changing in such a way that mutations that were once disadvantageous are favorable under the new conditions. For instance, mutations that endow houseflies with resistance to the pesticide DDT also reduce their growth rate. Before DDT was introduced, such mutations were a handicap to the flies that had them. But once DDT was part of the environment, the mutant alleles were advantageous, and natural selection increased their frequency in fly populations.

Chromosomal mutations that delete, disrupt, or rearrange many gene loci at once are almost certain to be harmful. But duplication of a gene or small pieces of DNA through errors in meiosis can provide an important source of genetic variation. If a repeated segment of DNA can persist over generations, mutations may accumulate in the duplicate copies without affecting the function of the original gene, eventually leading to new genes with novel functions. This process may have played a major role in evolution. For example, the remote ancestors of mammals carried a single gene for detecting odors that has since been duplicated repeatedly. As a result, mice have about 1,300 different genes that encode smell receptors. It is likely that such dramatic increases helped early mammals by enabling them to distinguish among many different smells. And repeated duplications of genes that control development are linked to the origin of vertebrate animals from an invertebrate ancestor.

In prokaryotes, mutations can quickly generate genetic variation in a population. Because prokaryotes multiply so rapidly, a beneficial mutation can increase in frequency in a matter of hours or days. And because prokaryotes have a single chromosome, with a single allele for each gene, a new allele can have an effect immediately. Mutation rates in animals and plants average about 1 in every 100,000 genes per generation. For these organisms, low mutation rates, long time spans between generations, and diploid genomes prevent most mutations from significantly affecting genetic variation from one generation to the next.

Sexual Reproduction

In organisms that reproduce sexually, most of the genetic variation in a population results from the unique combination of alleles that each individual inherits. (Of course, the origin of those allele variations is past mutations.)

Fresh assortments of existing alleles arise every generation from three random components of sexual reproduction: independent orientation of homologous chromosomes at metaphase I of meiosis (see Figure 8.16), crossing over (see Figure 8.18), and random fertilization. During meiosis, pairs of homologous chromosomes, one set inherited from each parent, trade some of their genes by crossing over. These homologous chromosomes separate into gametes independently of other chromosome pairs. Thus, gametes from any individual vary extensively in their genetic makeup. Finally, each zygote made by a mating pair has a unique assortment of alleles resulting from the random union of sperm and egg.

Populations as the Units of Evolution

One common misconception about evolution is that individual organisms evolve during their lifetimes. It is true that natural selection acts on individuals: Each individual's combination of traits affects its survival and reproductive success. But the evolutionary impact of natural selection is only apparent in the changes in a population of organisms over time.

A **population** is a group of individuals of the same species that live in the same area and interbreed. We can measure evolution as a change in the prevalence of certain heritable traits in a population over a span of generations. The increasing proportion of resistant insects in areas sprayed with pesticide is one example. Natural selection favored insects with alleles for pesticide resistance. As a result, these insects left more offspring than nonresistant individuals, changing the genetic makeup of the next generation's population.

Different populations of the same species may be geographically isolated from each other to such an extent that an exchange of genetic material never, or only rarely, occurs. Such isolation is common in populations confined to different lakes **(Figure 13.13)** or islands. For example, each population of Galápagos tortoise is restricted to its own island. Not all populations have such sharp boundaries, however; members of a population typically breed with one another and are therefore more closely related to each other than they are to members of a different population.

In studying evolution at the population level, biologists focus on the **gene pool**, which consists of all copies of every type of allele at every locus in all members of the population. For many loci, there are two or more alleles in the gene pool. For example, in a mosquito population, there may be two alleles relating to DDT breakdown, one that codes for an enzyme that breaks down DDT and one for a version of the enzyme that does not. In a mosquito population living in a village sprayed with DDT, the allele for the enzyme conferring resistance will increase in frequency and the other allele will decrease in frequency. When the relative frequencies of alleles in a population change like this over a number of generations, evolution is taking place.

Next, we'll explore how to test whether evolution is occurring in a population. ☑

▼ **Figure 13.13** **Lakes in Alaska containing isolated populations.**

Analyzing Gene Pools

Imagine a wildflower population with two varieties of blooms that are different colors **(Figure 13.14)**. An allele for red flowers, which we will symbolize by R, is dominant to an allele for white flowers, symbolized by r. These are the only two alleles for flower color in the gene pool of this hypothetical plant population. Now, let's say that 80%, or 0.8, of all flower-color loci in the gene pool have the R allele. We'll use the letter p to represent the relative frequency of the R allele in the population. Thus, $p = 0.8$. Because there are only two alleles in this example, the r allele must be present at the other 20% (0.2) of the gene pool's flower-color loci. (This accounts for 100% of the flower-color loci in the gene pool, or a relative frequency of 1.) Let's use the letter q for the frequency of the r allele in the population. For the wildflower population, $q = 0.2$. And because there are only two alleles for flower color, we can express their frequencies as follows:

$$p + q = 1$$

Frequency of one allele   Frequency of alternate allele

Notice that if we know the frequency of either allele in the gene pool, we can subtract it from 1 to calculate the frequency of the other allele.

From the frequencies of alleles, we can also calculate the frequencies of different genotypes in the population if the gene pool is completely stable (not evolving). In the wildflower population, what is the probability of producing an RR individual by "drawing" two R alleles from the pool of gametes? (Here we apply the rule of multiplication that you learned in Chapter 9; review Figure 9.11.) The probability of drawing an R sperm multiplied by the probability of drawing an R egg is $p \times p = p^2$, or $0.8 \times 0.8 = 0.64$. In other words, 64% of the plants in the population will have the RR genotype. Applying the same math, we also know the frequency of rr individuals in the population: $q^2 = 0.2 \times 0.2 = 0.04$. Thus, 4% of the plants are rr, giving them white flowers. Calculating the frequency of heterozygous individuals, Rr, is trickier. That's because the heterozygous genotype can form in two ways, depending on whether the sperm or egg supplies the dominant allele. So the frequency of the Rr genotype is $2pq$, which is $2 \times 0.8 \times 0.2 = 0.32$. In our imaginary wildflower population, 32% of the plants are Rr, with red flowers. **Figure 13.15** reviews these calculations.

Now we can write a general formula for calculating the frequencies of genotypes in a

▲ Figure 13.14 A population of wildflowers with two varieties of color.

gene pool from the frequencies of alleles, and vice versa:

$$p^2 \quad + \quad 2pq \quad + \quad q^2 \quad = \quad 1$$

Frequency of homozygotes for one allele

Frequency of heterozygotes

Frequency of homozygotes for alternate allele

Notice that the frequencies of all genotypes in the gene pool must add up to 1. This formula is called the Hardy-Weinberg equation, named for the two scientists who derived it.

▼ Figure 13.15 A mathematical swim in the gene pool. Each of the four boxes in the Punnett square corresponds to a probable "draw" of alleles from the gene pool.

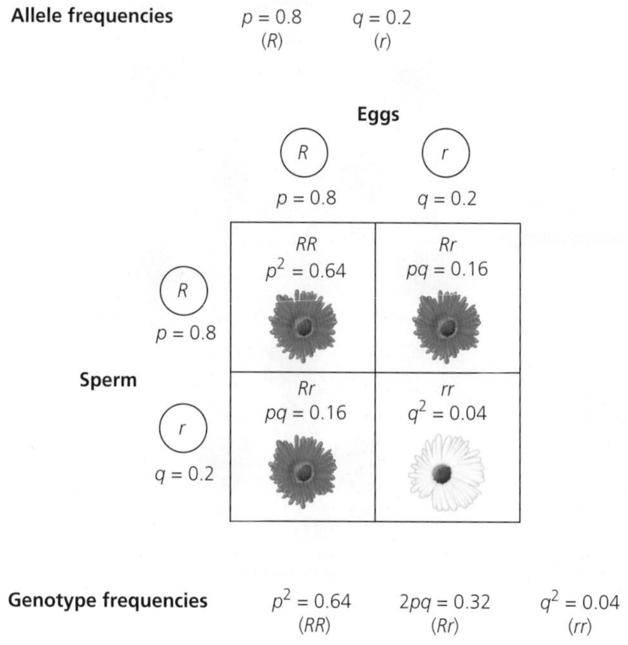

Allele frequencies $p = 0.8$ $q = 0.2$
                        (R)   (r)

Eggs

R   r
$p = 0.8$ $q = 0.2$

Sperm

R
$p = 0.8$

RR $p^2 = 0.64$	Rr $pq = 0.16$
Rr $pq = 0.16$	rr $q^2 = 0.04$

r
$q = 0.2$

Genotype frequencies $p^2 = 0.64$ $2pq = 0.32$ $q^2 = 0.04$
                        (RR)   (Rr)   (rr)

Population Genetics and Health Science

Public health scientists use the Hardy-Weinberg equation to calculate the percentage of a human population that carries the allele for certain inherited diseases. Consider phenylketonuria (PKU), which is an inherited inability to break down the amino acid phenylalanine. If untreated, the disorder has serious effects on brain development. PKU occurs in about 1 out of 10,000 babies born in the United States. Newborn babies are now routinely tested for PKU, and symptoms can be prevented if individuals living with the disease follow a strict diet that limits phenylalanine. In addition to occurring naturally, phenylalanine is found in the widely used artificial sweetener aspartame (Figure 13.16).

PKU is caused by a recessive allele (that is, one that must be present in two copies to produce the phenotype). Thus, we can represent the frequency of individuals in the U.S. population born with PKU with the q^2 term in the Hardy-Weinberg formula. For one PKU occurrence per 10,000 births, $q^2 = 0.0001$. Therefore, q, the frequency of the recessive allele in the population, equals the square root of 0.0001, or 0.01. And p, the frequency of the dominant allele, equals $1 - q$, or 0.99.

Now let's calculate the frequency of carriers, who are heterozygous individuals who carry the PKU allele in a single copy and may pass it on to offspring. Carriers are represented in the formula by $2pq$: $2 \times 0.99 \times 0.01$, or 0.0198. Thus, the Hardy-Weinberg formula tells us that about 2% of the U.S. population are carriers for the PKU allele. Estimating the frequency of a harmful allele is essential for any public health program dealing with genetic diseases.

▼ Figure 13.16 A warning to individuals with PKU.

INGREDIENTS: SORBITOL, MAGNESIUM STEARATE, ARTIFICIAL FLAVOR, **ASPARTAME†** (SWEETENER), ARTIFICIAL COLOR (YELLOW 5 LAKE, BLUE 1 LAKE), ZINC GLUCONATE.
†PHENYLKETONURICS: CONTAINS PHENYLALANINE

Microevolution as Change in a Gene Pool

As stated earlier, evolution can be measured as changes in the genetic composition of a population over time. It helps, as a basis of comparison, to know what to expect if a population is not evolving. A nonevolving population is in genetic equilibrium, which is also known as **Hardy-Weinberg equilibrium**.

☑ **CHECKPOINT**

1. Which term in the Hardy-Weinberg formula ($p^2 + 2pq + q^2 = 1$) corresponds to the frequency of individuals with *no* alleles for the recessive disease PKU?

2. Define microevolution.

■ Answers: 1. p^2 2. Microevolution is a change in a population's frequencies of alleles.

The population's gene pool remains constant. From generation to generation, the frequencies of alleles (p and q) and genotypes (p^2, $2pq$, and q^2) are unchanged. Sexual shuffling of genes cannot by itself change a large gene pool. Because a generation-to-generation change in allele frequencies of a population is evolution viewed on the smallest scale, it is sometimes referred to as **microevolution**. ☑

Mechanisms of Evolution

Now that we've defined microevolution as changes in a population's genetic makeup from generation to generation, we come to an obvious question: What mechanisms can change a gene pool? The three main causes of evolutionary change are natural selection, genetic drift, and gene flow.

Natural Selection

Natural selection is the most important mechanism of evolutionary change because it is the only process that promotes adaptation. Populations consist of varied individuals, and some variants leave more offspring than others. For example, rabbits might prefer to eat the red flower–producing plants in our hypothetical wildflower population. As a result, plants that produce white flowers (genotype *rr*) survive better and produce more offspring—they are better suited to an environment that includes rabbits. The frequency of the *r* allele on the gene pool would increase from one generation to the next. We'll examine natural selection in more detail shortly. But first, we look at two other mechanisms of evolutionary change: genetic drift, which is due to chance, and gene flow, the exchange of alleles between neighboring populations.

Genetic Drift

Flip a coin 1,000 times, and a result of 700 heads and 300 tails would make you very suspicious about that coin. But flip a coin 10 times, and an outcome of 7 heads and 3 tails would seem within reason. With a smaller sample, there is a greater chance of deviation from an idealized result—in this case, an equal number of heads and tails.

Let's apply this coin toss logic to a population's gene pool. If a new generation draws its alleles at random from the previous generation, then the larger the population (the sample size), the better the new generation will represent the gene pool of the previous generation. Thus, one requirement for a gene pool to maintain the status quo is a large population size. The gene pool of a small population may not be accurately represented in the next generation because of sampling error. The changed gene pool is analogous to the erratic outcome from a small sample of coin tosses.

Figure 13.17 applies this concept of sampling error to a small population of wildflowers. Chance causes

▼ **Figure 13.17 Genetic drift.** This hypothetical wildflower population consists of only ten plants. Due to random change over the generations, genetic drift can eliminate some alleles, as is the case for the *r* allele in generation 3 of this imaginary population.

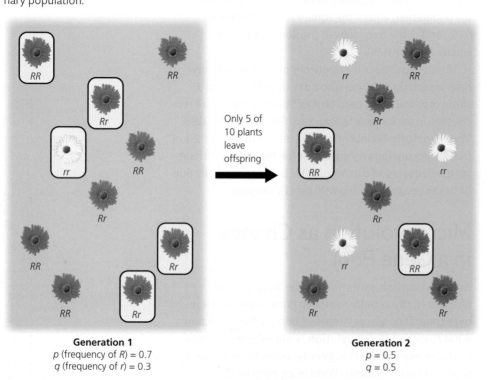

Generation 1
p (frequency of *R*) = 0.7
q (frequency of *r*) = 0.3

Only 5 of 10 plants leave offspring

Generation 2
$p = 0.5$
$q = 0.5$

Only 2 of 10 plants leave offspring

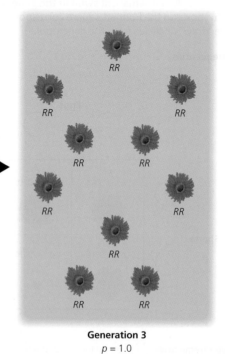

Generation 3
$p = 1.0$
$q = 0.0$

the frequencies of the alleles for red (R) and white (r) flowers to change over the generations. And that fits our definition of microevolution. This evolutionary mechanism, a change in the gene pool of a population due to chance, is called **genetic drift**. But what would cause a population to shrink down to a size where there is genetic drift? Two ways this can occur are the bottleneck effect and the founder effect, both of which we explore next.

The Bottleneck Effect

Disasters such as earthquakes, floods, and fires may kill large numbers of individuals, producing a small surviving population that is unlikely to have the same genetic makeup as the original population. Again, the gene pool of the surviving population is a small sample of the genetic diversity originally present. By chance, certain alleles may be overrepresented among the survivors. Other alleles may be underrepresented. And some alleles may be eliminated. Chance may continue to change the gene pool for many generations until the population is again large enough for sampling errors to be insignificant.

The analogy illustrated in **Figure 13.18** shows why genetic drift due to a drastic reduction in population size is called the **bottleneck effect**. Passing through a "bottleneck"—a severe reduction in population size—decreases the overall genetic variability in a population because at least some alleles are likely to be lost from the gene pool. We can see this concept at work in the potential loss of individual variation, and hence adaptability, in drastically reduced populations of endangered species.

▼ Figure 13.18 **The bottleneck effect.** The colored marbles in this analogy represent three alleles in an imaginary population. Shaking just a few of the marbles through the bottleneck is like an environmental disaster that drastically reduces the size of a population. Compared with the predisaster population, purple marbles are overrepresented in the new population, green marbles are underrepresented, and orange marbles are absent—all by chance. Similarly, a population that passes through a "bottleneck" event emerges with reduced variability.

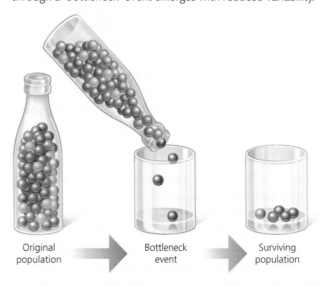

Original population → Bottleneck event → Surviving population

One such endangered species is the cheetah (**Figure 13.19**). The fastest of all running animals, cheetahs are magnificent cats that were once widespread in Africa and Asia. Like many African mammals, the number of cheetahs fell dramatically during the last ice age (around 10,000 years ago). At that time, the

MANY ENDANGERED SPECIES ARE DOOMED BY THEIR LACK OF GENETIC DIVERSITY.

▼ Figure 13.19 **Implications of the bottleneck effect in conservation biology.** Some endangered species, such as the cheetah, have low genetic variability. As a result, they are less adaptable to environmental changes, such as new diseases, than are species with a greater resource of genetic variation.

species suffered a severe bottleneck, possibly as a result of disease, human hunting, and periodic droughts. The South African cheetah population may have suffered a second bottleneck during the 1800s, when farmers hunted the animals to near extinction. Today, only a few small populations of cheetahs exist in the wild. Genetic variability in these populations is very low. In addition, the cheetahs remaining in Africa are being crowded into nature preserves and parks as human demands on the land increase. Along with crowding comes an increased potential for the spread of disease. With so little variability, the cheetah has a reduced capacity to adapt to such environmental challenges. Although captive breeding programs can boost cheetah population sizes, the species' pre-bottleneck genetic diversity can never be restored. ☑

The Founder Effect

When a few individuals colonize an isolated island, lake, or other new habitat, the genetic makeup of the colony is only a sample of the gene pool in the larger population. The smaller the colony (in other words, the smaller the sample size), the less likely it is to be representative of all the genetic diversity present in the population from which the colonists emigrated. If the colony succeeds, genetic drift will continue to change the frequency of alleles randomly until the population is large enough for genetic drift to be minimal. The type of genetic drift resulting from the establishment of a small, new population whose gene pool differs from that of the parent population is called the **founder effect**.

Numerous examples of the founder effect have been identified in geographically or socially isolated human populations. In such situations, disease-causing alleles that are rare in the larger population may become common in a small colony. For example, Amish and Mennonite communities in North America were founded by small numbers of European immigrants in the 1700s, and individuals within the community have since intermarried, remaining genetically separate from the larger population (Figure 13.20). Dozens of genetic diseases that are extremely rare elsewhere are relatively common in these communities. On the other hand, the high frequency of genetic diseases in populations has enabled genetic researchers to identify the mutation responsible for certain genetic disorders. In some cases, a disorder is treatable if detected early.

Gene Flow

Another source of evolutionary change is **gene flow**, which is genetic exchange with another population. A population may gain or lose alleles when fertile

▲ **Figure 13.20 The founder effect.** Small, isolated populations often have high frequencies of alleles that are rare in large populations.

individuals move into or out of the population or when gametes (such as plant pollen) are transferred between populations (Figure 13.21). For example, consider our

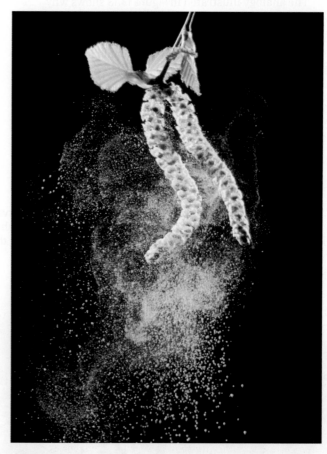

▲ **Figure 13.21 Gene flow.** The pollen of some plants can be carried by the wind for hundreds of miles, allowing gene flow to occur between distant populations.

hypothetical wildflower population in Figure 13.17. Suppose a neighboring population consists entirely of white-flowered individuals. A windstorm may blow pollen to our wildflowers from the neighboring population, resulting in a higher frequency of the white-flower allele in the next generation—a microevolutionary change.

Gene flow tends to reduce differences among populations. If it is extensive enough, gene flow can eventually join neighboring populations into a single population with a common gene pool. As people began to move about the world more freely, gene flow became an important agent of microevolutionary change in human populations that were previously isolated. ☑

▲ **Figure 13.22 A blue-footed booby.** Although the booby's large, webbed feet are highly advantageous in the water, they are clumsy for walking on land.

☑ **CHECKPOINT**

Which mechanism of microevolution tends to reduce differences between populations?

■ *Answer: gene flow*

Natural Selection: A Closer Look

Genetic drift, gene flow, and even mutation can cause microevolution. But only by rare chance would these events result in a population becoming better suited to life in its environment. In natural selection, on the other hand, only the events that produce genetic variation (mutation and sexual reproduction) are random. The process of natural selection, in which individuals better adapted to the environment are more likely to survive and reproduce, is *not* random. Consequently, only natural selection consistently leads to adaptive evolution—evolution that results in organisms better suited to their environment.

The evolutionary adaptations of organisms include many striking examples. **Consider these examples of structure and function that make the blue-footed booby suited to its home on the Galápagos Islands (Figure 13.22).** The bird's body and bill are streamlined like a torpedo, minimizing friction as it dives from heights up to 24 m (over 75 feet) into the shallow water below. To pull out of this high-speed dive once it hits the water, the booby uses its large tail as a brake. Those remarkable blue feet are an essential requirement for the male's reproductive success. Female boobies prefer males with the brightest blue feet. Thus, the male's courtship display is a dance that features frequent flashes of his colorful assets.

Such adaptations are the result of natural selection. By consistently favoring some alleles over others, natural selection results in populations of organisms becoming better adapted to their environment. However, the environment may change over time. As a result, what constitutes "better adapted" is a moving target, making adaptive evolution a continuous, dynamic process. You'll see an example of this in the next section, which examines Darwin's finches.

Evolutionary Fitness

The commonly used phrase "survival of the fittest" is misleading if we take it to mean head-to-head competition between individuals. Reproductive success, the key to evolutionary success, is generally more subtle and passive. In a varying population of moths, for example, certain individuals may produce more offspring than others because their wing colors hide them from predators better. Plants in a wildflower population may differ in reproductive success because some attract more pollinators, owing to slight variations in flower color, shape, or fragrance. In a given environment, such traits can lead to greater **relative fitness**, the contribution an individual makes to the gene pool of the next generation *relative to* the contributions of other individuals. The fittest individuals in the context of evolution are those that produce the largest number of viable, fertile offspring and thus pass on the most genes to the next generation. ☑

☑ **CHECKPOINT**

What is the best measure of relative fitness?

■ *Answer: the number of fertile off-spring an individual leaves*

THE PROCESS OF SCIENCE | Evolution in Action

Did Natural Selection Shape the Beaks of Darwin's Finches?

BACKGROUND

As you've learned in this chapter, Charles Darwin encountered many interesting organisms during his visit to the Galápagos Islands (see Figure 13.3). He was particularly intrigued by the 14 species of finches, which have come to be called Darwin's finches. These small birds are closely related and share many traits. However, they differ in their feeding habits and the size and shape of their beaks, which are specialized based on what they eat. Biologists hypothesize that Darwin's finches evolved from a small populations of ancestral birds that colonized one of the islands. Natural selection shaped the beaks of finches to make use of diverse foods in the new environment, a process that was repeated as birds migrated to neighboring islands with distinct environments. (You'll learn more about the evolution of Darwin's finches in Chapter 14.)

For 40 years, evolutionary ecologists Peter and Rosemary Grant and their students have studied Darwin's finches on Daphne (Figure 13.23a) The small size of the island, 100 acres, allowed the researchers to collect comprehensive data on its plant and animal inhabitants. In one investigation, the Grant team tested the hypothesis that the beak of the medium ground finch changed in response to changes in the available food resources. The Grants and their students captured every medium ground finch and recorded phenotypic variation in the population. One important character they measured was beak depth, which provides the crushing power to open the seeds that make up the bulk of the medium ground finch's diet (Figure 13.23b). The climate of Daphne alternates between wet and dry seasons. Seeds are abundant during the wet season, when plants bloom. Food becomes much scarcer during the desert-like dry season. The Grant team recorded the availability of edible seeds over time. They also quantified seed toughness—how difficult it is for the finches to crack open a seed to get to the food inside it.

METHOD

To test their hypothesis, the Grants used an observational study, a form of scientific inquiry in which scientists test hypotheses without manipulating the subjects

▼ Figure 13.23 An observational study on the effect of natural selection on beak shape in Darwin's finches

(a) Peter and Rosemary Grant collecting data on medium ground finches on Daphne, in the Galápagos Islands

Beak depth

(b) Medium ground finch

of the experiment. A few years after the Grants began their studies, the finch population experienced intense natural selection when Daphne experienced a severe drought. The Grants used data collected before, during, and after the drought to test their hypothesis.

RESULTS

Extremely low rainfall during the 1977 wet season caused seed production to plummet, resulting in a shortage of food to sustain the finches during the subsequent dry season **(Figure 13.23c)**. After the finches exhausted the supply of small, soft seeds, the toughest seeds with thick, hard-to-crack coats became the only food.

Of the 1,200 adult finches that lived on Daphne in June 1976, just 15% (180 birds) survived the drought. The proportion of birds with larger beaks—those able to crack the toughest seeds—increased in the population during this intense period of natural selection, evidence of evolution in action **(Figure 13.23d)**. As the Grants continued their work on Daphne, they observed additional droughts as well as other environmental changes that caused natural selection in the finch populations. Such long-term, observational field studies are an important method for testing hypotheses about evolution.

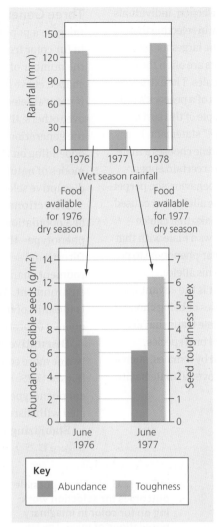

(c) Effect of rainfall on seed abundance and toughness

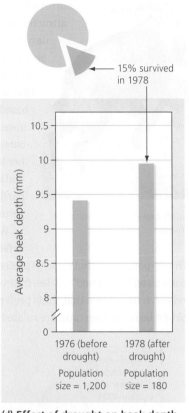

15% survived in 1978

(d) Effect of drought on beak depth in finch population

Thinking Like a Scientist

Why was an observational study better suited to testing the Grants' hypothesis than a controlled experiment?

For the answer, see Appendix D.

Sexual Selection

Sexual selection is a form of natural selection in which individuals with certain traits are more likely than other individuals to obtain mates. Because sexual selection has a direct impact on relative fitness, it is an especially powerful form of natural selection.

The males and females of an animal species obviously have different reproductive organs. But they may also have secondary sexual traits, noticeable differences not directly associated with reproduction or survival. This distinction in appearance, called **sexual dimorphism**, is often manifested in a size difference. Among male vertebrates, sexual dimorphism may also be evident in adornment, such as manes on lions, antlers on deer, and colorful plumage on peacocks and other birds **(Figure 13.24a)**.

In some species, secondary sex structures may be used to compete with members of the same sex (usually males) for mates. Contests may involve physical combat, but are more often ritualized displays **(Figure 13.24b)**. Such selection is common in species in which the winner acquires a harem of mates—an obvious boost to that male's evolutionary fitness.

▼ **Figure 13.24 Sexual dimorphism.**

(a) Sexual dimorphism in a finch species. Among vertebrates, including this pair of green-winged pytilia (native to Africa), males (right) are usually the showier sex.

(b) Competing for mates. Male Spanish ibex engage in nonlethal combat for the right to mate with females.

In a more common type of sexual selection, individuals of one sex (usually females) are choosy in selecting their mates. Males with the largest or most colorful adornments are often the most attractive to females. The extraordinary feathers of a peacock's tail are an example of this sort of "choose me!" statement. Every time a female chooses a mate based on a certain appearance or behavior, she perpetuates the alleles that caused her to make that choice and allows a male with that particular phenotype to perpetuate his alleles.

What is the advantage to females of being choosy? One hypothesis is that females prefer male traits that are correlated with "good" alleles. In several bird species, research has shown that traits preferred by females, such as bright beaks or long tails, are related to overall male health.

✓ CHECKPOINT

Beak depth in the population of medium ground finches on Daphne increased when easy-to-crack seeds became scarce. This is an example of which mode of natural selection: directional, disruptive, or stabilizing?

■ *Answer: directional*

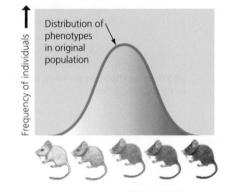

Distribution of phenotypes in original population

Frequency of individuals

Phenotypes (fur color)

◀ **Figure 13.25 Three possible outcomes for selection working on fur color in imaginary populations of mice.** The large downward arrows symbolize the pressure of natural selection working against certain phenotypes.

Three General Outcomes of Natural Selection

Imagine a population of mice with individuals ranging in fur color from very light to very dark gray. If we graph the number of mice in each color category, we get a bell-shaped curve like the one shown at the top of **Figure 13.25**. If natural selection favors certain fur-color phenotypes over others, then the population of mice will change over the generations. Three general outcomes are possible, depending on which phenotypes are favored. These three modes of natural selection are called directional selection, disruptive selection, and stabilizing selection.

Directional selection shifts the overall makeup of a population by selecting in favor of one extreme phenotype—the darkest mice, for example **(Figure 13.25a)**. Directional selection is most common when the local environment changes or when organisms migrate to a new environment. An actual example of directional selection is the shift of insect populations toward a greater frequency of pesticide-resistant individuals.

Disruptive selection can lead to a balance between two or more contrasting phenotypes in a population **(Figure 13.25b)**. A patchy environment, which favors different phenotypes in different patches, is one situation associated with disruptive selection.

Stabilizing selection favors intermediate phenotypes **(Figure 13.25c)**. Such selection typically occurs in relatively stable environments, where conditions tend to reduce physical variation. This evolutionary conservatism works by selecting against the more extreme phenotypes. For example, stabilizing selection keeps the majority of human birth weights between 3 and 4 kg (approximately 6.5 to 9 pounds). For babies much lighter or heavier than this, infant mortality is greater.

Of the three selection modes, stabilizing selection occurs in most situations, resisting change in well-adapted populations. Evolutionary spurts occur when a population is stressed by a change in the environment, such as happened with Darwin's finches on Daphne, or by migration to a new place. When challenged with a new set of environmental problems, a population either adapts through natural selection or dies off in that locale. The fossil record tells us that the population's extinction is the most common result. Those populations that do survive crises may change enough to become new species. (You'll learn more about this in Chapter 14.) ✓

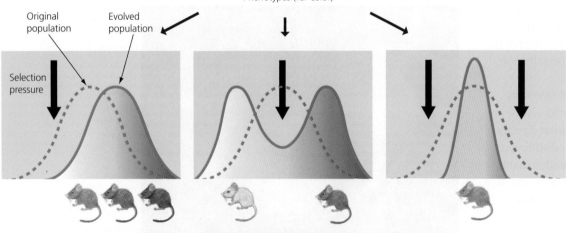

Original population

Evolved population

Selection pressure

(a) Directional selection shifts the overall makeup of the population by favoring variants at one extreme. In this case, the trend is toward darker color, perhaps because the landscape has been shaded by the growth of trees, making darker mice less noticeable to predators.

(b) Disruptive selection favors variants at opposite extremes over intermediate individuals. Here, the relative frequencies of very light and very dark mice have increased. Perhaps the mice have colonized a patchy habitat where a background of light soil is studded with dark rocks.

(c) Stabilizing selection removes extreme variants from the population, in this case eliminating individuals that are unusually light or dark. The trend is toward reduced phenotypic variation and increased frequency of an intermediate phenotype.

The Rising Threat of Antibiotic Resistance

As you probably know, antibiotics are drugs that kill infectious microorganisms. Before antibiotics, people often died from bacterial diseases such as whooping cough, and a minor wound—a razor nick or a scratch from a rose thorn—could result in a fatal infection. A new era in human health followed the introduction of penicillin, the first widely used antibiotic, in the 1940s. Suddenly, many diseases that had often been fatal could easily be cured. However, now medical experts fear that the process of evolution could end the era of antibiotics. In the same way that pesticides select for resistant insects, antibiotics select for resistant bacteria. A gene that codes for an enzyme that breaks down an antibiotic or a mutation that alters the binding site of an antibiotic can make a bacterium and its offspring resistant to that antibiotic. Again, we see both the random and nonrandom aspects of natural selection—the random genetic mutations in bacteria and the nonrandom selective effects as the environment favors the antibiotic-resistant phenotype. As mentioned in the chapter introduction, natural selection has been a problem in the efforts to eradicate malaria. Populations of the parasitic microbe that causes malaria have become resistant to drugs used to treat the disease.

The rapid evolution of antibiotic resistance has been fueled by their widespread use—and misuse. Livestock producers add antibiotics to animal feed as a growth promoter and to prevent illness, practices that may select for bacteria resistant to standard antibiotics. Doctors may prescribe antibiotics when they aren't warranted. And patients may stop taking the medication as soon as they feel better. This allows mutant bacteria that are killed more slowly by the drug to survive and multiply. Subsequent mutations in such bacteria may lead to full-blown antibiotic resistance.

A formidable "superbug" known as MRSA (methicillin-resistant *Staphylococcus aureus*) was the first sign that the power of antibiotics might be fading. *S. aureus* ("staph")

is common in health-care facilities, where natural selection for antibiotic resistance is strong because of the extensive use of antibiotics. Staph outbreaks also occur in community settings such as athletic facilities, schools, and military barracks. Some staph infections cause relatively minor skin disorders, but when bacteria invade the bloodstream, staph infections can be fatal. Although deaths from invasive MRSA acquired at health-care facilities have declined recently as a result of preventative measures, MRSA remains a serious threat to public health.

Drug-resistant microorganisms infect more than 2 million people and cause 23,000 deaths in the United States each year. The Centers for Disease Control (CDC) has identified 15 microorganisms that pose urgent or serious threats to public health. Some of the infections are associated with health-care facilities; others are passed on by contaminated food and water, sexual contact, or droplets exhaled from the respiratory tract of an infected person. **Figure 13.26** shows the estimated percentage of infections by antibiotic-resistant strains for several diseases.

Medical and pharmaceutical researchers are racing to develop new antibiotics and other drugs. However, experience suggests that our battle against the evolution of drug-resistant bacteria will continue into the future.

▶ **Figure 13.26 Urgent (red) and serious (orange) threats from antibiotic-resistant bacteria.** The area of the circles represents the total numbers of infections; slices show the percentage caused by antibiotic-resistant strains.

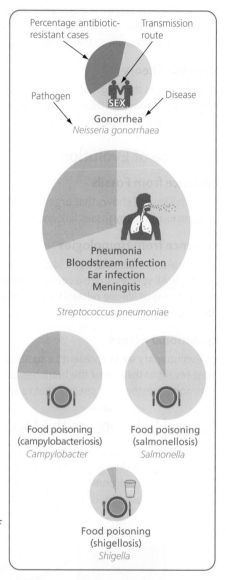

Chapter Review

SUMMARY OF KEY CONCEPTS

The Diversity of Life

Naming and Classifying the Diversity of Life

In the Linnaean system of classification, each species is assigned a two-part name. The first part is the genus, and the second part is unique for each species within the genus. In the taxonomic hierarchy, domain > kingdom > phylum > class > order > family > genus > species.

Explaining the Diversity of Life

Present-day biologists accept Darwin's theory of evolution by means of natural selection as the best explanation for the diversity of life. However, when Darwin published his theory in 1859, it was a radical departure from the prevailing views.

Charles Darwin and *The Origin of Species*

Darwin's Journey

During his voyage on the *Beagle*, Darwin observed adaptations of organisms that inhabited diverse environments, especially on the Galápagos Islands, off the South American coast. These observations, along with other insights, eventually led Darwin to formulate his theory of evolution.

Darwin's Theory

In his book *On the Origin of Species by Means of Natural Selection*, Darwin made two proposals: (1) Existing species descended from ancestral species and (2) natural selection is the mechanism of evolution.

Evidence of Evolution

Evidence from Fossils

The fossil record shows that organisms have appeared in a historical sequence, and many fossils link ancestral species with those living today.

Evidence from Homologies

Structural and molecular homologies reveal evolutionary relationships. All species share a common genetic code, suggesting that all forms of life are related through branching evolution from the earliest organisms.

Evolutionary Trees

An evolutionary tree represents a succession of related species, with the most recent at the tips of the branches. Each branch point represents a common ancestor of all species that radiate from it.

Natural Selection as the Mechanism for Evolution

Darwin proposed natural selection as the mechanism that produces adaptive evolutionary change. In a population that varies, individuals best suited for a particular environment are more likely to survive and reproduce than those that are less suited to that environment.

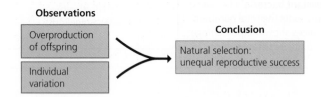

Natural Selection in Action

Natural selection has been observed in many scientific studies, including in the evolution of pesticide-resistant insects.

Key Points about Natural Selection

Individuals do not evolve. Only heritable traits, not those acquired during an individual's lifetime, can be amplified or diminished by natural selection. Natural selection only works on existing variation—new variation does not arise in response to an environmental change. Natural selection does not produce perfect organisms.

The Evolution of Populations

Sources of Genetic Variation

Mutation and sexual reproduction produce genetic variation. Mutation is the ultimate source of genetic variation.

Populations as the Units of Evolution

A population, members of the same species living in the same time and place, is the smallest biological unit that can evolve.

Analyzing Gene Pools

A gene pool consists of all the alleles in all the individuals making up a population. The Hardy-Weinberg formula can be used to calculate the frequencies of genotypes in a gene pool from the frequencies of alleles, and vice versa:

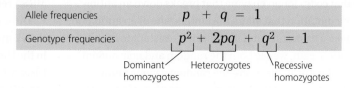

Allele frequencies	$p + q = 1$
Genotype frequencies	$p^2 + 2pq + q^2 = 1$

Dominant homozygotes · Heterozygotes · Recessive homozygotes

Population Genetics and Health Science

The Hardy-Weinberg formula can be used to estimate the frequency of a harmful allele, which is useful information for public health programs dealing with genetic diseases.

Microevolution as Change in a Gene Pool

Microevolution is generation-to-generation change in allele frequencies in a population.

Mechanisms of Evolution

Natural Selection

Natural selection is the most important mechanism of evolutionary change because it is the only process that promotes adaptation.

Genetic Drift

Genetic drift is a change in the gene pool of a small population due to chance. A bottleneck event (a drastic reduction in population size) and the founder effect (occurring in a new population started by a few individuals) are two situations that can lead to genetic drift.

Gene Flow

A population may gain or lose alleles by gene flow, which is genetic exchange with another population.

Natural Selection: A Closer Look

Of all the causes of evolution, only natural selection promotes evolutionary adaptations. Relative fitness is the contribution an individual makes to the gene pool of the next generation relative to the contributions of other individuals. The outcome of natural selection may be directional, disruptive, or stabilizing. Secondary sexual traits (such as sex-specific plumage or behaviors) can promote sexual selection, a type of natural selection in which mating preferences are determined by inherited traits.

Mastering Biology

For practice quizzes, BioFlix animations, MP3 tutorials, video tutors, and more study tools designed for this textbook, go to Mastering Biology™

SELF-QUIZ

1. Place these levels of classification in order from least inclusive to most inclusive: class, domain, family, genus, kingdom, order, phylum, species.

2. Which of the following is a true statement about Charles Darwin?
   a. He was the first to discover that living things can change, or evolve.
   b. He based his theory on the inheritance of acquired traits.
   c. He proposed natural selection as the mechanism of evolution.
   d. He was the first to realize that Earth is more than 6,000 years old.

3. How did the insights of Lyell and other geologists influence Darwin's thinking about evolution?

4. In a population with two alleles for a particular genetic locus, B and b, the allele frequency of B is 0.7. If this population is in Hardy-Weinberg equilibrium, what is the frequency of heterozygotes? What is the frequency of homozygous dominants? What is the frequency of homozygous recessives?

5. Define fitness from an evolutionary perspective.

6. Which of the following processes is the ultimate source of the genetic variation that serves as raw material for evolution?
   a. sexual reproduction
   b. mutation
   c. genetic drift
   d. natural selection

7. Which of the following is *not* a requirement of natural selection?
   a. genetic variation
   b. catastrophic events
   c. differential reproductive success
   d. overproduction of offspring

8. Compare and contrast how the bottleneck effect and the founder effect can lead to genetic drift.

9. Garter snakes with different color patterns behave differently when threatened. Of the three general outcomes of natural selection (directional, disruptive, or stabilizing), this example illustrates _____.

For answers to the Self Quiz, see Appendix D.

IDENTIFYING MAJOR THEMES

For each statement below, identify which major theme is evident (the relationship of structure to function, information flow, pathways that transform energy and matter, interactions within biological systems, or evolution) and explain how the statement relates to the theme. If necessary, review the themes (Chapter 1) and review the examples highlighted in blue in this chapter.

10. If two species have homologous genes with sequences that match closely, biologists conclude that these sequences must have been inherited from a relatively recent common ancestor.

11. The body and bill of a diving bird, the blue-footed booby, is streamlined like a torpedo.

12. Darwin hypothesized that as descendants of a remote ancestor spread into various habitats, natural selection resulted in diverse modifications that fit them to their environment.

For answers to Identifying Major Themes, see Appendix D.

THE PROCESS OF SCIENCE

13. **Interpreting Data** A population of snails has recently become established in a new region. The snails are preyed on by birds that break the snails open on rocks, eat the soft bodies, and leave the shells. The snails occur in both striped and unstriped forms. In one area, researchers counted both live snails and broken shells. Their data are summarized here:

	Striped Shells	Unstriped Shells
Number of live snails	264	296
Number of broken snail shells	486	377
Total	750	673

Based on these data, which snail form is subject to more predation by birds? Predict how the frequencies of striped and unstriped individuals might change over time.

14. Five years after the experiment described in the previous question, the researchers repeat the snail count in order to test their prediction. Surprisingly, they find that the ratio of broken shells to live snails has gone down for the striped and up for the unstriped snails. They notice that a weed with striped leaves has newly spread in the examined region. Formulate a hypothesis that explains this observation.

BIOLOGY AND SOCIETY

15. To what extent are people in a technological society exempt from natural selection? Explain your answer.

16. "Social Darwinism" suggests that the principle of "survival of the fittest" is applicable to human society. Is this theory a valid approach? Is it based on science?

14 How Biological Diversity Evolves

SEE A RESEMBLANCE BETWEEN THE NEANDERTHAL AND DARWIN? PIECING TOGETHER EVOLUTIONARY HISTORIES SHOWS WHO'S RELATED TO WHOM.

ENJOY CORN ON THE COB? ALLELES FROM TWO WILD GRASSES COULD BE USED TO PROTECT CORN CROPS FROM FUTURE DISASTERS.

Why Evolution Matters

The diversity of life on Earth is the product of evolutionary processes that have been happening for billions of years. During this time, the overwhelming majority of species that ever lived became extinct. The classification system used by biologists shows how all forms of life on Earth are related.

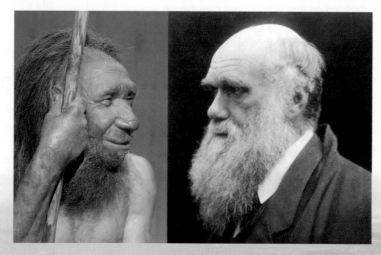

IF ROCKS COULD SPEAK, WHAT WOULD THEY TELL US? SEVERAL HUNDRED MILLION YEARS OF HISTORY IS "WRITTEN" IN THE LAYERS OF ROCK IN THE GRAND CANYON.

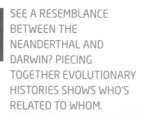

BIOLOGY AND SOCIETY Evolution in the Human-Dominated World

Humanity's Footprint

Humanity has had an extraordinary effect on the ecology and geology of Earth. Our indelible footprint includes the transport of organisms far from their natural homes, the prevalence of agriculture and domesticated animals, the existence of manufactured materials such as plastics and concrete, radioactivity from testing nuclear weapons, and climate-altering emissions from burning fossil fuels. Very little of the planet remains untouched by human activities. In 2017, for example, researchers reported that animals living 10 km (6.2 mi) deep in the Pacific Ocean were contaminated with toxic chemicals from industrial waste. The irreversibility of these changes has led some scientists to propose that a new epoch in Earth's history has begun: the Anthropocene (from the Greek *anthropos*, human). The Anthropocene signals a significant shift in the geologic record that includes a high rate of extinction and accelerating change to Earth. As you will learn in this chapter, scientists have divided the 4.6 billion years of Earth's history into a sequence of geologic eras, which are further subdivided into periods and epochs. This geologic time line is formally defined by the International Commission on Stratigraphy using long-established criteria. Naming a new epoch is no small matter, and experts are currently debating the proposal to acknowledge the Anthropocene.

Regardless of whether this new epoch is officially added to the geologic time line, it's a useful idea for nonscientists to consider. We are living in a time that is unique in Earth's 4.6-billion-year history. Humans are not the only organisms that have changed the environment on a global scale. More than 2 billion years ago, oxygen released by photosynthesis in single-celled prokaryotes eventually transformed the biosphere. But we are the only organisms with the ability to recognize our impact and understand its potential consequences for life on Earth. For example, human activities are modifying the global environment to such an extent that many species are disappearing. In the past 400 years—a very short time on a geologic scale—more than 1,000 species are known to have become extinct. Scientists estimate that this is 100 to 1,000 times the extinction rate seen in the past. Human-driven changes in the environment also bring about evolutionary change in populations of organisms, including pesticide-resistant insects and antibiotic-resistant bacteria (as you learned about in Chapter 13). Activities such as agriculture, hunting, and fishing apply selection pressures to specific populations of organisms. New species have arisen in recent geologic times, too, through transportation of plants, animals, and microbes to new environments.

We'll begin this chapter by discussing the birth of new species and then examine how biologists trace the evolution of biological diversity. We'll also take a closer look at how scientists classify living organisms.

Tailings (waste residue) dump from mining oil sands in Alberta, Canada. Oil sands, also called tar sands, contain a sticky, viscous form of petroleum. Mining these deposits leaves behind a large volume of tailings.

The Origin of Species

Natural selection, a microevolutionary mechanism, explains the striking ways in which organisms are suited to their environment. But what accounts for the tremendous diversity of life, the millions of species that have existed during Earth's history? This question intrigued Darwin, who referred to it in his diary as "that mystery of mysteries—the first appearance of new beings on this Earth."

When, as a young man, Darwin visited the Galápagos Islands (see Figure 13.3), he realized that he was visiting a place of origins. Though the volcanic islands were geologically young, they were already home to many plants and animals known nowhere else in the world. Among these unique inhabitants were marine iguanas (**Figure 14.1**), Galápagos tortoises (see Figure 13.4), and numerous species of small birds called finches, which you will learn more about in this chapter (see Figure 14.12). Surely, Darwin thought, not all of these species could have been among the original colonists. Some of them must have evolved later on, the myriad descendants of the original colonists, modified by natural selection from those original ancestors.

In the century and a half since the publication of Darwin's *On the Origin of Species by Means of Natural Selection*, new discoveries and technological advances—especially in molecular biology—have given scientists a wealth of new information about the evolution of life on Earth. For example, researchers have explained the genetic patterns underlying the homology of vertebrate limbs (see Figure 13.6). Hundreds of thousands more fossil discoveries have been cataloged since Darwin's time, including many of the transitional (intermediate) forms predicted by Darwin. In a fascinating convergence of old and new techniques, researchers have even been able to investigate the genetic material of certain fossils, including our ancient relatives, the Neanderthals (see Figure 14.23). New dating methods have confirmed that Earth is billions of years old, much older than even the most radical geologists of Darwin's time proposed. As you'll learn in this chapter, these dating methods have also enabled researchers to determine the ages of fossils and rocks, providing valuable insight into evolutionary relationships among groups of organisms. In addition, our enhanced understanding of geologic processes, such as the changing positions of continents, explains some of the geographic distributions of organisms and fossils that puzzled Darwin and his contemporaries.

In this chapter, you'll learn how evolution has woven the rich tapestry of life, beginning with **speciation**, the process in which one species splits into two or more species. Other topics include the origin of evolutionary novelty, such as the wings and feathers of birds and the large brains of humans, and the impact of mass extinctions, which clear the way for new adaptive explosions, such as the diversification of mammals following the disappearance of most of the dinosaurs.

▼ **Figure 14.1 A marine iguana (right), an example of the unique species inhabiting the Galápagos.** Darwin noticed that Galápagos marine iguanas—with a flattened tail that aids in swimming—are similar to, but distinct from, land-dwelling iguanas on the islands and on the South American mainland (left).

What Is a Species?

Species is a Latin word meaning "kind" or "appearance." Even as children we learn to distinguish between the kinds of plants and animals—between dogs and cats, for example, or between roses and dandelions—from differences in their appearance. Although the basic idea of species as distinct life-forms seems intuitive, devising a more formal definition is not so easy.

One way of defining a species (and the main definition used in this book) is the **biological species concept**. It defines a **species** as a group of populations whose members have the potential to interbreed with one another in nature and produce fertile offspring (offspring that can reproduce) **(Figure 14.2)**. Geography and culture may conspire to keep a Manhattan businesswoman and a Mongolian dairyman apart. But if the two did meet and mate, they could have viable babies who develop into fertile adults because all humans belong to the same species. In contrast, humans and chimpanzees, despite having a shared evolutionary history, are distinct species because they can't successfully interbreed.

We cannot apply the biological species concept to all situations. For example, basing the definition of species on reproductive compatibility excludes organisms that only reproduce asexually (producing offspring from a single parent), such as most prokaryotes. And because fossils are obviously not currently reproducing sexually, they cannot be evaluated by the biological species concept. In response to such challenges, biologists have developed other ways to define species. For example, most of the species named so far have been classified based on measurable physical traits such as number and type of teeth or flower structures. Another approach defines a species as the smallest group of individuals sharing a common ancestor and forming one branch on the tree of life. Yet another approach proposes defining a species solely on the basis of molecular data, a sort of bar code that identifies each species.

Each species concept is useful, depending on the situation and the questions being asked. The biological species concept, however, is particularly useful when focusing on how species originate—that is, when we ask: What prevents a member of one group from successfully interbreeding with a member of another group? You'll learn about the variety of answers to that question next. ☑

THE ORIGIN OF SPECIES

☑ CHECKPOINT

According to the biological species concept, what defines a species?

■ *Answer: the ability of its members to interbreed with one another and produce fertile offspring in a natural setting*

▼ Figure 14.2 **The biological species concept is based on reproductive compatibility rather than physical similarity.**

Similarity between different species. The eastern meadowlark (left) and the western meadowlark (right) are very similar in appearance, but they are separate species and do not interbreed.

Diversity within one species. Humans, as diverse in appearance as we are, belong to a single species (*Homo sapiens*) and can interbreed.

Reproductive Barriers between Species

Clearly, a fly will not mate with a frog or a fern. But what prevents closely related species from interbreeding? What, for example, maintains the species boundary between the eastern meadowlark and the western meadowlark (shown in Figure 14.2)? Their geographic ranges overlap in the Great Plains region, and they are so similar that only expert birders can tell them apart. And yet, these two bird species do not interbreed.

A **reproductive barrier** is anything that prevents individuals of closely related species from interbreeding. Let's examine the different kinds of reproductive barriers that isolate the gene pools of species (Figure 14.3). We can classify reproductive barriers as either prezygotic or postzygotic, depending on whether they block interbreeding before or after the formation of zygotes (fertilized eggs).

Prezygotic barriers prevent mating or fertilization between species (Figure 14.4). The barrier may be time-based (temporal isolation). For example, western spotted skunks breed in the fall, but the eastern species breeds in late winter. Temporal isolation keeps the species from mating even where they coexist on the Great Plains. In other cases, species live in the same region but not in the same habitats (habitat isolation). For example, one species of North American garter snake lives mainly in water, and a closely related species lives on land. Traits that enable individuals to recognize potential mates, such as a particular odor, coloration, or courtship ritual, can also function as reproductive barriers (behavioral isolation). In many bird species, for example, courtship behavior is so elaborate that individuals are unlikely to mistake a bird of

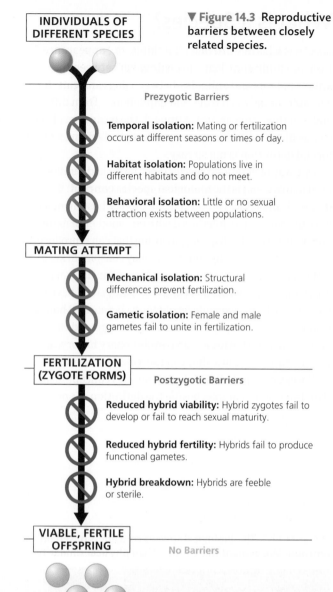

▼ **Figure 14.3** Reproductive barriers between closely related species.

INDIVIDUALS OF DIFFERENT SPECIES

Prezygotic Barriers

Temporal isolation: Mating or fertilization occurs at different seasons or times of day.

Habitat isolation: Populations live in different habitats and do not meet.

Behavioral isolation: Little or no sexual attraction exists between populations.

MATING ATTEMPT

Mechanical isolation: Structural differences prevent fertilization.

Gametic isolation: Female and male gametes fail to unite in fertilization.

FERTILIZATION (ZYGOTE FORMS)

Postzygotic Barriers

Reduced hybrid viability: Hybrid zygotes fail to develop or fail to reach sexual maturity.

Reduced hybrid fertility: Hybrids fail to produce functional gametes.

Hybrid breakdown: Hybrids are feeble or sterile.

VIABLE, FERTILE OFFSPRING

No Barriers

▶ **Figure 14.4 Prezygotic barriers.** Prezygotic barriers prevent mating or fertilization.

PREZYGOTIC BARRIERS

Temporal Isolation

These two closely related species of skunks mate at different times of the year.

Habitat Isolation

These two closely related species of garter snakes do not mate because one lives in the water and the other lives on land.

a different species as one of their kind. In still other cases, the reproductive structures of different species are physically incompatible (mechanical isolation). For example, pollinators such as the hummingbird in Figure 14.4 pick up pollen from the male parts of one flower and transfer it to the female parts of another flower. Floral structure determines the best fit between pollinator and flower. In still other cases, gametes (eggs and sperm) of different species are incompatible, preventing fertilization (gametic isolation). Gametic isolation is very important when fertilization is external. Male and female sea urchins of many species release eggs and sperm into the sea, but fertilization occurs only if species-specific molecules on the surface of egg and sperm attach to each other.

Postzygotic barriers operate if interspecies mating actually occurs and results in hybrid zygotes (**Figure 14.5**). (In this context, "hybrid" means that the egg comes from one species and the sperm from another species.) In some cases, hybrid offspring die before reaching reproductive maturity (reduced hybrid viability). For example, although certain closely related salamander species will hybridize, the offspring fail to develop normally because of genetic incompatibilities between the two species. In other cases of hybridization, offspring may become vigorous adults, but are infertile (reduced hybrid fertility). A mule, for example, is the hybrid offspring of a female horse and a male donkey. Mules are sterile—they cannot successfully breed with each other. Thus, horses and donkeys remain distinct species. In other cases, the first-generation hybrids are viable and fertile, but when these hybrids mate with one another or with either parent species, the offspring are feeble or sterile (hybrid breakdown). For example, different species of rice plants can

produce fertile hybrids, but the offspring of the hybrids do not survive.

In summary, reproductive barriers form the boundaries around closely related species. In most cases, it is not a single reproductive barrier but some combination of two or more that keeps species isolated. Next, we examine situations that make reproductive isolation and speciation possible. ☑

☑ **CHECKPOINT**

Why is behavioral isolation considered a prezygotic barrier?

■ *Answer: because it prevents mating and therefore the formation of a zygote*

▼ **Figure 14.5 Postzygotic barriers.** Postzygotic barriers prevent development of fertile adults.

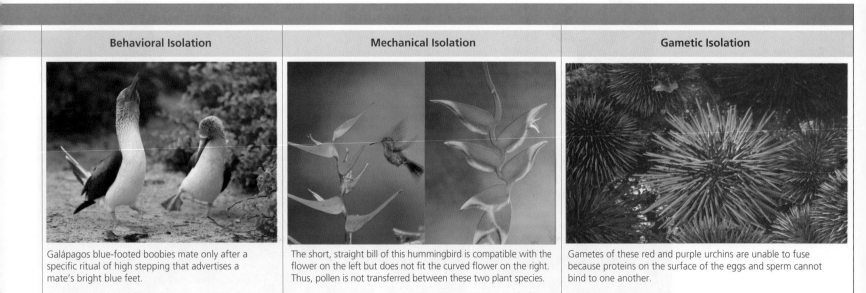

POSTZYGOTIC BARRIERS

Reduced Hybrid Viability

Some species of salamander can hybridize, but their offspring do not develop fully or, like this one, are frail and will not survive long enough to reproduce.

Reduced Hybrid Fertility

Horse

Donkey

Mule

The hybrid offspring of a horse and a donkey is a mule, which is sterile.

Hybrid Breakdown

The rice hybrids at the left and right are fertile, but plants of the next generation (middle) are small and sterile.

Behavioral Isolation

Galápagos blue-footed boobies mate only after a specific ritual of high stepping that advertises a mate's bright blue feet.

Mechanical Isolation

The short, straight bill of this hummingbird is compatible with the flower on the left but does not fit the curved flower on the right. Thus, pollen is not transferred between these two plant species.

Gametic Isolation

Gametes of these red and purple urchins are unable to fuse because proteins on the surface of the eggs and sperm cannot bind to one another.

Mechanisms of Speciation

A key event in the origin of many species occurs when a population is somehow cut off from other populations of the parent species. With its gene pool isolated, the splinter population can follow its own evolutionary course. Changes in its allele frequencies caused by genetic drift and natural selection will not be diluted by alleles entering from other populations (gene flow). Such reproductive isolation can result from two general scenarios: allopatric ("different country") speciation and sympatric ("same country") speciation. In **allopatric speciation**, the initial block to gene flow is a geographic barrier that physically isolates the splinter population. In contrast, **sympatric speciation** is the origin of a new species without geographic isolation. The splinter population becomes reproductively isolated even though it is in the midst of the parent population.

Allopatric Speciation

A variety of geologic processes can isolate populations. For example, the water level in a large lake may subside until there are several smaller lakes, each with a separate fish population. A stream may change course and divide populations of animals that cannot cross the water. Over time, a river flowing over rock may carve a deep canyon that separates the inhabitants on either side. On a larger scale, continents themselves can split and move apart, and the rise and fall of sea levels can submerge or expose land bridges between continents (see Figure 14.15).

How formidable must a geographic barrier be to interrupt gene flow between allopatric populations? The answer depends partly on the ability of the organisms to move about. Birds, mountain lions, and coyotes can cross mountain ranges, rivers, and canyons. Such barriers also do not hinder the windblown pollen of pine trees or the spread of seeds carried by animals capable of crossing the barrier. In contrast, small rodents may find a deep canyon or a wide river an impassable barrier **(Figure 14.6)**.

Speciation is more common for a small, isolated population because it is more likely than a large population to have its gene pool changed substantially by both genetic drift and natural selection. But for each small, isolated population that becomes a new species, many more simply perish in their new environment. Life on the frontier can be harsh, and most pioneer populations become extinct.

Even if a small, isolated population survives, it does not necessarily evolve into a new species. The population may adapt to its local environment and begin to look very different from the ancestral population, but that doesn't necessarily make it a new species. **Speciation occurs with the evolution of reproductive barriers between the isolated population and its parent population.**

▼ **Figure 14.6 Allopatric speciation of antelope squirrels on opposite rims of the Grand Canyon.** Harris's antelope squirrel (*Ammospermophilus harrisii*) is found on the south rim of the Grand Canyon. Just a few miles away on the north rim is the closely related white-tailed antelope squirrel (*Ammospermophilus leucurus*). Birds and other organisms that can disperse easily across the canyon have not diverged into different species on opposite rims.

Ammospermophilus harrisii

Ammospermophilus leucurus

▶ **Figure 14.7 Possible outcomes after geographic isolation of populations.** In this diagram, the orange and green arrows track populations over time. The mountain symbolizes a period of geographic isolation during which time genetic changes may occur in both populations. After a period of time, the populations are no longer separated by a geographic barrier (right side of diagram) and come back into contact. If the populations can interbreed freely (top diagram), speciation has not occurred. If the populations cannot interbreed (bottom), then speciation has occurred.

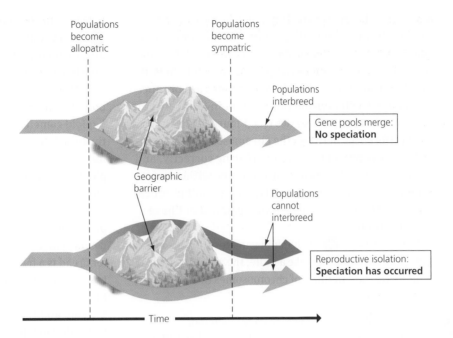

Populations become allopatric

Populations become sympatric

Populations interbreed

Gene pools merge: **No speciation**

Geographic barrier

Populations cannot interbreed

Reproductive isolation: **Speciation has occurred**

Time

In other words, if speciation occurs during geographic separation, the new species cannot breed with its ancestral population even if the two populations should come back into contact at some later time **(Figure 14.7)**. ☑

Sympatric Speciation

In sympatric speciation, a new species arises within the same geographic area as its parent species. Polyploidy, habitat complexity, and sexual selection are factors that can reduce gene flow in sympatric populations.

A species may originate from an accident during cell division that results in an extra set of chromosomes. New species formed in this way are **polyploid**, meaning their cells have more than two complete sets of chromosomes. Examples of polyploid speciation have been found in some animal species, especially fish and amphibians **(Figure 14.8)**.

However, it is most common in plants—an estimated 80% of present-day plant species are descended from ancestors that arose by polyploid speciation **(Figure 14.9)**.

Two distinct forms of polyploid speciation have been observed. In one form, polyploidy arises from a single parent species. For example, a failure of cell division might double the chromosome number from the original diploid number ($2n$) to tetraploid ($4n$). Because the polyploid individual cannot produce fertile hybrids with its parent species, immediate reproductive isolation results.

A second form of polyploid speciation can occur when two different species interbreed and produce hybrid offspring. Most instances of polyploid speciation in plants resulted from such hybridizations. How did these interspecies hybrids overcome the

☑ **CHECKPOINT**

What is necessary for allopatric speciation to occur?

■ *Answer: A population must be split into more than one group by a geographic barrier that interrupts gene flow between the two groups.*

▼ **Figure 14.8 Gray tree frog.** This amphibian is thought to have originated by polyploid speciation.

▼ **Figure 14.9 Chinese hibiscus.** Many ornamental varieties of this species have been produced through polyploidy, which may make the flowers larger or increase the number of petals.

☑ **CHECKPOINT**

What mechanism accounts
for most observed instances
of sympatric speciation? Why
might this be the case?

*Answer: Accidents of cell division
that result in polyploidy. Polyploidy
produces "instant" reproductive
isolation.*

postzygotic barrier of sterility (see Figure 14.5)? A mule is sterile because its parents' chromosomes don't match. A horse has 64 chromosomes (32 pairs); a donkey has 62 chromosomes (31 pairs). Therefore, a mule has 63 chromosomes. Recall that gametes are produced by meiosis, a cell division process that involves pairing of homologous chromosomes (see Figure 8.14). Structural differences between the chromosomes of horses and donkeys prevent them from pairing correctly, and the odd number leaves one chromosome without a potential partner. Consequently, mules don't produce viable gametes. Now, let's see why plants don't have the same problem.

In **Figure 14.10**, ❶ the haploid gametes from species A and species B combine. Like a mule, the resulting hybrid plant has an odd number of chromosomes ($n = 5$). The hybrid's chromosomes are not homologous; it is sterile. ❷ However, the hybrid may be able to reproduce asexually, as many plants can do. ❸ Subsequent errors in cell division may produce chromosome duplications that result in a diploid set of chromosomes ($2n = 10$). In this polyploid plant, chromosomes *can* pair in meiosis to form haploid gametes. The new polyploid species is reproductively isolated from both parent species. Biologists have identified several plant species that originated via polyploidy within the past 150 years. You will learn about one example next, in the Process of Science.

Many of the plant species we grow for food are polyploids, including oats, potatoes, bananas, strawberries, peanuts, apples, sugarcane, and wheat. The wheat used for bread is a hybrid of three different parent species and has six sets of chromosomes, two sets from each parent. Plant geneticists use chemicals to induce errors in cell division and generate new polyploids in the laboratory. By harnessing this evolutionary process, they can produce new hybrid species with desirable qualities.

Biologists have also identified cases in which subpopulations appear to be in the process of sympatric speciation. In some cases, subgroups of a population become adapted for exploiting food sources in different habitats, such as the shallow versus deep habitats of a lake. In another example, a type of sexual selection in which female fish choose mates based on color has contributed to rapid reproductive isolation. Because of its direct effect on reproductive success, sexual selection can interrupt gene flow within a population and may therefore be an important factor in sympatric speciation. But the most frequently observed mechanism of sympatric speciation involves large-scale genetic changes that occur in a single generation. ☑

▼ **Figure 14.10 Sympatric speciation in a plant.** The gametes from two different species result in a sterile hybrid, which may undergo asexual reproduction. Such a hybrid may eventually form a new species by polyploidy.

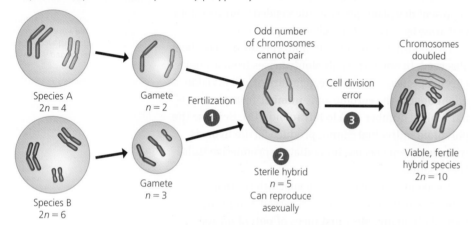

Species A
$2n = 4$

Species B
$2n = 6$

Gamete
$n = 2$

Gamete
$n = 3$

Fertilization

❶

Odd number
of chromosomes
cannot pair

❷

Sterile hybrid
$n = 5$
Can reproduce
asexually

Cell division
error

❸

Chromosomes
doubled

Viable, fertile
hybrid species
$2n = 10$

THE PROCESS OF SCIENCE　Evolution in the Human-Dominated World

Do Human Activities Facilitate Speciation?

BACKGROUND

Botanists estimate that more than 80% of the plant species alive today are descended from ancestors that formed by polyploidy speciation. As you saw in Figure 14.10, the first step in the process may be the hybridization of two "parent" species. The parent species are separated by a postzygotic barrier—the hybrid offspring may thrive, but they are unable to reproduce. This barrier can be overcome by an error in cell division that doubles the number of chromosomes, thus making the hybrid fertile.

As they travel the world, humans often transport plants, animals, and other organisms to regions beyond their natural range. These activities create opportunities for hybridization that would not otherwise exist.

One such example is a plant known as goatsbeard (genus *Tragopogon*, from the Greek words for "goat" and "beard"), a weedy member of the daisy family. Although goatsbeards are not native to North America, three species were brought from Europe by settlers in the early 1900s. Fifty years later, a botanist discovered that hybridization had produced two new species of goatsbeard. The new species were distinguished from their parental species by flower color and other phenotypic characters. More importantly, members of each new species could reproduce with each other but not with their parental species. The botanist hypothesized that one of the new species, *Tragopogon mirus* (**Figure 14.11a**), originated from *Tragopogon dubius* and *Tragopogon porrifolius* by

▼ **Figure 14.11** Testing a hypothesis about speciation.

(a) Researchers hypothesized that a new species of goatbeard, *Tragopogon mirus*, was formed by polyploidy speciation.

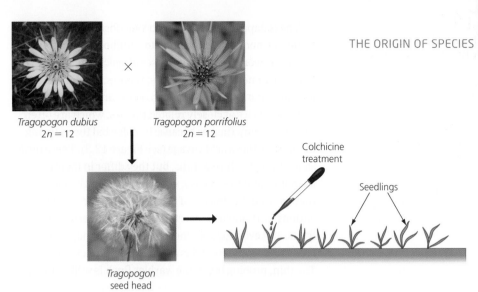

Tragopogon dubius
2n = 12

Tragopogon porrifolius
2n = 12

Colchicine treatment

Seedlings

Tragopogon
seed head

(b) After pollinating *Tragopogon porrifolius* with pollen from *T. dubius*, researchers collected and grew the seeds. Seedlings were treated with colchicine to encourage chromosome doubling.

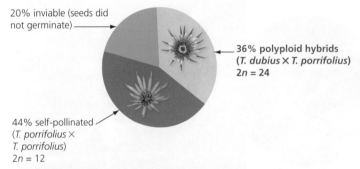

20% inviable (seeds did not germinate)

36% polyploid hybrids (*T. dubius* × *T. porrifolius*) 2n = 24

44% self-pollinated (*T. porrifolius* × *T. porrifolius*) 2n = 12

(c) Polyploid plants resulting from the hybridization of *T. porrifolius* and *T. dubius* were consistent with the new species, *T. mirus.*

polyploid speciation. In 2009, researchers combined traditional plant breeding methods with techniques in molecular genetics to test this hypothesis.

METHOD

The researchers grew *Tragopogon dubius* and *Tragopogon porrifolius* plants using seeds collected from naturally occurring populations in Washington state. *T. porrifolius* flowers were pollinated with pollen from *T. dubius* to produce hybrids (**Figure 14.11b**). The seeds resulting from these crosses were planted. In nature, polyploids arise infrequently by an error in cell division. For this experiment, the researchers greatly increased the chance of obtaining polyploids by treating the seedlings with a chemical, colchicine, that prevents chromosome separation during mitosis, thus doubling the number of chromosomes.

Like many plants, goatsbeards can self-pollinate. After 6 months, the scientists performed genetic tests to identify which plants resulted from self-pollination of *T. porrifolius* and which plants resulted from hybridization of *T. dubius* × *T. porrifolius.*

To determine whether a postzygotic barrier existed that would prevent the development of fertile adults, the hybrid plants were grown to maturity in a greenhouse (to isolate them from other pollen) and allowed to self-pollinate.

RESULTS

The researchers planted 134 seeds from 17 *T. dubius* × *T. porrifolius* crosses. Genetic tests showed that 36% of the seedlings were polyploid hybrids (**Figure 14.11c**). The hybrid plants were phenotypically indistinguishable from *T. mirus* plants found in natural populations.

When grown to maturity, hybrid plants formed by 13 of the 17 original crosses produced viable seeds. Plants grown from those seeds also proved to be fertile and produce viable offspring.

Results supported the hypothesis that *Tragopogon mirus* originated from *Tragopogon dubius* and *Tragopogon porrifolius* by polyploid speciation. In this instance, human activities (introducing plants to non-native regions) resulted in the emergence of a new species. Some biologists hypothesize that such human-mediated speciation may occur often in the Anthropocene.

Thinking Like a Scientist

Scientists have investigated polyploid evolution in several species that originated in the past several million years. What additional insights may be gained by studying *Tragopogon* hybrids?

For the answer, see Appendix D

Island Showcases of Speciation

Volcanic islands, such as the Galápagos and Hawaiian island chains, are initially devoid of life. Over time, colonists arrive via ocean currents or winds. Some of these organisms gain a foothold and establish new populations. In their new environment, these populations may diverge significantly from their distant parent populations. In addition, islands that have physically diverse habitats and that are far enough apart to permit populations to evolve in isolation but close enough to allow occasional dispersals to occur are often the sites of multiple speciation events.

The Galápagos Islands, which were formed by underwater volcanoes from 5 million to 1 million years ago, are one of the world's great showcases of speciation. They are home to numerous plants, snails, reptiles, and birds that are found nowhere else on Earth. For example, the islands have 14 species of closely related finches, which are often called Darwin's finches because he collected them during his around-the-world voyage (see Figure 13.3). These birds share many finch-like traits, but they differ in their feeding habits and their beaks, which are specialized for what they eat (see the Process of Science in Chapter 13). Their various foods include insects, large or small seeds, cactus fruits, and even eggs of other species. The woodpecker finch uses cactus spines or twigs as tools to pry insects from trees. The thin, probing bill of the warbler finch is well-suited for capturing small insects. **The distinctive beaks adapted for the specific diets of the different species of finches are an example of the correlation between structure and function (Figure 14.12).** The finches differ in their habitats as well as their beaks—some live in trees, and others spend most of their time on the ground.

How might Darwin's finch species have evolved from a small population of ancestral birds that colonized one of the islands? Completely isolated on the island, the founder population may have changed significantly as natural selection adapted it to the new environment, and thus it became a new species. Later, a few individuals of this new species may have migrated to a neighboring island, where, under different environmental conditions, this new founder population was changed enough through natural selection to become yet another new species. Some of these birds may then have recolonized the first island and coexisted there with the original ancestral species if reproductive barriers kept the species distinct. Multiple rounds of colonization and speciation on the many separate islands of the Galápagos probably followed. Today, each of the Galápagos Islands has several species of finches, with as many as ten on some islands. Reproductive isolation due to species-specific mating songs helps keep the species separate.

▼ Figure 14.12 **Galápagos finches with beaks adapted for specific diets.**

Cactus-seed-eater (cactus finch)

Tool-using insect-eater (woodpecker finch)

Insect-eater (warbler finch)

Observing Speciation in Progress

In contrast to microevolutionary change, which may be apparent in a population within a few generations, the process of speciation is often extremely slow. So you may be surprised to learn that we *can* see speciation occurring. Consider that life has been evolving over hundreds of millions of years and will continue to evolve. The species living today represent a snapshot, a brief instant in this vast span of time. The environment continues to change—sometimes rapidly due to human impact—and natural selection continues to act on affected populations. It is reasonable to assume that some of these populations are changing in ways that could eventually lead to speciation. Studying populations as they diverge gives biologists a window on the process of speciation.

Researchers have documented at least two dozen cases in which populations are currently diverging as they use different food resources or breed in different habitats. Numerous cases involve insects exploiting different food plants. In one well-studied example, a subpopulation of a fly that feeds on hawthorn fruits found a new resource when American colonists planted apple trees. Although the two fly populations are still regarded as subspecies, researchers have identified mechanisms that severely restrict gene flow between them. In other cases, biologists have identified animal populations that are diverging as a result of differences in male courtship behavior.

Although biologists are continually making observations and devising experiments to study evolution in progress, much of the evidence for evolution comes from the fossil record. So what does the fossil record say about the time frame for speciation—the length of time between when a new species forms and when its populations diverge enough to produce another new species? In one survey of 84 groups of plants and animals, the time for speciation ranged from 4,000 to 40 million years. Such long time frames tell us that it has taken vast spans of time for life on Earth to evolve.

As you've seen, speciation may begin with small differences. However, as speciation occurs again and again, these differences accumulate and may eventually lead to new groups that differ greatly from their ancestors. The cumulative effects of multiple speciations, as well as extinctions, have shaped the dramatic changes documented in the fossil record. We begin to examine such changes next.

Earth History and Macroevolution

Now let's turn our attention to macroevolution. **Macroevolution** is evolutionary change above the species level, for example, the origin of amphibians through a series of speciation events. Macroevolution also includes the impact of mass extinctions on the diversity of life and the origin of key adaptations such as flight. We begin with a look at the vast span of geologic time over which life's diversity has evolved.

The Fossil Record

Fossils are evidence of organisms that lived in the past (see Figure 13.5). The strata (layers) of sedimentary rocks provide a record of life on Earth—each rock layer contains a local sample of the organisms that existed at the time the sediment was deposited. Thus, the fossil record, the sequence in which fossils appear in rock strata, is an archive of macroevolution. For example, scan the wall of the Grand Canyon from rim to floor and you look back through hundreds of millions of years **(Figure 14.13)**. Younger strata formed atop older ones; correspondingly, younger fossils are found in layers closer to the surface, whereas the deepest strata contain the oldest

SEVERAL HUNDRED MILLION YEARS OF HISTORY IS "WRITTEN" IN THE LAYERS OF ROCK IN THE GRAND CANYON.

fossils. However, this only gives us the ages of fossils relative to each other. Like peeling off layers of wallpaper in an old house, we can infer the order in which the layers were applied but not the year that each layer was added.

By studying many sites, geologists have established a **geologic time scale** that divides Earth's history into a sequence of geologic periods. The time line presented in **Table 14.1** (on next page) is separated into four broad divisions: the Precambrian (a general term for the time before about 540 million years ago), followed by the Paleozoic, Mesozoic, and Cenozoic eras. Each of these divisions represents a distinct age in the history of Earth and its life. The boundaries between eras are marked by mass extinctions, when many forms of life disappeared from the fossil record and were replaced by species that diversified from the survivors.

The most common method geologists use to learn the ages of rocks and the fossils they contain is

▼ Figure 14.13 **Strata of sedimentary rock at the Grand Canyon.** The Colorado River has cut through more than a mile of rock, exposing sedimentary strata that are like huge pages from the book of life. Each stratum entombs fossils that represent some of the organisms from that period of Earth's history.

Figure Walkthrough

Mastering Biology
goo.gl/KEmNfa

Table 14.1				The Geologic Time Scale		
Geologic Time	**Period**	**Epoch**	**Age (millions of years ago)**	**Some Important Events in the History of Life**		**Relative Time Span**
Cenozoic era	Quaternary	Holocene	0.01	Historical time		Cenozoic
		Pleistocene	2.6	Ice ages; humans appear		Mesozoic
	Tertiary	Pliocene	5.3	Origin of genus *Homo*		Paleozoic
		Miocene	23	Continued speciation of mammals and angiosperms		
		Oligocene	34	Origins of many primate groups, including apes		
		Eocene	56	Angiosperm dominance increases; origins of most living mammalian orders		
		Paleocene	66	Major speciation of mammals, birds, and pollinating insects		
Mesozoic era	Cretaceous		145	Flowering plants (angiosperms) appear; many groups of organisms, including most dinosaur lineages, become extinct at end of period (Cretaceous extinctions)		
	Jurassic		201	Gymnosperms continue as dominant plants; dinosaurs become dominant		
	Triassic		252	Cone-bearing plants (gymnosperms) dominate landscape; speciation of dinosaurs, early mammals, and birds		
Paleozoic era	Permian		299	Extinction of many marine and terrestrial organisms (Permian extinctions); speciation of reptiles; origins of mammal-like reptiles and most living orders of insects		Pre-cambrian
	Carboniferous		359	Extensive forests of vascular plants; first seed plants; origin of reptiles; amphibians become dominant		
	Devonian		419	Diversification of bony fishes; first amphibians and insects		
	Silurian		444	Early vascular plants dominate land		
	Ordovician		488	Marine algae are abundant; colonization of land by diverse fungi, plants, and animals		
	Cambrian		541	Origin of most living animal phyla (Cambrian explosion)		
Precambrian			600	Diverse algae and soft-bodied invertebrate animals appear		
			635	Oldest animal fossils		
			1,200	Oldest known fossils of multicellular organisms		
			1,800	Oldest fossils of eukaryotic cells		
			2,700	Oxygen begins accumulating in atmosphere		
			3,500	Oldest fossils of cells (prokaryotes)		
			4,600	Approximate time of origin of Earth		

radiometric dating, a method based on the decay of radioactive isotopes (see Figure 2.17). For example, a living organism contains both the common isotope carbon-12 and the radioactive isotope carbon-14 in the same ratio as that present in the atmosphere. Once an organism dies, it stops accumulating carbon, and the stable carbon-12 in its tissues does not change. Carbon-14, however, spontaneously decays to another element. Carbon-14 has a half-life of 5,730 years, so half the carbon-14 in a specimen decays in about 5,730 years, half the remaining carbon-14 decays in the next 5,730 years, and so on. Scientists measure the ratio of carbon-14 to carbon-12 in a fossil to calculate its age.

Carbon-14 is useful for dating relatively young fossils—up to about 75,000 years old. There are isotopes with longer half-lives, such as uranium-235 (half-life 71.3 million years) and potassium-40 (half-life 1.3 billion years). However, organisms don't incorporate those elements into their bodies. Therefore, scientists use indirect methods to date older fossils. One commonly used method is to date layers of volcanic rock or ash above and below the sedimentary layer in which fossils are found. By inference, the age of the fossils is between those two dates. Potassium-argon dating is often used for volcanic rock, for example. Isotopes of uranium are useful for other types of ancient rock.

Plate Tectonics and Biogeography

If photographs of Earth were taken from space every 10,000 years and then spliced together, it would make a remarkable movie. The seemingly "rock solid" continents we live on drift about Earth's surface. According to the theory of **plate tectonics**, the continents and seafloors

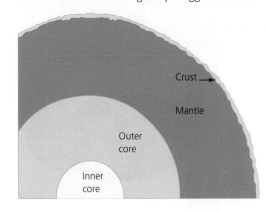

▼ Figure 14.14 **Cutaway view of Earth.** The thickness of the crust is greatly exaggerated here.

form a thin outer layer of solid rock, called the crust, which covers a mass of hot, viscous material called the mantle (**Figure 14.14**). The crust is not one continuous expanse, however. It is divided into giant, irregularly shaped plates that float atop the mantle (**Figure 14.15**). In a process called continental drift, movements in the mantle cause the plates to move. The boundaries of some plates are hotspots of geologic activity. In some cases, we feel an immediate, violent symptom of this activity, as when an earthquake signals that two plates are scraping past or colliding with each other. Although most movement is extremely slow, not much faster than the speed at which fingernails grow, continents have wandered thousands of miles over the long course of Earth's history.

By reshaping the physical features of the planet and altering the environments in which organisms live, continental drift has had a tremendous impact on the evolution of life's diversity. Two chapters in the continuing saga of continental drift had an especially strong influence on life.

Key

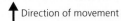 Zones of violent tectonic activity

↑ Direction of movement

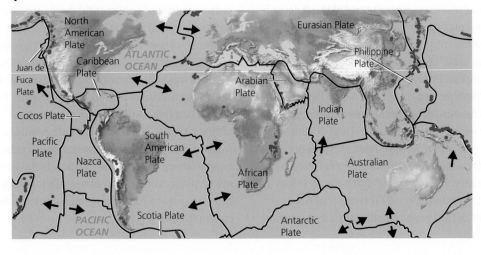

◀ Figure 14.15 **Earth's tectonic plates.** The red dots indicate zones where violent geologic activity, such as earthquakes and volcanic eruptions, takes place. Arrows indicate direction of continental drift.

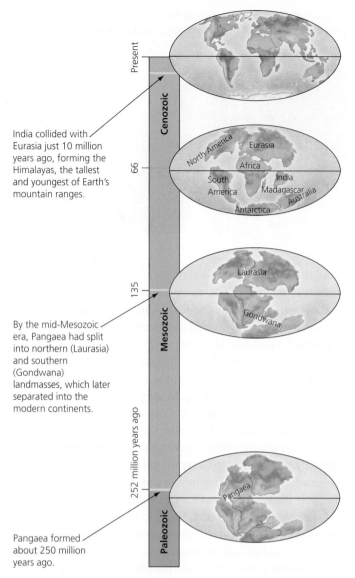

India collided with Eurasia just 10 million years ago, forming the Himalayas, the tallest and youngest of Earth's mountain ranges.

By the mid-Mesozoic era, Pangaea had split into northern (Laurasia) and southern (Gondwana) landmasses, which later separated into the modern continents.

Pangaea formed about 250 million years ago.

▲ Figure 14.16 **The history of plate tectonics.** The continents continue to drift, though not at a rate that's likely to cause any motion sickness for their passengers.

■ CHECKPOINT

If marsupials originated in Asia and reached Australia via South America, where else should paleontologists find fossil marsupials? (*Hint:* Look at Figure 14.16.)

■ *Answer: Antarctica*

dry. Changing ocean currents also undoubtedly affected land life as well as sea life. Thus, the formation of Pangaea had a tremendous impact on the physical environment and climate that reshaped biological diversity. Many species became extinct, and new opportunities arose for the survivors.

The second dramatic chapter in the history of continental drift began during the Mesozoic era. Pangaea started to break up, causing geographic isolation of colossal proportions. As the landmasses drifted apart, each continent became a separate evolutionary arena as its climates changed and its organisms diverged.

The history of continental mergers and separations explains many patterns of **biogeography**, the study of the past and present distribution of organisms. For example, almost all the animals and plants that live on Madagascar, a large island located off the southern coast of Africa, are unique—they diversified from ancestral populations after Madagascar was isolated from Africa and India. The more than 50 species of lemurs that currently inhabit Madagascar, for instance, evolved from a common ancestor over the past 40 million years.

Continental drift also explains the puzzling distribution of marsupials, mammals such as kangaroos, koalas, and wombats whose young complete their embryonic development in a pouch outside the mother's body. Australia and its neighboring islands are home to more than 200 species of marsupials, most of which are found nowhere else in the world **(Figure 14.17)**. The rest of the world is dominated by eutherian (placental) mammals, whose young complete their development in the mother's uterus. Looking at a current map of the world, you might hypothesize that marsupials evolved only on the island continent of Australia. But marsupials are not unique to Australia. More than a hundred species live in Central and South America. North America is also home to a few, including the Virginia opossum. The distribution of marsupials only makes sense in the context of continental drift—marsupials must have originated when the continents were joined. Fossil evidence suggests that marsupials originated in what is now Asia and later dispersed to the tip of South America while it was still connected to Antarctica. They made their way to Australia before continental drift separated Antarctica from Australia, setting "afloat" a great raft of marsupials. The few early eutherians that lived in Australia became extinct, whereas on other continents, most marsupials became extinct. Isolated on Australia, marsupials evolved and diversified, filling ecological roles analogous to those filled by eutherians on other continents. ☑

About 250 million years ago, near the end of the Paleozoic era, plate movements brought all the previously separated landmasses together into a supercontinent called Pangaea, which means "all land" **(Figure 14.16)**. Imagine some of the possible effects on life. Species that had been evolving in isolation came together and competed. As the landmasses joined, the total amount of shoreline was reduced. Ocean basins became deeper, lowering sea level and draining the shallow coastal seas. Then, as now, most marine species inhabited shallow waters, and the formation of Pangaea destroyed a considerable amount of that habitat. It was probably a long, traumatic period for terrestrial life as well. The interior of the vast new continent was cold and

Northern quoll, a carnivore

Sugar glider, an omnivore

Koala, an herbivore

▲ Figure 14.17 Australian
marsupials. The continent
of Australia is home to many
unique plants and animals,
including diverse marsupials,
mammals that have ecological roles filled by eutherians
on other continents.

Mass Extinctions and Explosive Diversifications of Life

The fossil record reveals that five mass extinctions have occurred over the last 540 million years. In each of these events, 50% or more of Earth's species died out. Of all the mass extinctions, those marking the ends of the Permian and Cretaceous periods have been the most intensively studied.

The Permian mass extinction, at about the time the merging continents formed Pangaea, claimed about 96% of marine species and took a tremendous toll on terrestrial life as well. Equally notable is the mass extinction at the end of the Cretaceous period. For 150 million years prior, dinosaurs dominated Earth's land and air, whereas mammals were few and small, resembling today's rodents. Then, about 66 million years ago, most of the dinosaurs became extinct, leaving behind only the descendants of one lineage, the birds. Remarkably, the massive die-off (which also included half of all other species) occurred in less than 10 million years—a brief period in geologic time.

But there is a flip side to the destruction. Each massive dip in species diversity was followed by explosive diversification of certain survivors. Extinctions seem to have provided the surviving organisms with new environmental opportunities. For example, mammals existed for at least 75 million years before undergoing an explosive increase in diversity just after the Cretaceous period (Figure 14.18). Their rise to prominence was undoubtedly associated with the void left by the extinction of the dinosaurs. The world would be a very different place today if many dinosaur lineages had escaped the Cretaceous extinctions or if none of the mammals had survived.

▼ Figure 14.18 **The diversification of mammals after the extinction of dinosaurs.** Widening lines reflect increasing numbers of species (not to scale).

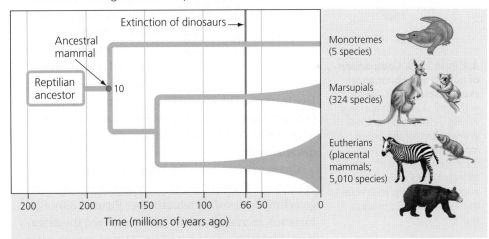

Mechanisms of Macroevolution

The fossil record can tell us what the great events in the history of life have been and when they occurred. Continental drift and mass extinctions followed by the diversification of survivors provide a big-picture view of how those changes came about. But now scientists are increasingly able to explain the basic biological mechanisms that underlie the macroevolutionary changes seen in the fossil record.

Large Effects from Small Genetic Changes

Scientists working at the interface of evolutionary biology and developmental biology—the research field abbreviated evo-devo—are studying how slight changes in the flow of genetic information can become magnified

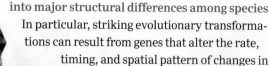

into major structural differences among species. In particular, striking evolutionary transformations can result from genes that alter the rate, timing, and spatial pattern of changes in an organism's form as it develops from a zygote into an adult.

An organism's shape depends in part on the relative growth rates of different body parts during development. For example, in the fetuses of both chimpanzees and humans, the skulls are rounded and the jaws are small, making the face rather flat. As development proceeds, accelerated growth in the jaw produces the elongated skull, sloping forehead, and massive jaws of an adult chimpanzee. In the human evolutionary lineage, mutations slowed the growth of the jaw relative to other parts of the skull. As a result, the skull of an adult human still resembles that of a child—and that of a baby chimpanzee (Figure 14.19).

Our large skull and complex brain are among our most distinctive features. The human brain is proportionately larger than the chimpanzee brain because growth of the organ is switched off much later in human development.

Changes in the rate of developmental events also explain the dramatic differences seen in the homologous limb bones of vertebrates (see Figure 13.6). For instance, increased growth rates produced the extra-long "finger" bones in bat wings. Slower growth rates of leg and pelvic bones led to the eventual loss of hind limbs in whales.

Evolutionary changes can also result from alterations in homeotic genes, the master control genes that determine such basic developmental events as where a pair of wings or legs will appear on a fruit fly (see Figure 11.9). A subtle change in the developmental program can have profound effects. Accordingly, changes in the number, nucleotide sequence, and regulation of homeotic genes have led to the huge diversity in body forms.

Next, we see how the process of evolution can produce new, complex structures.

Chimpanzee fetus

Chimpanzee adult

Human fetus

Human adult
(paedomorphic features)

▲ **Figure 14.19 Comparison of human and chimpanzee skull development.** Starting with fetal skulls that are very similar (left), the differential growth rates of the bones making up the skulls produce adult heads with very different proportions. The grid lines will help you relate the fetal skulls to the adult skulls.

The Evolution of Biological Novelty

The two squirrels in Figure 14.6 are different species, but they are very similar animals that live very much the same way. How do we account for the dramatic differences between dissimilar groups—squirrels and birds, for example? Let's see how the Darwinian theory of gradual change can explain the evolution of intricate structures such as eyes or of new (novel) structures such as feathers.

Adaptation of Old Structures for New Functions

The feathered flight of birds is a perfect marriage of structure and function. Consider the evolution of feathers, which are clearly essential to avian aeronautics. In a flight feather, separate filaments called barbs emerge from a central shaft that runs from base to tip. Each barb is linked to the next by tiny hooks that act much like the teeth of a zipper, forming a tightly connected sheet of barbs that is strong but flexible. In flight, the shapes and arrangements of various feathers produce lift, smooth airflow, and help with steering and balance. How did such a beautifully intricate structure evolve? Reptilian features apparent in fossils of *Archaeopteryx*, one of the earliest birds, offered clues in Darwin's time **(Figure 14.20)**, but the

▼ **Figure 14.20 An extinct bird.** Called *Archaeopteryx* ("ancient wing"), this animal lived near tropical lagoons in central Europe about 150 million years ago. Despite its feathers, *Archaeopteryx* has many features in common with reptiles. *Archaeopteryx* is not considered an ancestor of today's birds. Instead, it probably represents an extinct side branch of the bird lineage.

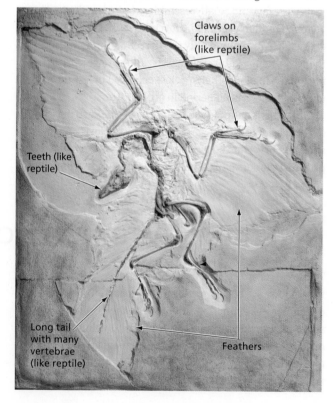

Claws on forelimbs (like reptile)

Teeth (like reptile)

Long tail with many vertebrae (like reptile)

Feathers

definitive answer came in 1996. Birds were not the first feathered animals on Earth—dinosaurs were.

The first feathered dinosaur to be discovered, a 130-million-year-old fossil found in northeastern China, was named *Sinosauropteryx* ("Chinese lizard-wing"). About the size of a turkey, it had short arms and ran on its hind legs, using its long tail for balance. Its unimpressive plumage consisted of a downy covering of hairlike feathers. Since the discovery of *Sinosauropteryx*, thousands of fossils of feathered dinosaurs have been found and classified into more than 30 different species. Although none was unequivocally capable of flying, many of these species had elaborate feathers that would be the envy of any modern bird. But the feathers seen in these fossils could not have been used for flight, nor would their reptilian anatomy have been suited to flying. So if feathers evolved before flight, what was their function? Their first utility may have been for insulation. It is possible that longer, winglike forelimbs and feathers, which increased the surface area of these forelimbs, were co-opted for flight after functioning in some other capacity, such as mating displays, thermoregulation, or camouflage (all functions that feathers also serve today). The first flights may have been only short glides to the ground or from branch to branch in tree-dwelling species. Once flight itself became an advantage, natural selection would have gradually adapted feathers and wings to fit their additional function.

Structures such as feathers that evolve in one context but become co-opted for another function are called **exaptations**. However, exaptation does not mean that a structure evolves in anticipation of future use. Natural selection cannot predict the future; it can only improve an existing structure in the context of its current use. ☑

From Simple to Complex Structures in Gradual Stages

Most complex structures have evolved in small steps from simpler versions having the same basic function—a process of refinement rather than the sudden appearance of complexity. Consider the amazing camera-like eyes of vertebrates and squids. Although these complex eyes evolved independently, the origin of both can be traced from a simple ancestral patch of photoreceptor cells through a series of incremental modifications that benefited their owners at each stage. Indeed, there appears to have been a single evolutionary origin of light-sensitive cells, and all animals with eyes—vertebrates and invertebrates alike—share the same master genes that regulate eye development.

Figure 14.21 illustrates the range of complexity in the structure of eyes among present-day molluscs, a large and diverse phylum of animals. Simple patches of pigmented cells enable limpets, single-shelled molluscs that cling to seaside rocks, to distinguish light from dark. When a shadow falls on them, they hold on more tightly—a behavioral adaptation that reduces the risk of being eaten. Other molluscs have eyecups that have no lenses or other

▼ Figure 14.21 **A range of eye complexity among molluscs.** The complex eye of the squid evolved in small steps. Even the simplest eye was useful to its owner.

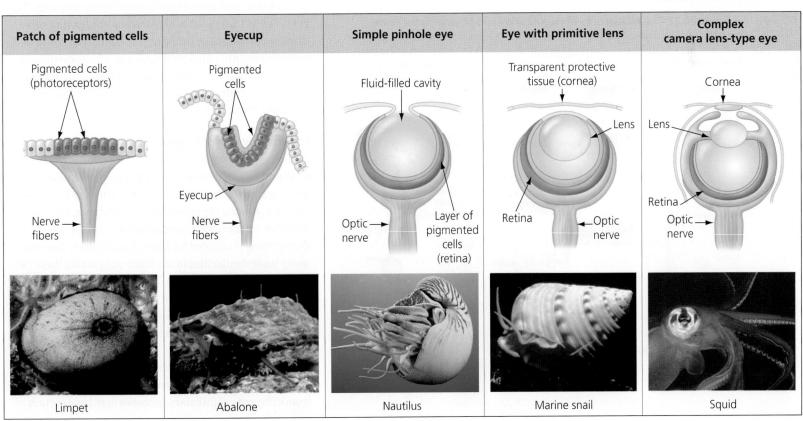

Patch of pigmented cells	Eyecup	Simple pinhole eye	Eye with primitive lens	Complex camera lens-type eye
Pigmented cells (photoreceptors); Nerve fibers	Pigmented cells; Eyecup; Nerve fibers	Fluid-filled cavity; Optic nerve; Layer of pigmented cells (retina)	Transparent protective tissue (cornea); Lens; Retina; Optic nerve	Cornea; Lens; Retina; Optic nerve
Limpet	Abalone	Nautilus	Marine snail	Squid

means of focusing images but can indicate light direction. In those molluscs that do have complex eyes, the organs probably evolved in small steps of adaptation. You can see examples of such small steps in Figure 14.21.

Classifying the Diversity of Life

The Linnaean system of taxonomy (see Figure 13.1) is quite a useful method of organizing life's diversity into groups. Ever since Darwin, however, biologists have had a goal beyond simple organization: to have classification reflect evolutionary relationships. In other words, how an organism is named and classified should reflect its place within the evolutionary tree of life. **Systematics**, which includes taxonomy, is a discipline of biology that focuses on classifying organisms and determining their evolutionary relationships.

Classification and Phylogeny

Biologists use **phylogenetic trees** to depict hypotheses about the evolutionary history, or **phylogeny**, of species. These branching diagrams reflect the hierarchical classification of groups nested within more inclusive groups. The tree in **Figure 14.22** shows the classification of some carnivores and their probable evolutionary relationships. Note

ALLELES FROM TWO WILD GRASSES COULD BE USED TO PROTECT CORN CROPS FROM FUTURE DISASTERS.

that each branch point represents the divergence of two lineages from a common ancestor. (You may recall Figure 13.8, which is a phylogenetic tree of tetrapods.)

Understanding phylogeny can have practical applications. For example, maize (corn) is an important food crop worldwide; it also provides us with snack favorites such as popcorn, tortilla chips, and corn dog batter. Thousands of years of artificial selection (selective breeding) transformed a scrawny grass with small ears of rock-hard kernels into the maize we know today. In the process, much of the plant's original genetic variation was stripped away. By constructing a phylogeny of maize, researchers have identified two species of wild grasses that may be maize's closest living relatives. The genomes of these plants may harbor alleles that offer disease resistance or other useful traits that could be transferred into cultivated maize by crossbreeding or genetic engineering—insurance against future disease outbreaks or other environmental changes that might threaten corn crops.

Identifying Homologous Characters

Homologous structures in different species may vary in form and function but exhibit fundamental similarities because they evolved from the same structure in a common ancestor. Among the vertebrates, for instance, the whale forelimb is adapted for steering in the water, whereas the bat wing is adapted for flight. Nonetheless, there are many basic similarities in the bones supporting these two structures (see Figure 13.6). Thus, homologous structures are one of the best sources of information for phylogenetic relationships. The greater the number of homologous structures between two species, the more closely the species are related.

There are pitfalls in the search for homology: Not all likeness is inherited from a common ancestor. Species from different evolutionary branches may have certain

▼ **Figure 14.22 The relationship of classification and phylogeny for some members of the order Carnivora.** The hierarchical classification is reflected in the finer and finer branching of the phylogenetic tree. Each branch point in the tree represents an ancestor common to species to the right of that branch point.

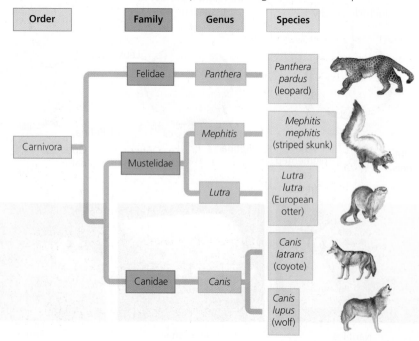

Order	Family	Genus	Species
Carnivora	Felidae	Panthera	Panthera pardus (leopard)
	Mustelidae	Mephitis	Mephitis mephitis (striped skunk)
		Lutra	Lutra lutra (European otter)
	Canidae	Canis	Canis latrans (coyote)
			Canis lupus (wolf)

▲ **Figure 14.23 Artist's reconstruction of Neanderthal.** DNA extracted from Neanderthals, extinct members of the human family, has allowed scientists to study their evolutionary relationship with modern humans.

structures that are superficially similar if natural selection has shaped analogous adaptations. This is called **convergent evolution**. Similarity due to convergence is called **analogy**, not homology. For example, the wings of insects and those of birds are analogous flight equipment: They evolved independently and are built from entirely different structures.

Comparing the embryonic development of two species can often reveal homology that is not apparent in the mature structures (for example, see Figure 13.7). There is another clue to distinguishing homology from analogy: The more complex two similar structures are, the less likely it is they evolved independently. For example, compare the skulls of a human and a chimpanzee (see Figure 14.19). Although each is a fusion of many bones, they match almost perfectly, bone for bone. It is highly improbable that such complex structures matching in so many details could have separate origins. Most likely, the genes required to build these skulls were inherited from a common ancestor.

If homology reflects common ancestry, then comparing the DNA sequences of organisms gets to the heart of their evolutionary relationships. The more recently two species have branched from a common ancestor, the more similar their DNA sequences should be. Scientists have sequenced the genomes of thousands of species. This enormous database has fueled a boom in the study of phylogeny and clarified many evolutionary relationships. In addition, some fossils are preserved in such a way that DNA fragments can be extracted for comparison with living organisms (**Figure 14.23**). ☑

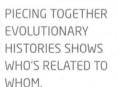

PIECING TOGETHER EVOLUTIONARY HISTORIES SHOWS WHO'S RELATED TO WHOM.

Inferring Phylogeny from Homologous Characters

Once homologous characters—characters that reflect an evolutionary relationship—have been identified for a group of organisms, how are these characters used to construct phylogenies? The most widely used approach is called cladistics. In **cladistics**, organisms are grouped by common ancestry. A **clade** (from the Greek word for "branch") consists of an ancestral species and all its evolutionary descendants—a distinct branch in the tree of life. Thus, identifying clades makes it possible to construct classification schemes that reflect the branching pattern of evolution.

Cladistics is based on the Darwinian concept of "descent with modification from a common ancestor"—species have some characters in common with their ancestors, but they also differ from them. To identify clades, scientists compare an ingroup with an outgroup (**Figure 14.24**). The ingroup (for example, the three mammals in Figure 14.24) is the group of species that is actually being analyzed. The outgroup (in Figure 14.24, the iguana, representing reptiles) is a species or group of species known to have diverged before the lineage that contains the groups being studied. By comparing members of the ingroup with each other and with the outgroup, we can determine what characters distinguish the ingroup from the outgroup. All the mammals in the ingroup have hair and mammary glands. These characters were present in the ancestral mammal, but not in the outgroup. Next, gestation, the carrying of offspring in the uterus within the female parent, is absent from the duck-billed platypus (which lays eggs with a shell). From this absence we might infer that the duck-billed platypus represents an early branch point in the mammalian clade. Proceeding in this manner, we can

☑ **CHECKPOINT**

Our forearms and a bat's wings are derived from the same ancestral prototype; thus, they are _____. In contrast, the wings of a bat and the wings of a bee are derived from totally unrelated structures; thus, they are _____.

■ *Answer: homologous; analogous*

▼ **Figure 14.24 Simplified example of cladistics.**

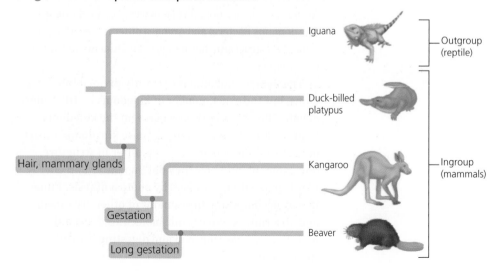

construct a phylogenetic tree. Each branch represents the divergence of two groups from a common ancestor, with the emergence of a lineage possessing one or more new features. The sequence of branching represents the order in which they evolved and when groups last shared a common ancestor. In other words, cladistics focuses on the changes that define the branch points in evolution.

The cladistics approach to phylogeny clarifies evolutionary relationships that were not always apparent in other taxonomic classifications. For instance, biologists traditionally placed birds and reptiles in separate classes of vertebrates (class Aves and class Reptilia, respectively). This classification, however, is inconsistent with cladistics. An inventory of homologies indicates that birds and crocodiles make up one clade, and lizards and snakes form another. If we go back as far as the ancestor that crocodiles share with lizards and snakes to make up a clade, then the class Reptilia must also include birds. The

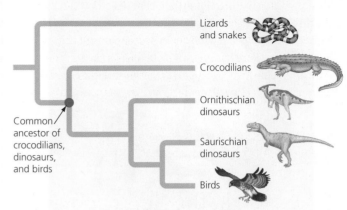

▲ **Figure 14.25 How cladistics is shaking phylogenetic trees.** Strict application of cladistics sometimes produces phylogenetic trees that conflict with classical taxonomy.

tree in **Figure 14.25** is thus more consistent with cladistics than with traditional classifications. ☑

Classification: A Work in Progress

Phylogenetic trees are hypotheses about evolutionary history. Like all hypotheses, they are revised (or in some cases rejected) based on new evidence. Molecular systematics and cladistics are combining to remodel phylogenetic trees and challenge traditional classifications.

Linnaeus divided all known forms of life between the plant and animal kingdoms, and the two-kingdom system prevailed in biology for more than 200 years. In the mid-1900s, the two-kingdom system was replaced by a five-kingdom system that placed all prokaryotes in one kingdom and divided the eukaryotes among four other kingdoms.

In the late 20th century, molecular studies and cladistics led to the development of a **three-domain system (Figure 14.26)**. This current scheme recognizes three basic groups: two domains of prokaryotes—Bacteria and Archaea— and one domain of eukaryotes, called Eukarya. The domains Bacteria and Archaea differ in a number of important structural, biochemical, and functional features (see Chapter 15).

The domain Eukarya is currently divided into kingdoms, but the exact number of kingdoms is still under debate. Biologists generally agree on the kingdoms Plantae, Fungi, and Animalia. These kingdoms consist of multicellular eukaryotes that differ in structure, development, and how they obtain matter and energy. Plants make their own food by photosynthesis. Fungi live by decomposing the remains of other organisms and absorbing small organic molecules. Most animals live by ingesting food and digesting it within their bodies.

The remaining eukaryotes, the protists, include all those eukaryotes that do not fit the definition of plant, fungus, or animal—effectively, a taxonomic grab bag.

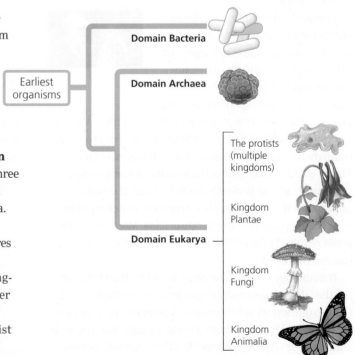

▲ **Figure 14.26 The three-domain classification system.** Molecular and cellular evidence supports the phylogenetic hypothesis that two lineages of prokaryotes, the domains Bacteria and Archaea, diverged very early in the history of life. Molecular evidence also suggests that domain Archaea is more closely related to domain Eukarya than to domain Bacteria.

Most protists are unicellular (amoebas, for example). But the protists also include certain large, multicellular organisms that are believed to be direct descendants of unicellular protists. For example, many biologists classify the seaweeds as protists because they are more closely related to some single-celled algae than they are to true plants.

It is important to understand that classifying Earth's diverse species is a work in progress as we learn more about organisms and their evolution. Charles Darwin envisioned the goals of modern systematics when he wrote in *The Origin of Species*, "Our classifications will come to be, as far as they can be so made, genealogies." ☑

☑ **CHECKPOINT**

What lines of evidence caused biologists to develop the three-domain system of classification?

■ *Answer: molecular studies and cladistics*

EVOLUTION CONNECTION Evolution in the Human-Dominated World

Evolution in the Anthropocene

In this chapter (and in Chapter 13), you've learned how the process of evolution gave rise to the vast diversity of organisms that inhabit Earth. The mechanism that produces adaptive evolutionary change is natural selection, which improves the fit between organisms and their environment. Because the environment may change over time, what constitutes a "good fit" is a moving target. The Anthropocene, as described in the Biology and Society section, is a time of environmental change on an epic scale. Consequently, habitats affected by human impacts are natural laboratories for studying evolutionary adaptation.

As you may know, many populations of pests and pathogens have become resistant to our efforts to kill them with pesticides and antibiotics (see Figures 13.11 and 13.26). On the flip side, other populations have become resistant to toxic pollutants that entered their environment from industrial processes. For example, an adaptation enables fish called tomcods to survive high levels of PCB (an industrial waste product) in the Hudson River.

In urban areas, natural vegetation and wildlife have been replaced by asphalt- and concrete-covered surfaces, buildings, and a high human population density. As a careful observer can verify, however, some organisms thrive in cities. Researchers are learning that urban populations have evolved adaptations to this new environment. For example, the toes of crested anole lizards **(Figure 14.27a)** have specialized scales that enable them to grip the surfaces of tree branches and leaves in their native forest habitat. Urban crested anoles in Puerto Rico have more of these sticky scales, an adaptation to slippery surfaces such as concrete and metal in the urban environment. The physiological stress response of European blackbirds **(Figure 14.27b)** that live in the hectic urban environment is lower than that of reclusive forest-dwelling members of the same species. Urban blackbirds also sing at a higher pitch, an adaptation that enables them to be heard over traffic noise. Scientists have even discovered a new species of mosquito adapted to life underground in the London subway system.

Climate change caused by global warming is the most sweeping environmental consequence of human activities. As a result of rising temperature, the distribution of many species that require cooler temperatures is shifting toward the poles or to higher elevations. These changes may result in zones where the ranges of closely related species overlap and hybridization occurs. In eastern North America, for example, the southern flying squirrel (*Glaucomys volans*) has expanded its range northward and is now hybridizing with the northern flying squirrel (*G. sabrinus*). (You will learn more about climate change in Chapter 18.)

▼ Figure 14.27 **Urban-adapted animals.**

(a) **Crested anole lizard**

(b) **European blackbird**

Chapter Review

SUMMARY OF KEY CONCEPTS

The Origin of Species

What Is a Species?

The diversity of life evolved through speciation, the process in which one species splits into two or more species. According to the biological species concept, a species is a group of populations whose members have the potential to interbreed in nature to produce fertile offspring. The biological species concept is just one of several possible ways to define species.

Reproductive Barriers between Species

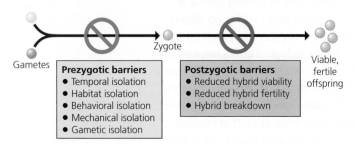

Gametes

Zygote

Viable, fertile offspring

Prezygotic barriers
- Temporal isolation
- Habitat isolation
- Behavioral isolation
- Mechanical isolation
- Gametic isolation

Postzygotic barriers
- Reduced hybrid viability
- Reduced hybrid fertility
- Hybrid breakdown

Mechanisms of Speciation

When the gene pool of a population is separated from other gene pools of the parent species, the splinter population can follow its own evolutionary course.

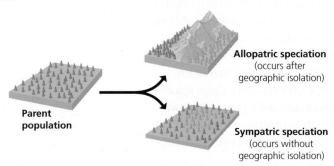

Allopatric speciation
(occurs after geographic isolation)

Parent population

Sympatric speciation
(occurs without geographic isolation)

Hybridization leading to polyploids is a common mechanism of sympatric speciation in plants.

Earth History and Macroevolution

Macroevolution refers to evolutionary change above the species level, for example, the origin of evolutionary novelty and new groups of species and the impact of mass extinctions on the diversity of life and its subsequent recovery.

The Fossil Record

Geologists have established a geologic time scale with four broad divisions: Precambrian, Paleozoic, Mesozoic, and Cenozoic. The most common method for determining the ages of fossils is radiometric dating.

Plate Tectonics and Biogeography

Earth's crust is divided into large, irregularly shaped plates that float atop a viscous mantle. About 250 million years ago, plate movements brought all the landmasses together into the supercontinent Pangaea, causing extinctions and providing new opportunities for the survivors to diversify. About 180 million years ago, Pangaea began to break up, causing geographic isolation. Continental drift accounts for patterns of biogeography, such as the diversity of unique marsupials in Australia.

Mass Extinctions and Explosive Diversifications of Life

The fossil record reveals long, relatively stable periods punctuated by mass extinctions followed in turn by explosive diversification of certain survivors. For example, during the Cretaceous extinctions, about 66 million years ago, the world lost an enormous number of species, including most of the dinosaurs. Mammals greatly increased in diversity after the Cretaceous period.

Mechanisms of Macroevolution

Large Effects from Small Genetic Changes

A subtle change in the genes that control a species' development can have profound effects. For example, the adult may retain body features that were strictly juvenile in ancestral species as a result of a change to the timing of developmental events.

The Evolution of Biological Novelty

An exaptation is a structure that evolves in one context and gradually becomes adapted for other functions. Most complex structures have evolved incrementally from simpler versions having the same function.

Classifying the Diversity of Life

Systematics, which includes taxonomy, focuses on classifying organisms and determining their evolutionary relationships.

Classification and Phylogeny

The goal of classification is to reflect phylogeny, the evolutionary history of species. Classification is based on the fossil record, homologous structures, and comparisons of DNA sequences. Homology (similarity based on shared ancestry) must be distinguished from analogy (similarity based on convergent evolution). Cladistics uses shared characters to group related organisms into clades—distinctive branches in the tree of life.

Classification: A Work in Progress

Biologists currently classify life into a three-domain system: Bacteria, Archaea, and Eukarya.

SELF-QUIZ

1. Distinguish between microevolution, speciation, and macroevolution.

2. Identify each of the following reproductive barriers as prezygotic or postzygotic.
   a. One lilac species lives on acidic soil, another on basic soil.
   b. Mallard and pintail ducks mate at different times of the year.
   c. Two species of leopard frogs have different mating calls.
   d. Hybrid offspring of two species of jimsonweed always die before reproducing.
   e. Pollen of one kind of pine tree cannot fertilize another kind.

3. Why is a small, isolated population more likely to undergo speciation than a large one?

4. Many species of plants and animals adapted to desert conditions probably did not arise there. Their success in living in deserts could be due to _____, structures that originally had one use but became adapted for different functions.

5. Mass extinctions
   a. cut the number of species to the few survivors left today.
   b. resulted mainly from the separation of the continents.
   c. occurred regularly, about every million years.
   d. were followed by diversification of the survivors.

6. The animals and plants of India are almost completely different from the species in nearby Southeast Asia. Why might this be true?
   a. They have become separated by convergent evolution.
   b. The climates of the two regions are completely different.
   c. India is in the process of separating from the rest of Asia.
   d. India was a separate continent until relatively recently.

7. A paleontologist estimates that when a particular rock formed, it contained 12 mg of the radioactive isotope potassium-40. The rock now contains 3 mg of potassium-40. The half-life of potassium-40 is 1.3 billion years. From this information, you can conclude that the rock is approximately _____ billion years old.

8. Why are biologists careful to distinguish similarities due to homology from similarities due to analogy when constructing phylogenetic trees?

9. In the three-domain system, which domain contains eukaryotic organisms?

For answers to the Self-Quiz, see Appendix D.

IDENTIFYING MAJOR THEMES

For each statement below, identify which major theme is evident (the relationship of structure to function, information flow, pathways that transform energy and matter, interactions within biological systems, or evolution) and explain how the statement relates to the theme. If necessary, review the themes (Chapter 1) and review the examples highlighted in blue in this chapter.

10. In the course of giraffe evolution, increased growth rates during development produced extra-long (11-in.) neck vertebrae.

11. The thin membrane of skin stretched between the long "finger" bones of bats is an adaptation for flight.

12. Plants make their own organic matter from inorganic nutrients by photosynthesis.

For answers to Identifying Major Themes, see Appendix D.

THE PROCESS OF SCIENCE

13. *Tragopogon dubius* and *Tragopogon porrifolius* are both common in Europe, but the hybrid species *Tragopogon mirus* has never been found there. Suggest a possible explanation for its absence.

14. **Interpreting Data** The carbon-14/carbon-12 ratio of a fossilized skull is about 6.25% of that of the skull of a present-day animal. Using the graph below, what is the approximate age of the fossil?

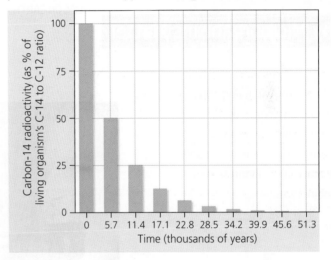

BIOLOGY AND SOCIETY

15. Can cloning be a solution to the alarming extinction rate of certain species? Justify your answer. Are you for or against reviving species that have been extinct for a long time out of scientific curiosity?

16. The red wolf (*Canis rufus*) was once widespread in the southeastern United States but was declared to be extinct in the wild. Biologists bred captive red wolf individuals and reintroduced them into areas of eastern North Carolina, where they are federally protected as endangered species. The current wild population is estimated to be about 50 individuals. However, a new threat to red wolves has arisen: hybridization with coyotes (*Canis latrans*), which have become more numerous in the areas inhabited by red wolves. Although red wolves and coyotes differ in morphology and DNA, they are capable of interbreeding and producing fertile offspring. Social behavior is the main reproductive barrier between the species and is more easily overcome when same-species mates are rare. For this reason, some people think that the endangered status of the red wolf should be withdrawn and resources should not be spent to protect what is not a "pure" species. Do you agree? Why or why not?

15 The Evolution of Microbial Life

CHAPTER CONTENTS

ARE WE THERE YET? IF YOU COUNT EACH MILE OF A ROAD TRIP AS 1 MILLION YEARS, YOU'D DRIVE FROM BANGOR, MAINE, TO MIAMI TO TRAVEL A DISTANCE EQUIVALENT TO THE HISTORY OF LIFE ON EARTH.

Why Microorganisms Matter

Microorganisms were the first life-forms on Earth and thus were the ancestors of all other organisms. Today, microorganisms are immensely important to all living things, including you.

SEAWEEDS AREN'T JUST USED FOR WRAPPING SUSHI— THEY'RE IN YOUR ICE CREAM, TOO.

THIS IS NOT A HOT TUB! SPECIALIZED ARCHAEANS ARE THE ONLY FORM OF LIFE THAT CAN SURVIVE IN THIS WATER.

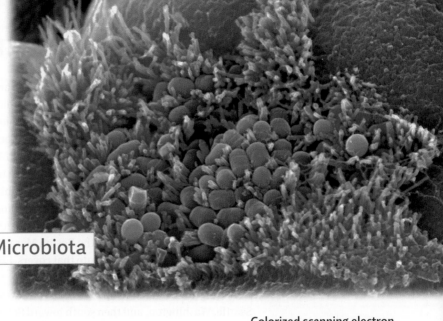

BIOLOGY AND SOCIETY | Human Microbiota

Our Invisible Inhabitants

You probably know that your body contains trillions of individual cells, but did you know that they aren't all "you"? In fact, microorganisms—prokaryotes, protists, and fungi—residing in and on your body are at least as numerous as your own cells. Your skin, mouth, nasal passages, and digestive and urogenital tracts are prime real estate for these microorganisms. Although each individual is so tiny that it would have to be magnified hundreds of times for you to see it, the weight of your microbial residents totals 2 to 5 pounds.

We acquire our microbial communities during the first two years of life, and they remain fairly stable thereafter. However, modern life is taking a toll on that stability. In our efforts to rid ourselves of disease-causing microbes, we don't discriminate between those that are harmful and those that are beneficial. We alter the balance of our microbial communities by taking antibiotics, purifying our water, sterilizing our food, attempting to germproof our surroundings, and scrubbing our skin and teeth. Scientists hypothesize that disrupting our microbial communities may increase our susceptibility to infectious diseases, predispose us to certain cancers, and contribute to conditions such as asthma and other allergies, irritable bowel syndrome, Crohn's disease, and autism. Researchers are even investigating whether having the wrong microbial community could make us fat. In addition, scientists are studying how our microbial communities have evolved over the course of human history. As you'll discover in the Evolution Connection section at the end of this chapter, for example, dietary changes invited decay-causing bacteria to make themselves at home on our teeth.

Throughout this chapter, you will learn about the benefits and drawbacks of human—microbe interactions. You will also sample a bit of the remarkable diversity of prokaryotes and protists. This chapter is the first of three that explore the magnificent diversity of life. And so it is fitting that we begin with the prokaryotes, Earth's first life-form, and the protists, the bridge between unicellular eukaryotes and multicellular plants, fungi, and animals.

Colorized scanning electron micrograph of bacteria (red) nestled among ciliated cells of a human small intestine (7000x).

Major Episodes in the History of Life

To put our survey of diversity into perspective and to understand the enormous span of time over which life has evolved, let's look at a brief overview of major events in the history of life on Earth. Our planet's history began 4.6 billion years ago—a period of time that is difficult to grasp. To visualize this immense scale, imagine taking a road trip across North America in which each mile traveled is the equivalent of passing through 1 million years. Our journey will take us from Kamloops, British Columbia, in Canada, to the finish line of the Boston Marathon in Boston, Massachusetts, on a route that covers 4,600 miles (Figure 15.1).

Setting out from Kamloops, we head southwest to Seattle, Washington, and then south toward San Francisco,

California. By the time we reach the California border, nearly 750 million years have passed, and the first rocks have formed on Earth's cooling surface. Our arrival at the Golden Gate Bridge coincides with the appearance of the first cells in the fossil record. After 1,100 million years, life on Earth has begun! Those earliest organisms were all **prokaryotes**, having cells that lack true nuclei. You'll learn more about the origin of these first cells in the next section.

Conditions on the young Earth were very different from conditions today. One difference that was critical to the origin and evolution of life was the lack of atmospheric O_2.

► **Figure 15.1 Some major episodes in the history of life.** On this 4,600-mile metaphorical road trip, each mile equals 1 million years in Earth's history.

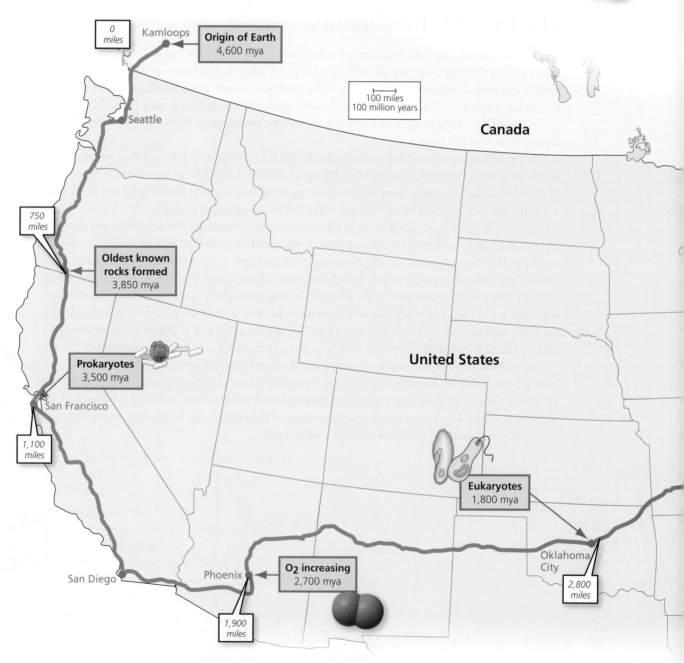

As we continue driving south on our metaphoric journey, diverse metabolic pathways are evolving among the prokaryotes. However, we won't reach our next milestone for another 800 million years. By that time we have passed through San Diego and turned east across the desert. At Phoenix, Arizona, Earth is 2,700 million years old, and atmospheric O_2 has begun to increase as a result of photosynthesis by autotrophic prokaryotes.

Nine hundred million years later, just past Oklahoma City, we find the first fossils of eukaryotic organisms. **Eukaryotes** are composed of one or more cells that contain nuclei and many other membrane-bound organelles absent in prokaryotic cells. Eukaryotic cells evolved from ancestral host cells that engulfed smaller prokaryotes. The mitochondria of our cells and those of every other eukaryote are descendants of those prokaryotes, as are the chloroplasts of plants and algae. Prokaryotes had been on Earth for 1.7 billion years before eukaryotes evolved. The appearance of the more complex eukaryotes, however, launched a period of tremendous diversification of eukaryotic forms. These new organisms were the protists. Protists are mostly microscopic and unicellular, and as you will learn in this chapter, they are represented today by a great diversity of species.

The next great event in the evolution of life was multicellularity. The oldest fossils that are clearly multicellular are 1.2 billion years old and fall on our metaphorical road trip about midway between St. Louis, Missouri, and Terre Haute, Indiana. The organisms that resulted in these fossils were tiny and not at all complex.

It isn't until 600 million years later (roughly 600 million years ago) that large, diverse, multicellular organisms appeared in the fossil record. At this point, we have traveled 4,000 miles on our road trip—or 4 billion years—and now find ourselves in Erie, near the western edge of Pennsylvania (coincidentally, where one of the authors was born). Less than 15% of our journey remains, and we still have not encountered much diversity of life. But that's about to change. A great diversification of animals, the so-called Cambrian explosion, marked the beginning of the Paleozoic era, about 541 million years ago (see Table 14.1). By the end of that period, all the major animal body plans, as well as all the major groups, had evolved.

The colonization of land by plants, fungi, and insects also occurred during the Paleozoic. This evolutionary transition began about 500 million years ago, a time that, in our road trip, corresponds to reaching Buffalo, New York.

By the time we get to Albany, New York, we're in the middle of the Mesozoic era, sometimes called the age of dinosaurs. At the end of the Mesozoic, 66 million years ago, we find ourselves about halfway across the state of Massachusetts. As we draw closer to Boston (and our modern time), more and more familiar organisms begin to dominate the landscape—flowering plants, birds, and mammals, including primates.

Modern humans, *Homo sapiens*, appeared roughly 195,000 years ago. At that point in our cross-country tour, we are less than two blocks shy of the finish line of our journey from the origin of Earth to the present day. This metaphorical road trip demonstrates that spans of time that seem lengthy in the context of our own lives are brief moments in the history of life on Earth.

Colonization of land
500 mya

Mid-Mesozoic
180 mya

Homo sapiens
0.195 mya

Large, complex multicellular organisms
600 mya

Oldest multicellular fossils
1,200 mya

Boston

Albany

4,600 miles

Buffalo

Erie

4,000 miles

Terre Haute

St Louis

3,400 miles

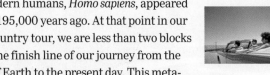

ARE WE
THERE YET?

The Origin of Life

Looking at the abundant diversity of life around us, it is difficult to imagine that Earth was not always lushly populated. But early in its 4.6-billion-year history, our planet had no oceans, no lakes, very little oxygen, and no life at all. It was bombarded by debris left over from the formation of the solar system—gigantic chunks of rock and ice whose impact with Earth generated heat so fierce that it vaporized all available water. Indeed, for the first several hundred million years of its existence, conditions on the young Earth were so harsh that it's doubtful life could have originated; if it did, it could not have survived.

Although conditions had become considerably calmer, the Earth of 4 billion years ago was still in violent turmoil. Water vapor had condensed into oceans on the planet's cooling surface, but volcanic eruptions belched gases such as carbon dioxide, methane, and ammonia and other nitrogen compounds into its atmosphere (Figure 15.2). Although lack of atmospheric O_2 would be deadly to most of Earth's present-day inhabitants, this O_2-free environment made the origin of life possible—O_2 is a corrosive agent that tends to disrupt chemical bonds and would have prevented the formation of complex molecules.

Life is an emergent property that arises from the specific arrangement and interactions of its molecular parts (see Figure 1.18). In this way, life is an example of interactions within systems. To learn how life originated from nonliving substances, biologists draw on research from the fields of chemistry, geology, and physics. In the next sections, we'll look at the attributes that must have arisen during the origin of life. Scientists agree on these properties, but which of several plausible scenarios led to life is still the subject of vigorous debate. ☑

A Four-Stage Hypothesis for the Origin of Life

According to one hypothesis for the origin of life, the first organisms were products of chemical evolution in four stages: (1) the synthesis of small organic molecules, such as amino acids and nucleotide monomers; (2) the joining of these small molecules into macromolecules, including proteins and nucleic acids; (3) the packaging of all these molecules into pre-cells, droplets with membranes that maintained an internal chemistry different from the surroundings; and (4) the origin of self-replicating molecules that eventually made inheritance of genetic material possible. Although we'll never know for sure how life on Earth began, this hypothesis leads to predictions that can be tested in the laboratory. Let's look at some of the observations and experimental evidence for each of these four stages.

Stage 1: Synthesis of Organic Compounds

The chemicals mentioned above that formed Earth's early atmosphere, such as water (H_2O), methane (CH_4), and ammonia (NH_3), are all small, inorganic molecules. In contrast, the structures and functions of life depend on more complex organic molecules, such as sugars, fatty acids, amino acids, and nucleotides, which are composed of the same elements. Could these complex molecules have arisen from the ingredients available?

This stage was the first to be extensively studied in the laboratory. In 1953, Stanley Miller, a graduate student of Nobel laureate Harold Urey, performed a now-classic experiment. He devised an apparatus to simulate conditions thought to prevail on early Earth (Figure 15.3). A flask of warmed water simulated the primordial sea. An "atmosphere"—in the form of gases added to a reaction chamber—contained hydrogen gas, methane, ammonia, and water vapor. To mimic the prevalent lightning of the early Earth, electrical sparks were discharged into the chamber. A condenser cooled the atmosphere, causing water and any dissolved compounds to "rain" into the miniature "sea."

The results of the Miller-Urey experiments were front-page news. After the apparatus had run for a week, an abundance of organic molecules essential for life, including amino acids, the monomers of

▼ Figure 15.2 An artist's rendition of conditions on early Earth.

▼ **Figure 15.3** Apparatus used to simulate early-Earth chemistry in Urey and Miller's experiments.

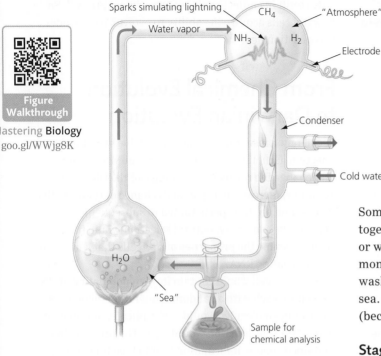

Sparks simulating lightning

CH_4

"Atmosphere"

Water vapor

NH_3 H_2

Electrode

Condenser

Cold water

H_2O

"Sea"

Sample for chemical analysis

organic molecules, including lipids, simple sugars, and nitrogenous bases such as uracil.

Stage 2: Abiotic Synthesis of Polymers

Once small organic molecules were present on Earth, how were they linked together to form polymers such as proteins and nucleic acids without the help of enzymes and other cellular equipment? Researchers have brought about such polymerization in the laboratory by dripping solutions of organic monomers onto hot sand, clay, or rock. The heat vaporizes the water in the solutions and concentrates the monomers on the underlying material. Some of the monomers then spontaneously bond together to form polymers. On early Earth, raindrops or waves may have splashed dilute solutions of organic monomers onto fresh lava or other hot rocks and then washed polypeptides and other polymers back into the sea. There they could accumulate in great quantities (because no life existed to consume them). ☑

Stage 3: Formation of Pre-Cells

The cell membrane forms a boundary that separates a living cell and its functions from the external environment. A key step in the origin of life would have been the isolation of a collection of organic molecules within a membrane. We'll call these molecular aggregates pre-cells—because they are not really cells but instead molecular packages with some of the properties of life. Within a confined space, certain combinations of molecules could be concentrated and interact more efficiently.

Researchers have demonstrated that pre-cells could have formed spontaneously from fatty acids (see Figure 3.11). The presence of a specific type of clay thought to be common on early Earth greatly speeds up the rate of spontaneous pre-cell formation. Unlike the plasma membranes of today's cells, these rudimentary membranes were quite porous, allowing organic monomers such as RNA nucleotides and amino acids to cross freely. However, polymers that formed inside the pre-cells were too large to exit. In addition to physically corralling molecules into a fluid-filled space, these pre-cells may have had certain properties of life. They may have had the ability to use chemical energy and grow. Most intriguingly, pre-cells made in laboratory experiments can divide to produce new pre-cells—a simple form of reproduction.

Stage 4: Origin of Self-Replicating Molecules

Life is defined partly by the process of inheritance, which is based on self-replicating molecules. Today's cells store their genetic information as DNA. They transcribe the

proteins, had collected in the "sea." Many laboratories have since repeated Miller's experiment using various atmospheric mixtures and have also produced organic compounds. In 2008, one of Miller's former graduate students discovered some samples from an experiment that Miller had designed with a different atmosphere, one that would mimic volcanic conditions. Reanalyzing these samples using modern equipment, he identified additional organic compounds that had been synthesized. Indeed, 22 amino acids had been produced under Miller's simulated volcanic conditions, compared with the 11 produced with the atmosphere in his original 1953 experiment.

Scientists are also testing other hypotheses for the origin of organic molecules on Earth. Some researchers are exploring the hypothesis that life may have begun in submerged volcanoes or deep-sea hydrothermal vents, gaps in Earth's crust where hot water and minerals gush into deep oceans. These environments, among the most extreme environments in which life exists today, could have provided the initial chemical resources for life.

Another intriguing hypothesis proposes that meteorites were the source of Earth's first organic molecules. Fragments of a 4.5-billion-year-old meteorite that fell to Earth in Australia in 1969 contain more than 80 types of amino acids, some in large amounts. Recent studies have shown that this meteorite also contains other key

information into RNA and then translate RNA messages
into specific enzymes and other proteins (see Figure
10.9). **This mechanism of information flow probably
emerged gradually through a series of small changes to
much simpler processes.**

What were the first genes like? One hypothesis is
that they were short strands of RNA that replicated
without the assistance of protein enzymes. Laboratory
experiments have shown that short RNA molecules can
assemble spontaneously from nucleotide monomers.
Furthermore, when RNA is added to a solution containing
a supply of RNA monomers, new RNA molecules complementary to parts of the starting RNA sometimes assemble.
So we can imagine a scenario on early Earth like the one
in **Figure 15.4**. RNA monomers adhere to clay particles
and become concentrated. Some monomers spontaneously join, forming the first small "genes." Then an RNA
chain complementary to one of these genes assembles. If
the new chain, in turn, serves as a template for another
round of RNA assembly, the result is a replica of the
original gene.

This replication process could have been aided by the
RNA molecules themselves, acting as catalysts for their
own replication. The discovery that some RNAs, which
scientists call **ribozymes**, can carry out enzyme-like
functions supports this hypothesis. Thus, the "chicken
and egg" paradox of which came first, enzymes or genes,
would be solved. Neither came first—both were RNA
molecules. Scientists use the term "RNA world" for the
period in the evolution of life when RNA may have served
as both simple genes and catalytic molecules. In 2013,
researchers demonstrated the plausibility of this scenario
by constructing a pre-cell that enclosed RNA capable of
replicating itself within the vesicle.

The pre-cells described above could assemble spontaneously and reproduce. These abilities, however, would
only result in endless copies of the original pre-cell. So
how did cells acquire the ability to evolve? We consider
this question next. ☑

From Chemical Evolution to Darwinian Evolution

Recall how present-day organisms evolve by natural
selection. Genetic variation arises by mutations, errors
that change the nucleotide sequence of the DNA. Some
variations increase an organism's chance of reproductive
success and are thus perpetuated in subsequent generations. In the same way, natural selection would have
begun to shape the properties of pre-cells that contained
self-replicating RNA.

Each pre-cell would contain its own population of RNA
molecules, each with a random sequence of monomers.
Some of the molecules would self-replicate, but their success at this reproduction would vary. The pre-cells that
contained genetic information that helped them reproduce more efficiently than others would have increased
in number, passing their abilities on to later generations.
Mutations would have resulted in additional variation
on which natural selection could work, and the most successful of these pre-cells would have continued to evolve.
At some point during millions of years of selection, DNA,
a more stable molecule, replaced RNA as the storehouse
of genetic information, and protocells passed a fuzzy
border to become true cells. The stage was then set for the
evolution of diverse life-forms, changes that we see documented in the fossil record.

▼ **Figure 15.4 A hypothesis for the first genes.**

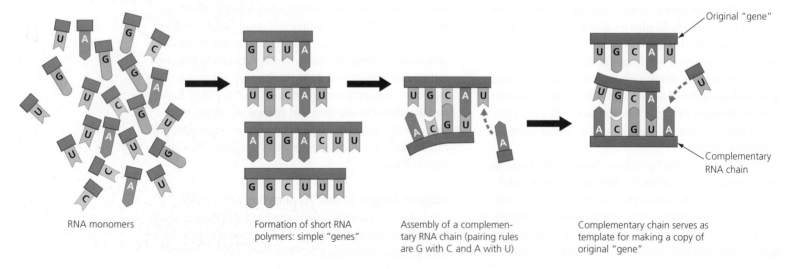

RNA monomers

Formation of short RNA
polymers: simple "genes"

Assembly of a complemen-
tary RNA chain (pairing rules
are G with C and A with U)

Complementary chain serves as
template for making a copy of
original "gene"

Original "gene"

Complementary
RNA chain

Prokaryotes

The history of prokaryotic life is a success story spanning billions of years. Prokaryotes lived and evolved all alone on Earth for about 2 billion years. They have continued to adapt and flourish on a changing Earth and in turn have helped to modify the planet. In this section, you will learn about prokaryotic structure and function, diversity, relationships, and ecological significance.

They're Everywhere!

Today, prokaryotes are found wherever there is life, including in and on the bodies of multicellular organisms. The collective biological mass (biomass) of prokaryotes is at least ten times that of all eukaryotes. Prokaryotes also thrive in habitats too cold, too hot, too salty, too acidic, or too alkaline for any eukaryote **(Figure 15.5)**. Biologists have even discovered prokaryotes living on the walls of a gold mine 3.3 km (2 miles) below Earth's surface. Molecular techniques have just begun to reveal the vast extent of prokaryote diversity. In a powerful new approach, researchers collect samples of prokaryotes from a particular environment, such as soil, water, or the human body, and sequence the DNA they contain. Computer software then assembles the collection of genomes of individual species present in that environment, known as a **microbiome**.

Though individual prokaryotes are small organisms **(Figure 15.6)**, they are giants in their collective impact on Earth and its life. We hear most about the relatively few species that cause illness. Bacterial infections are responsible for about half of all human diseases, including tuberculosis, cholera, many sexually transmitted infections, and certain types of food poisoning. However, prokaryotic life is much more than just a rogues' gallery. The Biology and Society section introduced our **microbiota**, the community of microorganisms that live in and on our bodies. Each of us harbors several hundred different species and genetic strains of prokaryotes, including a few whose positive effects are well studied. For example, some of our intestinal inhabitants supply essential vitamins and enable us to extract nutrition from food molecules that we can't otherwise digest. Many of the bacteria that live on our skin perform helpful housekeeping functions such as decomposing dead skin cells. Prokaryotes also guard the body against disease-causing intruders.

It would be hard to overstate the importance of prokaryotes to the health of the environment. Prokaryotes living in the soil and at the bottom of lakes, rivers, and oceans help to decompose dead organisms and other waste materials, returning vital chemical elements such as nitrogen to the environment. If prokaryotes were to disappear, the chemical cycles that sustain life would come to a halt, and all forms of eukaryotic life would also be doomed. In contrast, prokaryotic life would undoubtedly persist in the absence of eukaryotes, as it once did for 2 billion years on early Earth.

◀ **Figure 15.5** **A window to early life?** A hydrothermal vent 3.1 km (almost 2 miles) below the ocean's surface. Prokaryotes that live near the vent use minerals from the emitted gases as an energy source. This environment, which is very dark, hot, and under high pressure, is among the most extreme in which life is known to exist.

▼ **Figure 15.6** **Bacteria on the point of a pin.** The orange rods are individual bacteria, each about 5 μm long, on the point of a pin. Besides highlighting their tiny size, this micrograph will help you understand why a pin prick can cause infection.

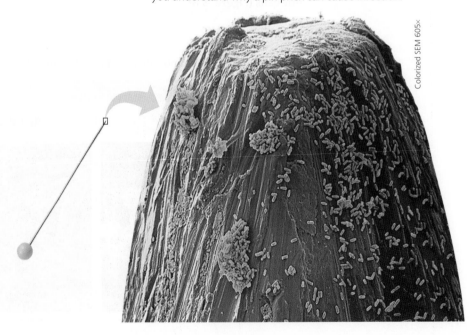

Colorized SEM 605×

The Structure and Function of Prokaryotes

Prokaryotes have a cellular organization fundamentally different from that of eukaryotes. Whereas eukaryotic cells have a membrane-enclosed nucleus and numerous other membrane-enclosed organelles, prokaryotic cells lack these structural features (see Figures 4.3 and 4.4). And nearly all species of prokaryotes have cell walls exterior to their plasma membranes. Despite the simplicity of the cell structures, though, prokaryotes display a remarkable range of diversity. In this section, you'll learn about aspects of their form, reproduction, and nutrition that help these organisms survive in their environments.

Prokaryotic Forms

Determining cell shape by microscopic examination is an important step in identifying a prokaryote. The micrographs in **Figure 15.7** show the three most common shapes. Spherical prokaryotic cells are called **cocci** (singular, *coccus*). Rod-shaped prokaryotes are called **bacilli** (singular, *bacillus*). Spiral-shaped prokaryotes include spirochetes, different species of which cause syphilis and Lyme disease.

Although all prokaryotes are unicellular, the cells of some species usually exist as groups of two or more cells. For example, cocci that occur in clusters are called staphylococci. Other cocci, including the bacterium that causes strep throat, occur in chains; they are called streptococci. Some prokaryotes grow branching chains of cells (**Figure 15.8a**). And some species even exhibit a simple division of labor among specialized types of cells (**Figure 15.8b**). Among unicellular species, moreover, there are some giants that actually dwarf most eukaryotic cells (**Figure 15.8c**).

▼ Figure 15.8 A diversity of prokaryotic shapes and sizes.

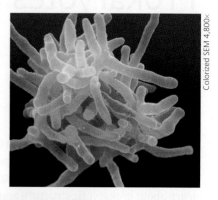

(a) Actinomycete. An actinomycete is a mass of branching chains of rod-shaped cells. These bacteria are common in soil, where they secrete antibiotics that inhibit the growth of other bacteria. Various antibiotic drugs, such as streptomycin, are obtained from actinomycetes.

Colorized SEM 4,800×

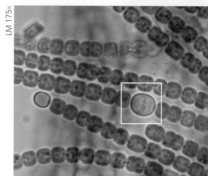

LM 175×

(b) Cyanobacteria. These photosynthetic cyanobacteria exhibit division of labor. The box highlights a cell that converts atmospheric nitrogen to ammonia, which can then be incorporated into amino acids and other organic compounds.

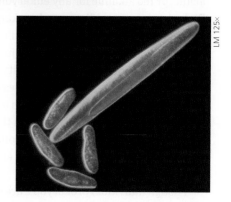

LM 125×

(c) Giant bacterium. *Epulopiscium fishelsoni*, which lives in the gut of a surgeonfish, is large enough to be seen with the unaided eye. It can grow to 10 times the size of a typical eukaryotic cell, such as the protists seen in the photo.

▼ Figure 15.7 Three common shapes of prokaryotic cells.

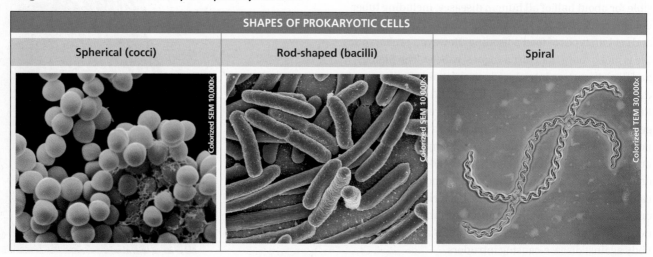

SHAPES OF PROKARYOTIC CELLS		
Spherical (cocci)	Rod-shaped (bacilli)	Spiral

Colorized SEM 10,000× | Colorized SEM 10,000× | Colorized TEM 30,000×

About half of all prokaryotic species are mobile. Many of those that travel have one or more flagella that propel the cells away from unfavorable places or toward more favorable places, such as nutrient-rich locales.

In many natural environments, prokaryotes attach to surfaces in a highly organized colony called a **biofilm**. A biofilm may consist of one or several species of prokaryotes, and it may include protists and fungi as well. As a biofilm becomes larger and more complex, it develops into a "city" of microbes. Communicating by chemical signals, members of the community engage in division of labor, including defense against invaders, among other activities.

Biofilms can form on almost any type of surface, such as rocks, organic material (including living tissue), metal, and plastic. A biofilm known as dental plaque (**Figure 15.9**) can cause tooth decay, a topic you'll learn more about in the Evolution Connection section at the end of this chapter. Biofilms are common among bacteria that cause disease in humans. For instance, ear infections and urinary tract infections are often the result of biofilm-forming bacteria. Biofilms of harmful bacteria can also form on implanted medical devices such as catheters, replacement joints, and pacemakers. The structure of biofilms makes these infections especially difficult to defeat. Antibiotics may not be able to penetrate beyond the outer layer of cells, leaving much of the community intact.

Prokaryotic Reproduction

Many prokaryotes can reproduce at a phenomenal rate if conditions are favorable. The cells copy their DNA almost continuously and divide again and again by the process of **binary fission**. Dividing by binary fission, a single cell becomes 2 cells, which then become 4, 8, 16, and so on. Some species can produce a new generation in only 20 minutes under optimal conditions. If reproduction continued unchecked at this rate, a single prokaryote could give rise to a colony outweighing Earth in only three days! Also, each time DNA is replicated prior to binary fission, spontaneous mutations occur. As a result, rapid reproduction generates a great deal of genetic variation in a prokaryotic population. If the environment changes, an individual that possesses a beneficial gene can quickly take advantage of the new conditions. For example, exposure to antibiotics may select for antibiotic resistance in a bacterial population (see the Evolution Connection section at the end of Chapter 13).

Fortunately for us, few prokaryotic populations can sustain exponential growth for long. Environments are usually limiting in resources such as food and space. Prokaryotes also produce metabolic waste products that may eventually pollute the colony's environment. Still, you can understand why kitchen sponges may harbor large numbers of bacteria (**Figure 15.10**) and why food can spoil so rapidly. Refrigeration retards food spoilage not because the cold kills bacteria but because most microorganisms reproduce very slowly at such low temperatures.

Some prokaryotes can survive during very harsh conditions by forming specialized cells called endospores. An **endospore** is a thick-coated, protective cell produced

▼ Figure 15.10 Household sponge contaminated with bacteria (red, green, yellow, and blue objects in this colorized micrograph).

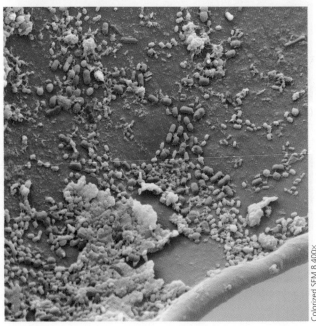

Colorized SEM 8,400×

▼ Figure 15.9 Dental plaque, a biofilm that forms on teeth.

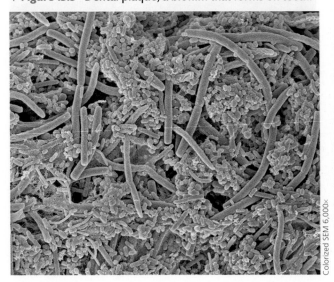

Colorized SEM 6,000×

within the prokaryotic cell when the prokaryote is exposed to unfavorable conditions. The endospore can survive all sorts of trauma and extreme temperatures—not even boiling water kills most of these resistant cells. And when the environment becomes more hospitable, the endospore can absorb water and resume growth. To ensure that all cells, including endospores, are killed when laboratory equipment is sterilized, microbiologists use an appliance called an autoclave, a pressure cooker that applies high-pressure steam at a temperature of 121°C (250°F). The food-canning industry uses similar methods to kill endospores of dangerous bacteria such as *Clostridium botulinum*, the cause of the potentially fatal disease botulism. ☑

Prokaryotic Nutrition

You are probably familiar with the methods by which multicellular organisms obtain energy and carbon, the two main resources needed for synthesizing organic compounds. Plants use carbon dioxide and the sun's energy in the process of photosynthesis; animals and fungi obtain both carbon and energy from organic matter. These modes of nutrition are both common among prokaryotes as well **(Figure 15.11)**. But the prokaryotic pathways that transform energy and matter are far more diverse than those of eukaryotes. Some species harvest energy from inorganic substances such as ammonia (NH_3) and hydrogen sulfide (H_2S). Soil bacteria that obtain energy from inorganic nitrogen compounds are essential to the chemical cycle that makes nitrogen available to plants (see Figure 20.34). Because they don't depend on sunlight for energy, these prokaryotes can thrive in conditions that seem totally inhospitable to life—even between layers of rocks buried hundreds of feet below Earth's surface! Near hydrothermal vents, where scalding water and hot gases surge into the sea more than a mile below the surface, bacteria that use sulfur compounds as a source of energy support diverse animal communities.

The metabolic talents of prokaryotes make them excellent symbiotic partners with animals, plants, and fungi. **Symbiosis** ("living together") is a close association between organisms of two or more species. In some cases of symbiosis, both organisms benefit from the partnership. For example, many of the animals that inhabit hydrothermal vent communities, including the giant tube worm shown in **Figure 15.12**, harbor sulfur bacteria within their bodies. The animals absorb sulfur compounds from the water. The bacteria use the compounds as an energy source to convert CO_2 from seawater into organic molecules that, in turn, provide food for their hosts. Similarly, photosynthesis by cyanobacteria provides food for a fungal partner in the symbiosis that forms some lichens.

▼ **Figure 15.11   Two methods of obtaining energy and carbon.**

LM 65x

(a) *Oscillatoria.* This is a species of cyanobacteria, a type of prokaryote that undergoes photosynthesis.

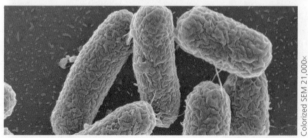

Colorized SEM 21,000x

(b) *Salmonella.* These bacteria, which cause a type of food poisoning, obtain energy and carbon from organic matter—in this case, living human cells.

▼ **Figure 15.12   Giant tube worm.** These animals, which can grow up to 2 m (over 6 ft) long, depend on symbiotic prokaryotes to supply them with food.

▲ **Figure 15.13** *Azolla* **(water fern, inset) floating in rice paddy.** These small plants grow rapidly and cover the water's surface, which helps exclude weeds. Decomposition of the short-lived plants releases nitrogen that fertilizes the rice.

In addition to photosynthesis, many cyanobacteria are also capable of nitrogen fixation, the process of converting atmospheric nitrogen (N_2) into a form usable by plants. Symbiosis with cyanobacteria gives plants such as the water fern *Azolla* an advantage in nitrogen-poor environments. This tiny, floating plant has been used to boost rice production for more than a thousand years (**Figure 15.13**). Other nitrogen-fixing bacteria live symbiotically in root nodules of plants in the legume family, a large group that includes many economically important species such as beans, soybeans, peas, and peanuts.

The Ecological Impact of Prokaryotes

As a result of their metabolic diversity, prokaryotes perform a variety of ecological services that are essential to our well-being. Let's turn our attention now to the vital role that prokaryotes play in sustaining the biosphere.

Prokaryotes and Chemical Recycling

Not too long ago, the atoms making up the organic molecules in your body were part of inorganic compounds found in soil, air, and water, as they will be again one day. **Life depends on the recycling of chemical elements between the biological and physical components of ecosystems, an example of interactions within biological systems.** Prokaryotes play essential roles in these chemical cycles. For example, nearly all the nitrogen that plants use to make proteins and nucleic acids comes from prokaryotic metabolism in the soil. In turn, animals get their nitrogen compounds from plants.

Another vital function of prokaryotes, mentioned earlier in the chapter, is the breakdown of organic wastes and dead organisms. Prokaryotes decompose organic matter and, in the process, return elements to the environment in inorganic forms that can be used by other organisms. If it were not for such decomposers, carbon, nitrogen, and other elements essential to life would become locked in the organic molecules of corpses and waste products. (You'll learn more about the role that prokaryotes play in chemical cycling in Chapter 20.)

Putting Prokaryotes to Work

People have put the metabolically diverse prokaryotes to work in cleaning up the environment. **Bioremediation** is the use of organisms to remove pollutants from water, air, or soil. One example of bioremediation is the use of prokaryotic decomposers to treat our sewage. Raw sewage is first passed through a series of screens and shredders, and solid matter settles out from the liquid waste. This solid matter, called sludge, is then gradually added to a culture of anaerobic prokaryotes, including both bacteria and archaea. The microbes decompose the organic matter in the sludge, converting it to material that can be used as landfill or fertilizer. The liquid wastes may then be passed through a trickling filter system consisting of a long horizontal bar that slowly rotates, spraying liquid wastes onto a bed of rocks (**Figure 15.14**). Aerobic prokaryotes and fungi growing on the rocks remove much of the organic matter from the liquid. The outflow from the rock bed is then sterilized and released back into the environment.

Bioremediation has also become an important tool for cleaning up toxic chemicals released into the soil and water by industrial processes. Naturally occurring prokaryotes capable of degrading pollutants such as oil, solvents, and pesticides are often already present in contaminated soil, but environmental workers may use methods to speed up their activity. An airplane is spraying chemical dispersants on oil from the disastrous

▼ **Figure 15.14** **Putting microbes to work in sewage treatment facilities.** This is a diagram of a trickling filter system, which uses bacteria, archaea, and fungi to treat liquid wastes after sludge is removed.

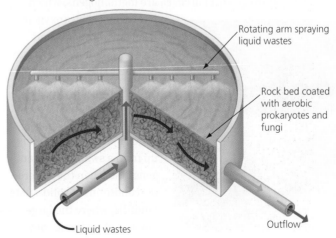

Rotating arm spraying liquid wastes

Rock bed coated with aerobic prokaryotes and fungi

Liquid wastes

Outflow

☑ CHECKPOINT

How do bacteria help
restore the atmospheric
CO_2 required by plants for
photosynthesis?

■ *Answer: By decomposing the
organic molecules of dead organisms and organic refuse such as
leaf litter, bacteria release carbon
from the organic matter in the form
of CO_2.*

Deepwater Horizon spill in the Gulf of Mexico (**Figure 15.15**). Working like detergents that help clean greasy dishes, these chemicals break oil into smaller droplets that offer more surface area for microbial attack.

We are just beginning to explore the great potential that prokaryotes offer for bioremediation. In 2016, research-ers discovered bacteria that can metabolize a plastic used to make products such as beverage bottles and synthetic fibers. These prokaryotes could offer a potential solution to removing the hundreds of millions of tons of plastics currently lingering in the environment. In the future, genetically engineered microbes may be put to work clean-ing up the wide variety of toxic waste products that con-tinue to accumulate in our landscapes and landfills. ☑

▲ **Figure 15.15** Spraying chemical dispersants on an oil spill in the Gulf of Mexico.

The Two Main Branches of Prokaryotic Evolution: Bacteria and Archaea

SPECIALIZED
ARCHAEANS ARE
THE ONLY FORM
OF LIFE THAT CAN
SURVIVE IN THIS
WATER.

By comparing diverse prokaryotes at the molecular level, biologists have identified two major branches of prokary-otic evolution: **bacteria** and **archaea**. Thus, life is organized into three domains—**Bacteria**, **Archaea**, and **Eukarya** (review Figure 14.26). Although bacteria and archaea have prokaryotic cell organization in common, they differ in many structural and physiological characteristics.

Archaea in Extreme Environments

Archaea are abundant in many habitats, including places where few other organisms can survive. One group of archaea, the extreme thermophiles ("heat lovers"), live in very hot water (**Figure 15.16**); some archaea even populate the deep-ocean vents that gush near-boiling water, such as the one shown in Figure 15.5. Another group is the extreme halophiles ("salt lovers"), archaea that thrive in such environments as Utah's Great Salt Lake, the Dead Sea, and seawater-evaporating ponds used to produce salt. A third group of archaea are methanogens, which live in anaerobic (oxygen-free) environments and give off methane as a waste product. They are abundant in the mud at the bottom of lakes and swamps. You may have seen methane, also called marsh gas, bubbling up from a swamp. Methano-gens flourish in the anaero-bic conditions of solid waste landfills, where the large amount of methane they

produce is a significant contributor to global warming (see Figure 18.43). Many municipalities collect this meth-ane and use it as a source of energy (**Figure 15.17**).

Great numbers of methanogens also inhabit the digestive tracts of animals. In humans, intestinal gas is largely the result of their metabolism. More important, methanogens aid digestion in cattle, deer, and other animals that depend heavily on cellulose for their nutrition. Normally, bloating does not occur in these animals because they regularly expel large volumes of gas produced by the methanogens. (And that may be more than you wanted to know about these gas-producing microbes!)

Archaea are also abundant in more moderate condi-tions, especially the oceans, where they can be found at all depths. They are a substantial fraction of the prokaryotes in waters 150 m (nearly 500 ft) below the surface and half of the prokaryotes that live below 1,500 m (0.9 mi).

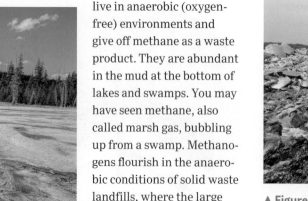

▼ **Figure 15.16** **Heat-loving archaea.** In this photo, you can see yellow and orange colonies of heat-loving prokaryotes growing in the Abyss Pool, West Thumb Geyser, in Yellowstone National Park, Wyoming.

▲ **Figure 15.17** Pipes for collecting gas generated by methanogenic archaea from a landfill.

Archaea are thus one of the most abundant cell types in Earth's largest habitat.

Bacteria That Cause Disease

Although most bacteria are harmless or even beneficial to us, a tiny minority of species cause a disproportionate amount of misery. Bacteria and other organisms that cause disease are called **pathogens**. We're healthy most of the time because our body's defenses check the growth of pathogen populations. Occasionally, the balance shifts in favor of a pathogen, and we become ill. Even some of the bacteria that are normal residents of the human body can make us sick when our defenses have been weakened by poor nutrition or a viral infection.

Most pathogenic bacteria cause disease by producing a poison—either an exotoxin or an endotoxin. **Exotoxins** are proteins that bacterial cells secrete into their environment. For example, *Staphylococcus aureus* (abbreviated *S. aureus*) produces several exotoxins. Although *S. aureus* is commonly found on the skin and in the nasal passages, if it enters the body through a wound, it can cause serious disease. One of its exotoxins causes layers of skin to slough off ("flesh-eating disease"); another can produce a potentially deadly disease called toxic shock syndrome, which has been associated with improper use of tampons. *S. aureus* exotoxins are also a leading cause of food poisoning. If food contaminated with *S. aureus* is not refrigerated, the bacteria reproduce and release exotoxins so potent that ingesting a millionth of a gram causes vomiting and diarrhea. Once the food has been contaminated, even boiling it will not destroy the exotoxins.

Endotoxins are chemical components of the outer membrane of certain bacteria. All endotoxins induce the same general symptoms: fever, aches, and sometimes a dangerous drop in blood pressure (septic shock). Septic shock triggered by an endotoxin of the pathogen that causes bacterial meningitis **(Figure 15.18)** can kill a healthy person in a matter of days, or even hours. Because the bacteria are easily transmitted among people living in close contact, many colleges require students to

be vaccinated against this disease. Other examples of endotoxin-producing bacteria include the species of *Salmonella* that cause food poisoning and typhoid fever.

Sanitation is generally the most effective way to prevent bacterial disease. The installation of water treatment and sewage systems continues to be a public health priority throughout the world. Antibiotics have been discovered that can cure most bacterial diseases. However, resistance to widely used antibiotics has evolved in many of these pathogens.

In addition to sanitation and antibiotics, a third defense against bacterial disease is education. A case in point is Lyme disease, which is caused by a spirochete bacterium carried by ticks. Disease-carrying ticks live on deer and field mice but also bite humans. Lyme disease usually starts as a red rash shaped like a bull's-eye around a tick bite **(Figure 15.19)**. Antibiotics can cure the disease if administered within a month of exposure. If untreated, Lyme disease can cause debilitating arthritis, heart disease, and nervous system disorders. Because there is no vaccine, the best defense against Lyme disease is public education about avoiding tick bites and the importance of seeking treatment if a rash develops.

The potential of some pathogens to cause serious harm has led to their use as biological weapons. One of the greatest threats is from endospores of the bacterium that causes anthrax. When anthrax endospores enter the lungs, they germinate, and the bacteria multiply, producing an exotoxin that eventually accumulates to lethal levels in the blood. Although the bacteria can be killed by certain antibiotics, the antibiotics don't eliminate the toxin already in the body. As a result, inhalation anthrax has a very high death rate. In a terrorist incident in 2001, five people died when anthrax spores were mailed to members of the news media and the U.S. Senate.

Another bacterium considered to have dangerous potential as a weapon is *Clostridium botulinum*. Unlike other biological agents, the weapon form of *C. botulinum* is the exotoxin it produces, botulinum, rather than the living

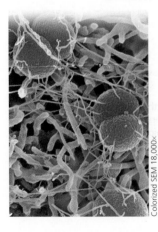

◀ **Figure 15.18 Bacteria that cause meningitis.** *Neisseria meningitidis*, an endotoxin-producing pathogen.

Colorized SEM 18,000×

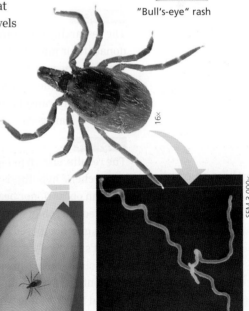

▶ **Figure 15.19 Lyme disease, a bacterial disease transmitted by ticks.** The bacterium that causes Lyme disease (shown in the micrograph at bottom right) is carried from deer to humans by ticks.

"Bull's-eye" rash

16×

Tick that carries the Lyme disease bacterium

SEM 3,000×

Spirochete that causes Lyme disease

microbes. Botulinum, the deadliest poison on Earth, blocks transmission of the nerve signals that cause muscle contraction, resulting in paralysis of the muscles required for breathing. Thirty grams of botulinum, a bit more than an ounce, could kill every person in the United States. On the other hand, the minute amount of botulinum in Botox is used for cosmetic purposes. When the toxin is injected under the skin, it relaxes the facial muscles that cause wrinkles. ☑

☑ **CHECKPOINT**

How can an exotoxin be harmful even after bacteria are killed?

■ *Answer: Exotoxins are secreted poisons that can remain harmful even when the bacteria that secrete them are gone.*

THE PROCESS OF SCIENCE Human Microbiota

Are Intestinal Microbiota to Blame for Obesity?

BACKGROUND

As you learned in the Biology and Society section, our bodies are home to trillions of bacteria that cause no harm or are even beneficial to our health. In the past decade, researchers have made enormous strides in characterizing our microbiota and have begun to investigate the specific effects of these residents on our physiological processes. Because our intestinal microbes are known to be involved in some aspects of food processing, researchers speculate that they might be involved in obesity. Let's examine how a team of scientists investigated the impact of microbiota on body composition—the amount of fat versus lean body mass.

Researchers routinely test hypotheses in animal models before using human subjects. Mice that have been raised in germ-free conditions have no microbiota **(Figure 15.20a)**, making them ideal subjects for this type of experiment. Therefore, rather than manipulate the microbiota of human subjects, the scientists tested the hypothesis that intestinal microbiota of an obese person would increase the amount of body fat in mice.

METHOD

The researchers recruited four pairs of female twins to donate their microbiota for the experiment. In each pair of twins, one was obese and the other was lean. Researchers extracted the microbes from the feces of each individual and then transplanted these microbiota into separate groups of lean, germ-free mice **(Figure 15.20b)**.

RESULTS

The results, shown in **Figure 15.20c**, supported the researchers' hypothesis. Mice that received microbiota from an obese donor became more obese; mice that received microbiota from a lean donor remained lean.

Is a microbe-based cure for obesity just around the corner? It's not likely. The experiment described here—and many similar experiments—represents an early stage of scientific investigation. A great deal more research is needed to determine whether our microbial residents are responsible for obesity. If that proves to be the case, the next challenge will be figuring out how to safely manipulate the complex ecosystem within our bodies.

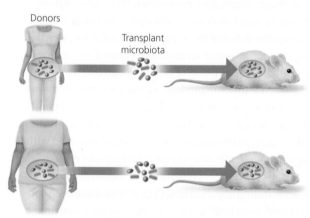

(a) Mice used in the experiment possessed no microbiota.

(b) Microbiota from lean and obese donors were transplanted into germ-free mice.

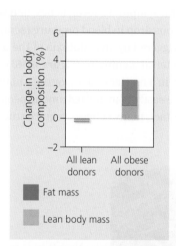

(c) The graph shows the change in body composition (lean versus fat mass) of mice that received microbiota from a lean donor (left) or an obese donor (right).

▲ **Figure 15.20 Experiment to investigate the effect of microbiota on body composition.**
Data from: V. K. Ridaura et al., Gut Microbiota from Twins Discordant for Obesity Modulate Metabolism in Mice. *Science* 341 (2013). DOI: 10.1126/science.1241214.

Thinking Like a Scientist

How would researchers test the hypothesis that the intestinal microbiota of an obese person would increase the amount of body fat in people rather than in mice? What are the drawbacks of such an experiment?

For the answer, see Appendix D.

Protists

The fossil record indicates that the first eukaryotes evolved from prokaryotes around 2 billion years ago. Extensive evidence supports the theory that mitochondria originated when small, oxygen-using prokaryotes began living symbiotically inside larger prokaryotic cells. The symbionts became interdependent and eventually became single organisms with inseparable parts—a eukaryotic cell with a mitochondrion. These primal eukaryotes were not only the predecessors of the great variety of modern protists; they were also ancestral to all other eukaryotes, the plants, fungi, and animals.

The term protists is not a taxonomic category. At one time, protists were classified in a fourth eukaryotic kingdom (kingdom Protista). However, recent genetic and structural studies shattered the notion of protists as a unified group—some protists are more closely related to fungi, plants, or animals than they are to each other. Hypotheses about protist phylogeny (and thus, classification) are changing rapidly as new information causes scientists to revise their ideas. Although there is general agreement about some relationships, others are hotly disputed. Thus, **protist** is a bit of a catch-all category that includes all eukaryotes that are not fungi, animals, or plants. Most, but not all, protists are unicellular. Because their cells are eukaryotic, even the simplest protists are much more complex than any prokaryote.

One mark of protist diversity is the variety of ways they obtain their nutrition. Some protists are autotrophs, producing their food by photosynthesis. Photosynthetic protists belong to an informal category called **algae** (singular, *alga*), which also includes cyanobacteria. Protistan algae may be unicellular, colonial, or multicellular, such as the one shown in **Figure 15.21a**. Other protists are heterotrophs, acquiring their food from other organisms. Some heterotrophic protists eat bacteria or other protists; some are fungus-like and obtain organic molecules by absorption. Still others are parasitic. A **parasite** derives its nutrition from a living host, which is harmed by the interaction. For example, the parasitic trypanosomes shown among human red blood cells in **Figure 15.21b** cause sleeping sickness, a debilitating disease common in parts of Africa. Yet other protists are mixotrophs, capable of both photosynthesis and heterotrophy. *Euglena* (**Figure 15.21c**), a common inhabitant of pond water, can change its mode of nutrition, depending on availability of light and nutrients.

Protist habitats are also diverse. Most protists are aquatic, living in oceans, lakes, and ponds, but they are found almost anywhere there is moisture, including terrestrial habitats such as damp soil and leaf litter. Others are symbionts that reside in the bodies of various host organisms.

Because the classification of protists remains a work in progress, our brief survey of protists is not organized to correspond with any hypothesis about phylogeny. Instead, we'll look at four informal categories of protists: protozoans, slime molds, unicellular and colonial algae, and seaweeds. ☑

▼ **Figure 15.21** **Protist modes of nutrition.**

(a) **An autotroph:** *Caulerpa*, a multicellular alga

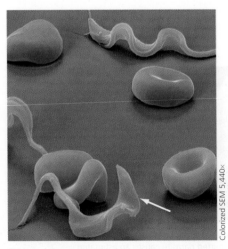

(b) **A heterotroph:** parasitic trypanosome (arrow)

Colorized SEM 5,440×

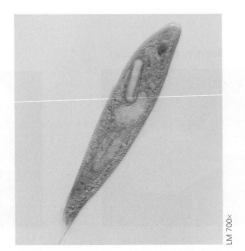

(c) **A mixotroph:** *Euglena*

LM 700×

Protozoans

Protists that live primarily by ingesting food are called **protozoans (Figure 15.22)**. Protozoans thrive in all types of aquatic environments. Most species eat bacteria or other protozoans, but some can absorb nutrients dissolved in the water. Protozoans that live as parasites in animals, though in the minority, cause some of the world's most harmful diseases.

Flagellates are protozoans that move by means of one or more flagella. Most species are free-living (nonparasitic). However, this group also includes some nasty parasites that make people sick. An example is *Giardia*, a common waterborne parasite that causes severe diarrhea. People most often pick up *Giardia* by drinking water contaminated with feces containing the parasite. For example, a swimmer in a lake or river might accidentally ingest water, or a hiker might drink contaminated water from a seemingly pristine stream. (Boiling the water first will kill *Giardia*.) Another flagellate is *Trichomonas*, a common sexually transmitted parasite. The parasite travels through the reproductive tract by moving its flagella and undulating part of its membrane. In women, these protozoans feed on white blood cells and bacteria living on the cells lining the vagina. *Trichomonas* also infects the cells lining the male reproductive tract, but limited availability of food results in very small population sizes. Consequently, males typically have no symptoms of infection.

Other flagellates live symbiotically in a relationship that benefits both partners. Termites are notoriously destructive to wooden structures, but they lack enzymes to digest the tough, complex cellulose molecules that make up wood. Flagellates that reside in the termite's digestive tract break down the cellulose into simpler molecules, sharing the bounty with their hosts.

Amoebas are characterized by great flexibility in their body shape and the absence of permanent organelles for locomotion. Most species move and feed by means of **pseudopodia** (singular, *pseudopodium*), temporary extensions of the cell. Amoebas can assume virtually any shape as they creep over rocks, sticks, or mud at the bottom of a pond or ocean. One species of parasitic amoeba causes amoebic dysentery, a disease that is responsible for an estimated 100,000 deaths worldwide every year. Other protozoans with pseudopodia include the **forams**, which have shells. Although they are single-celled, the largest forams grow to a diameter of several centimeters. Ninety percent of forams that have been identified are fossils. The fossilized shells, which are a component of sedimentary rock such as limestone, are excellent markers for relating the ages of rocks in different parts of the world.

Apicomplexans are all parasitic, and some cause serious human diseases. They are named for a structure at their apex (tip) that is specialized for penetrating host cells and tissues. This group of protozoans includes *Plasmodium*, the parasite that causes malaria. Another apicomplexan is *Toxoplasma*, which requires a feline host to complete its complex life cycle. Cats that eat wild birds or rodents may be infected and shed the parasite in their feces. People can become infected by handling cat litter of outdoor cats, but they don't become sick because the immune system keeps the parasite in check. However, a woman who is newly infected with *Toxoplasma* during pregnancy can pass the parasite to her unborn child, who may suffer damage to the nervous system as a result.

Ciliates are protozoans named for their hairlike structures called cilia, which provide movement of the protist and sweep food into the protist's "mouth." Nearly all ciliates are free-living (nonparasitic) and include both heterotrophs and mixotrophs. If you have had the opportunity to explore protist diversity in a droplet of pond water, you may have seen the common freshwater ciliate *Paramecium*. ☑

☑ CHECKPOINT

What three modes of locomotion occur among protozoans?

■ *Answer: movement using flagella, pseudopodia, and cilia*

▼ **Figure 15.22   A diversity of protozoans.**

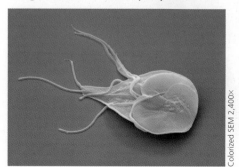

A flagellate: *Giardia*. This flagellated protozoan parasite can colonize and reproduce within the human intestine, causing disease.

Colorized SEM 2,400×

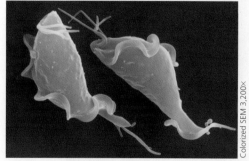

Another flagellate: *Trichomonas*. This common sexually transmitted parasite causes an estimated 5 million new infections each year.

Colorized SEM 3,200×

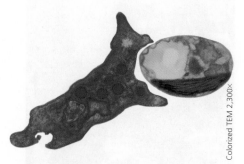

An amoeba. This amoeba is beginning to wrap pseudopodia around an algal cell in preparation for engulfing it.

Colorized TEM 2,300×

Slime Molds

Slime molds are multicellular protists related to amoebas. Although slime molds were once classified as fungi, DNA analysis showed that they arose from different evolutionary lineages. Like many fungi, slime molds feed on dead plant material. Two distinct types of slime molds have been identified.

In plasmodial slime molds, the feeding body is an amoeboid mass called a plasmodium (yellow structure in **Figure 15.23a**) that extends pseudopodia among the leaf litter and other decaying material on a forest floor. The weblike form of the plasmodium enlarges the organism's surface area, increasing its contact with food, water, and oxygen—an example of structure and function. A plasmodium can measure several centimeters across, with its network of fine filaments taking in bacteria and bits of dead organic matter amoeboid-style. Large as it is, though, the plasmodium is actually a single cell with many nuclei. Its mass of cytoplasm is not divided by plasma membranes. When the plasmodium runs out of food, or its environment dries up, the slime mold grows reproductive structures **(Figure 15.23b)**. Stalked bodies that arise from the plasmodium bear tough-coated spores that can survive harsh conditions. If you look closely at a decaying log or the mulch in landscaped areas, you may notice these spore-bearing structures. Like the endospores of anthrax bacteria, slime mold spores can absorb water and grow once favorable conditions return.

Cellular slime molds are also common on rotting logs and decaying organic matter. The feeding stage consists of solitary amoeboid cells that function independently of each other rather than a plasmodium. But when food is in short supply, the amoeboid cells swarm together to form a single unit, a slug-like colony that wanders around for a short time. Some of the cells then dry up

(a) The slime mold's feeding stage (yellow structure) extends pseudopodia over its food.

(b) The slime mold's reproductive structure produces and disperses spores.

▲ **Figure 15.23** **Plasmodial slime mold.**

and form a stalk supporting a reproductive structure in which other cells develop into spores.

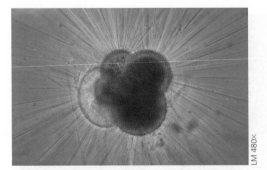

A foram. A foram cell secretes a shell made of organic material hardened with calcium carbonate. The pseudopodia, seen here as thin rays, extend through small pores in the shell.

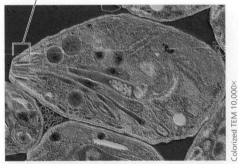

Apical complex

An apicomplexan. *Toxoplasma gondii* invades cells in the host's intestine. The parasite then multiplies and spreads via the bloodstream.

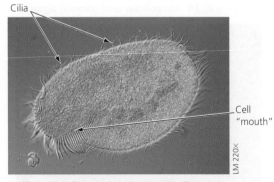

Cilia

Cell "mouth"

A ciliate. The ciliate *Paramecium* uses its cilia to move through pond water. Cilia lining the oral groove keep a current of water containing food moving toward the cell "mouth."

Unicellular and Colonial Algae

Algae are protists and cyanobacteria whose photosynthesis supports food chains in freshwater and marine ecosystems. Researchers are currently trying to harness their ability to convert light energy to chemical energy for another purpose—to produce biofuels. We'll look at four groups of protistan algae. Three of these groups—dinoflagellates, diatoms, and green algae—are unicellular, and one group is colonial.

Many unicellular algae are components of **phytoplankton**, the photosynthetic organisms, mostly microscopic, that drift near the surfaces of ponds, lakes, and oceans. Dinoflagellates are abundant in the vast aquatic pastures of phytoplankton. Each **dinoflagellate** species has a characteristic shape reinforced by external plates made of cellulose **(Figure 15.24a)**. The beating of two flagella in perpendicular grooves produces a spinning movement. Dinoflagellate blooms—population explosions—sometimes cause warm coastal waters to turn pinkish orange, a phenomenon known as a red tide. Toxins produced by some red-tide dinoflagellates have caused massive fish kills and are poisonous to humans as well. One group of dinoflagellates resides within the cells of reef-building corals. Without these algal partners, corals could not build and sustain the massive reefs that provide the food, living space, and shelter that support the splendid diversity of the reef community.

Diatoms have glassy cell walls containing silica, the mineral used to make glass **(Figure 15.24b)**. The cell wall consists of two halves that fit together like the bottom and lid of a shoe box. Diatoms store their food reserves in the form of an oil that provides buoyancy, keeping diatoms floating as plankton near the sunlit surface. The organic remains of diatoms that lived hundreds of millions of years ago are thought to be the main component of oil deposits. But why wait millions of years? Researchers are currently working on methods of growing diatoms and processing the oil into biodiesel.

Green algae are named for their grass-green chloroplasts. Unicellular green algae flourish in most freshwater lakes and ponds, as well as many home pools and aquariums. The green algal group also includes colonial forms, such as *Volvox*, shown in **Figure 15.24c**. Each *Volvox* colony is a hollow ball of flagellated cells (the small green dots in the photo) that are very similar to certain unicellular green algae. The balls within the balls in Figure 15.24c are daughter colonies that will be released when the parent colonies rupture. Of all photosynthetic protists, green algae are the most closely related to plants.

Seaweeds

Defined as large, multicellular marine algae, **seaweeds** grow on rocky shores and just offshore. Their cell walls have slimy and rubbery substances that cushion their bodies against the agitation of the waves. Some seaweeds are as large and complex as many plants. And although the word *seaweed* implies a plantlike appearance, the similarities between these algae and plants are a consequence of convergent evolution. In fact, the closest relatives of seaweeds are certain unicellular algae, which is why many biologists include seaweeds in the same category as protists. Seaweeds are classified into three different groups **(Figure 15.25)**, based partly on the types of pigments present in their chloroplasts: green algae, red algae, and brown algae (some of which are known as kelp).

Many coastal people, particularly in Asia, harvest seaweeds for food. For example, in Japan and Korea, some seaweed species, including brown algae called kombu, are ingredients in soups. Other seaweeds, such as red algae called nori, are used to wrap sushi. Marine algae are rich in iodine and other essential minerals, but much of their organic material consists of unusual polysaccharides

▼ **Figure 15.24 Unicellular and colonial algae.**

Colorized SEM 667×

(a) **Dinoflagellate.** Note the wall of protective plates.

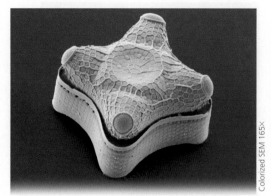

Colorized SEM 165×

(b) **Diatom.** Note the glassy walls.

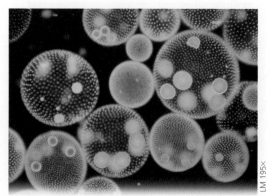

LM 195×

(c) *Volvox.* A colonial green alga

▼ Figure 15.25 The three major groups of seaweeds.

Green algae. This sea lettuce is an edible species that inhabits the intertidal zone, where the land meets the ocean.

Red algae. These seaweeds are most abundant in the warm coastal waters of the tropics.

Brown algae. This group includes the largest seaweeds, known as kelp, which grow as marine "forests."

that humans cannot digest. They are eaten mostly for their rich tastes and unusual textures. The gel-forming substances in the cell walls of seaweeds are widely used as thickeners for such processed foods as puddings, ice cream, and salad dressing. The seaweed extract called agar provides the gel-forming base for the media microbiologists use to culture bacteria in petri dishes. ☑

SEAWEEDS AREN'T JUST USED FOR WRAPPING SUSHI— THEY'RE IN YOUR ICE CREAM, TOO.

EVOLUTION CONNECTION | Human Microbiota

The Sweet Life of *Streptococcus mutans*

Did you ever wonder why eating candy leads to cavities? A biofilm-forming species of bacteria called *Streptococcus mutans* is the culprit. *S. mutans* thrives in the anaerobic environment found in the tiny crevices in tooth enamel. Using sucrose (table sugar) to make a sticky polysaccharide, the bacteria glue themselves in place and build up thick deposits of plaque. Unless you make an effort to remove it, the plaque becomes mineralized, turning into a crusty tartar that must be scraped off your teeth by a dental hygienist **(Figure 15.26)**. Within this fortress, *S. mutans* ferments sugars to obtain energy, releasing lactic acid as a by-product. The acid attacks tooth enamel and eventually eats through it. Other bacteria then use the entrance and infect the soft tissue in the interior of the tooth.

Early humans were hunter-gatherers, living on food foraged in the wild. In a major cultural shift, this lifestyle was replaced by agriculture, which provided a diet rich in carbohydrates from grain. An even later dietary shift brought processed flour and sugar to the table. Each change in diet altered the environment inhabited by our oral microbiota. Studies of prehistoric human remains have correlated dental disease with these changes in diet.

Recent research links *S. mutans* directly to these increases in tooth decay. In one study, researchers analyzed DNA from tartar on the teeth of prehistoric humans who lived in Europe from 7,500 to 400 years ago. The tartar of hunter-gatherers included many species of bacteria but few known to cause tooth decay. *S. mutans* first appeared as one member of a diverse bacterial community after the agricultural diet had been established. Diversity of the oral microbiota dropped dramatically about 400 years ago, around the time sugar was introduced into the diet, and *S. mutans* became the overwhelmingly dominant species. Thus, we can infer that natural selection in the high-sugar environment favored *S. mutans*.

What adaptations gave *S. mutans* an advantage over other species? Another research team investigated genetic changes that allowed *S. mutans* to thrive in a high-sugar environment. Their results turned up more than a dozen genes that improved the ability of *S. mutans* to metabolize sugars and survive increased acidity. They also discovered chemical weapons produced by *S. mutans* that kill harmless bacteria—their competitors for space in the limited terrain of the human oral cavity.

It appears that *S. mutans* took full advantage of an evolutionary opportunity afforded by the human sweet tooth. Having adapted to new environmental conditions and ousted its competitors, it is now firmly entrenched as the dominant member of the mouth's microbial community.

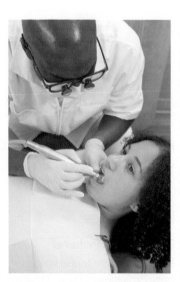

▲ **Figure 15.26 Checking the effects of *Streptococcus mutans*.**

Chapter Review

SUMMARY OF KEY CONCEPTS

Major Episodes in the History of Life

Major Episode	Millions of Years Ago
Plants and fungi colonize land	500
Fossils of large, diverse multicellular organisms	600
Oldest fossils of multicellular organisms	1,200
Oldest eukaryotic fossils	1,800
Beginning of atmospheric accumulation of O_2	2,700
Oldest prokaryotic fossils	3,500
Origin of Earth	4,600

The Origin of Life

A Four-Stage Hypothesis for the Origin of Life

One scenario suggests that the first organisms were products of chemical evolution in four stages:

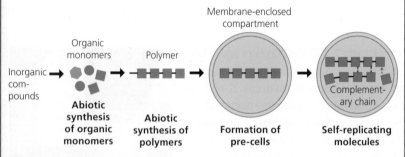

From Chemical Evolution to Darwinian Evolution

Over millions of years, natural selection favored the most efficient pre-cells, which evolved into the first prokaryotic cells.

Prokaryotes

They're Everywhere!

Prokaryotes are found wherever there is life and greatly outnumber eukaryotes. Prokaryotes thrive in habitats where eukaryotes cannot live. A few prokaryotic species cause serious disease, but most are either benign or beneficial to other forms of life.

Prokaryotes

Prokaryotic cells lack nuclei and other membrane-enclosed organelles. Most have cell walls. Prokaryotes exhibit three common shapes.

About half of all prokaryotic species are mobile, most of these using flagella to move. Some prokaryotes can survive extended periods of harsh conditions by forming endospores. Many prokaryotes can reproduce by binary fission at high rates if conditions are favorable, but growth is usually restricted by limited resources.

Prokaryotes include species that obtain energy from the sun (like plants) and from organic material (like animals and fungi). Some species derive energy from inorganic substances such as ammonia (NH_3) or hydrogen sulfide (H_2S). Some prokaryotes live in symbiotic associations with animals, plants, or fungi.

The Ecological Impact of Prokaryotes

Prokaryotes help recycle chemical elements between the biological and physical components of ecosystems. People use prokaryotes to remove pollutants from water, air, and soil in the process called bioremediation.

The Two Main Branches of Prokaryotic Evolution: Bacteria and Archaea

The prokaryotic lineage includes domains Bacteria and Archaea. Many archaea are "extremophiles" capable of surviving under conditions (such as high heat or salt concentrations) that would kill other forms of life; other archaea are found in more moderate environments. Some bacteria cause disease, mostly by producing exotoxins or endotoxins. Sanitation, antibiotics, and education are the best defenses against bacterial disease.

Protists

Protists are unicellular eukaryotes and their closest multicellular relatives.

Protozoans

Protozoans, including flagellates, amoebas, apicomplexans, and ciliates, primarily live in aquatic environments and ingest their food.

Slime Molds

Slime molds (including plasmodial slime molds and cellular slime molds) resemble fungi in appearance and lifestyle as decomposers but are not at all closely related.

Unicellular and Colonial Algae

Unicellular algae, including dinoflagellates, diatoms, and unicellular green algae, are photosynthetic protists that support food chains in freshwater and marine ecosystems.

Seaweeds

Seaweeds—which include green, red, and brown algae—are large, multicellular marine algae that grow on and near rocky shores.

Mastering Biology

For practice quizzes, BioFlix animations, MP3 tutorials, video tutors, and more study tools designed for this textbook, go to Mastering Biology™

SELF-QUIZ

1. Place these events in the history of life on Earth in order.
   a. accumulation of O_2 in Earth's atmosphere
   b. colonization of land by plants and fungi
   c. diversification of animals (Cambrian explosion)
   d. origin of eukaryotes
   e. origin of humans
   f. origin of multicellular organisms
   g. origin of prokaryotes

2. Why is the formation of membranes of fundamental importance in the evolution of cells?

3. What are some of the extreme environments that could have provided the organic chemicals crucial for the evolution of life?

4. Humans have symbiotic relationships with prokaryotes. *E. coli*, the main constituent of the gut flora, provides its host with vitamin K. What benefit does *E. coli* derive?

5. The bacteria that cause tetanus can be killed only by prolonged heating at temperatures considerably above boiling. What does this suggest about tetanus bacteria?

6. Name the three types of archaea found in extreme environments and identify the one that is a source of energy.

7. Besides algae, what other group of organisms can undergo photosynthesis like plants?

8. Which of the following can exist in both unicellular and multicellular forms?
   a. dinoflagellates
   b. *Volvox*
   c. *Paramecium*
   d. diatoms

9. Which protozoan group consists solely of parasitic forms?
   a. apicomplexans
   b. flagellates
   c. ciliates
   d. amoebas

For answers to the Self-Quiz, see Appendix D.

IDENTIFYING MAJOR THEMES

For each statement below, identify which major theme is evident (the relationship of structure to function, information flow, pathways that transform energy and matter, interactions within biological systems, or evolution) and explain how the statement relates to the theme. If necessary, review the themes (Chapter 1) and review the examples highlighted in blue in this chapter.

10. DNA in cells is transcribed into RNA and then translated to produce specific enzymes and other proteins.

11. Prokaryotic metabolic processes are far more diverse than those of eukaryotes.

12. The weblike form of the plasmodium enlarges the organism's surface area, increasing its contact with food, water, and oxygen.

For answers to the Identifying Major Themes, see Appendix D.

THE PROCESS OF SCIENCE

13. Imagine you are on a team designing a moon base that will be self-contained and self-sustaining. Once supplied with building materials, equipment, and organisms from Earth, the base will be expected to function indefinitely. One of the members of your team has suggested that everything sent to the base be chemically treated or irradiated so that no bacteria of any kind are present. Do you think this is a good idea? Predict some of the consequences of eliminating all bacteria from an environment.

14. **Interpreting Data** Because bacteria divide by binary fission, their population size doubles every generation. Suppose the generation time for bacteria that cause food poisoning (for example, *Staphylococcus aureus* or *Salmonella*) is 30 minutes at room temperature. The number of cells in a population can be calculated using this formula.

Initial number of cells $\times 2^{\text{(Number of generations)}}$ = Population size

For example, if a dish of potato salad is contaminated with 10 bacteria, the bacterial population after 1 hour (two generations) is $10 \times 2^2 = 40$. Fill in the table below to show how the bacterial population increases when a dish of potato salad is left on the kitchen counter after dinner.

Time (Hours)	Number of Generations	Number of Bacteria
0	0	10
1	2	40
2	4	
4	8	
6	12	
8	16	
10	20	
12	24	

Why does the rate of increase change over time? Describe how a graph of the data would look.

BIOLOGY AND SOCIETY

15. Allergies are caused by an excessive reaction of the immune system to normally harmless substances, such as certain foods, dust mites, and pollen. Surveys indicate that allergies are on the rise, with about one third of the population affected at some point in their lives. The rise of allergies can be noticed in all the countries that are undergoing industrial development. Propose an explanation for this.

16. Antibiotics are widely used to prevent infections in livestock, and studies have revealed that many physicians readily prescribe antibiotics to patients suffering from common cold or may even prescribe them in a preventative fashion. What are the possible negative consequences of the excessive usage of antibiotics?

16 The Evolution of Plants and Fungi

I'M EATING WHAT?!
IF YOU'VE EVER HAD
MUSHROOM PIZZA,
YOU'VE SNACKED
ON THE FUNGI'S
REPRODUCTIVE
STRUCTURE.

Why Plants and Fungi Matter

Whether you're a vegetarian or a meat-lover, most of the food you eat comes either directly or indirectly from the bounty provided by plants. The evolutionary history of plants stretches back hundreds of millions of years to the first tiny plants that gained a tentative toehold on land. Fungi made all of this plant life possible (and some are quite delicious, too).

HOW MANY DIFFERENT
FRUITS DO YOU SEE
IN THIS PHOTO?
ANSWER: ALL BUT THE
CARROTS AND GARLIC
ARE FRUITS.

WE'RE SURROUNDED
BY ANCIENT ATOMS.
BURNING COAL
RELEASES CO_2
CAPTURED BY PLANTS
THAT LIVED MORE
THAN 300 MILLION
YEARS AGO.

BIOLOGY AND SOCIETY Plant-Fungus Interactions

The Diamond of the Kitchen

Black truffles ready to be thinly sliced or grated for some lucky diner. Black truffles, which can be cultivated, are much less expensive than the white variety, which is found only in a small region of Italy.

To the untrained eye, a truffle (the fungus, not the chocolate) is an unappealing lump that you might not eat on a dare. But despite their unappetizing appearance, truffles are treasured by gourmets as "diamonds of the kitchen." White truffles, the rarest variety, can cost hundreds of dollars per ounce for the best specimens. What's the attraction? Rather than their taste, truffles are prized for their powerful scent, which has been described as earthy or moldy but is nonetheless enticing in small amounts. Chefs only need to add a few shavings from a truffle to infuse a dish with its delectable essence.

What is this weird-looking food? Truffles are actually the subterranean reproductive bodies of certain fungi. The job of these reproductive bodies is to produce spores, single cells that are capable of growing into a new fungus, just as a seed can grow into a new plant. It is generally advantageous for seeds or spores to start their lives in a new place, away from competition with their parents. And that's where the potent truffle smell comes in. Attracted by the heady aroma, certain animals excavate and eat the fungi and later deposit the hardy spores in their feces. Human noses are not sensitive enough to locate the buried treasures, so truffle hunters use pigs or trained dogs to sniff out their quarry.

Besides being culinary gems, truffles represent a mutually beneficial relationship between plants and fungi. The roots of most plants are surrounded—in some cases even permeated—by a finely woven web of fungal filaments. Truffles, for example, have such a relationship with certain species of trees, which is why skilled truffle hunters look for their treasures beneath oak and hazelnut trees. This association is an example of symbiosis, an interaction in which one species lives in or on another species. In this case, both species benefit from the association. The ultrathin fungal filaments of fungi reach into pockets between the soil grains that are too small for roots to enter, absorbing water and inorganic nutrients and passing them to the plant. The plant returns the favor by supplying the fungus with sugars and other organic molecules. This relationship between fungus and root is evident in some of the oldest plant fossils, suggesting that the mutually beneficial symbiosis was crucial to the colonization of land.

Colonizing Land

What exactly is a plant? A **plant** is a multicellular eukaryote that carries out photosynthesis and has a set of adaptations for living on land. This definition distinguishes plants from animals and fungi, which are also eukaryotic, multicellular organisms but are not photosynthetic. Large algae, including seaweeds, are photosynthetic as well as being eukaryotic and multicellular. However, algae lack terrestrial adaptations and thus are classified as protists rather than plants (see Figure 15.25). Some plants, such as water lilies, do live in the water, but these aquatic plants evolved from terrestrial ancestors (just as several species of aquatic mammals, such as whales, evolved from terrestrial mammals).

Terrestrial Adaptations of Plants

Why does life on land require a special set of adaptations? Consider how algae fare after being washed up on the beach. Bodies that were upright in the buoyant water go limp on land and soon shrivel in the drying air. In addition, algae are not equipped to obtain the carbon dioxide needed for photosynthesis from the air. Clearly, living on land poses different problems than living in water. In this section, we discuss some of the terrestrial adaptations that distinguished plants from algae and that allowed plants to colonize land. As you will see in later sections, it took more than 100 million years to complete the transition. The earliest land plants lacked some of the adaptations that made subsequent groups more successful in the terrestrial environment.

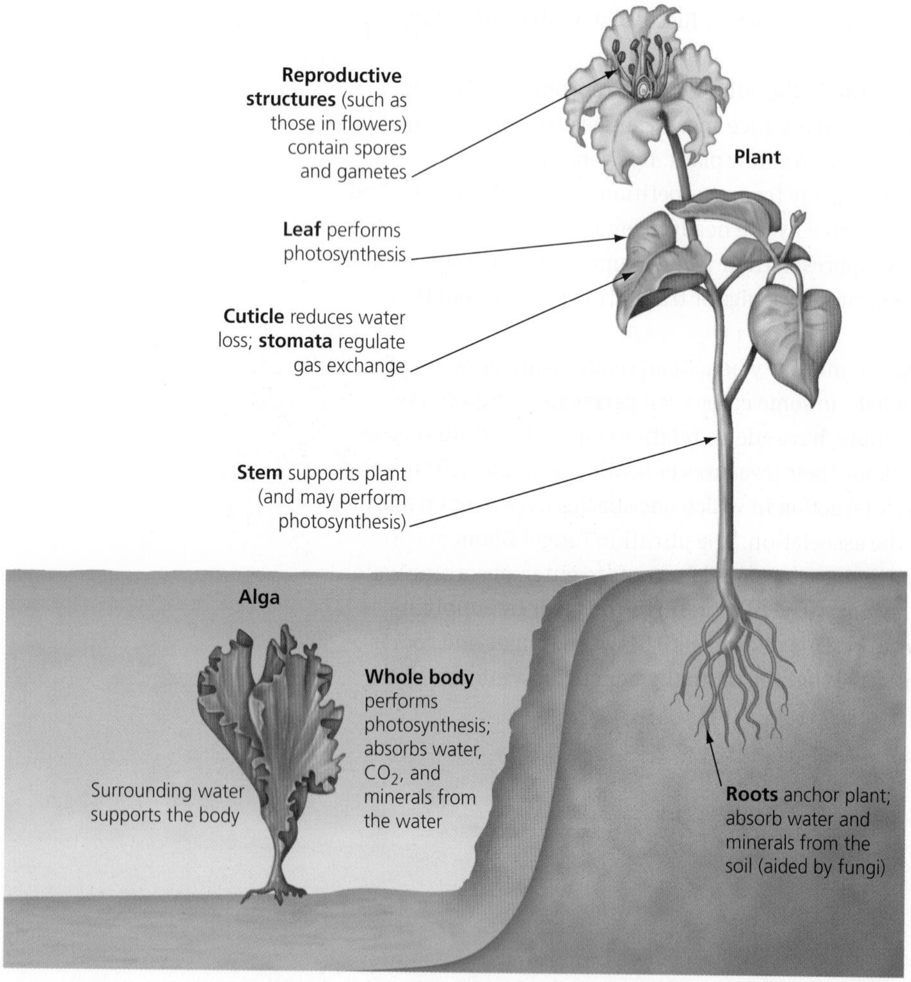

Adaptations of the Plant Body

Let's compare the terrestrial environment of plants with the aquatic environment of algae. For algae, carbon dioxide and minerals are available by diffusion from the surrounding water (Figure 16.1). But resources on land are found in two very different places: Carbon dioxide is mainly available in the air, whereas mineral nutrients and water are found mainly in the soil. Thus, the complex bodies of plants have organs specialized in different ways to function in these two environments. Subterranean organs called **roots** anchor the plant in soil and absorb minerals and water from the soil. Aboveground, **shoots** are organ systems that consist of photosynthetic leaves supported by stems.

Roots typically have many fine branches that thread among the grains of soil, providing a large surface area that maximizes contact with mineral-bearing water in the soil—just one example of how plant organ systems exemplify the relationship between structure and function. In addition, as you learned in the Biology and Society section at the beginning of this chapter, most plants have symbiotic fungi associated with their roots. These root-fungus combinations, called **mycorrhizae** ("fungus roots"), enlarge the root's functional surface area (Figure 16.2). For their part, the fungi absorb water and essential minerals from the soil and provide these materials to the plant. In turn, the sugars produced by the plant nourish the fungi. These mycorrhizal relationships are key adaptations that made it possible for plants to live on land.

Shoots also show structural adaptations to the terrestrial environment. Leaves are the main photosynthetic organs of most plants. Exchange of carbon dioxide (CO_2) and oxygen (O_2) between the atmosphere and the photosynthetic interior of a leaf occurs via **stomata** (singular, *stoma*), the microscopic pores found on a leaf's surface (see Figure 7.2). A waxy layer called the **cuticle** coats the

▼ Figure 16.1 Structural adaptations of algae and plants.

Reproductive **structures** (such as those in flowers) contain spores and gametes

Leaf performs photosynthesis

Cuticle reduces water loss; **stomata** regulate gas exchange

Stem supports plant (and may perform photosynthesis)

Plant

Alga

Surrounding water supports the body

Whole body performs photosynthesis; absorbs water, CO_2, and minerals from the water

Roots anchor plant; absorb water and minerals from the soil (aided by fungi)

▼ Figure 16.2 Mycorrhizae: symbiotic associations of fungi and roots. The finely branched filaments of the fungus (white in the photo) provide an extensive surface area for absorption of water and minerals from the soil.

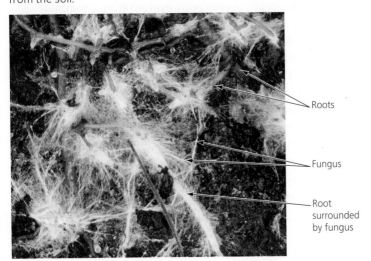

Roots

Fungus

Root surrounded by fungus

▼ Figure 16.3 Network of vascular tissue in a leaf. The vascular tissue is visible as yellow veins on the underside of the leaf in the photograph.

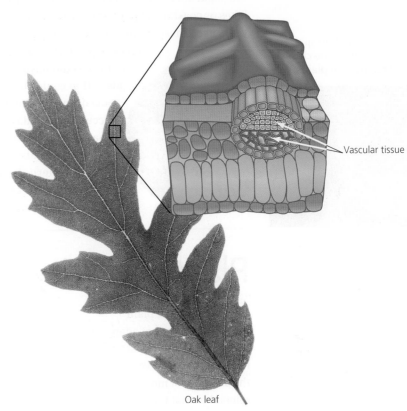

Vascular tissue

Oak leaf

leaves and other aerial parts of most plants, helping the plant body retain water (see Figure 18.8b).

To transport vital materials between roots and shoots, most plants have **vascular tissue**, a network of tube-shaped cells that branch throughout the plant **(Figure 16.3)**. There are two types of vascular tissue. One type is specialized for transporting water and minerals from roots to leaves, and the other distributes sugars from the leaves to the roots and other nonphotosynthetic parts of the plant.

Vascular tissue also solved the problem of structural support on land. The cell walls of many of the cells in vascular tissue are hardened by a chemical called **lignin**. The structural strength of lignified vascular tissue, otherwise known as wood, is amply demonstrated by its use as a building material.

Reproductive Adaptations

Adapting to land also required a new mode of reproduction. For the protist algae, the surrounding water ensures that gametes (sperm and eggs) and developing offspring stay moist. The aquatic environment also provides a means of dispersing the gametes and offspring. Plants, however, must keep their gametes and developing offspring from drying out in the air. Plants produce their gametes in a structure that allows gametes to develop without dehydrating. In addition, the egg remains within tissues of the mother plant and is fertilized there. In plants, but not algae, the zygote (fertilized egg) develops into an embryo while still contained within the female parent, which protects the embryo and keeps it from dehydrating **(Figure 16.4)**. Further adaptations in some plant groups allow sperm to travel via air and improve the survival of offspring during dispersal. ✓

▼ Figure 16.4 The protected embryo of a plant. Internal fertilization, with sperm and egg combining within a moist chamber on the female plant, is an adaptation for living on land. The female plant continues to nourish and protect the plant embryo, which develops from the zygote.

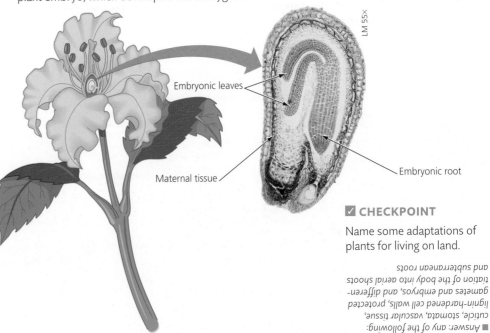

LM 55×

Embryonic leaves

Maternal tissue

Embryonic root

✓ CHECKPOINT

Name some adaptations of plants for living on land.

■ Answer: any of the following: cuticle, stomata, vascular tissue, lignin-hardened cell walls, protected gametes and embryos, and differentiation of the body into aerial shoots and subterranean roots

The Origin of Plants from Green Algae

The algal ancestors of plants carpeted moist fringes of lakes or coastal salt marshes more than 500 million years ago. These shallow-water habitats were subject to occasional drying, and natural selection would have favored algae that could survive periodic droughts. Some species accumulated adaptations that enabled them to live permanently above the water line. A modern-day lineage of green algae, the **charophytes** (Figure 16.5), may resemble

▼ Figure 16.5 **Two species of charophytes, the closest algal relatives of plants.**

LM 150×

one of these early plant ancestors. Plants and present-day charophytes probably evolved from a common ancestor.

Adaptations making life on dry land possible had accumulated by about 470 million years ago, the age of the oldest known plant fossils. The evolutionary novelties of these first land plants opened the new frontier of a terrestrial habitat. Early plant life would have thrived in the new environment. Bright sunlight was abundant on land, the atmosphere had a wealth of carbon dioxide, and at first there were relatively few pathogens and plant-eating animals. The stage was set for an explosive diversification of plant life.

Plant Diversity

The history of the plant kingdom is a story of adaptation to diverse terrestrial habitats. As we survey the diversity of modern plants, remember that the evolutionary past is the key to the present.

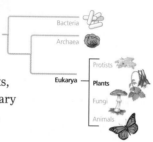

▼ Figure 16.6 **Highlights of plant evolution.** This phylogenetic tree highlights the evolution of structures that allowed plants to move onto land; these structures still exist in modern plants. As we survey the diversity of plants, miniature versions of this tree will help you place each plant group in its evolutionary context.

Highlights of Plant Evolution

The fossil record chronicles four major periods of plant evolution, which are also evident in the diversity of modern plants (Figure 16.6). Each stage is marked by the evolution of structures that opened new opportunities on land.

1 After plants originated from an algal ancestor approximately 470 million years ago, early diversification gave rise to nonvascular plants, including mosses, liverworts, and hornworts. These plants, called **bryophytes**, lack true roots and leaves. Bryophytes also lack lignin, the wall-hardening material that enables other plants to stand tall. Without lignified cell walls, bryophytes have weak upright support. The most familiar bryophytes are **mosses**. A mat of moss actually consists of many plants growing in a tight pack, holding one another up. Structures that protect the gametes and embryos are a terrestrial adaptation that originated in bryophytes.

2 The second period of plant evolution, which began about 425 million years ago, was the diversification of

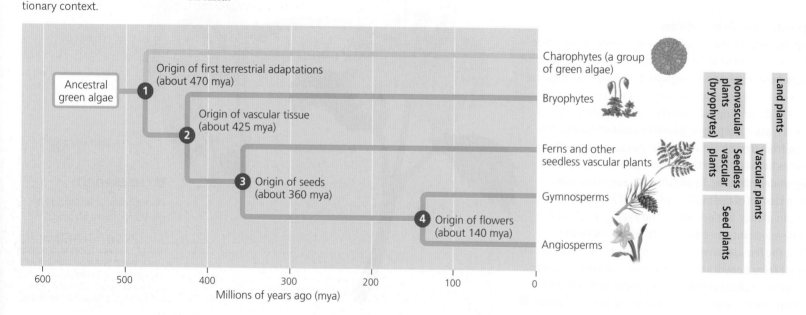

▼ **Figure 16.7** The major groups of plants.

PLANT DIVERSITY			
Bryophytes (nonvascular plants)	**Ferns** (seedless vascular plants)	**Gymnosperms** (naked-seed plants)	**Angiosperms** (flowering plants)

plants with vascular tissue. The presence of conducting tissues hardened with lignin allowed vascular plants to grow much taller, rising above the ground to achieve significant height. The earliest vascular plants lacked seeds. Today, this seedless condition is retained by **ferns** and a few other groups of vascular plants.

❸ The third major period of plant evolution began with the origin of the seed about 360 million years ago. Seeds advanced the colonization of land by further protecting plant embryos from drying and other hazards. A **seed** consists of an embryo packaged along with a store of food within a protective covering. The seeds of early seed plants were not enclosed in any specialized chambers. These plants gave rise to the **gymnosperms** ("naked seeds"). Today, the most widespread and diverse gymnosperms are the **conifers**, consisting mainly of cone-bearing trees, such as pines.

❹ The fourth major episode in the evolutionary history of plants was the emergence of flowering plants, or **angiosperms** ("contained seeds"), at least 140 million years ago. The **flower** is a complex reproductive structure that bears seeds within protective chambers called ovaries. This contrasts with the naked seeds of gymnosperms. The great majority of living plants—some 250,000 species—are angiosperms, including all our fruit and vegetable crops, grains and other grasses, and most trees.

With these highlights as our framework, we are now ready to survey the four major groups of modern plants: bryophytes, ferns, gymnosperms, and angiosperms **(Figure 16.7)**. ☑

Bryophytes

Mosses, which are bryophytes, may sprawl as low mats over acres of land **(Figure 16.8)**. Mosses display two of the key terrestrial adaptations that made the move onto land possible: (1) a waxy cuticle that helps prevent dehydration and (2) the retention of developing embryos within the female plant. However, mosses are not totally liberated from their ancestral aquatic habitat. Mosses need water to reproduce because their sperm need to swim to reach eggs located within the female plant. (A film of rainwater or dew is enough moisture for the sperm to travel.) In addition, because most mosses have no vascular tissue

☑ **CHECKPOINT**

Name the four major groups of plants. Name an example of each.

■ Answer: bryophytes (mosses), seedless vascular plants (ferns), gymnosperms (conifers), angiosperms (plants that produce fruits and vegetables)

▼ **Figure 16.8 A peat moss bog in Scotland.** Mosses are bryophytes, which are nonvascular plants. Sphagnum mosses, collectively called peat moss, carpet at least 3% of Earth's land surface. They are most commonly found in high northern latitudes. The ability of peat moss to absorb and retain water makes it an excellent addition to garden soil.

353

to carry water from soil to aerial parts of the plant, they need to live in damp places.

If you examine a mat of moss closely, you can see two distinct forms of the plant. The green, spongelike plant that is the more obvious is called the **gametophyte**. Careful examination will reveal the other form of the moss, called a **sporophyte**, growing out of a gametophyte as a stalk with a capsule at its tip **(Figure 16.9)**. The cells of the gametophyte are haploid—they have one set of chromosomes (see Figure 8.13). In contrast, the sporophyte is made up of diploid cells (with two chromosome sets).

These two different stages of the plant life cycle are named for the types of reproductive cells they produce. Gametophytes produce gametes (sperm and eggs), and sporophytes produce spores. A **spore** is a haploid cell that can develop into a new individual without fusing with another cell (two gametes must fuse to form a zygote). Spores usually have tough coats that enable them to survive in harsh environments. Seedless plants, including mosses and ferns, disperse their offspring as spores rather than as multicellular seeds.

This type of life cycle, in which the gametophyte and sporophyte take turns producing each other, is called **alternation of generations**. Gametophytes produce gametes that unite to form zygotes, which develop into new sporophytes. And sporophytes produce spores that give rise to new gametophytes. Thus, genetic information flows through alternating generations **(Figure 16.10)**. Among plants, mosses and other bryophytes are unique in having the gametophyte as the larger, more obvious plant. As we continue our survey of plants, we'll see an increasing dominance of the sporophyte as the more highly developed generation. ☑

▼ **Figure 16.9 The two forms of a moss.** The feathery plant we generally know as a moss is the gametophyte. The stalk with the capsule at its tip is the sporophyte.

Spore capsule

Sporophytes

Gametophytes

Figure Walkthrough

Mastering **Biology**
goo.gl/JeHa2v

▶ **Figure 16.10 Alternation of generations.** Plants have life cycles very different from ours. Each of us is a diploid individual; the only haploid stages in the human life cycle, as for nearly all animals, are sperm and eggs. By contrast, plants have alternating generations: Diploid (2*n*) individuals (sporophytes) and haploid (*n*) individuals (gametophytes) generate each other in the life cycle.

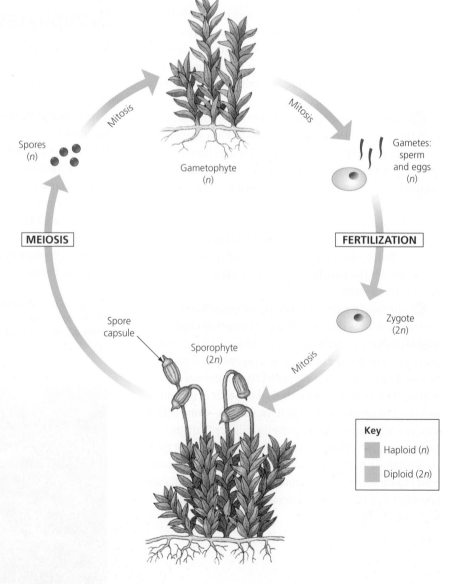

Spores
(*n*)

Gametophyte
(*n*)

Gametes: sperm and eggs
(*n*)

MEIOSIS

FERTILIZATION

Spore capsule

Sporophyte
(2*n*)

Zygote
(2*n*)

Mitosis

Key

Haploid (*n*)

Diploid (2*n*)

☑ **CHECKPOINT**

Bryophytes, like all plants, have a life cycle that involves an alternation of generations. What are the two generations called? Which generation dominates in bryophytes?

■ *Answer: gametophyte, sporophyte; gametophyte*

Ferns

The evolution of vascular tissue allowed ferns to colonize a greater variety of habitats than mosses. Ferns are by far the most diverse seedless vascular plants, with more than 12,000 known species. However, the sperm of ferns, like those of mosses, have flagella and must swim through a film of water to fertilize eggs. Most ferns inhabit the tropics, although many species are found in temperate forests, such as many woodlands in the United States (Figure 16.11).

During the Carboniferous period, from about 360 to 300 million years ago, ancient ferns were part of a much greater diversity of seedless plants that formed vast, swampy tropical forests over much of what is now Eurasia and North America (Figure 16.12). As the plants died, they fell into stagnant wetlands and did not decay completely. Their remains formed thick organic deposits. Later, seawater flooded the swamps, marine sediments covered the organic deposits, and pressure and heat gradually converted them to coal. Coal is black sedimentary rock made up of fossilized plant material. Like coal, oil and natural gas also formed from the remains of long-dead organisms; thus, all three are known as **fossil fuels**. Since the Industrial Revolution, coal has been a crucial source of energy for people. However, burning these fossil fuels releases CO_2 and other gases that contribute to global climate change (see Figure 18.46). ☑

BURNING COAL RELEASES CO_2 CAPTURED BY PLANTS THAT LIVED MORE THAN 300 MILLION YEARS AGO.

▼ Figure 16.11 **Ferns (seedless vascular plants).** The ferns in the foreground are growing on the forest floor in Redwood National Park, California. The fern generation familiar to us is the sporophyte generation. You would have to crawl on the forest floor and explore with careful hands and sharp eyes to find fern gametophytes (upper right), tiny plants growing on or just below the soil surface.

New sporophyte

Gametophyte

Cluster of spore capsules

"Fiddlehead" (young leaf ready to unfurl)

◀ Figure 16.12 **A "coal forest" of the Carboniferous period.** This painting, based on fossil evidence, reconstructs one of the great seedless forests. Most of the large trees belong to ancient groups of seedless vascular plants that are represented by just a few present-day species. The plants near the base of the trees are ferns.

☑ CHECKPOINT

Why are ferns able to grow taller than mosses?

Answer: Vascular tissue hardened with lignin allows ferns to stand taller and transport nutrients farther. ▪

Gymnosperms

"Coal forests" dominated the North American and Eurasian landscapes until near the end of the Carboniferous period. At that time, the global climate turned drier and colder, and the vast swamps began to disappear. This climatic change provided an opportunity for seed plants, which can complete their life cycles on dry land and withstand long, harsh winters. Of the earliest seed plants, the most successful were the gymnosperms, and several kinds grew along with the seedless plants in the Carboniferous swamps. Their descendants include the conifers, or cone-bearing plants.

Conifers

Perhaps you have had the fun of hiking or skiing through a forest of conifers, the most common gymnosperms. Pines, firs, spruces, junipers, cedars, and redwoods are all conifers. A broad band of coniferous forests covers much of northern Eurasia and North America and extends southward in mountainous regions (Figure 16.13). The 6,600-hectare Øvre Pasvik National Park in Norway is the largest undisturbed coniferous forest in northernmost Europe.

Conifers are among the tallest, largest, and oldest organisms on Earth. Coastal redwoods, native to the northern California coast in the United States, are the world's tallest trees—up to 110 m, the height of a 33-story building.

Giant sequoias, relatives of redwoods that grow in the Sierra Nevada mountains of California, are massive. One, known as the General Sherman tree, is about 84 m (275 feet) high and outweighs the combined weight of a dozen space shuttles. Bristlecone pines, another species of California conifer, are among the oldest organisms alive. A recently discovered specimen is more than 5,000 years old; it was a seedling when people invented writing.

Nearly all conifers are evergreens, meaning they retain leaves throughout the year. Even during winter, they perform a limited amount of photosynthesis on sunny days. And when spring comes, conifers already have fully developed leaves that can take advantage of the sunnier days. The needle-shaped leaves of pines and firs are also adapted to survive dry seasons. A thick cuticle covers the leaf, and the stomata are located in pits, further reducing water loss.

Coniferous forests are highly productive; you probably use products harvested from them every day. For example, conifers provide much of our timber for building and wood pulp for paper production. What we call wood is actually an accumulation of vascular tissue with lignin, which gives the tree structural support.

Terrestrial Adaptations of Seed Plants

Compared with ferns, most gymnosperms have three additional adaptations that make survival in diverse terrestrial habitats possible: (1) further reduction of the gametophyte, (2) pollen, and (3) seeds.

The first adaptation is an even greater development of the diploid sporophyte compared with the haploid gametophyte generation (Figure 16.14). A pine tree or other

▼ **Figure 16.13 A coniferous forest in Sweden, northern Europe.** Coniferous forests are widespread in northern Eurasia and North America; conifers also grow in the Southern Hemisphere, though they are less numerous there.

▶ **Figure 16.14 Three variations on alternation of generations in plants.**

Key
- Haploid (*n*)
- Diploid (*2n*)

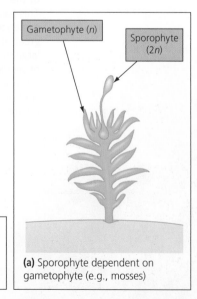

(a) Sporophyte dependent on gametophyte (e.g., mosses)

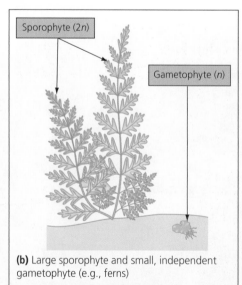

(b) Large sporophyte and small, independent gametophyte (e.g., ferns)

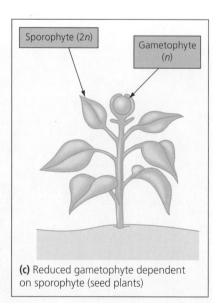

(c) Reduced gametophyte dependent on sporophyte (seed plants)

conifer is a sporophyte with tiny gametophytes living in its cones (Figure 16.15). In contrast to what is seen in bryophytes and ferns, gymnosperm gametophytes are totally dependent on and protected by the tissues of the parent sporophyte.

A second adaptation of seed plants to dry land came with the evolution of pollen. A **pollen grain** is actually the much-reduced male gametophyte; it houses cells that will develop into sperm. In the case of conifers, **pollination**, the delivery of pollen from the male parts of a plant to the female parts of a plant, occurs via wind. This mechanism for sperm transfer contrasts with the swimming sperm of mosses and ferns. In seed plants, the use of tough, airborne pollen that carries sperm-producing cells to the egg is a terrestrial adaptation that led to even greater success and diversity of plants on land.

The third important terrestrial adaptation of seed plants is the seed itself. A seed consists of a plant embryo packaged along with a food supply within a protective coat. Seeds develop from **ovules**, structures that contain the female gametophytes (Figure 16.16). In conifers, the ovules are located on the scales of female cones. Once released from the parent plant, the seed can remain

dormant for days, months, or even years. Under favorable conditions, the seed can then **germinate**, or sprout: Its embryo emerges through the seed coat as a seedling. Some seeds drop close to their parents, and others are carried far by the wind or animals. ☑

☑ **CHECKPOINT**

Contrast the modes of sperm delivery in ferns versus conifers.

■ *Answer: The flagellated sperm of ferns must swim through water to reach eggs. In contrast, the airborne pollen of conifers delivers sperm-producing cells to eggs in ovules without the need to go through water.*

▼ **Figure 16.16** **From ovule to seed.**

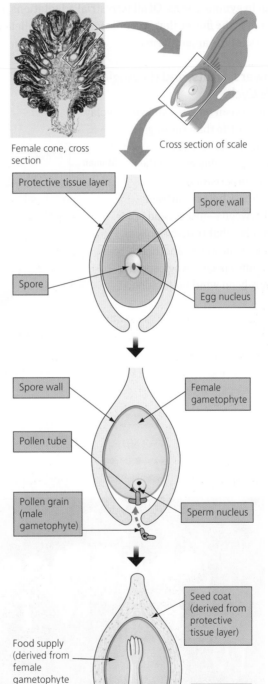

Female cone, cross section

Cross section of scale

Key

Haploid (*n*)

Diploid (2*n*)

Protective tissue layer

Spore wall

Spore

Egg nucleus

(a) Ovule. The sporophyte produces spores within a tissue surrounded by a protective tissue layer. The spore develops into a female gametophyte, which produces an egg nucleus.

Spore wall

Female gametophyte

Pollen tube

Pollen grain (male gametophyte)

Sperm nucleus

(b) Fertilized ovule. After pollination, the pollen grain grows a tiny tube that enters the ovule, where it releases a sperm nucleus that fertilizes the egg.

Seed coat (derived from protective tissue layer)

Food supply (derived from female gametophyte tissue)

Multicellular embryo (new sporophyte)

(c) Seed. Fertilization triggers the transformation of ovule to seed. The fertilized egg (zygote) develops into a multicellular embryo; the rest of the gametophyte forms a tissue that stockpiles food; and the protective tissue layer of the ovule hardens to become the seed coat.

▼ **Figure 16.15** **A pine tree, the sporophyte, bearing two types of cones containing gametophytes.** The leaf-like structures, or scales, of the female cone bears a structure called an ovule containing a female gametophyte. Male cones release clouds of millions of pollen grains, the male gametophytes. Some of these pollen grains land on female cones on trees of the same species. The sperm can fertilize eggs in the ovules of the female cones. The ovules eventually develop into seeds.

Scale

Ovule-producing cones; the scales contain female gametophytes

Pollen-producing cones; they produce male gametophytes

Ponderosa pine

Angiosperms

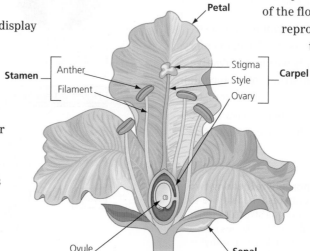

Angiosperms dominate the modern landscape. About 250,000 angiosperm species have been identified, compared to about 700 species of gymnosperms. Several unique adaptations account for the success of angiosperms. For example, refinements in vascular tissue make water transport even more efficient in angiosperms than in gymnosperms. Of all terrestrial adaptations, however, it is the flower that accounts for the unparalleled success of the angiosperms.

Flowers, Fruits, and the Angiosperm Life Cycle

No organism makes a showier display of its sex life than the angiosperm. From roses to dandelions, flowers are the site of procreation. This showiness helps to attract go-betweens—insects and other animals—that transfer pollen from one flower to another of the same species. Angiosperms that rely on wind pollination, including grasses and many trees, have much smaller, less flamboyant flowers. In those species, the plant's reproductive energy is allocated to making massive amounts of pollen for release into the wind.

A flower is a short stem bearing modified leaves that are attached in concentric circles at its base **(Figure 16.17)**. The outer layer consists of the **sepals**, which are usually green. They enclose the flower before it opens (think of the green "wrapping" on a rosebud). When the sepals are peeled away, the next layer is the **petals**, which are often colorful—these are the showy structures that attract pollinators. Plucking off the petals reveals the **stamens**, the male reproductive structures. Pollen grains develop in the **anther**, a sac at the top of each stamen. At the center of the flower is the **carpel**, the female reproductive structure. It includes the **ovary**, a protective chamber containing one or more ovules, in which eggs develop. The sticky tip of the carpel, called the **stigma**, traps pollen. As you can see in **Figure 16.18**, the basic structure of a flower can exist in many beautiful variations.

◀ **Figure 16.17** Structure of a flower.

▼ **Figure 16.18 A diversity of flowers.**

Prickly pear cactus

Bleeding heart

California poppy

Summer snapdragon

▼ **Figure 16.19** The angiosperm life cycle.

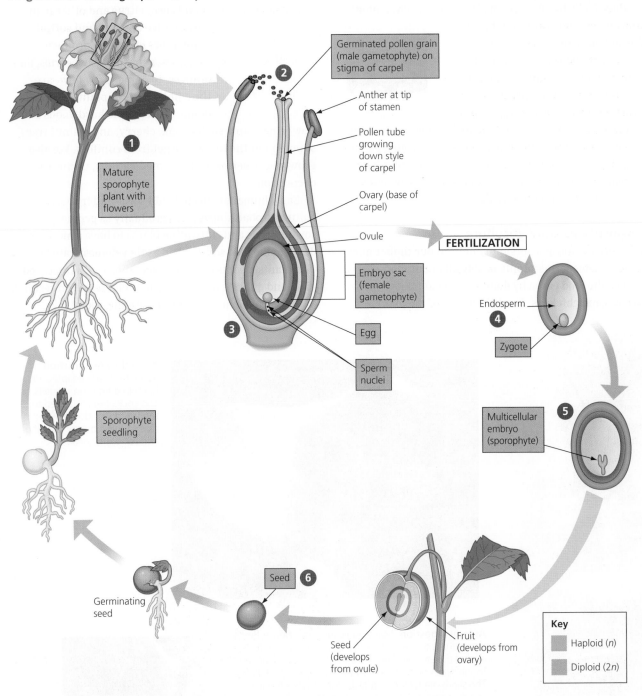

In angiosperms, as in gymnosperms, the sporophyte generation is dominant and produces the gametophyte generation within its body. **Figure 16.19** highlights key stages in the angiosperm life cycle. ❶ The flower is part of the sporophyte plant. As in gymnosperms, the pollen grain is the male gametophyte of angiosperms. The female gametophyte is located within an ovule, which in turn resides within a chamber of the ovary. ❷ After a pollen grain lands on the stigma, a pollen tube grows down to the ovule and ❸ releases a sperm nucleus that fertilizes an egg within the embryo sac. ❹ This produces a zygote, which ❺ develops into an embryo. The tissue surrounding the embryo develops into nutrient-rich **endosperm**, which will provide a food supply for the growing plant. ❻ The whole ovule develops into a seed, which can germinate and develop into a new sporophyte to begin the cycle anew. The seed's enclosure within an ovary is what distinguishes angiosperms from gymnosperms, which have a naked seed.

▼ **Figure 16.20 Fruits and
seed dispersal.** Different
types of fruits are adapted
for different methods of
dispersal.

A **fruit** is the ripened ovary of a flower. Thus, fruits are
produced only by angiosperms. As seeds are developing
from ovules, the ovary wall thickens,
forming the fruit that encloses the
seeds. A pea pod is an example of a
fruit with seeds (mature ovules, the
peas) encased in the ripened ovary
(the pod). Fruits protect and help
disperse seeds. As **Figure 16.20** demonstrates, many angiosperms depend on animals to disperse seeds, often passing
them through the digestive tract. Conversely, most land
animals, including humans, rely on angiosperms as a food
source, directly or indirectly. ☑

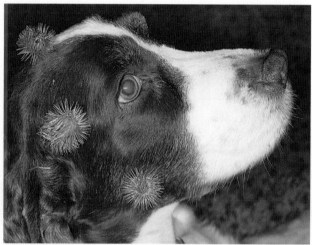

ALL BUT THE
CARROTS AND
GARLIC IN THIS
PHOTO ARE FRUITS.

Angiosperms and Agriculture

Whereas gymnosperms supply most of our timber and
paper, angiosperms supply nearly all of our food—as
well as the food eaten by domesticated animals, such
as cows and chickens. More than 90% of the plant
kingdom is made up of angiosperms, including cereal
grains such as wheat and corn, citrus and other fruit
trees, coffee and tea, and cotton.
Many types of garden produce—
tomatoes, squash, strawberries, and
oranges, to name just a few—are
the edible fruits of plants we have
domesticated. Fine hardwoods from
flowering plants such as oak, cherry, and walnut trees
supplement the timber we get from conifers. We also
grow angiosperms for fiber, medications, perfumes, and
decoration.

Early humans collected wild seeds and fruits. Agriculture probably developed gradually as people began
sowing seeds and cultivating plants to have a more
dependable food source. And as they domesticated certain plants, people began to select those with improved
yield and quality. In this way, artificial selection resulted
in the cornucopia of plant fare we enjoy today.

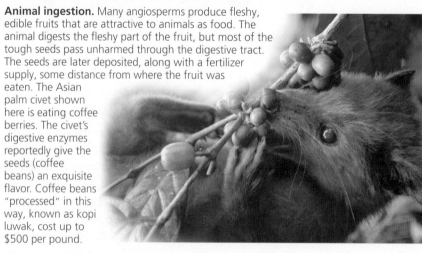

Animal transportation.
Some fruits are adapted to
hitch free rides on animals.
The cockleburs attached to
the fur of this dog may be
carried miles before opening
and releasing seeds.

Animal ingestion. Many angiosperms produce fleshy,
edible fruits that are attractive to animals as food. The
animal digests the fleshy part of the fruit, but most of the
tough seeds pass unharmed through the digestive tract.
The seeds are later deposited, along with a fertilizer
supply, some distance from where the fruit was
eaten. The Asian
palm civet shown
here is eating coffee
berries. The civet's
digestive enzymes
reportedly give the
seeds (coffee
beans) an exquisite
flavor. Coffee beans
"processed" in this
way, known as kopi
luwak, cost up to
$500 per pound.

Wind dispersal. Some
angiosperms depend on
wind for seed dispersal.
Here, milkweed pods open
to release masses of seeds
(brown) carried by silken
parachutes (part of the
seed coat).

Plant Diversity as a Nonrenewable Resource

The ever-increasing human population, with its demand for space and natural resources, is extinguishing plant species at an unprecedented rate. The problem is especially critical for forest ecosystems, which are home to as many as 80% of the world's terrestrial plant and animal species. Deforestation has been an ongoing human activity for centuries. Worldwide, fewer than 25% of the original forests remain; the figure in the contiguous United States is around 10%. Typically, forests are cut down to harvest timber or to clear land for housing or large-scale agriculture (Figure 16.21). Much of the forest that remains is tropical, and it is disappearing, too. From 2000 to 2015, an estimated 1,020,000 km^2 (393,825 mi^2, more than the combined areas of Texas and New Mexico) of tropical forest was lost, mostly to agriculture. However, conservation efforts and government policies in tropical countries have slowed the rate of loss from nearly 80,000 km^2 (30,888 mi^2) per year from 2000 to 2005 to 56,000 km^2 (21,622 mi$_2$) per year from 2010 to 2015.

Why does the loss of tropical forests matter? In addition to forests being centers of biodiversity, millions of people worldwide depend on these forests for their livelihood. There are other practical reasons to be concerned about the loss of plant diversity represented in tropical forests. More than 120 prescription drugs are made from substances derived from plants (Table 16.1). Pharmaceutical companies were led to most of these species by local peoples who use the plants in preparing their traditional medicines. Today, researchers are seeking to combine their scientific skills with the stores of local knowledge in partnerships that will develop new drugs and also benefit the local economies.

Scientists are working to slow the loss of plant diversity, in part by researching sustainable ways for people to benefit from forests. The goal of such efforts is to encourage management practices that use forests as resources without damaging them. The solutions we propose must be economically realistic; people who live where there are tropical rain forests must be able to make a living. But if the only goal is profit for the short term, then the destruction will continue until the forests are gone. We need to appreciate the rain forests and other ecosystems as living treasures that can regenerate only slowly. Only then will we learn to work with them in ways that preserve their biological diversity for the future.

Throughout our survey of plants in this chapter, we have seen how entangled the botanical world is with other terrestrial life. We switch our attention now to that other group of organisms that moved onto land with plants: the kingdom Fungi. ✓

☑ **CHECKPOINT**

In what way are forests renewable resources? In what way are they not?

■ Answer: Forests are renewable in the sense that new trees can grow where old growth has been removed by logging. Habitats that are permanently destroyed cannot be replaced, however, so forests must be harvested in a sustainable manner.

▼ **Figure 16.21 Cultivated land bordering a tropical forest in Uganda.** Bwindi Impenetrable National Park (on the right) is renowned for its biodiversity, which includes half the world's remaining mountain gorillas.

Table 16.1	A Sampling of Medicines Derived from Plants		
Compound	Source		Example of Use
Atropine	Belladonna plant		Pupil dilator in eye exams
Digitalin	Foxglove		Heart medication
Menthol	Wild mint		Ingredient in cough medicines, decongestants
Morphine	Opium poppy		Pain reliever
Quinine	Quinine tree		Malaria preventive
Paclitaxel (Taxol)	Pacific yew		Ovarian cancer drug
Tubocurarine	Curare tree		Muscle relaxant during surgery
Vinblastine	Periwinkle		Leukemia drug

Source: Adapted from R. Moore et al., *Botany*, 2nd ed. Dubuque, IA: Brown, 1998, Table 2.2, p. 37.

Fungi

Fungi are eukaryotes, and most are multicellular, but many have body structures and modes of reproduction unlike those of any other organism **(Figure 16.22).** Despite appearances, a mushroom is more closely related to you than it is to any plant! Molecular studies indicate that fungi and animals arose from a common ancestor more than 1 billion years ago. The oldest undisputed fossils of fungi, however, are only about 460 million years old, perhaps because the ancestors of terrestrial fungi were microscopic and fossilized poorly. Biologists who study fungi have described more than 100,000 species, but there may be as many as 1.5 million. Classifying fungi is an ongoing area of research.

The word *fungus* often evokes unpleasant images. Fungi rot timbers, spoil food, and afflict people with athlete's foot and worse. However, ecosystems would collapse without fungi to decompose dead organisms, fallen leaves, feces, and other organic materials. Fungi recycle vital chemical elements back to the environment in forms that other organisms can assimilate. And you have already learned that nearly all plants have mycorrhizae, fungus-root associations that help plants absorb minerals and water from the soil. In addition to these ecological roles, fungi have been used by people in various ways for centuries. You are probably familiar with some of these fungi, including mushrooms, mold, and yeast. We eat fungi (mushrooms and extremely expensive truffles, for instance), culture fungi to produce antibiotics and other drugs, add them to dough to make bread rise, culture them in milk to produce a variety of cheeses, and use them to ferment beer and wine. In this section, we'll discuss the characteristics common to all fungi and then survey their wide-ranging ecological impact. ☑

☑ **CHECKPOINT**

Name three ways that we benefit from fungi in our environment.

■ Answer: Fungi help recycle nutrients by decomposing dead organisms; mycorrhizae help plants absorb water and nutrients; some fungi serve us as food.

▼ **Figure 16.22   A gallery of diverse fungi.**

Bracket fungi. These are the reproductive structures of a fungus that absorbs nutrients as it decomposes a fallen tree on a forest floor.

A "fairy ring." Some mushroom-producing fungi poke up "fairy rings," which can appear on a lawn overnight. A ring develops at the edge of the main body of the fungus, which consists of an underground mass of tiny filaments (hyphae) within the ring. As the underground fungal mass grows outward from its center, the diameter of the fairy ring produced at its expanding perimeter increases annually.

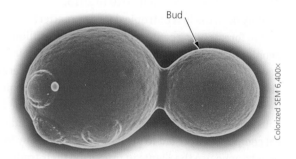

Bud

Colorized SEM 6,400×

Budding yeast. Yeasts are unicellular fungi. This yeast cell is reproducing asexually by a process called budding.

Colorized SEM 2,800×

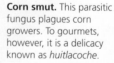

Mold. Molds grow rapidly on their food sources, which are often our food sources as well. The mold on this orange reproduces asexually by producing chains of microscopic spores (inset) that are dispersed via air currents.

Corn smut. This parasitic fungus plagues corn growers. To gourmets, however, it is a delicacy known as *huitlacoche.*

Characteristics of Fungi

We'll begin our look at the structure and function of fungi with an overview of how fungi obtain nutrients.

Fungal Nutrition

Fungi are heterotrophs that acquire their nutrients by **absorption**. A fungus digests food outside its body by secreting powerful digestive enzymes into the food. The enzymes decompose complex molecules to simpler compounds that the fungus can absorb. Fungi absorb nutrients from such nonliving organic material as fallen logs, animal corpses, and the wastes of live organisms.

Fungal Structure

The bodies of most fungi are constructed of thread-like filaments called **hyphae** (singular, *hypha*). Fungal hyphae are minute threads of cytoplasm surrounded by a plasma membrane and cell wall. The cell walls of fungi differ from the cellulose walls of plants. Fungal cell walls are usually built mainly of chitin, a strong but flexible polysaccharide that is also found in the external skeletons of insects. Most fungi have multicellular hyphae, which consist of chains of cells separated by cross-walls with pores. In many fungi, cell-to-cell channels allow ribosomes, mitochondria, and even nuclei to flow between cells.

Fungal hyphae branch repeatedly, forming an interwoven network called a **mycelium** (plural, *mycelia*), the feeding structure of the fungus **(Figure 16.23)**. Fungal mycelia usually escape our notice because they are often subterranean, but they can be huge. In fact, scientists have discovered that the mycelium of one humongous fungus in Oregon is 5.5 km—that's 3.4 miles!—in diameter and spreads through 2,200 acres of forest. This fungus is at least 2,600 years old and weighs hundreds of tons, qualifying it as one of Earth's oldest and largest organisms.

A mycelium maximizes contact with its food source by mingling with the organic matter it is decomposing and absorbing. A bucketful of rich organic soil may contain as much as a kilometer of hyphae. A fungal mycelium grows rapidly, adding hyphae as it branches within its food. The great majority of fungi are nonmotile; they cannot run, swim, or fly in search of food. But the mycelium makes up for the lack of mobility by swiftly extending the tips of its hyphae into new territory. ☑

Fungal Reproduction

The mushroom in Figure 16.23 is actually made up of tightly packed hyphae. Mushrooms arise from an underground mycelium. Although the mycelium obtains food from organic material via absorption, the function of the mushroom is reproduction. Unlike a truffle, which relies on animals to disperse its spores, a mushroom pops up above ground to disperse its spores on air currents.

Fungi typically reproduce by releasing haploid spores that are produced either sexually or asexually. The output of spores is mind-boggling. For example, puffballs, which are the reproductive structures of certain fungi, can spew clouds containing trillions of spores. Easily carried by wind or water, spores germinate to produce mycelia if they land in a moist place where there is food. Spores thus function in dispersal and account for the wide geographic distribution of many species of fungi. The airborne spores of fungi have been found more than 160 km (100 miles) above Earth. Closer to home, try leaving a slice of bread out for a week, and you will observe the furry mycelia that grow from the invisible spores raining down from the surrounding air.

IF YOU'VE EVER HAD MUSHROOM PIZZA, YOU'VE SNACKED ON THE FUNGI'S REPRODUCTIVE STRUCTURE.

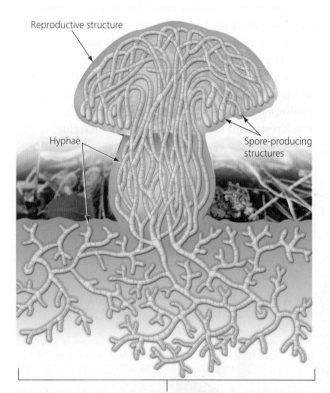

Reproductive structure

Hyphae

Spore-producing structures

Mycelium

◀ **Figure 16.23 The fungal mycelium.** A mushroom consists of tightly packed hyphae that extend upward from a much more massive mycelium of hyphae growing underground. The photo at the bottom shows a mycelium made up of the cottony threads that decompose organic litter.

☑ **CHECKPOINT**

How do nonmotile fungi colonize new territories that are both nearby and far away?

■ Answer: Nonmotile mycelia can extend the tips of their hyphae into nearby habitats, while spores released from the reproductive structures get carried by wind or water.

What Killed the Pines?

BACKGROUND

In 1932, Puerto Rico had many acres of worn-out farmland and little cash. The soil could no longer grow the usual crops and was in danger of eroding away. Growing trees seemed like the perfect solution. Trees would stabilize the soil and could be used for paper, furniture, and building materials. Because the native trees had a low market value, the U.S. Forest Service decided to introduce pine trees, which usually thrive in nutrient-poor soils, grow quickly, and require very little care. Pine species from the warm, wet southeastern United States were selected to match the conditions in Puerto Rico. Seeds from these trees sprouted, and the seedlings grew to about four inches high. Then they died. There were no signs of disease or insect damage, but none of the trees made it past the seedling stage, and no one knew why.

Researchers conducted more than 90 experiments over a period of 20 years, trying to solve the mystery. They tried many kinds of fertilizers, soils, and watering routines. They tested 27 different pine species from all over the world. All attempts failed—except one. In the successful laboratory experiment, researchers had treated the seedlings with mycorrhizae collected from other pine forests. This result gave scientists a vital clue.

They hypothesized that adding mycorrhizae to the soil would improve the growth of the pines in Puerto Rico.

METHOD

In 1955, researchers planted slash pine seedlings in an experimental field in Puerto Rico. They treated one group of pines with mycorrhizal fungi collected from soil in a North Carolina pine forest. The rest of the pines served as controls. Adding mycorrhizae disturbed the soil around the trees, so the soil around half of the control trees was disturbed in the same way. The other control trees were left undisturbed (**Figure 16.24a**).

RESULTS

Only 36% of the control trees survived, and none of the control trees grew much during the four-year experiment (**Figure 16.24b**). In contrast, 85% of the pines treated with mycorrhizae survived; these trees grew well. Today, pine forests grown with the help of mycorrhizae thrive in Puerto Rico, providing habitat for wildlife and protection from erosion and storms as well as economic benefits (**Figure 16.24c**). As a result of the experiment described here and subsequent research, it has become common practice to add mycorrhizae everywhere that pines are cultivated.

Thinking Like a Scientist

Why did the scientists use two controls, one with disturbed soil and one with undisturbed soil?

For the answer, see Appendix D.

▼ **Figure 16.24 An experiment to test the benefit of mycorrhizae on pine growth.**
Data from: C. B. Briscoe. Early Results of Mycorrhizal Inoculation of Pine in Puerto Rico, *Caribbean Forester* July-Dec.: 73–77 (1959).

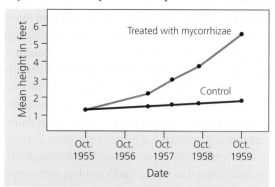

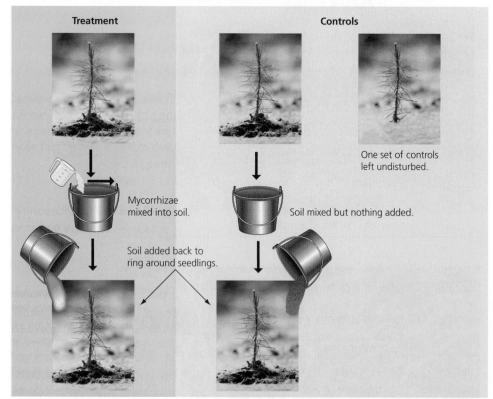

(b) Results. Pine trees treated with fungal mycorrhizae grew much better than did control trees. (The disturbed and undisturbed controls were combined.)

(a) Experimental groups. One group of seedlings was treated with mycorrhizae, a procedure that disturbed the soil. Control seedlings were not treated with mycorrhizae. The soil of half the controls was disturbed; the other half was undisturbed.

(c) Pine forest in Puerto Rico

The Ecological Impact of Fungi

Fungi have been major players in terrestrial communities ever since plants and fungi together moved onto land. Let's examine a few examples of how fungi continue to have an enormous ecological impact, including numerous interactions with people.

Fungi as Decomposers

Fungi and bacteria are the principal decomposers that keep ecosystems stocked with the inorganic nutrients essential for plant growth. **This vital role of decomposers is an example of interactions within biological systems.** Without decomposers, carbon, nitrogen, and other elements would accumulate in nonliving organic matter. Plants and the animals they feed would starve because elements taken from the soil would not be returned.

Fungi are well adapted as decomposers of organic refuse. Their invasive hyphae enter the tissues and cells of dead organisms and digest polymers, including the cellulose of plant cell walls. A succession of fungi, in concert with bacteria and, in some environments, invertebrate animals, is responsible for the complete breakdown of organic litter. The air is so loaded with fungal spores that as soon as a leaf falls or an insect dies, it is covered with spores and soon after infiltrated by fungal hyphae.

We may applaud fungi that decompose forest litter or dung, but it's a different story when molds attack our food or our shower curtains. A significant amount of the world's fruit harvest is lost each year to fungal attack. And a wood-digesting fungus does not distinguish between a fallen oak limb and the oak planks of a boat. During the Revolutionary War, the British lost more ships to fungal rot than to enemy attack.

Parasitic Fungi

Parasitism is a relationship in which two species live in contact and one organism benefits while the other is harmed. Parasitic fungi absorb nutrients from the cells or body fluids of living hosts. Of the 100,000 known species of fungi, about 30% make their living as parasites.

About 500 species of fungi are known to be parasitic in humans and other animals. A baffling disease known as coccidioidomycosis, or valley fever, causes devastating illness in some people; others suffer only mild flulike symptoms. People contract the disease when they inhale the spores of a fungus that lives in the soil of the southwestern United States. In recent years, researchers have noticed an uptick in reported cases of valley fever, perhaps due to changes in climate patterns or development of once-rural areas inhabited by the fungus. Less serious fungal diseases include vaginal yeast infections and ringworm, so named because it appears as circular red areas on the skin. The ringworm fungus can infect almost any skin surface, where

▼ **Figure 16.25 Parasitic fungi that cause plant disease.** The parasitic fungus that causes Dutch elm disease evolved with European species of elm trees, and it is relatively harmless to them. But it has been deadly to American elms since it was accidentally introduced in 1926.

it produces intense itching and sometimes blisters. One species attacks the feet, causing athlete's foot. Another species is responsible for the misery known as jock itch.

The great majority of fungal parasites infect plants. American chestnut and American elm trees, once common in forests, fields, and city streets, were devastated by fungal epidemics in the 1900s **(Figure 16.25)**. Fungi are also serious agricultural pests, and some of the fungi that attack food crops are toxic to humans.

Commercial Uses of Fungi

It would not be fair to fungi to end our discussion with an account of diseases. In addition to their positive global impact as decomposers, fungi also have a number of practical uses for people.

Most of us have eaten mushrooms, although we may not have realized that we were ingesting the reproductive extensions of subterranean fungi. Your grocery store probably stocks portobello, shiitake, and oyster mushrooms along with common button mushrooms. If you like to cook with mushrooms, you can buy a mushroom "garden"—mycelium embedded in a rich food source—that makes it easy to grow your own. Some enthusiasts gather edible fungi from fields and forests **(Figure 16.26)**, but only experts should dare to eat wild fungi. Some poisonous species resemble edible ones, and there are no simple rules to help the novice distinguish one from the other.

Other fungi are used in food production. The distinctive flavors of certain kinds of cheeses, including Roquefort and blue cheese, come from the fungi used to ripen them. And people have used yeasts (unicellular fungi) for thousands of years to produce alcoholic beverages and cause bread to rise (see Figure 6.15).

Fungi are medically valuable as well. Some fungi produce antibiotics that are used to treat bacterial diseases. In fact, the first antibiotic discovered was penicillin, made by the common mold *Penicillium* **(Figure 16.27)**. Researchers are also investigating fungal products that show potential as anticancer drugs.

As sources of medicines and food, as decomposers, and as partners with plants in mycorrhizae, fungi play vital roles in life on Earth. ☑

▼ **Figure 16.26 Fungi eaten by people.**

Chicken of the woods.
This fleshy shelf fungus is said to taste like chicken.

Chanterelle mushrooms. Gourmet mushrooms are highly prized by chefs for their earthy flavors and interesting textures.

Giant puffball. These humongous fungi can grow to more than 2 feet in diameter. They are only edible when immature, before the spores form. Beware: Deadly fungi often resemble small puffballs.

▼ **Figure 16.27 Fungal production of an antibiotic.** Penicillin is made by the common mold *Penicillium*. In this petri dish, the clear area between the mold and the growing *Staphylococcus* bacteria is where the antibiotic produced by the *Penicillium* inhibits the growth of the bacteria.

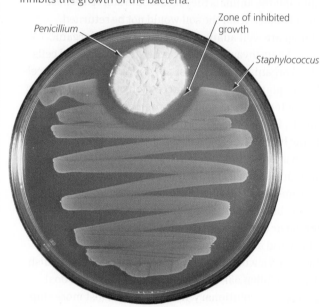

Penicillium

Zone of inhibited growth

Staphylococcus

☑ **CHECKPOINT**

1. What is athlete's foot?
2. What do you think is the natural function of the antibiotics that fungi produce in their native environments?

■ *Answers:* **1.** *Athlete's foot is infection of the foot's skin with ringworm fungus.* **2.** *The antibiotics block the growth of microorganisms, especially bacteria, that compete with the fungi for nutrients and other resources.*

EVOLUTION CONNECTION Plant-Fungus Interactions

A Pioneering Partnership

Evolution is not just about the origin and adaptation of individual species. Relationships between species are also an evolutionary product. Particularly relevant to this chapter is the symbiotic association of fungi and plant roots—mycorrhizae—that made life's move onto land possible.

Liverworts **(Figure 16.28)** are so inconspicuous that you may never have noticed one, but their ancestors were some of the first plants to colonize the land 470 million years ago. Like mosses, liverworts belong to the group of nonvascular plants known as bryophytes. Their name comes from the shape of the flat, leaf-like structures that act as solar collectors, which looked to an early botanist like the shape of a liver. ("Wort" is

from an old English word meaning "plant.") The simple root-like structures of liverworts consist of only one cell, limiting their ability to gather water and nutrients. As you learned earlier in this chapter, fungi greatly extend the area for absorbing resources. The fungi that teamed up with the liverworts were probably living in the soil, where there were few high-energy molecules to absorb. Conditions during the invasion of land were particularly well-suited to this partnership because, although the soil was poor in nutrients, the air was rich in CO_2. Water and CO_2 provide the raw materials for photosynthesis. Just as a farmer might provide a chef with the ingredients to make food for both of them, the fungi and liverworts make a winning team. And as the plants and fungi died and decomposed, they helped to create the nutrient-rich soil that made the land more hospitable for the next waves of immigrants.

After protists and plants, fungi is the third group of eukaryotes that we have surveyed so far. Strong evidence suggests that they evolved from protist ancestors that also gave rise to the fourth and most diverse group of eukaryotes: the animals, the topic of the next chapter.

◄ **Figure 16.28 Liverworts.** Symbiotic relationships with fungi helped early nonvascular plants colonize land. The mycorrhizal fungus receives food from its photosynthetic partner. The fungus in turn, helps the liverwort absorb water and minerals.

Chapter Review

SUMMARY OF KEY CONCEPTS

Colonizing Land

Terrestrial Adaptations of Plants

Plants are multicellular photosynthetic eukaryotes with adaptations for living on land.

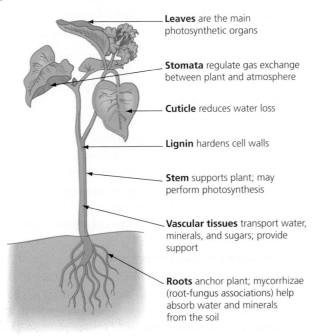

Leaves are the main photosynthetic organs

Stomata regulate gas exchange between plant and atmosphere

Cuticle reduces water loss

Lignin hardens cell walls

Stem supports plant; may perform photosynthesis

Vascular tissues transport water, minerals, and sugars; provide support

Roots anchor plant; mycorrhizae (root-fungus associations) help absorb water and minerals from the soil

The Origin of Plants from Green Algae

Plants evolved from a group of multicellular green algae called charophytes.

Plant Diversity

Highlights of Plant Evolution

Four periods of plant evolution are marked by terrestrial adaptations in major plant groups.

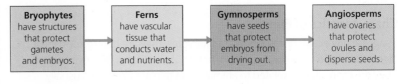

| **Bryophytes** have structures that protect gametes and embryos. | **Ferns** have vascular tissue that conducts water and nutrients. | **Gymnosperms** have seeds that protect embryos from drying out. | **Angiosperms** have ovaries that protect ovules and disperse seeds. |

Bryophytes

The most familiar bryophytes are mosses. Mosses display two key terrestrial adaptations: a waxy cuticle that prevents dehydration and the retention of developing embryos within the female plant's body. Mosses are most common in moist environments because their sperm must swim to the eggs and because they lack lignin in their cell walls and thus cannot stand tall. Bryophytes are unique among plants in having the gametophyte as the dominant generation in the life cycle.

Ferns

Ferns are seedless plants that have vascular tissues but still use flagellated sperm to fertilize eggs. During the Carboniferous period, giant ferns were among the plants that decayed to thick deposits of organic matter, which were gradually converted to coal.

Gymnosperms

A drier and colder global climate near the end of the Carboniferous period favored the evolution of the first seed plants. The most successful were the gymnosperms, represented by conifers. Needle-shaped leaves with thick cuticles and sunken stomata are adaptations to dry conditions. Conifers and most other gymnosperms have three additional terrestrial adaptations: (1) further reduction of the haploid gametophyte and greater development of the diploid sporophyte; (2) sperm-bearing pollen, which doesn't require water for transport; and (3) seeds, which consist of a plant embryo packaged along with a food supply inside a protective coat.

Angiosperms

Angiosperms supply nearly all our food and much of our fiber for textiles. The evolution of the flower and more efficient water transport help account for the success of the angiosperms. The dominant stage is a sporophyte with gametophytes in its flowers. The female gametophyte is located within an ovule, which in turn resides within a chamber of the ovary. Fertilization of an egg in the female gametophyte produces a zygote, which develops into an embryo. The whole ovule develops into a seed. The seed's enclosure within an ovary is what distinguishes angiosperms from gymnosperms, which have naked seeds. A fruit is the ripened ovary of a flower. Fruits protect and help disperse seeds. Angiosperms are a major food source for animals, and animals aid plants in pollination and seed dispersal. Agriculture constitutes a unique kind of evolutionary relationship among plants, people, and other animals.

Plant Diversity as a Nonrenewable Resource

Deforestation to meet the demand of human activities for space and natural resources is causing the extinction of plant species at an unprecedented rate. The problem is especially critical for tropical forests.

Fungi

Characteristics of Fungi

Fungi are unicellular or multicellular eukaryotes; they are heterotrophs that digest their food externally and absorb the nutrients from the environment. They are more closely related to animals than to plants. A fungus usually consists of a mass of threadlike hyphae, forming a mycelium. The cell walls of fungi are mainly composed of chitin. Although most fungi are nonmotile, a mycelium can grow very quickly, extending the tips of its hyphae into new territory. Mushrooms are reproductive structures that extend from the underground mycelium. Fungi reproduce and disperse by releasing spores that are produced either sexually or asexually.

The Ecological Impact of Fungi

Fungi and bacteria are the principal decomposers of ecosystems. Many molds destroy fruit, wood, and human-made materials. About 500 species of fungi are known to be parasites of people and other animals.

Commercial Uses of Fungi

A variety of fungi are eaten or used in food production. In addition, some fungi are medically valuable.

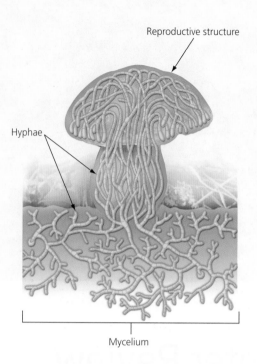

Reproductive structure

Hyphae

Mycelium

Mastering **Biology**

For practice quizzes, BioFlix animations, MP3 tutorials, video tutors, and more study tools designed for this textbook, go to Mastering Biology™

SELF-QUIZ

1. Which of the following structures is common to all four major plant groups? vascular tissue, flowers, seeds, cuticle, pollen

2. Multicellular eukaryotic photosynthetic algae are classified as protists rather than plants because most of them lack _____ adaptations.

3. Complete the following analogies:
   a. Gametophyte is to haploid as _____ is to diploid.
   b. _____ are to conifers as flowers are to _____.
   c. Ovule is to seed as ovary is to _____.

4. The carpel of the flower of an angiosperm does not contain
   a. a stigma.   c. an ovary.
   b. a style.   d. an anther.

5. Of the following, which is the earliest step in the formation of fossil fuels?
   a. subjecting organic matter to extreme pressure
   b. fungi and bacteria converting the remains of organisms to inorganic nutrients
   c. incomplete decomposition of organic matter
   d. subjecting organic matter to extreme heat

6. You discover a new species of plant. Under the microscope, you find that it produces flagellated sperm. A genetic analysis shows that its dominant generation has diploid cells. What kind of plant do you have?

7. Why are many fruits green when their seeds are immature?

8. Which of the following terms includes all others in the list? angiosperm, fern, vascular plant, gymnosperm, seed plant

9. A nonvascular plant with no true roots or lignified walls is a
   a. fern.
   b. bryophyte.
   c. charophyte.
   d. green alga.

10. What is a pollen grain?

11. Mycorrhizae are symbiotic associations between _____ and _____.

12. Contrast the heterotrophic nutrition of a fungus with your own heterotrophic nutrition.

For answers to the Self-Quiz, see Appendix D.

IDENTIFYING MAJOR THEMES

For each statement below, identify which major theme is evident (the relationship of structure to function, information flow, pathways that transform energy and matter, interactions within biological systems, or evolution) and explain how the statement relates to the theme. If necessary, review themes (Chapter 1) and review the examples highlighted in blue in this chapter.

13. Gametophytes produce gametes that unite to form zygotes, which develop into new sporophytes. And sporophytes produce spores that give rise to new gametophytes, transmitting DNA through an alternation of generations.

14. Roots typically have many fine branches that thread among the grains of soil, providing a large surface area that maximizes contact with mineral-bearing water in the soil.

15. Vascular tissue was an adaptation that allowed ferns to colonize a greater variety of habitats than mosses.

For answers to Identifying Major Themes, see Appendix D.

THE PROCESS OF SCIENCE

16. Quorn, the major meat substitute product in the UK, is based on a mycoprotein extracted from the fungus *Fusarium venenatum*. What characteristics of this fungus do you think make it ideal for food production? From an ecological point of view, what may be the advantages of eating a fungal protein over meat? When Quorn was released in the US in 2002, the Center for Science in the Public Interest (CSPI) had a few concerns about the product. One of those was that Quorn was marketed as a "mushroom product." Why would the CSPI take offense at this labeling? Speculate why the company might have chosen this labeling.

17. **Interpreting Data** Airborne pollen of wind-pollinated plants such as pines, oaks, weeds, and grasses causes seasonal allergy symptoms in many people. As global warming lengthens the growing season for plants,

scientists predict longer periods of misery for allergy sufferers. However, global warming does not affect all regions equally (see Figure 18.44). The table below shows the length of the average season in nine locations for ragweed pollen, an allergen that affects millions of people. Calculate the change in length of the pollen season from 1995 to 2009 for each location and graph this information against latitude. Is there a latitudinal trend in the length of ragweed season? You may want to record the data on a map to help you visualize the geographic locations at which samples were taken.

Average Length of Ragweed Pollen Season in Nine Locations (averages obtained from at least 15 years of data)				
Location	Latitude (°N)	Length of Pollen Season in 1995 (days)	Length of Pollen Season in 2009 (days)	Change in Length of Pollen Season (days)
Oklahoma City, OK	35.47	88	89	
Rogers, AR	36.33	64	69	
Papillion, WI	41.15	69	80	
Madison, WI	43.00	64	76	
La Crosse, WI	43.80	58	71	
Minneapolis, MN	45.00	62	78	
Fargo, ND	46.88	36	52	
Winnipeg, MB, Canada	50.07	57	82	
Saskatoon, SK, Canada	52.07	44	71	

Data from: L. Ziska et al., Recent Warming by Latitude Associated with Increased Length of Ragweed Pollen Season in Central North America. *Proceedings of the National Academy of Sciences* 108: 4248–4251 (2011).

BIOLOGY AND SOCIETY

18. Why are tropical forests being destroyed so rapidly? What kinds of social, technological, and economic factors are responsible? Most forests in more industrialized Northern Hemisphere countries have already been cut. Do the more industrialized nations have a right to pressure the less industrialized nations in the Southern Hemisphere to slow or stop the destruction of their forests? Defend your answer. What kinds of benefits, incentives, or programs might slow the destruction of tropical forests?

19. Many prescription drugs are derived from natural plant products. Numerous other plant substances, including caffeine and nicotine, have effects in the human body as well. There is also a wide array of plant products, in the form of pills, powders, or teas, marketed as herbal medicines. Some people prefer taking these "natural" products to pharmaceuticals. Others use herbal supplements to boost energy, promote weight loss, strengthen the immune system, relieve stress, and more. The U.S. Federal Drug Administration, which approves pharmaceuticals, is also responsible for regulating herbal remedies. What does the label "FDA-approved" on an herbal remedy mean? How does that compare to FDA approval of a drug? The FDA website is a good place to start your research. (Note that the FDA classifies herbal remedies as dietary supplements.)

17 The Evolution of Animals

Why Animal Diversity Matters

We are members of the animal kingdom, a group that encompasses a vast diversity of species ranging in size from microscopic to gargantuan. The evolutionary history of animals spans more than 600 million years, including several million years of human evolution.

IS THAT AN ANIMAL OR A PLANT IN YOUR BATHTUB? A LOOFAH IS THE SKELETON OF A PLANT RELATED TO CUCUMBERS. A SEA SPONGE IS THE SKELETON OF AN ANIMAL.

BEWARE OF RAW FISH! FISH TAPEWORM, FOUND IN SPECIES THAT LIVE OR BREED IN FRESH WATER, IS THE LARGEST PARASITE THAT INFECTS HUMANS.

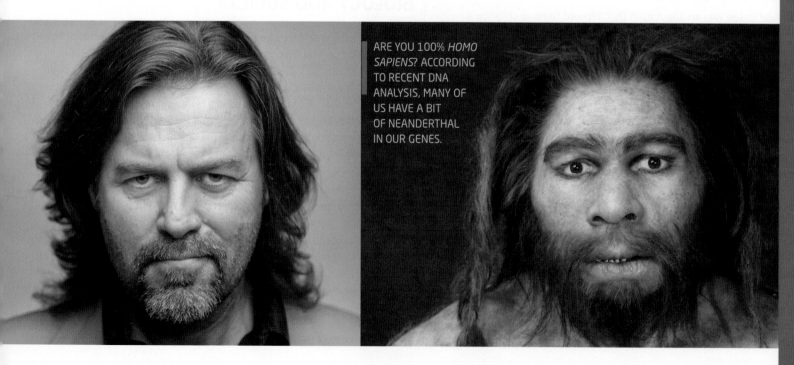

ARE YOU 100% *HOMO SAPIENS*? ACCORDING TO RECENT DNA ANALYSIS, MANY OF US HAVE A BIT OF NEANDERTHAL IN OUR GENES.

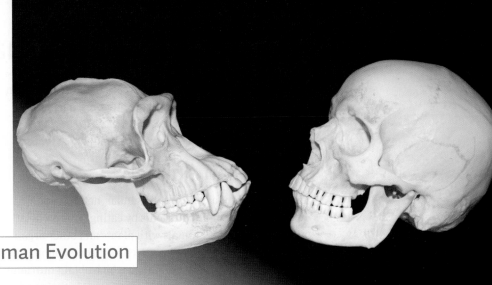

BIOLOGY AND SOCIETY Human Evolution

Comparison of human and chimpanzee skulls, showing a large difference in brain size.

Evolving Adaptability

Humans (*Homo sapiens*) have managed to colonize almost every major habitat on Earth. What makes humans such successful animals? Much of our success is due to brain power. Species larger than us, such as elephants and whales, do have larger brains, but our brains are much larger than you might predict based on our body size. For example, the ratio of brain volume to body mass in humans is roughly 2.5 times the brain volume to body mass ratio in chimpanzees, our closest primate relatives. Brain size is not the only difference. Our brain cells make more connections with each other, and those connections can be adjusted as needed, making our brains exceptionally flexible. In addition, the configuration of the human brain is distinctive. The part of our brain that deals with problem solving, language, logic, and understanding other people is particularly well-developed.

Although body size has remained roughly the same for about the last 1.5 million years of human evolution, brain size has increased by about 40%. There is evidence that our ancestors' brain size increased as the environment became less predictable. In a changing world, individuals who can plan, imagine, evaluate, and change their behavior have an advantage over individuals who rely more on instinct. Our large, flexible brains have allowed us to create a global society that can share information, culture, and technology. Other species can fly, breathe underwater, or produce millions of offspring in their lifetimes, but none can match our ability to learn and change our behavior.

Humans are just one of the 1.3 million species of animals that have been named and described by biologists. This amazing diversity arose through hundreds of millions of years of evolution as natural selection shaped animal adaptations to Earth's many environments. In this chapter, we'll look at the 9 most abundant and widespread of the roughly 35 phyla (major groups) in the kingdom Animalia. We'll give special attention to the major milestones in animal evolution and conclude by reconnecting with the fascinating subject of human evolution.

The Origins of Animal Diversity

Animal life began in Precambrian seas with the evolution of multicellular creatures that ate other organisms. We are among their descendants.

What Is an Animal?

Animals are eukaryotic, multicellular, heterotrophic organisms that obtain nutrients by eating. This mode of nutrition contrasts animals with plants and other organisms that construct organic molecules through photosynthesis. It also contrasts with fungi, which obtain nutrients by absorption after digesting the food outside their bodies (see Figure 16.23). Most animals digest food within their bodies after ingesting other organisms, dead or alive, whole or by the piece **(Figure 17.1)**.

In addition, animal cells lack the cell walls that provide strong support in the bodies of plants and fungi. And most animals have muscle cells for movement and nerve cells that control the muscles. The most complex animals can use their muscular and nervous systems for many functions other than eating. Some species even use massive networks of nerve cells called brains to interpret complex sensory information and control the body's activities.

Most animals are diploid and reproduce sexually; eggs and sperm are the only haploid cells. The life cycle of a sea star **(Figure 17.2)** includes basic stages found in most animal life cycles. ❶ Male and female adult animals make haploid gametes by meiosis, and ❷ an egg and a sperm

fuse, producing a zygote. ❸ The zygote divides by mitosis, forming ❹ an early embryonic stage called a **blastula**, which is usually a hollow ball of cells. ❺ In most animals, one side of the blastula folds inward, forming a stage called a **gastrula**. ❻ The gastrula develops into a saclike embryo with inner, outer, and middle cell layers and an opening at one end. After the gastrula stage, many animals develop directly into adults. Others, such as the sea star, develop into a **larva** ❼, an immature individual that looks different from the adult animal. (A tadpole, a larval frog, is another example.) ❽ The larva undergoes a major change of body form, called **metamorphosis**, in becoming an adult capable of reproducing sexually. ☑

✅ **CHECKPOINT**

What mode of nutrition distinguishes animals from fungi, both of which are heterotrophs?

■ *Answer: ingestion (eating)*

▼ **Figure 17.1 Nutrition by ingestion, the animal way of life.** Few animals ingest a piece of food as large as the gazelle being eaten by this rock python. The snake will spend two weeks or more digesting its meal.

▼ **Figure 17.2 The life cycle of a sea star as an example of animal development.**

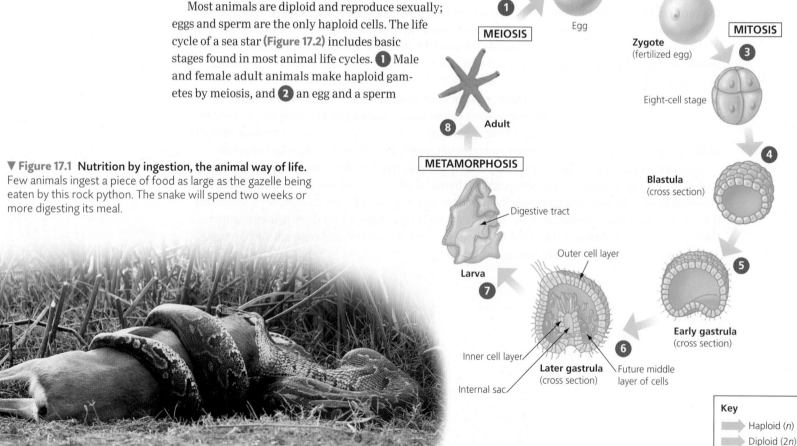

Sperm

❷ **FERTILIZATION**

❶

MEIOSIS

Egg

Zygote
(fertilized egg)

MITOSIS

❸

Eight-cell stage

❽ **Adult**

METAMORPHOSIS

❹

Blastula
(cross section)

Digestive tract

Larva
❼

Outer cell layer

❺

Inner cell layer

Later gastrula
(cross section)

Future middle
layer of cells

Early gastrula
(cross section)

❻

Internal sac

Key

Haploid (*n*)

Diploid (2*n*)

Early Animals and the Cambrian Explosion

Scientists hypothesize that animals evolved from a colonial flagellated protist (Figure 17.3). Although molecular data point to a much earlier origin, the oldest animal fossils that have been found are about 560 million years old. Animal evolution must have been under way already for some time prior to that—the fossils reveal a variety of shapes, and sizes range from 1 cm to 1 m in length (Figure 17.4).

Animal diversification appears to have accelerated rapidly from 535 to 525 million years ago, during the Cambrian period. Because so many animal body plans and new phyla appear in the fossils from such an evolutionarily short time span, biologists call this episode the Cambrian explosion. The most celebrated source of Cambrian fossils is located in the mountains of British Columbia, Canada. The Burgess Shale, as it is known, provided a cornucopia of perfectly preserved animal fossils. In contrast to the Precambrian animals, many Cambrian animals had hard parts such as shells, and many are clearly related to existing animal groups. For example, scientists have classified more than a third of the species found in the Burgess Shale as arthropods, the group that includes present-day crabs, shrimps, and insects (Figure 17.5). Other fossils are more difficult to place. Some are downright weird, like the spiky creature near the center of the drawing, known as *Hallucigenia*, and *Opabinia*, the five-eyed predator grasping a worm with the long, flexible appendage that protrudes in front of its mouth.

What ignited the Cambrian explosion? Scientists have proposed several hypotheses, including increasingly complex predator-prey relationships and an increase in atmospheric oxygen. **But whatever the cause of the rapid diversification, it is likely that the set of "master control" genes—the genetic framework of information flow for building complex bodies—was already in place.** Much of the diversity in body form among the animal phyla is associated with variations in where and when these genes are expressed within developing embryos.

In the last half billion years, animal evolution has to a large degree merely generated variations of the animal forms that originated in the Cambrian seas. Continuing research will help test hypotheses about the Cambrian explosion. But even as the explosion becomes less mysterious, it will seem no less wondrous. ✓

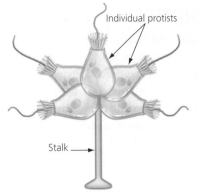

◀ **Figure 17.3 Hypothetical common ancestor of animals.** As you'll learn shortly, the individual cells of this colonial flagellated protist resemble the feeding cells of sponges.

Individual protists

Stalk

▶ **Figure 17.4 Fossils of Precambrian animals.** All of the oldest animal fossils are impressions of soft-bodied animals. Most of them do not appear to be related to any living group of animals.

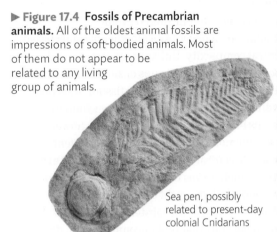

Sea pen, possibly related to present-day colonial Cnidarians

Impression of upper surface of *Tribrachidium heraldicum*, which had a hemispheric shape and three-part symmetry unlike any living animal (up to 5 cm across)

▼ **Figure 17.5 A Cambrian seascape.** This drawing is based on fossils from the Burgess Shale. The flat-bodied animals are extinct arthropods called trilobites. A photo of a fossil trilobite is shown at the right.

☑ **CHECKPOINT**

Why is animal evolution during the early Cambrian referred to as an "explosion"?

■ Answer: because a great diversity of animals evolved in a relatively short time span

Animal Phylogeny

Historically, biologists have categorized animals by "body plan"—general features of body structure. Distinctions between body plans were used to construct phylogenetic trees showing the evolutionary relationships among animal groups. More recently, a wealth of genetic data has allowed evolutionary biologists to modify and refine groups. **Figure 17.6** represents a revised set of hypotheses about the evolutionary relationships among nine major animal phyla based on both structural and genetic similarities.

A major branch point in animal evolution distinguishes sponges from all other animals based on structural complexity. Unlike more complex animals, sponges lack tissues, groups of similar cells that perform a function (such as nervous tissue). A second major evolutionary split is based on body symmetry: radial versus bilateral (**Figure 17.7**). Like the flowerpot, the sea anemone has **radial symmetry**, identical all around a central axis. The shovel has **bilateral symmetry**, which means there's only one way to split it into two equal halves—right down the midline. A bilateral animal, such as the lobster in Figure 17.7, has a definite "head end" that first encounters food, danger, and other stimuli when traveling. In most bilateral animals, a nerve center in the form of a brain is at the head end, near a concentration of sense organs such as eyes. Thus, bilateral symmetry is an adaptation that aids movement, such as crawling, burrowing, or swimming. Indeed, many radial animals are stationary, whereas most bilateral animals are mobile.

Among bilateral animals, analysis of embryonic development sets the echinoderms and chordates apart from the evolutionary branch that includes molluscs, flatworms, annelids, roundworms, and arthropods. Molluscs, flatworms, and annelids have genetic similarities that are not shared by roundworms and arthropods.

The evolution of body cavities also helped lead to more complex animals. A **body cavity (Figure 17.8)** is a fluid-filled space separating the digestive tract from the outer body wall. The cavity enables the animal's internal organs to grow and move independently of the outer body wall, and the fluid cushions them from injury. In soft-bodied animals such as earthworms, the fluid is under pressure and functions as a hydrostatic skeleton. Of the phyla shown in Figure 17.6, only sponges, cnidarians, and flatworms lack a body cavity.

With the overview of animal evolution in Figure 17.6 as our guide, we're ready to take a closer look at the nine most numerous animal phyla. ✓

✓ CHECKPOINT

1. In the phylogeny shown in Figure 17.6, chordates (the phylum that includes humans) are most closely related to which other animal phylum?

2. A round pizza displays _____ symmetry, whereas a slice of pizza displays _____ symmetry.

■ *Answers: 1. echinoderms 2. radial; bilateral*

▼ **Figure 17.6 An overview of animal phylogeny.** Only 9 of the more than 30 animal phyla (the exact number is not agreed upon) are included in the tree and in the text. The branching takes into account the body plan, embryonic development, and genetic relationships of the organisms.

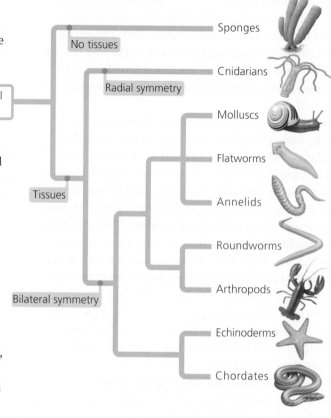

Sponges

No tissues

Cnidarians

Radial symmetry

Ancestral protist

Molluscs

Flatworms

Tissues

Annelids

Roundworms

Arthropods

Bilateral symmetry

Echinoderms

Chordates

▼ **Figure 17.7 Body symmetry.**

Radial symmetry. Parts radiate from the center, so any slice through the central axis divides into mirror images.

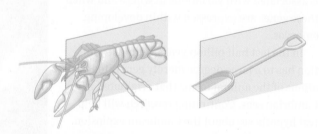

Bilateral symmetry. Only one slice can divide left and right sides into mirror-image halves.

▼ **Figure 17.8 Body plans of bilateral animals.** The various organ systems of these animals develop from the three tissue layers that form in the embryo.

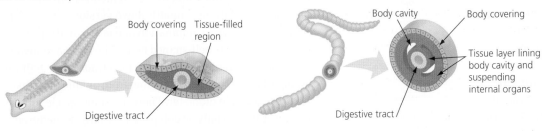

(a) No body cavity: for example, flatworm

(b) Body cavity: for example, earthworm

Major Invertebrate Phyla

Living on land as we do, our sense of animal diversity is biased in favor of vertebrates, animals with a backbone, such as amphibians, reptiles, and mammals. However, vertebrates make up less than 5% of all animal species. If we were to sample the animals in an aquatic habitat, such as a pond, tide pool, or coral reef, or if we were to consider the millions of insects that share our terrestrial world, we would find ourselves in the realm of **invertebrates**, animals without backbones. We give special attention to the vertebrates simply because we humans are among the backboned ones. However, by exploring the other 95% of the animal kingdom—the invertebrates—we'll discover an astonishing diversity of beautiful creatures that too often escape our notice.

Sponges

Sponges (phylum Porifera) are stationary animals that appear so immobile that you might mistake them for plants. The simplest of all animals, sponges probably evolved very early from colonial protists. They have no nerves or muscles, but their individual cells can sense and react to changes in the environment. The cell layers of sponges are loose associations that are not considered tissues. Sponges range in height from 1 cm (about half an inch) to 2 m (more than 6 ft). Although some live in fresh water, the majority of the 5,500 or so species of sponges are marine.

Sponges are examples of suspension feeders, animals that collect food

A LOOFAH IS THE SKELETON OF A PLANT RELATED TO CUCUMBERS. A SEA SPONGE IS THE SKELETON OF AN ANIMAL.

particles from water passed through some type of food-trapping equipment. The body of a sponge resembles a sac perforated with holes. Water is drawn through the pores into a central cavity and then flows out of the sponge through a larger opening **(Figure 17.9)**. Flagella on specialized cells called choanocytes sweep water through the sponge's porous body. Tiny nets encircling the flagella trap bacteria and other food particles, which the choanocytes then engulf. Specialized cells called amoebocytes pick up food from the choanocytes, digest it, and carry the nutrients to other cells. Amoebocytes also manufacture the fibers that make up a sponge's skeleton. In some sponges, these fibers are sharp and spur-like, like the ones shown in Figure 17.9. Other sponges have softer, more flexible skeletons; these pliable, honeycombed skeletons are often used as natural sponges in the bath or to wash cars. ☑

☑ CHECKPOINT

In what fundamental way does the structure of a sponge differ from that of all other animals?

Answer: A sponge has no tissues.

▼ **Figure 17.9 Anatomy of a sponge.** To obtain enough food to grow by 3 ounces, a sponge must filter roughly 275 gallons of water through its body, enough to fill 3.5 typical-size bathtubs.

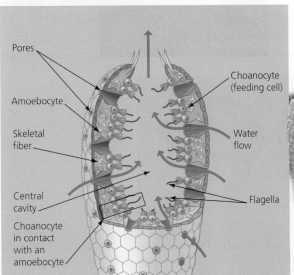

Sponges
Cnidarians
Molluscs
Flatworms
Annelids
Roundworms
Arthropods
Echinoderms
Chordates

Pores
Amoebocyte
Skeletal fiber
Central cavity
Choanocyte in contact with an amoebocyte
Choanocyte (feeding cell)
Water flow
Flagella

Cnidarians

Cnidarians (phylum Cnidaria) are characterized by the presence of body tissues—as are all the remaining animals we will discuss—as well as by radial symmetry and tentacles with stinging cells. Cnidarians include sea anemones, hydras, corals, and jellies (sometimes called jellyfish, although they are not fish). Most of the 10,000 cnidarian species are marine.

The basic body plan of a cnidarian is a sac with a central digestive compartment, the **gastrovascular cavity**. A single opening to this cavity functions as both mouth and anus. This basic body plan has two variations: the stationary **polyp** and the floating **medusa (Figure 17.10)**. Polyps adhere to larger objects and extend their tentacles,

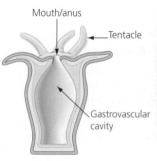

Sponges
Cnidarians
Molluscs
Flatworms
Annelids
Roundworms
Arthropods
Echinoderms
Chordates

waiting for prey. Examples of the polyp body plan are corals, sea anemones, and hydras. A medusa (plural, *medusae*) is a flattened, mouth-down version of the polyp. It moves freely by a combination of passive drifting and contractions of its bell-shaped body. The largest jellies are medusae with tentacles 60–70 m long (more than half the length of a football field) dangling from an umbrella-like body up to 2 m in diameter. There are some species of cnidarians that live only as polyps, others only as medusae, and still others that pass through both a medusa stage and a polyp stage in their life cycle.

Cnidarians are carnivores that use tentacles arranged in a ring around the mouth to capture prey and push the food into the gastrovascular cavity, where digestion begins. The undigested remains are eliminated through the mouth/anus. The tentacles are armed with batteries of cnidocytes ("stinging cells") that function in defense and in the capture of prey **(Figure 17.11)**. The phylum Cnidaria is named for these stinging cells. ☑

☑ CHECKPOINT

In what fundamental way does the body plan of a cnidarian differ from that of other animals?

■ Answer: *The body of a cnidarian is radially symmetric.*

▼ **Figure 17.10 Polyp and medusa forms of cnidarians.** Note that cnidarians have two tissue layers, distinguished in the diagrams by blue and yellow. The gastrovascular cavity has only one opening, which functions as both mouth and anus.

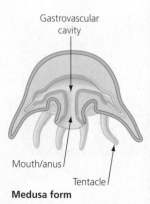

Mouth/anus
Tentacle
Gastrovascular cavity

Polyp form

Sea anemone
Coral
Hydra

Gastrovascular cavity
Mouth/anus
Tentacle

Medusa form

Jelly

▼ **Figure 17.11 Cnidocyte action.** When a trigger on a tentacle is stimulated by touch, a fine thread shoots out from a capsule. Some cnidocyte threads entangle prey, while others puncture the prey and inject a poison.

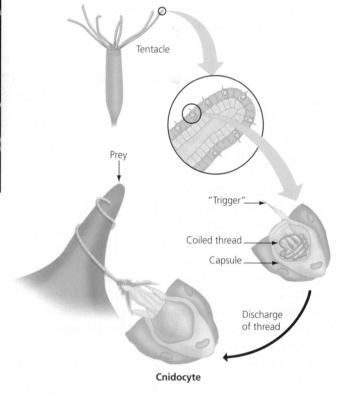

Tentacle
Prey
"Trigger"
Coiled thread
Capsule
Discharge of thread
Cnidocyte

Molluscs

Snails and slugs, oysters and clams, and octopuses and squids are all **molluscs** (phylum Mollusca). Molluscs are soft-bodied animals, but most are protected by a hard shell. Many molluscs feed by extending a file-like organ called a **radula** to scrape up food. For example, the radula of some aquatic snails slides back and forth like a backhoe, scraping and scooping algae off rocks. You can observe a radula in action by watching a snail graze on the glass wall of an aquarium. In cone snails, a group of predatory marine molluscs, the radula is modified to inject venom into prey. The sting of some cone snails is painful, or even fatal, to people.

There are 100,000 known species of molluscs, with most being marine animals. All molluscs have a similar body plan **(Figure 17.12)**. The body has three main parts: a muscular foot, usually used for movement; a visceral mass containing most of the internal organs; and a fold of tissue called the mantle. The **mantle** drapes over the visceral mass and secretes the shell if one is present. The three major groups of molluscs are gastropods, bivalves, and cephalopods **(Figure 17.13)**.

Most **gastropods**, including snails, are protected by a single spiraled shell into which the animal can retreat when threatened. Slugs and sea slugs lack shells. Many gastropods have a distinct head with eyes at the tips of tentacles (think of a garden snail). Marine, freshwater, and terrestrial gastropods make up about three-quarters of the living mollusc species.

The **bivalves**, including clams, oysters, mussels, and scallops, have shells divided into two halves hinged

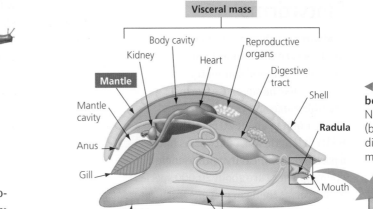

together. None of the bivalves have a radula. There are both marine and freshwater species, with most being sedentary, using their muscular foot for digging and anchoring in sand or mud.

Cephalopods are all marine animals and generally differ from gastropods and sedentary bivalves in that their bodies are fast and agile. A few have large, heavy shells, but in most the shell is small and internal (as in squids) or missing (as in octopuses). Cephalopods have large brains and sophisticated sense organs, which contribute to their success as mobile predators. They use beak-like jaws and a radula to crush or rip apart prey. The mouth is at the base of the foot, which is drawn out into several long tentacles for catching and holding prey. The colossal squid, discovered in the ocean depths near Antarctica, is the largest living invertebrate. Scientists estimate that this massive cephalopod grows to an average length of 13 m—as long as a school bus. ☑

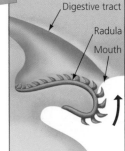

◄ **Figure 17.12 The general body plan of a mollusc.** Note the small body cavity (brown) and the complete digestive tract, with both mouth and anus (pink).

☑ **CHECKPOINT**

Classify these molluscs: A garden snail is an example of a _____; a clam is an example of a _____; a squid is an example of a _____.

■ Answer: gastropod, bivalve, cephalopod

▼ **Figure 17.13 Mollusc diversity.**

MAJOR GROUPS OF MOLLUSCS			
Gastropods	**Bivalves** (hinged shell)	**Cephalopods** (large brain and tentacles)	
Snail (spiraled shell) **Sea slug** (no shell)	**Scallop.** This scallop has many eyes (small round structures) peering out between the two halves of the hinged shell.	**Octopus.** Octopuses live on the seafloor, where they search for crabs and other food. They have no shell. The brain of an octopus is larger and more complex, proportionate to body size, than that of any other invertebrate.	**Nautilus.** The shell of the nautilus is a coiled series of chambers. The animal inhabits only the outermost chamber; the other chambers contain gas and fluid that enable the nautilus to regulate its buoyancy.

Flatworms

Flatworms (phylum Platyhelminthes) are the simplest animals with bilateral symmetry. True to their name, these worms are ribbonlike and range from about 1 mm to about 20 m (about 65 ft) in length. Most flatworms have a gastrovascular cavity with a single opening. There are about 20,000 species of flatworms living in marine, freshwater, and damp terrestrial habitats **(Figure 17.14)**.

The gastrovascular cavity of free-living flatworms called planarians is highly branched, providing an extensive surface area for the absorption of nutrients. When the animal feeds, a muscular tube projects through the mouth and sucks food in. Planarians live on the undersurfaces of rocks in freshwater ponds and streams.

Parasitic flatworms include blood flukes called schistosomes, which are a major health problem in the tropics. These worms have suckers that attach to the inside of the blood vessels near the human host's intestines. Infection by these flatworms causes a long-lasting disease called schistosomiasis, with such symptoms as severe abdominal pain, anemia, and dysentery. Although schistosomes are not found in the United States, more than 200 million people around the world are infected by these parasites each year.

Tapeworms parasitize many vertebrates, including people. Most tapeworms have a very long, ribbonlike body with repeated parts. There is no mouth and no gastrovascular cavity. The head of a tapeworm is equipped with suckers and hooks that lock the worm to the intestinal lining of the host—a gruesomely effective example of structure fitting function. Bathed in partially digested food in the intestines of its host, the tapeworm simply absorbs nutrients across its body surface. Behind the head is a long ribbon of units that are little more than sacs of sex organs. At the back of the worm, mature units containing thousands of eggs break off and leave the host's body with the feces. People can become infected with tapeworms by eating undercooked beef, pork, or fish containing tapeworm larvae. The larvae are microscopic, but the adults can reach lengths of 2 m (6.5 ft) in the intestine. Such large tapeworms can cause intestinal blockage and rob enough nutrients from the host to cause nutritional deficiencies. Fortunately, an orally administered drug can kill the adult worms. ✅

FISH TAPEWORM LARVAE ARE FOUND IN SPECIES THAT LIVE OR BREED IN FRESH WATER.

✅ CHECKPOINT

Flatworms are the simplest animals to display a body plan that is _____.

▼ Figure 17.14 **Flatworm diversity.**

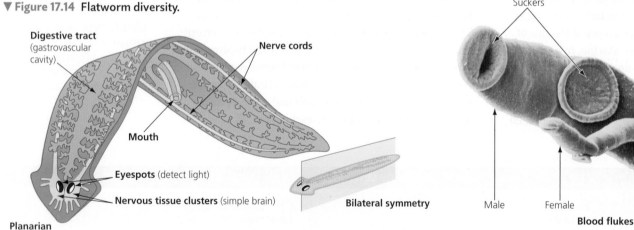

Planarian

- Digestive tract (gastrovascular cavity)
- Nerve cords
- Mouth
- Eyespots (detect light)
- Nervous tissue clusters (simple brain)
- Bilateral symmetry

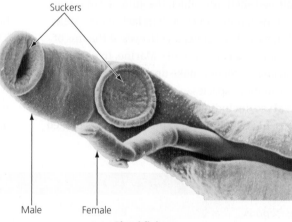

Suckers

Male   Female

Blood flukes

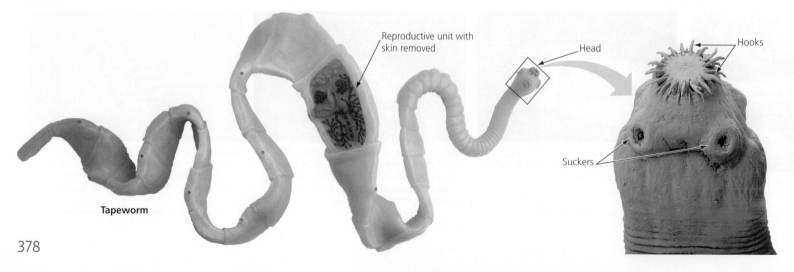

Tapeworm

Reproductive unit with skin removed

Head

Hooks

Suckers

Annelids

Annelids (phylum Annelida) are worms that have **body segmentation**, which is the subdivision of the body along its length into a series of repeated parts called segments. In annelids, the segments look like a set of fused rings. There are about 16,500 annelid species, ranging in length from less than 1 mm to the giant Australian earthworm, which can grow up to 3 m (nearly 10 ft) long. Annelids live in damp soil, the sea, and most freshwater habitats. Like all other bilateral animals except flatworms, annelids have a **complete digestive tract**, which is a digestive tube with two openings: a mouth and an anus. A complete digestive tract can process food and absorb nutrients as a meal moves in one direction from one specialized digestive organ to the next. In people, for example, the mouth, stomach, and intestines act as digestive organs.

Earthworms, like all annelids, are segmented both externally and internally **(Figure 17.15)**. The body cavity is partitioned by walls (only two segment walls are fully shown here). Many of the internal structures, such as the nervous system (yellow in the figure) and organs that dispose of fluid wastes (green) are repeated in each segment. Segmental blood vessels include one main heart and five pairs of accessory hearts. The digestive tract, however, is not segmented; it passes through the segment walls from the mouth to the anus.

Earthworms eat their way through the soil, extracting nutrients from organic matter as the soil passes through the digestive tract. Undigested material is eliminated as castings through the anus. Farmers and gardeners value earthworms because the animals aerate the soil and

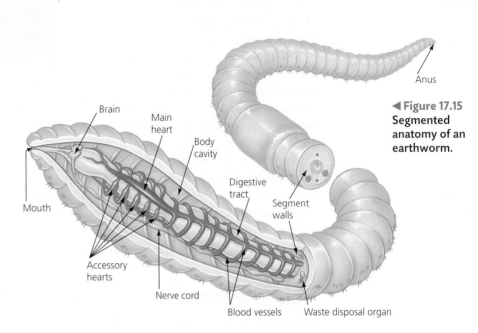

Sponges
Cnidarians
Molluscs
Flatworms
Annelids
Roundworms
Arthropods
Echinoderms
Chordates

Brain
Main heart
Body cavity
Digestive tract
Segment walls
Mouth
Accessory hearts
Nerve cord
Blood vessels
Waste disposal organ
Anus

◄ **Figure 17.15** Segmented anatomy of an earthworm.

increase the amount of mineral nutrients available to plants. Because crops provide the earthworms' food, land use and farming practices have a huge effect on earthworm populations. In one study, for example, corn fields that were plowed yearly averaged 39,000 earthworms per acre, while populations in unplowed dairy pastures numbered at least 1,333,000 per acre.

Recent molecular evidence has identified two major groups of annelids. Their names, Errantia (from the old French *errant*, traveling) and Sedentaria (from the Latin *sedere*, sit), reflect very different ways of life **(Figure 17.16)**.

Most **errantians** are marine and many have active, mobile ways of life. Some errantians, including the ragworm in Figure 17.16, crawl or burrow in the sediments;

▼ **Figure 17.16** Annelid diversity.

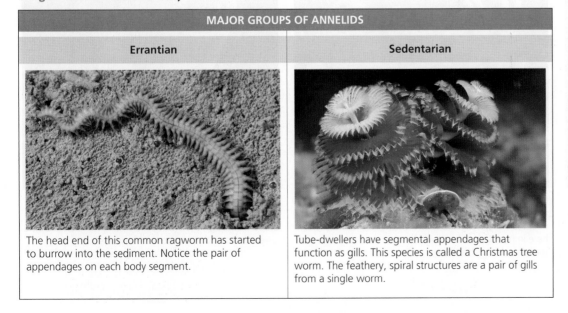

MAJOR GROUPS OF ANNELIDS	
Errantian	**Sedentarian**
The head end of this common ragworm has started to burrow into the sediment. Notice the pair of appendages on each body segment.	Tube-dwellers have segmental appendages that function as gills. This species is called a Christmas tree worm. The feathery, spiral structures are a pair of gills from a single worm.

others are free-swimming. Their heads are well-equipped with sensory organs for moving about in search of small invertebrates to eat. Segmental appendages aid locomotion in many species.

Sedentarians, which include earthworms, many tube-dwellers, and leeches, tend to be less mobile than errantians. Tube-dwellers, such as the one shown in Figure 17.16, build tubes by secreting calcium carbonate or by mixing mucus with bits of sand and broken shells. The circlet of feathery tentacles seen at the mouth of each tube extends from the head of the worm inside. The tentacles are coated with mucus that traps suspended food particles. The tentacles also function in gas exchange. Leeches are notorious for the bloodsucking

Medicinal leech

habits of some species. However, most species are free-living carnivores that eat small invertebrates such as snails and insects. A few terrestrial species inhabit damp vegetation in the tropics, but the majority of leeches live in fresh water. Leeches that feed on blood have razor-like jaws with hundreds of tiny teeth that cut through the skin. They secrete saliva containing an anesthetic and an anticoagulant into the wound. The anesthetic makes the bite virtually painless, and the anticoagulant prevents clotting as the leech drains excess blood from the wound. A European freshwater species called *Hirudo medicinalis* is occasionally used to drain blood from bruised tissues and to help relieve swelling in fingers or toes that have been sewn back on after accidents. ☑

☑ **CHECKPOINT**

The body plan of an annelid displays _____, meaning that the body is divided into a series of repeated regions.

■ *Answer: segmentation*

Roundworms

Roundworms (also called **nematodes**, members of the phylum Nematoda) get their common name from their cylindrical body, which is usually tapered at both ends **(Figure 17.17)**. Roundworms are among the most numerous and widespread of all animals. About 25,000 species of roundworms are known, and perhaps ten times that number actually exist. Roundworms range in length from about 1 mm to 1 m. They are found in most aquatic habitats, in wet soil,

Sponges
Cnidarians
Molluscs
Flatworms
Annelids
Roundworms
Arthropods
Echinoderms
Chordates

and as parasites in the body fluids and tissues of plants and animals.

Free-living roundworms are important decomposers. They live virtually everywhere there is decaying organic matter, and their numbers are huge. Ninety thousand nematodes have been found in a single rotting apple. Recently, researchers even found nematodes living 2 miles underground, where they survive by grazing on microbes. Other species of nematodes thrive as parasites in plants and animals. Some are major agricultural pests that attack the roots of plants. At least 50 parasitic roundworm species infect people; they include pinworms, hookworms, and the parasite that causes trichinosis. ☑

☑ **CHECKPOINT**

Which phylum is most closely related to the roundworms? (*Hint:* Refer to the phylogenetic tree.)

■ *Answer: arthropods*

▼ **Figure 17.17 Roundworm diversity.**

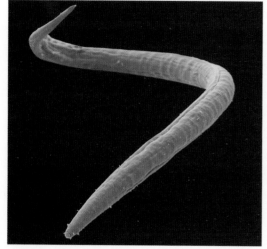

(a) A free-living roundworm. This species has the classic roundworm shape: cylindrical with tapered ends. The ridges indicate muscles that run the length of the body.

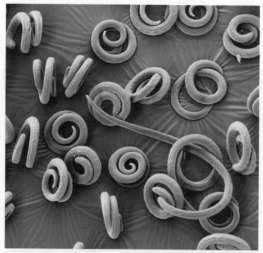

(b) Parasitic roundworms in pork. The potentially fatal disease trichinosis is caused by eating undercooked pork infected with *Trichinella* roundworms. The worms (shown here in pork tissue) burrow into a person's intestine and then invade muscle tissue.

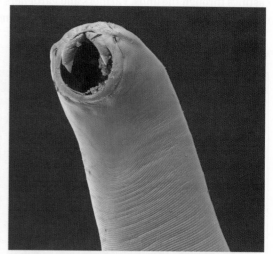

(c) Head of hookworm. Hookworms sink their hooks into the wall of the host's small intestine and feed on blood. Although the worms are small (less than 1 cm), a severe infestation can cause serious anemia.

Arthropods

Arthropods (phylum Arthropoda) are named for their jointed appendages. Crustaceans (such as crabs and lobsters), arachnids (such as spiders and scorpions), and insects (such as grasshoppers and moths) are examples of arthropods **(Figure 17.18)**. Zoologists estimate that the total arthropod population numbers about a billion billion (10^{18}) individuals. Researchers have identified more than a million arthropod species, mostly insects. In fact, two out of every three species of life that have been scientifically described are arthropods. And arthropods are represented in nearly all habitats of the biosphere. In species diversity, distribution, and sheer numbers, arthropods must be regarded as the most successful animal phylum.

Sponges
Cnidarians
Molluscs
Flatworms
Annelids
Roundworms
Arthropods
Echinoderms
Chordates

◀ **Figure 17.18** Arthropod diversity.

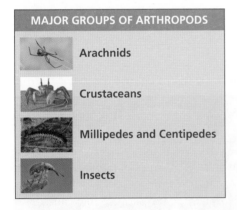

MAJOR GROUPS OF ARTHROPODS

Arachnids

Crustaceans

Millipedes and Centipedes

Insects

General Characteristics of Arthropods

Arthropods are segmented animals. In contrast with the repeating similar segments of annelids, however, arthropod segments and their appendages have become specialized for a great variety of functions. This evolutionary flexibility contributed to the great diversification of arthropods. Specialization of segments (or of fused groups of segments) provides for an efficient division of labor among body regions. For example, the appendages of different segments may be adapted for walking, feeding, sensory reception, swimming, or defense **(Figure 17.19)**.

The body of an arthropod is completely covered by an **exoskeleton**, an external skeleton. This coat is constructed from layers of protein and a tough polysaccharide called chitin. The exoskeleton can be a thick, hard armor over some parts of the body (such as the head), yet be paper-thin and flexible in other locations (such as the joints). The exoskeleton protects the animal and provides points of attachment for the muscles that move the appendages. There are, of course, advantages to wearing hard parts on the outside. Our own skeleton is interior to most of our soft tissues, an arrangement that doesn't provide much protection from injury. But our skeleton does offer the advantage of being able to grow along with the rest of our body. In contrast, a growing arthropod must occasionally shed its old exoskeleton and secrete a larger one. This process, called molting, leaves the animal temporarily vulnerable to predators and other dangers. The following sections explore the major groups of arthropods. ☑

☑ CHECKPOINT

Seafood lovers look forward to soft-shell crab season, when almost the entire animal can be eaten without cracking a hard shell to get at the meat. An individual crab is only a "soft-shell" for a matter of hours; harvesters therefore capture crabs and keep them in holding tanks until they are ready to be cooked. What process makes soft-shell crabs possible?

■ *Answer: Crabs have a thick exoskeleton that must be shed (molted) to allow the crab's body to grow. Soft-shell crabs have molted but have not yet secreted a new hard shell.*

◀ **Figure 17.19 Anatomy of a lobster, a crustacean.** The whole body, including the appendages, is covered by an exoskeleton. The body is segmented, but this characteristic is obvious only in the abdomen.

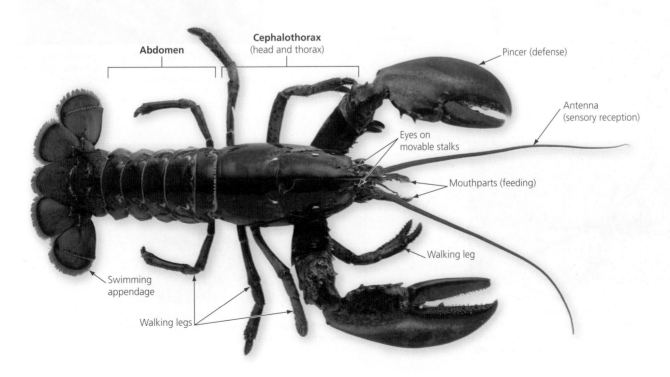

Abdomen

Cephalothorax (head and thorax)

Pincer (defense)

Antenna (sensory reception)

Eyes on movable stalks

Mouthparts (feeding)

Walking leg

Swimming appendage

Walking legs

Arachnids

Arachnids include scorpions, spiders, ticks, and mites (Figure 17.20). Most arachnids live on land. Members of this arthropod group usually have four pairs of walking legs and a specialized pair of feeding appendages.

In spiders, these feeding appendages are fang-like and equipped with poison glands. As a spider uses these appendages to immobilize and dismantle its prey, it spills digestive juices onto the torn tissues and sucks up its liquid meal.

▼ **Figure 17.20 Arachnid characteristics and diversity.**

Pair of feeding appendages

Leg (four pairs)

Pair of silk-spinning appendages

Scorpion. Scorpions have a pair of large pincers that function in defense and food capture. The tip of the tail bears a poisonous stinger. Scorpions sting people only when prodded or stepped on.

Dust mite. This microscopic house dust mite is a ubiquitous scavenger in our homes. Dust mites are harmless except to people who are allergic to the mites' feces.

Spider. Like most spiders, including the tarantula in the large photo above, this black widow spins a web of liquid silk, which solidifies as it comes out of specialized glands. A black widow's venom can kill small prey but is rarely fatal to humans.

Wood tick. Wood ticks and other species carry bacteria that cause Rocky Mountain spotted fever. Lyme disease is carried by several different species of ticks.

Crustaceans

Crustaceans, which are nearly all aquatic, include delectable species of crabs, lobsters, crayfish, and shrimps (Figure 17.21). Barnacles, which anchor themselves to rocks, boat hulls, and even whales, are also crustaceans. One group of crustaceans, the isopods, is represented on land by pill bugs. All of these animals exhibit the arthropod characteristic of multiple pairs of specialized appendages.

▼ **Figure 17.21 Crustacean characteristics and diversity.**

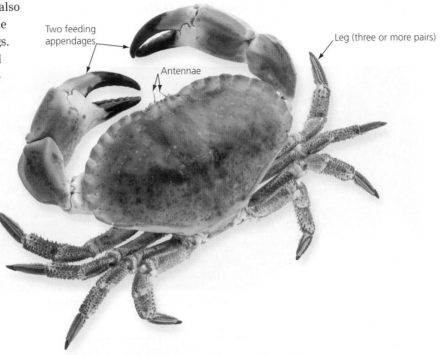

Two feeding appendages

Antennae

Leg (three or more pairs)

Crab. Ghost crabs are common along shorelines throughout the world. They scurry along the surf's edge and then quickly bury themselves in sand.

Shrimp. Naturally found in Pacific waters from Africa to Asia, giant prawns are widely cultivated as food.

Pill bug. Commonly found in moist locations with decaying leaves, such as under logs, pill bugs get their name from their tendency to roll up into a tight ball when they sense danger.

Crayfish. The red swamp crayfish, cultivated worldwide for food, is native to the southeastern United States. When released outside its native range, it competes aggressively with other species, endangering the local ecosystem.

Barnacles. Barnacles are stationary crustaceans with exoskeletons hardened into shells by calcium carbonate (lime). The jointed appendages projecting from the shell capture small plankton.

Millipedes and Centipedes

Millipedes and **centipedes** are terrestrial arthropods that have similar segments over most of the body. Although they superficially resemble annelids, their jointed legs reveal they are arthropods **(Figure 17.22)**.

Millipedes eat decaying plant matter. They have two pairs of short legs per body segment. Centipedes are carnivores, with a pair of poison claws used in defense and to paralyze prey, such as cockroaches and flies. Each of a centipede's body segments bears a single pair of legs.

▶ **Figure 17.22** Millipedes and centipedes.

Two pairs of legs per segment

One pair of legs per segment

Millipede. Like most millipedes, this one has an elongated body with two pairs of legs per trunk segment.

Centipede. Centipedes can be found in dirt and leaf litter. Their venomous claws can harm cockroaches and spiders but not people.

Insect Anatomy

Like the grasshopper in **Figure 17.23**, most **insects** have a three-part body: head, thorax, and abdomen. The head usually bears a pair of sensory antennae and a pair of eyes. The mouthparts of insects are adapted for particular kinds of eating—for example, for biting and chewing plant material in grasshoppers, for lapping up fluids in houseflies, and for piercing skin and sucking blood in mosquitoes. Most adult insects have three pairs of legs and one or two pairs of wings, all extending from the thorax.

Flight is obviously one key to the great success of insects. An animal that can fly can escape many predators, more readily find food and mates, and disperse to new habitats much faster than an animal that must crawl on the ground. Because their wings are extensions of the exoskeleton and not true appendages, insects can fly without sacrificing legs. By contrast, the flying vertebrates—birds and bats—have one of their two pairs of legs modified for wings, which explains why these vertebrates are generally not very swift on the ground.

▼ **Figure 17.23** Anatomy of a grasshopper.

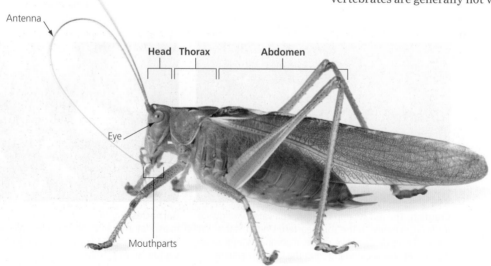

Antenna

Head   Thorax   Abdomen

Eye

Mouthparts

Insect Diversity

In species diversity, insects outnumber all other forms of life combined (Figure 17.24). They live in almost every terrestrial habitat and in fresh water, and flying insects fill the air. Insects are rare in the seas, where crustaceans are the dominant arthropods. The oldest insect fossils date back to about 400 million years ago. Later, the evolution of flight sparked an explosion in insect variety.

▼ Figure 17.24 Insect diversity.

Rhinoceros beetle. Only males have a "horn," which is used for fighting other males and for digging.

Rainbow shield bugs. The smaller insects are juveniles of the same species.

Greater arid-land katydid. Often called the red-eyed devil, this scary-looking predator displays its wings and spiny legs when threatened.

Robber fly. This predatory insect injects prey with enzymes that liquefy the tissues and then sucks in the resulting fluid.

Praying mantis. There are more then 2,000 species of praying mantises throughout the world.

Weevil. Light reflected from varying thicknesses of chitin produces the brilliant colors of this weevil.

Buckeye butterfly. The eyespots on the wings of this butterfly may startle predators.

Blue dasher dragonfly. The huge multifaceted eyes of dragonflies wrap around the head for nearly 360° vision. Each of the four wings works independently, giving dragonflies excellent maneuverability as they pursue prey.

Many insects undergo metamorphosis in their development. In the case of grasshoppers and some other insect groups, the young resemble adults but are smaller and have different body proportions. The animal goes through a series of molts, each time looking more like an adult, until it reaches full size. In other cases, insects have distinctive larval stages specialized for eating and growing that are known by such names as maggots (fly larvae), grubs (beetle larvae), or caterpillars (larvae of moths and butterflies). The larval stage looks entirely different from the adult stage, which is specialized for dispersal and reproduction. Metamorphosis from the larva to the adult occurs during a pupal stage **(Figure 17.25)**.

Forensic entomologists (forensic scientists who study insects) use their knowledge of insect life cycles to help solve criminal cases. For example, blowfly maggots feed on decaying flesh. Female blowflies, which can smell a dead body from up to a mile away, typically arrive and lay their eggs in a fresh corpse within minutes. By knowing the length of each stage in the life cycle of a blowfly, an entomologist can determine how much time has passed since death occurred.

Animals so numerous, diverse, and widespread as insects are bound to affect the lives of all other terrestrial organisms, including people, in many ways. Thus, **insects provide many examples of interactions within biological systems.** The bees, flies, and other insects that pollinate our crops and orchards exemplify a beneficial interaction. Other interactions are harmful to people. For example, insects are carriers of the microbes that cause many human diseases, such as malaria and West Nile disease. Insects also compete with people for food by eating our field crops. In an effort to minimize their losses, farmers in the United States spend billions of dollars each year on pesticides, spraying crops with massive doses of insecticide poisons. But try as they might, not even humans have significantly challenged the preeminence of insects and their arthropod kin. Rather, the evolution of pesticide resistance has caused humans to change their pest-control methods (see Figure 13.11). ☑

☑ **CHECKPOINT**

Which major arthropod group is mainly aquatic? Which is the most numerous?

■ *Answer: crustaceans; insects*

▼ **Figure 17.25 Metamorphosis of a monarch butterfly.**

The **larva (caterpillar)** spends its time eating and growing, molting as it grows.

After several molts, the larva becomes a **pupa** encased in a cocoon.

Within the pupa, the larval organs break down and adult organs develop from cells that were dormant in the larva.

Finally, the **adult** emerges from the cocoon.

The butterfly flies off and reproduces, nourished mainly by calories stored when it was a caterpillar.

Echinoderms

The **echinoderms** (phylum Echinodermata) are named for their spiny surfaces (*echin* is Greek for "spiny"). Among the echinoderms are sea stars, sea urchins, sea cucumbers, and sand dollars (Figure 17.26).

Echinoderms include about 7,000 species, all of them marine. Most move slowly, if at all. Echinoderms lack body segments, and most have radial symmetry as adults. Both the external and the internal parts of a sea star, for instance, radiate from the center like the spokes of a wheel. In contrast to the adult, the larval stage of echinoderms is bilaterally symmetrical. This supports other evidence that echinoderms are not closely related to other radial animals, such as cnidarians, that never show bilateral symmetry. Most echinoderms have an **endoskeleton** (interior skeleton) constructed from hard plates just beneath the skin.

Bumps and spines of this endoskeleton account for the animal's rough or prickly surface. Unique to echinoderms is the **water vascular system**, a network of water-filled canals that circulate water throughout the echinoderm's body, facilitating gas exchange (the entry of O_2 and the removal of CO_2) and waste disposal. The water vascular system also branches into extensions called tube feet. A sea star or sea urchin pulls itself slowly over the seafloor using its suction-cup-like tube feet. Sea stars also use their tube feet to grip prey during feeding.

Looking at sea stars and other adult echinoderms, you may think they have little in common with humans and other vertebrates. But as shown by the phylogenetic tree at the left, echinoderms share an evolutionary branch with chordates, the phylum that includes vertebrates. Analysis of embryonic development can differentiate the echinoderms and chordates from the evolutionary branch that includes molluscs, flatworms, annelids, roundworms, and arthropods. With this context in mind, we're now ready to make the transition in our discussion from invertebrates to vertebrates. ☑

Sponges
Cnidarians
Molluscs
Flatworms
Annelids
Roundworms
Arthropods
Echinoderms
Chordates

☑ CHECKPOINT

Contrast the skeleton of an echinoderm with that of an arthropod.

■ Answer: An echinoderm has an endoskeleton; an arthropod has an exoskeleton.

▼ Figure 17.26 Echinoderm diversity.

Sea star. When a sea star encounters an oyster or clam, it grips the mollusc's shell with its tube feet (see inset) and positions its mouth next to the narrow opening between the two halves of the prey's shell. The sea star then pushes its stomach out through its mouth and the crack in the mollusc's shell.

Sea urchin. In contrast to sea stars, sea urchins are spherical and have no arms. If you look closely, you can see the long tube feet projecting among the spines. Unlike sea stars, which are mostly carnivorous, sea urchins mainly graze on seaweed and other algae.

Tube feet

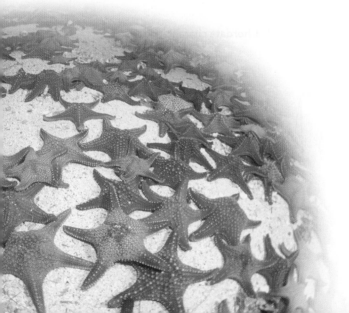

Sea cucumber. On casual inspection, this sea cucumber does not look much like other echinoderms. However, a closer look would reveal many echinoderm traits, including five rows of tube feet.

Sand dollar. Live sand dollars have a skin of short, movable spines covering a rigid skeleton. A set of five pores (arranged in a star pattern) allows seawater to be drawn into the sand dollar's body.

Vertebrate Evolution and Diversity

Most of us are curious about our family ancestry. Biologists are also interested in the larger question of tracing human ancestry within the animal kingdom. In this section, we trace the evolution of the vertebrates, the group that includes humans and our closest relatives. All vertebrates have endoskeletons, a characteristic shared with most echinoderms. However, vertebrate endoskeletons are unique in having a skull and a backbone, a series of bones called vertebrae (singular, *vertebra*), for which the group is named (Figure 17.27). Our first step in tracing the vertebrate lineage is to determine where vertebrates fit in the animal kingdom.

▼ Figure 17.27 **A vertebrate endoskeleton.** This snake skeleton, like those of all vertebrates, has a skull and a backbone consisting of vertebrae.

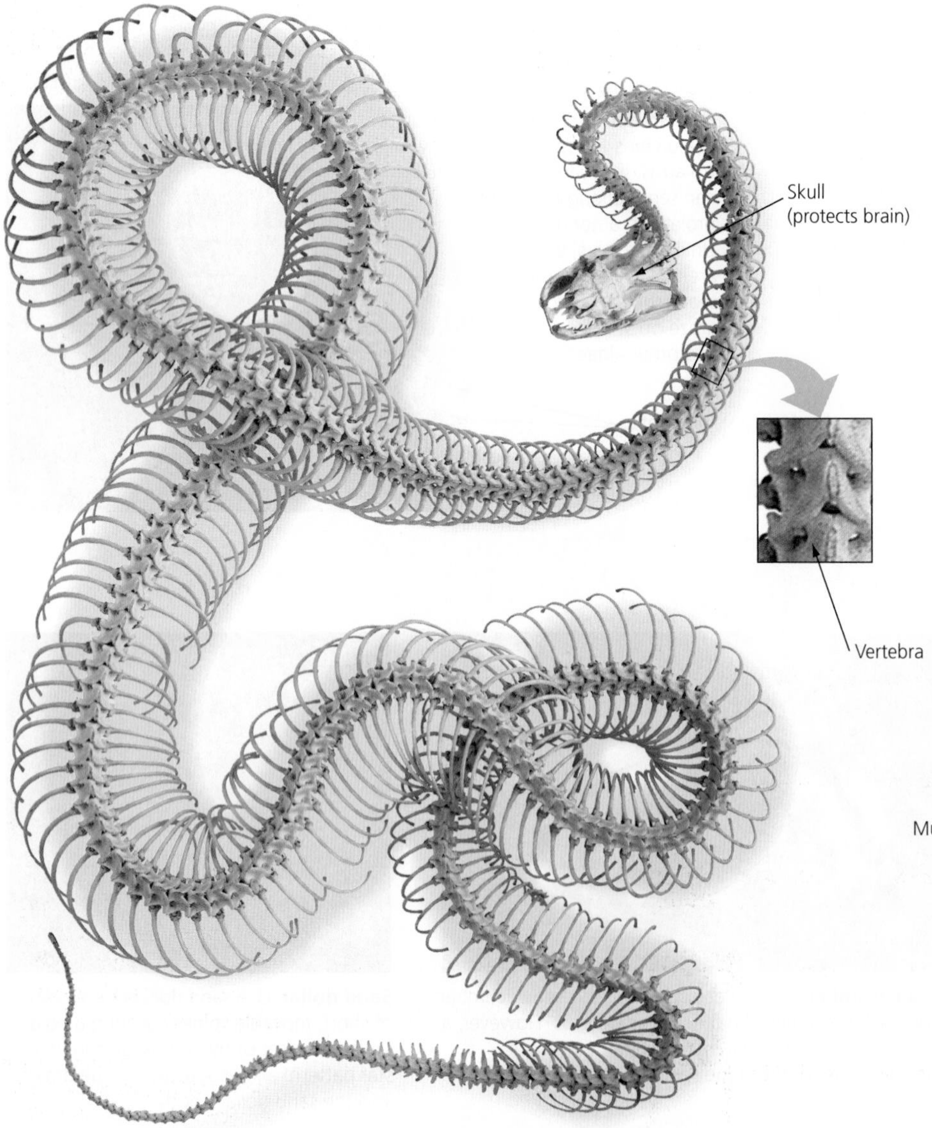

Skull
(protects brain)

Vertebra

Characteristics of Chordates

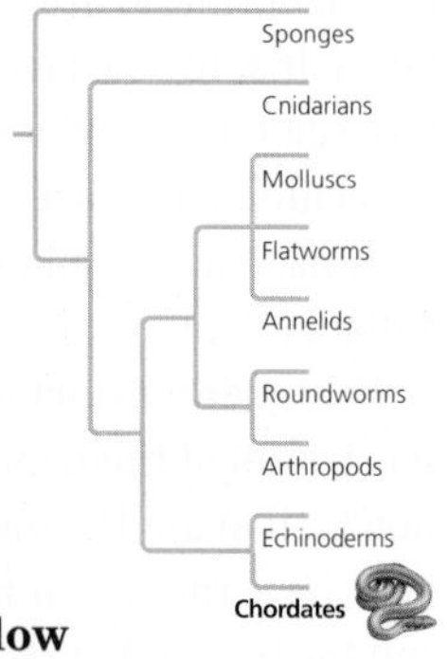

The last phylum in our survey of the animal kingdom is the phylum Chordata. **Chordates** share four key features that appear in the embryo and sometimes in the adult (Figure 17.28). These four chordate characteristics are (1) a **dorsal, hollow nerve cord**; (2) a **notochord**, which is a flexible, longitudinal rod located between the digestive tract and the nerve cord; (3) **pharyngeal slits**, which are grooves in the pharynx, the region of the digestive tube just behind the mouth; and (4) a **post-anal tail**, which is a tail to the rear of the anus. Although these chordate characteristics are often difficult to recognize in the adult animal, they are always present in chordate embryos. For example, the notochord, for which our phylum is named, persists in adult humans only in the form of the cartilage disks that function as cushions between the vertebrae. Back injuries described as "ruptured disks" or "slipped disks" refer to these notochord remnants.

Body segmentation is another chordate characteristic. Chordate segmentation is apparent in the backbone of vertebrates (see Figure 17.27) and is also evident in the segmental muscles of all chordates (see the chevron-shaped—<<<<—muscles in the lancelet in Figure 17.29). Segmental musculature is not so obvious in adult humans unless one is motivated to sculpt "washboard abs."

Sponges

Cnidarians

Molluscs

Flatworms

Annelids

Roundworms

Arthropods

Echinoderms

Chordates

▼ Figure 17.28 **Chordate characteristics.**

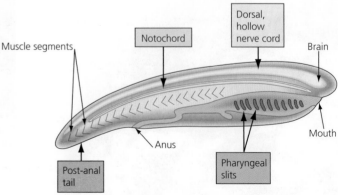

Muscle segments

Notochord

Dorsal, hollow nerve cord

Brain

Mouth

Pharyngeal slits

Post-anal tail

Anus

Two groups of chordates, **tunicates** and **lancelets** (Figure 17.29), are invertebrates. All other chordates are **vertebrates**, which retain the basic chordate characteristics but have additional features that are unique—including, of course, the backbone. **Figure 17.30** is an overview of chordate and vertebrate evolution that will provide a context for our survey. ☑

▼ **Figure 17.29 Invertebrate chordates.**

Lancelet. This marine invertebrate owes its name to its bladelike shape. Only a few centimeters long, lancelets wiggle backward into the gravel, leaving their mouth exposed, and filter tiny food particles from the seawater.

Tunicates. The tunicate, or sea squirt, is a stationary animal that filters food from the water. These pastel sea squirts get their nickname from their coloration and the fact that they can quickly expel water to startle intruders.

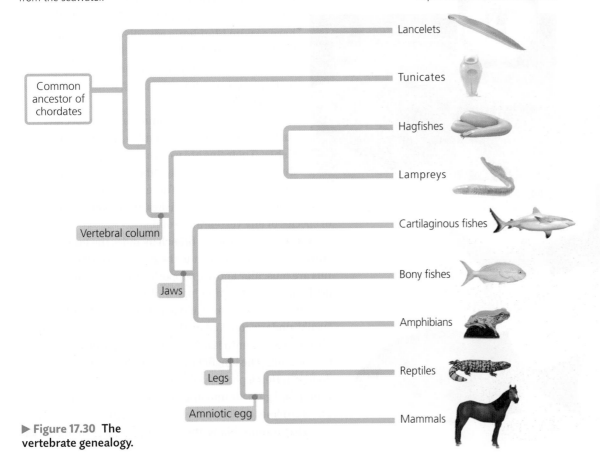

▶ **Figure 17.30 The vertebrate genealogy.**

☑ **CHECKPOINT**

During our early embryonic development, what four features do we share with invertebrate chordates such as lancelets?

■ *Answers: (1) dorsal, hollow nerve cord; (2) notochord; (3) pharyngeal slits; (4) post-anal tail*

Fishes

The first vertebrates were aquatic and probably evolved during the early Cambrian period about 542 million years ago. In contrast with most living vertebrates, they lacked jaws, hinged bone structures that work the mouth.

Two types of jawless fishes survive today: hagfishes and lampreys. Present-day hagfishes scavenge dead or dying animals on the cold, dark seafloor (**Figure 17.31a**). When threatened, a hagfish exudes an enormous amount of slime from special glands on the sides of its body. Recently, some hagfishes have become endangered because their skin is used to make "eel-skin" belts, purses, and boots. Most species of lampreys are parasites that use their jawless mouths as suckers to attach to the sides of large fish (**Figure 17.31b**). The rasping tongue then penetrates the skin, allowing the lamprey to feed on its victim's blood and tissues.

We know from the fossil record that the first jawed vertebrates were fishes that evolved about 440 million years ago. They had two pairs of fins, making them agile swimmers. Some early fishes were active predators up to 10 m (33 ft) in length that could chase prey and bite off chunks of flesh. Even today, most fishes are carnivores.

Cartilaginous fishes, such as sharks and rays, have a flexible skeleton made of cartilage (**Figure 17.31c**). Most sharks are adept predators because they are fast swimmers with streamlined bodies, acute senses, and powerful jaws. A shark does not have keen eyesight, but its sense of smell is very sharp. In addition, special electrosensors on the head can detect minute electrical fields produced by muscle contractions in nearby animals. Sharks also have a **lateral line system**, a row of sensory organs running along each side of the body. Sensitive to changes in water pressure, the lateral line system enables a shark to detect minor vibrations caused by animals swimming in its neighborhood. There are about 1,000 living species of cartilaginous fishes, nearly all of them marine.

The skeletons of **bony fishes** are reinforced by calcium (**Figure 17.31d**). Bony fishes have a lateral line system, a keen sense of smell, and excellent eyesight. On each side of the head, a protective flap called the **operculum** (plural, *opercula*) covers a chamber housing the gills, feathery external organs that extract oxygen from water. Movement of the operculum allows the fish to breathe without swimming. By contrast, sharks lack opercula and must swim to pass water over their gills. The need to move water over the gills is why a shark must keep moving to stay alive. Also unlike sharks, bony fishes have a **swim bladder**, a gas-filled sac that enables the fish to control its buoyancy. Thus, many bony fishes can conserve energy by remaining almost motionless, in contrast to sharks, which sink if they stop swimming.

Most bony fishes, including familiar species such as tuna, trout, and goldfish, are **ray-finned fishes**. Their fins are webs of skin supported by thin, flexible skeletal rays, the feature for which the group was named. There are approximately 27,000 species of ray-finned fishes, the greatest number of species of any vertebrate group.

A second evolutionary branch includes the **lobe-finned fishes**. In contrast to the ray-finned fishes, their fins are muscular and are supported by stout bones that are homologous to amphibian limb bones. Early lobe-fins lived in coastal wetlands and may have used their fins to "walk" underwater. Today, three lineages of lobe-fins survive. The coelacanth is a deep-sea dweller once thought to be extinct. The lungfishes are represented by several Southern Hemisphere species that inhabit stagnant waters and gulp air into lungs connected to the pharynx. The third lineage of lobe-fins adapted to life on land and gave rise to amphibians, the first terrestrial vertebrates. ✓

Lancelets
Tunicates
Hagfishes
Lampreys
Cartilaginous fishes
Bony fishes
Amphibians
Reptiles
Mammals

▼ Figure 17.31 **Fish diversity.**

(a) Hagfish

(b) Lamprey (inset: mouth)

(c) Shark, a cartilaginous fish

Lateral line

Operculum

(d) Bony fish

✓ **CHECKPOINT**

A shark has a _____ skeleton, whereas a tuna has a _____ skeleton.

■ *Answer: cartilaginous; bony*

Amphibians

In Greek, the word *amphibios* means "living a double life." Most **amphibians**, which include frogs and salamanders, exhibit a mixture of aquatic and terrestrial adaptations **(Figure 17.32a)**. Most species are tied to water because their eggs, lacking shells, dry out quickly in the air. A frog may spend much of its time on land, but it lays its eggs in water **(Figure 17.32b)**. An egg develops into a larva called a tadpole, a legless, aquatic algae-eater with gills, a lateral line system resembling that of fishes, and a long-finned tail **(Figure 17.32c)**. The tadpole undergoes a radical metamorphosis when changing into a frog **(Figure 17.32d)**. When a

young frog crawls onto shore and begins life as a terrestrial insect-eater, it has four legs, air-breathing lungs instead of gills, external eardrums, and no lateral line system. But even as adults, amphibians are most abundant in damp habitats, such as swamps and rain forests. This is partly because amphibians depend on their moist skin to supplement lung function in exchanging gases with the environment. Thus, even those frogs that are adapted to relatively dry habitats spend much of their time in humid burrows or under piles of damp leaves. With their thin, permeable skin and susceptibility to dehydration, amphibians are especially vulnerable to environmental threats such as water pollution and climate change.

Amphibians were the first vertebrates to colonize land. They descended from fishes that had lungs and fins with muscles and skeletal supports strong enough to enable some movement, however clumsy, on land **(Figure 17.33)**. The fossil record chronicles the evolution of four-limbed amphibians from fish-like ancestors. Terrestrial vertebrates—amphibians, reptiles, and mammals—are collectively called **tetrapods**, which means "four feet." ☑

Lancelets
Tunicates
Hagfishes
Lampreys
Cartilaginous fishes
Bony fishes
Amphibians
Reptiles
Mammals

☑ CHECKPOINT

Amphibians were the first _____, four-footed terrestrial vertebrates.

■ *Answer: tetrapods*

▼ Figure 17.32 **Amphibian diversity.**

Malayan horned frog

Texas barred tiger salamander

(a) Frogs and salamanders: the two major groups of amphibians

(b) Frog eggs

(c) Tadpole

(d) Adult tree frog

▼ Figure 17.33 **The origin of tetrapods.**

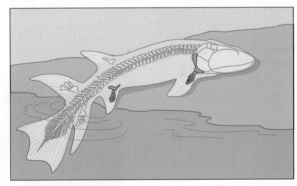

Lobe-finned fish. Fossils of some lobe-finned fishes have skeletal supports extending into their fins.

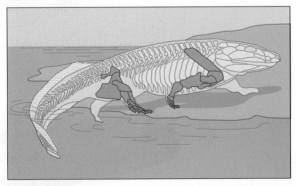

Early amphibian. Fossils of early amphibians have limb skeletons that probably functioned in helping them move on land.

391

Reptiles

Reptiles (including birds) and mammals are **amniotes**. The evolution of amniotes from an amphibian ancestor included many adaptations for living on land. The adaptation that gives the group its name is the **amniotic egg**, a fluid-filled egg with a water-proof shell that encloses the developing embryo **(Figure 17.34)**. The amniotic egg functions as a self-contained "pond" that enables amniotes to complete their life cycle on land.

The **reptiles** include snakes, lizards, turtles, crocodiles, alligators, and birds, along with a number of extinct groups, including most of the dinosaurs. The European grass snake in Figure 17.34 displays two reptilian adaptations to living on land: scaled waterproof skin, preventing dehydration in dry air, and amniotic eggs with shells, providing a watery, nutritious internal environment where the embryo can develop. These adaptations allowed reptiles to break their ancestral ties to aquatic habitats. Reptiles cannot breathe through their dry skin and so obtain most of their oxygen through their lungs.

Nonbird Reptiles

Nonbird reptiles are sometimes referred to as "cold-blooded" animals because they do not use their metabolism extensively to control body temperature. Reptiles do regulate body temperature, but largely through behavioral adaptations. For example, many lizards regulate their internal temperature by basking in the sun when the air is cool and seeking shade when the air is too warm. Because lizards and other nonbird reptiles absorb external heat rather than generating much of their own, they are said to be **ectotherms**, a term more accurate than "cold-blooded." By heating directly with solar energy rather than through the metabolic breakdown of food, a nonbird reptile can survive on less than 10% of the calories required by a mammal of equivalent size.

As successful as reptiles are today, they were far more widespread, numerous, and diverse during the Mesozoic era, which is sometimes known as the "age of reptiles." Reptiles diversified extensively during that era, producing a dynasty that lasted until about 66 million years ago. Dinosaurs, the most diverse reptile group, included the largest animals ever to inhabit land. Some were gentle giants that lumbered about while browsing vegetation. Others were voracious carnivores that chased their larger prey on two legs.

The age of reptiles began to fade about 70 million years ago. Around that time, the global climate became cooler and more variable. This was a period of mass extinctions that claimed all the dinosaurs by about 66 million years ago, except for one lineage (see Table 14.1). That lone surviving lineage is represented today by the reptilian group we know as birds. ☑

☑ **CHECKPOINT**

What is an amniotic egg?

■ *Answer: a shelled egg surrounding fluid that contains an embryo*

▶ **Figure 17.34 Reptile diversity.**

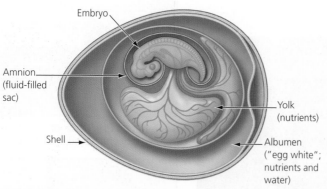

Aminiotic egg. The embryo and its life-support system are enclosed in a waterproof shell.

Embryo

Amnion (fluid-filled sac)

Shell

Yolk (nutrients)

Albumen ("egg white"; nutrients and water)

Snake. The shells of nonbird reptile eggs are leathery and flexible. This European grass snake is nonvenomous. When threatened, it may hiss and strike. If the bluff fails, it goes limp, pretending to be dead.

Lizard. The Gila monster, a desert-dweller of the Southwest, is the only venomous lizard native to the United States. Although large (up to 2 feet long), it moves too slowly to pose any danger to people.

Lancelets
Tunicates
Hagfishes
Lampreys
Cartilaginous fishes
Bony fishes
Amphibians
Reptiles
Mammals

Birds

You may have noticed that the reptilian egg resembles the more familiar chicken egg. Birds have scaly skin on their legs and feet; even feathers—their signature feature—are modified scales. Genetic and fossil evidence shows that **birds** are indeed reptiles, having evolved from a lineage of small, two-legged dinosaurs called theropods. Today, birds look quite different from reptiles because of their distinctive flight equipment—almost all of the 10,000 living bird species are airborne. The few flightless species, including the ostrich and the penguin, evolved from flying ancestors.

Almost every element of bird anatomy is adapted in some way that enhances flight. The bones have a honeycombed structure that makes them strong but light. (The wings of airplanes have the same basic construction.) For example, a huge seagoing species called the frigate bird has a wingspan of more than 2 m (6.6 ft), but its whole skeleton weighs only about 113 g (a mere 4 ounces, less than an iPhone). Another adaptation that reduces the weight of birds is the absence of some internal organs found in other vertebrates. Female birds, for instance, have only one ovary instead of a pair. Also, today's birds are toothless, an adaptation that trims the weight of the head, preventing uncontrolled nosedives. Birds do not chew food in the mouth but grind it in the gizzard, a muscular chamber of the digestive tract near the stomach.

Flying requires a great expenditure of energy and an active metabolism. Unlike other reptiles, birds are **endotherms**, meaning they use their own metabolic heat to maintain a warm, constant body temperature.

A bird's most obvious flight equipment is its wings. Bird wings are airfoils that illustrate the same principles of aerodynamics as the wings of an airplane **(Figure 17.35)**. A bird's flight motors are its powerful breast muscles, which are anchored to a keel-like breastbone. It is mainly these flight muscles that we call "white meat" on chicken and turkey breasts. Some birds, such as eagles and hawks, have wings adapted for soaring on air currents and flap their wings only occasionally. Other birds, including hummingbirds, excel at maneuvering but must flap continuously to stay aloft. Feathers are made of the same protein that forms the scales of reptiles. Feathers may have functioned first as insulation, helping birds retain body heat, or for courtship displays. Only later were they adapted as flight gear. ☑

☑ **CHECKPOINT**

Birds differ from other reptiles in their main source of body heat, with birds being _____ and other reptiles being _____.

◄ Answer: endotherms; ectotherms

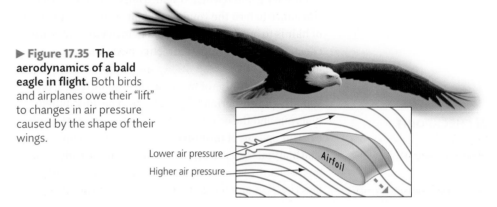

▶ **Figure 17.35 The aerodynamics of a bald eagle in flight.** Both birds and airplanes owe their "lift" to changes in air pressure caused by the shape of their wings.

Lower air pressure

Higher air pressure

Airfoil

Crocodile. Found throughout central and southern Africa, the Nile crocodile can grow up to 20 feet long and weigh nearly 2,000 pounds.

Birds. These red-crowned cranes, native to China, are performing an elaborate courtship dance.

Dinosaur. This *Herrerasaurus* skeleton is from a carnivorous theropod discovered in Argentina.

393

Mammals

There are two major lineages of amniotes: one that led to the reptiles and one that produced the mammals. The first **mammals** arose about 200 million years ago and were probably small, nocturnal insect-eaters. Mammals became much more diverse after the downfall of the dinosaurs. Most mammals are terrestrial. These include nearly 1,000 species of winged mammals, the bats. And the roughly 80 species of dolphins, porpoises, and whales are totally aquatic. The blue whale—an endangered mammal that grows to lengths of nearly 30 m (about as long as a basketball court)—is the largest animal that has ever lived. Mammals have two unique characteristics: mammary glands (which produce milk, a nutrient-rich substance to feed the young) and hair. The main function of hair is to insulate the body and help maintain a warm, constant internal temperature; like birds, mammals are endotherms.

The major groups of mammals are monotremes, marsupials, and eutherians **(Figure 17.36)**. The duck-billed platypus and the echidna, or spiny anteater, are the only existing species of **monotremes**, egg-laying mammals. The platypus lives along rivers in eastern Australia and on the nearby island of Tasmania. The female usually lays two eggs and incubates them in a leaf nest. After hatching, the young nurse by licking up milk secreted by rudimentary mammary glands onto the mother's fur.

Most mammals are born rather than hatched. During pregnancy in marsupials and eutherians, the embryos are nurtured inside the mother by an organ called the **placenta**. Consisting of both embryonic and maternal tissues, the placenta joins the embryo to the mother within the uterus. The embryo receives oxygen and nutrients from maternal blood that flows close to the embryonic blood system in the placenta.

Marsupials, the so-called pouched mammals, include kangaroos, koalas, and opossums. These mammals have a brief pregnancy and give birth to embryonic offspring. After birth, the tiny offspring make their way to an external pouch on the mother's abdomen, where they attach to mammary glands. Nearly all marsupials live in Australia, New Zealand, and North and South America. Australian marsupials have diversified extensively, filling terrestrial habitats that on the other continents are occupied by eutherian mammals (see Figure 14.17).

Eutherians are also called **placental mammals** because their placentas provide a more intimate and longer-lasting association between the mother and her developing young than do marsupial placentas. Eutherians make up almost 95% of the 5,300 species of living mammals. Dogs, cats, cows, rodents, rabbits, bats, and whales are all examples of eutherian mammals. One of the eutherian groups is the primates, which include monkeys, apes, and humans. ☑

Lancelets

Tunicates

Hagfishes

Lampreys

Cartilaginous fishes

Bony fishes

Amphibians

Reptiles

Mammals

☑ **CHECKPOINT**

What are two defining characteristics of mammals?

■ *Answer: mammary glands and hair*

▼ **Figure 17.36** Mammalian diversity.

MAJOR GROUPS OF MAMMALS		
Monotremes (hatched from eggs)	**Marsupials** (embryonic at birth)	**Eutherians** (fully developed at birth)

Monotremes, such as this duck-billed platypus, are the only mammals that lay eggs. Like other mammals, platypus mothers nourish their young with milk.

The young of marsupials are born very early in their development. The newborn kangaroo will finish its growth while nursing from a nipple in its mother's pouch.

In eutherians (placental mammals), young develop within the uterus of the mother. There they are nurtured by the flow of blood through the dense network of vessels in the placenta. This newborn foal is coated by remnants of the placenta.

The Human Ancestry

We have now traced animal phylogeny to the **primates**, the mammalian group that includes us—*Homo sapiens*—and our closest kin. To understand what that means, we must follow our ancestry back to the trees, where some of our most treasured traits originated.

The Evolution of Primates

Primate evolution provides a context for understanding human origins. The fossil record supports the hypothesis that primates evolved from insect-eating mammals during the late Cretaceous period, about 65 million years ago. Those early primates were small, arboreal (tree-dwelling) mammals. Thus, primates were first distinguished by characteristics that were shaped, through natural selection, by the demands of living in the trees. For example, primates have limber shoulder joints, which make it possible to swing from branch to branch. The agile hands of primates can hang on to branches and manipulate food. Nails have replaced claws in many primate species, and the fingertips are very sensitive. The eyes of primates are close together on the front of the face. The overlapping fields of vision of the two eyes enhance depth perception, an obvious advantage when swinging in trees. Excellent eye-hand coordination is also important for arboreal maneuvering. Parental care is essential for young animals in the trees. Mammals devote more energy to caring for their young than most other vertebrates, and primates are among the most attentive parents of all mammals. Most primates have single births and nurture their offspring for a long time. Although humans never lived in trees, we retain in modified form many traits that originated there.

Taxonomists divide the primates into three main groups **(Figure 17.37)**. The first includes lemurs, lorises, and bush babies. These primates live in Madagascar, southern Asia, and Africa. Tarsiers, small nocturnal tree-dwellers found only in Southeast Asia, form the second group of primates. The third group of primates, **anthropoids**, includes monkeys and apes. All monkeys in the New World (the Americas) are arboreal and are distinguished by prehensile (grasping) tails that function as an extra appendage for swinging. If you see a monkey in a zoo swinging by its tail, you know it's from the New World. Although some Old World (African and Asian) monkeys are also arboreal, their tails are not prehensile. And many Old World monkeys, including baboons, macaques, and mandrills, are mainly ground-dwellers. Anthropoids also have a fully

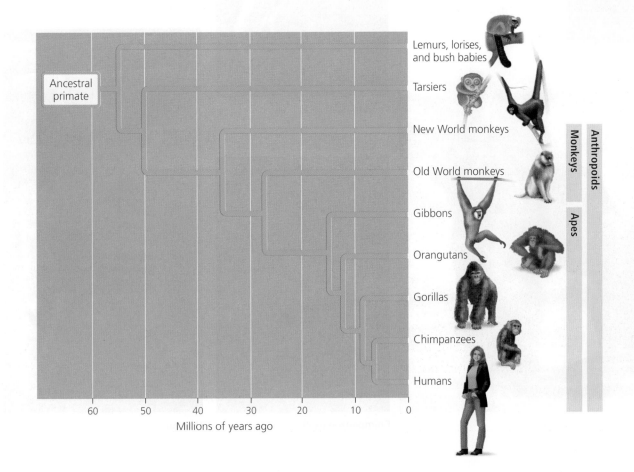

◀ Figure 17.37 Primate phylogeny.

Figure Walkthrough

Mastering **Biology**
goo.gl/qFiJkL

Ancestral primate

Lemurs, lorises, and bush babies

Tarsiers

New World monkeys

Old World monkeys

Gibbons

Orangutans

Gorillas

Chimpanzees

Humans

Monkeys

Apes

Anthropoids

60 50 40 30 20 10 0
Millions of years ago

opposable thumb; that is, they can touch the tips of all four fingers with their thumb.

Our closest anthropoid relatives are the nonhuman apes: gibbons, orangutans, gorillas, and chimpanzees. They live only in tropical regions of the Old World. Except for some gibbons, apes are larger than monkeys, with relatively long arms and short legs, and most have no tail. Although all apes are capable of living in trees, only gibbons and orangutans are primarily arboreal. Gorillas and chimpanzees are highly social. Apes have larger brains proportionate to body size than monkeys, and their behavior is more adaptable. And, of course, the apes include humans. **Figure 17.38** shows examples of primates.

▼ **Figure 17.38 Primate diversity.**

Red ruffed lemur

Tarsier

Black spider monkey
(New World monkey)

Orangutan (ape)

Patas monkey (Old World monkey)

Gibbon (ape)

Gorilla (ape)

Chimpanzee (ape)

Human

The Emergence of Humankind

Humanity is one very young twig on the tree of life. In the continuum of life spanning 3.5 billion years, the fossil record and molecular systematics indicate that humans and chimpanzees have shared a common African ancestry for all but the last 6–7 million years (see Figure 17.37). Put another way, if we compressed the history of life to a year, the human branch has existed for only 18 hours.

Some Common Misconceptions

Certain misconceptions about human evolution persist in the minds of many, long after these myths have been debunked by the fossil evidence. One of these myths is expressed in the question "If chimpanzees were our ancestors, then why do they still exist?" In fact, scientists do not think that humans evolved from chimpanzees. Rather, as illustrated in Figure 17.37, the lineages that led to present-day humans and chimpanzees diverged from a common ancestor several million years ago. Each branch then evolved separately. As an analogy, consider a large family reunion attended by the descendants of a man who was born in 1830. Although the attendees share this common ancestor who lived several generations ago, they may

be only distantly related to each other, as sixth or seventh cousins. Similarly, chimpanzees are not our parent species, but more like our very distant phylogenetic cousins, related through a common ancestor that lived hundreds of thousands of generations in the past.

Another myth envisions human evolution as a ladder with a series of steps leading directly from an ancestral anthropoid to *Homo sapiens*. This is often illustrated as a parade of fossil **hominins** (members of the human family) becoming progressively more modern as they march across the page. In fact, as **Figure 17.39** shows, there were times in hominin history when several human species coexisted. Scientists have identified about 20 species of fossil hominins, demonstrating that human phylogeny is more like a multibranched bush than a ladder, with our species being the tip of the only twig that still lives.

Although evidence from hundreds of thousands of hominin fossils debunks these and other myths about human evolution, many fascinating questions about our ancestry remain. The discovery of each new hominin fossil brings scientists a little closer to solving the puzzle of how we became human. In the following pages, you'll learn about some of the important clues that have been discovered so far.

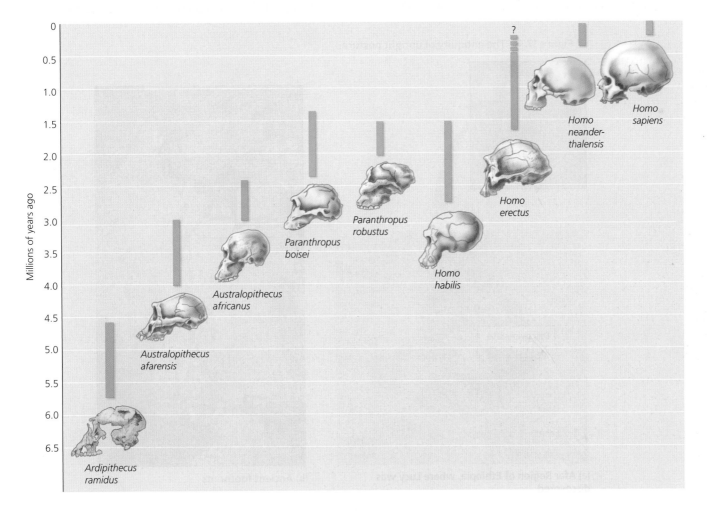

◀ **Figure 17.39 A timeline of human evolution.** The orange bars indicate the time span during which each species lived. Notice that there have been times when two or more hominin species coexisted. The skulls are all drawn to the same scale so you can compare brain sizes.

Australopithecus and the Antiquity of Bipedalism

Present-day humans and chimpanzees clearly differ in two major physical features: Humans are bipedal (walk upright) and have much larger brains. When did these features emerge? In the early 1900s, scientists hypothesized that increased brain size was the initial change that separated hominins from other apes. That hypothesis was overturned when a team of researchers in Ethiopia **(Figure 17.40a)** unearthed a stunning 3.24-million-year-old female hominin that had a small brain and walked on two legs. Officially named *Australopithecus afarensis*, but nicknamed Lucy by her discoverers, the individual was only about 3 feet tall and with a head about the size of a softball. Corroborating evidence of early bipedalism was found soon after—the footprints of two upright-walking hominins preserved in a 3.6-million-year-old layer of volcanic ash **(Figure 17.40b)**. Since these initial discoveries, many more *Australopithecus afarensis* fossils have been found, including most of the skeleton of a 3-year-old member of the species. Other species of *Australopithecus* have also been discovered.

Scientists are now certain that bipedalism is a very old trait. Another lineage known as "robust" australopiths—including *Paranthropus boisei* and *Paranthropus robustus* in Figure 17.39—were also small-brained bipeds. So was *Ardipithecus ramidus*, the oldest hominin shown in Figure 17.39.

Homo habilis and the Evolution of Inventive Minds

Enlargement of the human brain is first evident in fossils from East Africa dating to about 2.4 million years ago. Thus, the fundamental human trait of an enlarged brain evolved a few million years after bipedalism.

Anthropologists have found skulls with brain capacities intermediate in size between those of the latest *Australopithecus* species and those of *Homo sapiens*. Simple handmade stone tools are sometimes found with the larger-brained fossils, which have been dubbed *Homo habilis* ("handy man"). After walking upright for about 2 million years, humans were finally beginning to use their manual dexterity and big brains to invent tools that enhanced their hunting, gathering, and scavenging on the African savanna.

Homo erectus and the Global Dispersal of Humanity

The first species to extend humanity's range from Africa to other continents was *Homo erectus*. Skeletons of *Homo erectus* dating to 1.8 million years ago found in the Republic of Georgia represent the oldest known

▼ **Figure 17.40 The antiquity of upright posture.**

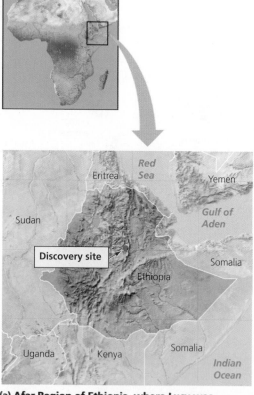

(a) Afar Region of Ethiopia, where Lucy was discovered

(b) Ancient footprints

fossils of hominins outside Africa. *Homo erectus* was taller than *Homo habilis* and had a larger brain capacity. Intelligence enabled this species to continue succeeding in Africa and also to survive in the colder climates of the north. *Homo erectus* resided in huts or caves, built fires, made clothes from animal skins, and designed stone tools. In anatomical and physiological adaptations, *Homo erectus* was poorly equipped for life outside the tropics, but made up for the deficiencies with cleverness and social cooperation.

Eventually, *Homo erectus* migrated to populate many regions of Asia and Europe, moving as far as Indonesia. This species gave rise to regionally diverse descendants, including, as you'll learn in the next section, the Neanderthals. ☑

Homo neanderthalensis

One of the most intriguing members of our family tree is *Homo neanderthalensis*. Fossilized remains of Neanderthals, as this species is commonly called, were first discovered in the Neander Valley in Germany 150 years ago. This hominin had a large brain and hunted big game with tools made from stone and wood. Neanderthals were living in Europe as far back as 350,000 years ago and spread to the Near East, central Asia, and southern Siberia. By 28,000 years ago, the species was extinct.

Analysis of DNA extracted from Neanderthal fossils proved that humans are not the descendants of Neanderthals, as was once thought. Rather, humans and Neanderthals shared a common ancestor, and their lineages diverged about 400,000 years ago. You may be surprised to learn that sequencing of Neanderthal genomes suggests that interbreeding between Neanderthals and some populations of *Homo sapiens* left a genetic legacy in our species. Roughly 2% of the genomes of most present-day humans came from Neanderthals. Africans are the exception—their DNA carries no detectable trace of Neanderthal ancestry. Scientists also learned that at least some Neanderthals had pale skin and red hair (see Figure 14.23). By comparing the genomes of more than 28,000 Europeans with Neanderthal genomes, researchers have identified a dozen Neanderthal alleles that make modern humans more susceptible to conditions such as depression, tobacco addiction, and autoimmune diseases. Clues from the Neanderthal genome will also help scientists understand the genetic differences that were important in the evolution of *Homo sapiens*.

MANY OF US HAVE A BIT OF NEANDERTHAL IN OUR GENES.

The Origin and Dispersal of *Homo sapiens*

Evidence from fossils and DNA studies is coming together to support a compelling hypothesis about how our own species, *Homo sapiens*, emerged and spread around the world. The oldest known fossils of *Homo sapiens* were discovered in Ethiopia and date from 160,000 to 195,000 years ago. The Ethiopian fossils support molecular evidence about the origin of humans: DNA studies strongly suggest that all living humans can trace their ancestry back to a single African *Homo sapiens* lineage that began 160,000 to 200,000 years ago.

The oldest fossils of *Homo sapiens* outside Africa are from the Middle East and date back about 115,000 years. Evidence suggests that our species emerged from Africa in one or more waves, spreading first into Asia and then to Europe, Southeast Asia, Australia, and finally to the New World (North and South America) **(Figure 17.41)**. The date of the first arrival of humans in the New World is uncertain, although the generally accepted evidence suggests a minimum of 15,000 years ago.

Certain uniquely human traits have allowed for the development of human societies. The primate brain continues to grow after birth, and the period of growth is longer for a human than for any other primate. The extended period of human development also lengthens the time parents care for their offspring, which contributes to the child's ability to benefit from the experiences of earlier generations. This is the basis of human culture—social transmission of accumulated knowledge, customs, beliefs, and art over generations. The major means of this transmission is language, spoken and written. Humans have evolved culturally as well as biologically.

Nothing has had a greater impact on life on Earth than *Homo sapiens*. The global consequences of human evolution have been enormous. Cultural evolution made modern *Homo sapiens* a new force in the history of life—a species that could defy its physical limitations. We do not have to wait to adapt to an environment through natural selection; we simply change the environment to meet our needs. ☑

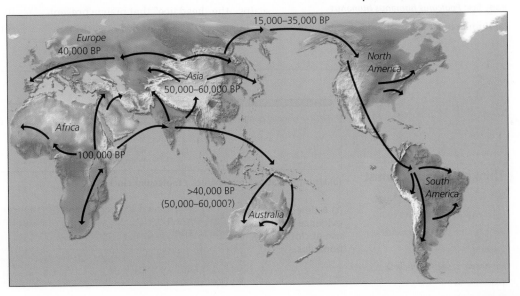

▼ **Figure 17.41 The spread of *Homo sapiens* (dates given as years before present, BP).**

Europe
40,000 BP

15,000–35,000 BP

North America

Asia
50,000–60,000 BP

Africa

100,000 BP

>40,000 BP
(50,000–60,000?)

Australia

South America

What Can Lice Tell Us About Ancient Humans?

BACKGROUND

Imagine standing naked in a snowstorm. You probably wouldn't last long. When did humans first start wearing clothes? Archeological evidence provides a very rough time frame. Ancient members of genus *Homo* in Europe had tools for scraping animal hides about 780,000 years ago. Animal hides could have been used for clothing but would have had other uses as well. Needles, which indicate the construction of more complex clothing, were in use by 50,000 years ago. To zero in on a narrower date range for the origin of clothing, researchers turned to some of our most unwelcome guests: lice.

Lice are tiny blood-sucking parasites **(Figure 17.42a)** that attach their eggs to fur or similar fibers. Our distant ancestors, which were fur-covered like present-day primates, were parasitized by one species of louse. (Pubic lice, or "crabs," have a different species evolutionary history and are members of a different genus.) How did lice persist in modern humans when, with the exception of a few patches of hair, we are furless?

Around 3 million years ago, environmental changes required hominins to exert more energy in search of food and water. Muscle activity generates a great deal of heat; evaporation of sweat from the skin cools the body. However, fur prevents sweat from evaporating efficiently. By about 1.2 million years ago, hominins had lost most of their body hair, stranding lice on our scalps—the "head lice" that plague present-day humans. We can also be parasitized by clothing lice, which belong to the same species as head lice but have adaptations for grasping cloth fibers rather than hair or fur. Researchers hypothesized that analyzing the evolutionary divergence of lice into two distinct types would allow them to estimate when our ancestors first wore clothes.

METHOD

As you may recall, populations that don't interbreed because they occupy different habitats may accumulate genetic differences (see Figure 14.7). The longer two populations have been separated, the greater the divergence of their genes. Thus, the number of genetic differences between populations can be used to estimate how long the populations have been separated, a method known as a molecular clock **(Figure 17.42b)**. Imagine that a population of people all sang one song. As groups formed new isolated communities, differences would creep into their lyrics. Knowing the average number of word changes per year would let you estimate how long the populations had been singing separately. To construct a molecular clock for the divergence of head lice and clothing lice, researchers compared four DNA sequences for which the mutation rate—the average number of changes—was known.

RESULTS

The molecular clock shows that clothing offered a new habitat for the lice between 83,000 and 170,000 years ago, when populations of head lice and clothing lice began to diverge **(Figure 17.42c)**. Thus, modern humans originated the use of clothing, probably in Africa. Humans spreading north (see Figure 17.41) would have encountered the exceptionally cold climate that resulted from a series of ice ages, making clothing particularly useful.

▼ **Figure 17.42 Using lice to date the origin of clothing.**

Head lice: *Pediculus humanus capitis*

Body lice: *Pediculus humanus humanus*

(a) Human head lice (top) and clothing lice.

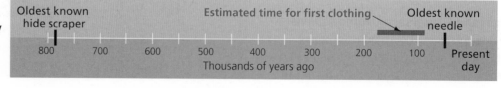

DNA in original population

After populations separate, each has unique mutations

Mutations accumulate over time

(b) Using mutations to contruct a molecular clock.

(c) A timeline of evidence for the use of clothing by hominins.

Oldest known hide scraper

Estimated time for first clothing

Oldest known needle

| 800 | 700 | 600 | 500 | 400 | 300 | 200 | 100 | Present day |

Thousands of years ago

Thinking Like a Scientist

Why is a molecular clock more likely to be accurate when it is based on more genes?

For answers, see Appendix D.

EVOLUTION CONNECTION | Human Evolution

Are We Still Evolving?

Imagine that you could take a time machine 100,000 years into the past and bring back a *Homo sapiens* man. If you dressed him in jeans and a T-shirt and took him for a stroll around campus, chances are that no one would look twice. Did we stop evolving after becoming *Homo sapiens*? In some ways, yes. The human body has not changed much in the past 100,000 years. And by the time *Homo sapiens* began to travel out of Africa, all of the complex characteristics that define our humanity, including our big-brained intelligence and our capacity for language and symbolic thought, had already evolved.

But as humans wandered far from their site of origin and settled in diverse environments, populations encountered different selective forces. Some traits of people today reflect evolutionary responses of ancient ancestors to their physical and cultural environment. For example, the high frequency of sickle hemoglobin (see Figure 9.21) in certain populations is an adaptation that protects against the deadly disease malaria. In other malarial regions, a group of inherited blood disorders called thalassemia serves the same adaptive function.

Diet has had a significant effect on human evolution, too. For example, the ability to digest lactose as adults (see the Evolution Connection section of Chapter 3) evolved in populations that kept dairy herds. Reliance by early farmers on starchy crops such as rice or tubers also left a genetic trace: extra copies of the gene that encodes the starch-digesting enzyme amylase.

One of the most striking differences among people is skin color (Figure 17.43). The loss of skin pigmentation in humans who migrated north from Africa is thought to be an adaptation to low levels of ultraviolet (UV) radiation in northern latitudes. Dark pigment blocks the UV radiation necessary for synthesizing vitamin D—essential for proper bone development—in the skin. Recent research has turned up numerous other examples of adaptations that enabled us to colonize Earth's varied environments. For instance, indigenous people in the Andes Mountains in South America live at altitudes up to 12,000 feet (2.3 miles), where the air has 40% less oxygen than at sea level (Figure 17.44). Researchers have identified genes that have undergone evolutionary changes in response to this challenging environment. High-altitude adaptations have also been discovered in people who are native to the Tibetan Plateau in Asia. Despite evolutionary tweaks such as these, however, we remain a single species.

▼ Figure 17.44 Quechuans, indigenous people adapted to living at high altitude.

▼ Figure 17.43 People with different adaptations to UV radiation.

Chapter Review

SUMMARY OF KEY CONCEPTS

The Origins of Animal Diversity

What Is an Animal?

Animals are eukaryotic, multicellular, heterotrophic organisms that obtain nutrients by ingestion. Most animals reproduce sexually and develop from a zygote to a blastula and then to a gastrula. After the gastrula stage, some develop directly into adults, whereas others pass through a larval stage.

Early Animals and the Cambrian Explosion

Animals probably evolved from a colonial flagellated protist. Precambrian animals were soft-bodied. Animals with hard parts appeared during the Cambrian period. Between 535 and 525 million years ago, animal diversity increased rapidly.

Animal Phylogeny

Major branches of animal evolution are defined by two key evolutionary differences: the presence or absence of tissues and radial versus bilateral body symmetry. A tissue-lined body cavity evolved in a number of later branches.

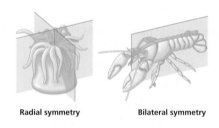

Radial symmetry   **Bilateral symmetry**

Major Invertebrate Phyla

This tree shows the eight major invertebrate phyla, as well as chordates, which include a few invertebrates.

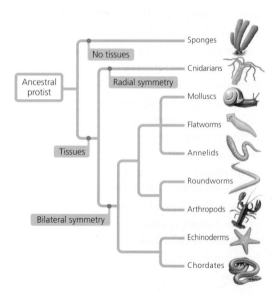

Sponges

Sponges (phylum Porifera) are stationary animals with porous bodies but no tissues. Specialized cells draw water through pores in the sides of the body and trap food particles.

Cnidarians

Cnidarians (phylum Cnidaria) have radial symmetry, a gastrovascular cavity with a single opening, and tentacles with stinging cnidocytes. The body is either a stationary polyp or a floating medusa.

Molluscs

Molluscs (phylum Mollusca) are soft-bodied animals often protected by a hard shell. The body has three main parts: a muscular foot, a visceral mass, and a fold of tissue called the mantle.

MOLLUSCS		
Gastropods	Bivalves	Cephalopods

Flatworms

Flatworms (phylum Platyhelminthes) are the simplest bilateral animals. They may be free-living (such as planarians) or parasitic (such as tapeworms).

Annelids

Annelids (phylum Annelida) are segmented worms with complete digestive tracts. They may be free-living or parasitic.

Roundworms

Roundworms, also called nematodes (phylum Nematoda), are unsegmented and cylindrical with tapered ends. They may be free-living or parasitic.

Arthropods

Arthropods (phylum Arthropoda) are segmented animals with an exoskeleton and specialized, jointed appendages.

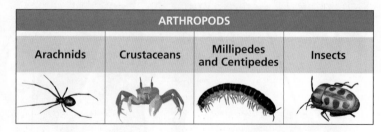

ARTHROPODS			
Arachnids	Crustaceans	Millipedes and Centipedes	Insects

Echinoderms

Echinoderms (phylum Echinodermata) are stationary or slow-moving marine animals that lack body segments and possess a unique water vascular system. Bilaterally symmetric larvae usually change to radially symmetric adults. Echinoderms have a bumpy endoskeleton.

Vertebrate Evolution and Diversity

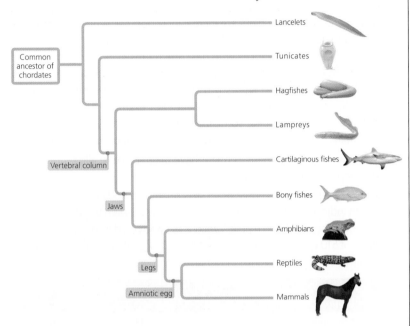

Characteristics of Chordates

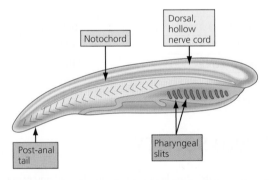

Tunicates and lancelets are invertebrate chordates. The vast majority of chordates are vertebrates, possessing a skull and backbone.

Fishes

Hagfishes and lampreys are jawless vertebrates. Cartilaginous fishes, such as sharks, are mostly predators with powerful jaws and a flexible skeleton made of cartilage. Bony fishes have a stiff skeleton reinforced by calcium. Bony fishes are further classified into ray-finned fishes and lobe-finned fishes (including lungfishes).

Amphibians

Amphibians are tetrapod vertebrates that usually deposit their eggs (lacking shells) in water. Aquatic larvae typically undergo a radical metamorphosis into the adult stage. Their moist skin requires that amphibians spend much of their adult life in humid environments.

Reptiles

Reptiles are amniotes, vertebrates that develop in a fluid-filled egg enclosed by a shell. Reptiles include terrestrial ectotherms with lungs and waterproof skin covered by scales. Scales and amniotic eggs enhanced reproduction on land. Birds are endothermic reptiles with wings, feathers, and other adaptations for flight.

Mammals

Mammals are endothermic vertebrates with mammary glands and hair. There are three major groups of mammals: Monotremes lay eggs; marsupials use a placenta but give birth to tiny embryonic offspring that usually complete development while attached to nipples inside the mother's pouch; and eutherians, or placental mammals, use their placenta in a longer-lasting association between the mother and her developing young.

MAMMALS		
Monotremes	**Marsupials**	**Eutherians**

The Human Ancestry

The Evolution of Primates

The first primates were small, arboreal mammals that evolved from insect-eating mammals about 65 million years ago. Anthropoids consist of New World monkeys (with prehensile tails), Old World monkeys (without prehensile tails), apes, and humans.

The Emergence of Humankind

Chimpanzees and humans evolved from a common ancestor about 6–7 million years ago. Species of the genus *Australopithecus*, which lived at least 4 million years ago, walked upright and had a small brain. Enlargement of the human brain in *Homo habilis* came later, about 2.4 million years ago. *Homo erectus* was the first species to extend humanity's range from its birthplace in Africa to other continents. *Homo erectus* gave rise to regionally diverse descendants, including the Neanderthals (*Homo neanderthalensis*). Current data indicate a relatively recent dispersal of modern Africans that gave rise to today's human diversity.

Mastering Biology

For practice quizzes, BioFlix animations, MP3 tutorials, video tutors, and more study tools designed for this textbook, go to Mastering Biology™

SELF-QUIZ

1. Bilateral symmetry in the animal kingdom is best correlated with
   a. an ability to sense equally in all directions.
   b. the presence of a skeleton.
   c. motility and active predation and escape.
   d. development of a body cavity.

2. Identify which of the following categories includes all others in the list: arthropod, arachnid, insect, butterfly, crustacean, millipede.

3. The oldest group of tetrapods is the _____.

4. Reptiles are much more extensively adapted to life on land than amphibians because reptiles
   a. have a complete digestive tract.
   b. lay eggs that are enclosed in shells.
   c. are endothermic.
   d. go through a larval stage.

5. What is the name of the phylum to which humans belong? For what anatomical structure is the phylum named? Where in your body is a derivative of this anatomical structure found?

6. Which of the following types of animals is not included in the human ancestry? (*Hint:* See Figure 17.30.)
   a. a bird
   b. a bony fish
   c. an amphibian
   d. a primate

7. Put the following list of species in order, from the oldest to the most recent: *Homo erectus, Australopithecus* species, *Homo habilis, Homo sapiens.*

8. Match each of the following animals to its phylum:
   a. lancelet   1. Platyhelminthes
   b. clam   2. Arthropoda
   c. tapeworm   3. Porifera
   d. dust mite   4. Chordata
   e. sponge   5. Mollusca

For answers to the Self-Quiz, see Appendix D.

IDENTIFYING MAJOR THEMES

For each statement below, identify which major theme is evident (the relationship of structure to function, information flow, pathways that transform energy and matter, interactions within biological systems, or evolution) and explain how the statement relates to the theme. If necessary, review the themes (Chapter 1) and review the examples highlighted in blue in this chapter.

9. It is likely that the set of "master control" genes that allow building complex bodies was in place before rapid diversification.

10. The head of a tapeworm is equipped with suckers and hooks that lock the worm to the intestinal lining of the host.

11. Bees and flies pollinate our crops and orchards. Other insects are carriers of the microbes that cause many human diseases. Insects also compete with people for food by eating our crops.

For answers to Identifying Major Themes, see Appendix D.

THE PROCESS OF SCIENCE

12. Imagine that you are a marine biologist. As part of your exploration, you dredge up an unknown animal from the seafloor. Describe some of the characteristics you should look at to determine the phylum to which the creature should be assigned.

13. Vegan and vegetarian diets are increasingly popular. While vegans consume no animals or animal products, many vegetarians are less strict. Talk to acquaintances who describe themselves as vegetarians, or who follow a meat-free diet, and determine which taxonomic groups they avoid eating (see Figures 17.6 and 17.30). Try to generalize about their diet. What do they consider "meat"? For example, do they avoid eating vertebrates but eat some invertebrates? Do they avoid only birds and mammals? What about fish? Do they eat dairy products or eggs?

14. Which evolutionary changes might have taken place to allow humans to develop a complex language?

15. **Interpreting Data** Average brain size, relative to body mass, gives a rough indication of the intelligence of a species. Graph the data below for anthropoids. *Homo naledi*, which was discovered in 2015, had a mean brain of 510 cm^3; its mean body mass was 37 kg. What do these data imply about the intelligence of *H. naledi* compared to other hominids?

Anthropoid Species	Mean Brain Volume (cm³)	Mean Body Mass (kg)
Australopithecus afarensis	440	37
Chimpanzee	405	46
Gorilla	500	105
Homo erectus	940	58
Homo habilis	610	34
Homo neanderthalensis	1480	65
Homo sapiens	1330	64
Paranthropus boisei	490	41

BIOLOGY AND SOCIETY

16. Coral reefs harbor a greater diversity of animals than any other environment in the sea. Australia's Great Barrier Reef has been protected as a marine reserve and is a mecca for scientists and nature enthusiasts. Elsewhere, such as in Indonesia and the Philippines, coral reefs are in danger. Many reefs have been depleted of fish, and runoff from the shore has covered coral with sediment. Nearly all the changes in the reefs can be traced back to human activities. What kinds of activities do you think might be contributing to the decline of the reefs? What are some reasons to be concerned about this decline? Do you think the situation is likely to improve or worsen in the future? Why? What might the local people do to halt the decline? Should the more industrialized countries help? Why or why not?

17. Many of us, including many anthropologists and zoologists, think that humans are at a special position among all animals due to a range of characteristics, such as an upright posture, high intelligence, handling of complex tools, development of a complex language system, a prolonged childhood, and an extended lifespan. However, many of these characteristics can be observed in other species as well. Are we really unique among animals? Justify your answer.

Ecology

18 An Introduction to Ecology and the Biosphere

CHAPTER CONTENTS

Why Ecology Matters

From the icy polar regions to the lush tropics, from the vibrant coral reefs to the deepest ocean, Earth's biosphere is a treasure trove of life that is still being explored. The study of ecology helps us understand the interactions of organisms with these diverse environments and offers insight into our own impact on the natural world.

VAMPIRE ALERT! YOUR HOME MAY BE INFESTED WITH "VAMPIRE" DEVICES THAT CONSUME ELECTRICITY EVEN WHILE YOU SLEEP.

POLLUTION DOESN'T JUST VANISH INTO THIN AIR. AIRBORNE POLLUTANTS FROM CAR EXHAUST CAN COMBINE WITH WATER AND RETURN TO EARTH AS ACID PRECIPITATION FAR AWAY.

WHAT HAPPENS TO ALL THE TRASH WE DISCARD? LANDFILLS AREN'T THE ONLY PLACE WHERE GARBAGE ACCUMULATES.

BIOLOGY AND SOCIETY	Climate Change

A polar bear with her two cubs on Arctic ice. Climate change is shrinking the permanent sea ice that polar bears use as hunting platforms.

Penguins, Polar Bears, and People in Peril

Ninety-seven percent of climate scientists agree: The global climate is changing due to human activities. The change is driven by a rapid rise in temperatures—the ten hottest years on record have occurred since 1998. The current rate of warming is *ten times* faster than the average rate during the warm-up that followed the last ice age. So what do we know about climate change now, and what can we expect for the future?

The northernmost regions of the Northern Hemisphere and the Antarctic Peninsula have heated up the most. In parts of Alaska, for example, winter temperatures have risen by 5–6°F. The permanent Arctic sea ice is shrinking; each summer brings thinner ice and more open water. Polar bears, which stalk their prey on ice and need to store up body fat for the warmer months when there is no ice, are showing signs of starvation as their winter hunting grounds melt away. At the other end of the planet, diminishing sea ice near the Antarctic Peninsula limits the access of Adélie penguins to their food supply, and spring blizzards of unprecedented frequency and severity are taking a heavy toll on their eggs and chicks. But these charismatic animals are just a distress signal that should alert us to our own danger. We are already feeling the effects of climate change in more frequent and larger wildfires, increased coastal flooding, deadly heat waves, and altered precipitation patterns that bring drought to some regions and torrential downpours to others.

What does climate change mean for the future of life on Earth? Any predictions that scientists make now about future impacts of global climate change are based on incomplete information. Much remains to be discovered about species diversity and about the complex interactions of organisms (living things) with each other and with their environments. There is overwhelming evidence that human enterprises are responsible for the changes that are occurring. How we respond to this crisis will determine whether circumstances improve or worsen. And the process begins with understanding the basic concepts of ecology, which we start to explore in this chapter.

An Overview of Ecology

In your study of biology so far, you have learned about the diversity of life on Earth and about the molecular and cellular structures and processes that make life tick. **Ecology**, the scientific study of the interactions between organisms and their environments, offers a different perspective on life—biology from the skin out, so to speak.

Humans have always had an interest in other organisms and their environments. As hunters and gatherers, prehistoric people had to learn where and when game and edible plants could be found in greatest abundance. Naturalists, from Aristotle to Darwin and beyond, made the process of observing and describing organisms in their natural habitats an end in itself rather than simply a means of survival. We can still gain valuable insight from watching nature and recording its structure and processes **(Figure 18.1)**. As you might expect, hypothesis-driven science performed in natural environments is fundamental to ecology. But ecologists also test hypotheses using laboratory experiments, where conditions can be simplified and controlled. And some ecologists take a theoretical approach, devising mathematical and computer models, which enable them to simulate large-scale experiments that are impossible to conduct in the field.

▼ Figure 18.1 **Discovery science in a rain forest canopy.** A biologist collects insects in a rain forest on the eastern slope of the Andes mountain range in Argentina.

Ecology and Environmentalism

Technological innovations have enabled people to colonize just about every environment on Earth. Even so, our survival depends on Earth's resources, which have been profoundly altered by human activities **(Figure 18.2)**. Climate change is just one of the many environmental issues that have stirred public concern in recent decades. Some of our industrial and agricultural practices have contaminated the air, soil, and water. Our relentless quest for land and other resources has endangered a lengthy list of plant and animal species and has even driven some to extinction.

The science of ecology can provide the understanding needed to solve environmental problems. But these problems cannot be solved by ecologists alone, because they require making decisions based on values and ethics. On a personal level, each of us makes daily choices that affect our ecological impact. And legislators and corporations, motivated by environmentally aware voters and consumers, must address questions that have wider implications: How should the use of land and water be regulated? Should we try to save all species or just certain ones? What alternatives to environmentally destructive practices can be developed? How can we balance environmental impacts with economic needs?

▼ Figure 18.2 **Human impact on the environment.** Cleanup crews burn oil collected from the surface of the Gulf of Mexico after the Deepwater Horizon disaster in 2010.

A Hierarchy of Interactions

Many different factors can potentially affect an organism's interactions with its environment. **Biotic factors**—all of the organisms in the area—make up the living component of the environment. Other organisms may compete with an individual for food and other resources, prey upon it, or change its physical and chemical environment. **Abiotic factors** make up the environment's nonliving component and include chemical and physical factors, such as temperature, light, water, minerals, and air. An organism's **habitat**, the specific environment it lives in, includes the biotic and abiotic factors of its surroundings.

When we study the interactions between organisms and their environments, it is convenient to divide ecology into four increasingly comprehensive levels: organismal ecology, population ecology, community ecology, and ecosystem ecology.

An **organism** is an individual living thing. **Organismal ecology** is concerned with the evolutionary adaptations that enable organisms to meet the challenges posed by their abiotic environments. The distribution of organisms is limited by the abiotic conditions they can tolerate. For example, amphibians such as the salamander in **Figure 18.3a** cannot live in cold climates because they gain most of their body warmth by absorbing heat from their surroundings. Temperature and precipitation shifts due to global climate change have already affected the distributions of some salamander species, and many more will feel the impact in the coming decades.

The next level of organization in ecology is the **population**, a group of individuals of the same species living in a particular geographic area. **Population ecology** concentrates mainly on factors that affect population density and growth **(Figure 18.3b)**. Biologists who study endangered species are especially interested in this level of ecology.

A **community** consists of all the organisms that inhabit a particular area; it is an assemblage of populations of different species. Questions in **community ecology** focus on how interactions between species, such as predation and competition, affect community structure and organization **(Figure 18.3c)**.

An **ecosystem** includes all the abiotic factors in addition to the community of species in a certain area. For example, a savanna ecosystem includes not only the organisms, such as diverse plants and animals, but also the soil, water sources, sunlight, and other abiotic factors of the environment. Thus, an ecosystem encompasses a wide variety of interactions within and between biological systems. In **ecosystem ecology**, questions concern energy flow and the cycling of chemicals among the various biotic and abiotic factors **(Figure 18.3d)**.

▼ **Figure 18.3** Examples of questions at different levels of ecology.

(a) **Organismal ecology.** What range of temperatures can a red salamander tolerate?

(b) **Population ecology.** What factors affect the survival of emperor penguin chicks?

(c) **Community ecology.** How do predators such as this beech marten affect the diversity of rodents in a community?

(d) **Ecosystem ecology.** What processes recycle vital chemical elements such as nitrogen within a savanna ecosystem in Africa?

The **biosphere** is the global ecosystem—the sum of all the planet's ecosystems, or all of life and where it lives. The most complex level in ecology, the biosphere includes the atmosphere to an altitude of several kilometers, the land down to water-bearing rocks about 1,500 m (almost a mile) deep, lakes and streams, caves, and the oceans to a depth of several kilometers. But despite its grand scale, organisms within the biosphere are linked; events in one part may have far-reaching effects. ☑

☑ CHECKPOINT

1. What does the ecosystem level of classification have in common with the community level of classification?
2. What does the ecosystem level include that the community level does not?

Answers: 1. all the biotic factors of the area. 2. the abiotic factors of the area.

Living in Earth's Diverse Environments

Whether you have seen the world by traveling or through television and movies, you have probably noticed that there are striking regional patterns in the distribution of life. For example, some terrestrial areas, such as the tropical forests of South America and Africa, are home to plentiful plant life, whereas other areas, such as deserts, are relatively barren. Coral reefs are alive with vibrantly colored organisms; other parts of the ocean appear empty by comparison.

The distribution of life varies on a local scale, too. In the aerial view of a New Zealand wilderness in **Figure 18.4**, we can see a forest, a large lake, a meandering river, and mountains. Within these different environments, variation occurs on an even smaller scale. For example, we would find that the lake has several different habitats, and each habitat has a characteristic community of organisms.

Abiotic Factors of the Biosphere

Patterns in the distribution of life mainly reflect differences in the abiotic factors of the environment. Let's look at several major abiotic factors that influence where organisms live.

Energy Source

Life depends on pathways that transform energy and matter. Thus, all organisms require a usable source of energy to live. Solar energy from sunlight, captured by chlorophyll during the process of photosynthesis, powers most ecosystems. In the image shown in **Figure 18.5**, colors are keyed to the relative abundance of chlorophyll. Green areas on land indicate high densities of plant life. Orange areas on land, including the Sahara region of Africa and much of the western United States, are much less productive. Green regions of the ocean contain an abundance of algae and photosynthetic bacteria compared to darker regions.

Lack of sunlight is seldom the most important factor limiting plant growth for terrestrial ecosystems, although shading by trees does create intense competition for light among plants growing on forest floors. In many aquatic environments, however, light cannot penetrate beyond certain depths. As a result, most photosynthesis in a body of water occurs near the surface.

Surprisingly, life also thrives in environments that are completely dark. A mile or more below the ocean's surface lies the remarkable world of hydrothermal vents, sites near the adjoining edges of giant plates of Earth's crust where molten rock and hot gases surge upward from Earth's interior. Towering chimneys, some as tall as a nine-story building, emit scalding water and hot gases **(Figure 18.6)**. These ecosystems are powered by bacteria that derive energy from the oxidation of inorganic chemicals such as hydrogen sulfide. Bacteria with similar metabolic talents support communities of cave-dwelling organisms on land.

▼ Figure 18.4 Local variation of the environment in a New Zealand wilderness.

▼ Figure 18.5 **Distribution of life in the biosphere.** In this image of Earth, colors are keyed to the relative abundance of chlorophyll, which correlates with the regional densities of photosynthetic organisms.

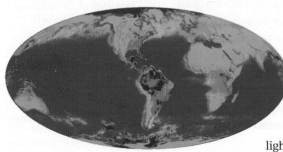

▼ Figure 18.6 **A deep-sea hydrothermal vent.** "Black smokers" in the Pacific Ocean west of Washington spew plumes of hot gases from Earth's interior. Giant tube worms (inset), annelids that may grow to 2 m long, are members of the vent community.

Temperature

Temperature is an important abiotic factor because of its effect on metabolism. Few organisms can maintain a sufficiently active metabolism at temperatures close to 0°C (32°F), and temperatures above 45°C (113°F) destroy the enzymes of most organisms. Most organisms function best within a specific range of environmental temperatures. For example, the American pika (Figure 18.7) has a high body temperature well suited to the chilly climate of its mountain habitat. On warm days, however, pikas must take refuge in crevices where pockets of cold air prevent fatal overheating. In winter, pikas depend on a blanket of snow to insulate their shelters from the perilous cold.

▲ Figure 18.7 **An American pika.** This diminutive relative of rabbits lives at high elevations in the western United States and Canada.

Water

Water is essential to all life. Aquatic organisms are surrounded by water, but they face problems of water balance if their own solute concentration does not match that of their surroundings (see Figure 5.14). For terrestrial organisms, the primary threat is drying out in the air. Many land animals have watertight coverings that reduce water loss. **For example, scales in reptiles, such as snakes and lizards, are structures that perform the function of sealing in moisture (Figure 18.8a).** Most plants have waxy coatings on their leaves and other aerial parts (Figure 18.8b). Wax from carnauba palm leaves is valued for the glossy, waterproof coat it adds to polishing products for cars, surfboards, furniture, and shoes. Carnauba wax is an ingredient of many other products as well, including cosmetics such as lipstick and mascara.

Inorganic Nutrients

The distribution and abundance of photosynthetic organisms, including plants, algae, and photosynthetic bacteria, depend on the availability of inorganic nutrients such as compounds of nitrogen and phosphorus. Plants obtain these nutrients from the soil. Soil structure, pH, and nutrient content often play major roles in determining the distribution of plants. In many aquatic ecosystems, low levels of nitrogen and phosphorus limit the growth of algae and photosynthetic bacteria.

Other Aquatic Factors

Several abiotic factors are important in aquatic, but not terrestrial, ecosystems. Although terrestrial organisms have a plentiful supply of oxygen from the air, aquatic organisms must depend on oxygen dissolved in water. This is a critical factor for many species of fish. Cold, fast-moving water has a higher oxygen content than warm or stagnant water. Salinity (saltiness), currents, and tides also play a role in many aquatic ecosystems.

Other Terrestrial Factors

Some abiotic factors affect terrestrial, but not aquatic, ecosystems. For example, wind is often an important factor on land. Wind increases an organism's rate of water loss by evaporation. The resulting increase in evaporative cooling can be advantageous on a hot summer day, but it can cause dangerous wind chill on a cold day. In some ecosystems, frequent occurrences of natural disturbances such as storms or fire play a role in the distribution of organisms. ☑

☑ CHECKPOINT

Why is hydrogen sulphide vital for deep-sea hydrothermal vent ecosystems?

■ *Answer: In the absence of solar energy, these ecosystems are powered by bacteria that derive energy from the oxidation of inorganic chemicals like hydrogen sulphide.*

(a) Scales on a Carolina anole lizard.

(b) Beaded water droplets showing the water repellency of the leaf's waxy coating.

▲ Figure 18.8 **Watertight coverings.**

The Evolutionary Adaptations of Organisms

The ability of organisms to live in Earth's diverse environments demonstrates the close relationship between the fields of ecology and evolutionary biology. Charles Darwin was an ecologist, although he predated the word *ecology*. It was the geographic distribution of organisms and their exquisite adaptations to specific environments that provided Darwin with evidence for evolution. Evolutionary adaptation via natural selection results from the interactions of organisms with their environments, which brings us back to our definition of ecology. Thus, events that occur in the short term, during the course of an individual's lifetime, may translate into effects over the longer scale of evolutionary time. For instance, because the availability of water affects a plant's growth and ultimately its reproductive success, precipitation has an impact on the gene pool of a plant population. **After a period of lower-than-average rainfall, drought-resistant individuals may be more prevalent in a plant population—an example of evolution.** Organisms also evolve in response to biotic interactions, such as predation and competition. ☑

☑ CHECKPOINT

How are the fields of ecology and evolution linked?

■ *Answer: The process of evolutionary adaptation via natural selection results from the interactions of organisms with their environments (ecology).*

▼ **Figure 18.9** A mourning dove demonstrating its physiological response to cold weather.

Adjusting to Environmental Variability

The abiotic factors in a habitat may vary from year to year, seasonally, or over the course of a day. An individual's abilities to adjust to environmental changes that occur during its lifetime are themselves adaptations refined by natural selection. For instance, if you see a bird on a cold day, it may look unusually fluffy (**Figure 18.9**). Small muscles in the skin raise the bird's feathers, a physiological response that traps insulating pockets of air. Some species of birds adjust to seasonal cold by growing heavier feathers. And some bird species respond to the onset of cold weather by migrating to warmer regions—a behavioral response. Note that these responses occur during the lifetime of an individual, so they are not examples of evolution, which is change in a population over time.

Physiological Responses

Like birds, mammals can adjust to a cold day by contracting skin muscles—in this case attached to hairs—to create a temporary layer of insulation. (Our own muscles do this, too, but we just get "goose bumps" instead of a furry insulation.) The blood vessels in the skin also constrict, which slows the loss of body heat. In both cases, the adjustment occurs in just seconds.

A gradual, though still reversible, physiological adjustment that occurs in response to an environmental change is called **acclimation**. For example, suppose you moved from Boston, which is essentially at sea level, to the mile-high city of Denver, where there is less oxygen. One physiological response to your new environment would be a gradual increase in the number of your red blood cells, which transport O_2 from your lungs to other parts of your body. Acclimation can take days or weeks. This is why high-altitude climbers, such as those attempting to scale Mount Everest, need extended stays at a high-elevation base camp before proceeding to the summit.

The ability to acclimate is generally related to the range of environmental conditions a species naturally experiences. Species that live in very warm climates, for example, usually cannot acclimate to extreme cold. Among vertebrates, birds and mammals can generally tolerate the greatest temperature extremes because, as endotherms, they use their metabolism to regulate internal temperature. In contrast, ectothermic reptiles can tolerate only a more limited range of temperatures (**Figure 18.10**).

▼ **Figure 18.10** **The number of lizard species in different regions of the contiguous United States.** Notice that there are fewer and fewer lizard species in more northern regions. This reflects lizards' ectothermic physiology, which depends on environmental heat for keeping the body warm enough for the animal to be active.

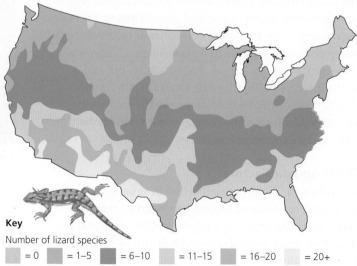

Key
Number of lizard species
■ = 0 ■ = 1–5 ■ = 6–10 ■ = 11–15 ■ = 16–20 ■ = 20+

Anatomical Responses

Many organisms respond to environmental challenge with some type of change in body shape or structure. When the change is reversible, the response is an example of acclimation. Many mammals, for example, grow a heavier coat of fur before the winter cold sets in and shed it when summer comes. In some animals, fur or feather color changes seasonally as well, camouflaging the animal against winter snow and summer vegetation (Figure 18.11).

Other anatomical changes are irreversible over the lifetime of an individual. Environmental variation can affect growth and development so much that there may be remarkable differences in body shape within a population. You can see an example in Figure 18.12, which shows the "flagging" that wind causes in certain trees. In general, plants are more anatomically changeable than animals. Rooted and unable to move to a better location, plants rely entirely on their anatomical and physiological responses to survive environmental fluctuations.

▲ Figure 18.11 The arctic fox in winter and summer coats.

▼ Figure 18.12 Wind as an abiotic factor that shapes trees. The mechanical disturbance of the prevailing wind hinders limb growth on the windward side of this fir tree near the timberline in the Rocky Mountains, while limbs on the other side grow normally. This anatomical response is an evolutionary adaptation that reduces the number of limbs that are broken during strong winds.

Behavioral Responses

In contrast to plants, most animals can respond to an unfavorable change in the environment by moving to a new location. Such movement may be fairly localized. For example, many desert ectotherms, including reptiles, maintain a reasonably constant body temperature by shuttling between sun and shade. Some animals are capable of migrating great distances in response to such environmental cues as the changing seasons. Many migratory birds overwinter in Africa, returning to the northern latitudes of Europe to breed during the summer. And we humans, with our large brains and available technology, have an especially rich range of behavioral responses available to us (Figure 18.13). ☑

CHECKPOINT

What is acclimation?

■ *Answer: a gradual, reversible change in anatomy or physiology in response to an environmental change*

▼ Figure 18.13 Behavioral responses have expanded the geographic range of humans. Dressing for the weather is a thermoregulatory behavior unique to people.

Biomes

The abiotic factors you learned about in the previous section are largely responsible for the distribution of life on Earth. (You'll learn about the role of biotic factors in species distribution in Chapter 20.) Using various combinations of these factors, ecologists have categorized Earth's environments into biomes. A **biome** is a major terrestrial or aquatic life zone, characterized by vegetation type in terrestrial biomes and the physical environment in aquatic biomes. In this section, we'll briefly survey the aquatic biomes, followed by the terrestrial biomes.

Aquatic biomes, which occupy roughly 75% of Earth's surface, are determined by their salinity and other physical factors. Freshwater biomes (lakes, streams and rivers, and wetlands) typically have salt concentrations of less than 1%. The salt concentrations of marine biomes (oceans, intertidal zones, and coral reefs) are generally around 3%.

☑ **CHECKPOINT**

Why does sewage cause heavy algal growth in lakes?

■ *Answer: Sewage adds mineral nutrients that stimulate growth of the algae.*

Freshwater Biomes

Freshwater biomes cover less than 1% of Earth, and they contain a mere 0.01% of its water. But they harbor a disproportionate share of biodiversity—an estimated 6% of all described species. Moreover, we depend on freshwater biomes for drinking water, crop irrigation, sanitation, and industry.

Freshwater biomes fall into two broad groups: standing water, which includes lakes and ponds, and flowing water, such as rivers and streams. The difference in water movement results in profound differences in ecosystem structure.

Lakes and Ponds

Standing bodies of water range from small ponds only a few square meters in area to large lakes, such as North America's Great Lakes, that are thousands of square kilometers (**Figure 18.14**).

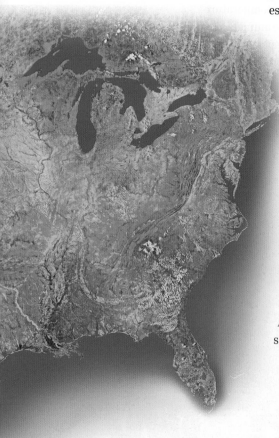

◀ **Figure 18.14 A satellite view of the Great Lakes.**

In lakes and large ponds, communities of plants, algae, and animals are distributed according to the depth of the water and the distance from shore (**Figure 18.15**). Shallow water near shore and the upper layer of water away from shore make up the **photic zone**, so named because light is available for photosynthesis. Microscopic algae and cyanobacteria grow in the photic zone, joined by rooted plants and floating plants such as water lilies in the photic area near shore. If a lake or pond is deep enough or murky enough, it has an **aphotic zone**, where light levels are too low to support photosynthesis.

The **benthic realm** is at the bottom of all aquatic biomes. Made up of sand and organic and inorganic sediments, the benthic realm is occupied by communities of organisms that may include algae, aquatic plants, worms, insect larvae, molluscs, and microorganisms. Dead material that "rains" down from the productive surface waters of the photic zone is a major source of food for animals of the benthic realm.

The mineral nutrients nitrogen and phosphorus typically regulate the growth of **phytoplankton**, the collective name for microscopic algae and cyanobacteria that drift near the surfaces of aquatic biomes. Many lakes and ponds are affected by large inputs of nitrogen and phosphorus from sewage and runoff from fertilized lawns and farms. These nutrients often produce heavy growth of algae, which reduces light penetration. When the algae die and decompose, a pond or lake can suffer serious oxygen depletion, killing fish that are adapted to high-oxygen conditions. ☑

▼ **Figure 18.15 Zones in a lake.**

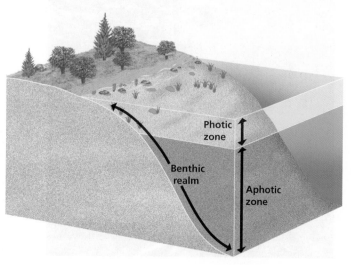

Photic zone

Benthic realm

Aphotic zone

▲ Figure 18.16 A stream in the Appalachian Mountains.

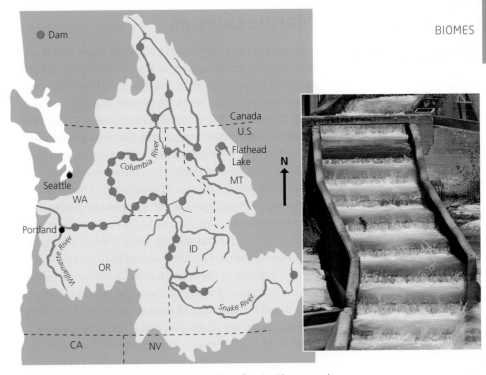

▲ Figure 18.17 Damming the Columbia River basin. This map shows only the largest of the 250 dams that have altered freshwater ecosystems throughout the Pacific Northwest. These great concrete obstacles make it difficult for salmon to swim upriver to their breeding streams, though many dams now have "fish ladders" that provide detours (inset).

Rivers and Streams

Rivers and streams, which are bodies of flowing water, generally support communities of organisms quite different from those of lakes and ponds (Figure 18.16). A river or stream changes greatly between its source (perhaps a spring or snowmelt in the mountains) and the point at which it empties into a lake or the ocean. Near a source, the water is usually cold, low in nutrients, and clear. The channel is often narrow, with a swift current that does not allow much silt to accumulate on the bottom. The current also inhibits the growth of phytoplankton; most of the organisms found here are supported by the photosynthesis of algae attached to rocks or by organic material (such as leaves) carried into the stream from the surrounding land. The most abundant benthic animals are usually insects that eat algae, leaves, or one another. Trout are often the predominant fishes, locating their food, including insects, mainly by sight in the clear water.

Downstream, a river or stream typically widens and slows. There the water is usually warmer and may be murkier because of sediments and phytoplankton suspended in it. Worms and insects that burrow into mud are often abundant, as are waterfowl, frogs, and catfish and other fishes that find food more by scent and taste than by sight.

People have altered rivers by constructing dams to control flooding, to provide reservoirs of drinking water, or to generate hydroelectric power. In many cases, dams have completely changed the downstream ecosystems, altering the rate and volume of water flow and affecting fish and invertebrate populations (Figure 18.17). Many streams and rivers have also been affected by pollution from human activities.

Wetlands

A **wetland** is a transitional biome between an aquatic ecosystem and a terrestrial one. Freshwater wetlands include swamps, bogs, and marshes (Figure 18.18). Covered with water either permanently or periodically, wetlands support the growth of aquatic plants and are rich in species diversity. Migrating waterfowl and many other birds depend on wetland "pit stops" for food and shelter during their journeys. In addition, wetlands provide water storage areas that reduce flooding. Wetlands also improve water quality by trapping pollutants such as metals and organic compounds in their sediments.

▼ Figure 18.18 A wetland near Kent, Ohio.

Marine Biomes

Gazing out over a vast ocean, you might think that it is the most uniform environment on Earth. But marine habitats can be as different as night and day. The deepest ocean, where hydrothermal vents are located, is perpetually dark. In contrast, the vivid coral reefs nearer the surface are utterly dependent on sunlight. Habitats near shore are different from those in mid-ocean, and the seafloor hosts different communities than the open waters.

As in freshwater biomes, the seafloor is known as the benthic realm (Figure 18.19). The **pelagic realm** of the oceans includes all open water. In shallow areas, such as the continental shelves (the submerged parts of continents), the photic zone includes both pelagic and benthic regions. In these sunlit areas, photosynthesis by phytoplankton and multicellular algae provides energy for a diverse community of animals. Sponges, burrowing worms, clams, sea anemones, crabs, and echinoderms inhabit the benthic realm. **Zooplankton** (free-floating animals, including many microscopic ones), fishes, marine mammals, and many other types of animals are abundant in the pelagic photic zone.

The **coral reef** biome occurs in the photic zone of warm tropical waters in scattered locations around the globe. A coral reef is built up slowly by successive generations of coral animals—a diverse group of cnidarians that secrete a hard external skeleton—and by multicellular algae encrusted with limestone (Figure 18.20). Unicellular algae live within the coral's cells, providing the coral with food. The physical structure and productivity of coral reefs support a huge variety of invertebrates and fishes.

The photic zone extends down a maximum of 200 m (about 656 feet) in the ocean. Although there is not enough light for photosynthesis between 200 and 1,000 m (down a little more than half a mile), some light does reach these depths of the aphotic zone. This dimly lit world, sometimes called the twilight zone, is dominated by a fascinating variety of small fishes and crustaceans. Food sinking from the photic zone provides some sustenance for these animals. In addition, many of them migrate to the surface at night to feed. Some fishes in the twilight zone have enlarged eyes, enabling them to see in the very dim light, and light-emitting organs that attract mates and prey.

Below 1,000 m—a depth greater than the height of two Empire State Buildings—the ocean is completely and permanently dark. Adaptation to this environment has produced many bizarre-looking creatures. Most of the benthic organisms here are deposit feeders, animals that consume dead organic material in the sediments on the seafloor. Crustaceans, annelid worms, sea anemones, and echinoderms such as sea cucumbers, sea stars, and sea urchins are common. Food is scarce, however. The

▼ Figure 18.19 **Ocean life.** (Zone depths and organisms not drawn to scale.)

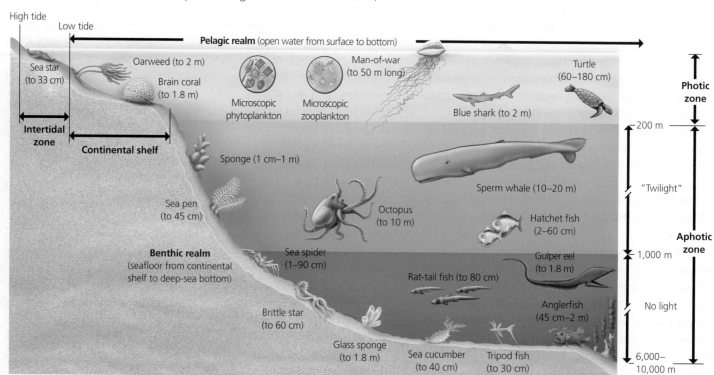

▼ Figure 18.21 Organisms clinging to the rocks of an intertidal zone on the Pacific coast of Washington.

▲ Figure 18.20 A coral reef in the Red Sea off the coast of Egypt.

density of animals is low except at hydrothermal vents, the prokaryote-powered ecosystems mentioned earlier (see Figure 18.6).

The marine environment also includes distinctive biomes, such as the intertidal zone and estuaries, where the ocean interfaces with land or with fresh water. In the **intertidal zone**, where the ocean meets land, the shore is pounded by waves during high tide and exposed to the sun and drying winds during low tide. The rocky intertidal zone is home to many sedentary organisms, such as sea stars, barnacles, and mussels, which attach to rocks and thus are prevented from being washed away **(Figure 18.21)**. On sandy beaches, suspension-feeding worms, clams, and predatory crustaceans bury themselves in the ground.

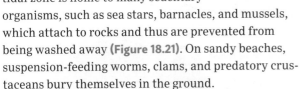

LANDFILLS AREN'T THE ONLY PLACE WHERE GARBAGE ACCUMULATES.

Figure 18.22 shows an **estuary**, a transition area between a river and the ocean. The saltiness of estuaries ranges from nearly that of fresh water to that of the ocean. With their waters enriched by nutrients from rivers, estuaries, like freshwater wetlands, are among the most productive areas on Earth. Oysters, crabs, and many fishes live or reproduce in estuaries. Estuaries are also crucial nesting and feeding areas for waterfowl. Mudflats and salt marshes are extensive coastal wetlands that often border estuaries.

For centuries, people viewed the ocean as a limitless resource, harvesting its bounty with increasingly effective and indiscriminate technologies and using it as a dumping ground for wastes. The negative effects of these practices are now becoming clear. Populations of commercial fish species are declining. Trillions of small bits of plastic debris float beneath the surface of vast swaths of ocean, concentrated by converging currents in several regions dubbed "garbage patches." Many marine habitats are polluted by nutrients or toxic chemicals; it will be

years before the full extent of the damage from the massive Deepwater Horizon oil spill in the Gulf of Mexico in 2010 (see Figure 18.2) is known. Because of their nearness to land, estuaries are especially vulnerable. Many have been completely replaced by development on landfill; other threats include pollution and alteration of freshwater inflow. Coral reefs are imperiled by ocean acidification and rising sea surface temperatures due to global warming.

Meanwhile, our knowledge of marine biomes is woefully incomplete. A recent decade-long scientific census of marine life announced the discovery of more than 6,000 new species. ☑

☑ **CHECKPOINT**

What are phytoplankton? Why are they essential to other oceanic life?

■ *Answer: Phytoplankton are photosynthetic algae and bacteria. They are food for animals in the photic zone; those animals in turn may become food for animals in the aphotic zone.*

▼ Figure 18.22 Waterfowl in an estuary on the southeast coast of England.

How Climate Affects Terrestrial Biome Distribution

Terrestrial biomes are determined primarily by climate, especially temperature and rainfall. Before we survey these biomes, let's look at the broad patterns of global climate that help explain their locations.

Earth's global climate patterns are largely the result of the input of solar energy—which warms the atmosphere, land, and water—and of the planet's movement in space. Because of Earth's curvature, the intensity of sunlight varies according to latitude (**Figure 18.23**). The equator receives the greatest intensity of solar radiation and thus has the highest temperatures, which in turn evaporate water from Earth's surface. As this warm, moist air rises, it cools, diminishing its ability to hold moisture. The water vapor condenses into clouds and eventually falls as rain (**Figure 18.24**). This process largely explains why rain forests are concentrated in the **tropics**—the region from the Tropic of Cancer to the Tropic of Capricorn.

After losing moisture over equatorial zones, dry high-altitude air masses spread away from the equator until they cool and descend at latitudes of about 30° north and south. Many of the world's great deserts—the Sahara in North Africa and the Arabian on the Arabian Peninsula, for example—are centered at these latitudes because of the dry air they receive.

Latitudes between the tropics and the Arctic Circle and the Antarctic Circle are called **temperate zones**. Generally, these regions have milder climates than the tropics or the polar regions. Notice in Figure 18.24 that some of the descending dry air moves into the latitudes above 30°. At first these air masses pick up moisture, but they tend to drop it as they cool at higher latitudes. This is why the north and south temperate zones tend to be relatively wet. Coniferous forests dominate the landscape at the wet but cool latitudes around 60° north.

Proximity to large bodies of water and the presence of landforms such as mountain ranges also affect climate. Oceans and large lakes moderate climate by absorbing heat when the air is warm and releasing heat to cold air. Mountains affect climate in two major ways. First, air temperature drops as elevation increases. As a result, driving up a tall mountain offers a quick tour of several biomes. **Figure 18.25** shows the scenery you might encounter on a journey from the scorching lowlands of the Sonoran Desert to a cool coniferous forest at an elevation of 11,000 feet above sea level.

▼ **Figure 18.23** **Uneven heating of Earth.**

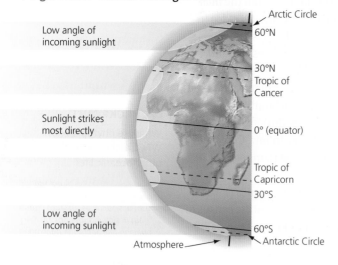

▼ **Figure 18.24** **How uneven heating of Earth produces various climates.**

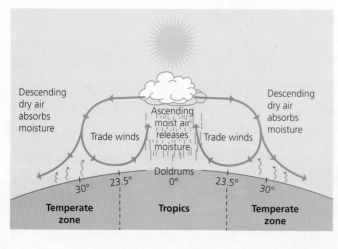

▶ **Figure 18.25** **The effect of altitude on vegetation.** The zones shown are typical of the Sonoran Desert region in southwestern North America.

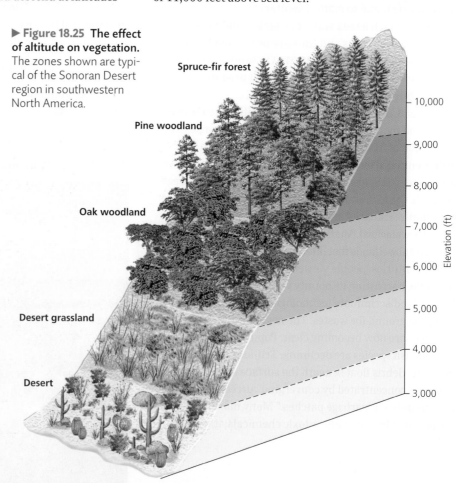

Second, mountains can block the flow of moist air from a coast, causing radically different climates on opposite sides of a mountain range. In the example shown in **Figure 18.26**, moist air moves in off the Pacific Ocean and encounters the Coast Range in California. Air flows upward, cools at higher altitudes, and drops a large amount of rainfall. The world's tallest trees, the coastal redwoods, thrive here. Precipitation increases again farther inland as the air moves up and over higher mountains (the Sierra Nevada). By the time it reaches the eastern side of the Sierra, the air contains little moisture; as this dry air descends, it absorbs moisture. As a result, there is little precipitation on the eastern side of the mountains. This effect, called a rain shadow, is responsible for the desert that covers much of central Nevada. ☑

▼ Figure 18.26 **How mountains affect rainfall.**

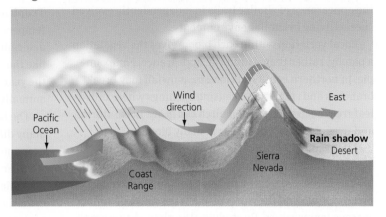

☑ **CHECKPOINT**

Why is there so much rainfall in the tropics?

■ *Answer: Air at the equator rises as it is warmed by direct sunlight. As the air rises, it cools. This causes cloud formation and rainfall because cool air holds less moisture than warm air.*

Terrestrial Biomes

Terrestrial ecosystems are grouped into biomes primarily on the basis of their vegetation type (**Figure 18.27**). By providing food, shelter, and nesting sites for animals, as well as much of the organic material for the decomposers that recycle mineral nutrients, plants build the foundation for the communities of organisms typical of each biome. The geographic distribution of plants, and thus of biomes, largely depends on climate, with temperature and rainfall often the key factors determining the kind of biome that exists in a particular region. If the climate in two geographically separate areas is similar, the same type of biome may occur in both. Coniferous forests, for instance, extend in a broad band across North America, Europe, and Asia.

Each biome is characterized by a type of biological community rather than an assemblage of particular species. For example, the groups of species living in the deserts of southwestern North America and in the Sahara Desert of Africa are different, but both groups are adapted to desert conditions. Organisms in widely separated

▼ Figure 18.27 **A map of the major terrestrial biomes.** Although this map has sharp boundaries, biomes actually grade into one another. We'll use smaller versions of this map, highlighted by color coding, during our closer look at the terrestrial biomes in the next several pages.

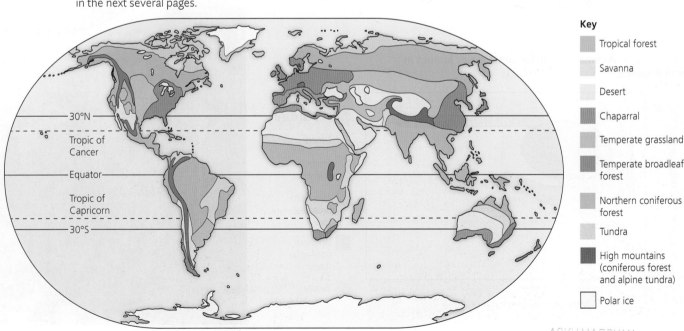

Key

- Tropical forest
- Savanna
- Desert
- Chaparral
- Temperate grassland
- Temperate broadleaf forest
- Northern coniferous forest
- Tundra
- High mountains (coniferous forest and alpine tundra)
- Polar ice

biomes may look alike because of convergent evolution, the appearance of similar traits in independently evolved species living in similar environments.

Local variation within each biome gives the vegetation a patchy, rather than a uniform, appearance. For example, in northern coniferous forests, snowfall may break branches and small trees, causing openings where broadleaf trees such as aspen and birch can grow. Local storms and fires also create openings in many biomes.

The graph in **Figure 18.28** shows the ranges of precipitation and temperature that characterize terrestrial biomes. The *x*-axis shows the range of annual average precipitation, and the *y*-axis displays the range of annual average temperature. By studying the plots on this graph, we can compare these abiotic factors in different biomes. For example, although the range of precipitation in temperate broadleaf forests is similar to that of northern coniferous forests, the lower range of temperatures in northern coniferous forests reveals a significant difference in the abiotic environments of these two biomes. Grasslands are typically drier than forests, and deserts are drier still.

Today, concern about global warming is generating intense interest in the effect of climate on vegetation patterns. Using powerful new tools such as satellite imagery, scientists are documenting shifts in latitude of biome borders, decreases in snow and ice coverage, and changes in the length of the growing season. At the same time, many natural biomes have been fragmented and altered by human activity. We'll discuss both of these issues after we survey the major terrestrial biomes, beginning near the equator and generally approaching the poles.

CHECKPOINT

Why are climbing plants common in tropical rain forests?

■ *Answer: Climbing is a plant adaptation for reaching sunlight in a closed canopy, where little sunlight reaches the forest floor.*

Tropical Forest

Tropical forests occur in equatorial areas where the temperature is warm year-round. The type of vegetation is determined primarily by rainfall. In tropical rain forests, like the one shown in **Figure 18.29**, rain falls throughout the year, totaling 200 to 400 cm (6.6 to 13 *feet*!) of rain annually. Rainfall is less plentiful in other tropical forests.

The layered structure of tropical rain forests provides many different habitats. Treetops form a closed canopy over one or two layers of smaller trees and a shrub understory. Few plants grow in the deep shade of the forest floor. Many trees are covered by woody vines growing toward the light. Other plants, such as orchids, gain access to sunlight by growing on the branches or trunks of tall trees. Scattered trees reach full sunlight by towering above the canopy. Many of the animals also dwell in trees, where food is abundant. Monkeys, birds, insects, snakes, bats, and frogs find food and shelter many meters above the ground.

Tropical dry forests predominate in lowland areas that have a prolonged dry season or scarce rainfall at any time. The plants found there are a mixture of thorny shrubs and trees and succulents. In regions with distinct wet and dry seasons, deciduous trees that conserve water by shedding their leaves during the dry season are common. ☑

▼ **Figure 18.29 Tropical rain forest in Borneo.**

▼ **Figure 18.28 A climate graph for some major biomes in North America.**

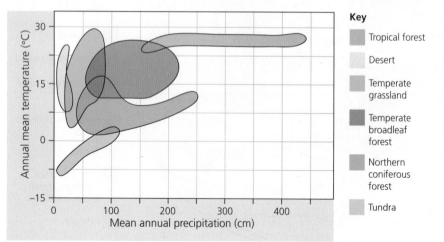

Key

Tropical forest

Desert

Temperate grassland

Temperate broadleaf forest

Northern coniferous forest

Tundra

Savanna

Savannas, such as the one shown in **Figure 18.30**, are dominated by grasses and scattered trees. The temperature is warm year-round. Rainfall averages 30 to 50 cm (roughly 12 to 20 inches) per year, with dramatic seasonal variation.

Fire, caused by lightning or human activity, is an important abiotic factor in the savanna. The grasses survive burning because the growing points of their shoots are below ground. Other plants have seeds that sprout rapidly after a fire. Poor soil and lack of moisture, along with fire and grazing animals, prevent the establishment of most trees in the first place. The luxuriant growth of grasses and small broadleaf plants during the rainy season provides a rich food source for plant-eating animals.

Many of the world's large grazing mammals and their predators inhabit savannas. African savannas are home to zebras and many species of antelope, as well as to lions and cheetahs. Several species of kangaroo are the dominant grazers of Australian savannas. Oddly, though, the large grazers are not the dominant plant-eaters in savannas. That distinction belongs to insects, especially ants and termites. Other animals include burrowers such as mice, moles, gophers, and ground squirrels.

Desert

Deserts are the driest of all biomes, characterized by low and unpredictable rainfall—less than 30 cm (about 12 inches) per year. Some deserts are very hot, with daytime soil surface temperatures above 60°C (140°F) and large daily temperature fluctuations. Other deserts, such as those west of the Rocky Mountains and the Gobi Desert, spanning northern China and southern Mongolia, are relatively cold. Air temperatures in cold deserts may fall below −30°C (−22°F).

Desert vegetation typically includes water-storing plants, such as cacti, and deeply rooted shrubs. Various snakes, lizards, and seed-eating rodents are common inhabitants. Arthropods such as scorpions and insects also thrive in the desert. Evolutionary adaptations of desert plants and animals include a remarkable array of mechanisms that conserve water. For example, the "pleated" stem of saguaro cacti **(Figure 18.31)** enables the plants to expand when they absorb water during wet periods. Some desert mice *never* drink, deriving all their water from the metabolic breakdown of the seeds they eat. Protective adaptations that deter feeding by mammals and insects, such as spines on cacti and poisons in the leaves of shrubs, are common in desert plants. ☑

☑ **CHECKPOINT**

1. How does the savanna climate vary seasonally?

2. What abiotic factor characterizes deserts?

Answers: 1. Temperature stays about the same year-round, but rainfall varies dramatically. 2. Rainfall is low and unpredictable.

▼ **Figure 18.30 Savanna in the Serengeti Plain in Tanzania.**

▼ **Figure 18.31 Sonoran Desert.**

Chaparral

The climate that supports **chaparral** vegetation results mainly from cool ocean currents circulating offshore, producing mild, rainy winters. Summers are hot and dry. This biome is limited to small coastal areas, some in California (Figure 18.32). The largest region of chaparral surrounds the Mediterranean Sea; in fact, Mediterranean is another name for this biome. Dense, spiny, evergreen shrubs dominate chaparral. Annual plants are also common during the wet winter and spring months. Animals characteristic of the chaparral are deer, fruit-eating birds, seed-eating rodents, and lizards and snakes.

Chaparral vegetation is adapted to periodic fires caused by lightning. Many plants contain flammable chemicals and burn fiercely, especially where dead brush has accumulated. After a fire, shrubs use food reserves stored in the surviving roots to support rapid shoot regeneration. Some chaparral plants produce seeds that will germinate only after a hot fire. The ashes of burned vegetation fertilize the soil with mineral nutrients, promoting regrowth of the plant community. Houses do not fare as well. The firestorms that race through the densely populated canyons of Southern California can be devastating to the human inhabitants.

Temperate Grassland

Temperate grasslands have some of the characteristics of tropical savannas, but they are mostly treeless, except along rivers or streams, and are found in regions of relatively cold winter temperatures. Rainfall, averaging between 25 and 75 cm per year (approximately 10 to 30 inches), with frequent severe droughts, is too low to support forest growth. Periodic fires and grazing by large mammals also prevent invasion by woody plants. These grazers include the bison and pronghorn in North America, the wild horses and sheep of the Asian steppes, and kangaroos in Australia. As in the savanna, however, the dominant plant-eaters are invertebrates, especially grasshoppers and soil-dwelling nematodes.

Without trees, many birds nest on the ground. Many small mammals, such as rabbits, voles, ground squirrels, prairie dogs, and pocket gophers, dig burrows to escape predators. Temperate grasslands like the one shown in Figure 18.33 once covered much of central North America.

Because grassland soil is both deep and rich in nutrients, these habitats provide fertile land for agriculture. Most grassland in the United States has been converted to cropland or pasture, and very little natural prairie exists today. ☑

☑ CHECKPOINT

1. What is one way that homeowners in chaparral areas can protect their neighborhoods from fire?
2. How do people now use most of the North American land that was once temperate grassland?

■ *Answers: 1. by keeping the area clear of dead brush, which is flammable. 2. for farming.*

▼ **Figure 18.32 Chaparral in California.**

▼ **Figure 18.33 Temperate grassland in Saskatchewan, Canada.**

Temperate Broadleaf Forest

Temperate broadleaf forests occur throughout midlatitudes where there is sufficient moisture to support the growth of large trees. Annual precipitation is relatively high at 75 to 150 cm (30 to 60 inches) and typically distributed evenly around the year. Temperature varies seasonally over a wide range, with hot summers and cold winters. In the Northern Hemisphere, dense stands of deciduous trees are trademarks of temperate forests, such as the one pictured in **Figure 18.34**. Deciduous trees drop their leaves before winter, when temperatures are too low for effective photosynthesis and water lost by evaporation is not easily replaced from frozen soil.

Numerous invertebrates live in the soil and the thick layer of leaf litter that accumulates on the forest floor. Some vertebrates, such as mice, shrews, and ground squirrels, burrow for shelter and food, while others, including many species of birds, live in the trees. Predators include bobcats, foxes, black bears, and mountain lions. Many mammals that inhabit these forests enter a dormant winter state called hibernation, and some bird species migrate to warmer climates.

Virtually all the original temperate broadleaf forests in North America were cut for timber or cleared for agriculture or development. These forests tend to recover after disturbance, however, and today we see deciduous trees growing in undeveloped areas over much of their former range.

Coniferous Forest

Cone-bearing evergreen trees such as pine, spruce, fir, and hemlock dominate **coniferous forests** in the Northern Hemisphere. (Other kinds of conifers grow in parts of South America, Africa, and Australia.) The northern coniferous forest, also known as the boreal forest or **taiga** (**Figure 18.35**), is the largest terrestrial biome on Earth, stretching in a broad band across North America and Asia south of the Arctic Circle. Taiga is also found north of the Arctic Circle in Finland due to the warming action of the Gulf Stream and at cool, high elevations in more temperate latitudes. The taiga is characterized by long, snowy winters and short, wet summers that are sometimes warm. The slow decomposition of conifer needles in the thin, acidic soil makes few nutrients available for plant growth. The conical shape of many conifers prevents too much snow from accumulating on their branches and breaking them. Animals of the taiga include moose, elk, hares, bears, wolves, grouse, and migratory birds. The Asian taiga is home to the dwindling number of Siberian tigers that remain in the wild.

The **temperate rain forests** of coastal North America (from Alaska to Oregon) are also coniferous forests. Warm, moist air from the Pacific Ocean supports this unique biome, which, like most coniferous forests, is dominated by a few tree species, typically hemlock, Douglas fir, and redwood. These forests are heavily logged, and the old-growth stands of trees are rapidly disappearing. ☑

☑ CHECKPOINT

1. How does the loss of leaves function as an adaptation of deciduous trees to cold winters?

2. What type of trees are characteristic of the taiga?

■ Answers: **1.** by reducing loss of water from the trees when that water cannot be replaced because of frozen soil. **2.** conifers such as pine, spruce, fir, and hemlock.

▼ **Figure 18.34**
Temperate broadleaf forest in Maine in autumn.

▼ **Figure 18.35**
Northern coniferous forest in Finland, with the sky lit by the northern lights.

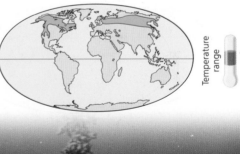

Tundra

Tundra covers expansive areas of the Arctic between the taiga and polar ice. **Permafrost** (permanently frozen subsoil), bitterly cold temperatures, and high winds are responsible for the absence of trees and other tall plants in the arctic tundra shown in **Figure 18.36**. The arctic tundra receives very little annual precipitation. However, water cannot penetrate the underlying permafrost, so melted snow and ice accumulate in pools on the shallow topsoil during the short summer.

Tundra vegetation includes small shrubs, grasses, mosses, and lichens. When summer arrives, flowering plants grow quickly and bloom in a rapid burst. Caribou, musk oxen, wolves, and small rodents called lemmings are among the mammals found in the arctic tundra. Many migratory birds use the tundra as a summer breeding ground. During the brief but productive warm season, the marshy ground supports the aquatic larvae of insects, providing food for migratory waterfowl, and clouds of mosquitoes often fill the tundra air.

On very high mountaintops at all latitudes, including the tropics, high winds and cold temperatures create plant communities called alpine tundra. Although these communities are similar to arctic tundra, there is no permafrost beneath alpine tundra.

Polar Ice

Polar ice covers the land at high latitudes north of the arctic tundra in the Northern Hemisphere and in Antarctica in the Southern Hemisphere (**Figure 18.37**). The temperature in these regions is extremely cold year-round, and precipitation is very low. Only a small portion of these landmasses is free of ice or snow, even during the summer. Nevertheless, small plants, such as mosses and lichens, eke out a living, and invertebrates such as nematodes, mites, and wingless insects called springtails inhabit the frigid soil. Nearby sea ice provides feeding platforms for large animals such as polar bears (in the Northern Hemisphere), penguins (in the Southern Hemisphere), and seals. Seals, penguins, and other marine birds visit the land to rest and breed. The polar marine biome provides the food that sustains these birds and mammals. In the Antarctic, penguins feed at sea, eating a variety of fish, squids, and small shrimplike crustaceans known as krill. Antarctic krill, an important food source for many species of fish, seals, squids, seabirds, and filter-feeding whales as well as penguins, depend on sea ice for breeding and as a refuge from predators. As the amount of sea ice declines and the ice season becomes shorter—consequences of global climate change—krill habitat is shrinking. ☑

▼ **Figure 18.36** Arctic tundra in Yukon Territory, Canada.

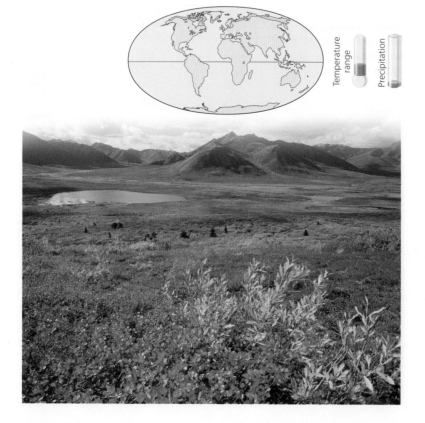

Temperature range   Precipitation

▼ **Figure 18.37** Polar ice in Antarctica.

Temperature range   Precipitation

The Water Cycle

Biomes are not self-contained units. Rather, all parts of the biosphere are linked by the global water cycle, illustrated in **Figure 18.38**, and by nutrient cycles (see Chapter 20). Consequently, events in one biome may reverberate throughout the biosphere.

As you learned earlier in this chapter, water and air move in global patterns driven by solar energy. Precipitation and evaporation continuously move water between the land, oceans, and the atmosphere. Water also evaporates from plants, which pull it from the soil in a process called transpiration.

Over the oceans, evaporation exceeds precipitation. The result is a net movement of water vapor to clouds that are carried by winds from the oceans across the land. On land, precipitation exceeds evaporation and transpiration. The excess precipitation may stay on the surface or it may trickle through the soil to become groundwater. Both surface water and groundwater eventually flow back to the sea, completing the water cycle.

Just as the water draining from your shower carries dead skin cells from your body along with the day's grime, the water washing over and through the ground carries traces of the land and its history. For example, water flowing from land to the sea carries with it silt (fine soil particles) and chemicals such as fertilizers and pesticides.

Erosion from coastal development has caused silt to muddy the waters of some coral reefs, dimming the light available to the photosynthetic algae that power the reef community. Chemicals in surface water may travel hundreds of miles by stream and river to the ocean, where currents then carry them even farther from their point of origin. For instance, traces of pesticides and chemicals from industrial wastes have been found in marine mammals in the Arctic and in deep-sea octopuses and squids. Airborne pollutants such as nitrogen oxides and sulfur oxides, which combine with water to form acid precipitation, are distributed by the water cycle, too.

Human activity also affects the global water cycle itself in a number of important ways. One of the main sources of atmospheric water is transpiration from the dense vegetation making up tropical rain forests. The destruction of these forests changes the amount of water vapor in the air. Pumping large amounts of groundwater to the surface for irrigation increases the rate of evaporation over land and may deplete groundwater supplies. In addition, global warming affects the water cycle in complex ways that will have far-reaching effects on precipitation patterns. We'll consider some of these environmental impacts in the following sections. ☑

POLLUTANTS FROM CAR EXHAUST CAN RETURN TO EARTH AS ACID PRECIPITATION.

☑ **CHECKPOINT**

What is the main way that living organisms contribute to the water cycle?

■ *Answer: Plants move water from the ground to the air via transpiration.*

Figure Walkthrough

Mastering **Biology**
goo.gl/FaPcb2

▼ Figure 18.38 **The global water cycle.**

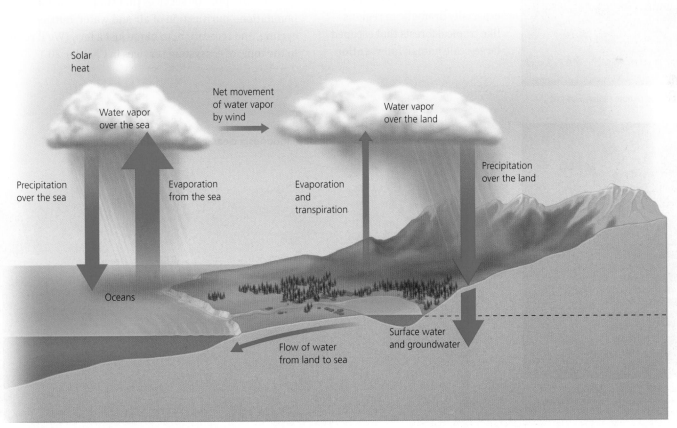

Solar heat

Water vapor over the sea

Net movement of water vapor by wind

Water vapor over the land

Precipitation over the sea

Evaporation from the sea

Evaporation and transpiration

Precipitation over the land

Oceans

Flow of water from land to sea

Surface water and groundwater

Human Impact on Biomes

For hundreds of years, people have been using increasingly effective technologies to capture or produce food, to extract resources from the environment, and to build cities. It is now clear that the environmental costs of these enterprises are staggering. In this section, you'll see some examples of how human activities are affecting forest and freshwater resources. Throughout the remainder of this unit, you'll learn about the role of ecological knowledge in achieving **sustainability**, the goal of developing, managing, and conserving Earth's resources in ways that meet the needs of people today without compromising the ability of future generations to meet their needs.

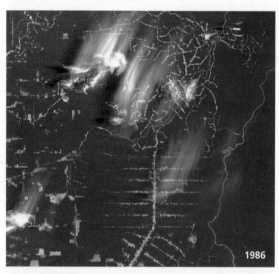

▼ **Figure 18.39** Satellite photos of the Rondonia area of the Brazilian rain forest.

1986. In 1986, development in this remote region was just beginning.

2001. Same area 15 years later, after loggers and farmers have moved in. The "fishbone" pattern marks the new network of roads carved through the forest.

Forests

The map in Figure 18.27 shows the terrestrial biomes that would be expected to flourish under the prevailing climatic conditions. However, about three-quarters of Earth's land surface has been altered by thousands of years of human occupation. Most of the land that we've appropriated is used for agriculture; another hefty chunk is covered by the asphalt and concrete of development. Changes in vegetation are especially dramatic in regions like tropical forests that escaped large-scale human intervention until recently. Satellite photos of a small area in Brazil show how thoroughly a landscape can be altered in a short amount of time (**Figure 18.39**).

Every year, more and more forested land is cleared for agriculture. You might think that this land is needed to feed new mouths as the human population continues to grow, but that's not entirely the case. Unsustainable agricultural practices have degraded much of the world's cropland so severely that it is unusable. Researchers estimate that replacing worn-out farmland accounts for up to 80% of the deforestation occurring today. Tropical forests, such as the one shown in **Figure 18.40**, are also being cleared to grow palm oil for products such as cosmetics and a long list of packaged foods, including cookies, crackers, potato chips, chocolate products, and soups. Other forests are being lost to logging, mining, and air pollution, problems that are hitting coniferous forests especially hard. (As we mentioned in the previous section, most temperate broadleaf forests were replaced by human enterprises long ago.) Land that hasn't been directly converted to food production and living space also bears the imprint of our presence. Roads penetrate regions that are otherwise unaltered, bringing pollution to the wilderness, providing avenues for new diseases to emerge, and slicing vast tracts of biome into segments that are too small to support a full array of species.

Land uses that provide resources such as food, fuel, and shelter are clearly beneficial to us. But natural ecosystems also provide services that support the human population—purification of air and water, nutrient cycling, and recreation, to name just a few. (We'll return to the topic of ecosystem services in Chapter 20.)

▼ **Figure 18.40** A tropical forest in Indonesia clear-cut for a palm oil plantation.

Fresh Water

The impact of human activities on freshwater ecosystems may pose an even greater threat to life on Earth—including ourselves—than the damage to terrestrial ecosystems. Freshwater ecosystems are being polluted by large amounts of nitrogen and phosphorus compounds that run off from heavily fertilized farms or from livestock feedlots. A wide variety of other pollutants, such as industrial wastes, also contaminate freshwater habitats, drinking water, and groundwater. Some regions of the world face dire shortages of water as a result of the overuse of groundwater for irrigation, extended droughts (partially caused by global climate change), or poor water management practices.

Las Vegas, the population center of Clark County, Nevada, is one example of a city whose water resources are increasingly stressed by drought and overconsumption. **Figure 18.41a** is a satellite photo of Las Vegas in 1972, when the population of Clark County was 307,421. **Figure 18.41b** shows the same area in 2013, when the population had swelled to more than 2 million. In contrast to the disappearance of greenery in the photos of Brazilian rain forest, the mark of human activities in Figure 18.41b is the notable expansion of greenery—the result of watering lawns and golf courses. Las Vegas is situated in a high valley in the Mojave Desert. Where does it get the water to turn barren desert into green fields?

Las Vegas taps underground aquifers for some water, but its main water supply is Lake Mead. Lake Mead is an enormous reservoir formed by the Hoover Dam on the Colorado River, which in turn receives almost all of its water from snowmelt in the Rocky Mountains. With decreased annual snowfall, attributable largely to climate change, the flow of the Colorado has greatly diminished.

The water level in Lake Mead has dropped drastically (**Figure 18.42**), and parched cities and farms farther downstream are pleading for more water.

To ensure an adequate water supply for the future, Las Vegas is looking for new sources of water. Among other options, Las Vegas is eyeing the abundant supply of groundwater in the northern end of the valley where it lies. Although sparsely populated, that area is home to many ranchers whose livelihoods depend on the groundwater. It is also home to numerous endangered species. Not surprisingly, environmentalists and residents of the north valley are resisting efforts to pipe its groundwater to Las Vegas.

Nevada is just one of many places where the hard realities of climate change are beginning to affect daily life. Battles over water resources are shaping up throughout the arid West and Southwest of the United States, where changing precipitation patterns due to global warming are projected to continue the drought for many years to come. In other regions of the world, including China, India, and North Africa, the increasing demands of economic, agricultural, and population growth are straining water resources that are already scarce.

While policymakers are dealing with current crises and planning how to manage resources in the future, researchers are seeking methods of sustainable agriculture and water use. Basic ecological research is an essential component of ensuring that sufficient food and water will be available for people now—and for the generations to come. Next, we take a closer look at a major threat to sustainability: climate change. ✓

▼ **Figure 18.42 Low water level in Lake Mead.** The white "bathtub ring" is caused by mineral deposits on rocks that were once submerged.

▼ **Figure 18.41 Satellite photos of Las Vegas, Nevada.**

(a) 1972

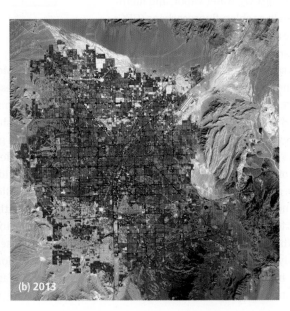

(b) 2013

✓ **CHECKPOINT**

Why is decreased snowfall in the Rocky Mountains a concern for people who live in Las Vegas?

■ *Answer: Snowmelt from the Rockies flows into the Colorado River, which supplies water for Las Vegas residents.*

Climate Change

Rising concentrations of carbon dioxide (CO_2) and certain other gases in the atmosphere are changing global climate patterns. This was the conclusion of the assessment report released by the Intergovernmental Panel on Climate Change (IPCC) in 2014. Thousands of scientists and policymakers from more than 100 countries participated in producing the report, which is based on data published in thousands of scientific papers. Thus, there is no debate among scientists about whether climate change is occurring.

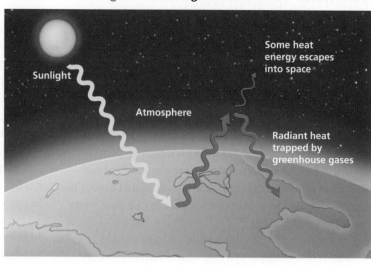

▼ Figure 18.43 The greenhouse effect.

The Greenhouse Effect and Global Warming

Why is Earth's atmosphere becoming warmer? A familiar and useful analogy is a greenhouse, which is used to grow plants when the weather outside is too cold. Its transparent glass or plastic walls allow solar radiation to pass through and trap some of the heat that accumulates inside the building. Similarly, certain gases in Earth's atmosphere are transparent to solar radiation but absorb or reflect heat. Some of these **greenhouse gases** are natural, including CO_2, water vapor, and methane. Others, such as chlorofluorocarbons (CFCs, found in some aerosol sprays and refrigerants), are synthetic. As **Figure 18.43** shows, greenhouse gases act as a blanket that traps heat in the atmosphere. This heating, often called the **greenhouse effect**, is highly beneficial. Without it, the average air temperature on Earth would be a frigid −18°C (−0.4°F), far too cold for most life as we know it. However, increasing the insulation that the blanket provides is making Earth overly warm.

The effect of increasing greenhouse gases is the increase in the average global temperature, which has risen about 1°C (1.8°F) since 1900 at an accelerating pace. Further increases of 2 to 4.5°C are likely by the end of this century, according to the IPPC. Temperature increases are not distributed evenly around the world. The largest increases are in the northernmost regions of the Northern Hemisphere and parts of Antarctica (**Figure 18.44**).

▼ Figure 18.44 **Differences in average temperatures during 2006–2016 compared with long-term averages during 1951–1980, in °C.** The largest temperature increases are shown in red. Gray indicates regions for which no data are available.

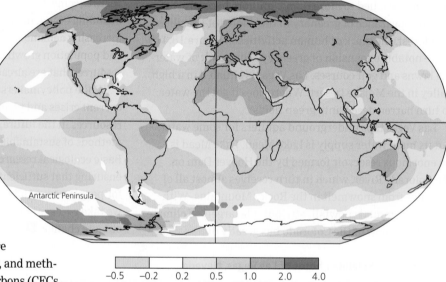

Antarctic Peninsula

| −0.5 | −0.2 | 0.2 | 0.5 | 1.0 | 2.0 | 4.0 |

More than 90% of the heat trapped by greenhouse gases is being stored in the ocean. Water expands as it warms, causing sea levels to rise. Melting of the massive ice sheets of Greenland and Antarctica, as well as mountain glaciers, is also contributing to sea level rise. Rising sea levels will cause catastrophic flooding of coastal areas worldwide.

Global warming is causing an increase in extreme weather events. In some regions, a greater proportion of the total precipitation falls in torrential downpours that cause flooding; other regions undergo intense drought. Prolonged heat waves occur more frequently. Higher sea surface temperatures fuel more powerful hurricanes. In addition, warm weather begins earlier each year. ✓

☑ CHECKPOINT

Why are gases such as CO_2 and methane called greenhouse gases?

■ Answer: They allow solar radiation to pass through the atmosphere but prevent the heat from reflecting back out, much as the glass of a greenhouse retains the sun's heat inside the building.

The Accumulation of Greenhouse Gases

After many years of data collection and debate, the vast majority of scientists are confident that human activities have caused the rising concentrations of greenhouse gases. Major sources of emissions include agriculture, landfills, and the burning of wood and fossil fuels (oil, coal, and natural gas).

Let's take a closer look at rising concentrations of CO_2. For 650,000 years, the atmospheric concentration of CO_2 did not exceed 300 parts per million (ppm); the concentration before the Industrial Revolution was 280 ppm. In 2016, the average atmospheric CO_2 was 404 ppm and continuing to rise (Figure 18.45). The levels of other greenhouse gases have increased dramatically, too. Remember that CO_2 is removed from the atmosphere by the process of photosynthesis and stored in organic molecules such as carbohydrates (see Figure 6.2). These molecules are eventually broken down by cellular respiration, releasing CO_2. Overall, uptake of CO_2 by photosynthesis roughly equals the release of CO_2 by cellular respiration (Figure 18.46). However, extensive deforestation has significantly decreased the incorporation of CO_2 into organic material. At the same time, CO_2 is flooding into the atmosphere from the burning of fossil fuels and wood, a process that releases CO_2 from organic material much more rapidly than cellular respiration.

CO_2 is also exchanged between the atmosphere and the surface waters of the oceans. For decades, the oceans have acted as massive sponges, soaking up considerably more CO_2 than they have released. But now, the excess CO_2 has made the oceans more acidic, a change that could have a profound effect on marine communities. As ocean acidification worsens, many species of plankton and marine animals such as corals and molluscs will be unable to build their shells or exoskeletons. Their demise will remove critical links from marine food webs and ultimately damage marine ecosystems around the world. ☑

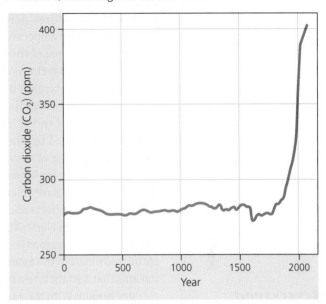

▼ Figure 18.45 **Atmospheric concentration of CO_2.** Notice that the concentration was relatively stable until the Industrial Revolution, which began in the late 1700s.

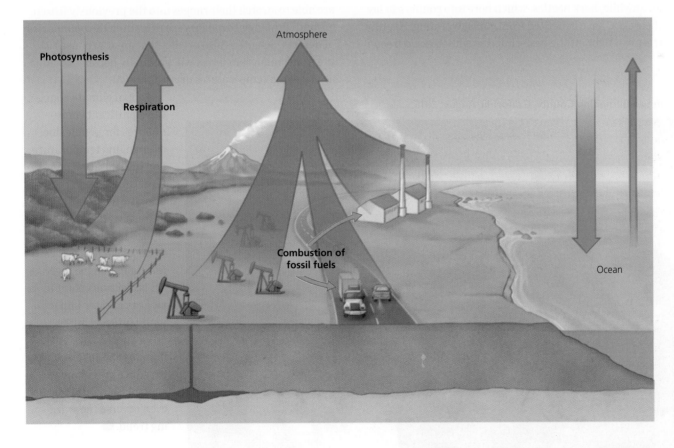

◀ Figure 18.46 **How CO_2 enters and leaves the atmosphere.**

☑ **CHECKPOINT**

What is the major source of CO_2 released by human activities?

■ *Answer: burning fossil fuels*

429

Effects of Climate Change on Ecosystems

The sentiment expressed by the poet John Donne that "no man is an island" is equally true for every other species in nature: Every species needs others to survive. Climate change is knocking some of these interactions out of sync. In temperate and polar climates, life cycle events of many plants and animals are triggered by warming temperatures. Across the Northern Hemisphere, the warm temperatures of spring are arriving earlier. Satellite images show earlier greening of the landscape, and flowering occurs sooner. A variety of species, including birds and frogs, have begun their breeding seasons earlier. But for other species, the environmental cue that spring has arrived is day length, which is not affected by climate change. Consequently, the winter white fur of snowshoe hares may be conspicuous against a greening landscape, or plants may bloom before pollinators have emerged.

The combined effects of climate change on forest ecosystems in western North America have spawned catastrophic wildfire seasons (Figure 18.47). In these regions, spring snowmelt in the mountains releases water into streams that sustain forest moisture levels over the summer dry season. With the earlier arrival of spring, snowmelt begins earlier and dwindles away before the dry season ends. As a result, the fire season is lasting longer. Meanwhile, bark beetles, which bore into conifers to lay their eggs, have benefited from global warming. Healthy trees can fight off the pests, but drought-stressed trees

✓ **CHECKPOINT**

How have bark beetles benefited from global warming?

◼ *Answer: Drought-stressed trees are less resistant to the beetles; with a longer warm season, beetles can reproduce twice in a year rather than once.*

▼ **Figure 18.47 Roaring wildfire in San Bernadino County, California, August 2016.**

▼ **Figure 18.48 Pines infested by bark beetles in British Columbia, Canada.** Brown or reddish foliage indicates dead or dying trees. Green trees are still healthy.

are too weak to resist (Figure 18.48). As a bonus for the beetles, the warmer weather allows them to reproduce twice a year rather than once. In turn, vast numbers of dead trees add fuel to a fire. Wildfires burn longer, and the number of acres burned has increased dramatically.

The map of terrestrial biomes (see Figure 18.27), which is primarily determined by temperature and rainfall, is also changing. Melting permafrost is shifting the boundary of the tundra northward as shrubs and conifers are able to stretch their ranges into the previously frozen ground. Prolonged droughts are extending the boundaries of deserts. Scientists also predict that great expanses of the Amazonian rain forest will gradually become savanna as increased temperatures dry out the soil.

Global climate change has significant consequences for people, too, as changing temperature and precipitation patterns affect food production, the availability of fresh water, and the spread of mosquito-borne diseases such as yellow fever and dengue fever. All projections point to an even greater impact in the future. Unlike other species, however, humans can take action to reduce greenhouse gas emissions and maybe even reverse the warming trend. ✓

How Does Climate Change Affect Species Distribution?

BACKGROUND

As you've learned in this chapter, climate is a critical determinant of where organisms live. With rising temperatures caused by global warming, the ranges of many species have already shifted toward the poles. For example, robins have been sighted north of the Arctic Circle for the first time. Ranges can also shift to higher elevations, where temperatures are cooler. Landforms narrow as elevation increases, so less habitat is available higher up.

Diana fritillary butterflies (Figure 18.49a) live in forests and meadows in the midwestern and southeastern United States. The larvae hatch in the fall and spend the winter under forest leaf litter near the violets that are their food source. Adult butterflies emerge in spring and feed on plant nectar. Conservation biologists have observed that Diana fritillary butterflies have recently become rarer and several populations have completely disappeared. A team of ecologists hypothesized that the butterfly's range has been reduced as a result of temperature increases caused by global warming.

METHOD

To test their hypothesis, the researchers compiled records of where Diana fritillary populations were located. They visualized the data by plotting the locations of Diana fritillary populations during two time periods, 1777 to 1960 and 1961 to 2010 on a map of the Eastern United States (Figure 18.49b). They gathered this information from museum specimens, government reports, private collections, and online databases, as well as from their own field work. They analyzed the data, by entering the date, latitude, longitude, and elevation of each Diana fritillary record into a mapping program.

RESULTS

The results showed that the range of Diana fritillary had become significantly smaller and had divided into two geographically separated groups of populations. The lower-elevation populations were the most likely to disappear, while mountain populations were expanding somewhat (though not enough to balance low elevation losses). The range of the species is retreating about 60 feet uphill every 10 years. Thus, the results support the hypothesis that increasing temperatures have reduced the range of Diana fritillary butterflies. In addition, the researchers found that rising temperatures are affecting the butterfly's life cycle. Females are now emerging from their cocoons earlier in the spring, an event triggered by temperature cues. This shift could threaten future Diana fritillary populations if the timing of their life cycle gets out of sync with the availability of violets, their food source.

▼ Figure 18.49 Investigating the effect of climate change on the range of a butterfly.

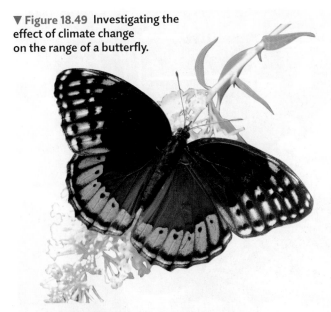

(a) Female Diana fritillary butterfly

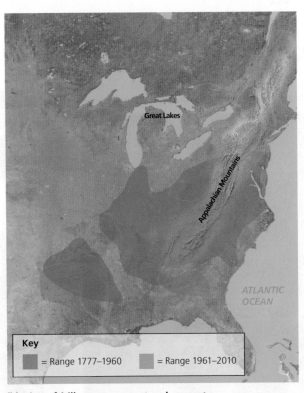

Great Lakes

Appalachian Mountains

ATLANTIC OCEAN

Key

■ = Range 1777–1960   ■ = Range 1961–2010

(b) Diana fritillary ranges past and present

Data from: C. N. Wells and D.W. Tonkyn. Range Collapse in the Diana Fritillary, *Speyeria diana* (Nymphalidae), *Insect Conservation and Diversity 7*: 365–80 (2014).

Thinking Like a Scientist

Explain how these researchers tested their hypothesis without using a controlled experiment.

For the answer, see Appendix D.

Looking to Our Future

Despite the attempts of many nations to curb carbon emissions, atmospheric CO_2 is increasing at an accelerating pace. At this rate, further climate change is inevitable. However, with effort, ingenuity, and international cooperation, we may be able to begin reducing emissions. Individuals can take action by working to influence politicians and large corporations, which wield the most power to effect large-scale change. But your personal habits have an impact, too. The amount of greenhouse gas emitted as a result of the actions of a single individual is that person's **carbon footprint** (from the fact that the most important greenhouse gas is CO_2). A carbon footprint can be estimated using a set of rough calculations; several different calculators are available online.

Home energy use is one major contributor to the carbon footprint. It's easy to reduce your energy consumption by turning off the lights, TV, and other electrical appliances when they aren't in use. Unplug "vampire" devices—electronics such as cell phone chargers, videogame consoles, and other computer, video, and audio equipment that draw electricity even when they aren't being used.

Transportation is another significant part of the carbon footprint. If you have a car, keep it well maintained,

"VAMPIRE" DEVICES CONSUME ELECTRICITY EVEN WHILE YOU SLEEP.

consolidate trips, share rides with friends, and use alternative means of transportation whenever possible.

Manufactured goods are a third category in the carbon footprint. Every item you purchase generated its own carbon footprint in the process of going from raw materials to store shelves. You can reduce your carbon emissions by not buying unnecessary goods and by recycling—or better yet, reusing—items instead of putting them in the trash (Figure 18.50). Landfills are the largest human-related source of methane, a greenhouse gas more potent than CO_2. The methane is released by prokaryotes that decompose (break down) landfill waste.

Changes in your eating habits can also shrink your carbon footprint. Like landfills, the digestive system of cattle depends on methane-producing bacteria. The methane released by cattle and by the bacteria that decompose their manure accounts for about 20% of methane emissions in the United States. Thus, replacing beef and dairy products in your diet with fish, chicken, eggs, and vegetables reduces your carbon footprint. In addition, eating locally grown fresh foods may lower the greenhouse gas emissions that result from food processing and transportation (Figure 18.51). Many websites offer additional suggestions for reducing your carbon footprint. ☑

☑ **CHECKPOINT**

What is a personal carbon footprint?

■ Answer: the amount of greenhouse gas that a person is responsible for emitting

▼ **Figure 18.50 Don't trash it, recycle!** Students at Duke University in Durham, North Carolina, used 3,500 cardboard boxes left over from move-in day to build Fort Duke. The structure, which was 16 feet high and covered 490 square feet, was soon dismantled, and the cardboard was hauled off to a recycling facility.

▼ **Figure 18.51 Shopping at a farmers' market.** Eating locally grown foods may reduce your carbon footprint—and they taste good, too!

EVOLUTION CONNECTION | Climate Change

Climate Change as an Agent of Natural Selection

Environmental change has always been a part of life; in fact, it is a key ingredient of evolutionary change. Will evolutionary adaptation counteract the negative effects of climate change on organisms? Researchers have documented microevolutionary shifts in a few populations, including red squirrels, a few bird species, and alpine bumblebees (Figure 18.52a). It appears that some populations, especially those with high genetic variability and short life spans, may avoid extinction by means of evolutionary adaptation. However, evolutionary adaptation is unlikely to save long-lived species, such as polar bears and penguins, that are experiencing rapid habitat loss (Figure 18.52b). The rate of climate change is incredibly fast compared with major climate shifts in evolutionary history. If climate change continues on its present course, thousands of species—the IPCC estimates as many as 30% of all plants and animals—will face extinction by midcentury.

(a) **Alpine bumblebee.** Alpine bumblebees may evolve quickly enough to survive climate change.

(b) **Adelie penguin.** Adelie penguins, which live on the rapidly warming Antarctic Peninsula, are not likely to make it.

▲ Figure 18.52 **Which species will survive climate change?**

Chapter Review

SUMMARY OF KEY CONCEPTS

An Overview of Ecology

Ecology is the scientific study of interactions between organisms and their environments. The environment includes abiotic (nonliving) and biotic (living) components. Ecologists use observation, experiments, and computer models to test hypothetical explanations of these interactions.

Ecology and Environmentalism

Human activities have had an impact on all parts of the biosphere. Ecology provides the basis for understanding and addressing these environmental problems.

A Hierarchy of Interactions

Ecologists study interactions at four increasingly complex levels.

Organismal ecology (individual)

Population ecology (group of individuals)

Community ecology (all organisms in a particular area)

Ecosystem ecology (all organisms and abiotic factors)

Living in Earth's Diverse Environments

The biosphere is an environmental patchwork in which abiotic factors affect the distribution and abundance of organisms.

Abiotic Factors of the Biosphere

Abiotic factors include the availability of sunlight, water, nutrients, and temperature. In aquatic habitats, dissolved oxygen, salinity, current, and tides are also important. Additional factors in terrestrial environments include wind and fire.

The Evolutionary Adaptations of Organisms

Adaptation via natural selection results from the interactions of organisms with their environments.

Adjusting to Environmental Variability

Organisms also have adaptations that enable them to cope with environmental variability, including physiological, behavioral, and anatomical responses to changing conditions.

Biomes

A biome is a major terrestrial or aquatic life zone.

Freshwater Biomes

Freshwater biomes include lakes, ponds, rivers, streams, and wetlands. Lakes vary, depending on depth, with regard to light penetration (photic and aphotic zones), temperature, nutrients, oxygen levels, and community structure. Rivers change greatly from their source to the point at which they empty into a lake or ocean. The bottom of an aquatic biome is its benthic realm.

Marine Biomes

Marine life is distributed into distinct realms (benthic and pelagic) and zones (photic, aphotic, and intertidal) according to the depth of the water, degree of light penetration, distance from shore, and open water versus deep-sea bottom. Marine biomes include the pelagic realm and the benthic realm of the oceans, coral reefs, intertidal zones, and estuaries. Coral reefs, which occur in warm tropical waters above the continental shelf, have an abundance of biological diversity. An ecosystem found near hydrothermal vents in the deep ocean is powered by chemical energy from Earth's interior instead of sunlight. Estuaries, located where a freshwater river or stream merges with the ocean, are some of the most biologically productive environments on Earth.

How Climate Affects Terrestrial Biome Distribution

The geographic distribution of terrestrial biomes is based mainly on regional variations in climate. Climate is largely determined by the uneven distribution of solar energy on Earth. Proximity to large bodies of water and the presence of landforms such as mountains also affect climate.

Terrestrial Biomes

Most terrestrial biomes are named for their climate and predominant vegetation. The major terrestrial biomes include tropical forest, savanna, desert, chaparral, temperate grassland, temperate broadleaf forest, coniferous forest, tundra, and polar ice. If the climate in two geographically separate areas is similar, the same type of biome may occur in both.

The Water Cycle

The global water cycle links aquatic and terrestrial biomes. Human activities are disrupting the water cycle.

Human Impact on Biomes

Land use by humans has altered vast tracts of forest and degraded the services provided by natural ecosystems. Unsustainable agricultural practices have depleted cropland fertility. Human activities have polluted freshwater ecosystems, which are vital for life. Agriculture, population growth, drought, and declining snowfall are all factors in the rapid depletion of freshwater resources in some regions.

Climate Change

The Greenhouse Effect and Global Warming

So-called greenhouse gases, including CO_2 and methane, increase the amount of heat retained in Earth's atmosphere. The accumulation of these gases has caused increases in the average global temperature.

The Accumulation of Greenhouse Gases

Human activities, especially the burning of fossil fuels, are responsible for the rise in greenhouse gases over the past century. Release of CO_2 has exceeded the amount that can be absorbed by natural processes.

Effects of Climate Change on Ecosystems

Climate change is disrupting interactions between species. Devastating wildfires are among the effects of climate change in certain ecosystems. Climate change is also shifting biome boundaries.

Looking to Our Future

Each person has a carbon footprint—that person's responsibility for a portion of global greenhouse gas emissions. We can take action to reduce our carbon footprints.

Mastering Biology

For practice quizzes, BioFlix animations, MP3 tutorials, video tutors, and more study tools designed for this textbook, go to Mastering Biology™

SELF-QUIZ

1. Place these levels of ecological study in order from the least to the most comprehensive: community ecology, ecosystem ecology, organismal ecology, population ecology.

2. Name several abiotic factors that might affect the community of organisms living inside a home fish tank.

3. Arctic fox changes the color of fur to camouflage against winter snow. This anatomical response is known as _____ which is a _____ change.

4. Which of the following sea creatures might be described as a pelagic animal of the aphotic zone?

a. a coral reef fish

b. a giant clam near a deep-sea hydrothermal vent

c. an intertidal snail

d. a deep-sea squid

5. Identify the following biomes on the graph below: tundra, northern coniferous forest, desert, temperate grassland, temperate broadleaf forest, and tropical forest.

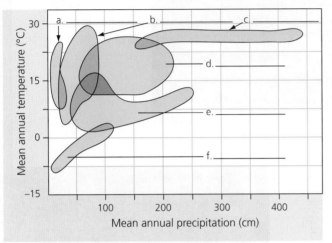

6. You board a ten-hour circumpolar flight from Finland to Russia. The flight starts and ends in the same biome. You are most likely in a _____ biome.

7. What three abiotic factors account for the rarity of trees in arctic tundra?

8. What free-floating microscopic animals inhabit the pelagic photic zone of oceans?

9. What is the greenhouse effect? How is the greenhouse effect related to global warming?

10. The recent increase in atmospheric CO_2 concentration is mainly a result of an increase in

a. plant growth.

b. the absorption of heat radiating from Earth.

c. the burning of fossil fuels and wood.

d. cellular respiration by the increasing human population.

11. How are changes in temperature affecting the map of terrestrial biomes?

For answers to the Self-Quiz, see Appendix D.

IDENTIFYING MAJOR THEMES

For each statement below, identify which major theme is evident (the relationship of structure to function, information flow, pathways that transform energy and matter, interactions within biological systems, or evolution) and explain how the statement relates to the theme. If necessary, review the themes (Chapter 1) and review the examples highlighted in blue in this chapter.

12. Reptilian scales and the waxy coating on many leaves reduce water loss.

13. Other organisms may compete with an individual for food and other resources, prey upon it, or change its physical and chemical environment.

14. Solar energy from sunlight captured by chlorophyll during the process of photosynthesis powers most ecosystems.

15. After a period of lower-than-average rainfall, drought-resistant individuals may be more prevalent in a plant population.

For answers to Identifying Major Themes, see Appendix D.

THE PROCESS OF SCIENCE

16. Design a laboratory experiment to measure the effect of water temperature on the population growth of a certain phytoplankton species from a pond.

17. **Interpreting Data** This graph shows average monthly temperature and precipitation for a city in the Northern Hemisphere. Based on the biome descriptions on pages 385–390, in which biome is this city located? Explain your answer.

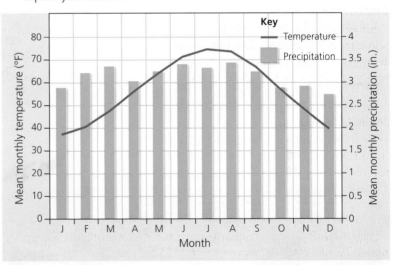

BIOLOGY AND SOCIETY

18. How will climate change affect the planet's ability to renew resources in the future? Is it safe to assume that technological advances in the future may solve the challenges posed by the supposed shortage of resources? Instead of worrying about the future, should we instead focus on providing for the society's current needs?

19. The concept of carbon footprint can be applied to a single person, an organization, a product, a city, or a country. What are the practical limitations of this estimate of greenhouse gas emissions?

19 Population Ecology

DO YOU BUY TOO MUCH STUFF? THE AVERAGE AMERICAN GENERATES ABOUT 40 POUNDS OF TRASH PER WEEK, TWICE AS MUCH AS THE AVERAGE MEXICAN.

Why Population Ecology Matters

Population ecology has many practical applications in such diverse fields as conservation, agriculture, and combating the spread of infectious diseases. In addition, understanding population ecology provides insight into one of the most important environmental issues we currently face—the ever-increasing human population.

IS THE ABUNDANCE OF NATURE UNLIMITED? THE SAYING "THERE ARE PLENTY OF FISH IN THE SEA" COULD BECOME MEANINGLESS IF OVERHARVESTING CONTINUES.

HOW LARGE IS THE HUMAN POPULATION? COUNTING ONE PERSON PER SECOND, IT WOULD TAKE MORE THAN 238 YEARS TO COUNT ALL 7.5 BILLION PEOPLE ALIVE TODAY.

BIOLOGY AND SOCIETY | Biological Invasions

Invasion of the Lionfish

Lionfish, with their graceful, flowing fins, bold stripes, and eye-catching array of spines, are striking members of tropical reef communities. They are also favorites of saltwater aquarium enthusiasts—especially the red lionfish, a native to the coral reefs of the South Pacific and Indian Oceans. There are a few drawbacks to owning a red lionfish, however. The spines are venomous and can inflict an intensely painful sting. Lionfish are merciless predators, so any tankmates must be chosen with care. And they are large. A 2-inch juvenile can rapidly become an 18-inch adult that requires, at minimum, a 120-gallon tank. Apparently, some aquarium owners who regretted their purchase released their lionfish into the wild.

Freed from the competitors and predators of their native reefs as well as their tanks, red lionfish have multiplied exponentially. Within a few years of the first sightings off the southeastern coast of Florida, lionfish populations had spread up the East Coast. They have since invaded islands and coastlines throughout the Atlantic and Caribbean regions and the Gulf of Mexico. The speed of the onslaught has stunned scientists, who are just beginning to document its devastating effects on native ecosystems. Lionfish consume prodigious numbers of other fish, including species that are key to maintaining the legendary diversity of reef communities and juveniles of economically important fishes such as grouper and snapper. Some biologists think our best hope of stopping the lionfish invasion is for *us* to consume *them*. The National Oceanic and Atmospheric Administration (NOAA) has launched an "Eat Lionfish" campaign to encourage human predation on the tasty fish.

Lionfish are not the only invasive species that have had devastating effects. For as long as people have traveled the world, they have carried—intentionally or accidentally—thousands of species to new habitats. Many of these non-native species have established populations that spread far and wide, leaving environmental havoc in their wake. We humans, too, have multiplied and spread far from our point of origin, radically changing our environment in the process. As you explore population ecology in this chapter, you'll also learn about trends in human population growth and other applications of this area of ecological research.

The red lionfish, a beautiful but deadly invader, is a threat to coral reef communities. This lionfish was photographed in the Caribbean, halfway around the world from its native habitat.

An Overview of Population Ecology

Ecologists usually define a **population** as a group of individuals of a single species that occupy the same general area at the same time. These individuals rely on the same resources, are influenced by the same environmental factors, and are likely to interact and breed with one another. For example, the red lionfish that live in the vicinity of a specific reef are a population.

Population ecology is concerned with changes in population size and the factors that regulate populations over time. A population ecologist might describe a population in terms of its size (number of individuals), age structure (proportion of individuals of different ages), or density (number of individuals per unit area or volume). Population ecologists also study population dynamics, the interactions between biotic and abiotic factors that cause variation in population size. One important aspect of population dynamics—and a major topic for this chapter—is population growth.

Population ecology plays a key role in applied research. For example, it provides critical information for conservation and restoration projects (**Figure 19.1**). Population ecology is used to develop sustainable fisheries and to manage wildlife populations. Studying the population ecology of pests provides insight into controlling how they spread. Population ecologists also study human population growth, one of the most critical environmental issues of our time.

Let's consider what a snapshot of a population might look like. The first question is, which individuals are included in this population? A population's geographic boundaries may be natural, as with lionfish inhabiting a particular coral reef. But ecologists often define a population's boundaries in more arbitrary ways that fit their research questions. For example, an ecologist studying the contribution of asexual reproduction to the population growth of sea anemones might define a population as all the anemones of one species in a particular tide pool. Another researcher studying the effects of hunting on deer might define a population as all the deer within a particular state. Yet another researcher, attempting to understand the spread of HIV/AIDS, might study the HIV infection rate of the human population in one nation or throughout the world.

▼ **Figure 19.1 Ecologists getting up close and personal with members of the populations they study.**

A biologist in Oregon releases a 5-week-old northern spotted owl, which has been banded to monitor the population of this threatened species.

Researchers in Namibia collect blood samples from a cheetah that has been sedated.

A researcher marks meerkats in the Kuruman River Reserve in Northern Cape, South Africa.

Population Density

Our snapshot of a population includes **population density**, the number of individuals of a species per unit area or volume of the habitat: the number of largemouth bass per cubic kilometer (km^3) of a lake, for example, or the number of oak trees per square kilometer (km^2) in a forest, or the number of nematodes per cubic meter (m^3) in the forest's soil. In rare cases, an ecologist can actually count all the individuals within the boundaries of the population. For example, we could count the total number of oak trees (say, 200) in a forest covering 50 km^2 (about 19 square miles). The population density would be the total number of trees divided by the area, or 4 trees per square kilometer (4/km^2).

In most cases, however, it is impractical or impossible to count all individuals in a population. Instead, ecologists use a variety of sampling techniques to estimate population density. For example, they might estimate the density of alligators in the Florida Everglades based on a count of individuals in a few sample plots of 1 km^2 each. Generally speaking, the larger the number and size of sample plots, the more accurate the estimates. Population densities may also be estimated by indicators such as number of bird nests or rodent burrows **(Figure 19.2)** rather than by actual counts of organisms.

Keep in mind that population density is not a constant number. It changes when individuals are born or die and when new individuals enter the population (immigration) or leave it (emigration).

Population Age Structure

The **age structure** of a population—the distribution of individuals in different age-groups—reveals information that is not apparent from population density. For instance, age structure can provide insight into the history of a population's survival or reproductive success and how it relates to environmental factors. **Figure 19.3** shows the age structure of males in a population of cactus finches on the Galápagos Islands in 1987. (See Figure 14.12 for other examples of Galápagos finches.) Four-year-old birds, born in 1983, made up almost half the population, but there were not any 2- or 3-year-olds. Why was there such dramatic variation? For their food, cactus finches depend on plants, which in turn depend on rainfall. The 1983 baby boom resulted from unusually wet weather that produced abundant plant growth and food for the finches. Severe droughts in 1984 and 1985 limited the food supply, preventing reproduction and causing many deaths. As we'll see later in this chapter, age structure is also a useful tool for predicting future changes in a population. ☑

☑ **CHECKPOINT**

What does an age structure show?

Answer: the distribution of individuals in different age-groups ■

▼ **Figure 19.2 An indirect census of a prairie dog population.** We could get a rough estimate of the number of prairie dogs in this colony in Saskatchewan, Canada, by counting the number of burrows constructed by the rodents and then multiplying by the number of animals that use a typical burrow.

▼ **Figure 19.3 Age structure for the males in a population of large cactus finches (inset) on one of the Galápagos Islands in 1987.**

Life Tables and Survivorship Curves

Life tables track survivorship, the chance of an individual in a given population surviving to various ages. The life insurance industry uses life tables to predict how long, on average, a person of a given age will live. Starting with a population of 100,000 people, **Table 19.1** shows the number of people who are expected to be alive at the beginning of each age interval, based on death rates in 2012. For example, 94,351 out of 100,000 people are expected to live to age 50. Their chance of surviving to age 60, shown in the last column of the same row, is 0.941; 94% of 50-year-olds will reach the age of 60. The chance of 80-year-olds surviving to age 90, however, is only 0.418. Population ecologists have adopted this technique and constructed life tables to help them understand the structure and dynamics of various plant and animal species. By identifying the most vulnerable stage of the life cycle, life table data may also help conservationists develop effective measures to protect species whose populations are declining.

Ecologists represent life table data graphically in a **survivorship curve**, a plot of the number of individuals still alive at each age in the maximum life span **(Figure 19.4)**. By using a percentage scale instead of actual ages on the *x*-axis, we can compare species with widely varying life spans, such as humans and squirrels, on the same graph. The curve for the human population (red) shows that most people survive to the older age intervals. Ecologists refer to the shape of this curve as Type I survivorship. Species that exhibit a Type I curve—humans

CHECKPOINT

Sea turtles lay their eggs in nests on sandy beaches. Although a single nest may contain as many as 200 eggs, only a small proportion of the hatchlings survive the journey from the beach to the open ocean. Once sea turtles have matured, however, their death rate is low. What type of survivorship do sea turtles exhibit?

■ *Answer: Type III*

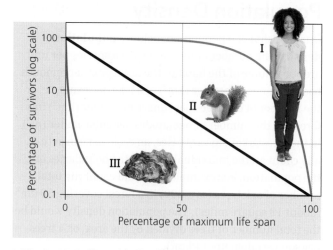

▲ Figure 19.4 **Three idealized types of survivorship curves.**

and many other large mammals—usually produce few offspring but give them good care, increasing the likelihood that they will survive to maturity.

In contrast, a Type III curve (blue) indicates low survivorship for the very young, followed by a period when survivorship is high for those few individuals who live to a certain age. Species with this type of survivorship curve usually produce very large numbers of offspring but provide little or no care for them. Some species of fish, for example, produce millions of eggs at a time, but most of these offspring die as larvae from predation or other causes. Many invertebrates, including oysters, also have Type III survivorship curves.

A Type II curve (black) is intermediate, with survivorship constant over the life span. That is, individuals are no more vulnerable at one stage of the life cycle than another. This type of survivorship has been observed in some invertebrates, lizards, and rodents. ✔

Life History Traits as Adaptations

A population's pattern of survivorship is an important feature of its **life history**, the set of traits that affect an organism's schedule of reproduction and survival. Some key life history traits are the age at first reproduction, the frequency of reproduction, the number of offspring, and the amount of parental care given. As you might expect from the different types of survivorship curves, life history traits vary among organisms. Let's take a closer look at how natural selection shapes these traits.

As you may recall, reproductive success is key to evolutionary success (see Chapter 13). Accordingly,

Table 19.1	Life Table for the U.S. Population in 2012		
	Number Living at Start of Age Interval	Number Dying During Interval	Chance of Surviving Interval
Age Interval	(N)	(D)	1 − (D/N)
0–10	100,000	759	0.992
10–20	99,241	301	0.997
20–30	98,940	900	0.991
30–40	98,040	1,234	0.987
40–50	96,805	2,454	0.975
50–60	94,351	5,546	0.941
60–70	88,805	10,465	0.882
70–80	78,340	20,485	0.739
80–90	57,855	33,653	0.418
90+	24,202	24,202	0.000

you may wonder why all organisms don't simply produce a large number of offspring. One reason is that reproduction is expensive in terms of time, energy, and nutrients—resources that are available in limited amounts. An organism that gives birth to a large number of offspring will not be able to provide a great deal of parental care. Consequently, the combination of life history traits represents trade-offs that balance the demands of reproduction and survival. **In other words, life history traits, like anatomical features, are shaped by evolution.**

Because selective pressures vary, life histories are very diverse. Nevertheless, ecologists have observed some patterns that are useful for understanding how natural selection influences life history characteristics.

One life history pattern is typified by small-bodied, short-lived species (for example, insects, small rodents, and dandelions) that develop and reach sexual maturity rapidly, have a large number of offspring, and offer little or no parental care. In plants, "parental care" is measured by the amount of nutritional material stocked in each seed. Many small, nonwoody plants produce thousands of tiny seeds. Such organisms have an **opportunistic life history**, one that enables the plant or animal to take immediate advantage of favorable conditions. In general, populations with this life history pattern exhibit a Type III survivorship curve.

In contrast, organisms with an **equilibrial life history** develop and reach sexual maturity slowly and produce a few well-cared-for offspring. Organisms that have an equilibrial life history are typically larger-bodied, longer-lived species (for example, bears and elephants). Populations with this life history pattern exhibit a Type I survivorship curve. Plants with comparable life history traits include certain trees. For example, coconut palms produce relatively few seeds, but those seeds are well stocked with nutrient-rich material. **Table 19.2** compares key traits of opportunistic and equilibrial life history patterns.

What accounts for the differences in life history patterns?

Some ecologists hypothesize that the potential survival rate of the offspring and the likelihood that the adult will live to reproduce again are the critical factors. In a harsh, unpredictable environment, an adult may have just one good shot at reproduction, so it may be an advantage to invest in quantity rather than quality. On the other hand, in an environment where favorable conditions are more dependable, an adult is more likely to survive to reproduce again. Seeds are more likely to fall on fertile ground, and newly emerged animals are more likely to survive to adulthood. In that case, it may be more advantageous for the adult to invest its energy in producing a few well-cared-for offspring at a time.

Of course, there is much more diversity in life history patterns than the two extremes described here. Nevertheless, the contrasting patterns are useful for understanding the interactions between life history traits and our next topic, population growth. ✓

Dandelions have an opportunistic life history.

Elephants have an equilibrial life history.

Table 19.2	Some Life History Characteristics of Opportunistic and Equilibrial Populations	
Characteristic	Opportunistic Populations (such as many wildflowers)	Equilibrial Populations (such as many large mammals)
Climate	Relatively unpredictable	Relatively predictable
Maturation time	Short	Long
Life span	Short	Long
Death rate	Often high	Usually low
Number of offspring per reproductive episode	Many	Few
Number of reproductions per lifetime	Usually one	Often several
Timing of first reproduction	Early in life	Later in life
Size of offspring or eggs	Small	Large
Parental care	Little or none	Often extensive

441

Population Growth Models

Population size fluctuates as new individuals are born or move into an area and others die or move out of it. Some populations—for example, trees in a mature forest—are relatively constant over time. Other populations change rapidly, even explosively. Consider a single bacterium that divides every 20 minutes. There would be two bacteria after 20 minutes, four after 40 minutes, eight after 60 minutes, and so on. In just 12 hours, the population would approach 70 billion cells. If reproduction continued at this rate for a day and a half—a mere 36 hours—there would be enough bacteria to form a layer a foot deep over the entire Earth! Population ecologists use idealized models to investigate how the size of a particular population may change over time under different conditions. We'll describe two basic mathematical models that illustrate fundamental concepts of population growth.

The Exponential Population Growth Model: The Ideal of an Unlimited Environment

The first model, known as exponential growth, works like compound interest on a savings account: The principal (population size) grows faster with each interest payment (the individuals added to the population). **Exponential population growth** describes the expansion of a population in an ideal, unlimited environment. In this model, the population size of each new generation is calculated by multiplying the current population size by a constant factor that represents the birth rate minus the death rate. Let's look at how such a population grows. In **Figure 19.5**, we begin with a population of 20 rabbits, indicated on the *y*-axis. Each month, there are more rabbit births than deaths; as a result, the population size increases each month.

Notice in Figure 19.5 that each increase is larger than the previous one. In other words, the larger the population, the faster it grows. The increasing speed of population growth produces a J-shaped curve that is typical of exponential growth. The slope of the curve shows how rapidly the population is growing. At the outset, when the population is small, the curve is almost flat: Over the first 4 months, the population increases by a total of only 37 individuals, an average of 9.25 births per month. By the end of 7 months, the growth rate has increased to an average of 15 births per month. The largest increase is seen in the period from 10 to 12 months, when an average of 85 rabbits are born each month.

Exponential population growth is common in certain situations. For example, a disturbance such as a fire, flood, hurricane, drought, or cold snap may suddenly reduce the size of a population. Organisms that have opportunistic life history patterns can rapidly take advantage of the lack of competition and quickly recolonize the habitat by exponential population growth. Human activity can also be a major cause of disturbance, and plants and animals with opportunistic life history traits commonly occupy road cuts, freshly cleared fields and woodlots, and poorly maintained lawns. However, no natural environment can sustain exponential growth indefinitely. ☑

▼ Figure 19.5 **Exponential growth of a rabbit population.**

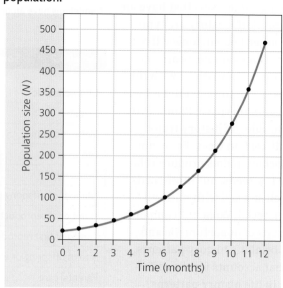

The Logistic Population Growth Model: The Reality of a Limited Environment

Most natural environments do not have an unlimited supply of the resources needed to sustain population growth. Environmental factors that restrict population growth are called **limiting factors**. Limiting factors ultimately control the number of individuals that can occupy a habitat. Ecologists define **carrying capacity** as the maximum population size that a particular environment can sustain. In **logistic population growth**, the growth rate decreases as the population size approaches carrying capacity. When the population is at carrying capacity, the growth rate is zero.

You can see the effect of limiting factors in the graph in **Figure 19.6**, which shows the growth of a population of fur seals on St. Paul Island, off the coast of Alaska. (For simplicity, only the mated bulls were counted. Each has a harem of females, as shown in the photograph.) Before 1925, the seal population on the island remained

low—between 1,000 and 4,500 mated bulls—because of uncontrolled hunting. After hunting was controlled, the population increased rapidly until about 1935, when it began to level off and started fluctuating around a population size of about 10,000 bull seals—the carrying capacity for St. Paul Island. In this instance, the main limiting factor was the amount of space suitable for breeding territories.

The carrying capacity for a population varies, depending on the species and the resources available in the habitat. For example, carrying capacity might be considerably less than 10,000 for a fur seal population on a smaller island with fewer breeding sites. Even in one location, it is not a fixed number. Organisms interact with other organisms in their communities, including predators, pathogens, and food sources, and these interactions may affect carrying capacity. Changes in abiotic factors may also increase or decrease carrying capacity. In any case, the concept of carrying capacity expresses an essential fact of nature: Resources are finite.

Ecologists hypothesize that selection for organisms exhibiting equilibrial life history patterns occurs in environments where the population size is at or near carrying capacity. Because competition for resources is keen under these circumstances, organisms gain an advantage by allocating energy to their own survival and to the survival of their descendants.

Figure 19.7 compares logistic growth (shown in blue) with exponential growth (shown in red). As you can see, the logistic curve is J-shaped at first, but gradually levels off to resemble an S shape as carrying capacity is reached. Both the logistic model and the exponential model of population growth are theoretical ideals. No natural population fits either one perfectly. However, these models are useful starting points for studying population growth. Ecologists use them to predict how populations will grow in certain environments and as a basis for constructing more complex models. ✓

▼ Figure 19.6 **Logistic growth of a seal population.**

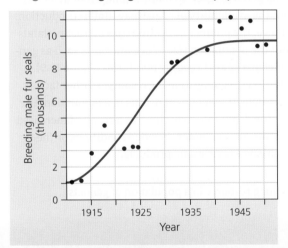

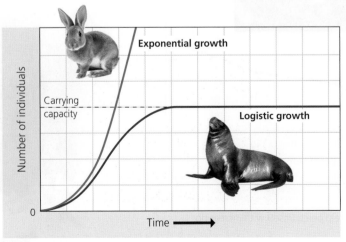

Exponential growth

Logistic growth

Carrying capacity

Number of individuals

Time ➡

◀ Figure 19.7 **Comparison of exponential and logistic growth.**

☑ **CHECKPOINT**

What happens when a population reaches its carrying capacity?

■ *Answer: Enough resources are available to sustain that population size, but the population does not continue to increase.*

Regulation of Population Growth

Now let's take a closer look at how population growth is regulated in nature. What stops a population from continuing to increase after reaching carrying capacity?

Density-Dependent Factors

Several **density-dependent factors**—limiting factors whose intensity is related to population density—can limit growth in natural populations. The most obvious is **intraspecific competition**, the competition between individuals of the same species for the same limited resources. As a limited food supply is divided among more and more individuals, birth rates may decline as

individuals have less energy available for reproduction. Density-dependent factors may also depress a population's growth by increasing the death rate. For example, in a population of song sparrows, both factors reduced the number of offspring that survived and left the nest **(Figure 19.8a)**. As the number of competitors for food increased, female song sparrows laid fewer eggs. In addition, the death rate of eggs and nestlings increased with increasing population density.

Plants that grow close together may experience an increased death rate as intraspecific competition for resources increases. And those that do survive will produce fewer flowers, fruits, and seeds than uncrowded individuals. After seeds sprout, gardeners often pull out some of the seedlings to allow sufficient resources for the remaining plants. Intraspecific competition is also the reason that plants purchased from a nursery come with instructions to space the plants a certain distance apart.

A limited resource may be something other than food or nutrients. Like a game of musical chairs, the number of safe hiding places may limit a prey population by exposing some individuals to a greater risk of predation. For example, young kelp perch hide from predators in "forests" of the large seaweed known as kelp (see Figure 15.25). In the experiment shown in **Figure 19.8b**, the proportion of perch eaten by a predator increased with increasing perch density. In many animals that defend a territory, the availability of space may limit reproduction. For instance, the number of nesting sites on rocky islands may limit the population size of oceanic birds such as gannets, which maintain breeding territories **(Figure 19.9)**.

In addition to competition for resources, other factors may cause density-dependent deaths in a population. For example, the death rate may climb as a result of increased disease transmission under crowded conditions or the accumulation of toxic waste products.

▼ Figure 19.8 **Density-dependent regulation of population growth.**

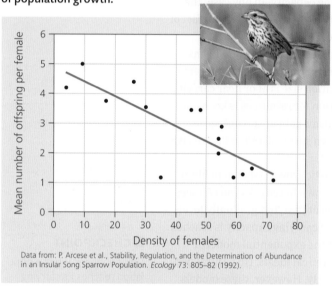

Data from: P. Arcese et al., Stability, Regulation, and the Determination of Abundance in an Insular Song Sparrow Population. *Ecology* 73: 805–82 (1992).

(a) Declining reproductive success of song sparrows (inset) with increasing population density.

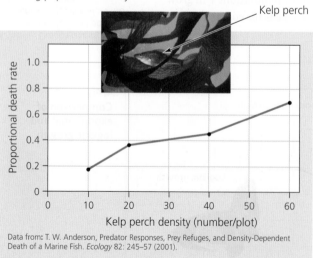

Data from: T. W. Anderson, Predator Responses, Prey Refuges, and Density-Dependent Death of a Marine Fish. *Ecology* 82: 245–57 (2001).

(b) Increasing death rate of kelp perch (inset) with increasing population density.

▼ Figure 19.9 **Space as a limiting resource in a population of gannets.**

Density-Independent Factors

In many natural populations, abiotic factors such as weather may limit or reduce population size well before other limiting factors become important. A population-limiting factor whose intensity is unrelated to population density is called a **density-independent factor**. If we look at the growth curve of such a population, we see something like exponential growth followed by a rapid decline rather than a leveling off. **Figure 19.10** shows this effect for a population of aphids, insects that feed on the sugary sap of plants. These and many other insects undergo virtually exponential growth in the spring and then rapidly die off when the weather turns hot and dry in the summer. A few individuals may remain, allowing population growth to resume if favorable conditions return. In some populations of insects—many mosquitoes and grasshoppers, for instance—the adults die off entirely, leaving behind eggs that will initiate population growth the following year. In addition to seasonal changes in the weather, environmental disturbances such as fire, floods, and storms can affect a population's size regardless of its density.

Over the long term, most populations are probably regulated by a complex interaction of density-dependent and density-independent factors. Although some populations remain fairly stable in size and are presumably close to a carrying capacity that is determined by biotic factors such as competition or predation, most populations for which we have long-term data do fluctuate.

Population Cycles

Some populations of insects, birds, and mammals undergo dramatic fluctuations in density with remarkable regularity. "Booms" characterized by rapid exponential growth are followed by "busts," during which the population falls back to a minimal level. Lemmings, small rodents that live in the tundra, are a striking example. In lemming populations, boom-and-bust growth cycles occur every three to four years. Some researchers hypothesize that natural changes in the lemmings' food supply may be the underlying cause. Another hypothesis is that stress from crowding during the "boom" triggers hormonal changes that may cause the "bust" by reducing birth rates.

Population cycles of the snowshoe hare and the lynx illustrate interconnections within biological systems (Figure 19.11). The lynx is one of the main predators of the snowshoe hare in the far northern forests of Canada and Alaska. About every ten years, both hare and lynx populations show a rapid increase followed by a sharp decline. What causes these boom-and-bust cycles? Since ups and downs in the two populations seem to almost match each other on the graph, does this mean that changes in one directly affect the other? For the hare cycles, there are three main hypotheses. First, cycles may be caused by winter food shortages that result from overgrazing. Second, cycles may be due to predator-prey interactions. Many predators other than lynx, such as coyotes, foxes, and great-horned owls, eat hares, and together these predators might overexploit their prey. Third, cycles may be affected by a combination of food resource limitation and excessive predation. Recent field studies support the hypothesis that the ten-year cycles of the snowshoe hare are largely driven by excessive predation but are also influenced by fluctuations in the hare's food supplies. Long-term studies are the key to unraveling the complex causes of such population cycles. ✓

✓ CHECKPOINT

List some density-dependent factors that limit population growth.

■ Answer: food and nutrient limitations, insufficient space for territories or nests, increase in disease and predation, accumulation of toxins

▼ Figure 19.10 **Weather change as a density-independent factor limiting growth of an aphid population.**

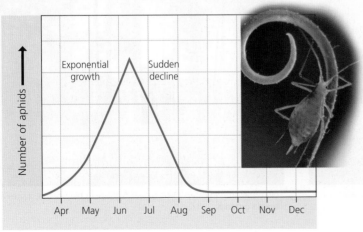

▼ Figure 19.11 **Population cycles of the snowshoe hare and the lynx.**

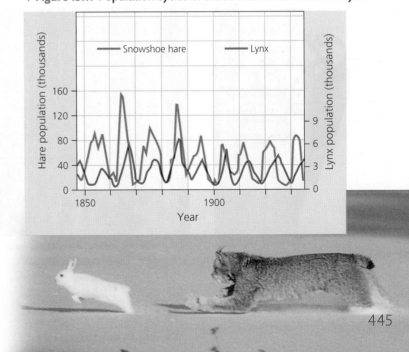

Applications of Population Ecology

To a great extent, we humans have converted Earth's natural ecosystems to ecosystems that produce goods and services for our own benefit. In some cases, we try to increase populations of organisms that we wish to harvest and decrease populations of organisms that we consider pests. Other efforts are aimed at saving populations that are perilously close to extinction. Principles of population ecology help guide us toward these various resource management goals.

☑ **CHECKPOINT**

What is a key factor in the recovery of red-cockaded woodpecker populations?

■ *Answer: removal of understory vegetation by controlled burns*

Conservation of Endangered Species

The U.S. Endangered Species Act, signed into law in 1973, defines an **endangered species** as one that is "in danger of extinction throughout all or a significant portion of its range." **Threatened species** are defined as those that are likely to become endangered in the near future. Endangered and threatened species are characterized by population sizes that are highly reduced or steadily declining. The challenge for conservationists is to determine the circumstances that threaten a species with extinction and try to remedy the situation.

The red-cockaded woodpecker was one of the first species to be listed as endangered **(Figure 19.12)**. The bird requires longleaf pine forests, where it drills its nest holes in mature, living pine trees. Originally found throughout the southeastern United States, the woodpeckers declined in number as suitable habitats were lost to logging and agriculture. Moreover, we have altered the composition of many of the remaining forests by suppressing the fires that are a natural occurrence in these ecosystems. Research revealed that breeding birds tend to abandon nests when vegetation among the pines is thick and higher than about 4.5 m (15 feet). Apparently, the birds require a clear flight path between their home trees and the neighboring feeding grounds. Armed with an understanding of the factors that regulate population growth of this species, wildlife managers protected critical habitat and began a maintenance program that included controlled fires to reduce forest undergrowth. As a result of such measures, populations of red-cockaded woodpeckers are beginning to recover. ☑

Sustainable Resource Management

Principles of population ecology can help guide us toward resource management goals, such as increasing populations we wish to harvest or save from extinction or decreasing populations we consider pests. Wildlife managers, fishery biologists, and foresters use **sustainable resource management**: practices that allow use of a natural resource without damaging it. This means maintaining a high population growth rate to replenish the population. According to the logistic growth model, the fastest growth rate occurs when the population size is at

▼ Figure 19.12 **A red-cockaded woodpecker and its habitat.**

A red-cockaded woodpecker perches at the entrance to its nest in a longleaf pine tree.

High, dense undergrowth impedes the woodpeckers' access to feeding grounds.

Low undergrowth offers birds a clear flight path between nest sites and feeding grounds.

roughly half the carrying capacity of the habitat. Theoretically, a resource manager should achieve the best results by harvesting the population down to this level. However, the logistic model assumes that growth rate and carrying capacity are stable over time, assumptions that are not realistic for some populations. In many cases, the amount of scientific information available is insufficient. In addition, human economic and political pressures often outweigh ecological concerns. The result may be unsustainably high harvest levels that ultimately deplete the resource.

Fish, the only wild animals still hunted on a large scale, are particularly vulnerable to overharvesting. For example, in the northern Atlantic cod fishery, estimates of cod stocks were too high, and the practice of discarding young cod (not of legal size) at sea caused a higher death rate than was predicted. The fishery collapsed in 1992 and has not recovered (Figure 19.13).

Until the 1970s, marine fisheries concentrated on species such as cod that inhabit the continental shelves (see Figure 18.19). As these resources dwindled, attention turned to deeper waters, most commonly the continental slopes below 600 m. In many of these new locations, however, catches are initially high but then quickly fall off as stocks are depleted. Deeper waters are colder, and food is relatively scarce. Fishes that are adapted to this environment, such as Chilean sea bass and orange roughy, typically grow more slowly, take longer to reach maturity, and have a lower reproductive rate than continental shelf species. Sustainable catch rates can't be estimated without knowing these essential life history traits for the target species. In addition, knowledge of population ecology alone is not sufficient; sustainable fisheries also require knowledge of community and ecosystem characteristics. ☑

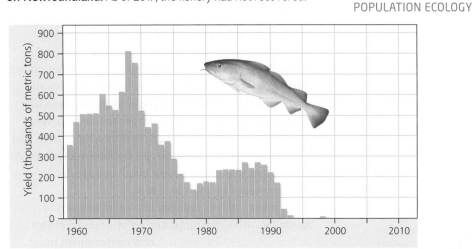

▼ Figure 19.13 The collapse of the northern cod fishery off Newfoundland. As of 2017, the fishery had not recovered.

THE SAYING "THERE ARE PLENTY OF FISH IN THE SEA" COULD SOON BECOME MEANINGLESS.

Invasive Species

Like the lionfish featured in the Biology and Society section, organisms that are introduced into non-native habitats can have a devastating effect on the ecosystem. An **invasive species** is a non-native species that has spread far beyond its original point of introduction and causes environmental or economic damage by colonizing and dominating suitable habitats. Throughout the world, there are thousands of invasive species, including plants, mammals, birds, fishes, arthropods, and molluscs. Regardless of where you live, an invasive plant or animal is probably living nearby. Invasive species are a leading cause of local extinctions. And the economic costs of invasive species are enormous.

Not every organism that is introduced to a new habitat is successful, and not every species that survives in its new habitat becomes invasive. There is no single explanation for why any non-native species turns into a damaging pest, but invasive species typically exhibit an opportunistic life history pattern. A female lionfish, for example, is sexually mature at a year old and can produce 2 million eggs per year.

The life history traits of zebra mussels (Figure 19.14), fingernail-size molluscs native to western Asia, also aided their spectacular success. Female zebra mussels begin reproducing early in life and can lay a million eggs in a year. Fertilized eggs develop into free-swimming larvae, which eventually settle and attach themselves to any available object. The structures with which zebra mussels glue themselves to surfaces are another factor in their success. The mussel secretes a combination of molecules that hardens into exceptionally sticky threads—an example of the relationship between structure and function.

Zebra mussels were first discovered in the Great Lakes in 1988, probably carried in the ballast water of ocean-going ships passing through

☑ CHECKPOINT

Why do managers try to maintain populations of fish and game species at about half their carrying capacity?

■ *Answer: to prevent overharvesting yet maintain lower population levels so that growth rate is high*

▼ Figure 19.14 Hundreds of zebra mussels, an invasive mollusc, attached to rock from a Texas lake.

the Saint Lawrence Seaway. They quickly spread to rivers and lakes beyond the Great Lakes region and are now widely distributed. The economic damage caused by zebra mussels results from their astronomically large populations. Adult zebra mussels form thick layers that clog pipes and the water intakes of cities, power plants, and factories. As many as 70,000 individuals can be found in 1 square meter! In addition, these tiny molluscs frequently attach to larger native bivalves, interfering with their ability to feed and reproduce, and they compete with native species for food and spaces. As a result, populations of native bivalves have declined in areas colonized by zebra mussels.

For a non-native organism like zebra mussels to become invasive, the biotic and abiotic factors of the new environment must be compatible with the organism's needs and tolerances. For example, Burmese pythons set loose in South Florida—either accidently released by damaging storms or deliberately freed by disenchanted pet owners—found a hot, humid climate similar to their native area. Prey such as birds, mammals, reptiles, and amphibians are readily available, especially in the Everglades. As a result, South Florida is now home to a burgeoning population of the giant reptiles **(Figure 19.15)**. Burmese pythons released in a less favorable environment might survive for a short time but would not be able to establish a population. ✓

▼ **Figure 19.15 A Burmese python.** Researchers from the U.S. Geological Survey captured this enormous snake in the Everglades in 2017.

✅ **CHECKPOINT**

What distinguishes invasive species from organisms that are introduced to non-native habitats but do not become invasive?

■ *Answer: Invasive species spread far from where they are introduced, and they cause environmental or economic damage.*

Biological Control of Pests

The absence of biotic factors that limit population growth, such as pathogens, predators, or herbivores, may contribute to the success of invasive species. Accordingly, efforts to eliminate or control these troublesome organisms often focus on **biological control**, the intentional release of a natural enemy to attack a pest population. Agricultural researchers have long been interested in identifying potential biological agents to control insects, weeds, and other organisms that reduce crop yield.

Biological control has been effective in numerous instances, especially with invasive insects and plants. In Britain, field trials are currently underway to investigate whether a sap-sucking psyllid can combat Japanese knotweed, a garden plant brought from Japan in the nineteenth century. It can grow over a meter a month and its stout rhizomes can push their way through concrete, making it one of the most damaging invasive weeds in Europe and North America. The psyllids feed exclusively on knotweed in Japan and is seen as a potential alternative to expensive and unsustainable chemical control methods.

One potential pitfall of biological control is the danger that an imported control agent may be as invasive as its target. One cautionary tale comes from introducing the mongoose **(Figure 19.16)** to control rats. Rats that originated in India and northern Asia were accidentally transported around the world and became invasive in many places. For sugarcane growers, rat infestation meant massive crop damage. Cane planters imported the small Indian mongoose, a fierce little carnivore, to deal with the problem. In time, mongooses were introduced to dozens of natural habitats, including all of the largest Caribbean and Hawaiian islands—and became invasive themselves. Mongooses are not picky eaters, and they have voracious appetites. On island after island, populations of reptiles, amphibians, and ground-nesting birds have declined or vanished as mongoose populations have grown and spread. They also prey on domestic poultry and ruin crops, costing millions of dollars a year. Clearly, rigorous research is needed to assess the safety and effectiveness of potential biological control agents.

◀ **Figure 19.16 A small Indian mongoose.**

Can Fences Stop Cane Toads?

BACKGROUND

In 1935, Australian agricultural experts imported the cane toad **(Figure 19.17a)** to control invasive beetles that were destroying sugar cane, a major crop. Unfortunately, the beetles were usually underground or in the air, safe from being eaten by cane toads. Instead of recruiting an ally in the battle against beetles, the agriculturalists had introduced a super-invader. From the original introduction of 102 cane toads, the amphibians spread rapidly. They now occupy more than 1 million km^2 (over 386,000 mi^2) of Australia, an area larger than Texas and Arizona, with a total population estimated at 1.5 billion. The toads' voracious appetite threatens many native species. However, the greatest danger is to predators, such as snakes, lizards, and marsupials, which risk death when they try to eat the toxic toads.

Humans have unwittingly aided the cane toad invasion. To survive the dry season on the semi-arid plains of Australia's Northern Territory, the toads gather near permanent water sources. Cattle farmers in this region have expanded suitable cane toad habitat by providing additional watering holes—artificial stock ponds built to sustain their livestock through the dry season. The stock ponds are rings of packed dirt holding water that can be directed to troughs. Researchers hypothesized that blocking access to the ponds could limit the cane toad invasion.

METHOD

During the dry season, researchers selected nine stock ponds that were at least 40 yards apart. Three of the ponds were left as unfenced controls **(Figure 19.17b)**. Three ponds were surrounded by closed, toad-proof fences made of stakes and mesh from the ground to several feet above water. Three ponds had open fences (raised up so that toads could hop under the fence to access water, but otherwise identical to the closed fences). Researchers removed the toads from all of the ponds. Then, the researchers waited to see if the ponds in the immediate vicinity of each would be reinvaded by toads. On day 70 (early wet season), the researchers counted live and dead toads in and around each pond. They repeated this on day 190 (early dry season) and day 365 (late dry season).

RESULTS

There were almost no dead toads at either the unfenced or open-fence ponds at any time in the year, but toad mortality was high near the closed-fenced ponds late in the dry season **(Figure 19.17c)**. Almost no live toads were found at the closed-fenced ponds, while there were 30–100 toads per survey at the open-fence and unfenced ponds. The closed-fenced ponds acted as ecological traps: Toads were attracted to the scent of water but died when they could not get inside fences to access the ponds. Enclosing stock ponds, the researchers concluded, could protect the dry regions of Australia from cane toad invasion.

▼ **Figure 19.17 An experiment to test a strategy for limiting the spread of cane toads.**
Data from: M. Letnic et al. Restricting Access to Invasion Hubs Enables Sustained Control of an Invasive Vertebrate. Journal of Applied Ecology 52(2): 341– 47 (2015).

(a) The cane toad has spread across large parts of Australia.

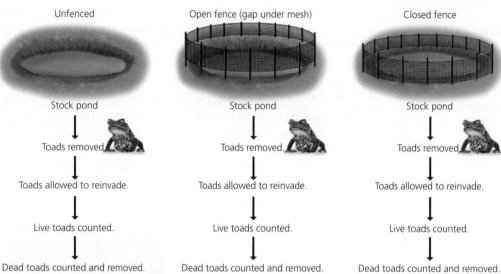

(b) Experimental design testing the effectiveness of toad barriers

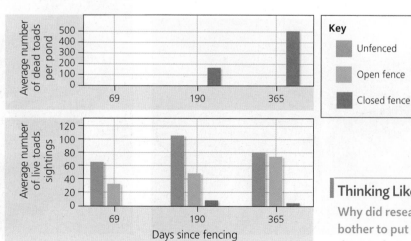

(c) Influence of fencing on toad reinvasions of ponds

Thinking Like a Scientist

Why did researchers bother to put up fences that toads could go under?

For the answer, see Appendix D.

Integrated Pest Management

In contrast to enterprises such as fisheries, which harvest resources from natural ecosystems, agricultural operations create their own highly managed ecosystems. A typical crop population consists of genetically similar individuals (a monoculture) planted in close proximity to each other— a banquet laid out for the many plant-eating animals and pathogenic bacteria, viruses, and fungi in the community. The tilled, fertile ground nurtures weeds as well as crops. Thus, farmers wage an eternal war against pests that compete with their crop for soil minerals, water, and light; that siphon nutrients from the growing plants; or that consume their leaves, roots, fruits, or seeds. At home, you may be engaged in combat against pests on a smaller scale as you attempt to eradicate the weeds, insects, fungi, and bacteria that attack your lawn and garden or the mosquitoes that make your summer evenings miserable.

Like invasive species, most crop pests have an opportunistic life history pattern that enables them to rapidly take advantage of a favorable habitat. The history of agriculture abounds with examples of devastating pest outbreaks. For example, the boll weevil **(Figure 19.18)** is an insect that feeds on cotton plants both as larvae and as adults. Its unstoppable spread across the southern United States in the early 1900s severely damaged local economies and had a lasting impact on the region. The folklore of the invasion includes the song "Boll Weevil Blues," which has been recorded by many artists, including The White Stripes. Viruses, fungi, bacteria, nematodes, and other plant-eating insects can cause massive damage as well.

When synthetic herbicides and insecticides such as DDT were developed in the 1940s, they quickly became the method of choice in agriculture. However, chemical solutions to pest problems bring numerous problems of their own. These chemicals are pollutants that can be carried great distances by air or water currents. In addition, natural selection may result in populations that are not affected by a pesticide (see Figure 13.11). Furthermore, most insecticides kill both the pest and its natural predators. Because prey species often have a higher reproductive rate than predators, pest populations rapidly rebound before their predators can reproduce. There may be other unintended damage as well, such as killing pollinators that are essential for both agricultural and natural ecosystems.

Integrated pest management (IPM) uses a combination of biological, chemical, and cultural methods for sustainable control of agricultural pests. Researchers are also investigating IPM approaches to invasive species. IPM relies on knowledge of the population ecology of the pest and its associated predators and parasites, as well as plant growth dynamics. In contrast to traditional methods of pest control, IPM advocates tolerating a low level of pests rather than attempting total eradication. Thus, many pest control measures are aimed at lowering the habitat's carrying capacity for the pest population by using pest-resistant varieties of crops, mixed-species plantings, and crop rotation to deprive the pest of a dependable food source. Biological control is also used when possible. For example, many gardeners release ladybird beetles to control aphid infestations **(Figure 19.19)**. Pesticides are applied when necessary, but adherence to the principles of IPM prevents the overuse of chemicals. ☑

▲ Figure 19.18 **A boll weevil on a damaged boll (seed pod) of a cotton plant.**

▲ Figure 19.19 **Ladybird beetles feeding on aphids.** Each of these voracious predators can eat as many as 50 aphids per day.

Human Population Growth

Now that we have examined the regulation of population growth in other organisms, what about our own species? Let's begin by looking at the history of the human population and then consider some current and future trends in population growth.

The History of Human Population Growth

In the few seconds it takes you to read this sentence, approximately 30 babies will be born somewhere in the world and 13 people will die. An imbalance between births and deaths is the cause of population growth (or decline), and as the line graph in **Figure 19.20** shows, the human population is expected to continue increasing for at least the next several decades. The bar graph in Figure 19.20 tells a different part of the story. The number of people added to the population each year has been declining since the 1980s. How do we explain these patterns of human population growth?

Let's begin with the rise in world population from approximately 480 million people in 1500 to the current population of more than 7.5 billion. In the exponential population growth model introduced earlier in this chapter, we assumed that the net rate of increase (birth rate minus death rate) was constant—births and deaths were roughly equal. As a result, population growth depended only on the size of the existing population. Throughout most of human history, this assumption held true. Although parents had many children, the death rate was also high, resulting in a rate of increase only slightly

higher than 0. Consequently, human population growth was initially very slow. (If we extended the *x*-axis of Figure 19.20 back in time to year 1, when the population was roughly 300 million, the line would be almost flat for 1,500 years.) The 1 billion mark was not reached until the early 1800s. As economic development in Europe and the United States led to advances in nutrition and sanitation, and, later, medical care, people took control of their population's growth rate. At first, the death rate decreased while the birth rate remained the same. The net rate of increase rose, and population growth began to pick up steam by the beginning of the 1900s. By midcentury, improvements in nutrition, sanitation, and health care had spread to the developing world, spurring growth at a breakneck pace as birth rates far outstripped death rates.

As the world population skyrocketed from 2 billion in 1927 to 3 billion just 33 years later, some scientists became alarmed. They feared that Earth's carrying capacity would be reached and that density-dependent factors would maintain that population size through human suffering and death. But the overall growth rate peaked in 1962. In the more developed nations, advanced medical care continued to improve survivorship, but effective contraceptives held down the birth rate. As a result, the overall growth rate of the world's population began a downward trend as the difference between birth rate and death rate decreased. In the most developed nations, the overall rate of increase is near zero **(Table 19.3)**. In the developing world, on the other hand, death rates have dropped, but high birth rates persist. As a result, these populations are growing rapidly—of the 77.8 million people added to the world in 2016, nearly 74 million were in developing nations. Thus, the total world population continues to increase. ✓

COUNTING ONE PERSON PER SECOND, IT WOULD TAKE MORE THAN 238 YEARS TO COUNT ALL 7.5 BILLION PEOPLE ALIVE TODAY.

✅ CHECKPOINT

Why was the growth rate of the world's population so high during most of the 1900s? What accounts for the recent decrease in the growth rate of the world's population?

Answer: In the 1900s, the death rate decreased dramatically due to improved nutrition, sanitation, and health care, but the birth rate remained high. As a result, the overall population growth rate was high. The recent decrease in growth rate is the result of lower birth rates in some regions.

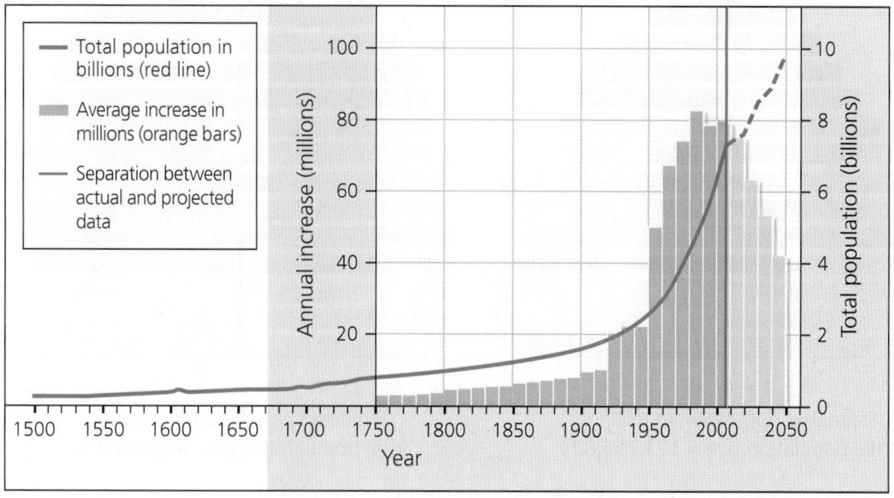

▲ Figure 19.20 Five centuries of human population growth, projected to 2050.

Table 19.3	Population Trends in 2016		
Population	Birth Rate per 1,000	Death Rate per 1,000	Growth Rate (%)
World	18.5	7.8	1.1
More developed countries	10.7	10.1	0.06
Less developed countries	20.1	7.3	1.3

Age Structures

Age structures, which were introduced at the beginning of this chapter, are helpful for predicting a population's future growth. **Figure 19.21** shows the estimated and projected age structures of Mexico's population in 1990, 2015, and 2040. In these diagrams, the area to the left of each vertical line represents the number of males in each age-group; females are represented on the right side of the line. The three different colors represent the portion of the population in their prereproductive years (0–14), prime reproductive years (15–44), and postreproductive years (45 and older). Within each of these broader groups, each horizontal bar represents the population in a 5-year age-group.

In 1990, each age-group was larger than the one above it, indicating a high birth rate. The pyramidal shape of this age structure is typical of a population that is growing rapidly. In 2015, the population's growth rate was lower; notice that the three youngest (bottommost) age-groups are roughly the same size. However, the population continues to be affected by its earlier expansion. This situation, which results from the increased proportion of women of childbearing age in the population, is known as **population momentum**. Girls who were 0–14 years old in the 1990 age structure (outlined in pink) were in their reproductive prime in 2015, and girls who were 0–14 years old in 2015 (outlined in blue) will carry the legacy of rapid growth forward to 2040. Putting the brakes on

a rapidly expanding population is like stopping a freight train; the actual event takes place long after the decision to do it was made. Even when fertility—the number of live births over a woman's lifetime—is reduced to replacement rate (an average of two children per female), the total population size will continue to increase for several decades. Thus, the percentage of individuals under the age of 15 gives a rough idea of future growth. In less developed countries, about 28% of the population is in this age-group. In contrast, 16% of the population of more developed nations is under the age of 15. Population momentum also explains why the total population size worldwide (line graph in Figure 19.20) continues to increase even though fewer people are added to the population each year, as shown by the bar graph in Figure 19.20.

Age structure diagrams may also indicate social conditions. For instance, an expanding population has an increasing need for schools, employment, and infrastructure. A large elderly population requires that extensive resources be allotted to health care. Let's look at trends in the age structure of the United States from 1990 to 2040 **(Figure 19.22)**. The noticeable bulge in the 1990 population (highlighted with the yellow screen) corresponds to the "baby boom" that lasted for about two decades after World War II ended in 1945. The large number of children swelled school enrollments, prompting construction of new schools and creating a demand for teachers. On the other hand, graduates who were born near the end of the

Figure
Walkthrough

Mastering **Biology**
goo.gl/pWmvvv

▼ **Figure 19.21 Population momentum in Mexico.**

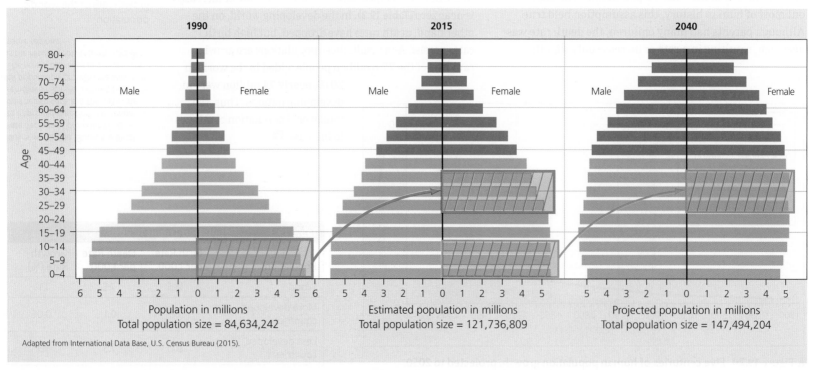

Adapted from International Data Base, U.S. Census Bureau (2015).

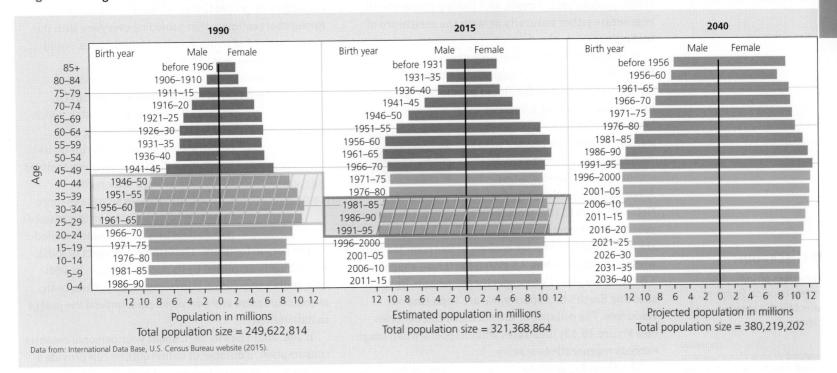

▼ **Figure 19.22** Age structures for the United States in 1990, 2015 (estimated), and 2040 (projected).

Data from: International Data Base, U.S. Census Bureau website (2015).

boom faced stiff competition for jobs. Because they make up such a large segment of the population, boomers have had an enormous influence on social, economic, and political trends. They also produced a boomlet of their own, seen in the 0–4 age-group in 1990 and the bump (pink screen) in the 2015 age structure.

Where are the baby boomers now? The leading edge has reached retirement age, which will place pressure on programs such as Medicare and Social Security. In 2015, roughly 60% of the U.S. population was between 20 and 64, the ages most likely to be in the workforce, and 13.5% was over 65. In 2040, these age groups are projected to make up 55% and 20% of the population, respectively. In part, the increase in the elderly population results from people living longer. The percentage of the population over 80, which was 2.8% in 1990, is projected to rise to nearly 7.5%—more than 28.5 million people—in 2040. ☑

Our Ecological Footprint

How large a human population can Earth support? Figure 19.20 shows that the world's population is increasing rapidly, though at a slower rate than it did in the past century. The rate of increase, as well as population momentum, indicates that the populations of most developing nations will continue to increase for the foreseeable future. The U.S. Census Bureau projects a global population of almost 8 billion by 2025 and 9.8 billion by 2050. But these numbers are only part of the story. Trillions

of bacteria can live in a petri dish *if* they have sufficient resources. Do we have sufficient resources to sustain 8 or 9 billion people?

To accommodate all the people expected to live on our planet in the coming decades and improve the diets of those who are currently malnourished or undernourished, world food production must increase dramatically. But agricultural lands are already under pressure. Overgrazing by the world's growing herds of livestock is turning vast areas of grassland into desert. Water use, which increased sixfold from 1900 to 2000, continues to rise, causing rivers to run dry and levels of groundwater to drop. The effects of climate change on precipitation patterns are likely to have a devastating impact on agriculture in some regions. And because so much open space will be needed to support the expanding human population, many other species are expected to become extinct.

The concept of an ecological footprint is one approach to understanding resource availability and usage. An **ecological footprint** is an estimate of the land and water area required to provide the resources an individual or a nation consumes—for example, food, fuel, and housing—and to absorb the waste it generates. Our carbon footprint, the emission of carbon dioxide and other greenhouse gases (see Figure 18.45), is the largest component of humanity's ecological footprint.

Comparing our demand for resources with Earth's capacity to renew these resources, or **biocapacity**, gives us a broad view of the sustainability of human

☑ **CHECKPOINT**

Why is the percentage of individuals under the age of 15 a good indication of future population growth?

■ Answer: These individuals have not yet entered their reproductive years. If they make up a large percentage of the population (a bottom-heavy age structure), future population growth will be high.

activities. When used sustainably, resources such as crops, pastureland, forests, and fishing grounds will regenerate either naturally or with the assistance of technology.

Is humanity's current resource usage sustainable? When the total area of ecologically productive land on Earth is divided by the global population, we each have a share of about 1.7 global hectares (1 hectare = 2.47 acres; a global hectare is a hectare with world-average ability to produce resources and absorb wastes). According to the Global Footprint Network, in 2012 (the most recent year for which complete data are available), the average ecological footprint for the world's population was 2.6 global hectares—roughly 1.5 times the planet's biocapacity per person. By overshooting Earth's biocapacity, we are depleting our resources. The collapse of the northern cod fisheries (see Figure 19.13) illustrates what happens when usage exceeds regenerative capacity.

The green footprints in **Figure 19.23** compare the average ecological footprint per person for several countries to the "fair share" (1.7 gha) and world average (2.6 gha) footprints. As the giant footprints of the United States

THE AVERAGE AMERICAN GENERATES ABOUT 40 POUNDS OF TRASH PER WEEK.

and Australia illustrate, individuals in affluent nations consume a disproportionate amount of resources. Researchers estimate that providing everyone with the same standard of living as in the United States would require the resources of 3.9 planet Earths.

The impact of population size on sustainability is seen in the national total ecological footprints in Figure 19.23 (blue footprints), which is calculated by multiplying each country's per person footprint by its population. Earth's total biocapacity is estimated to be 12 billion gha.

The United States alone uses more than 17% of that. China, with a massive population, uses 29%, despite its modest per person footprint. The ecological impact of India, too, is hugely amplified by its population size. In addition, India's population is growing rapidly. Thus, both overconsumption and overpopulation imperil the goal of sustainability.

If you would like to learn about your personal resource consumption, a number of online quizzes can provide a rough estimate of your ecological footprint. Like carbon footprint calculators (described in Chapter 18), these tools are useful for learning how to reduce your environmental impact. ✓

✓ CHECKPOINT

How does an individual's large ecological footprint affect Earth's carrying capacity?

■ *Answer: The more resources required to sustain an individual, the lower Earth's carrying capacity will be. (That is, Earth can sustain fewer people if each of those people consumes a large share of available resources.)*

▼ Figure 19.23 **Personal and national ecological footprints of several countries.**

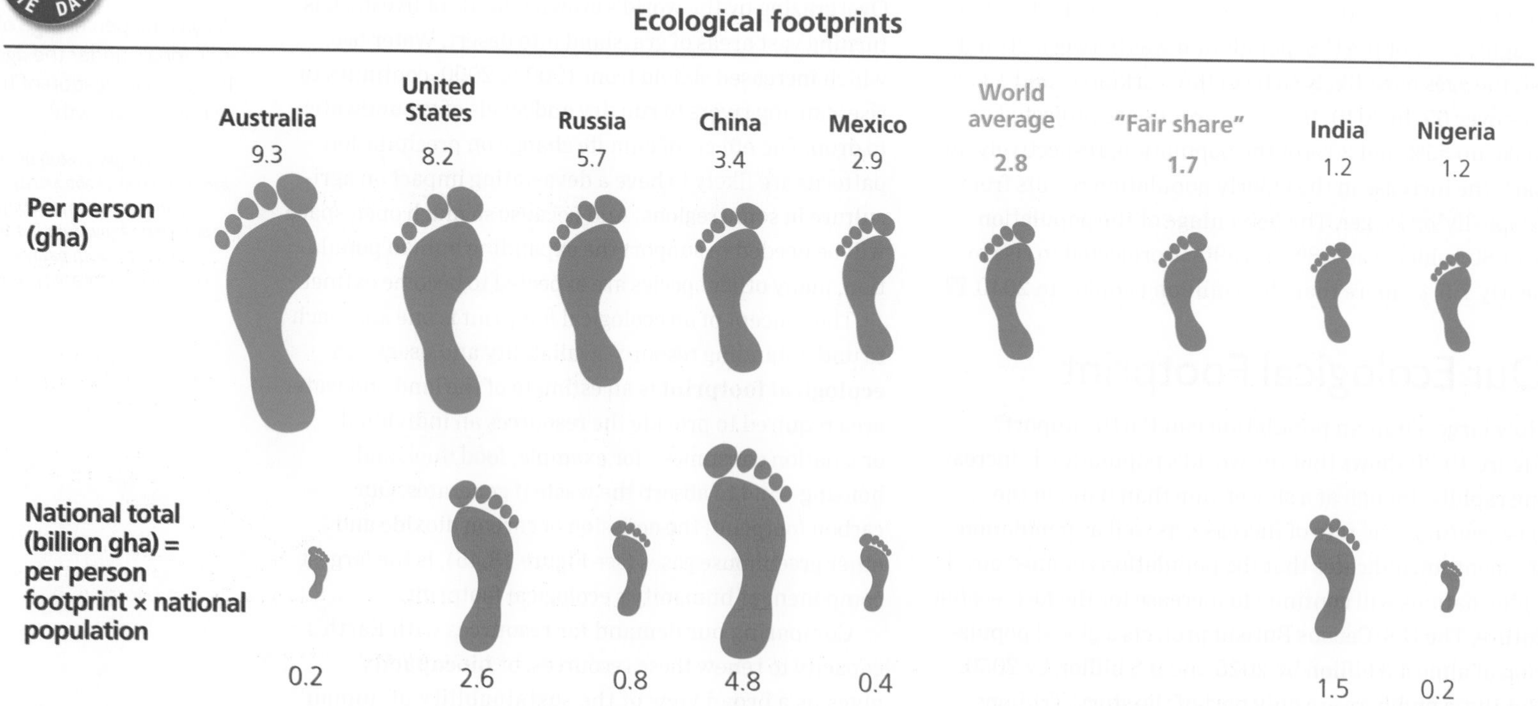

Ecological footprints

	Australia	United States	Russia	China	Mexico	World average	"Fair share"	India	Nigeria
	9.3	8.2	5.7	3.4	2.9	2.8	1.7	1.2	1.2
Per person (gha)									
National total (billion gha) = per person footprint × national population	0.2	2.6	0.8	4.8	0.4			1.5	0.2

Humans as an Invasive Species

The magnificent pronghorn antelope (*Antilocapra americana*) is the descendant of ancestors that roamed the open plains and shrub deserts of North America millions of years ago (Figure 19.24). With strides that cover 6 m (20 feet) or more at its top speed of 97 km/h (60 mph), it is easily the fastest mammal on the continent. The pronghorn's speed is more than a match for its major predator, the wolf, which typically takes adults that have been weakened by age or illness. What selection pressure promoted such extravagant speed? Ecologists hypothesize that the pronghorn's ancestors were running from the now-extinct American cheetah, a fleet-footed predator that bore some similarities to the more familiar African cheetah.

Cheetahs were not the only danger in the pronghorn's environment. During the Pleistocene epoch, which lasted from 1.8 million to 10,000 years ago, North America was also home to an intimidating list of other predators: lions, jaguars, saber-toothed cats with canine teeth up to 7 inches long, and towering short-faced bears, which stood 11 feet tall and weighed three-quarters of a ton. There were plenty of potential prey for these fearsome predators, including massive ground sloths, bison with horns that spread 10 feet, elephant-like mammoths, a variety of horses and camels, and several species of pronghorns.

Of all these species of large mammals, only *Antilocapra americana* remained at the end of the Pleistocene. The others went extinct during a relatively brief period of time that coincided with the spread of humans throughout North America. Although the cause of the extinctions has been hotly disputed, many scientists think that the human invasion, combined with climate change at the end of the last ice age, was responsible. Taken together, changes in the biotic and abiotic environments happened too rapidly for an evolutionary response from these large mammals.

The role of humans in the Pleistocene extinctions was merely a preview of things to come. The human population continues to increase, colonizing almost every corner of the world. Like other invasive species, we change the environment of the other organisms that share our habitats. As the scope and speed of human-induced environmental changes increase, extinctions are occurring at an accelerating pace. (This rapid loss of biodiversity will be our unifying focus in the next chapter.)

▼ Figure 19.24 A pronghorn antelope racing across the North American plains.

Chapter Review

SUMMARY OF KEY CONCEPTS

An Overview of Population Ecology

A population consists of all the members of a species living in the same place at the same time. Population ecology focuses on the factors that influence a population's size, density, age structure, and growth rate.

Population Density

Population density, the number of individuals of a species per unit area or volume, can be estimated by a variety of sampling techniques.

Population Age Structure

A graph showing the distribution of individuals in different age-groups often provides useful information about the population.

Life Tables and Survivorship Curves

A life table tracks the chance of an individual in a population surviving to various ages. Survivorship curves can be classified into three general types.

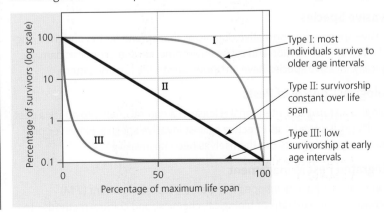

Type I: most individuals survive to older age intervals

Type II: survivorship constant over life span

Type III: low survivorship at early age intervals

Life History Traits as Adaptations

Life history traits are shaped by evolutionary adaptation. Most populations probably fall between the extreme opportunistic life histories (reach sexual maturity rapidly; produce many offspring; little or no parental care) of many insects and the equilibrial life histories (develop slowly and produce few, well-cared-for offspring) of many larger-bodied species.

Population Growth Models

The Exponential Population Growth Model: The Ideal of an Unlimited Environment

Exponential population growth is the accelerating increase that occurs when growth is unlimited. The exponential model predicts that the larger a population becomes, the faster it grows.

The Logistic Population Growth Model: The Reality of a Limited Environment

Logistic population growth occurs when growth is slowed by limiting factors. The logistic model predicts that a population's growth rate will be low when the population size is either small or large and highest when the population is at an intermediate level relative to the carrying capacity.

Regulation of Population Growth

Over the long term, most population growth is limited by a mixture of density-independent factors, which affect the same percentage of individuals regardless of population size, and density-dependent factors, which intensify as a population increases in density. Some populations have regular boom-and-bust cycles.

Applications of Population Ecology

Conservation of Endangered Species

Endangered and threatened species are characterized by very small population sizes. One approach to conservation is identifying and attempting to supply the critical combination of habitat factors needed by the population.

Sustainable Resource Management

Resource managers apply principles of population ecology to help determine sustainable harvesting practices.

Invasive Species

Invasive species are non-native organisms that spread far beyond their original point of introduction and cause environmental and economic damage. Typically, invasive species have an opportunistic life history pattern.

Biological Control of Pests

Biological control, the intentional release of a natural enemy to attack a pest population, is sometimes effective against invasive species. However, prospective control agents can themselves become invasive.

Integrated Pest Management

Crop scientists have developed integrated pest management (IPM) strategies—combinations of biological, chemical, and cultural methods—to deal with agricultural pests.

Human Population Growth

The History of Human Population Growth

The human population grew rapidly during the 1900s and is currently more than 7.5 billion. A shift from high birth and death rates to low birth and death rates has lowered the rate of growth in more developed countries. In developing nations, death rates have dropped, but birth rates are still high.

Age Structures

The age structure of a population affects its future growth. The wide base of the age structure of Mexico in 1990—the 0–14 age-group—predicts continued population growth in the next generation. Population momentum is the continued growth that occurs after a population's high fertility rate has been reduced to replacement rate; it is a result of girls in the 0–14 age-group reaching their childbearing years. Age structures may also indicate social and economic trends, as in the age structure on the right below.

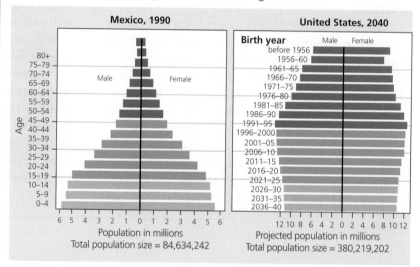

Mexico, 1990
Total population size = 84,634,242

United States, 2040
Projected population size = 380,219,202

Our Ecological Footprint

An ecological footprint represents the amount of land and water needed to produce the resources used by an individual or nation. There is a huge disparity between resource consumption by individuals in more developed and less developed nations.

Mastering Biology

For practice quizzes, BioFlix animations, MP3 tutorials, video tutors, and more study tools designed for this textbook, go to Mastering Biology™

SELF-QUIZ

1. What two values would you need to know to figure out the human population density of your community?

2. If members of a species produce a large number of offspring but provide minimal parental care, then a Type _____ survivorship curve is expected. In contrast, if members of a species produce few offspring and provide them with long-standing care, then a Type _____ survivorship curve is expected.

3. Use this graph of the idealized exponential and logistic growth curves to complete the following.

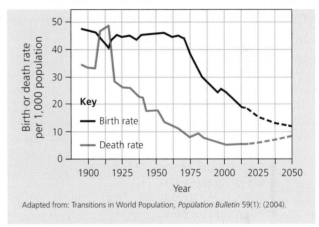

   a. Label the axes and curves on the graph.

   b. What does the dotted line represent?

   c. For each curve, indicate and explain where population growth is the most rapid.

   d. Which of these curves better represents global human population growth?

4. Which of the following describes the effects of a density-dependent limiting factor?

   a. A forest fire kills all the pine trees in a patch of forest.

   b. Early rainfall triggers the explosion of a locust population.

   c. Drought decimates a wheat crop.

   d. Rabbits multiply, and their food supply begins to dwindle.

5. What type of life history do elephants typically have?

6. Which of the following statements is false when taking an integrated pest management (IPM) approach to control an invasive species?

   a. IPM advocates complete eradication of the pest.

   b. IPM lowers the habitat's carrying capacity for the pest population.

   c. IPM uses a combination of biological, chemical and cultural methods.

   d. IPM permits use of pesticides.

7. A study of the human ecological footprint shows that

   a. we have already overshot the planet's capacity to sustain us.

   b. Earth can sustain the current population, but not much more.

   c. Earth can sustain a population about double the current population.

   d. the size of the human population will soon crash.

For answers to the Self-Quiz, see Appendix D.

IDENTIFYING MAJOR THEMES

For each statement below, identify which major theme is evident (the relationship of structure to function, information flow, pathways that transform energy and matter, interactions within biological systems, or evolution) and explain how the statement relates to the theme. If necessary, review the themes (Chapter 1) and review the examples highlighted in blue in this chapter.

8. The mussel secretes a combination of molecules that hardens into exceptionally sticky threads.

9. The lynx is one of the main predators of the snowshoe hare in the far northern forests of Canada and Alaska. About every ten years, both lynx and hare populations show a rapid increase followed by a sharp decline.

10. Life history traits, like anatomical features, are shaped by trade-offs that balance the demands of reproduction and survival, both of which determine fitness.

For answers to Identifying Major Themes, see Appendix D.

THE PROCESS OF SCIENCE

11. What are the limiting factors for a population of tropical fish in an aquarium? How would you test if there is any interspecies competition with another type of fish?

12. Interpreting Data The graph below shows data for population trends in Mexico from 1890 to 2016, with projected trends for 2016–2050. How has Mexico's rate of population growth changed over this time period? How is it expected to change by midcentury? Describe Mexico's projected age structure in 2050.

Adapted from: Transitions in World Population, *Population Bulletin* 59(1): (2004).

BIOLOGY AND SOCIETY

13. Wars and epidemics can limit the growth of the human population. How big has the influence of wars and epidemics been on the human population in the past?

14. The ever-increasing size of the human population is responsible for several environmental problems. In 1979, the Chinese government introduced the one-child policy to control population growth in the country. Various dispensations followed, and the policy was scrapped in 2015, allowing couples to have two children for the first time in three decades. What impact do you think this had on the country's age structure pyramid? How could this have affected the population balance in China?

20 Communities and Ecosystems

Why Ecological Interactions Matter

Humans are a part of the ecosystems that sustain life on Earth. Our world is shaped by interactions between species and between individual organisms and their environment.

BEWARE. TUNA MAY BE CONTAMINATED WITH MERCURY AND OTHER TOXINS.

THERE'S SAFETY IN NUMBERS. BIODIVERSITY CAN PROTECT YOU FROM CATCHING LYME DISEASE FROM THIS TICK.

DO THE MATH. RAISING BEEF REQUIRES EIGHT TIMES AS MUCH LAND AS RAISING SOYBEANS.

BIOLOGY AND SOCIETY | Importance of Biodiversity

A protein in Tasmanian devil milk may kill "superbugs" in humans someday.

Why Biodiversity Matters

Tasmanian devils aren't just cartoon characters. Real Tasmanian devils are fierce meat-eating marsupials that live up to their name. They can run incredibly fast, have an exceptionally strong bite, and can emit a loud screech. They are also an endangered species. Why not leave these strange animals to become extinct? One reason: They may save your life one day. A protein recently found in Tasmanian devil milk kills antibiotic-resistant bacteria. Unfortunately, Tasmanian devil populations have dropped drastically in the last 20 years due to habitat loss and a new, incurable facial cancer that spreads when the devils bite each other, which happens often. (They are called "devils" for a reason.) Their entire habitat is an island off the coast of Australia. They breed slowly and have low genetic diversity. It may be a relief to learn that steps have been taken to protect the species.

Like all species, Tasmanian devils play a unique role in their community. They kill small mammals, including destructive introduced species, but they specialize on scavenging dead animals. Loss of this species will have ripple effects on the entire community.

Human activities such as habitat destruction, pollution, and overharvesting threaten species around the world. Some people would conserve biodiversity for its own sake, but there are practical reasons as well. Humans depend on diverse ecological communities. Every day, our lives are improved by species that manufacture oxygen, provide food, eat pests, moderate the climate, absorb flood waters, reduce erosion, and provide essential products, even unexpected cures for diseases. It is impossible to put an exact dollar value on these ecosystem services, but they may be worth over $30 trillion a year.

Each species is integral in the complex set of interactions that create biological communities. When one species is lost, the results are unpredictable and can be devastating. Without pollinators, some plants go extinct. Species without predators or competitors may over-consume their resources, permanently damaging the environment.

In this chapter, we will explore how species interact with each other and with their environment. We will see how communities recover from disturbances and how we can protect and restore the ecosystems that provide us with vital services.

Biodiversity

Biodiversity is short for biological diversity, the variety of living things. It includes genetic diversity, species diversity, and ecosystem diversity. Thus, the loss of biodiversity encompasses more than just the fate of individual species.

▼ **Figure 20.1 Cavendish bananas.** All are clones and are threatened by a new fungal disease.

Genetic Diversity

The genetic diversity (variation in alleles) within a population is the raw material of natural selection, which allows organisms to adapt to a changing environment (Chapter 13). As populations are lost, so are their genetic resources, genes that produce structures or functions that are actually or potentially beneficial. Reduction in genetic diversity, therefore, threatens the survival of species. When an entire species is lost, so are all of its genetic resources.

Genetic diversity has economic value. Many researchers and biotechnology leaders are enthusiastic about the potential that genetic "bioprospecting" holds for development of new medicines, industrial chemicals, and other products. Bioprospecting may also protect the world's food supply. Cavendish bananas account for virtually all the banana market, generating over $10 billion in revenue per year. All Cavendish bananas are clones, with no genetic diversity, so one pathogen could wipe out every individual. One such pathogen has emerged in Southeast Asia, triggering a search for wild species of bananas that were previously thought to have no value. If wild bananas have information in their genes that makes them resistant to the pathogen, that information could be transferred to the Cavendish.

☑ **CHECKPOINT**

How does the loss of genetic diversity endanger a population?

■ Answer: A population with decreased genetic diversity has less ability to evolve in response to environmental change.

Gene prospectors are racing against the destruction of wild bananas, as rain forest is replaced by palm oil plantations (**Figure 20.1**). ☑

Species Diversity

Ecologists believe that we are pushing species toward extinction at an alarming rate. The present rate of species loss may be as much as 100 times higher than at any time in the past 100,000 years. Some researchers estimate that at the current rate of destruction, over half of all currently living plant and animal species will be gone by the end of this century. **Figure 20.2** shows two species, one extinct and the other endangered. The International Union for Conservation of Nature (IUCN) compiles scientific assessments of the conservation status of species worldwide. Here are some examples of where things stand:

- Approximately 13% of the known 10,004 bird species and a quarter of the identified 5,488 mammalian species assessed are threatened with extinction.

- More than 20% of the discovered freshwater fishes in the world either have become extinct during human history or are seriously threatened.

- Roughly 41% of all assessed amphibian species are in danger of extinction.

- In Britain, one in five species of plants is threatened with extinction, and 20 species are currently recognized as extinct. More than 10,000 plant species worldwide are in danger of extinction.

▼ **Figure 20.2 Recent additions to the list of species eradicated or endangered by humans.**

Clouded leopard. In 2013, scientists gave up hope that the Formosan clouded leopard, a subspecies found only on the island of Taiwan, still exists. This photo shows a similar subspecies living in a zoo.

Hector's dolphin. This New Zealand dolphin is the smallest in the world. It is now endangered.

Ecosystem Diversity

Ecosystem diversity is the third component of biological diversity. Recall that an ecosystem includes both the organisms and the abiotic factors in a particular area. Because of the network of interactions among populations of different species, the loss of one species can have a negative effect on the entire ecosystem. As discussed in the Biology and Society section, the disappearance of natural ecosystems results in the loss of **ecosystem services**, functions performed by an ecosystem that directly or indirectly benefit people. These vital services include air and water purification, climate regulation, and erosion control. For example, forests absorb and store carbon from the atmosphere, a service that vanishes when forests are destroyed or degraded (see Figure 18.39). Coral reefs not only are rich in species diversity **(Figure 20.3)** but also provide a wealth of benefits to people, including food, storm protection, and recreation. An estimated 20% of the world's coral reefs have already been destroyed by human activities. A recent study found that 93% of Australia's Great Barrier Reef has suffered from a severe form of damage called bleaching. Later in this chapter, you'll learn how declining ecosystem diversity can directly influence human health. ☑

▼ **Figure 20.3 A coral reef, a colorful display of biodiversity.**

Causes of Declining Biodiversity

Ecologists have identified four main factors responsible for the loss of biodiversity: habitat destruction, including fragmentation; invasive species; overexploitation; and pollution. The ever-expanding size and dominance of the human population are at the root of all four factors. In addition, scientists expect global climate change to become a leading cause of extinctions in the near future (see Chapter 18).

Habitat Destruction
The massive destruction and fragmentation of habitats caused by agriculture, urban development, forestry, and mining pose the single greatest threat to biodiversity **(Figure 20.4)**. According to the IUCN, habitat destruction affects more than 85% of all birds, mammals, and amphibians that are threatened with extinction. The destruction of its forest habitat, along with trade in its gorgeous pelt, doomed the Formosan clouded leopard. The remaining subspecies of clouded leopard, which inhabit the forests of Southeast Asia, are also vulnerable to extinction as a result of deforestation.

Invasive Species
Ranking second behind habitat destruction as a cause of biodiversity loss is the introduction of invasive species. Uncontrolled population growth of human-introduced species has caused havoc when the introduced species have competed with, preyed on, or parasitized native species (see Chapter 19). The lack of interactions with other species that could keep non-native species in check is often a key factor in a non-native species becoming invasive.

☑ **CHECKPOINT**

When ecosystems are destroyed, the services they provide are lost. What are some examples of ecosystem services?

■ *Answer: Possible answers are mentioned in the Biology and Society section. Services include providing medicines, air and water purification, climate regulation, and resources such as water, food, and wood.*

▼ **Figure 20.4 Habitat destruction.** In a controversial method known as mountaintop removal, mining companies blast the tops off of mountains and then scoop out coal. The earth removed from the mountain is dumped into a neighboring valley.

461

Overexploitation

Unsustainable marine fisheries (see Figure 19.13) demonstrate how people can overexploit wildlife by harvesting at rates that exceed the ability of populations to rebound. American bison, Galápagos tortoises, and tigers are among the many terrestrial species whose numbers have been drastically reduced by excessive commercial harvesting, poaching, and sport hunting. Overharvesting also threatens some plants, including rare trees such as mahogany and rosewood that produce valuable wood.

Pollution

Air and water pollution (**Figure 20.5**) contribute to declining populations of hundreds of species worldwide. Pollutants emitted into the atmosphere may be carried aloft for thousands of miles before falling to earth in the form of acid precipitation. The global water cycle can transport pollutants from terrestrial to aquatic ecosystems hundreds of miles away. ☑

▲ **Figure 20.5   A pelican stuck in oil from the 2010 Gulf of Mexico disaster.** Wildlife is often the most visible casualty of pollution, but the impact extends throughout the ecosystem.

☑ **CHECKPOINT**

What are the four main causes of declining biodiversity?

■ *Answer: habitat destruction, invasive species, overexploitation, and pollution*

Community Ecology

On your next walk through a field or woodland, or your own neighborhood, observe the variety of species present. You may see birds in trees, butterflies on flowers, dandelions in the grass of a lawn, or lizards darting for cover as you approach. Each of these organisms interacts with other organisms as it carries out its life activities. An organism's biotic environment includes not just individuals from its own population, but also populations of other species living in the same area. Ecologists call such an assemblage of species living close enough together for potential interaction a **community**. In **Figure 20.6**, the lion, the zebra, the vultures, the plants, and the unseen microbes are all members of an ecological community in Kenya.

Interspecific Interactions

Our study of communities begins with interactions between species. These **interspecific interactions** can be classified according to the net effect on the individuals concerned, which may be helpful (+), harmful (−), or neutral (0). In some cases, two populations in a community compete for a resource such as food or space. The effect of this interaction is generally negative for individuals of both species (−/−) because neither species has access to the full range of resources in the habitat. On the other hand, some interspecific interactions benefit both parties (+/+). For example, the interactions between flowers and their pollinators are mutually beneficial. In a third type of interspecific interaction, one species exploits another species as a source of food. This interaction is beneficial to individuals of one population and harmful to the other (+/−). Sometimes, individuals of one species can benefit from another species without harming or helping the individuals of that other species. A lizard may benefit

▼ **Figure 20.6   Diverse species in a Kenyan savanna community.**

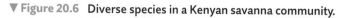

from a tree's shade, but not influence the tree's fitness (+/0). In the next several pages, you will learn more about these interspecific interactions and how they affect communities. You will also see that interspecific interactions can be powerful agents of natural selection.

Interspecific Competition (–/–)

In the logistic model of population growth (see Figure 19.6), increasing population density reduces the resources available for each individual. This intraspecific (within-species) competition for limited resources ultimately limits population growth. In **interspecific competition** (between-species competition), the population growth of a species may be limited by the population densities of competing species as well as by the density of its own population (intraspecific competition).

What determines whether populations in a community compete with each other? Each species has an **ecological niche**, defined as its total use of the biotic and abiotic resources in its environment. For example, the ecological niches of a barnacle in the genus *Chthamalus* **(Figure 20.7a)** includes space on rocks in the intertidal

zone where it lives, the food it filters from the water, and climatic conditions such as the water temperature, air temperature, and humidity that enable it to survive. In other words, the ecological niche encompasses everything this barnacle needs for its existence. The ecological niche of a barnacle in the genus *Balanus* includes some of the same resources used by the *Chthamalus* barnacle, but *Balanus* can't survive as high above the low tide limit. Consequently, when these two species inhabit the same area, they are competitors for resources such as space on rocks near the low tide level.

Ecologist Joseph Connell investigated the effects of interspecific competition between populations of these two barnacles in a community along the Scottish coast. When he removed the *Balanus* barnacles from the study site, *Chthamalus* barnacles spread into the area formerly occupied by *Balanus* barnacles **(Figure 20.7b)**. This study showed that interspecific competition can have a direct, negative effect on fitness by reducing available habitat.

If the ecological niches of two species are too similar, they cannot coexist in the same place. Ecologists call this the **competitive exclusion principle**, a concept introduced by Russian ecologist G. F. Gause, who demonstrated this effect with an elegant series of experiments. Gause used two closely related species of protists, *Paramecium caudatum* and *P. aurelia*. First, he established the carrying capacity for each species separately under the conditions used to grow them in the laboratory **(Figure 20.8**, top graph). Then he grew the two species in the same habitat. Within two weeks, the *P. caudatum* population had crashed (bottom graph). Gause concluded that the requirements of these two species were so similar that the superior competitor—in this case, *P. aurelia*—deprived *P. caudatum* of essential resources. ✓

▼ **Figure 20.7** Similar barnacle species competing for habitat.

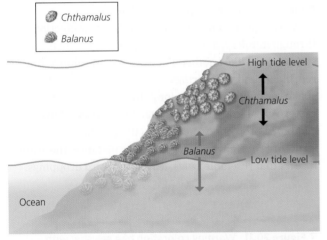

- 🪨 *Chthamalus*
- 🐚 *Balanus*

High tide level

Chthamalus

Balanus

Low tide level

Ocean

(a) *Chthamalus* range is restricted when *Balanus* is present.

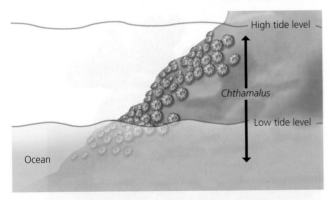

High tide level

Chthamalus

Low tide level

Ocean

(b) *Chthamalus* range expands when *Balanus* is absent.

▼ **Figure 20.8** Competitive exclusion in laboratory populations of *Paramecium*.

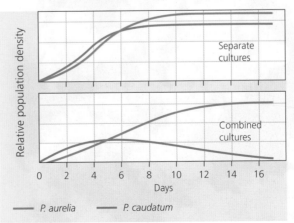

Relative population density

Separate cultures

Combined cultures

0 2 4 6 8 10 12 14 16

Days

— *P. aurelia*   — *P. caudatum*

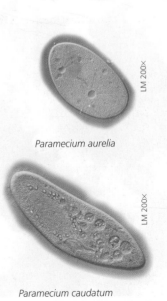

LM 200×

Paramecium aurelia

LM 200×

Paramecium caudatum

Mutualism (+/+)

In **mutualism**, both species benefit from an interaction. Some mutualisms occur between symbiotic species—those in which the organisms have a close physical association with each other. In the symbiotic root-fungus associations known as mycorrhizae (see Figure 16.2), the fungus delivers mineral nutrients to the plant and receives organic nutrients in return. Coral reef ecosystems depend on the symbiotic mutualism between certain species of coral animals and millions of unicellular algae that live in the cells of each coral polyp **(Figure 20.9)**. Reefs are constructed by successive generations of colonial corals that secrete an external calcium carbonate skeleton. The sugars that the algae produce by photosynthesis provide at least half of the energy used by the coral animals. This energy input enables the coral to form new skeleton rapidly enough to outpace erosion and competition for space from fast-growing seaweeds. In return, the algae gain a secure shelter that allows access to light. They also use the coral's waste products, including CO_2 and ammonia, a valuable source of nitrogen. Mutualism can also occur between species that are not symbiotic, such as flowers and their pollinators.

▲ **Figure 20.9 Mutualism.** The cells of coral polyps are inhabited by unicellular algae.

▼ **Figure 20.10 Cryptic coloration.** Camouflage conceals the pygmy seahorse from predators.

Predation (+/−)

Predation refers to an interaction in which one species (the predator) kills and eats another (the prey). Because predation has such a negative impact on the reproductive success of the prey, numerous adaptations for predator avoidance have evolved in prey populations through natural selection. For example, some prey species, such as the pronghorn antelope, run fast enough to escape their predators (see the Evolution Connection section in Chapter 19). Others, such as rabbits, flee into shelters. Still other prey species rely on mechanical structures for protection. The porcupine's sharp quills and the hard shell of the clam function as defenses.

Adaptive coloration is a type of defense that has evolved in many species of animals. Camouflage, called **cryptic coloration**, makes potential prey difficult to spot against its background **(Figure 20.10)**. **Warning coloration**, bright patterns of yellow, red, or orange in combination with black, often marks animals with effective chemical defenses. Predators learn to associate these color patterns with undesirable consequences, such as a noxious taste or painful sting, and avoid potential prey with similar markings. Learned responses are not inherited; each individual predator must learn them, so signals must be clear. The vivid colors of some sea slugs from the southern Pacific and Indian Oceans **(Figure 20.11)**, warn of their ability to sting or poison predators.

A prey species may also gain significant protection through mimicry, a "copycat" adaptation in which one species looks like another. For example, the alternating red, black, and yellow rings of the harmless scarlet king snake resemble the bold color pattern of the venomous eastern coral snake **(Figure 20.12)**. Some insects have combined protective coloration with adaptations of body structures in elaborate disguises. For instance, there are insects that resemble twigs, leaves, and bird droppings.

▼ **Figure 20.11 Warning coloration of a sea slug with chemical defenses.**

▲ **Figure 20.12 Mimicry in snakes.** The color pattern of the nonvenomous scarlet king snake (left) is similar to that of the venomous eastern coral snake (right).

Some even do a passable imitation of a vertebrate. For example, the colors on the dorsal side of certain caterpillars are an effective camouflage, but when disturbed, the caterpillars flip over to reveal the snakelike eyespots of their ventral side (**Figure 20.13.**). Eyespots that resemble vertebrate eyes are common in several groups of moths and butterflies. A flash of these large "eyes" startles would-be predators. In other species, an eyespot may deflect a predator's attack away from vital body parts.

Herbivory (+/–)

Herbivory is the consumption of plant parts or algae by an animal. Although herbivory is not usually fatal to plants, a plant whose body parts have been partially eaten by an animal must expend energy to replace the loss. Consequently, numerous defenses against herbivores have evolved in plants. Thorns and spines are obvious anti-herbivore devices, as anyone who has plucked a rose from a thorny rosebush or brushed against a spiky cactus knows. Chemical toxins are also very common in plants. Like the chemical defenses of animals, toxins in plants are distasteful, and

herbivores learn to avoid them. Among such chemical weapons are the poison strychnine, produced by a tropical vine called *Strychnos toxifera*; morphine, from the opium poppy; nicotine, produced by the tobacco plant; mescaline, from peyote cactus; and tannins, from a variety of plant species. Other defensive compounds that are not toxic but may be distasteful are responsible for the familiar flavors of peppermint, cloves, and cinnamon (**Figure 20.14**). Some plants even produce chemicals that cause abnormal development in insects that eat them. Neem trees produce chemicals that mimic developmental hormones, leaving insects with curled wings and twisted abdomens. Chemical companies have taken advantage of the poisonous properties of certain plants to produce pesticides. For example, nicotine is used as an insecticide. ☑

▼ **Figure 20.14 Flavorful plants.**

Peppermint. Parts of the peppermint plant yield a pungent oil.

Cloves. The cloves used in cooking are the flower buds of this plant.

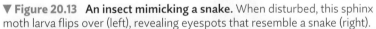

▼ **Figure 20.13 An insect mimicking a snake.** When disturbed, this sphinx moth larva flips over (left), revealing eyespots that resemble a snake (right).

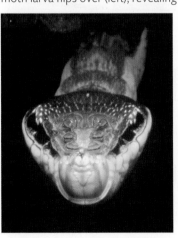

Cinnamon. Cinnamon comes from the inner bark of this tree.

Parasites and Pathogens (+/−)

Both plants and animals may be victimized by parasites or pathogens. These interactions are beneficial to one species (the parasite or pathogen) and harmful to the other, known as the host. A **parasite** lives on or in a **host** from which it obtains nourishment. Invertebrate parasites include tapeworms and a variety of roundworms, which live inside a host's body. External parasites, such as ticks, lice, and mosquitoes, attach to their victims temporarily to feed on blood or other body fluids (see the Process of Science section). Plants are also attacked by parasites, including roundworms and aphids, tiny insects that tap into the phloem to suck plant sap (see Figure 19.10). In any parasite population, reproductive success is greatest for individuals that are best at locating and feeding on their hosts. For example, some aquatic leeches first locate a host by detecting movement in the water and then confirm its identity based on the host's body temperature and chemical cues on its skin.

Pathogens are disease-causing bacteria, viruses, fungi, or protists that can be thought of as microscopic parasites. Non-native pathogens have provided opportunities to investigate the effects of pathogens on communities. For example, ecologists studied the consequences of the epidemic in North America of chestnut blight, a disease caused by a protist from Asia. The loss of chestnuts, massive canopy trees that once dominated many forest communities in North America, had a significant impact. Trees such as oaks and hickories that had formerly competed with chestnuts became more numerous; overall, the diversity of tree species increased. Dead chestnut trees also furnished niches for other organisms, such as insects, cavity-nesting birds, and, eventually, decomposers. ✓

Commensalism (+/0)

Commensalism is an interaction in which the individual of one species benefits and the individual of the other species is neither helped nor harmed. For example, the sea horse in Figure 20.10 benefits by matching the coral to hide from predators, but there is no cost or benefit to the coral to have the sea horse near it.

Trophic Structure

Now that we have looked at how populations in a community interact with one another, let's consider the community as a whole. The feeding relationships among the various species in a community are referred to as its **trophic structure**. A community's trophic structure determines the pathways that transform energy and matter as they move from photosynthetic organisms to herbivores and then to predators.

☑ **CHECKPOINT**

How is the interaction between a parasite and its host similar to predator-prey and herbivore-plant interactions?

■ *Answer: In all three interactions, individuals of one species benefit and individuals of the other species are harmed (+/−).*

The sequence of food transfer between trophic levels is called a **food chain**.

Figure 20.15 shows two food chains, one terrestrial and one aquatic. At the bottom of both chains is the trophic level that supports all others. This level consists of autotrophs, which ecologists call **producers** (see Figure 6.1). Photosynthetic producers transform light energy to chemical energy stored in the bonds of organic compounds like sugar. Plants are the main producers on land. In water, the main producers are photosynthetic protists and cyanobacteria, collectively called phytoplankton. Multicellular algae and aquatic plants are also important producers in shallow waters. In a few communities, such as those surrounding hydrothermal vents, the producers are prokaryotes that get their energy from inorganic chemicals.

Organisms in trophic levels above the producers are called consumers because they consume other organisms. **Herbivores** eat producers, such as plants and algae. Herbivores are called **primary consumers** because they are the first level of consumers on the food chain. Primary

▼ **Figure 20.15 Examples of food chains.** The arrows trace the transfer of food from one trophic level to the next in terrestrial and aquatic communities.

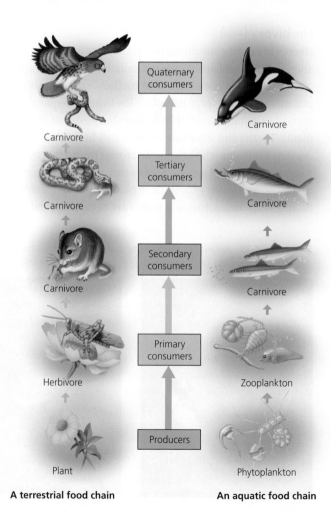

A terrestrial food chain   An aquatic food chain

▲ **Figure 20.16 Compost bin in a garden.** As it decays, the rich organic material of compost provides a slow-release source of inorganic nutrients for plants.

consumers on land include many insects, snails, and certain vertebrates, such as grazing mammals and birds that eat seeds and fruits. In aquatic environments, primary consumers include a variety of zooplankton (mainly protists and microscopic animals such as small shrimps) that eat phytoplankton.

Above the primary consumers, the trophic levels are made up of **carnivores**, which eat the consumers from the level below. On land, **secondary consumers** include many small mammals, such as the mouse shown in Figure 20.15 eating an herbivorous insect, and a great variety of birds, frogs, and spiders, as well as lions and other large carnivores that eat grazers. In aquatic communities, secondary consumers are mainly small fishes that eat zooplankton. Higher trophic levels include **tertiary consumers** (third-level consumers), such as snakes that eat mice and other secondary consumers. Most communities have secondary and tertiary consumers. As the figure indicates, some also have a higher level—**quaternary consumers** (fourth-level consumers). These include hawks in terrestrial communities and killer whales in the marine environment. High-level consumers are rare—you'll find out why shortly.

Figure 20.15 shows only those consumers that eat living organisms. Some consumers derive their energy from **detritus**, the dead material left by all trophic levels, including animal wastes, plant litter, and dead bodies. Different organisms consume detritus in different stages of decay. We refer to those organisms that consume detritus in its final stages of breakdown as **decomposers**. Enormous numbers of decomposers, mainly prokaryotes and fungi, live in soil or the mud at the bottom of lakes and oceans, where they secrete enzymes that digest molecules in organic material and convert them to inorganic forms. These inorganic nutrients are the raw materials used by plants and phytoplankton to make new organic materials that may eventually become food for consumers. Many gardeners keep a compost pile, using the services of decomposers to break down organic material from kitchen scraps and yard trimmings (Figure 20.16).

Biological Magnification

Organisms can't metabolize many of the toxins produced by industrial wastes or applied as pesticides; after consumption, the chemicals remain in the body. These toxins become concentrated as they pass through a food chain, a process called **biological magnification**.

Figure 20.17 shows the biological magnification of chemicals called PCBs (organic compounds used in electrical equipment until 1977) in an Arctic food chain, as zooplankton—the base of the pyramid—feed on phytoplankton contaminated by PCBs in the water. Cod feed on contaminated zooplankton. Because each cod consumes many zooplankton, the concentration of PCBs is higher in cod than in zooplankton. For the same reason, the concentration of PCBs in salmon is higher than in cod. The top-level predators—humans in this example—have the highest concentrations of PCBs in the food chain and are the organisms most severely affected by any toxic compounds in the environment. Because PCBs are fat-soluble, they can be concentrated in milk. Many other synthetic chemicals that cannot be degraded by microorganisms, including DDT (a chemical used in insecticides) and mercury, can also become concentrated through biological magnification.

TUNA MAY BE CONTAMINATED WITH MERCURY AND OTHER TOXINS.

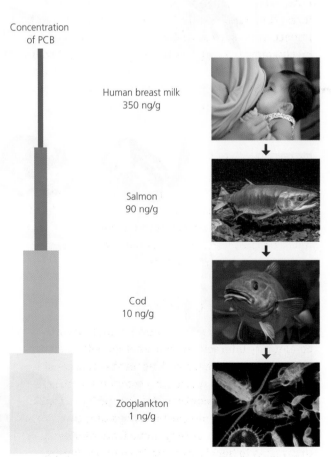

Concentration of PCB

Human breast milk
350 ng/g

Salmon
90 ng/g

Cod
10 ng/g

Zooplankton
1 ng/g

◀ **Figure 20.17 Biological magnification of PCBs in an Arctic food chain.** Eating top-level consumers, such as salmon from contaminated waters, can cause health problems for people, especially for pregnant women. (Note: ng/g = nanograms PCB per gram of fatty tissue.)

Food Webs

Few, if any, communities are so simple that they are
characterized by a single, unbranched food chain. Several types of primary consumers usually feed on the
same plant species, and one species of primary consumer
may eat several different plants. Such branching of food
chains occurs at the other trophic levels as well. Thus,
the feeding relationships in a community are woven into
elaborate **food webs**.

Consider the grasshopper mouse in **Figure 20.18**, the
secondary consumer shown crunching on a grasshopper
in Figure 20.15. Its diet also includes plants, making it a
primary consumer, too. It is an **omnivore**, an animal that
eats producers as well as consumers of different levels.
The rattlesnake that eats the mouse also feeds on more
than one trophic level. The blue arrow leading to the
rattlesnake indicates that it eats primary consumers—the
rattlesnake is a secondary consumer. The purple arrow
shows that the rattlesnake eats secondary consumers,
too, so it is also a tertiary consumer. Sound complicated?
Actually, Figure 20.18 shows a simplified food web. An
actual food web would involve many more organisms at
each trophic level, and most of the animals would have a
more diverse diet than shown in the figure. ✓

▶ **Figure 20.18   A simplified
food web for a Sonoran
desert community.** As in the
food chains of Figure 20.15,
the arrows in this web indicate
"who eats whom," the direction of nutrient transfers. We
also continue the color-coding
introduced in Figure 20.15 for
the trophic levels and food
transfers.

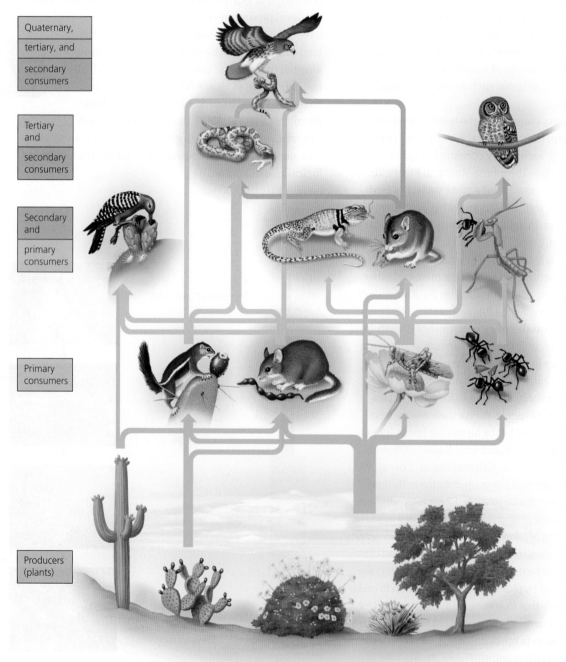

Quaternary,
tertiary, and
secondary
consumers

Tertiary
and
secondary
consumers

Secondary
and
primary
consumers

Primary
consumers

Producers
(plants)

Species Diversity in Communities

The **species diversity** of a community—the variety of species that make up the community—has two components. The first component is **species richness**, or the number of different species in the community. The other component is the **relative abundance** of the different species, the proportional representation of each species in a community. To understand why both components are important for describing species diversity, imagine walking through the woodlands shown in **Figure 20.19**. On the path through woodland A, you would pass by four different species of trees, but most of the trees you encounter would be the same species. Now imagine walking on a path through woodland B. You would see the same four species of trees that you saw in woodland A—the species richness of the two woodlands is the same. However, woodland B might seem more diverse to you because no single species predominates. As **Figure 20.20** shows, the relative abundance of one species in woodland A is much higher than the relative abundances of the other three species. In woodland B, all four species are equally abundant. As a result, species diversity is greater in woodland B. Because plants provide food and shelter for many animals, a diverse plant community promotes animal diversity.

Although the abundance of a dominant species such as a forest tree can have an impact on the diversity of other species in the community, a nondominant species may also exert control over community composition. A **keystone species** is a species whose impact on its community is much larger than its total mass or abundance indicates. The term *keystone species* was derived from the wedge-shaped stone at the top of an arch that locks the other pieces in place. If the keystone is removed, the arch collapses. A keystone species occupies an ecological niche that holds the rest of its community in place.

To investigate the role of a potential keystone species in a community, ecologists compare diversity when the species is present to that when it is absent. Experiments by Robert Paine in the 1960s were among the first to provide evidence of the keystone species effect. Paine manually removed a predator, a sea star of the genus *Pisaster* (**Figure 20.21**), from experimental areas within the intertidal zone of the Washington coast. The result was that *Pisaster*'s main prey, a mussel, outcompeted many of the other shoreline organisms (algae, barnacles, and snails, for instance) for the important resource of space on the

Woodland A

Woodland B

▲ **Figure 20.19** Which woodland is more diverse?

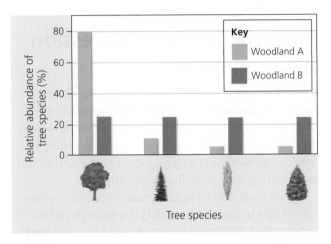

◀**Figure 20.20** Relative abundance of tree species in woodlands A and B.

 VISUALIZING THE DATA

▼ **Figure 20.21** Communities before and after the removal of *Pisaster* sea star.

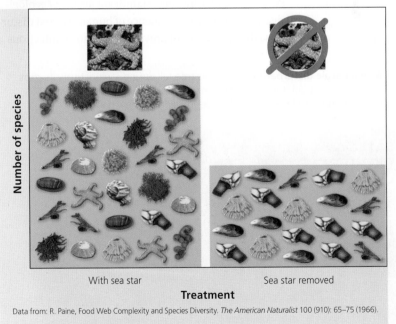

Data from: R. Paine, Food Web Complexity and Species Diversity. *The American Naturalist* 100 (910): 65–75 (1966).

rocks. The number of different organisms present in experimental areas dropped from more than 15 species to fewer than 8 species.

Ecologists have identified other species that play a key role in ecosystem structure. For instance, the decline of sea otters off the western coast of Alaska allowed populations of sea urchins, their main prey, to increase. The greater abundance of sea urchins, which consume seaweeds such as kelp, has resulted in the loss of many of the kelp "forests" (see Figure 15.25) and the diversity of marine life that they support. In many ecosystems, however, ecologists are just beginning to understand the complex relationships among species; the value of an individual species may not be apparent until it is gone. We will further explore the value of biodiversity in the Process of Science section. ☑

Disturbances and Succession in Communities

Most communities are constantly changed by disturbances. **Disturbances** are episodes that alter biological communities, at least temporarily, by destroying organisms and changing the availability of resources such as mineral nutrients and water. Examples of natural disturbances are storms, fires, floods, and droughts.

Small-scale natural disturbances often have positive effects on a biological community. For example, when a large tree falls in a windstorm, it creates new habitats (Figure 20.22). More light may now reach the forest floor, giving small seedlings the chance to grow. The depression left by the tree's roots may fill with water and be used as egg-laying sites by frogs, salamanders, and insects.

People are by far the most significant agents of ecological disturbance. One consequence of human-caused disturbance is the emergence of previously unknown infectious diseases. Three-quarters of emerging diseases have jumped to humans from another vertebrate species. In many cases, people come into contact with these pathogens through activities such as clearing land for agriculture, building roads, or hunting in previously isolated ecosystems. Habitat destruction may also cause pathogen-carrying animals to venture closer to human dwellings in search of food. ☑

Ecological Succession

After a disturbance strips away vegetation, the area may be colonized by a variety of species, which are gradually replaced by other species, in a process called **ecological succession**. Ecological succession that begins in an area with no soil is called **primary succession** (Figure 20.23). Examples include cooled lava flows and the rubble left by a melting glacier. Often the only life forms initially present are autotrophic bacteria. Lichens and mosses, which grow from windblown spores, are commonly the first multicellular producers to colonize the area. Soil can be deposited by wind and water or develop gradually as rocks weather and organic matter accumulates from the decomposing remains of the early colonizers. Lichens and mosses are eventually overgrown by weedy plants that are hardy and grow fast. These pioneers are replaced by competitors, such as grasses and shrubs, that sprout from seeds blown in from nearby areas or carried in by animals. These later successional species store energy and nutrients underground, giving them a competitive

☑ CHECKPOINT

How could a community appear to have relatively little diversity even though it is rich in species?

Answer: if one or a few of the diverse species accounted for almost all the organisms in the community, with the other species being rare

☑ CHECKPOINT

Why might the effect of a small-scale natural disturbance, such as a forest tree felled by wind, be considered positive?

Answer: A fallen tree alters abiotic factors such as water and sunlight in a small patch of the forest and provides new habitats for other organisms.

▼ Figure 20.23 **Primary succession underway on a lava flow in Volcanoes National Park, Hawaii.**

▼ Figure 20.22 **A small-scale disturbance.** When this tree fell during a windstorm, its root system and the surrounding soil uplifted, resulting in a depression that filled with water. The dead tree, the root mound, and the water-filled depression are new habitats.

advantage over weedy plants that grow from seeds every year. Finally, the area is colonized by plants that become the community's dominant form of vegetation. Primary succession can take hundreds or thousands of years.

In **secondary succession**, the soil remains after a disturbance. For example, fires lead to secondary succession (Figure 20.24). Human activities can also lead to secondary succession. In the tropics, people clear rainforests through slash-and-burn farming for agriculture and settlements. Once the soil is depleted of its chemical nutrients, the site is abandoned and the residents move to a new clearing. Whenever human intervention stops, secondary succession begins. ☑

▼ Figure 20.24 Secondary succession after a fire.

What is the main abiotic factor that distinguishes primary from secondary succession?

■ *Answer: absence of soil (primary succession) versus presence of soil (secondary succession) at the onset of succession*

Ecosystem Ecology

In addition to the community of species in a given area, an **ecosystem** includes all the abiotic factors, such as energy, soil, atmosphere, and water. Let's look at a small-scale ecosystem—a terrarium—to see how the community interacts with these abiotic factors (Figure 20.25). A terrarium microcosm demonstrates that matter, including water and nutrients, is recycled in an ecosystem, while energy dissipates as it flows through the trophic levels (Figure 5.2).

Energy flow is the passage of energy through the components of the ecosystem. Energy enters the terrarium in the form of sunlight (yellow arrows). Plants (producers) convert light energy to chemical energy through the process of photosynthesis. Animals (consumers) take in some of this chemical energy in the form of organic compounds when they eat plants. Decomposers in the soil obtain chemical energy when they feed on the wastes and remains of other organisms. Every use of chemical energy involves a loss of some energy to the surroundings in the form of heat (red arrows). Energy is not destroyed, but it dissipates, so that it is not available to do work. Imagine that your backpack was leaking pennies as you walked around campus. The pennies wouldn't disappear, but gathering them up would be more energy than they're worth. Because so much of the energy captured by photosynthesis is lost as heat, this ecosystem would run out of energy if it were not powered by a continuous inflow of energy from the sun.

Chemical cycling is the use and reuse of chemical elements such as carbon and nitrogen within the ecosystem. Chemical cycling (blue arrows in Figure 20.25) involves the transfer of matter within the ecosystem. Chemical elements are not created or destroyed. They are cycled between the abiotic components of the ecosystem (including air, water, and soil) and the biotic components of the ecosystem (the community). Plants acquire these

elements in inorganic form from the air and soil and use them to construct organic molecules. Animals, such as the snail in Figure 20.25, consume some of these organic molecules. Decomposers feeding on dead plant and animal tissue produce CO_2 and mineral wastes, returning most of the elements to the soil and air in inorganic form. Some elements are also returned to the air and soil as the by-products of plant and animal metabolism. If nutrients are lost due to erosion or because crops are removed from where they are produced, nutrients may become limited.

In summary, energy flow and chemical cycling involve the transfer of energy and matter through the trophic levels of the ecosystem. However, energy flows through, and ultimately out of, ecosystems, whereas chemicals are recycled within and between ecosystems.

▼ Figure 20.25 **A terrarium ecosystem.** Though it is small and artificial, this sealed terrarium illustrates the two major ecosystem processes: energy flow and chemical cycling.

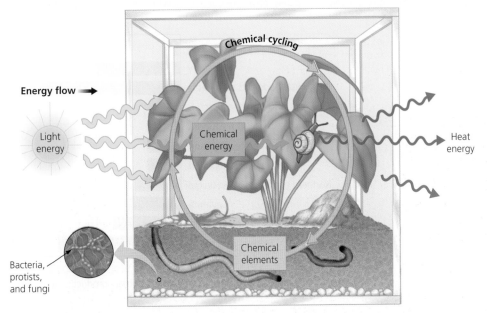

Chemical cycling

Energy flow ➡

Light energy

Chemical energy

Heat energy

Chemical elements

Bacteria, protists, and fungi

Energy Flow in Ecosystems

All organisms require energy for growth, maintenance, reproduction, and, in many species, locomotion. In this section, we take a closer look at energy flow through ecosystems. As we do, consider why there are so few top-level carnivores and how we can apply energy flow concepts to use our resources efficiently.

Primary Production and the Energy Flow in Ecosystems

Each day, Earth receives about 10^{19} kcal of solar energy, the equivalent of about 100 million atomic bombs. Most of this energy is absorbed, scattered, or reflected by the atmosphere or by Earth's surface. Of the visible light that reaches producers, only about 1% is converted to chemical energy by photosynthesis.

Ecologists call the amount, or mass, of living organic material in an ecosystem the **biomass**. The rate at which an ecosystem's producers convert solar energy to the chemical energy stored in organic compounds is called **primary production**. The primary production of the entire biosphere is roughly 165 billion tons of biomass per year.

Different ecosystems vary considerably in their primary production **(Figure 20.26)** as well as in their contribution to the total production of the biosphere. Primary production on land is mostly limited by sunlight, water, and nutrients. Primary production in water is limited mostly by sunlight and nutrients. Tropical rain forests are among the most

productive terrestrial ecosystems and contribute a large portion of the planet's overall production of biomass. Coral reefs also have very high production, but their contribution to global production is small because they cover such a small area. Interestingly, even though the open ocean has very low production, it contributes the most to Earth's total primary production because of its huge size—it covers 65% of Earth's surface area. Whatever the ecosystem, primary production sets the limit of the energy available for the entire ecosystem because consumers must acquire their organic fuels from producers. Now let's see how this energy budget is divided among the different trophic levels in an ecosystem's food web. ☑

Ecological Pyramids

When energy flows through the trophic levels of an ecosystem, much of it is lost at each link in the food chain. Consider the transfer of organic matter from producers to primary consumers. In most ecosystems, herbivores eat only a fraction of the plant material produced, and they can't digest all of what they do consume. For example, a caterpillar feeding on leaves will defecate undigested material as feces, which still contain energy. **(Figure 20.27)**. Another 35% of the energy is expended in cellular respiration (Figure 5.2), which powers the caterpillar's metabolism. Only about 15% of the energy in the caterpillar's food is transformed into caterpillar biomass. Only this biomass (and the energy it contains) is available to the consumer that eats the caterpillar.

▼ **Figure 20.26 Primary production of different ecosystems.** Primary production is the amount of biomass created by the producers of an ecosystem over a unit of time, in this case a year. Aquatic ecosystems are color-coded blue in these histograms; terrestrial ecosystems are green.

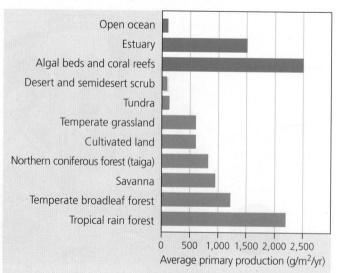

▼ **Figure 20.27 What becomes of a caterpillar's food?** Only about 15% of the calories of plant material this herbivore consumes will be stored as biomass available to the next link in the food chain.

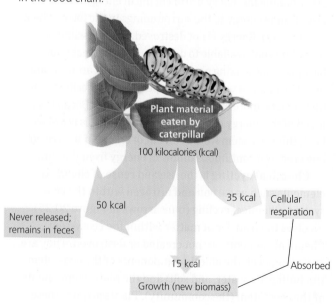

▼ **Figure 20.28** An idealized energy pyramid.

As energy is transferred, some is lost as heat.

10 kcal

100 kcal

1,000 kcal

10,000 kcal

1,000,000 kcal of sunlight

Tertiary consumers

Secondary consumers

Primary consumers

Producers

Figure 20.28, called an **energy pyramid**, illustrates the cumulative loss of energy with each transfer in a food chain. Each tier of the pyramid represents all of the organisms in one trophic level, and the width of each tier indicates how much of the chemical energy of the tier below is actually incorporated into the organic matter of that trophic level. Note that producers convert only about 1% of the energy in the sunlight available to them to primary production. In this generalized pyramid, 10% of the energy available at each trophic level becomes incorporated into the next higher level. Actual efficiencies of energy transfer are usually in the range of 5–20%. In other words, 80–95% of the energy at one trophic level never reaches the next.

An important implication of this stepwise decline of energy in a trophic structure is that the amount of energy available to top-level consumers is small compared with that available to lower-level consumers. Only a tiny fraction of the energy stored by photosynthesis flows through a food chain to a tertiary consumer, such as a snake feeding on a mouse. This explains why top-level consumers such as lions and hawks require so much geographic territory: It takes a lot of vegetation to support trophic levels so many steps removed from photosynthetic production. You can also understand why most food chains are limited to three to five levels; there is simply not enough energy at the very top of an ecological pyramid to support another trophic level. There are, for example, no nonhuman predators of lions, eagles, and killer whales; the biomass in populations of

these top-level consumers is insufficient to supply yet another trophic level with a reliable source of nutrition.

Ecosystem Energetics and Human Resource Use

The dynamics of energy flow apply to the human population as much as to other organisms. The two energy pyramids in **Figure 20.29** are based on the same generalized model used to construct Figure 20.28, with roughly 10% of the energy in each trophic level available for consumption by the next trophic level. The pyramid on the left shows energy flow from producers (represented by corn) to people as primary consumers—vegetarians. The pyramid on the right illustrates energy flow from the same corn crop, with people as secondary consumers, eating beef. Clearly, the human population has less energy available to it when people eat at higher trophic levels than as primary consumers.

Worldwide, only about 20% of agricultural land is used to produce plants for direct human consumption. The rest of the land produces food for livestock. In either case, large-scale agriculture is environmentally expensive. Land is cleared of its native vegetation, fossil fuels are burned, chemical fertilizers and pesticides are applied, and in many regions, water is used for irrigation. Currently, people in many countries cannot afford to buy meat and are vegetarians by necessity. As nations become more affluent, the demand for meat increases—and so do the environmental costs of food production. ☑

RAISING BEEF REQUIRES EIGHT TIMES AS MUCH LAND AS RAISING SOYBEANS.

☑ **CHECKPOINT**

Why is the energy at different trophic levels represented as a pyramid in Figure 20.28?

■ Answer: Energy is lost with each transfer from one trophic level to the next. Thus, the collective amount of energy at the producer level (the base of the pyramid) is greater than the collective amount of energy at the level of primary consumers, and so on.

▼ **Figure 20.29** Food energy available to the human population at different trophic levels.

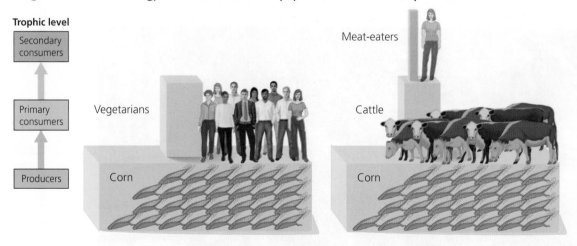

Trophic level

Secondary consumers

Primary consumers

Producers

Vegetarians

Corn

Meat-eaters

Cattle

Corn

Chemical Cycling in Ecosystems

The sun (or, in some cases, Earth's interior) supplies ecosystems with a continual input of energy, but aside from an occasional meteorite, there are no extraterrestrial sources of chemical elements. Life, therefore, depends on the recycling of chemicals. While an organism is alive, much of its chemical stock changes continuously, as nutrients are acquired and waste products are released. Atoms present in the complex molecules of an organism at the time of its death are returned to the environment by the action of decomposers, replenishing the pool of inorganic nutrients that plants and other producers use to build new organic matter (**Figure 20.30**). In a sense, each living thing only borrows an ecosystem's chemical elements, returning what is left in its body after it dies. Let's take a closer look at how chemicals cycle between organisms and the abiotic components of ecosystems.

The General Scheme of Chemical Cycling

Because chemical cycles in an ecosystem involve both biotic components (organisms and nonliving organic material) and abiotic (geologic and atmospheric) components, they are called **biogeochemical cycles**. **Figure 20.31** is a general scheme for the cycling of a mineral nutrient within an ecosystem. Note that the cycle has an **abiotic reservoir** (white box) where a chemical accumulates outside of living organisms. The atmosphere, for example, is an abiotic reservoir for carbon. The water of aquatic ecosystems contains dissolved carbon, nitrogen, and phosphorus compounds.

Let's trace our way around our general biogeochemical cycle. **1** Producers incorporate chemicals from the abiotic reservoir into organic compounds. **2** Consumers feed on the producers, incorporating some of the chemicals into their own bodies. **3** Both producers and consumers release some chemicals back to the environment in waste products. **4** Decomposers play a central role by breaking down the complex organic molecules in detritus such as plant litter, animal wastes, and dead organisms. The products of this metabolism are inorganic molecules that replenish the abiotic reservoirs. Geologic processes such as erosion and the weathering of rock also contribute to the abiotic reservoirs. Producers use the inorganic molecules from abiotic reservoirs as raw materials for synthesizing new organic molecules (carbohydrates and proteins, for example), and the cycle continues.

Biogeochemical cycles can be local or global. Soil is the main reservoir for nutrients in a local cycle, such as phosphorus. In contrast, for those chemicals that spend part of their time in gaseous form—carbon and nitrogen are examples—the cycling is fueled by atmospheric processes, and the atmospheric reservoir is essentially global. For instance, some of the carbon a plant acquires from the air may have been released into the atmosphere by the respiration of a plant or animal on another continent.

Now let's examine three important biogeochemical cycles more closely: the cycles for carbon, phosphorus, and nitrogen. As you study the cycles, look for the four basic steps we described, as well as the geologic and atmospheric processes that move chemicals around and between ecosystems. In all the diagrams, the main abiotic reservoirs appear in white boxes. ☑

The Carbon Cycle

Carbon, the major ingredient of all organic molecules, has an atmospheric reservoir and cycles globally. Other abiotic

☑ **CHECKPOINT**

What is the role of abiotic reservoirs in biogeochemical cycles? What are the three major abiotic reservoirs?

■ *Answer: Abiotic reservoirs are sources of the nutrients used by producers for synthesizing organic molecules. The atmosphere, soil, and water are the three major abiotic reservoirs.*

▼ **Figure 20.30 Plant growth on fallen tree.** In the temperate rain forest of Olympic National Park, in Washington, plants—including other trees—quickly take advantage of the mineral nutrients supplied by decomposing "nurse logs."

▼ **Figure 20.31 General scheme for biogeochemical cycles.**

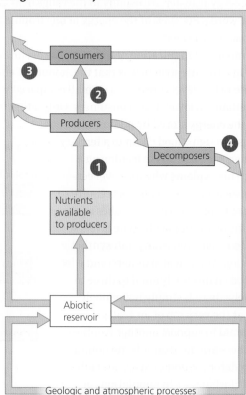

reservoirs of carbon include fossil fuels and dissolved carbon compounds in the oceans. The transformation of carbon dioxide (plus water) to sugar (plus oxygen) and back in photosynthesis and cellular respiration (see Figure 6.2) is mainly responsible for the cycling of carbon between the biotic and abiotic worlds (**Figure 20.32**). **1** Photosynthesis removes CO_2 from the atmosphere and incorporates it into organic molecules, which are **2** passed along the food chain by consumers. **3** Cellular respiration returns CO_2 to the atmosphere. **4** Decomposers break down the carbon compounds in detritus; that carbon, too, is eventually released as CO_2. On a global scale, the return of CO_2 to the atmosphere by respiration closely balances its removal by photosynthesis. However, increasing levels of CO_2 caused by **5** the burning of wood and fossil fuels (coal and petroleum) are contributing to climate change (see Figure 18.46). ☑

The Phosphorus Cycle

Organisms require phosphorus as an ingredient of nucleic acids, phospholipids, and ATP and (in vertebrates) as a mineral component of bones and teeth. In contrast to the carbon cycle and the other major biogeochemical cycles, the phosphorus cycle does not have an atmospheric component. Phosphorus for terrestrial ecosystems can come from the soil, water, or out of rocks; in fact, rocks that have high phosphorus content are mined for fertilizer.

At the center of **Figure 20.33**, **1** the weathering (breakdown) of rock gradually adds inorganic phosphate (PO_4^{3-}) to the soil. **2** Plants absorb dissolved phosphate from the soil and assimilate it by building the phosphorus atoms into organic compounds. **3** Consumers obtain phosphorus in organic form by eating plants. **4** Phosphates are returned to the soil by the action of decomposers on animal waste and the remains of dead plants and animals. **5** Some of the phosphates drain from terrestrial ecosystems into the sea, where they may settle and eventually become part of new rocks. Phosphorus removed from the cycle in this way will not be available to living organisms until **6** geologic processes uplift the rocks and expose them to weathering.

Phosphates move from land to aquatic ecosystems much more rapidly than they are replaced, and soil characteristics may also decrease the amount of phosphate available to plants. As a result, phosphate is a limiting factor in many terrestrial ecosystems. Farmers and gardeners often use phosphate fertilizer, such as crushed phosphate rock or bone meal (finely ground bones from slaughtered livestock or fish), to boost plant growth.

▼ Figure 20.32 **The carbon cycle.**

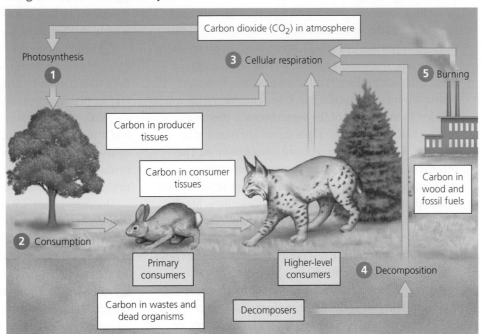

Figure Walkthrough
Mastering **Biology**
goo.gl/Hg88JN.qr

▼ Figure 20.33 **The phosphorus cycle.**

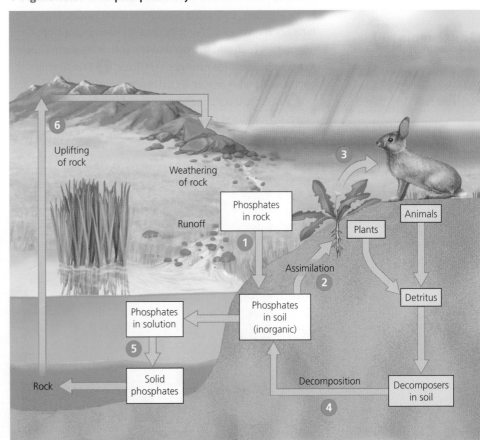

☑ **CHECKPOINT**

How does carbon get into terrestrial plants?

Answer: CO_2 from the atmosphere enters the leaves.

The Nitrogen Cycle

As an ingredient of proteins and nucleic acids, nitrogen is essential to the structure and functioning of all organisms. Nitrogen has two abiotic reservoirs, the atmosphere and the soil. The atmospheric reservoir is huge; almost 80% of the atmosphere is nitrogen gas (N_2). However, plants cannot use nitrogen gas. Like someone dying of thirst in the middle of the ocean because they can't remove salt from the water, plants are surrounded by nitrogen, a critical resource, in an unusable form. The process of **nitrogen fixation** converts gaseous N_2 to ammonia (NH_3). NH_3 then picks up another H^+ to become ammonium (NH_4^+), which plants can assimilate. Most of the nitrogen available in natural ecosystems comes from biological fixation performed by certain bacteria. Without these organisms, the reservoir of usable soil nitrogen would be extremely limited.

Figure 20.34 illustrates the actions of two types of nitrogen-fixing bacteria. ❶ Some bacteria live symbiotically in the roots of certain species of plants, supplying their hosts with a direct source of usable nitrogen. The largest group of plants with this mutualistic relationship is the legumes, including peanuts and soybeans. Many farmers improve soil fertility by alternating crops of legumes, which add nitrogen to the soil, with plants such as corn that require nitrogen fertilizer. ❷ Free-living bacteria in soil or water fix nitrogen, resulting in NH_4^+.

❸ After nitrogen is fixed, some of the ammonium is taken up and used by plants. ❹ Nitrifying bacteria in the soil also convert some of the ammonium to nitrate (NO_3^-),

☑ **CHECKPOINT**

What are the abiotic reservoirs of nitrogen? In what form does nitrogen occur in each reservoir?

■ *Answer: atmosphere: N_2; soil:*
NH_4^+ and NO_3^-

❺ which is more readily acquired by plants. Plants use this nitrogen to make molecules such as amino acids, which are then incorporated into proteins.

❻ When an herbivore (represented here by a rabbit) eats a plant, it digests the proteins into amino acids and then uses the amino acids to build the proteins it needs. Higher-order consumers get nitrogen from the organic molecules of their prey. Because animals form nitrogen-containing waste products during protein metabolism, consumers excrete some nitrogen into soil or water. The urine that rabbits and other mammals excrete contains urea, a nitrogen compound that is widely used as fertilizer.

Organisms that are not consumed eventually die and become detritus, which is decomposed by bacteria and fungi. ❼ The decomposition of organic compounds releases ammonium into the soil, replenishing that abiotic reservoir. Under low-oxygen conditions, however, ❽ soil bacteria known as denitrifying bacteria strip the oxygen atoms from nitrates, releasing N_2 back into the atmosphere and depleting the soil of usable nitrogen.

Human activities are disrupting the nitrogen cycle by adding extra nitrogen to the biosphere. Combustion of fossil fuels and modern agricultural practices are two major sources of nitrogen. Many farmers apply enormous amounts of synthetic nitrogen fertilizer. However, less than half that fertilizer is actually used by the crop plants. Some nitrogen escapes to the atmosphere, where it forms nitrous oxide (N_2O), a gas that contributes to global warming. And as you'll learn next, nitrogen fertilizers also pollute aquatic systems. ☑

▼ **Figure 20.34**   **The nitrogen cycle.**

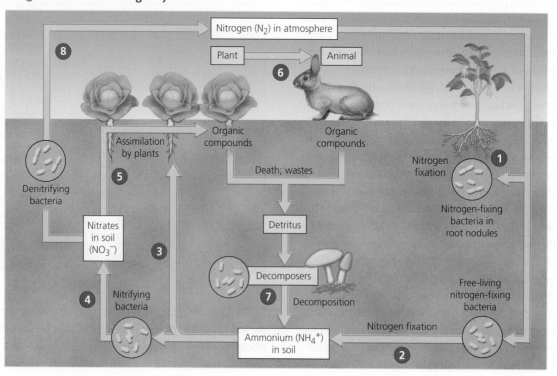

Nutrient Pollution

Low nutrient levels, especially of phosphorus and nitrogen, often limit the growth of producers, such as algae and cyanobacteria in aquatic ecosystems. Nutrient pollution occurs when human activities add excess amounts of these chemicals to aquatic ecosystems, increasing populations of producers and disrupting the ecosystem in which they live.

In many areas, phosphate and nitrogen pollution comes from the large amount of inorganic fertilizers routinely applied to crops, lawns, and golf courses and runoff of animal waste from livestock feedlots (where hundreds of animals are penned together). Phosphates are also a common ingredient in dishwasher detergents, making the outflow from sewage treatment facilities—which also contains phosphorus from human waste—a major source of phosphate pollution. Pollution may also come from sewage treatment facilities when extreme conditions (such as unusual storms) or malfunctioning equipment prevents them from meeting water quality standards. Nutrient pollution of lakes and rivers results in heavy growth of algae and cyanobacteria **(Figure 20.35)**.

In an example of how far-reaching this problem can be, nutrient runoff from Midwestern farm fields has been linked to an annual summer "dead zone" in the Gulf of Mexico that is nearly devoid of animals **(Figure 20.36)**. Vast algal blooms extend outward from where the Mississippi River deposits its nutrient-laden waters. It might seem that producers would add oxygen to the water through photosynthesis, allowing consumers to thrive. However, producers consume oxygen when they do not get sufficient light, such as at night or when they are at the bottom of thick mats of algae. Algae have a short life span. As the algae die, bacteria decompose the huge quantities of biomass. Cellular respiration by bacteria diminishes the supply of dissolved oxygen over an area that ranges from 13,000 km² to 22,000 km² (from roughly the size of Connecticut to the size of New Jersey). Oxygen depletion disrupts benthic (aquatic, bottom dwelling) communities, displacing fish and invertebrates that move along or near the substrate and killing organisms that are attached to it. More than 400 recurring or permanent coastal dead zones totaling approximately 245,000 km² (about the area of Michigan) have been documented worldwide. Some algal blooms are caused by toxic algae, greatly increasing the ecological and economic damage. ☑

▼ **Figure 20.35** Producer growth in Lake Erie resulting from nutrient pollution.

▼ **Figure 20.36** The Gulf of Mexico dead zone.

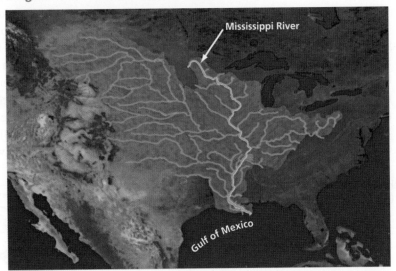

Light blue lines represent rivers draining into the Mississippi River (shown in bright blue). Nitrogen runoff carried by these rivers ends up in the Gulf of Mexico. In the images below, red and orange indicate high concentrations of phytoplankton. Bacteria feeding on dead phytoplankton deplete the water of oxygen, creating a "dead zone."

Summer

Winter

☑ **CHECKPOINT**

How does the excessive addition of mineral nutrients to a pond eventually result in the loss of most fish in the pond?

Answer: Excessive mineral nutrients initially cause population explosions of producers and the organisms that feed on them. The respiration of so much life, especially of the microbes decomposing all the organic refuse, consumes most of the pond's oxygen, which the fish require.

Conservation and Restoration Biology

As we have seen in this unit, many of the environmental problems facing us today have been caused by human enterprises. But the science of ecology is not just useful for telling us how things have gone wrong. Ecological research is also the foundation for finding solutions to these problems and for reversing the negative consequences of ecosystem alteration. Thus, we end the ecology unit by highlighting these beneficial applications of ecological research.

Conservation biology is a goal-oriented science that seeks to understand and counter the loss of biodiversity. Conservation biologists recognize that biodiversity can be sustained only if we maintain the ecosystems and processes that shaped the evolution of the species. Thus, the goal is not simply to preserve individual species but to sustain ecosystems, where natural selection can continue to function, and to maintain the genetic variability on which natural selection acts. The expanding field of **restoration ecology** uses ecological principles to develop methods of returning degraded areas to their natural state.

Biodiversity "Hot Spots"

CHECKPOINT

What is a biodiversity hot spot?

Answer: a relatively small area with a disproportionately large number of species, including endangered species

Conservation biologists are applying their understanding of population, community, and ecosystem dynamics in establishing parks, wilderness areas, and other legally protected nature reserves. Choosing locations for these protected zones often focuses on **biodiversity hot spots**. These relatively small areas have a large number of endangered and threatened species and an exceptional

concentration of **endemic species**, species that are found nowhere else. Together, the "hottest" of Earth's biodiversity hot spots, shown in **Figure 20.37**, total less than 1.5% of Earth's land surface but are home to a third of all species of plants and vertebrates. For example, all of the many species of lemurs on Earth—more than 50 species—are endemic to Madagascar, a large island off the eastern coast of Africa. In fact, almost all of the mammals, reptiles, amphibians, and plants that inhabit Madagascar are endemic. There are also hot spots in aquatic ecosystems, such as certain river systems and coral reefs. Because biodiversity hot spots can also be hot spots of extinction, they rank high on the list of areas demanding strong global conservation efforts.

Concentrations of species provide an opportunity to protect many species in very limited areas. We will explore this more in the Evolution Connection essay. However, preserving biodiversity requires more than protecting the hot spots, which are usually defined using only the most noticeable organisms, especially vertebrates and plants. Invertebrates and microorganisms are often overlooked. Furthermore, species endangerment is a global problem, and focusing on hot spots should not detract from efforts to conserve habitats and species diversity in other areas. Finally, even the protection of a nature reserve does not shield organisms from the effects of climate change or other threats, such as invasive species or infectious disease. To stem the tide of biodiversity loss, we will have to address environmental problems globally as well as locally. ✅

▶ **Figure 20.37** Earth's terrestrial (purple) and marine (red) biodiversity hot spots.

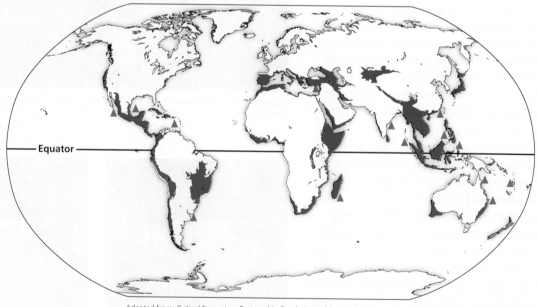

Adapted from: Critical Ecosystem Partnership Fund, *Annual Report 2014*, Conservation International.

Conservation at the Ecosystem Level

In the past, most conservation efforts focused on saving individual species, and this work continues. (You have already learned about one example, the red-cockaded woodpecker; see Figure 19.12.) More and more, however, conservation biology aims at sustaining the biodiversity of entire communities and ecosystems. On an even broader scale, conservation biology considers the biodiversity of whole landscapes. Ecologically, a **landscape** is a regional assemblage of interacting ecosystems, such as an area with forest, adjacent fields, wetlands, streams, and streamside habitats. **Landscape ecology** is the application of ecological principles to the study of land-use patterns. Its goal is to make ecosystem conservation a functional part of the planning for land use.

▼ **Figure 20.38** **Edges between ecosystems within a landscape.**

Natural edges. Forests border grassland ecosystems in Lake Clark National Park, Alaska.

Edges created by human activity. Forest edges surround farmland in the Cotswolds region of south central England.

Edges between ecosystems are prominent features of landscapes, whether natural or altered by human activity **(Figure 20.38)**. Edges have their own sets of physical conditions—such as soil type and surface features—that differ from the ecosystems on either side of them. Edges also may have their own type and amount of disturbance. For instance, the edge of a forest often has more blown-down trees than a forest interior because the edge is less protected from strong winds. Because of their specific physical features, edges also have their own communities of organisms. Some organisms thrive in edges because those organisms require resources found only there. For instance, white-tailed deer browse on woody shrubs found in edges between woods and fields, so deer populations often expand as edges do when forests are logged or fragmented by development.

Edges can have both positive and negative effects on biodiversity. A recent study in a tropical rain forest in western Africa indicated that natural edge communities are important sites of speciation. On the other hand, landscapes where human activities have produced edges often have fewer species.

Another important landscape feature, especially where habitats have been severely fragmented, is the **movement corridor**, a narrow strip or series of small clumps of suitable habitat connecting otherwise isolated patches. In places where there is extremely heavy human impact, artificial corridors are sometimes constructed **(Figure 20.39)**. Corridors can promote dispersal of individuals from a population and thereby help sustain populations. Corridors are especially important to species that migrate between different habitats seasonally. But a corridor can also be harmful—as, for example, in the spread of disease, especially among small subpopulations in closely situated habitat patches. ☑

☑ **CHECKPOINT**

How is a landscape different from an ecosystem?

Answer: A landscape is more inclusive in that it consists of several interacting ecosystems in the same region.

▼ **Figure 20.39** **An artificial corridor.** This bridge over a road provides an artificial corridor for animals in Canada.

Does Biodiversity Protect Human Health?

BACKGROUND

As humans extract natural resources, build roads, and convert forests to agricultural land, they are fragmenting many landscapes and decreasing biodiversity in those communities. Species that require large areas to survive are disappearing as their habitat is destroyed. Species that thrive in small fragments are increasing in numbers. The loss of biodiversity due to fragmentation can be seen in the northern forests of the United States, where populations of the white-footed mouse (*Peromyscus leucopus*) are exploding (**Figure 20.40a**). White-footed mice carry Lyme disease.

In humans, Lyme disease can be fatal if it spreads to the heart or nervous system. The bacterium that causes Lyme disease (*Borrelia burgdorferi*) is carried by black-legged ticks (*Ixodes scapularis*) in the northern forests of the United States and in similar biomes worldwide. The tick, a parasite, requires more than one host species to complete its life cycle (**Figure 20.40b**). Adult ticks feed on the blood of deer and humans, then females lay eggs. Larvae hatch and feed on the blood of an intermediate host. Then the larvae grow into nymphs. Once a nymph, a tick attaches to another intermediate host. As the tick feeds, the Lyme disease bacteria is spread, either from the tick to the host or vice versa. Host species differ in their ability to infect ticks with bacteria. White-footed mice are particularly likely to infect ticks. The higher biodiversity found in large forest fragments would potentially help break the chain of disease transmission because more biologically diverse communities include many intermediate hosts, for example, birds and lizards, that do not spread the bacteria, even if they are infected.

In recent years, the number of cases of Lyme disease in the United States has spiked. To find out if fragmentation is increasing the risk of Lyme disease, researchers conducted a study to test the hypothesis that smaller forest fragments would have an increased density of infected ticks.

METHOD

Researchers studied 14 maple forest fragments of different sizes in southeastern New York. They chose a county with an extremely high number of Lyme disease cases. They identified fragments from aerial photographs and collected ticks by dragging cloths along the ground in a standard pattern. A subset of all the ticks collected was analyzed for presence of Lyme disease bacteria.

RESULTS

The smallest forest fragments had much higher densities of infected black-legged tick nymphs (**Figure 20.40c**). In the small fragments, white-footed mice are extremely common, and other intermediate hosts may be rare or absent. White-footed mice are also the most likely intermediate host in the landscapes studied to infect ticks with Lyme disease. These findings suggest that high biodiversity provides people some protection from Lyme disease. Future studies conducted on other parasite-transmitted diseases may show a similar pattern.

BIODIVERSITY CAN PROTECT YOU FROM CATCHING LYME DISEASE FROM THIS TICK.

▼ **Figure 20.40 Habitat fragmentation, decreased biodiversity, and Lyme disease transmission by ticks.** Data from: B. F. Allan et al., Effect of Forest Fragmentation on Lyme Disease Risk. *Conservation Biology* 17(1): 267–72 (2003), Figure 1c, p. 271.

(a) The white-footed mouse spreads Lyme disease in forest communities.

Thinking Like a Scientist

Researchers ensured that the fragments they studied were a minimum distance apart. Why was this important?

For answer, see Appendix D.

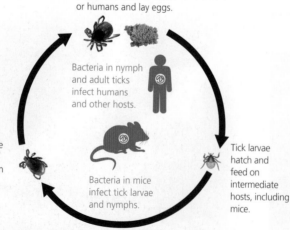

Adult ticks feed on deer or humans and lay eggs.

Bacteria in nymph and adult ticks infect humans and other hosts.

Larvae mature into nymphs, which feed on a variety of hosts.

Tick larvae hatch and feed on intermediate hosts, including mice.

Bacteria in mice infect tick larvae and nymphs.

(b) Black-legged ticks can spread Lyme disease at different stages of their life cycle.

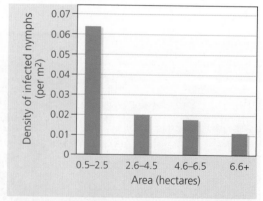

(c) Larger forest fragments have a lower density of infected tick nymphs. (A hectare is an area of 100 × 100 m.)

Restoring Ecosystems

One of the major strategies in restoration ecology is **bioremediation**, the use of living organisms to detoxify polluted ecosystems. For example, bacteria have been used to clean up old mining sites and oil spills. Researchers are also investigating the potential of using plants to remove toxic substances such as heavy metals and organic pollutants (for example, PCBs) from contaminated soil (**Figure 20.41**).

Some restoration projects have the broader goal of returning ecosystems to their natural state. Such projects may involve replanting vegetation, fencing out non-native animals, or removing dams that restrict water flow. Hundreds of restoration projects are currently under way in the United States. One of the most ambitious endeavors is the Kissimmee River Restoration Project in south central Florida.

The Kissimmee River was once a shallow, meandering river that wound its way from Lake Kissimmee southward into Lake Okeechobee. During about half of the year, the river flooded into a wide floodplain, creating wetlands that provided habitat for large numbers of birds, fishes, and invertebrates. And as the floods deposited the river's load of nutrient-rich silt on the floodplain, they boosted soil fertility and maintained the water quality of the river.

Between 1962 and 1971, the U.S. Army Corps of Engineers converted the 166-km wandering river to a straight canal 9 m deep, 100 m wide, and 90 km long. This project, designed to allow development on the floodplain, drained approximately 31,000 acres of wetlands, with significant negative impacts on fish and wetland bird populations. Without the marshes to help filter and reduce agricultural runoff, the river transported phosphates and other excess nutrients from Lake Okeechobee farther south to the Everglades.

The restoration project involves removing water-control structures such as dams, reservoirs, and channel modifications and filling in about 35 km of the canal (**Figure 20.42**). The first phase of the restoration project was completed in 2004. The time line for completion has been extended repeatedly, but the latest end date is 2019. The photo in Figure 20.42 shows a section of the Kissimmee canal that has been plugged, diverting flow into the remnant river channels. Birds and other wildlife have returned in unexpected numbers to the 11,000 acres of wetlands that have been restored. The marshes are filled with native vegetation, and game fishes again swim in the river channels. ☑

☑ **CHECKPOINT**

The water in the Kissimmee River eventually flows into the Everglades. How will the Kissimmee River Restoration Project affect water quality in the Everglades ecosystem?

■ *Answer: Wetlands filter agricultural runoff, which prevents nutrient pollution from flowing downstream. By restoring this ecosystem service, the project will improve water quality in the Everglades.*

▼ Figure 20.41 **Bioremediation using plants.**
A researcher from the U.S. Department of Agriculture investigates the use of canola plants to reduce toxic levels of selenium in contaminated soil.

▼ Figure 20.42 **The Kissimmee River Restoration Project.**

The Goal of Sustainable Development

As the world population grows and becomes more affluent, the demand for the "provisioning" services of ecosystems, such as food, wood, and water, is increasing. Although these demands are currently being met for much of the world, they are satisfied at the expense of other critical ecosystem services, such as climate regulation and protection against natural disasters. Clearly, we have set ourselves and the rest of the biosphere on a precarious path into the future. How can we achieve **sustainable development**—development that meets the needs of people today without limiting the ability of future generations to meet their needs?

Many nations, scientific associations, corporations, and private foundations have embraced the concept of sustainable development. The Ecological Society of America, the world's largest organization of ecologists, endorses a research agenda called the Sustainable Biosphere Initiative. The goal of this initiative is to acquire the ecological information necessary for the responsible development, management, and conservation of Earth's resources. The research agenda includes the search for ways to sustain the productivity of natural and artificial ecosystems and studies of the relationship between biological diversity, global climate change, and ecological processes.

Sustainable development depends on more than continued research and application of ecological knowledge. It also requires that we connect the life sciences with the social sciences, economics, and humanities. Conservation of biodiversity is only one side of sustainable development; the other side is improving the human condition. Public education and the political commitment and cooperation of nations are essential to the success of this endeavor.

An awareness of our unique ability to alter the biosphere and jeopardize the existence of other species, as well as our own, may help us choose a path toward a sustainable future. The risk of a world without adequate natural resources for all its people is not a vision of the distant future. It is a prospect for your children's lifetime, or perhaps even your own. But although the current state of the biosphere is grim, the situation is far from hopeless. Now is the time to take action by aggressively pursuing greater knowledge about the diversity of life on our planet and by joining with others in working toward long-term sustainability (**Figure 20.43**). ☑

☑ CHECKPOINT

What is meant by sustainable development?

■ *Answer: development that meets current needs while ensuring an adequate supply of natural and economic resources for future generations*

▼ Figure 20.43 **Working toward sustainability.**

Students at the University of Virginia sorted trash from dumpsters to promote recycling. Recycling just one aluminum can saves enough energy to power a laptop for five hours.

A student at California State University, Fresno, pulls weeds in a plot of mustard and kale plants at the university's organic farm. Weeding by hand is more sustainable than herbicide use.

Saving the Hot Spots

As we have learned in this chapter, islands can be hot spots of both diversity and extinction. For example, the island of Guam was once home to several endemic bird species that are now extinct in the wild (Figure 20.44). Guam's story is not unique. Geographical separation often leads to evolution of new species (see Figure 14.7). Different conditions and random events shape species in ways that are specific to their islands. Eventually, endemic species and subspecies evolve.

On an island with no predators, investing in mechanical, physiological, or behavioral strategies to avoid predation is a waste. Random mutations that decrease genes for anti-predation traits actually leave individuals on the island better adapted to their environment. These individuals produce more offspring than others in their population. In fact, on such islands, there is no selection pressure to even maintain fear responses to predators. However, when predators arrive, native species have no defenses and may go extinct before any evolve. With no native competitors or predators to control their numbers, invasive predators can devastate island communities. This happened on Guam, which has no native snakes.

When one or more brown treesnakes hitchhiked to Guam in cargo soon after World War II, no one noticed. The snake is nocturnal and tends to stay hidden. At first, people found a few shed snake skins and reported the occasional snake sighting. Then native species started disappearing. Eventually, about half of the native birds and several native species of bats and lizards became extinct in the wild. Forests that once rang with birdsong fell silent. Guam's biodiversity continues to be affected. The island has seen a decline in trees dependent on pollination by the missing bird species. Brown treesnakes also cause economic damage, eating poultry and causing frequent power outages by shorting out high-voltage wires.

The isolation of islands provides humans a unique opportunity to protect them. On Guam, government agencies use trained dogs to prevent snakes from leaving to infest other areas. Snake traps are placed along fences. Brown treesnakes are unusually vulnerable to the common painkiller acetaminophen (Tylenol), so mouse carcasses laced with the drug are attached to tiny parachutes dropped from helicopters. A 2012 study showed that Guam's efforts resulted in treesnake reductions of over 80%. Someday, it may be possible to reintroduce into the wild some endemic bird species that are currently maintained in captive populations. Worldwide, over 230 island species have benefited from the eradication of pests.

Understanding the structure and function of communities and ecosystems provides us with tools for reversing some of the damage done by human activities. With this knowledge, we may ensure that we leave a diverse and healthy world for future generations.

Guam bridled white-eye

Guam kingfisher

Guam rail

(a) Some of Guam's endemic birds that are now extinct in the wild.

◀ Figure 20.44 Brown treesnakes as a threat to Guam's biodiversity.

(b) Guam uses detection dogs, traps, and poisoned bait to control brown treesnakes.

Chapter Review

SUMMARY OF KEY CONCEPTS

Biodiversity

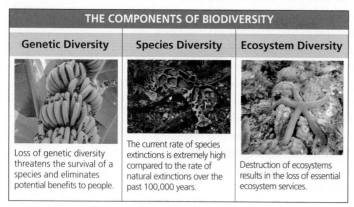

THE COMPONENTS OF BIODIVERSITY		
Genetic Diversity	**Species Diversity**	**Ecosystem Diversity**
Loss of genetic diversity threatens the survival of a species and eliminates potential benefits to people.	The current rate of species extinctions is extremely high compared to the rate of natural extinctions over the past 100,000 years.	Destruction of ecosystems results in the loss of essential ecosystem services.

Causes of Declining Biodiversity

Habitat destruction is the leading cause of extinctions. Invasive species, overexploitation, and pollution are also significant factors.

Community Ecology

Interspecific Interactions

Populations in a community interact in a variety of ways that can be generally categorized as being beneficial (+), harmful (–), or neutral (0) to individuals of different species. Because +/– interactions (exploitation of one species by another species) may have such a negative impact on the individual that is harmed, defensive evolutionary adaptations are common.

INTERACTIONS BETWEEN SPECIES IN A COMMUNITY					
Interspecific Interaction	**Effect on Species 1**	**Effect on Species 2**	**Interspecific Interaction**	**Effect on Species 1**	**Effect on Species 2**
Competition	–	–	**Exploitation**		
			Predation	+	–
Mutualism	+	+			
			Herbivory	+	–
Commensalism	+	0			
			Parasites and pathogens	+	–

Trophic Structure

The trophic structure of a community defines the feeding relationships among organisms. In the process of biological magnification, toxins become more concentrated as they are passed up a food chain to the top predators.

Increasing PCB concentration

Species Diversity in Communities

Diversity within a community includes species richness and relative abundance of different species. A keystone species is a species that has a great impact on the composition of the community despite a relatively low abundance or biomass.

Disturbances and Succession in Communities

Disturbances are episodes that damage communities, at least temporarily, by destroying organisms or altering the availability of resources such as mineral nutrients and water. People are the most significant cause of disturbances today.

Ecological Succession

The sequence of changes in a community after a disturbance is called ecological succession. Primary succession occurs where a community arises in a virtually lifeless area with no soil. Secondary succession occurs where a disturbance has destroyed an existing community but left the soil intact.

Ecosystem Ecology

Energy Flow in Ecosystems

An ecosystem is a biological community and the abiotic factors with which the community interacts. Energy must be added continuously to an ecosystem, because as it flows from producers to consumers and

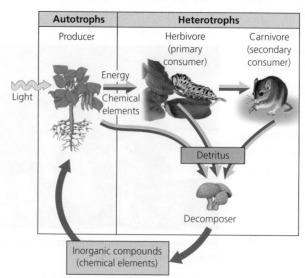

decomposers, it is constantly being lost. Chemical elements can be recycled between an ecosystem's living community and the abiotic environment. Trophic relationships determine an ecosystem's routes of energy flow and chemical cycling.

Primary production is the rate at which plants and other producers build biomass. Ecosystems vary considerably in their productivity. Primary production sets the spending limit for the energy budget of the entire ecosystem because consumers must acquire their organic fuels from producers. In a food chain, only about 10% of the biomass at one trophic level is available to the next, resulting in an energy pyramid.

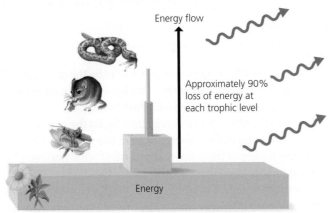

Energy flow

Approximately 90% loss of energy at each trophic level

Energy

When people eat producers instead of consumers, less photosynthetic production is required, which reduces the impact on the environment.

Chemical Cycling in Ecosystems

Biogeochemical cycles involve biotic and abiotic components. Each circuit has an abiotic reservoir through which the chemical cycles. Some chemical elements require "processing" by certain microorganisms before those chemical elements are available to plants as inorganic nutrients. A chemical's specific route through an ecosystem varies, depending on the element and the trophic structure of the ecosystem. Phosphorus is not very mobile and is cycled locally. Carbon and nitrogen spend part of their time in gaseous form and are cycled globally. Runoff of nitrogen and phosphorus, especially from agricultural land, causes algal blooms in aquatic ecosystems, lowering water quality and sometimes depleting the water of oxygen.

Conservation and Restoration Biology

Biodiversity "Hot Spots"

Conservation biology is a goal-oriented science that seeks to counter the loss of biodiversity. The front lines for conservation biology are biodiversity "hot spots," relatively small geographic areas that are especially rich in endangered species.

Conservation at the Ecosystem Level

Increasingly, conservation biology aims at sustaining the biodiversity of entire communities, ecosystems, and landscapes. Edges between ecosystems are prominent features of landscapes, with positive and negative effects on biodiversity. Corridors can promote dispersal and help sustain populations.

Restoring Ecosystems

In some cases, ecologists use microbes or plants to remove toxic substances, such as heavy metals, from ecosystems. Ecologists are working to revitalize some ecosystems by planting native vegetation, removing barriers to wildlife, and other means. The Kissimmee River Restoration Project is an attempt to undo the ecological damage done when the river was engineered into straight channels.

The Goal of Sustainable Development

Balancing the needs of people with the health of the biosphere, sustainable development has the goal of long-term prosperity of human societies and the ecosystems that support them.

Mastering **Biology**

For practice quizzes, BioFlix animations, MP3 tutorials, video tutors, and more study tools designed for this textbook, go to Mastering Biology™

SELF-QUIZ

1. Currently, the number one cause of biodiversity loss is _____.
2. According to the concept of competitive exclusion,
   a. two species cannot coexist in the same habitat.
   b. extinction or emigration is the only possible result of competitive interactions.
   c. intraspecific competition results in the success of the best-adapted individuals.
   d. two species cannot share exactly the same niche in a community.
3. Which of the following best describes a commensal interspecific interaction?
   a. +/0
   b. +/+
   c. +/−
   d. −/−
4. Match each organism with its trophic level (you may choose a level more than once).
   a. alga   1. decomposer
   b. grasshopper   2. producer
   c. zooplankton   3. tertiary consumer
   d. eagle   4. secondary consumer
   e. fungus   5. primary consumer
5. Why are the top predators in food chains most severely affected by pesticides such as DDT?
6. An episode that alters biological communities by destroying organisms and changing the availability of resources is known as a _____.
7. According to the energy pyramid, why is eating grain-fed beef a relatively inefficient means of obtaining the energy trapped by photosynthesis?

8. Local conditions, such as heavy rainfall or the removal of plants, may limit the amount of nitrogen, phosphorus, or calcium available to a particular terrestrial ecosystem. Why is the amount of carbon available to the ecosystem seldom a problem?

9. _____ is the use of living organisms to detoxify polluted ecosystems.

10. At what hierarchical level is conservation dealing with edges or boundaries between ecosystems aimed at?
    a. species
    b. communities
    c. biomes
    d. landscapes

For answers to the Self-Quiz, see Appendix D.

IDENTIFYING MAJOR THEMES

For each statement below, identify which major theme is evident (the relationship of structure to function, information flow, pathways that transform energy and matter, interactions within biological systems, or evolution) and explain how the statement relates to the theme. If necessary, review the themes (Chapter 1) and review the examples highlighted in blue in this chapter.

11. The porcupine's sharp quills and the hard shells of clams provide defenses against predators.

12. In some cases, two populations in a community compete for a resource such as food or space.

13. A community's trophic structure determines the pathways that transform resources as they move from photosynthetic organisms to herbivores and then to predators.

For answers to Identifying Major Themes, see Appendix D.

THE PROCESS OF SCIENCE

14. An ecologist studying desert plants performed the following experiment. She staked out several similar plots that included a few sagebrush plants and numerous small annual wildflowers. She found the same five wildflower species in similar numbers in all plots. Then she randomly selected half of the plots and enclosed them with fences to keep out kangaroo rats, the most common herbivores in the area. After two years, four species of wildflowers were no longer present in the fenced plots, but one wildflower species had increased dramatically. The unfenced control plots had not changed significantly in species composition. Using the concepts discussed in the chapter, what do you think happened?

15. Beavers are considered to be "ecosystem engineers"—keystone species that actively transform the environment. They cut down old trees and use the wood to build river dams. What are the likely impacts of the beavers' activities on the riparian (river-based) environment? What would be the consequences if beavers disappeared?

16. **Interpreting Data** In a classic study, John Teal measured energy flow in a salt marsh ecosystem. The table below shows some of his results.

Form of energy	Kcal/m²/yr	Efficiency of energy transfer (%)
Sunlight	600,000	n/a
Chemical energy in producers	6,585	
Chemical energy in primary consumers	81	

Data from: J. M. Teal, Energy Flow in the Salt Marsh Ecosystem of Georgia. *Ecology* 43: 614–24 (1962).

a. Calculate the efficiency of energy transfer by the producers. That is, what percentage of the energy in sunlight was converted into chemical energy and incorporated into plant biomass?

b. Calculate the efficiency of energy transfer by the primary consumers. What percentage of the energy in plant biomass was incorporated into the bodies of the primary consumers? What became of the rest of the energy (see Figure 20.27)?

c. How much energy is available for secondary consumers? Based on the efficiency of energy transfer by primary consumers, estimate how much energy will be available to tertiary consumers.

d. Draw an energy pyramid for the producers, primary consumers, and secondary consumers for this ecosystem (see Figure 20.28).

BIOLOGY AND SOCIETY

17. Produce a list of the different ways in which society benefits from the maintenance of biodiversity at the gene, species, and ecosystem levels. What role does a society have in maintaining biodiversity? How can individuals respond to the biodiversity crisis and ensure the conservation of endangered species? Can one person really make a difference? If there was one thing you could do to contribute towards biodiversity, what would it be?

18. Citizen scientists help in the massive task of documenting species. Bioblitz teams of experts, hobbyists, and even children identify and count all of the species they can in a specific area. Often, bioblitzes are held around the same time of year for multiple years. How would you convince your local government or school to hold a bioblitz in your community?

19. One reason that Tasmanian devils are important to preserving Tasmania's biodiversity is that they kill feral cats, which are a threat to native species, such as birds. Tasmanian devils are a natural control of an invasive species that involves lethal measures. When nonlethal methods, such as trapping and relocating, are not a feasible way to control non-native species that threaten native species, should lethal measures be used?

APPENDIX A Metric Conversion Table

Measurement	Unit and Abbreviation	Metric Equivalent	Approximate Metric-to-English Conversion Factor	Approximate English-to-Metric Conversion Factor
Length	1 kilometer (km)	= 1,000 (10^3) meters	1 km = 0.6 mile	1 mile = 1.6 km
	1 meter (m)	= 100 (10^2) centimeters	1 m = 1.1 yards	1 yard = 0.9 m
		= 1,000 millimeters	1 m = 3.3 feet	1 foot = 0.3 m
			1 m = 39.4 inches	
	1 centimeter (cm)	= 0.01 (10^{-2}) meter	1 cm = 0.4 inch	1 foot = 30.5 cm
				1 inch = 2.5 cm
	1 millimeter (mm)	= 0.001 (10^{-3}) meter	1 mm = 0.04 inch	
	1 micrometer (μm)	= 10^{-6} meter (10^{-3} mm)		
	1 nanometer (nm)	= 10^{-9} meter (10^{-3} μm)		
	1 angstrom (Å)	= 10^{-10} meter (10^{-4} μm)		
Area	1 hectare (ha)	= 10,000 square meters	1 ha = 2.5 acres	1 acre = 0.4 ha
	1 square meter (m^2)	= 10,000 square centimeters	1 m^2 = 1.2 square yards	1 square yard = 0.8 m^2
			1 m^2 = 10.8 square feet	1 square foot = 0.09 m^2
	1 square centimeter (cm^2)	= 100 square millimeters	1 cm^2 = 0.16 square inch	1 square inch = 6.5 cm^2
Mass	1 metric ton (t)	= 1,000 kilograms	1 t = 1.1 tons	1 ton = 0.91 t
	1 kilogram (kg)	= 1,000 grams	1 kg = 2.2 pounds	1 pound = 0.45 kg
	1 gram (g)	= 1,000 milligrams	1 g = 0.04 ounce	1 ounce = 28.35 g
			1 g = 15.4 grains	
	1 milligram (mg)	= 10^{-3} gram	1 mg = 0.02 grain	
	1 microgram (μg)	= 10^{-6} gram		
Volume (solids)	1 cubic meter (m^3)	= 1,000,000 cubic centimeters	1 m^3 = 1.3 cubic yards	1 cubic yard = 0.8 m^3
			1 m^3 = 35.3 cubic feet	1 cubic foot = 0.03 m^3
	1 cubic centimeter (cm^3 or cc)	= 10^{-6} cubic meter	1 cm^3 = 0.06 cubic inch	1 cubic inch = 16.4 cm^3
	1 cubic millimeter (mm^3)	= 10^{-9} cubic meter (10^{-3} cubic centimeter)		
Volume (liquids and gases)	1 kiloliter (kL or kl)	= 1,000 liters	1 kL = 264.2 gallons	1 gallon = 3.79 L
	1 liter (L)	= 1,000 milliliters	1 L = 0.26 gallon	1 quart = 0.95 L
			1 L = 1.06 quarts	
	1 milliliter (mL or ml)	= 10^{-3} liter	1 mL = 0.03 fluid ounce	1 quart = 946 mL
		= 1 cubic centimeter	1 mL = approx. $\frac{1}{4}$ teaspoon	1 pint = 473 mL
			1 mL = approx. 15–16 drops	1 fluid ounce = 29.6 mL
	1 microliter (μL or μl)	= 10^{-6} liter (10^{-3} milliliters)		1 teaspoon = approx. 5 mL
Time	1 second (s)	= $\frac{1}{60}$ minute		
	1 millisecond (ms)	= 10^{-3} second		
Temperature	Degrees Celsius (°C)		°F = $\frac{9}{5}$°C + 32	°C = $\frac{5}{9}$(°F − 32)

APPENDIX B The Periodic Table

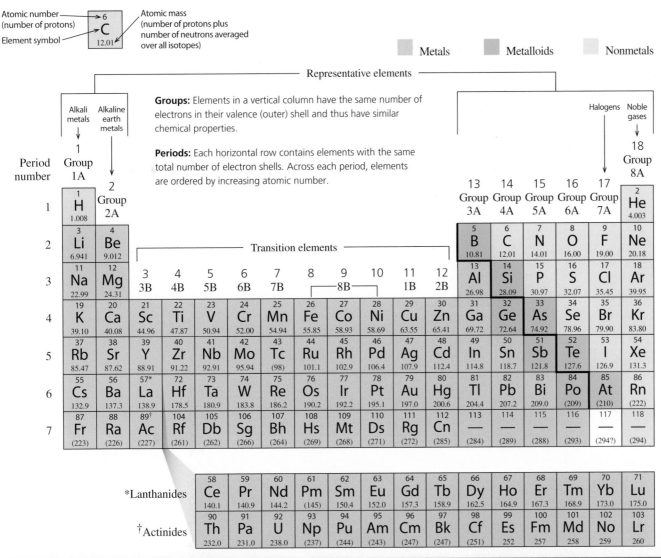

Atomic number (number of protons)
Element symbol
Atomic mass (number of protons plus number of neutrons averaged over all isotopes)

6
C
12.01

Metals   Metalloids   Nonmetals

Representative elements

Groups: Elements in a vertical column have the same number of electrons in their valence (outer) shell and thus have similar chemical properties.

Periods: Each horizontal row contains elements with the same total number of electron shells. Across each period, elements are ordered by increasing atomic number.

Name (Symbol)	Atomic Number	Name (Symbol)	Atomic Number	Name (Symbol)	Atomic Number	Name (Symbol)	Atomic Number	Name (Symbol)	Atomic Number
Actinium (Ac)	89	Copper (Cu)	29	Iron (Fe)	26	Osmium (Os)	76	Silicon (Si)	14
Aluminum (Al)	13	Curium (Cm)	96	Krypton (Kr)	36	Oxygen (O)	8	Silver (Ag)	47
Americium (Am)	95	Darmstadtium (Ds)	110	Lanthanum (La)	57	Palladium (Pd)	46	Sodium (Na)	11
Antimony (Sb)	51	Dubnium (Db)	105	Lawrencium (Lr)	103	Phosphorus (P)	15	Strontium (Sr)	38
Argon (Ar)	18	Dysprosium (Dy)	66	Lead (Pb)	82	Platinum (Pt)	78	Sulfur (S)	16
Arsenic (As)	33	Einsteinium (Es)	99	Lithium (Li)	3	Plutonium (Pu)	94	Tantalum (Ta)	73
Astatine (At)	85	Erbium (Er)	68	Livermorium (Lv)	116	Polonium (Po)	84	Technetium (Tc)	43
Barium (Ba)	56	Europium (Eu)	63	Lutetium (Lu)	71	Potassium (K)	19	Tellurium (Te)	52
Berkelium (Bk)	97	Fermium (Fm)	100	Magnesium (Mg)	12	Praseod`ymium (Pr)	59	Tennessine (Ts)	117
Beryllium (Be)	4	Flerovium (Fl)	114	Manganese (Mn)	25	Promethium (Pm)	61	Terbium (Tb)	65
Bismuth (Bi)	83	Fluorine (F)	9	Meitnerium (Mt)	109	Protactinium (Pa)	91	Thallium (Tl)	81
Bohrium (Bh)	107	Francium (Fr)	87	Mendelevium (Md)	101	Radium (Ra)	88	Thorium (Th)	90
Boron (B)	5	Gadolinium (Gd)	64	Mercury (Hg)	80	Radon (Rn)	86	Thulium (Tm)	69
Bromine (Br)	35	Gallium (Ga)	31	Molybdenum (Mo)	42	Rhenium (Re)	75	Tin (Sn)	50
Cadmium (Cd)	48	Germanium (Ge)	32	Moscovium (Mc)	115	Rhodium (Rh)	45	Titanium (Ti)	22
Calcium (Ca)	20	Gold (Au)	79	Neodymium (Nd)	60	Roentgenium (Rg)	111	Tungsten (W)	74
Californium (Cf)	98	Hafnium (Hf)	72	Neon (Ne)	10	Rubidium (Rb)	37	Uranium (U)	92
Carbon (C)	6	Hassium (Hs)	108	Neptunium (Np)	93	Ruthenium (Ru)	44	Vanadium (V)	23
Cerium (Ce)	58	Helium (He)	2	Nickel (Ni)	28	Rutherfordium (Rf)	104	Xenon (Xe)	54
Cesium (Cs)	55	Holmium (Ho)	67	Nihonium (Nh)	113	Samarium (Sm)	62	Ytterbium (Yb)	70
Chlorine (Cl)	17	Hydrogen (H)	1	Niobium (Nb)	41	Scandium (Sc)	21	Yttrium (Y)	39
Chromium (Cr)	24	Indium (In)	49	Nitrogen (N)	7	Seaborgium (Sg)	106	Zinc (Zn)	30
Cobalt (Co)	27	Iodine (I)	53	Nobelium (No)	102	Selenium (Se)	34	Zirconium (Zr)	40
Copernicium (Cn)	112	Iridium (Ir)	77	Oganesson (Og)	118				

PHOTO CREDITS

UNIT OPENERS:

Unit 1 clockwise from left Robert S. McNeil/Baylor College of Medicine/Science Source; Living Art Enterprises/Science Source; Bo1982/E+/Getty Images; Science Source; Thomas Deerinck/NCMIR/Science Source; Kris Mendoza/Moment/Getty Images; Chandlerphoto/Getty Images; **unit 2 clockwise from left** Erik Lam/Shutterstock; Jordi Chias/Nature Picture Library; Eric J. Simon; Kateryna Kon/Science Photo Library/Getty Images; SPL/Science Source; Isak55/Shutterstock; **unit 3 clockwise from left** Huayang/Moment/Getty Images; George Mulala/REUTERS; Michael Collier; Stephanie Schuller/Science Source; Valentyn Volkov/Shutterstock; Sabena Jane Blackbird/Alamy; **unit 4 clockwise from left** Anthony Mercieca/Science Source; Sylvain Cordier/Photodisc/Getty Images; Richard Whitcombe/Shutterstock; Bruce Miller/Alamy.

CHAPTER 1:

Chapter opening photo top right Torwai Studio/Shutterstock; **chapter opening photo left** Feng Wei Photography/Getty Images; **chapter opening photo bottom right** NASA; **p. 37** Eric J. Simon; **1.1 left** Michael Nichols/National Geographic/Getty Images; **1.1 right** Tim Ridley/Dorling Kindersley, Ltd.; **1.2** Adrian Sherratt/Alamy; **1.3** Eric J. Simon; **1.4** Will Vragovic/ZUMA Press/Newscom; **1.6a** Aqua Image/Alamy; **1.6b, 1.7** University of Central Florida; All marine turtle images taken in the Gulf of Mexico were obtained with the approval of the National Marine Fisheries Service (NMFS) under conditions not harmful to the turtles. Images were acquired while conducting authorized research activities under NMFS Research Permit 16733; **1.9 left** Gregor Bister/E+/Getty Images; **1.9 right** Adrian Arbib/Alamy; **1.10** SSS615/Shutterstock; **1.11** Isabelle Kuehn/Shutterstock; **1.12** Paula French/iStock/Getty Images; **1.13** Martin Dohrn/Royal College of Surgeons/Science Source; **1.14** Susumu Nishinaga/Science Source; **1.16** Ian Hooton/Science Source; **1.17** Sergey Novikov/Shutterstock; **1.18 left to right** NASA Goddard Institute for Space Studies and Surface Temperature Analysis; Prasit Chansareekorn/Moment Collection/Getty Images; Quinn Aikens/Shutterstock; Biology Pics/Science Source; **1.19 top left** Mike Hayward/Alamy; **1.19 top right** Title page from On The Origin of Species by Charles Darwin, M.A., London: John Murray, Albemarle Street, 1859; **1.19 bottom** Michael Nolan/Robert Harding World Imagery; **1.21** Ron Masessa/E+/Getty Images.

CHAPTER 2:

Chapter opening photo top right LP2 Studio/Shutterstock; **chapter opening photo center** VasiliyBudarin/Shutterstock; **chapter opening photo bottom** Steffen Binke/Alamy; **p. 57** Living Art Enterprises/Science Source; **2.1 top** Trevor Clifford/Pearson Education, Inc.; **2.1 bottom** Geoff Dann/Dorling Kindersley, Ltd.; **2.3 left to right** Anne Dowie/Pearson Education, Inc.; Jiri Hera/Fotolia; Howard Shooter/Dorling Kindersley, Ltd.; Alessio Cola/Shutterstock; Ulkastudio/Shutterstock; Ivan Polunin/Avalon; **2.5a** Rene Johnston/Toronto Star/Getty Images; **2.9** Fiona McAllister Photography/Moment/Getty Images; **2.10 left** Stephen Alvarez/National Geographic/Getty Images; **2.10 right** Andrew Syred/Science Source; **2.11** Alasdair James/E+/Getty Images; **2.12** Werayuth Tes/Shutterstock; **2.13** Nexus 7/Shutterstock; **2.14** Kristin Piljay/Pearson Education, Inc.; **p. 65** Doug Allan/Science Source; **2.15 top to bottom** Beth Van Trees/Shutterstock; Steve Gschmeissner/Science Source; lukovic photograpy/Shutterstock; Terekhov Igor/Shutterstock; Ian 2010/Fotolia; **2.16** Luiz A. Rocha/Shutterstock.

CHAPTER 3:

Chapter opening photo top Andras Csontos/Fotolia; **chapter opening photo right** Sunabesyou/Fotolia; **chapter opening photo bottom** khlungcenter/Shutterstock; **3.3 left** James Marvin Phelps/Shutterstock; **3.3 right** Ilolab/Fotolia; **p. 73** Arnulf Hettrich/ImageBroker/Alamy; **3.7** M. Unal Ozmen/Shutterstock; **3.9 top left** Dougal Waters/Photodisc/Getty Images; **3.9 top right** Biophoto Associates/Science Source; **3.9 center right** C. F. Armstrong/Science Source; **3.9 bottom** Biophoto Associates/Science Source; **3.10** Science Source; **3.12 left to right** Scanrail/Fotolia; Scanrail/Fotolia; Vincent Noel/Shutterstock; **3.13 clockwise from left** Thomas M Perkins/Shutterstock; Hannamariah/Shutterstock; Kayros Studio/Fotolia; Maksim Shebeko/Fotolia; Subbotina Anna/Fotolia; Bigacis/Shutterstock; Vladm/Shutterstock; Kellis/Fotolia; Valentin Mosichev/Shutterstock; **3.14 left** EcoPrint/Shutterstock; **3.14 right** Stockbyte/Getty Images; **3.15** Ron Vesely/MLB Photos/Getty Images; **3.16 left to right** Victoria Andreas/Shutterstock; Geoff Dann/Dorling Kindersley, Ltd.; Dave King/Dorling Kindersley, Ltd.; Tim Parmenter/Dorling Kindersley, Ltd.; Chris Ryan/OJO Images/Getty Images; Steve Gschmeissner/Science Source; Gts/Shutterstock; **3.21** Stefan Wackerhagen/Image Broker/Newscom; **3.28 top** Hemis/Alamy; **3.28 bottom** Friedrich Stark/Alamy.

CHAPTER 4:

Chapter opening photo top right Bhandol/Alamy; **chapter opening photo bottom left** Looker Studio/Shutterstock; **chapter opening photo center right** Robert S. McNeil/Baylor College of Medicine/Science Source; **chapter opening photo bottom right** Trevor Reeves/Shutterstock; **p. 89** Science Source; **4.1 top to bottom** M. I. Walker/Science Source; SPL/Science Source; Francine Iftode and Anne Aubusson; **4.3** NIBSC/Science Source; **4.6b** Slava Epstein, Northeastern University; **4.7 left** Biophoto Associates/Science Source; **4.7 right** Don W. Fawcett/Science Source; **4.10** MedImage/Science Source; **4.14** SPL/Science Source; **4.15c** Mary Martin/Biophoto Associates/Science Source; **4.16a** Michael Abbey/Science Source; **4.16b** Jeremy Burgess/Science Source; **4.18** Biology Pics/Science Source; **4.19** Don W. Fawcett/Science Source; **4.20a** Torsten Wittmann/Science Source; **4.20b** Roland Birke/Photolibrary/Getty Images; **4.21a** Eye of Science/Science Source; **4.21b** Charles Daghlian/Science Source; **4.22** Cavallini James/BSIP/Alamy.

CHAPTER 5:

Chapter opening photo top right Brent Hofacker/123RF; **chapter opening photo center** Valentyn Volkov/Shutterstock; **chapter opening photo bottom left** Tyler Olson/Shutterstock; **chapter opening photo bottom right** FotografiaBasica/Getty Images; **p. 109** Thomas Deerinck/NCMIR/Science Source; **5.1** Stephen Simpson/Photolibrary/Getty Images; **5.8** Laguna Design/Science Source; **5.15** David Cook/blueshiftstudios/Alamy; **5.19** SPL/Science Source.

CHAPTER 6:

Chapter opening photo bottom left Pung/Shutterstock; **chapter opening photo center** GJLP/CNRI/Science Source; **chapter opening photo top right** MaraZe/Shutterstock; **p. 125** Kris Mendoza/Moment/Getty Images; **6.1** Nikolay 007/Shutterstock; **6.3** Dmitrimaruta/Fotolia; **p. 132** Egd/Shutterstock; **p. 133** Hotwalkn/123RF; **6.12** Alex Staroseltsev/Shutterstock; **6.13** Maridav/Shutterstock; **6.16** Kristin Piljay/Alamy.

CHAPTER 7:

Chapter opening photo top right CSP_katia26/AGE Fotostock; **chapter opening photo center** John Kelly/Getty Images; **chapter opening photo bottom** Konradlew/Getty Images; **p. 141** Chandlerphoto/Getty Images; **7.1 left to right** Mark Bauer/Loop Images Ltd/Alamy; Tobias Friedrich/F1online digitale Bildagentur GmbH/Alamy; M. I. Walker/Science Source; **7.2 left to right** Pearson Education, Inc.; John Durham/Science Source; Biophoto Associates/Science Source; **7.6** Joyce/Fotolia; **7.7** Gabriela Tulian/Getty Images; **7.8b** Photos by LQ/Alamy; **7.14** Pascal Goetgheluck/Science Source.

CHAPTER 8:

Chapter opening photo left Biophoto Associates/Science Source; **chapter opening photo bottom** Du Cane Medical Imaging Ltd/Science Source; **chapter opening photo right** Szefei/123RF; **p. 155** Jordi Chias/Nature Picture Library; **8.1 left to right** Dr. Torsten Wittmann/Science Source; Dr. Yorgos Nikas/Science Source; Biophoto Associates/Science Source; Image Quest Marine; John Beedle/Photolibrary/Getty Images; **8.2 top to bottom** Eric Isselee/Shutterstock; Christian Musat/Shutterstock; Michaeljung/Fotolia;

Eric Isselee/Shutterstock; Ashley Toone/Alamy; **8.3** Ed Reschke/Photolibrary/Getty Images; **8.4** Biophoto Associates/Science Source; **8.7** Conly L. Rieder; **8.8a** Don W. Fawcett/Science Source; **8.8b** Kent Wood/Science Source; **8.10** Portra/Digital Vision/ Getty Images; **8.11** CNRI/Science Source; **8.12** Iofoto/ Shutterstock; **p. 167** Ed Reschke/Photolibrary/Getty Images; **8.17** David M. Phillips/Science Source; **8.19a** Scenics & Science/Alamy; **8.22 top** Lauren Shear/Science Source; **8.22 bottom** CNRI/Science Source; **8.23** Nhpa/Superstock; **p. 177** Ed Reschke/Photolibrary/ Getty Images.

CHAPTER 9:
Chapter opening photo top right Petrenko Andriy/ Shutterstock; **chapter opening photo left** Africa Studio/Shutterstock; **chapter opening photo bottom** KidStock/Getty Images; **p. 179** Eric Simon; **9.1** Science Source; **9.4** Patrick Lynch/Alamy; **9.5** James King-Holmes/Science Source; **9.8** Martin Shields/Science Source; **9.9 left to right** Tracy Morgan/Dorling Kindersley, Ltd.; Tracy Morgan/Dorling Kindersley, Ltd.; Eric Isselee/Shutterstock; Victoria Rak/Shutterstock; **9.10 left** Tracy Morgan/Dorling Kindersley, Ltd.; **9.10 right** Eric Isselee/Shutterstock; **9.12 left** James Woodson/Getty Images; **9.12 right** Cookie Studio/Shutterstock; **Table 9.1 top to bottom** David Terrazas Morales/Corbis; Editorial Image, LLC/Alamy; Eye of Science/Science Source; Science Source; **9.13 left** Ostill/Shutterstock; **9.13 right** Image Source/Getty Images; **9.16a left to right** Kuznetsov Alexey/Shutterstock; Erik Lam/123RF; Irina Oxilixo Danilova/Shutterstock; Purplequeue/Shutterstock; Markos86/Shutterstock; Eric Isselee/Shutterstock; **9.17** Saturn Stills/Science Source; **9.20** Mauro Fermariello/Science Source; **9.21** Eye of Science/Science Source; **9.23** Jean Dickey; **p. 195** Rootstock/Shutterstock; **9.25 clockwise from left** In Green/Shutterstock; Rido/Shutterstock; Andrew Syred/Science Source; Dave King/Dorling Kindersley, Ltd.; Jo Foord/ Dorling Kindersley, Ltd.; **9.28** Archive Pics/Alamy; **9.29 top left to bottom right** Jerry Young/Dorling Kindersley, Ltd.; Gelpi/Fotolia; Dave King/Dorling Kindersley, Ltd.; Dave King/ Dorling Kindersley, Ltd.; Tracy Morgan/Dorling Kindersley, Ltd.; Dave King/Dorling Kindersley, Ltd.; Jerry Young/Dorling Kindersley, Ltd.; Dave King/ Dorling Kindersley, Ltd.; Tracy Morgan/Dorling Kindersley, Ltd.; Irina Oxilixo Danilova/Shutterstock; **p. 199** Eric J. Simon; **p. 202** Arco/G. Lacz/GmbH/Alamy; **p. 203** Eric J. Simon.

CHAPTER 10:
Chapter opening photo top right JLPH/Cultura/ Getty Images; **chapter opening photo center** Volker Steger/Science Source; **chapter opening photo bottom left** Eye of Science/Science Source; **chapter opening photo bottom right** Thomas Deerinck/

NCMIR/Science Source; **p. 205** Kateryna Kon/ Science Photo Library/Getty Images; **10.3 left** Barrington Brown/Science Source; **10.3 right** Library of Congress Print and photographs Division; **10.11** Eye of Science/Science Source; **10.22** Georgette Apol/ Steve Bloom Images/Alamy; **10.23** Koji Niino/Mixa/ Alamy; **10.24** Oliver Meckes/Science Source; **10.26** Norm Thomas/Science Source; **10.28** Hazel Appleton/ Health Protection Agency Centre for Infections/ Science Source; **10.31** NIBSC/Science Photo Library/ Science Source; **10.32** Will & Deni McIntyre/Science Source; **10.33 clockwise from left** Scott Camazine/ Alamy; Cynthia Goldsmith/CDC; Cynthia Goldsmith/CDC; Cultura Creative (RF)/Alamy; Cynthia Goldsmith/CDC; National Institute of Allergy and Infectious Diseases (NIAD); Cynthia Goldsmith/CDC; Phanie/Alamy.

CHAPTER 11:
Chapter opening photo top Peathegee Inc/Blend Image/Getty Images; **chapter opening photo center** Alila Medical Media/Shutterstock; **chapter opening photo center right** S4svisuals/Shutterstock; **chapter opening photo bottom left** Akkharat Jarusilawong/Shutterstock; **p. 231** SPL/Science Source; **11.1 left to right** Steve Gschmeissner/Science Source; Steve Gschmeissner/Science Photo Library/Alamy; Ed Reschke/Oxford Scientific/Getty Images; **11.4** Iuliia Lodia/Fotolia; **11.9** Dr. Thomas Kaufman Dept. of Biology Indiana University; **11.10** Alila Medical Media/Shutterstock; **11.11** Videowokart/Shutterstock; **p. 239** Joseph T. Collins/Science Source; **11.13a** Courtesy of the Roslin Institute, Edinburgh; **11.13b** Randall S. Prather, PhD/Courtesy of Jim Curley; **11.13c** Robert Lanza; **11.15 left** Mauro Fermariello/Science Source; **11.15 right** Klaus Guldbrandsen/Science Source; **p. 244** Eric Isselee/Shutterstock; **11.19** Geo Martinez/Shutterstock; **11.21** Alastair Grant/Pool/AP Images; **11.23** Biophoto Associates/ Science Source.

CHAPTER 12:
Chapter opening photo top right Photolinc/Shutterstock; **chapter opening photo left** Volker Steger/ Science Source; **chapter opening photo bottom right** Dinodia Photos/Alamy; **p. 251** Isak55/Shutterstock; **12.1** STR/AP Images; **12.2 left** Prof. S. Cohen/Science Source; **12.2 right** Huntington Potter/ University of South Florida College of Medicine; **12.3** Dmitry Lobanov/Fotolia; **12.6** Eric Carr/Alamy; **12.7** Volker Steger/Science Source; **12.8** Inga Spence/ Alamy; **12.9 top** Christopher Gable and Sally Gable/ Dorling Kindersley, Ltd.; **12.9 bottom** U.S. Department of Agriculture (USDA); **12.10 left** International Rice Research Institute (IRRI); **12.10 right** Fotosearch RM/AGE Fotostock; **p. 257** Andy Manis/ Bloomberg/Getty Images; **12.13** Eric J. Simon; **21.15** Andrew Brookes/National Physical Laboratory/Science Source; **12.17** Steve Helber/AP Images; **12.18**

Fine Art Images/Heritage Image Partnership Ltd/ Alamy; **12.19** Volker Steger/Science Source; **12.21 top** Pikselstock/Shutterstock; **12.21 bottom** David Parker/Science Photo Library/Science Source; **12.24 left** Scott Camazine/Science Source; **12.24 right** FBI UPI Photo Service/Newscom; **12.25** James King-Holmes/Science Source; **12.26** Science Source; **12.27** Alex Milan Tracy/NurPhoto/Sipa U/Newscom; **12.28** Isak55/Shutterstock; **12.29** Image Point Fr/Shutterstock; **12.30** Pictures From History/Newscom; **12.31** Jekesai Njikizana/AFP/GettyImages.

CHAPTER 13:
Chapter opening photo top BW Folsom/Shutterstock; **chapter opening photo bottom left** Lapina/ Shutterstock; **chapter opening photo bottom right** John Bryant/Gallo Images/Getty Images; **p. 277** George Mulala/Reuters; **13.1** Aditya Singh/Moment/ Getty Images; **13.2a** Chrispo/Shutterstock; **13.2b** Sabena Jane Blackbird/Alamy; **13.3 left** Science Source; **13.3 right** Classic Image/Alamy; **13.4a** Celso Diniz/Shutterstock; **13.4b** Tim Laman/National Geographic/Getty Images; **13.5 clockwise from left** Francois Gohier/Science Source; Francois Gohier/ Science Source; Science Source; Sergei Cherkashin/ Reuters; Pixtal/SuperStock; **13.7 left** Dr.Keith Wheeler/Science Source; **13.7 right** Lennart Nilsson/ TT Nyhetsbyrån; **13.9** Blickwinkel/Hecker/Alamy; **13.10** Matthew Oldfield Underwater Photography/ Alamy; **13.11** Philippe Wojazer/Reuters; **13.12** Our Wild Life Photography/Alamy; **13.13** Adam Jones/ The Image Bank/Getty Images; **13.14** Andy Levin/ Science Source; **13.19** Steve Bloom Images/Alamy; **13.20** Planetpix/Alamy; **13.21** Heather Angel/Natural Visions/Alamy; **13.22** Mariko Yuki/Shutterstock; **13.23a** Peter Grant; **13.23b** Gabrielle Therin-Weise/ Photographer's Choice RF/Getty Images; **13.24a** Reinhard/ARCO/Nature Picture Library; **13.24b** Paolo-manzi/Shutterstock.

CHAPTER 14:
Chapter opening photo top right Sappington Todd/ Getty Images; **chapter opening photo center left** Oberhaeuser/Agencja Fotograficzna Caro/Alamy; **chapter opening photo center right** GL Archive/ Alamy; **chapter opening photo bottom** Keneva Photography/Shutterstock; **p. 303** Michael Collier; **14.1 left** Cathleen A Clapper/Shutterstock; **14.1 right** Peter Scoones/Nature Picture Library; **14.2 clockwise from left** Bill Draker/Rolf Nussbaumer Photography/Alamy; David Kjaer/Nature Picture Library; Jupiterimages/Stockbyte/Getty images; Photos/Getty Images; Phil Date/Shutterstock; Blvdone/Shutterstock; Comstock/Stockbyte/Getty Images; blvdone/Shutterstock; **14.4 left to right** APHIS Animal and Plant health inspection Service/ USDA; Jared Hobbs/SuperStock; McDonald/Photoshot; Joe McDonald/Corbis Documentary/Getty Images; J & C Sohns/Tier und Naturfotografie/

AGE Fotostock; Oyvind Martinsen Wildlife Collection/ Alamy; Jon G. Fuller/VWPics/Alamy; Danita Delimont/Gallo Images/AGE Fotostock; **14.5 left to right** Chuck Brown/Science Source; Dogist/Shutterstock; Alistair Duncan/Dorling Kindersley, Ltd.; Dorling Kindersley, Ltd.; Dr. Kazutoshi Okuno; **14.6 left to right** John Shaw/Photoshot; Morey Milbradt/ Stockbyte/Getty Images; Ron Niebrugge/Alamy; **14.8** Michelle Gilders/Alamy; **14.9** Artemyeva/Shutterstock; **14.11a** Florida Museum photo by Jeff Gage; **14.11b clockwise from left** Bill Brooks/Alamy; Arco Images GmbH/Alamy; David Chapman/Alamy; **14.12 top to bottom** Interfoto/Alamy; Mary Plage/Oxford Scientific/Getty Images; Ralph Lee Hopkins/National Geographic RF/Getty Images; **14.13** Barbara Rich/ Moment/Getty Images; **14.17 left to right** Jean-Paul Ferrero/Auscape International Pty Ltd/Alamy; Eugene Sergeev/Shutterstock; Kamonrat/Shutterstock; **14.19** George Atsametakis/Alamy; **14.20** Chris Hellier/Science Source; **14.21 left to right** Image Quest Marine; Christophe Courteau/Science Source; Reinhard Discherl/Alamy; Reinhard Dirscherl/Alamy Stock Photo; James Watt/Image Quest Marine; **p. 320** Florida Stock/Shutterstock; **14.23** Mark Thiessen/National Geographic Creative/Alamy; **14.27a** John Sullivan/ Alamy; **14.27b** MikeLane45/iStock/Getty Images.

CHAPTER 15:

Chapter opening photo top right MintImages/ Shutterstock; **chapter opening photo center left** Harry Vorsteher/Cultura Creative (RF)/Alamy; **chapter opening photo center right** Odilon Dimier/PhotoAlto sas/Alamy; **chapter opening photo bottom** Floris van Breugel/Nature Picture Library/Alamy; **p. 327** Stephanie Schuller/Science Source; **15.2** Mark Garlick/Science Photo Library/Corbis; **15.5** B. Murton/Southampton Oceanography Centre/Science Source; **15.6 left** Imagedb/Shutterstock; **15.6 right** Dr. Tony Brain and David Parker/Science Photo Library/Science Source; **15.7 left to right** Scimat/ Science Source; Niaid/CDC/Science Source; CNRI/ SPL/Science Source; **15.8a** David M. Phillips/Science Source; **15.8b** John Walsh/Science Source; **15.8c** Esther R. Angert; **15.9** Science Photo Library - Steve Gschmeissner/Brand X Pictures/Getty Images; **15.10** Eye of Science/Science Source; **15.11a** Sinclair Stammers/Science Source; **15.11b** Dr. Gary Gaugler/ Science Source; **15.12** Peter Batson/Image Quest Marine; **15.13 top** Huetter, C./Arco Images/Alamy; **15.13 bottom** Nigel Cattlin/Alamy; **15.15** Sipa USA/ Newscom; **15.16** Eric J. Simon; **15.17** Jim West/Alamy; **15.18** SPL/Science Source; **15.19 top** Centers for Disease Control and Prevention (CDC); **15.19 center** Scott Camazine/Science Source; **15.19 bottom left** Sarah2/Shutterstock; **15.19 bottom right** David M. Phillips/Science Source; **15.21a** Carol Buchanan/ F1online/AGE Fotostock; **15.21b** Eye of Science/ Science Source; **15.21c** Blickwinkel/NaturimBild/ Alamy; **15.22 left to right** Eye of Science/Science

Source; David M. Phillips/Science Source; Biophoto Associates/Science Source; Claude Carre/Science Source; Eye of Science/Science Source; Michael Abbey/Science Source; **15.23a** Nigel Downer/ Science Source; **15.23b** Eye of Science/Science Source; **15.24a** Eye of Science/Science Source; **15.24b** Steve Gschmeissner/Science Source; **15.24c** Manfred Kage/Science Source; **15.25 left to right** Marevision/AGE Fotostock; Marevision/AGE Fotostock; David Hall/Science Source; **15.26** Lucidio Studio, Inc/Moment/Getty Images.

CHAPTER 16:

Chapter opening photo top right Webphotographeer/E+/Getty Images; **chapter opening photo center** ITS AL Dente/Shutterstock; **chapter opening photo bottom** National Geographic Creative/Alamy; **p. 349** Valentyn Volkov/Shutterstock; **16.2** Science Source; **16.3** Steve Gorton/Dorling Kindersley, Ltd.; **16.4** Garry DeLong/Oxford Scientific/Getty Images; **16.5 left** Bob Gibbons/Alamy; **16.5 right** Gerd Guenther/Science Source; **16.7 left to right** Kristin Piljay; James Randklev/Photographer's Choice RF/Getty Images; V. J. Matthew/Shutterstock; Dale Wagler/ Shutterstock; **16.8** Duncan Shaw/Science Source; **16.9** John Serrao/Science Source; **16.11 clockwise from left** Jon Bilous/Shutterstock; Biophoto Associates/Science Source; Mothy20/Getty Images; Ed Reschke/Photolibrary/Getty Images; **16.12** Field Museum Library/Premium Archive/Getty Images; **16.13** Guillaume Hullin/Shutterstock; **16.15 left to right** Stephen P. Parker/Science Source; Morales/ AGE Fotostock; Gunter Marx/Alamy; **16.16** Gene Cox/Science Source; **16.18 left to right** Jean Dickey; Tyler Boyes/Shutterstock; Christopher Marin/Shutterstock; Jean Dickey; **16.20 clockwise from left** Jean Dickey; Scott Camazine/Science Source; Sonny Tumbelaka/AFP Creative/Getty Images; **16.21** Prill/ Shutterstock; **Table 16.1 top to bottom** Steve Gorton/Dorling Kindersley, Ltd.; Richard Griffin/Shutterstock; Photogal/Shutterstock; Dionisvera/Fotolia; Radu Razvan/Shutterstock; Dorling Kindersley/Getty Images; Photogal/Shutterstock; National Tropical Botanical Garden; Alle/Shutterstock; **16.22 clockwise from left** Jean Dickey; Stan Rohrer/Alamy; Science Source; Astrid & Hanns-Frieder Michler/Science Source; VEM/Science Source; Pabkov/Shutterstock; **16.23 top** Jupiterimages/Photos/Getty Images; **16.23 bottom** Blickwinkel/Alamy; **16.24a** Wolfness72/ Shutterstock; **16.24c** Raymond Louis/Alamy; **16.25** Bedrich Grunzweig/Science Source; **16.26 top left** Mikeledray/Shutterstock; **16.26 bottom left** Imagebroker/Food-Drinks/SuperStock; **16.26 right** Will Heap/Dorling Kindersley, Ltd.; **16.27** Christine Case; **16.28** Wildlife GmbH/Alamy.

CHAPTER 17:

Chapter opening photo top right Gazimal/Getty Images; **chapter opening photo center** Dropu/

Shutterstock; **chapter opening photo bottom** Sebastien Plailly/Science Source; **p. 371** Sabena Jane Blackbird/Alamy; **17.1** Gunter Ziesler/Photolibrary/Getty Images; **17.4** Sinclair Stammers/Science Source; **17.5 left** Christian Jegou/Publiphoto/Science Source; **17.5 right** Lorraine Hudgins/Shutterstock; **17.9** James Watt/Image Quest Marine; **17.10 top left** Joe Belanger/Shutterstock; **17.10 top right** Sue Daly/Nature Picture Library/Alamy; **17.10 center** Marek Mis/Science Source; **17.10 bottom** Pavlo Vakhrushev/Fotolia; **17.13 left to right** Georgette Douwman/Nature Picture Library; Jez Tryner/ Image Quest Marine; Christophe Courteau/Nature Picture Library; Marevision/AGE Fotostock; Reinhard Dirscherl/Alamy; **17.14 top right** CMB/AGE Fotostock; **17.14 bottom left** Geoff Brightling/Gary Stabb/Dorling Kindersley Ltd.; **17.14 bottom right** Eye of Science/Science Source; **17.16 left** Schulz, H./Juniors Bildarchiv GmbH/Alamy; **17.16 right** Blue-Sea.cz/Shutterstock; **17.17a** Steve Gschmeissner/Science Source; **17.17b** Eye of Science/Science Source; **17.17c** Eye of Science/Science Source; **p. 380** Astrid & Hanns-Frieder Michler/Science Source; **17.18 top to bottom** Snowleopard1/Getty Images; Maximilian Weinzierl/Alamy; Jean L. Dickey; **17.19** Dave King/Dorling Kindersley, Ltd.; **17.20 clockwise from left** Nenad Druzic/Moment Open/Getty Images; Dave King/Dorling Kindersley Ltd.; Andrew Syred/Science Source; Larry West/Science Source; Snowleopard1/Getty Images; **17.21 top** Dave King/ Dorling Kindersley Ltd.; **17.21 center left** Maximilian Weinzierl/Alamy; **17.21 center** Tom McHugh/ Science Source; **17.21 center right** Nature's Images/ Science Source; **17.21 bottom left** Dietmar Nill/ Nature Picture Library; **17.21 bottom right** Nancy Sefton/Science Source; **17.22 left** Jean L. Dickey; **17.22 right** Tom McHugh/Science Source; **17.23** Radius Images/Alamy; **17.24 clockwise from left** NH/Shutterstock; Stuart Wilson/Science Source; Huayang/Moment/Getty Images; Jean L. Dickey; Jean L. Dickey; Lkpro/Shutterstock; David Oberholzer/ Prisma by Dukas Presseagentur GmbH/Alamy; Stuart Wilson/Science Source; **17.25 top** Thomas Kitchin & Victoria Hurst/Design Pics Inc./Alamy; **17.25 bottom** Keith Dannemiller/Alamy; **17.26 clockwise from left** Roger Steene/Image Quest Marine; Andrew J. Martinez/Science Source; Jose B. Ruiz/Nature Picture Library; Image Quest Marine; Roger Steene/Image Quest Marine; Tbkmedia.de/ Alamy Alamy; **17.27** Colin Keates/Courtesy of the Natural History Museum/Dorling Kindersley Ltd.; **17.29 left** Heather Angel/Natural Visions/Alamy; **17.29 right** Image Quest Marine; **17.31a** Tom McHugh/Science Source; **17.31b** Marevision/AGE Fotostock; **17.31b inset** A Hartl/Blickwinkel/AGE Fotostock; **17.31c** George Grall/National Geographic/ Getty Images; **17.31d** Christian Vinces/Shutterstock; **17.32a left** Tom McHugh/Science Source; **17.32a right** Jack Goldfarb/AGE Fotostock; **17.32b** LeChatMachine/Fotolia; **17.32c left** Gary Meszaros/Science

Source; **17.32c right** Bill Brooks/Alamy; **17.34 left to right** Encyclopaedia Britannica/Universal Images Group North America LLC/Alamy; Sylvain Cordier/Photolibrary/Getty Images; Jerry Young/Dorling Kindersley Ltd.; Dlillc/Corbis; Miguel Periera/Courtesy of the Instituto Fundacion Miguel Lillo, Argentina/Dorling Kindersley Ltd.; **17.35** Adam Jones/Photodisc/Getty Images; **17.36 left to right** Jean-Philippe Varin/Science Source; Rebecca Jackrel/AGE Fotostock; Barry Lewis/Alamy; **17.38 clockwise from left** Creativ Studio Heinem/Westend61/AGE Fotostock; Siegfried Grassegger/ImageBroker/AGE Fotostock; Arco Images Gmbh/Tuns/Alamy; Juan Carlos Muñoz/AGE Fotostock; Anup Shah/Nature Picture Library; John P Kelly/Photographer's Choice RF/Getty Images; Anup Shah/Nature Picture Library; Lexan/123RF; P. Wegner/Arco Images/AGE Fotostock; **17.40** John Reader/SPL/Science Source; **17.42 top** Oxford Scientific/Getty Images; **17.42 bottom** Darlyne A. Murawski/National Geographic/Getty Images; **17.43** Kablonk/Superstock; **17.44** Kristin Piljay/Danita Delimont/Alamy.

CHAPTER 18:
Chapter opening photo top right Olegusk/Shutterstock; **chapter opening photo center** Vince Clements/Shutterstock; **chapter opening photo bottom** Olena Serditova/Alamy; **p. 407** Sylvain Cordier/Photodisc/Getty Images; **18.1** Philippe Psaila/Science Source; **18.2** US Coast Guard Photo/Alamy; **18.3a** Barry Mansell/Nature Picture Library; **18.3b** Sue Flood/Alamy; **18.3c** Juniors Bildarchiv/AGE Fotostock; **18.3d** Jeremy Woodhouse/Stockbyte/Getty Images; **18.4** David Wall/Alamy; **18.5** NASA Earth Observing System; **18.6 left** Science Source; **18.6 right** Peter Batson/Image Quest Marine; **18.7** NHPA/Photoshot; **18.8a** Jean Dickey; **18.8b** AGE Fotostock/SuperStock; **18.9** Jean Dickey; **18.11 left** Enn Li Photography/Getty Images; **18.11 right** J & C Sohns/Tier und Naturfotografie/AGE Fotostock; **18.12** Ed Reschke/Photolibrary/Getty Images; **18.13** Robert Stainforth/Alamy; **18.14** WorldSat International Inc./Science Source; **18.16** Ishbukar Yalilfatar/Shutterstock; **18.17** Kevin Schafer/Alamy; **18.18** Jean Dickey; **18.20** Digital Vision/Photodisc/Getty Images; **18.21** Ron Watts/All Canada Photos/Getty Images; **18.22** George McCarthy/Nature Picture Library; **18.29** AGE Fotostock/SuperStock; **18.30** Eric J. Simon; **18.31** Juan Carlos Muñoz/AGE Fotostock; **18.32** Earl Scott/Science Source; **18.33** Mark Coffey/All Canada Photos/SuperStock; **18.34** Flashbacknyc/Shutterstock; **18.35** Jorma Luhta/Nature Picture Library; **18.36** Paul Nicklen/National Geographic/Getty Images; **18.37** Gordon Wiltsie/National Geographic/Getty Images; **18.39** Planet Observer/Universal Images Group North America LLC/Alamy; **18.40** Crack Palinggi/Reuters; **18.41** NASA/Goddard Space Flight Center/US Geological Survey (USGS); **18.42** Jim West/Alamy; **18.47** Patrick T Fallon/Reuters; **18.48** Tracy Ferrero/Alamy; **18.49a** Roger Givens/

Dembinsky Photo Associates/Alamy; **18.50** CB2/ZOB/Supplied by WENN.com/Newscom; **18.51** Chris Cheadle/All Canada Photos/Getty Images; **18.52a** Alekcey/Shutterstock; **18.52b** Andreanita/123RF; **p. 433** Digital Vision/Photodisc/Getty Images.

CHAPTER 19:
Chapter opening photo top Rob245/Fotolia; **chapter opening photo center right** James Watt/Perspectives/Getty Images; **chapter opening photo bottom** Atlantide Phototravel/Corbis Documentary/Getty Images; **p. 437** Richard Whitcombe/Shutterstock; **19.1a** Avalon/Photoshot License/Alamy; **19.1b** Zuma Press, Inc./Alamy; **19.1c** Flpa/Alamy; **19.2** Wave Royalty Free/Design Pics Inc/Alamy; **19.3** WorldFoto/Alamy; **19.4 left to right** Roger Phillips/Dorling Kindersley, Ltd.; Jane Burton/Dorling Kindersley, Ltd.; Wavebreakmedia/Shutterstock; **Table 19.2 left** Prill/Shutterstock; **Table 19.2 right** Anke Van Wyk/Shutterstock; **19.5** Juniors Bildarchiv GmbH/Alamy; **19.6** Accent Alaska.com/Alamy; **19.7 left** Joshua Lewis/Shutterstock; **19.7 right** Wizdata/Shutterstock; **19.8a** Rick & Nora Bowers/Alamy; **19.8b** Mauricio Handler/National Geographic Creative/Alamy; **19.9** Design Pics/SuperStock; **19.10** Meul/ARCO/Nature Picture Library; **19.11** Alan & Sandy Carey/Science Source; **19.12 left to right** William Leaman/Alamy; USDA Forest Service; Gilbert S. Grant/Science Source; **19.13** Dan Burton/Nature Picture Library; **19.14** Max Faulkner/Fort Worth Star-Telegram/MCT/Getty Images; **19.15** U.S. Geological Survey; **19.16** Chris Johns/National Geographic/Getty Images; **19.17** Pete Evans/FOAP/Getty Images; **19.18** Nigel Cattlin/Alamy; **19.19** Andre Skonieczny/Image Broker/Alamy; **19.24** Franzfoto/Alamy.

CHAPTER 20:
Chapter opening photo top left Ming-Hsiang Chuang/Shutterstock; **chapter opening photo top right** Lauree Feldman/Photolibrary/Getty Images; **chapter opening photo bottom left** Kseniia Ivanova/EyeEm/Getty Images; **chapter opening photo bottom right** Bill Barksdale/Design Pics Inc/Alamy; **20.1** Sopotnicki/Shutterstock; **20.2 left** Gerard Lacz/AGE Fotostock; **20.2 right** Tobias Bernhard/Oxford Scientific/Getty Images; **20.3** Andre Seale/AGE Fotostock; **20.4** American Folklife Center/Library of Congress Print and photographs division [CRF-LE-C039-07]; **20.5** Gerald Herbert/AP Images; **20.6** Richard D. Estes/Science Source; **20.8** M. I. Walker/Science Source; **20.9** Little Dinosaur/Alamy; **20.10** Eric Lemar/Shutterstock; **20.11** WaterFrame/Alamy; **20.12 left** Robert Hamilton/Alamy; **20.12 right** Barry Mansell/Nature Picture Library; **20.13 left** Dante Fenolio/Science Source; **20.13 right** Andrew M. Snyder/Moment Open/Getty Images; **20.14 top left** Bildagentur-online/TH Foto-Werbung/Science Source; **20.14 top right** Luca Invernizzi Tetto/AGE Fotostock; **20.14 bottom** Tony Camacho/Science

Source; **20.16** Piotr Malczyk/Alamy; **20.17 top to bottom** Marlon Lopez MMG1 Design/Shutterstock; Thomas Kline/Design Pics/Getty Images; WaterFrame/Alamy; FLPA/Alamy; **20.21** Mark Conlin/Vwpics/Visual&Written SL/Alamy; **20.22** Matthijs Wetterauw/Alamy; **20.23** HiloFoto/Moment/Getty Images; **20.24** AdstockRF/Universal Images Group Limited/Alamy; **20.30** Don Paulson Photography/Purestock/Alamy; **20.35** Eric Albrecht/The Columbus Dispatch/AP Images; **20.36** Goddard Space Flight Center/NASA; **20.38 top** Lee Foster/Alamy; **20.38 bottom** Matthew Dixon/Shutterstock; **20.39** Tony Clevenger/AP Images; **20.40a** Szasz-Fabian Jozsef/Shutterstock; **20.40b top** Centers for Disease Control and Prevention; **20.40b bottom left** Scott Camazine/Science Source; **20.40b bottom right** Jonathan Oliver; **20.41** United States Department of Agriculture; **20.42** South Florida Water Management District; **20.43 left** Andrew_Shurtleff/The Daily Progress/AP Images; **20.43 right** ZUMA Press, Inc/Alamy; **20.44a left** Anthony Mercieca/Science Source; **20.44a top right** Joel Sartore/National Geographic Photo Ark/National Geographic/Getty Images; **20.44a bottom right** G. Ronald Austing/Science Source; **20.44b left to right** U.S Air Force photo by Senior Airman Benjamin Wiseman; Nature and Science/Alamy; Smith Collection/Gado/Archive Photos/Getty Images.

ILLUSTRATION AND TEXT CREDITS

CHAPTER 1:
1.6c: http://www.cell.com/current-biology/fulltext/S0960-9822(15)00328-0; **1.22:** http://www.savetheseaturtle.org/The-Evolution-of-Sea-Turtles.html; **page 54:** https://www.nps.gov/pais/learn/nature/kridley.htm.

CHAPTER 3:
3.19: Based on Protein Databank: http://www.pdb.org/pdb/explore/explore.do?structureId=1a00; **3.20:** Based on The Core module 8.11 and *Campbell Biology* 10e Fig. 19.10.

CHAPTER 7:
7.5: T. W. Engelmann, Bacterium photometricum, Ein Beitrag zur vergleichenden Physiologie des Licht- und Farbensinnes, Archiv. Fur Physiologie 30:95-124 (1883); **7.12:** Adapted from Richard and David Walker, *Energy, Plants and Man*, fig. 4.1, p. 69. Oxygraphics. Copyright Richard Walker. Used courtesy of Richard Walker, http://www.oxygraphics.co.uk.

CHAPTER 10:
Page 226: Text quotation by Joshua Lederberg, from Barbara J. Culliton, "Emerging Viruses, Emerging Threat," *Science*, 247, p. 279, 35/19/1990; **10.29:** Data

taken from Karaca, K., Bowen, R., Austgen, L. E., Teehee, M., Siger, L., Grosenbaugh, D., ... & Minke, J. M. (2005). "Recombinant canarypox vectored West Nile virus (WNV) vaccine protects dogs and cats against a mosquito WNV challenge." *Vaccine*, 23(29), 3808-3813; **page 229:** Dirlikov E, Ryff KR, Torres-Aponte J, et al. Update: Ongoing Zika Virus Transmission — Puerto Rico, November 1, 2015–April 14, 2016. *MMWR Morb Mortal Wkly Rep* 2016;65:451–455. DOI: http://dx.doi.org/10.15585/mmwr.mm6517e2 https://www.cdc.gov/mmwr/volumes/65/wr/mm6517e2.htm.

CHAPTER 12:
12.23: "Nic Volker and XIAP - The Impact of a Single Nucleotide," http://cbm.msoe.edu/markMyweb/genomicJmols/xiap.html MSOE CBM: Conservation of Amino Acids.; **Page 270:** MARYLAND v. KING CERTIORARI TO THE COURT OF APPEALS OF MARYLAND No. 12–207. Argued February 26, 2013–Decided June 3, 2013. SUPREME COURT OF THE UNITED STATES.

CHAPTER 14:
Page 304: Charles Darwin, *The Voyage of the Beagle* (Auckland: Floating Press, 1839); **14.15:** Adapted from "Active Volcanoes and Plate Tectonics, 'Hot Spots' and the 'Ring of Fire'" by Lyn Topinka, U.S. Geological Survey website, January 2, 2003; **page 323:** Charles Darwin in *The Origin of Species* (London: Murray, 1859).

CHAPTER 15:
15.1: Maps of North America. Published by Vidiani.com. http://www.vidiani.com/maps/maps_of_north_america/maps_of_usa/large_detailed_administrative_and_road_map_of_USA.jpg; **15.14:** Tortora, Gerard J.; Funke, Berdell R.; Case, Christine L., *Microbiology: an introduction*, 9th Ed., ©2007. Reprinted and Electronically reproduced by permission of Pearson Education, Inc.,

Upper Saddle River, New Jersey; **15.20:** Data from V. K. Ridaura et al., "Gut microbiota from twins discordant for obesity modulate metabolism in mice." *Science* 341 (2013). DOI: 10.1126/science.1241214

CHAPTER 16:
Table 16.1: Data from Randy Moore et al., *Botany*, 2nd ed. Dubuque, IA: Brown, 1998, Table 2.2, p. 37; **16.24b:** Briscoe, Charles B. 1959. "Early Results of Mycorrhizal Inoculation of Pine in Puerto Rico." *Carribean Forester*, July-December:73-77; **page 369:** Data from: L. Ziska et al., "Recent warming by latitude associated with increased length of ragweed pollen season in central North America." *Proceedings of the National Academy of Sciences* 108: 4248–4251 (2011).

CHAPTER 17:
17.39: Drawn from photos of fossils: *A. Ramidus* adapted from www.age-of-the-sage.org/evolution/ardi_fossilized_skeleton.html. *H. neanderthalensis* adapted from *The Human Evolution Coloring Book*. *P. boisei* drawn from a photo by David Bill.

CHAPTER 18:
18.25: J. H. Withgott and S. R. Brennan, *Environment: The Science Behind the Stories*, 3rd Ed., © 2008. Reprinted and electronically reproduced by permission of Pearson Education, Inc., Upper Saddle River, New Jersey; **18.44:** NASA.gov website GISS Surface Temperature Analysis (global map generator), http://data.giss.nasa.gov/gistemp/maps/ (settings for generating this specific map are shown below the map); **18.45:** Based on Climate Change 2013: The Physical Science Basis. Working Group I Contribution to the Fourth Assessment Report of the Intergovernmental Panel on Climate Change; **page 435:** Based on data from http://www.ncdc.noaa.gov/land-based-station-data/climate-normals/1981-2010-normals-data.

CHAPTER 19:
Table 19.1: Data from Centers for Disease Control and Prevention website; **19.8a:** Data from P. Arcese et al., "Stability, Regulation and the Determination of Abundance in an Insular Song Sparrow Population," *Ecology*, 73: 805–882 (1992); **19.8b:** Data from T. W. Anderson, "Predator Responses, Prey Refuges, and Density Dependent Mortality of a Marine Fish," *Ecology* 82: 245–257 (2001); **19.13:** Data from Fisheries and Oceans, Canada, 1999; **19.17b:** Data from Letnic, M., et al. "Restricting access to invasion hubs enables sustained control of an invasive vertebrate." *Journal of Applied Ecology* 52(2), 341-347 (2015); **19.17c:** Modified from: Letnic, M., Webb, J. K., Jessop, T. S., & Dempster, T. (2015). "Restricting access to invasion hubs enables sustained control of an invasive vertebrate." *Journal of Applied Ecology*, 52(2), 341-347. **19.20:** Data from United Nations, "The World at 6 Billion," 2007; **Table 19.3:** Data from Population Reference Bureau; **19.21:** Data from U.S. Census Bureau; **19.22:** Data from International Data Base, U.S. Census Bureau website (2015); **19.23:** Data from *Living Planet Report 2012: Biodiversity, Biocapacity and Better Choices*, World Wildlife Fund (2012); **page 456:** Data from "Transitions in World Population," *Population Bulletin*, 59: 1 (2004).

CHAPTER 20:
20.17: Data extracted from Figure 1, graph C in Kelly, B. C., Ikonomou, M. G., Blair, J. D., Morin, A. E., & Gobas, F. A. (2007). "Food web–specific biomagnification of persistent organic pollutants." *Science*, 317(5835), 236-239; **20.21:** Based on data from Paine, R. "Food Web Complexity and Species Diversity," *The American Naturalist*, 100 (910), 65-75 (1966). Retrieved from http://www.jstor.org/stable/2459379; **20.40c:** Data from Allan, B. F. et al. "Effect of forest fragmentation on Lyme disease risk," *Conservation Biology*, 17(1), 267-272. P. 271 Fig. 1c (2003); **page 486:** Data from J. M. Teal, "Energy Flow in the Salt Marsh Ecosystem of Georgia," *Ecology*, 43:614–624 (1962).

APPENDIX D Selected Answers

CHAPTER 1

1. b (some living organisms are single-celled)
2. atom, molecule, cell, tissue, organ, organism, population, ecosystem, biosphere; the cell
3. Photosynthesis cycles nutrients by converting the carbon in carbon dioxide to sugar, which is then consumed by other organisms. Additionally, the oxygen in water is released as oxygen gas. Photosynthesis contributes to energy flow by converting sunlight to chemical energy, which is then also consumed by other organisms, and by producing heat.
4. On average, those individuals with heritable traits best suited to the local environment produce the greatest number of offspring that survive and reproduce. This increases the frequency of those traits in the population over time. The result is the accumulation of evolutionary adaptations.
5. c
6. c
7. evolution
8. A fact should be confirmed to be the same by all observers, whereas opinions can differ from one observer to the next.
9. a3, b2, c1, d4
10. Information flow: Genes encode information to make proteins, which then produce physical traits.
11. Pathways that transform energy and matter: Sea turtles obtain both energy and molecular building blocks from the grass they consume.
12. Interactions within biological systems: Chemicals released in one part of the world can interact with ocean water in another part, affecting the life found there.

Thinking Like a Scientist

The buckets acted as a control group. By comparing the movement of the sea turtles with the movement of the floating buckets, the researchers could assign any difference to be the result of swimming by the turtles.

CHAPTER 2

1. electrons; neutrons
2. protons
3. Sodium-23 has an atomic number of 11 and a mass number of 23. The radioactive isotope, sodium-22, has an atomic number of 11 and a mass number of 22.
4. Organisms incorporate radioactive isotopes of an element into their molecules just as they do nonradioactive isotopes, and researchers can detect the presence of the radioactive isotopes.
5. Each carbon atom has only three covalent bonds instead of the required four.
6. The positively charged hydrogen regions would repel each other.
7. d
8. c
9. The positive and negative poles cause adjacent water molecules to become attracted to each other, forming hydrogen bonds. The properties of water such as cohesion, temperature regulation, and water's ability to act as a solvent all arise from this atomic "stickiness."
10. Because a nonpolar molecule cannot form hydrogen bonds, it would not have the properties that allow water to act as the basis of life, such as the ability to dissolve substances and water's cohesive properties.
11. The cola is an aqueous solution, with water as the solvent, sugar as the main solute, and the CO_2 making the solution acidic.
12. Pathways that transform energy and matter: The rearrangement of molecules within cells is an example of matter being transformed.
13. The relationship between structure and function: The polar structure of a water molecule helps enable it to carry out the function of supporting life.
14. Interactions within biological systems: Since there are interactions between the atmosphere and ocean, release of CO_2 into the atmosphere can affect ocean life.

Thinking Like a Scientist

The rat study could include a placebo-controlled group (rats who receive nonradioactive seed implantation). That would be unethical in a human trial.

CHAPTER 3

1. Isomers have different structures, or shapes, and the shape of a molecule usually helps determine the way it functions in the body.
2. dehydration reactions; water
3. hydrolysis
4. d
5. $C_6H_{12}O_6 + C_6H_{12}O_6 \rightarrow C_{12}H_{22}O_{11} + H_2O$
6. fatty acid; glycerol; triglyceride
7. c
8. c
9. If the change does not affect the shape of the protein in any way, then that change would not affect the function of the protein.
10. Hydrophobic amino acids are most likely to be found within the interior of a protein, far from the watery environment.
11. a
12. nucleotide
13. Both DNA and RNA are polynucleotides; both have the same phosphate group along the backbone; and both use A, C, and G bases. But DNA uses T while RNA uses U as a base; the sugar differs between them; and DNA is usually double-stranded, while RNA is usually single-stranded.
14. Structurally, a gene is a long stretch of DNA. Functionally, a gene contains the information needed to produce a protein.
15. Pathways that transform energy and matter: The energy within a molecule of glucose can be transformed to promote cellular work, and the matter within glucose can be used to build large molecules such as starch.
16. The relationship of structure to function: The specific structure of a polysaccharide (how the monosaccharide monomers are joined together) affects the function of that polysaccharide.
17. Information flow: Nucleic acids are information storage molecules, and the information of genes is stored as a precise nucleotide sequence.

Thinking Like a Scientist

A change in the nucleotide sequence of a chromosome outside of a gene may affect a protein that interacts with that gene, such as a protein that turns a gene on or off.

CHAPTER 4

1. d
2. A membrane is fluid because its components are not locked into place. A membrane is mosaic because it contains a variety of suspended proteins.
3. endomembrane system
4. smooth; circulating drugs
5. because many bacterial cells have walls, but no human cells do
6. Both organelles use membranes to organize enzymes and both provide energy to the cell. But chloroplasts capture energy from sunlight during photosynthesis, whereas mitochondria release energy from glucose during cellular respiration. Chloroplasts are only in photosynthetic plants and protists, whereas mitochondria are in almost all eukaryotic cells.
7. a3, b1, c5, d2, e4
8. nucleus, nuclear pores, ribosomes, rough ER, Golgi apparatus
9. Both are appendages that aid in movement and that extend from the surface of a cell. Cells with flagella typically have one long flagellum that propels the cell in a whiplike motion; cilia are usually shorter, are more numerous, and beat in a coordinated fashion.
10. Information flow: The precise DNA nucleotide sequence of a gene contains the information necessary to build a protein.
11. Interactions within biological systems: Only by acting together can organelles such as the nucleus, ER, and ribosomes express a genetic message.
12. Pathways that transform energy and matter: The energy in sunlight is transformed into the chemical energy of sugar molecules.

Thinking Like a Scientist

Most antibiotics were discovered in nature. Therefore, we need to preserve natural habitats because we depend upon such areas for new drug discoveries.

CHAPTER 5

1. The dog has potential energy when it crouches before jumping.

When it leaps onto the truck, the potential energy is converted to kinetic energy to drive muscle movement.

2. Energy; entropy
3. 3,733 g (or 3.733 kg); remember that 1 Calorie on a food label equals 1,000 Calories of heat energy.
4. The three phosphate groups store chemical energy, a form of potential energy. The release of a phosphate group makes some of this potential energy available to cells to perform work.
5. Hydrolases are enzymes that participate in hydrolysis reactions, breaking down large molecules into the smaller molecules that make them up. Enzymes often have names that end in -ase, so a hydrolase is an enzyme that performs hydrolysis reactions.
6. An inhibitor's binding to another site on the enzyme can cause the enzyme's active site to change shape.
7. b
8. *Hypertonic* and *hypotonic* are relative terms. A solution that is hypertonic to tap water could be hypotonic to seawater. When using these terms, you must provide a comparison, as in "The solution is hypertonic to the cell's cytoplasm."
9. Passive transport moves atoms or molecules along their concentration gradient (from higher to lower concentration), and active transport moves them against their concentration gradient.
10. a
11. Pathways that transform energy and matter: You are burning food energy to change the position of the water, which is gaining potential energy.
12. Interactions within biological systems: Many different parts of the human body must interact in order to produce an overall outcome (such as muscle movement).
13. The relationship of structure to function: What an enzyme does (its function) depends upon it having a particular shape (its structure).

Thinking Like a Scientist

Directed evolution, as with natural selection, involves reproduction and variation. It differs in that scientists perform tests to determine which variations are most fit.

CHAPTER 6

1. d
2. Plants produce organic molecules by photosynthesis. Consumers must acquire organic material by consuming it rather than making it.
3. In breathing, your lungs exchange CO_2 and O_2 between your body and the atmosphere. In cellular respiration, your cells consume the O_2 in extracting energy from food and release CO_2 as a waste product.
4. the electron transport chain
5. NAD^+
6. The majority of the energy provided by cellular respiration is generated during the electron transport chain. Shutting down that pathway will deprive cells of energy very quickly.
7. b
8. Glycolysis
9. a
10. Because fermentation supplies only 2 ATP per glucose molecule compared with about 32 from cellular respiration, the yeast will have to consume 16 times as much glucose to produce the same amount of ATP.
11. The relationship of structure to function: The structure of a mitochondrion, with its highly folded membranes, correlates with the function of promoting chemical reactions using membrane-bound proteins.
12. Pathways that transform energy and matter: The linked processes of cellular respiration and photosynthesis both involve transformations of energy and matter.
13. Interactions within biological systems: Your metabolism consists of many chemical reactions that interact with each other.

Thinking Like a Scientist

No. The process of science calls for a continuous reevaluation of all ideas as new evidence comes to light. Such skepticism about currently accepted explanations is a benefit of the process of science, not a weakness.

CHAPTER 7

1. thylakoids; stroma
2. Because NADPH and ATP are produced by the light reactions on the stroma side, they are more readily available for the Calvin cycle, which consumes the NADPH and ATP and occurs in the stroma.
3. b
4. "Photo" refers to the light required for photosynthesis to proceed, and "synthesis" refers to the fact that it makes sugar. Putting it together, the word "photosynthesis" means "to make, using light."
5. Green light is reflected by chlorophyll, not absorbed, and therefore cannot drive photosynthesis.
6. H_2O
7. c
8. glucose
9. c
10. Pathways that transform energy and matter: Photosynthetic organisms transform the energy of sunlight into the energy stored in biological molecules. These molecules then provide energy to organisms that consume them.
11. The relationship of structure to function: The highly folded structure of the thylakoids corresponds to its function of promoting the many chemical reactions of photosynthesis.
12. Interactions within biological systems: Deforestation reduces carbon fixation, which allows more carbon dioxide to remain in the atmosphere, which in turn affects climate around the world.

Thinking Like a Scientist

Natural sunlight is a combination of many wavelengths of light. By using a prism, Engelmann was able to visualize which of these colors was responsible for driving photosynthesis.

CHAPTER 8

1. c
2. They have identical genes (DNA).
3. They are in the form of very long, thin strands.
4. d
5. prophase and telophase
6. a. 1, 1; b. 1, 2; c. 2, 4; d. $2n$, n; e. individually, by homologous pair; f. identical, unique; g. repair/growth/asexual reproduction, gamete formation
7. 39
8. Prophase II or metaphase II; it cannot be during meiosis I because then you would see an even number of chromosomes; it cannot be during a later stage of meiosis II because then you would see the sister chromatids separated.
9. benign; malignant
10. 16 ($2n = 8$, so $n = 4$ and $2^n = 2^4 = 16$)
11. Nondisjunction would create just as many gametes with an extra copy of chromosome 3 or 16, but extra copies of chromosome 3 or 16 are probably fatal.
12. Information flow: Genetic information is passed from one generation to the next by the duplication of chromosomes during mitosis and meiosis.
13. Interactions within biological systems: Through their interactions, a number of different proteins all contribute to signal a cell to divide or not.
14. The relationship of structure to function: Examining the structure of the region gives insight into its current function. A chromosome region that is active will be loosely coiled, whereas a region that is inactive will be compact.

Thinking Like a Scientist

Both navel oranges and bdelloid rotifers illustrate that studying a single rare case can provide a result that is generally useful.

CHAPTER 9

1. genotype; phenotype
2. a. the law of independent assortment; b. the law of segregation
3. c
4. c
5. d
6. d
7. d
8. Rudy must be $X^D Y^0$. Carla must be $X^D X^d$ (because she had a son with the disease). There is a ¼ chance that their second child will be a male with the disease.
9. Height appears to result from polygenic inheritance, like human skin color. See Figure 9.22
10. The brown allele appears to be dominant, the white allele

recessive. The brown parent appears to be homozygous dominant, *BB*, and the white mouse is homozygous recessive, *bb*. The F_1 mice are all heterozygous, *Bb*. If two of the F_1 mice are mated, ¾ of the F_2 mice will be brown.

11. The best way to find out whether a brown F_2 mouse is homozygous dominant or heterozygous is to do a testcross: Mate the brown mouse with a white mouse. If the brown mouse is homozygous, all the offspring will be brown. If the brown mouse is heterozygous, you would expect half the offspring to be brown and half to be white.

12. Freckles are dominant, so Tim and Jan must both be heterozygous. There is a ¾ chance that they will produce a child with freckles and a ¼ chance that they will produce a child without freckles. The probability that the next two children will have freckles is $\frac{3}{4} \times \frac{3}{4} = \frac{9}{16}$.

13. According to the odds, half their children will be heterozygous and have elevated cholesterol levels. There is a ¼ chance that their next child will be homozygous, *hh*, and have an extremely high cholesterol level, like Katerina.

14. Because it is the father's sperm (that could carry either an X or a Y chromosome) that determines the sex of the offspring, not the mother's egg (which always carries an X).

15. The mother is a heterozygous carrier, and the father is normal. See the brown boxed area at the bottom of Figure 9.28 for a pedigree. One-quarter of their children will be boys suffering from hemophilia; ¼ will be female carriers.

16. For a woman to be colorblind, she must inherit X chromosomes bearing the colorblindness allele from both parents. Her father has only one X chromosome, which he passes on to all his daughters, so he must be colorblind. A male only needs to inherit the colorblindness allele from a carrier mother; both his parents are usually phenotypically normal.

17. The genotype of the black short-haired parent rabbit is *BBSS*. The genotype of the brown long-haired parent is *bbss*. The F_1 rabbits will

all be black and short-haired, *BbSs*. The F_2 rabbits will be black short-haired, black long-haired, brown short-haired, and brown long-haired, in a proportion of 9:3:3:1.

18. The relationship of structure to function: a mutation in the sickle-cell gene changes the shape of the protein hemoglobin, which changes how red blood cells function.

19. Interactions within biological systems: Skin color and other polygenic traits arise from the interactions of many different gene products.

20. Information flow: Genes carry genetic information from one generation to the next.

Thinking Like a Scientist

Dogs come in a much wider variety of physical forms than humans, which makes it easier to associate specific genetic mutations with specific phenotypes.

CHAPTER 10

1. polynucleotides; nucleotides
2. sugar (deoxyribose), phosphate, nitrogenous base
3. chromosome, gene, codon, nucleotide
4. Each daughter DNA molecule will have half the radioactivity of the parent molecule because one polynucleotide from the original parental DNA molecule winds up in each daughter DNA molecule.
5. UGG; ACC; TGG
6. A gene is the polynucleotide sequence with information for making one polypeptide. Each codon—a triplet of bases in DNA or RNA—codes for one amino acid. Transcription occurs when RNA polymerase produces mRNA using one strand of DNA as a template. A ribosome is the site of translation, or polypeptide synthesis, and tRNA molecules serve as interpreters of the genetic code. Each tRNA molecule has an amino acid attached at one end and a three-base anticodon at the other end. Beginning at the start codon, mRNA moves relative to the ribosome a codon at a time. A tRNA with a complementary anticodon pairs with each codon,

adding its amino acid to the polypeptide chain. The amino acids are linked by peptide bonds. Translation stops at a stop codon, and the finished polypeptide is released. The polypeptide folds to form a functional protein, sometimes in combination with other polypeptides.

7. a3, b3, c1, d2, e2 and e3
8. d
9. d
10. The genetic material of these viruses is RNA, which is replicated inside the infected cell by special enzymes encoded by the virus. The viral genome (or its complement) serves as mRNA for the synthesis of viral proteins.
11. reverse transcriptase; The process of reverse transcription occurs only in infections by RNA-containing retroviruses like HIV. Cells do not require reverse transcriptase (their RNA molecules do not undergo reverse transcription), so reverse transcriptase can be knocked out without harming the human host.
12. Evolution: The universal nature of the genetic code indicates that it arose very early in the history of life on Earth.
13. The relationship of structure to function: The structure of a tRNA provides insight into how it manages to connect nucleic acids (via the anticodon) and amino acids (via the attachment site).
14. Information flow: Nucleic acids store and carry genetic information used by the cell to produce proteins.

Thinking Like a Scientist

Just giving an injection could have a physical effect on the animal and thereby influence the results. Placebos allow researchers to keep all variables the same except for the vaccine itself.

CHAPTER 11

1. c
2. operon
3. b
4. a
5. DNA polymerase and other proteins required for transcription do not have access to tightly packed DNA.

6. No, the mitochondrial DNA is derived from the acceptor cell.
7. nuclear transplantation
8. which genes are active in a particular sample of cells
9. b
10. embryonic tissue (embryonic stem cells), umbilical cord blood, and bone marrow (adult stem cells)
11. The production of genetically identical animals for experimentation, the production of organs in pigs for transplant into humans, and restocking populations of endangered animals are a few possible uses of reproductive cloning.
12. Master control genes, called homeotic genes, regulate many other genes during development.
13. The relationship of structure to function: The ability of any protein (such as the repressor) to act (function) depends upon its precise shape (form).
14. Interactions within biological systems: The endocrine system secretes hormones that interact with nearly every cell in the body.
15. Information flow: The proteins produced by master control genes communicate information to other cells about their developmental fate.

Thinking Like a Scientist

Each cancer could be caused by unique mutations that require unique treatments. Also, testing the drugs on human cancer cells makes the results more relevant for treating humans.

CHAPTER 12

1. b
2. DNA ligase
3. Such an enzyme creates DNA fragments with "sticky ends," single-stranded regions whose unpaired bases can hydrogen-bond to the complementary sticky ends of other fragments created by the same enzyme.
4. PCR
5. Different people tend to have different numbers of repeats at each STR site. DNA fragments prepared from the STR sites of different people will thus have different lengths, causing them to migrate to different locations on a gel.
6. b
7. b

8. Chop the genome into fragments using restriction enzymes, clone and sequence each fragment, and reassemble the short sequences into a continuous sequence for every chromosome.

9. c, b, a, d

10. Evolution: Analysis of DNA sequences can provide insight into evolutionary history.

11. Interactions within biological systems: New functions arise through the interactions of many smaller parts, such as the many proteins that are involved in metabolism.

12. Information flow: Genes are the units of genetic information that are conveyed from one generation to the next.

Thinking Like a Scientist

Random mutations happen all the time. If a sequence is the same in very different species, it is evidence that the exact sequence is essential to survival.

CHAPTER 13

1. species, genus, family, order, class, phylum, kingdom, domain

2. c

3. Lyell and other geologists presented evidence for the gradual change of geologic features over millions of years. Darwin applied this idea to suggest that species evolve through the slow accumulation of small changes over long periods of time.

4. *Bb*: 0.42; *BB*: 0.49; *bb*: 0.09

5. The fitness of an individual (or of a particular genotype) is measured by the relative number of alleles that it contributes to the gene pool of the next generation compared with the contribution of others. Thus, the number of fertile offspring produced determines an individual's fitness.

6. b

7. b

8. Both effects result in populations small enough for significant sampling error in the gene pool for the first few generations. A bottleneck event reduces the size of an existing population in a given location. The founder effect occurs when a new, small

population colonizes a new territory.

9. disruptive selection

10. Information flow: The evolutionary history of a species is documented in the genetic information inherited from its ancestral species.

11. The relationship between structure and function: The bird's shape minimizes friction as it dives into water from a height of 75 feet.

12. Evolution: Natural selection is the process by which life evolves.

Thinking Like a Scientist

To perform a controlled experiment, the Grants would have to artificially manipulate food resources for a subset of the finch population. Such an experiment would be extremely difficult to carry out in the field, and capturing birds for a laboratory experiment would be disruptive to the island's ecology. An observational study allowed the Grants to test their hypothesis under natural conditions with the entire medium ground finch population of Daphne.

CHAPTER 14

1. Microevolution is a change in the gene pool of a population, often associated with adaptation. Speciation is an evolutionary process in which one species splits into two or more species. Macroevolution is evolutionary change above the species level, for example, the origin of evolutionary novelty and new taxonomic groups and the impact of mass extinctions on the diversity of life and its subsequent recovery. Macroevolution is marked by major changes in the history of life, and these changes are often noticeable enough to be evident in the fossil record.

2. prezygotic: a, b, c, e; postzygotic: d

3. because a small gene pool is more likely to be changed substantially by genetic drift and natural selection

4. exaptations

5. d

6. d

7. 2.6

8. Homologies reflected a shared evolutionary history, whereas analogies do not. Analogies result from convergent evolution.

9. Eukarya

10. Information flow: Genetic changes that affect the rate of developmental events can have a profound effect on body form.

11. The relationship of structure to function: The structure of the skin between the "finger" bones of the wing provides the surface area needed for flight.

12. Pathways that transform energy and matter: In the process of photosynthesis, plants convert the sun's energy into chemical energy that is stored in the bonds of organic molecules.

Thinking Like a Scientist

New species of *Tragopogon* originated very recently (within the past several decades). This gives researchers an opportunity to study the adaptive and genetic changes that occur early in a species' existence.

CHAPTER 15

1. g, a, d, f, c, b, e

2. The membrane forms a boundary to separate the living cell and its functions from the external environment. A key step in the origin of life would have been the isolation of organic monomers like RNA nucleotides and amino acids within the fluid surrounded by the membrane.

3. deep sea hydrothermal vents, submerged volcanoes, and meteorites

4. a large array of nutrients from the host's intestinal tract

5. They can form endospores.

6. Halophiles, thermophiles and methanogens; the latter generates methane, which is a source of energy.

7. They are eukaryotes that are not plants, animals, or fungi.

8. b

9. a

10. Information flow: Cells store information in DNA, which is used to make RNA, which is used to make proteins.

11. Pathways that transform energy and matter: When prokaryotes break down complex molecules to simpler ones, they transform energy and matter.

12. Relationship of structure to function: The web form increases the surface area, which facilitates absorption of necessary resources.

Thinking Like a Scientist

Extract microbiota from an obese "donor," then introduce equal amounts of these microbes into the intestinal tracts of lean and obese recipients. Measure the body composition of recipients before microbial transplant and measure again at a pre-determined amount of time after the transplant. Recipients would adhere to a prescribed diet during the experiment. Drawbacks: In practical terms, human subjects are more variable than lab-raised mice, which can be bred and raised in standardized conditions. In addition, each person has his own unique microbiota that will interact with the transplanted microbes. Solving these problems creates ethical issues. For example, it would be unethical for the researchers to treat subjects with large doses of antibiotics to kill off their existing microbiota. Also, the transplanted microbes could result in health problems for the recipients.

CHAPTER 16

1. cuticle

2. terrestrial

3. a. sporophyte b. cones; angiosperms c. fruit

4. c

5. d

6. a fern

7. Green fruits are harder to be spotted and thus less likely to be eaten than other fruits.

8. vascular plant

9. b

10. a much-reduced male gametophyte that houses cells that will develop into sperm

11. plant roots; fungi

12. A fungus digests its food externally by secreting digestive juices into the food and then absorbing the small nutrients that result from digestion. In contrast, humans and most other animals ingest relatively large pieces of food and digest the food within their bodies.

13. Information Flow: DNA from different gametophytes combines and is passed to the sporophyte generation. The spores contain the DNA for the next generation of gametophytes.
14. The Relationship of Structure to Function: The fine branches of roots increase the surface area, which facilitates absorption of necessary resources.
15. Evolution: Natural selection led to the development of vascular systems in ferns, giving them an advantage in a variety of habitats. This could also be an example of the relationship of structure to function because vascular tissue functions to move materials throughout the plant.

Thinking Like a Scientist

So they could tell whether disturbing the soil influenced pine growth, even if mycorrhizae were not added.

CHAPTER 17

1. c
2. arthropod
3. amphibians
4. b
5. chordata; notochord; cartilage disks between your vertebrae
6. a
7. *Australopithecus* species, *Homo habilis, Homo erectus, Homo sapiens*
8. a4, b5, c1, d2, e3
9. Information flow: The information needed to direct the development of diverse body forms is contained in DNA, which is transmitted between generations.
10. Relationship of structure to function: The suckers and hooks function as effective tools that allow tapeworms to attach to hosts.
11. Interactions within biological systems: Insects have multiple types of interactions with people. Some interactions are beneficial to both humans and insects. Others are harmful to humans, insects, or both.

Thinking Like a Scientist

Mutations result in genetic differences between populations that don't interbreed. Analyzing more genes provides multiple, independent lines of evidence.

CHAPTER 18

1. organismal ecology, population ecology, community ecology, ecosystem ecology
2. light, water temperature, chemicals added
3. acclimation; reversible
4. d
5. a. desert; b. temperate grassland; c. tropical forest; d. temperate broadleaf forest; e. northern coniferous forest; f. tundra
6. taiga
7. permafrost, very cold winters, and high winds
8. zooplanktons
9. Carbon dioxide and other gases in the atmosphere absorb heat energy radiating from Earth and reflect it back toward Earth. This is called the greenhouse effect. As the carbon dioxide concentration in the atmosphere increases, more heat is retained, causing global warming.
10. c
11. terrestrial biomes are extending to previously frozen ground; the boundary of the tundra is shifting northward
12. The relationship of structure to function: The structures of molecules like the keratin in reptilian scales and the wax that coats plant leaves create a water-tight barrier.
13. Interactions within biological systems: Organisms in an ecosystem are connected in many ways, including competition and predation.
14. Pathways that transform energy and matter: Photosynthesis is a pathway that uses light energy to produce energy-rich molecules from carbon dioxide and water.
15. Evolution: As a result of random mutation, plants in a population vary in their resistance to drought. During dry periods, individuals

that are more drought resistant leave more offspring than other individuals in the population. Therefore, the drought-resistant individuals pass along their genes. This process of natural selection leads to evolution of increased drought resistance.

Thinking Like a Scientist

The researchers tested their hypothesis with an observational study: They collected data from historical and present-day population records and compared these observations to determine changes in the butterfly's range.

CHAPTER 19

1. the number of people and the land area in which they live
2. III; I
3. a. The *x*-axis is time; the *y*-axis is the number of individuals; the red curve represents exponential growth; the blue curve represents logistic growth.
   b. carrying capacity
   c. In exponential growth, the size of the population increases more and more rapidly. In logistic growth, the population grows fastest when it is about one-half the carrying capacity.
   d. exponential growth curve, though the worldwide growth rate is slowing
4. d
5. equilibrial
6. a
7. a
8. The Relationship of Structure to Function: The structure of the molecules in this combination causes them to function as effective anchors.
9. Interactions within Biological Systems: Changes in population size can be caused by interactions between species (such as predation and herbivory) as well as between organisms and their environment.
10. Evolution: Individuals have limited resources to invest in reproduction and survival. Alleles that lead to a better balance between survival and reproduction will lead to greater

fitness, and those alleles will become more common in the population.

Thinking Like a Scientist

A valid control must match the experimental treatment in every way except for the variable in question. Researchers were testing whether blocking access would limit the toads. The open-fence treatment allowed researchers to determine whether some other aspect of the fence (such as giving predators a place to perch) was responsible for any reduction in toad numbers.

CHAPTER 20

1. habitat destruction
2. d
3. a
4. a2, b5, c5, d3 or d4, e1
5. because the pesticides become concentrated in their prey
6. disturbance
7. Only about 10% of the energy trapped by photosynthesis is turned into biomass by the plant, and only about 10% of that energy is turned into the meat of a grazing animal. Therefore, grain-fed beef provides only about 1% of the energy captured by photosynthesis.
8. Many nutrients come from the soil, but carbon comes from CO_2 in the air.
9. Bioremediation
10. d
11. The relationship of structure to function: The nature of the structures (sharp or hard) allows them to function for defense.
12. Interactions within biological systems: Competition between different species is a type of interaction.
13. Pathways that transform energy and matter: As energy flows through an ecosystem, some is lost as heat. Matter is conserved.

Thinking Like a Scientist

If fragments are too close together, they may act as one large fragment for species that can move between them.

Glossary

A

abiotic factor (ā'bī-ot'-ik)
A nonliving component of an ecosystem, such as air, water, light, minerals, or temperature.

abiotic reservoir
The part of an ecosystem where a chemical, such as carbon or nitrogen, accumulates or is stockpiled outside of living organisms.

ABO blood groups
Genetically determined classes of human blood that are based on the presence or absence of carbohydrates A and B on the surface of red blood cells. The ABO blood group phenotypes, also called blood types, are A, B, AB, and O.

absorption
The uptake of small nutrient molecules by an organism's own body. In animals, absorption is the third main stage of food processing, following digestion; in fungi, it is acquisition of nutrients from the surrounding medium.

acclimation (ak'-li-mā'-shun)
Physiological adjustment that occurs gradually, though still reversibly, in response to an environmental change.

acid
A substance that increases the hydrogen ion (H⁺) concentration in a solution.

activation energy
The amount of energy that reactants must absorb before a chemical reaction will start. An enzyme lowers the activation energy of a chemical reaction, allowing it to proceed faster.

activator
A protein that switches on a gene or group of genes by binding to DNA.

active site
The part of an enzyme molecule where a substrate molecule attaches—typically, a pocket or groove on the enzyme's surface.

active transport
The movement of a substance across a biological membrane against its concentration gradient, aided by specific transport proteins and requiring the input of energy (often as ATP).

adenine (A) (ad'-uh-nēn)
A double-ring nitrogenous base found in DNA and RNA.

ADP (adenosine diphosphate) (a-den'-ō-sēn dī-fos'-fāt)
A molecule composed of adenosine and two phosphate groups. The molecule ATP is made by combining a molecule of ADP with a third phosphate in an energy-consuming reaction.

adult stem cell
A cell present in adult tissues that generates replacements for nondividing differentiated cells.

aerobic (ār-ō'-bik)
Containing or requiring molecular oxygen (O_2).

age structure
The relative number of individuals of each age in a population.

AIDS
Acquired immunodeficiency syndrome; the late stages of HIV infection, characterized by a reduced number of T cells; usually results in death caused by infections that would be defeated by a properly functioning immune system.

alga (al'-guh)
(plural, **algae**) An informal term that describes a great variety of photosynthetic protists, including unicellular, colonial, and multicellular forms. Prokaryotes that are photosynthetic autotrophs are also regarded as algae.

allele (uh-lē'-ul)
An alternative version of a gene.

allopatric speciation
The formation of a new species in populations that are geographically isolated from one another. *See also* sympatric speciation.

alternation of generations
A life cycle in which there is both a multicellular diploid form, the sporophyte, and a multicellular haploid form, the gametophyte; a characteristic of plants and multicellular green algae.

alternative RNA splicing
A type of regulation at the RNA-processing level in which different mRNA molecules are produced from the same primary transcript, depending on which RNA segments are treated as exons and which as introns.

amino acid (uh-mēn'-ō)
An organic molecule containing a carboxyl group, an amino group, a hydrogen atom, and a variable side chain (also called a radical group, or R group); serves as the monomer of proteins.

amniote (am'-nē-ōt)
Member of a clade of tetrapods that has an amniotic egg containing specialized membranes that protect the embryo. Amniotes include mammals and reptiles (including birds).

amniotic egg (am'-nē-ot'-ik)
A shelled egg in which an embryo develops within a fluid-filled amniotic sac and is nourished by yolk. Produced by reptiles (including birds) and egg-laying mammals, it enables them to complete their life cycles on dry land.

amoeba (uh-mē'-buh)
A general term for a protozoan (animal-like protist) characterized by great structural flexibility and the presence of pseudopodia.

amphibian
Member of a class of vertebrate animals that includes frogs and salamanders.

anaerobic (*an'-ār-ō'-bik*)
Lacking or not requiring molecular oxygen (O_2).

analogy
The similarity between two species that is due to convergent evolution rather than to descent from a common ancestor with the same trait.

anaphase
The third stage of mitosis, beginning when sister chromatids separate from each other and ending when a complete set of daughter chromosomes has arrived at each of the two poles of the cell.

anecdotal evidence
An assertion based on a single or just a few examples. Anecdotal evidence is not considered valid proof of a generalized conclusion.

angiosperm (*an'-jē-ō-sperm*)
A flowering plant, which forms seeds inside a protective chamber called an ovary.

animal
A eukaryotic, multicellular, heterotrophic organism that obtains nutrients by ingestion.

annelid (*an'-uh-lid*)
A segmented worm. Annelids include earthworms, polychaetes, and leeches.

anther
A sac in which pollen grains develop, located at the tip of a flower's stamen.

anthropoid (*an'-thruh-poyd*)
Member of a primate group made up of the apes (gibbons, orangutans, gorillas, chimpanzees, and bonobos), monkeys, and humans.

anticodon (*an'-tī-kō'-don*)
On a tRNA molecule, a specific sequence of three nucleotides that is complementary to a codon triplet on mRNA.

aphotic zone (*ā-fō'-tik*)
The region of an aquatic ecosystem beneath the photic zone, where light levels are too low for photosynthesis to take place.

apicomplexan (*ap'-ē-kom-pleks'-un*)
A type of parasitic protozoan (animal-like protist). Some apicomplexans cause serious human disease.

aqueous solution
A solution in which water is the solvent.

arachnid
Member of a major arthropod group that includes spiders, scorpions, ticks, and mites.

Archaea (*ar-kē'-uh*)
One of two prokaryotic domains of life, the other being Bacteria.

archaean (*ar-kē'-uhn*)
(plural, **archaea**) An organism that is a member of the domain Archaea.

arthropod (*ar'-thruh-pod*)
Member of the most diverse phylum in the animal kingdom; includes the horseshoe crab, arachnids (for example, spiders, ticks, scorpions, and mites), crustaceans (for example, crayfish, lobsters, crabs, and barnacles), millipedes, centipedes, and insects. Arthropods are characterized by a chitinous exoskeleton, molting, jointed appendages, and a body formed of distinct groups of segments.

artificial selection
The selective breeding of domesticated plants and animals to promote the occurrence of desirable traits in the offspring.

asexual reproduction
The creation of genetically identical offspring by a single parent, without the participation of gametes (sperm and egg).

atom
The smallest unit of matter that retains the properties of an element.

atomic mass
The total mass of an atom.

atomic number
The number of protons in each atom of a particular element. Elements are ordered by atomic number in the periodic table of the elements.

ATP (adenosine triphosphate)
(*a-den'-ō-sēn trī-fos'-fāt*)
A molecule composed of adenosine and three phosphate groups; the main energy source for cells. A molecule of ATP can be broken down to a molecule of ADP (adenosine diphosphate) and a free phosphate; this reaction releases energy that can be used for cellular work.

ATP synthase
A protein cluster, found in a cellular membrane (including the inner membrane of mitochondria, the thylakoid membrane of chloroplasts, and the plasma membrane of prokaryotes), that uses the energy of a hydrogen ion concentration gradient to make ATP from ADP. An ATP synthase provides a port through which hydrogen ions (H^+) diffuse.

autosome
A chromosome not directly involved in determining the sex of an organism; in mammals, for example, any chromosome other than X or Y.

autotroph (*ot'-ō-trōf*)
An organism that makes its own food from inorganic ingredients, thereby sustaining itself without eating other organisms or their molecules. Plants, algae, and photosynthetic bacteria are autotrophs.

B

bacillus (*buh-sil'-us*)
(plural, **bacilli**) A rod-shaped prokaryotic cell.

Bacteria
One of two prokaryotic domains of life, the other being Archaea.

bacteriophage (*bak-tēr'-ē-ō-fāj*)
A virus that infects bacteria; also called a phage.

bacterium
(plural, **bacteria**) An organism that is a member of the domain Bacteria.

base
A substance that decreases the hydrogen ion (H^+) concentration in a solution.

benign tumor
An abnormal mass of cells that remains at its original site in the body.

benthic realm
A seafloor or the bottom of a freshwater lake, pond, river, or stream. The benthic realm is occupied by communities of organisms known as benthos.

bilateral symmetry
An arrangement of body parts such that an organism can be divided equally by a single cut passing longitudinally through it. A bilaterally symmetric organism has mirror-image right and left sides.

binary fission
A means of asexual reproduction in which a parent organism, often a single cell, divides into two individuals of about equal size.

binomial
The two-part format for naming a species; for example, *Homo sapiens*.

biocapacity
Earth's capacity to produce the resources such as food, water, and fuel consumed by humans and to absorb human-generated waste.

biodiversity
The variety of living things; includes genetic diversity, species diversity, and ecosystem diversity.

biodiversity hot spot
A small geographic area that contains a large number of threatened or endangered species and an exceptional concentration of endemic species (those found nowhere else).

biofilm
A surface-coating cooperative colony of prokaryotes.

biogeochemical cycle
Any of the various chemical circuits occurring in an ecosystem, involving both biotic and abiotic components of the ecosystem.

biogeography
The study of the past and present distribution of organisms.

bioinformatics
A scientific field of study that uses mathematics to develop methods for organizing and analyzing large sets of biological data.

biological control
The intentional release of a natural enemy to attack a pest population.

biological magnification
The accumulation of persistent chemicals in the living tissues of consumers in food chains.

biological species concept
The definition of a species as a population or group of populations the members of which have the potential in nature to interbreed and produce fertile offspring.

biology
The scientific study of life.

biomass
The amount, or mass, of living organic material in an ecosystem.

biome (*bī-ōm*)
A major terrestrial or aquatic life zone, characterized by vegetation type in terrestrial biomes and the physical environment in aquatic biomes.

bioremediation
The use of living organisms to detoxify and restore polluted and degraded ecosystems.

biosphere
The global ecosystem; the entire portion of Earth inhabited by life; all of life and where it lives.

biotechnology
The manipulation of living organisms to perform useful tasks.

biotic factor (bī-ot'-ik)
A living component of a biological community; any organism that is part of an individual's environment.

bird
Member of a group of reptiles with feathers and adaptations for flight.

bivalve
Member of a group of molluscs that includes clams, mussels, scallops, and oysters.

blastula (blas'-tyū-luh)
An embryonic stage that marks the end of cleavage during animal development; a hollow ball of cells in many species.

body cavity
A fluid-filled space separating the digestive tract from the outer body wall.

body segmentation
Subdivision of an animal's body into a series of repeated parts called segments.

bony fish
A fish that has a stiff skeleton reinforced by calcium salts.

bottleneck effect
Genetic drift resulting from a drastic reduction in population size. Typically, the surviving population is no longer genetically representative of the original population.

bryophyte (brī'-uh-fīt)
A type of plant that lacks xylem and phloem; a nonvascular plant. Bryophytes include mosses and their close relatives.

buffer
A chemical substance that decreases the hydrogen ion (H^+) concentration in a solution.

C

calorie
The amount of energy that raises the temperature of 1 g of water by 1°C. Commonly reported as Calories, which are kilocalories (1,000 calories).

Calvin cycle
The second of two stages of photosynthesis; a cyclic series of chemical reactions that occur in the stroma of a chloroplast, using the carbon in CO_2 and the ATP and NADPH produced by the light reactions to make the energy-rich sugar molecule G3P, which is later used to produce glucose.

cancer
A malignant growth or tumor caused by abnormal and uncontrolled cell division.

carbohydrate (kar'-bō-hī'-drāt)
A biological molecule consisting of a simple sugar (a monosaccharide), two monosaccharides joined into a double sugar (a disaccharide), or a chain of monosaccharides (a polysaccharide).

carbon fixation
The initial incorporation of carbon from CO_2 into organic compounds by autotrophic organisms such as photosynthetic plants, algae, or bacteria.

carbon footprint
The amount of greenhouse gas emitted as a result of the actions of a person, nation, or other entity.

carcinogen (kar-sin'-uh-jin)
A cancer-causing agent, either high-energy radiation (such as X-rays or UV light) or a chemical.

carnivore
An animal that mainly eats other animals. *See also* herbivore; omnivore.

carpel (kar'-pul)
The egg-producing part of a flower, consisting of a stalk with an ovary at the base and a stigma, which traps pollen, at the tip.

carrier
An individual who is heterozygous for a recessively inherited disorder and who therefore does not show symptoms of that disorder.

carrying capacity
The maximum population size that a particular environment can sustain.

cartilaginous fish (kar-ti-laj'-uh-nus)
A fish that has a flexible skeleton made of cartilage.

cell cycle
An ordered sequence of events (including interphase and the mitotic phase) that extends from the time a eukaryotic cell is first formed from a dividing parent cell until its own division into two cells.

cell cycle control system
A cyclically operating set of proteins that triggers and coordinates events in the eukaryotic cell cycle.

cell division
The reproduction of a cell.

cell plate
A membranous disk that forms across the midline of a dividing plant cell. During cytokinesis, the cell plate grows outward, accumulating more cell wall material and eventually fusing into a new cell wall.

cell theory
The theory that all living things are composed of cells and that all cells come from earlier cells.

cellular respiration
The aerobic harvesting of energy from food molecules; the energy-releasing chemical breakdown of food molecules, such as glucose, and the storage of potential energy in a form that cells can use to perform work; involves glycolysis, the citric acid cycle, the electron transport chain, and chemiosmosis.

cellulose (sel'-yū-lōs)
A large polysaccharide composed of many glucose monomers linked into cable-like fibrils that provide structural support in plant cell walls. Because cellulose cannot be digested by animals, it acts as fiber, or roughage, in the diet.

centipede
A carnivorous terrestrial arthropod that has one pair of long legs for each of its numerous body segments, with the front pair modified as poisonous claws.

central vacuole (vak'-yū-ōl)
A membrane-enclosed sac occupying most of the interior of a mature plant cell, having diverse roles in reproduction, growth, and development.

centromere (sen'-trō-mer)
The region of a chromosome where two sister chromatids are joined and where spindle microtubules attach during mitosis and meiosis. The centromere divides at the onset of anaphase during mitosis and anaphase II of meiosis.

cephalopod
Member of a group of molluscs that includes squids and octopuses.

chaparral (shap-uh-ral')
A terrestrial biome limited to coastal regions where cold ocean currents circulate offshore, creating mild, rainy winters and long, hot, dry summers; also known as the Mediterranean biome. Chaparral vegetation is adapted to fire.

character
A heritable feature that varies among individuals within a population, such as flower color in pea plants or eye color in humans.

charophyte (kar'-uh-fīt')
A member of the green algal group that shares features with land plants. Charophytes are considered the closest relatives of land plants; modern charophytes and modern plants likely evolved from a common ancestor.

chemical bond
An attraction between two atoms resulting from a sharing of outer-shell electrons or the presence of opposite charges on the atoms.

chemical cycling
The use and reuse of chemical elements such as carbon within an ecosystem.

chemical energy
Energy stored in the chemical bonds of molecules; a form of potential energy.

chemical reaction
A process leading to chemical changes in matter, involving the making and/or breaking of chemical bonds. A chemical reaction involves rearranging atoms, but no atoms are created or destroyed.

chemotherapy (kē'-mō-ther'-uh-pē)
Treatment for cancer in which drugs are administered to disrupt cell division of the cancer cells.

chlorophyll (klor'-ō-fil)
A light-absorbing pigment in chloroplasts that plays a central role in converting solar energy to chemical energy.

chloroplast (klō'-rō-plast)
An organelle found in plants and photosynthetic protists. Enclosed by two membranes, a chloroplast absorbs sunlight and uses it to power the synthesis of organic food molecules (sugars).

chordate (kor'-dāt)
An animal that at some point during its development has a dorsal, hollow nerve cord, a notochord, pharyngeal slits, and a post-anal tail. Chordates include lancelets, tunicates, and vertebrates.

chromatin (krō'-muh-tin)
The combination of DNA and proteins that constitutes chromosomes; often used to refer to the diffuse, very extended form taken by the chromosomes when a eukaryotic cell is not dividing.

chromosome (krō'-muh-sōm)
A gene-carrying structure found in the nucleus of a eukaryotic cell and most visible when compacted during mitosis and meiosis; also, the main gene-carrying structure of a prokaryotic cell. Each chromosome consists of one very long threadlike DNA molecule and associated proteins. *See also* chromatin.

chromosome theory of inheritance
A basic principle in biology stating that genes are located on chromosomes and that the behavior of chromosomes during meiosis accounts for inheritance patterns.

cilia (*sil'-ē-a*)
Extensions from a eukaryotic cell that are generally shorter and more numerous than flagella and that propel the cell by moving in a coordinated back-and-forth motion.

ciliate (*sil'-ē-it*)
A type of protozoan (animal-like protist) that moves and feeds by means of cilia.

citric acid cycle
The metabolic cycle that is fueled by acetyl CoA formed after glycolysis in cellular respiration. Chemical reactions in the cycle complete the metabolic breakdown of glucose molecules to carbon dioxide. The cycle occurs in the matrix of mitochondria and supplies most of the NADH molecules that carry energy to the electron transport chains. Also referred to as the Krebs cycle.

clade
An ancestral species and all its descendants—a distinctive branch in the tree of life.

cladistics (*kluh-dis'-tiks*)
The study of evolutionary history; specifically, an approach to systematics in which organisms are grouped by common ancestry.

class
In classification, the taxonomic category above order.

cleavage
The process of cytokinesis in animal cells, characterized by pinching of the plasma membrane.

clone
As a verb, to produce genetically identical copies of a cell, organism, or DNA molecule. As a noun, the collection of cells, organisms, or molecules resulting from cloning; also (colloquially), a single organism that is genetically identical to another because it arose from the cloning of a somatic cell.

cnidarian (*nī-dār'-ē-an*)
An animal characterized by cnidocytes, radial symmetry, a gastrovascular cavity, and a polyp or medusa body form. Cnidarians include hydras, jellies, sea anemones, and corals.

coccus (*kok'-us*)
(plural, **cocci**) A spherical prokaryotic cell.

codominant
Expressing two different alleles of a gene in a heterozygote.

codon (*kō'-don*)
A three-nucleotide sequence of DNA or mRNA that specifies a particular amino acid or termination signal; the basic unit of the genetic code.

cohesion (*kō-hē'-zhun*)
The attraction between molecules of the same kind.

commensalism
An interaction in which the individual of one species benefits and the individual of the other species is neither helped nor harmed.

community ecology
The study of how interactions between species affect community structure and organization.

community
All the organisms inhabiting and potentially interacting in a particular area; an assemblage of populations of different species.

competitive exclusion principle
The concept that populations of two species cannot coexist in a community if their niches are nearly identical. Using resources more efficiently and having a reproductive advantage, one of the populations will eventually outcompete and eliminate the other.

complementary DNA (cDNA)
A DNA molecule made in vitro using mRNA as a template and the enzyme reverse transcriptase. A cDNA molecule therefore corresponds to a gene but lacks the introns present in the DNA of the genome.

complete digestive tract
A digestive tube with two openings, a mouth and an anus.

compound
A substance containing two or more different elements in a fixed ratio; for example, table salt (NaCl) consists of one atom of the element sodium (Na) for every atom of chlorine (Cl).

concentration gradient
An increase or decrease in the density of a chemical substance within a given region. Cells often maintain concentration gradients of hydrogen ions across their membranes. When a gradient exists, the ions or other chemical substances involved tend to move from where they are more concentrated to where they are less concentrated.

conifer (*kon'-uh-fer*)
A gymnosperm, or naked-seed plant, most of which produce cones.

coniferous forest (*kō-nif'-rus*)
A terrestrial biome characterized by conifers, cone-bearing evergreen trees.

conservation biology
A goal-oriented science that seeks to understand and counter the loss of biodiversity.

conservation of energy
The principle that energy can be neither created nor destroyed.

consumer
An organism that obtains its food by eating plants or by eating animals that have eaten plants.

control group
In a controlled experiment, a set of subjects that lacks (or does not receive) the specific factor being tested. Ideally, the control group should be identical to the experimental group in all other respects.

controlled experiment
A component of the process of science whereby a scientist carries out two parallel tests, an experimental test and a control test. The experimental test differs from the control by one factor, the variable.

convergent evolution
The evolution of similar features in different evolutionary lineages, which can result from living in very similar environments.

coral reef
Tropical marine biome characterized by hard skeletal structures secreted primarily by the resident cnidarians.

covalent bond
An attraction between atoms that share one or more pairs of electrons.

CRISPR-Cas9 system
A technique for editing genes in living cells, involving a bacterial protein called Cas9 associated with a guide RNA complementary to a gene sequence of interest.

cross
The cross-fertilization of two different varieties of an organism or of two different species; also called hybridization.

crossing over
The exchange of segments between chromatids of homologous chromosomes during prophase I of meiosis.

crustacean
Member of a major arthropod group that includes lobsters, crayfish, crabs, shrimps, and barnacles.

cryptic coloration
Adaptive coloration that makes an organism difficult to spot against its background.

cuticle (*kyū'-tuh-kul*)
(1) In animals, a tough, nonliving outer layer of the skin. (2) In plants, a waxy coating on the surface of stems and leaves that helps retain water.

cytokinesis (*sī'-tō-kuh-nē'-sis*)
The division of the cytoplasm to form two separate daughter cells. Cytokinesis usually occurs during telophase of mitosis, and the two processes (mitosis and cytokinesis) make up the mitotic (M) phase of the cell cycle.

cytoplasm (*sī'-tō-plaz'-um*)
Everything within a eukaryotic cell inside the plasma membrane and outside the nucleus; consists of a semifluid medium (cytosol) and organelles; can also refer to the interior of a prokaryotic cell.

cytosine (C) (*sī'-tuh-sēn*)
A single-ring nitrogenous base found in DNA and RNA.

cytoskeleton
A meshwork of fine fibers in the cytoplasm of a eukaryotic cell; includes microfilaments, intermediate filaments, and microtubules.

cytosol (*sī'-tuh-sol*)
The fluid part of the cytoplasm, in which organelles are suspended.

D

data
Recorded verifiable observations.

decomposer
An organism that absorbs nutrients from nonliving organic material, such as dead organisms, fallen plant material, and the wastes of living organisms, and converts them to inorganic forms.

dehydration reaction (*dē-hī-drā'-shun*)
A chemical process in which a polymer forms when monomers are linked by the removal of water

molecules. One molecule of water is removed for each pair of monomers linked. The atoms in the water molecule are provided by the two monomers involved in the reaction. A dehydration reaction is essentially the reverse of a hydrolysis reaction.

density-dependent factor
A limiting factor whose effects intensify with increasing population density.

density-independent factor
A limiting factor whose occurrence and effects are not related to population density.

dependent variable
A factor whose value is measured in an experiment to see whether it is influenced by changes in another factor (the independent variable).

desert
A terrestrial biome characterized by low and unpredictable rainfall (less than 30 cm per year).

detritus (di-trī′-tus)
Dead organic matter.

diatom (dī′-uh-tom)
A unicellular photosynthetic alga with a unique glassy cell wall containing silica.

diffusion
The spontaneous movement of particles of any kind down a concentration gradient, that is, movement of particles from where they are more concentrated to where they are less concentrated.

dihybrid cross (dī′-hī′-brid)
A mating of individuals differing at two genetic loci.

dinoflagellate (dī′-nō-flaj′-uh-let)
A unicellular photosynthetic alga with two flagella situated in perpendicular grooves in cellulose plates covering the cell.

diploid (dip′-loid)
Containing two sets of chromosomes (pairs of homologous chromosomes) in each cell, one set inherited from each parent; referring to a 2n cell.

directional selection
Natural selection that acts in favor of the individuals at one end of a phenotypic range.

disaccharide (dī-sak′-uh-rīd)
A sugar molecule consisting of two monosaccharides (simple sugars) linked by a dehydration reaction.

disruptive selection
Natural selection that favors extreme over intermediate phenotypes.

disturbance
In an ecological sense, a force that damages a biological community, at least temporarily, by destroying organisms and altering the availability of resources needed by organisms in the community. Examples include fires, floods, and droughts.

DNA (deoxyribonucleic acid)
(dē-ok′-sē-rī′-bō-nū-klā′-ik)
The genetic material that organisms inherit from their parents; a double-stranded helical macromolecule consisting of nucleotide monomers with deoxyribose sugar, a phosphate group, and the nitrogenous bases adenine (A), cytosine (C), guanine (G), and thymine (T). *See also* gene.

DNA cloning
The production of many identical copies of a specific segment of DNA.

DNA ligase (lī′-gās)
An enzyme, essential for DNA replication, that creates new chemical bonds between adjacent DNA nucleotides; used in genetic engineering to paste a specific piece of DNA containing a gene of interest into a bacterial plasmid or other vector.

DNA microarray
A glass slide containing thousands of different kinds of single-stranded DNA fragments arranged in an array (grid). Tiny amounts of DNA fragments, representing different genes, are attached to the glass slide. These fragments are tested for hybridization with various samples of cDNA molecules, thereby measuring the expression of thousands of genes at one time.

DNA polymerase (puh-lim′-er-ās)
An enzyme that assembles DNA nucleotides into polynucleotides using a preexisting strand of DNA as a template.

DNA profiling
A procedure that analyzes an individual's unique collection of genetic markers using PCR and gel electrophoresis. DNA profiling can be used to determine whether two samples of genetic material came from the same individual.

DNA sequencing
Determining the complete nucleotide sequence of a gene or DNA segment.

domain
A taxonomic category above the kingdom level. The three domains of life are Archaea, Bacteria, and Eukarya.

dominant allele
In a heterozygote, the allele that determines the phenotype with respect to a particular gene; the dominant version of a gene is usually represented with a capital italic letter (e.g., *F*).

dorsal, hollow nerve cord
One of the four hallmarks of chordates; the chordate brain and spinal cord.

double-blind experiment
A scientific experiment in which some information is withheld from both the test subject and the experimenter. A double-blind experiment is meant to eliminate bias on the part of the participants and the researchers.

double helix
The form assumed by DNA in living cells, referring to its two adjacent polynucleotide strands wound into a spiral shape.

Down syndrome
A human genetic disorder resulting from a condition called trisomy 21, the presence of an extra chromosome 21; characterized by heart and respiratory defects and varying degrees of developmental disability.

E

echinoderm (ih-kī′-nuh-derm)
Member of a group of slow-moving or stationary marine animals characterized by a rough or spiny skin, a water vascular system, typically an endoskeleton, and radial symmetry in adults. Echinoderms include sea stars, sea urchins, and sand dollars.

ecological footprint
An estimate of the amount of land required to provide the resources, such as food, water, fuel, and housing, that an individual or a nation consumes and to absorb the waste it generates.

ecological niche
The sum of a species' use of the biotic and abiotic resources in its environment.

ecological succession
The process of biological community change resulting from disturbance; transition in the species composition of a biological community, often following a flood, fire, or volcanic eruption. *See also* primary succession; secondary succession.

ecology
The scientific study of the interactions between organisms and their environments.

ecosystem (ē′-kō-sis-tem)
All the organisms in a given area, along with the nonliving (abiotic) factors with which they interact; a biological community and its physical environment.

ecosystem ecology
The study of energy flow and the cycling of chemicals among the various biotic and abiotic factors in an ecosystem.

ecosystem service
Function performed by an ecosystem that directly or indirectly benefits people.

ectotherm (ek′-tō-therm)
An animal that warms itself mainly by absorbing heat from its surroundings.

electromagnetic spectrum
The full range of radiation, from the very short wavelengths of gamma rays to the very long wavelengths of radio signals.

electron
A subatomic particle with a single unit of negative electrical charge. One or more electrons move around the nucleus of an atom.

electron transport
A reaction in which one or more electrons are transferred to carrier molecules. A series of such reactions, called an electron transport chain, can release the energy stored in high-energy molecules such as glucose. *See also* electron transport chain.

electron transport chain
A series of electron carrier molecules that shuttle electrons during a series of chemical reactions. These reactions release energy that is used to make ATP; located in the inner membrane of mitochondria, the thylakoid membrane of chloroplasts, and the plasma membrane of prokaryotes.

element
A substance that cannot be broken down into other substances by chemical means. Scientists recognize 92 chemical elements that occur naturally and several dozen more that have been created in the laboratory.

embryonic stem cell (ES cell)
Any of the cells in the early animal embryo that differentiate during development to give rise to all the kinds of specialized cells in the body.

emerging virus
A virus that has appeared suddenly or has recently come to the attention of medical scientists.

endangered species
As defined in the U.S. Endangered Species Act, a species that is in danger of extinction throughout all or a significant portion of its range.

endemic species
A species whose distribution is limited to a specific geographic area.

endocytosis (en'-dō-sī-tō'-sis)
The movement of materials from the external environment into the cytoplasm of a cell via vesicles or vacuoles.

endomembrane system
A network of organelles that partitions the cytoplasm of eukaryotic cells into functional compartments. Some of the organelles are structurally connected to each other, whereas others are structurally separate but functionally connected by the traffic of vesicles among them.

endoplasmic reticulum (ER)
(reh-tik'-yuh-lum)
An extensive membranous network in a eukaryotic cell, continuous with the outer nuclear membrane and composed of ribosome-studded (rough) and ribosome-free (smooth) regions. See also rough ER; smooth ER.

endoskeleton
A hard interior skeleton located within the soft tissues of an animal; found in all vertebrates and a few invertebrates (such as echinoderms).

endosperm
In flowering plants, a nutrient-rich mass formed by the union of a sperm cell with the diploid central cell of the embryo sac during double fertilization; provides nourishment to the developing embryo in the seed.

endospore
A thick-coated, protective cell produced within a prokaryotic cell exposed to harsh conditions.

endotherm
An animal that derives most of its body heat from its own metabolism.

endotoxin
A poisonous component of the outer membrane of certain bacteria.

energy
The capacity to cause change, or to move matter in a direction it would not move if left alone.

energy flow
The passage of energy through the components of an ecosystem.

energy pyramid
A diagram depicting the cumulative loss of energy with each transfer in a food chain.

enhancer
A eukaryotic DNA sequence that helps stimulate the transcription of a gene at some distance from it. An enhancer functions by means of a transcription factor called an activator, which binds to it and then to the rest of the transcription apparatus.

entropy (en'-truh-pē)
A measure of disorder, or randomness. One form of disorder is heat, which is random molecular motion.

enzyme (en'-zīm)
A molecule (usually a protein, but sometimes RNA) that serves as a biological catalyst, changing the rate of a chemical reaction without itself being changed in the process.

enzyme inhibitor
A chemical that interferes with an enzyme's activity by changing the enzyme's shape, either by plugging up the active site or by binding to another site on the enzyme.

epigenetic inheritance
Inheritance of traits transmitted by mechanisms not directly involving the nucleotide sequence of a genome; frequently involves chemical modification of DNA bases and/or histone proteins.

equilibrial life history
(ē-kwi-lib'-rē-ul)
The pattern of reaching sexual maturity slowly and producing few offspring but caring for the young; often seen in long-lived, large-bodied species.

errantian
A member of a major annelid lineage that includes mostly marine worms with an active lifestyle.

estuary (es'-chuh-wār-ē)
The area where a freshwater stream or river merges with seawater.

Eukarya (yū-kār'-yuh)
The domain of eukaryotes, organisms made up of eukaryotic cells; includes all of the protists, plants, fungi, and animals.

eukaryote (yū-kār'-ē-ōt)
An organism characterized by eukaryotic cells. See also eukaryotic cell.

eukaryotic cell (yū-kār'-ē-ot'-ik)
A type of cell that has a membrane-enclosed nucleus and other membrane-enclosed organelles. All organisms except bacteria and archaea (including protists, plants, fungi, and animals) are composed of eukaryotic cells.

eutherian (yū-thēr'-ē-un)
Mammal whose young complete their embryonic development in the uterus, nourished via the mother's blood vessels in the placenta; also called a placental mammal.

evaporative cooling
A property of water whereby a body becomes cooler as water evaporates from it.

evolution
Descent with modification; the idea that living species are descendants of ancestral species that were different from present-day ones; also, defined more narrowly as the change in the genetic composition of a population from generation to generation. Evolution is the central unifying theme of biology.

evolutionary adaptation
An inherited characteristic that enhances an organism's ability to survive and reproduce in a particular environment.

evolutionary tree
A branching diagram that reflects a hypothesis about evolutionary relationships among groups of organisms.

exaptation
A structure that evolves in one context and gradually becomes adapted for other functions.

exocytosis (ek'-sō-sī-tō'-sis)
The movement of materials out of the cytoplasm of a cell via membranous vesicles or vacuoles.

exon (ek'-son)
In eukaryotes, a coding portion of a gene. See also intron.

exoskeleton
A hard, external skeleton that protects an animal and provides points of attachment for muscles.

exotoxin
A poisonous protein secreted by certain bacteria.

experiment
A scientific test, often carried out under controlled conditions, that

involves changing just one factor (the variable) at a time.

experimental group
A set of subjects that has (or receives) the specific factor being tested in a controlled experiment. Ideally, the experimental group should be identical to the control group for all other factors.

exponential population growth
A model that describes the expansion of a population in an ideal, unlimited environment.

extracellular matrix
The meshwork that surrounds animal cells, consisting of a web of protein and polysaccharide fibers embedded in a liquid, jelly, or solid.

F

F$_1$ generation
The offspring of two parental (P generation) individuals. F$_1$ stands for first filial.

F$_2$ generation
The offspring of the F$_1$ generation. F$_2$ stands for second filial.

facilitated diffusion
The passage of a substance across a biological membrane down its concentration gradient aided by specific transport proteins.

fact
A piece of information that is correct based on all current information. A fact is not subject to opinion and can be independently verified.

family
In classification, the taxonomic category above genus.

fat
A large lipid molecule made from an alcohol called glycerol and three fatty acids; a triglyceride. Most fats function as energy-storage molecules.

fermentation
The anaerobic harvest of energy from food by some cells. Different pathways of fermentation can produce different end products, including ethanol and lactic acid.

fern
Any of a group of seedless vascular plants.

fertilization
The union of a haploid sperm cell with a haploid egg cell, producing a zygote.

flagella (*fla-jel´-uh*)
Extensions from a eukaryotic cell that propel the cell with an undulating, whiplike motion.

flagellate (*flaj´-uh-lit*)
A protozoan (animal-like protist) that moves by means of one or more flagella.

flatworm
A bilateral animal with a thin, flat body form, a gastrovascular cavity with a single opening, and no body cavity. Flatworms include planarians, flukes, and tapeworms.

flower
In an angiosperm, a short stem with four sets of modified leaves, bearing structures that function in sexual reproduction.

fluid mosaic
A description of membrane structure, depicting a cellular membrane as a mosaic of diverse protein molecules suspended in a fluid bilayer of phospholipid molecules.

food chain
The sequence of food transfers between the trophic levels of a community, beginning with the producers.

food web
A network of interconnecting food chains.

foram
A marine protozoan (animal-like protist) that secretes a shell and extends pseudopodia through pores in its shell.

forensics
The scientific analysis of evidence for crime scene investigations and other legal proceedings.

fossil
A preserved imprint or remains of an organism that lived in the past.

fossil fuel
An energy deposit formed from the fossilized remains of long-dead plants and animals.

fossil record
The ordered sequence of fossils as they appear in rock layers, marking the passing of geologic time.

founder effect
Genetic drift resulting from the establishment of a new, small population whose gene pool represents only a sample of the genetic variation present in the original population.

fruit
A ripened, thickened ovary of a flower, which protects dormant seeds and aids in their dispersal.

functional group
A group of atoms that form the chemically reactive part of an organic molecule. A particular functional group usually behaves similarly in different chemical reactions.

fungus
(plural, **fungi**) A heterotrophic eukaryote that digests its food externally and absorbs the resulting small nutrient molecules. Most fungi consist of a netlike mass of filaments called hyphae. Molds, mushrooms, and yeasts are examples of fungi.

G

gamete (*gam´-ēt*)
A sex cell; a haploid egg or sperm. The union of two gametes of opposite sex (fertilization) produces a zygote.

gametophyte (*guh-mē´-tō-fīt*)
The multicellular haploid form in the life cycle of organisms undergoing alternation of generations; results from a union of spores and mitotically produces haploid gametes that unite and grow into the sporophyte generation.

gastropod
Member of the largest group of molluscs, including snails and slugs.

gastrovascular cavity
A digestive compartment with a single opening that serves as both the entrance for food and the exit for undigested wastes; may also function in circulation, body support, and gas exchange. Jellies and hydras are examples of animals with a gastrovascular cavity.

gastrula (*gas´-trū-luh*)
The embryonic stage resulting from gastrulation in animal development. Most animals have a gastrula made up of three layers of cells: ectoderm, endoderm, and mesoderm.

gel electrophoresis
(*jel e-lek´-trō-fōr-ē´-sis*)
A technique for sorting macromolecules. A mixture of molecules is placed on a gel between a positively charged electrode and a negatively charged one; negatively charged molecules migrate toward the positive electrode. The molecules separate in the gel according to their rates of migration.

gene
A unit of inheritance in DNA (or RNA, in some viruses) consisting of a specific nucleotide sequence that programs the amino acid sequence of a polypeptide. Most of the genes of a eukaryote are located in its chromosomal DNA; a few are carried by the DNA of mitochondria and chloroplasts.

gene cloning
The production of multiple copies of a gene.

gene expression
The process whereby genetic information flows from genes to proteins; the flow of genetic information from the genotype to the phenotype: DNA → RNA → protein.

gene flow
The gain or loss of alleles from a population by the movement of individuals or gametes into or out of the population.

gene pool
All copies of every type of allele at every locus in all members of a population at any one time.

gene regulation
The turning on and off of specific genes within a living organism.

genetically modified (GM) organism
An organism that has acquired one or more genes by artificial means. If the gene is from another organism, typically of another species, the recombinant organism is also known as a transgenic organism. *See also* transgenic organism.

genetic code
The set of rules giving the correspondence between nucleotide triplets (codons) in mRNA and amino acids in protein.

genetic drift
A change in the gene pool of a population due to chance. Effects of genetic drift are most pronounced in small populations.

genetic engineering
The direct manipulation of genes for practical purposes.

genetics
The scientific study of heredity (inheritance).

genomics
The study of whole sets of genes and their interactions.

genotype (*jē´-nō-tīp*)
The genetic makeup of an organism.

genus (*jē´-nus*)
(plural, **genera**) In classification, the taxonomic category above species; the first part of a species' binomial; for example, *Homo*.

geologic time scale
A time scale established by geologists that divides Earth's history into a sequence of geologic periods, grouped into four divisions: Precambrian, Paleozoic, Mesozoic, and Cenozoic.

germinate
To initiate growth, as in a plant seed or a plant or fungal spore.

glycogen (*glī´-kō-jen*)
A complex, extensively branched polysaccharide made up of many glucose monomers; serves as a temporary energy-storage molecule in liver and muscle cells.

glycolysis (*glī-kol´-uh-sis*)
The multistep chemical breakdown of a molecule of glucose into two molecules of pyruvic acid; the first stage of cellular respiration in all organisms; occurs in the cytoplasmic fluid.

Golgi apparatus (*gol´-jē*)
An organelle in eukaryotic cells consisting of stacks of membranous sacs that modify, store, and ship products of the endoplasmic reticulum.

granum (*gran´-um*)
(plural, **grana**) A stack of hollow disks formed of thylakoid membrane in a chloroplast. Grana are the sites where light energy is trapped by chlorophyll and converted to chemical energy during the light reactions of photosynthesis.

green alga (*al´-guh*)
One of a group of photosynthetic protists that includes unicellular, colonial, and multicellular species. Green algae are the photosynthetic protists most closely related to plants.

greenhouse effect
The warming of the atmosphere caused by CO_2, CH_4, and other gases that absorb heat radiation and slow its escape from Earth's surface.

greenhouse gas
Any of the gases in the atmosphere that absorb heat radiation, including CO_2, methane, water vapor, and synthetic chlorofluorocarbons.

growth factor
A protein secreted by certain body cells that stimulates other cells to divide.

guanine (G) (*gwa´-nēn*)
A double-ring nitrogenous base found in DNA and RNA.

gymnosperm (*jim´-nō-sperm*)
A naked-seed plant. Its seed is said to be naked because it is not enclosed in an ovary.

H

habitat
A place where an organism lives; a specific environment in which an organism lives.

half-life
The amount of time it takes for 50% of a sample of a radioactive isotope to decay.

haploid
Containing a single set of chromosomes; referring to an *n* cell.

Hardy-Weinberg equilibrium
The condition describing a nonevolving population (one that is in genetic equilibrium).

heat
The amount of kinetic energy contained in the movement of the atoms and molecules in a body of matter. Heat is energy in its most random form.

herbivore
An animal that eats mainly plants, algae, or phytoplankton. *See also* carnivore; omnivore.

herbivory
The consumption of plant parts or algae by an animal.

heredity
The transmission of traits from one generation to the next.

heterotroph (*het'-er-ō-trōf*)
An organism that cannot make its own organic food molecules from inorganic ingredients and must obtain them by consuming other organisms or their organic products; a consumer (such as an animal) or a decomposer (such as a fungus) in a food chain.

heterozygous (*het'-er-ō-zī'-gus*)
Having two different alleles for a given gene.

histone (*his'-tōn*)
A small protein molecule associated with DNA and important in DNA packing in the eukaryotic chromosome.

HIV
Human immunodeficiency virus; the retrovirus that attacks the human immune system and causes AIDS.

homeotic gene (*hō'-mē-ot'-ik*)
A master control gene that determines the identity of a body structure of a developing organism, presumably by controlling the developmental fate of groups of cells. (In plants, such genes are called organ identity genes.)

hominin (*hah'-mi-nin*)
Any anthropoid on the human branch of the evolutionary tree, more closely related to humans than to chimpanzees.

homologous chromosomes (*hō-mol'-uh-gus*)
The two chromosomes that make up a matched pair in a diploid cell. Homologous chromosomes are of the same length, centromere position, and staining pattern and possess genes for the same characteristics at corresponding loci. One homologous chromosome is inherited from the organism's father, the other from the mother.

homology (*hō-mol'-uh-jē*)
Similarity in characteristics resulting from a shared ancestry.

homozygous (*hō'-mō-zī'-gus*)
Having two identical alleles for a given gene.

host
An organism that is exploited by a parasite or pathogen.

human gene therapy
A recombinant DNA procedure intended to treat disease by altering an afflicted person's genes.

Human Genome Project
An international collaborative effort that sequenced the DNA of the entire human genome.

hybrid
The offspring of parents of two different species or of two different varieties of one species; the offspring of two parents that differ in one or more inherited traits; an individual that is heterozygous for one or more pairs of genes.

hydrogen bond
A type of weak chemical bond formed when a partially positive hydrogen atom from one polar molecule is attracted to the partially negative atom in another molecule (or in another part of the same molecule).

hydrogenation
The artificial process of converting unsaturated fats to saturated fats by adding hydrogen.

hydrolysis (*hī-drol'-uh-sis*)
A chemical process in which macromolecules are broken down by the chemical addition of water molecules to the bonds linking their monomers; an essential part of digestion. A hydrolysis reaction is essentially the opposite of a dehydration reaction.

hydrophilic (*hī'-drō-fil'-ik*)
"Water-loving"; pertaining to polar, or charged, molecules (or parts of molecules), which are soluble in water.

hydrophobic (*hī'-drō-fō'-bik*)
"Water-fearing"; pertaining to nonpolar molecules (or parts of molecules), which do not dissolve in water.

hypertonic
In comparing two solutions, referring to the one with the greater concentration of solutes.

hypha (*hī'-fuh*)
(plural, **hyphae**) One of many filaments making up the body of a fungus.

hypothesis (*hī-poth'-uh-sis*)
(plural, **hypotheses**) A tentative explanation that a scientist proposes for a specific phenomenon that has been observed.

hypotonic
In comparing two solutions, referring to the one with the lower concentration of solutes.

I

incomplete dominance
A type of inheritance in which the phenotype of a heterozygote (*Aa*) is intermediate between the phenotypes of the two types of homozygotes (*AA* and *aa*).

independent variable
A factor whose value is manipulated or changed during an experiment to reveal possible effects on another factor (the dependent variable).

induced fit
The interaction between a substrate molecule and the active site of an enzyme, which changes shape slightly to embrace the substrate and catalyze the reaction.

insect
An arthropod that usually has three body segments (head, thorax, and abdomen), three pairs of legs, and one or two pairs of wings.

interphase
The phase in the eukaryotic cell cycle when the cell is not actually dividing. During interphase, cellular metabolic activity is high, chromosomes and organelles are duplicated, and cell size may increase. Interphase accounts for 90% of the cell cycle. *See also* mitosis.

interspecific competition
Competition between individuals of two or more species that require similar limited resources.

interspecific interaction
Any interaction between members of different species.

intertidal zone (*in'-ter-tīd'-ul*)
A shallow zone where the waters of an estuary or ocean meet land.

intraspecific competition
Competition between individuals of the same species for the same limited resources.

intron (*in'-tron*)
In eukaryotes, a nonexpressed (noncoding) portion of a gene that is excised from the RNA transcript. *See also* exon.

invasive species
A non-native species that has spread far beyond the original point of introduction and causes environmental or economic damage by colonizing and dominating suitable habitats.

invertebrate
An animal that does not have a backbone.

ion
An atom or molecule that has gained or lost one or more electrons, thus acquiring an electrical charge.

ionic bond
An attraction between two ions with opposite electrical charges. The electrical attraction of the opposite charges holds the ions together.

isomer (*ī'-sō-mer*)
One of two or more molecules with the same molecular formula but different structures and thus different properties.

isotonic (*ī-sō-ton'-ik*)
Having the same solute concentration as another solution.

isotope (*ī'-sō-tōp*)
A variant form of an atom. Different isotopes of an element have the same number of protons and electrons but different numbers of neutrons.

K

karyotype (*kār'-ē-ō-tīp*)
A display of micrographs of the metaphase chromosomes of a cell, arranged by size and centromere position.

keystone species
A species whose impact on its community is much larger than its biomass or abundance indicates.

kinetic energy (*kuh-net'-ik*)
Energy of motion. Moving matter performs work by transferring its motion to other matter, such as leg muscles pushing bicycle pedals.

kingdom
In classification, the broad taxonomic category above phylum.

L

lancelet
One of a group of bladelike invertebrate chordates.

landscape
A regional assemblage of interacting ecosystems.

landscape ecology
The application of ecological principles to the study of land-use patterns; the scientific study of the biodiversity of interacting ecosystems.

larva
An immature individual that looks different from the adult animal.

lateral line system
A row of sensory organs along each side of a fish's body. Sensitive to changes in water pressure, it enables a fish to detect minor vibrations in the water.

law of independent assortment
A general rule of inheritance, first proposed by Gregor Mendel, that states that when gametes form during meiosis, each pair of alleles for a particular character segregates (separates) independently of each other pair.

law of segregation
A general rule of inheritance, first proposed by Gregor Mendel, that states that the two alleles in a pair segregate (separate) into different gametes during meiosis.

life cycle
The entire sequence of generation-to-generation stages in the life of an organism, from fertilization to the production of its own offspring

life history
The traits that affect an organism's schedule of reproduction and survival.

life table
A listing of survivals and deaths in a population in a particular time period and predictions of how long, on average, an individual of a given age will live.

light reactions
The first of two stages in photosynthesis; the steps in which solar energy is absorbed and converted to chemical energy in the form of ATP and NADPH. The light reactions power the sugar-producing Calvin cycle but produce no sugar themselves.

lignin (*lig'-nin*)
A chemical that hardens the cell walls of plants. Lignin makes up most of what we call wood.

limiting factor
An environmental factor that restricts the number of individuals that can occupy a particular habitat, thus holding population growth in check.

linked genes
Genes located close enough together on a chromosome that they are usually inherited together.

lipid
An organic compound consisting mainly of carbon and hydrogen atoms linked by nonpolar covalent bonds and therefore mostly hydrophobic and insoluble in water. Lipids include fats, waxes, phospholipids, and steroids.

lobe-finned fish
A bony fish with strong, muscular fins supported by bones. *See also* ray-finned fish.

locus
(plural, **loci**) The particular site where a gene is found on a chromosome. Homologous chromosomes have corresponding gene loci.

logistic population growth
A model that describes population growth that decreases as population size approaches carrying capacity.

lysogenic cycle (*lī-sō-jen'-ik*)
A bacteriophage reproductive cycle in which the viral genome is incorporated into the bacterial host chromosome as a prophage. New phages are not produced, and the host cell is not killed or lysed unless the viral genome leaves the host chromosome.

lysosome (*lī'-sō-sōm*)
A digestive organelle in eukaryotic cells; contains enzymes that digest the cell's food and wastes.

lytic cycle (*lit'-ik*)
A viral reproductive cycle resulting in the release of new viruses by lysis (breaking open) of the host cell.

M

macroevolution
Evolutionary change above the species level. Examples of macroevolutionary change include the origin of a new group of organisms through a series of speciation events, the impact of mass extinctions on the diversity of life, and the origin of key adaptations.

macromolecule
A giant molecule formed by joining smaller molecules. Examples of macromolecules include proteins, polysaccharides, and nucleic acids.

malignant tumor
An abnormal tissue mass that spreads into neighboring tissue and to other parts of the body; a cancerous tumor.

mammal
Member of a class of endothermic amniotes that possesses mammary glands and hair.

mantle
In molluscs, the outgrowth of the body surface that drapes over the animal. The mantle produces the shell and forms the mantle cavity.

marsupial (*mar-sū'-pē-ul*)
A pouched mammal, such as a kangaroo, opossum, or koala. Marsupials give birth to embryonic offspring that complete development while housed in a pouch and attached to nipples on the mother's abdomen.

mass
A measure of the amount of matter in an object.

mass number
The sum of the number of protons and neutrons in an atom's nucleus.

matter
Anything that occupies space and has mass.

medusa (*med-ū'-suh*)
(plural, **medusae**) One of two types of cnidarian body forms; a floating, umbrella-like body form; also called a jelly.

meiosis (*mī-ō'-sis*)
In a sexually reproducing organism, the process of cell division that produces haploid gametes from diploid cells within the reproductive organs.

messenger RNA (mRNA)
The type of ribonucleic acid that encodes genetic information from DNA and conveys it to ribosomes, where the information is translated into amino acid sequences.

metabolism (*muh-tab'-uh-liz-um*)
The total of all the chemical reactions in an organism.

metamorphosis
(*met'-uh-mōr'-fuh-sis*)
The transformation of a larva into an adult.

metaphase (*met'-eh-fāz*)
The second stage of mitosis. During metaphase, the centromeres of all the cell's duplicated chromosomes are lined up along the center line of the cell.

metastasis (*muh-tas'-tuh-sis*)
The spread of cancer cells beyond their original site.

microbiome
The collection of genomes of individual microbial species present in a particular environment, such as the human intestinal tract.

microbiota
The community of microorganisms that live in and on the body of an animal.

microevolution
A change in a population's gene pool over a succession of generations.

microRNA (miRNA)
A small, single-stranded RNA molecule that associates with one or more proteins in a complex that can degrade or prevent translation of an mRNA with a complementary sequence.

microtubule
The thickest of the three main kinds of fibers making up the cytoskeleton of a eukaryotic cell; a straight, hollow tube made of globular proteins called tubulins. Microtubules form the basis of the structure and movement of cilia and flagella.

millipede
A terrestrial arthropod that has two pairs of short legs for each of its numerous body segments and that eats decaying plant matter.

mitochondrion (*mī'-tō-kon'-drē-on*)
(plural, **mitochondria**) An organelle in eukaryotic cells where cellular respiration occurs. Enclosed by two concentric membranes, it is where most of the cell's ATP is made.

mitosis (*mī-tō'-sis*)
The division of a single nucleus into two genetically identical daughter nuclei. Mitosis and cytokinesis make up the mitotic (M) phase of the cell cycle.

mitotic (M) phase
The phase of the cell cycle when mitosis divides the nucleus and distributes its chromosomes to the daughter nuclei and cytokinesis divides the cytoplasm, producing two daughter cells.

mitotic spindle
A spindle-shaped structure formed of microtubules and associated proteins that is involved in the movement of chromosomes during mitosis and meiosis. (A spindle is shaped roughly like a football.)

molecular biology
The study of biological structures, functions, and heredity at the molecular level.

molecule
A group of two or more atoms held together by covalent bonds.

mollusc (mol'-lusk)
A soft-bodied animal characterized by a muscular foot, mantle, mantle cavity, and radula. Molluscs include gastropods (snails and slugs), bivalves (clams, oysters, and scallops), and cephalopods (squids and octopuses).

monohybrid cross
A mating of individuals that are heterozygous for the character being followed.

monomer (mon'-uh-mer)
A chemical subunit that serves as a building block of a polymer.

monosaccharide (mon'-uh-sak'-uh-rīd)
The smallest kind of sugar molecule; a single-unit sugar; also known as a simple sugar.

monotreme (mon'-uh-trēm)
An egg-laying mammal, such as the duck-billed platypus.

moss
Any of a group of seedless nonvascular plants.

movement corridor
A series of small clumps or a narrow strip of quality habitat (usable by organisms) that connects otherwise isolated patches of quality habitat.

mutagen (myū'-tuh-jen)
A chemical or physical agent that interacts with DNA and causes a mutation.

mutation
Any change to the genetic information of a cell or virus.

mutualism
An interspecific interaction in which both partners benefit.

mycelium (mī-sē'-lē-um)
(plural, **mycelia**) The densely branched network of hyphae in a fungus.

mycorrhiza (mī'-kō-rī'-zuh)
(plural, **mycorrhizae**) A mutually beneficial symbiotic association of a plant root and fungus.

N

NADH
An electron carrier (a molecule that carries electrons) involved in cellular respiration and photosynthesis. NADH carries electrons from glucose and other fuel molecules and deposits them at the top of an electron transport chain. NADH is generated during glycolysis and the citric acid cycle.

NADPH
An electron carrier (a molecule that carries electrons) involved in photosynthesis. Light drives electrons from chlorophyll to $NADP^+$, forming NADPH, which provides the high-energy electrons for the reduction of carbon dioxide to sugar in the Calvin cycle.

natural selection
A process in which individuals with certain inherited traits are more likely to survive and reproduce than are individuals that do not have those traits.

nematode (nēm'-uh-tōd)
See roundworm.

neutron
An electrically neutral particle (a particle having no electrical charge) found in the nucleus of an atom.

nitrogen fixation
The conversion of atmospheric nitrogen (N_2) to ammonia (NH_3). NH_3 then picks up another H^+ to become NH_4^+ (ammonium), which plants can absorb and use.

nondisjunction
An accident of meiosis or mitosis in which a pair of homologous chromosomes or a pair of sister chromatids fails to separate at anaphase.

notochord (nō'-tuh-kord)
A flexible, cartilage-like, longitudinal rod located between the digestive tract and nerve cord in chordate animals, present only in embryos in many species.

nuclear envelope
A double membrane, perforated with pores, that encloses the nucleus and separates it from the rest of the eukaryotic cell.

nuclear transplantation
A technique in which the nucleus of one cell is placed into another cell that already has a nucleus or in which the nucleus has been previously destroyed. The cell is then stimulated to grow, producing an embryo that is a genetic copy of the nucleus donor.

nucleic acid (nū-klā'-ik)
A polymer consisting of many nucleotide monomers; serves as a blueprint for proteins and, through the actions of proteins, for all cellular structures and activities. The two types of nucleic acids are DNA and RNA.

nucleoid
A non—membrane-enclosed region in a prokaryotic cell where the DNA is concentrated.

nucleolus (nū-klē'-ō-lus)
A structure within the nucleus of a eukaryotic cell where ribosomal RNA is made and assembled with proteins to make ribosomal subunits; consists of parts of the chromatin DNA, RNA transcribed from the DNA, and proteins imported from the cytoplasm.

nucleosome (nū'-klē-ō-sōm)
The bead-like unit of DNA packing in a eukaryotic cell; consists of DNA wound around a protein core made up of eight histone molecules.

nucleotide (nū'-klē-ō-tīd)
An organic monomer consisting of a five-carbon sugar covalently bonded to a nitrogenous base and a phosphate group. Nucleotides are the building blocks of nucleic acids, including DNA and RNA.

nucleus
(plural, **nuclei**) (1) An atom's central core, containing protons and neutrons. (2) The genetic control center of a eukaryotic cell.

O

omnivore
An animal that eats both plants and animals. *See also* carnivore; herbivore.

oncogene (on'-kō-jēn)
A cancer-causing gene; usually contributes to malignancy by abnormally enhancing the amount or activity of a growth factor made by the cell.

operator
In prokaryotic DNA, a sequence of nucleotides near the start of an operon to which an active repressor can attach. The binding of repressor prevents RNA polymerase from attaching to the promoter and transcribing the genes of the operon.

operculum (ō-per'-kyū-lum)
(plural, **opercula**) A protective flap on each side of a bony fish's head that covers a chamber housing the gills.

operon (op'-er-on)
A unit of genetic regulation common in prokaryotes; a cluster of genes with related functions, along with the promoter and operator that control their transcription.

opportunistic life history
The pattern of reproducing when young and producing many offspring that receive little or no parental care; often seen in short-lived, small-bodied species.

order
In classification, the taxonomic category above family.

organelle (ōr-guh-nel')
A membrane-enclosed structure with a specialized function within a eukaryotic cell.

organic compound
A chemical compound containing the element carbon.

organism
An individual living thing, such as a bacterium, fungus, protist, plant, or animal.

organismal ecology
The study of the evolutionary adaptations that enable individual organisms to meet the challenges posed by their abiotic environments.

osmoregulation
The control of the gain or loss of water and dissolved solutes in an organism.

osmosis (oz-mō'-sis)
The diffusion of water across a selectively permeable membrane.

ovary
(1) In animals, the female gonad, which produces egg cells and reproductive hormones. (2) In flowering plants, the base of a carpel in which the egg-containing ovules develop.

ovule (ō'-vyūl)
In a seed plant, a reproductive structure that contains the female gametophyte and the developing egg. An ovule develops into a seed.

P

P generation
The parent individuals from which offspring are derived in studies of inheritance. P stands for parental.

paleontologist
A scientist who studies fossils.

parasite
An organism that exploits another organism (the host) from which it obtains nourishment or other benefits; an organism that benefits at the expense of another organism, which is harmed in the process.

passive transport
The diffusion of a substance across a biological membrane without any input of energy.

pathogen
A disease-causing virus or organism.

pedigree
A family tree representing the occurrence of heritable traits in parents and offspring across a number of generations.

peer review
The evaluation of scientific work by impartial, often anonymous, experts in that same field. Peer review is considered a good means of recognizing valid scientific sources.

pelagic realm (*puh-laj'-ik*)
The open-water region of an ocean.

peptide bond
The covalent linkage between two amino acid units in a polypeptide, formed by a dehydration reaction between two amino acids.

periodic table of the elements
A table listing all of the chemical elements (both natural and human-made) ordered by atomic number (the number of protons in the nucleus of a single atom of that element).

permafrost
Continuously frozen subsoil found in the arctic tundra.

petal
A modified leaf of a flowering plant. Petals are the often colorful parts of a flower that advertise it to insects and other pollinators.

pH scale
A measure of the relative acidity of a solution, ranging in value from 0 (most acidic) to 14 (most basic).

phage (*fāj*)
See bacteriophage.

phagocytosis (*fag'-ō-sī-tō'-sis*)
Cellular "eating"; a type of endocytosis whereby a cell engulfs large molecules, other cells, or particles into its cytoplasm.

pharyngeal slit (*fuh-rin'-jē-ul*)
A gill structure in the pharynx, found in chordate embryos and some adult chordates.

phenotype (*fē'-nō-tīp*)
The expressed traits of an organism.

phospholipid (*fos'-fō-lip'-id*)
A molecule that is a part of the inner bilayer of biological membranes, having a hydrophilic head and a hydrophobic tail.

phospholipid bilayer
A double layer of phospholipid molecules (each molecule consisting of a phosphate group bonded to two fatty acids) that is the primary component of all cellular membranes.

photic zone (*fō'-tik*)
Shallow water near the shore or the upper layer of water away from the shore; region of an aquatic ecosystem where sufficient light is available for photosynthesis.

photon (*fō'-ton*)
A fixed quantity of light energy. The shorter the wavelength of light, the greater the energy of a photon.

photosynthesis (*fō'-tō-sin'-thuh-sis*)
The process by which plants, algae, and some bacteria transform light energy to chemical energy stored in the bonds of sugars. This process requires an input of carbon dioxide (CO_2) and water (H_2O) and produces oxygen gas (O_2) as a waste product.

photosystem
A light-harvesting unit of a chloroplast's thylakoid membrane; consists of several hundred molecules, a reaction-center chlorophyll, and a primary electron acceptor.

phylogenetic tree (*fī'-lō-juh-net'-ik*)
A branching diagram that represents a hypothesis about evolutionary relationships between organisms.

phylogeny (*fī-loj'-uh-nē*)
The evolutionary history of a species or group of related species.

phylum (*fī'-lum*)
(plural, **phyla**) In classification, the taxonomic category above class.

phytoplankton
Photosynthetic organisms, mostly microscopic, that drift near the surfaces of ponds, lakes, and oceans.

placebo
A harmless but ineffective procedure or treatment that is given purely for psychological reasons or to act as a control in a blind experiment.

placenta (*pluh-sen'-tuh*)
In most mammals, the organ that provides nutrients and oxygen to the embryo and helps dispose of its metabolic wastes. The placenta is formed from embryonic tissue and the mother's endometrial blood vessels.

placental mammal (*pluh-sen'-tul*)
Mammal whose young complete their embryonic development in the uterus, nourished via the mother's blood vessels in the placenta; also called a eutherian.

plant
A multicellular eukaryote that carries out photosynthesis and has a set of structural and reproductive terrestrial adaptations, including a multicellular, dependent embryo.

plasma membrane
The thin double layer of lipids and proteins that sets a cell off from its surroundings and acts as a selective barrier to the passage of ions and molecules into and out of the cell; consists of a phospholipid bilayer in which proteins are embedded.

plasmid
A small ring of self-replicating DNA separate from the larger chromosome(s). Plasmids are most frequently derived from bacteria.

plate tectonics (*tek-tahn'-iks*)
The theory that the continents are part of great plates of Earth's crust that float on the hot, underlying portion of the mantle. Movements in the mantle cause the continents to move slowly over time.

pleiotropy (*plī'-uh-trō-pē*)
The control of more than one phenotypic character by a single gene.

polar ice
A terrestrial biome that includes regions of extremely cold temperature and low precipitation located at high latitudes north of the arctic tundra and in Antarctica.

polar molecule
A molecule containing an uneven distribution of charge due to the presence of polar covalent bonds (bonds having opposite charges on opposite ends). A polar molecule will have a slightly positive pole (end) and a slightly negative pole.

pollen grain
In a seed plant, the male gametophyte that develops within the anther of a stamen. It houses cells that will develop into sperm.

pollination
In seed plants, the delivery, by wind or animals, of pollen from the male (pollen-producing) parts of a plant to the stigma of a carpel on the female part of a plant.

polygenic inheritance (*pol'-ē-jen'-ik*)
The additive effect of two or more genes on a single phenotypic character.

polymer (*pol'-uh-mer*)
A large molecule consisting of many identical or similar molecular units, called monomers, covalently joined together in a chain.

polymerase chain reaction (PCR) (*puh-lim'-uh-rās*)
A technique used to obtain many copies of a DNA molecule or many copies of part of a DNA molecule. A small amount of DNA mixed with the enzyme DNA polymerase, DNA nucleotides, and a few other ingredients replicates repeatedly in a test tube.

polynucleotide (*pol'-ē-nū'-klē-ō-tīd*)
A polymer made up of many nucleotides covalently bonded together.

polyp (*pol'-ip*)
One of two types of cnidarian body forms; a stationary (sedentary), columnar, hydra-like body.

polypeptide
A chain of amino acids linked by peptide bonds.

polyploid
An organism that has more than two complete sets of chromosomes as a result of an accident of cell division.

polysaccharide (*pol'-ē-sak'-uh-rīd*)
A carbohydrate polymer consisting of many monosaccharides (simple sugars) linked by covalent bonds.

population
A group of interacting individuals belonging to one species and living in the same geographic area at the same time.

population density
The number of individuals of a species per unit area or volume of the habitat.

population ecology
The study of how members of a population interact with their environment, focusing on factors that influence population density and growth.

population momentum
In a population in which fertility (the number of live births over a woman's lifetime) averages two children (replacement rate), the continuation of population growth as girls reach their reproductive years.

post-anal tail
A tail posterior to the anus, found in chordate embryos and most adult chordates.

postzygotic barrier (*pōst'-zī-got'-ik*)
A reproductive barrier that prevents development of fertile adults if hybridization occurs.

potential energy
Stored energy; the energy that an object has due to its location and/or arrangement. Water behind a dam and chemical bonds both possess potential energy.

predation
An interaction in which an individual of one species, the predator, kills and eats an individual of the other species, the prey.

prezygotic barrier (*prē'-zī-got'-ik*)
A reproductive barrier that prevents mating between species or hinders fertilization of eggs if members of different species attempt to mate.

primary consumer
An organism that eats only producers (autotrophs); an herbivore.

primary production
The amount of solar energy converted to the chemical energy stored in organic compounds by producers in an ecosystem during a given time period.

primary succession
A type of ecological succession in which a biological community begins in an area without soil. *See also* secondary succession.

primate
Member of the mammalian group that includes lorises, bush babies, lemurs, tarsiers, monkeys, apes, and humans.

primer
A short stretch of nucleic acid bound by complementary base pairing to a DNA sequence and elongated with DNA nucleotides. During PCR, primers flank the desired sequence to be copied.

prion (*prī'-on*)
An infectious form of protein that may multiply by converting related proteins to more prions. Prions cause several related diseases in different animals, including scrapie in sheep, mad cow disease, and Creutzfeldt-Jakob disease in humans.

producer
An organism that makes organic food molecules from carbon dioxide, water, and other inorganic raw materials: a plant, alga, or autotrophic bacterium; the trophic level that supports all others in a food chain or food web.

product
An ending material in a chemical reaction.

prokaryote (*prō-kār'-ē-ōt*)
An organism characterized by prokaryotic cells. *See also* prokaryotic cell.

prokaryotic cell (*prō-kār'-ē-ot'-ik*)
A type of cell lacking a nucleus and other membrane-bound organelles. Prokaryotic cells are found only among organisms of the domains Bacteria and Archaea.

promoter
A specific nucleotide sequence in DNA, located at the start of a gene, that is the binding site for RNA polymerase and the place where transcription begins.

prophage (*prō'-fāj*)
Phage DNA that has inserted into the DNA of a prokaryotic chromosome.

prophase
The first stage of mitosis. During prophase, duplicated chromosomes condense to form structures visible with a light microscope, and the mitotic spindle forms and begins moving the chromosomes toward the center of the cell.

protein
A biological polymer constructed from hundreds to thousands of amino acid monomers. Proteins perform many functions within living cells, including providing structure, transport, and acting as enzymes.

proteomics
The systematic study of the full protein sets (proteomes) encoded by genomes.

protist (*prō'-tist*)
Any eukaryote that is not a plant, animal, or fungus.

proton
A subatomic particle with a single unit of positive electrical charge, found in the nucleus of an atom.

proto-oncogene (*prō'-tō-on'-kō-jēn*)
A normal gene that can be converted to a cancer-causing gene.

protozoan (*prō'-tō-zō'-un*)
A protist that lives primarily by ingesting food; a heterotrophic, animal-like protist.

provirus
Viral DNA that inserts into a host genome.

pseudopodium (*sū'-dō-pō'-dē-um*)
(plural, **pseudopodia**) A temporary extension of an amoeboid cell. Pseudopodia function in moving cells and engulfing food.

pseudoscience
A field of study that is falsely presented or mistakenly regarded as having a scientific basis when it does not.

Punnett square
A diagram used in the study of inheritance to show the results of random fertilization.

Q

quaternary consumer (*kwot'-er-nār-ē*)
An organism that eats tertiary consumers.

R

radial symmetry
An arrangement of the body parts of an organism like pieces of a pie around an imaginary central axis. Any slice passing longitudinally through a radially symmetric organism's central axis divides the organism into mirror-image halves.

radiation therapy
Treatment for cancer in which parts of the body that have cancerous tumors are exposed to high-energy radiation to disrupt cell division of the cancer cells.

radioactive isotope
An isotope whose nucleus decays spontaneously, giving off particles and energy.

radiometric dating
A method for determining the ages of fossils and rocks from the ratio of a radioactive isotope to the nonradioactive isotope(s) of the same element in the sample.

radula (*rad'-yū-luh*)
A file-like organ found in many molluscs, typically used to scrape up or shred food.

ray-finned fish
A bony fish in which fins are webs of skin supported by thin, flexible skeletal rays. All but one living species of bony fishes are ray-fins. *See also* lobe-finned fish.

reactant
A starting material in a chemical reaction.

recessive allele
In heterozygotes, the allele that has no noticeable effect on the phenotype; the recessive version of a gene is usually represented with a lowercase italic letter (e.g., *f*).

recombinant DNA
A DNA molecule carrying genes derived from two or more sources, often from different species.

regeneration
The regrowth of body parts from pieces of an organism.

relative abundance
The proportional representation of a species in a biological community; one component of species diversity.

relative fitness
The contribution an individual makes to the gene pool of the next generation relative to the contribution of other individuals in the population.

repetitive DNA
Nucleotide sequences that are present in many copies in the DNA of a genome. The repeated sequences may be long or short and may be located next to each other or dispersed in the DNA.

repressor
A protein that blocks the transcription of a gene or operon.

reproductive barrier
A factor that prevents individuals of closely related species from interbreeding

reproductive cloning
Using a body cell from a multicellular organism to make one or more genetically identical individuals. *See also* therapeutic cloning.

reptile
Member of the clade of amniotes that includes snakes, lizards, turtles, crocodiles, alligators, birds, and a number of extinct groups (including most of the dinosaurs).

restoration ecology
A field of ecology that develops methods of returning degraded ecosystems to their natural state.

restriction enzyme
A bacterial enzyme that cuts up foreign DNA at one very specific nucleotide sequence. Restriction enzymes are used in DNA technology to cut DNA molecules in reproducible ways.

restriction fragment
A molecule of DNA produced from a longer DNA molecule cut up by a restriction enzyme.

restriction site
A specific sequence on a DNA strand that is recognized and cut by a restriction enzyme.

retrovirus
An RNA virus that reproduces by means of a DNA molecule. It reverse-transcribes its RNA into DNA, inserts the DNA into a cellular chromosome, and then transcribes more copies of the RNA from the viral DNA. HIV and a number of cancer-causing viruses are retroviruses.

reverse transcriptase (*tran-skrip'-tās*)
An enzyme that catalyzes the synthesis of DNA on an RNA template.

ribosomal RNA (rRNA)
(*rī'-buh-sōm'-ul*)
The type of ribonucleic acid that, together with proteins, makes up ribosomes.

ribosome (*rī'-buh-sōm*)
A cellular structure consisting of RNA and protein organized into two subunits and functioning as the site of protein synthesis in the cytoplasm. The ribosomal subunits are constructed in the nucleolus and then transported to the cytoplasm where they act.

ribozyme (*rī'-bō-zīm*)
An RNA molecule that functions as an enzyme.

RNA (ribonucleic acid)
(*rī'-bō-nū-klā'-ik*)
A type of nucleic acid consisting of nucleotide monomers, with a ribose sugar, a phosphate group, and the nitrogenous bases adenine (A), cytosine (C), guanine (G), and uracil (U); usually single-stranded; functions in protein synthesis and as the genome of some viruses.

RNA interference (RNAi)
A biotechnology technique used to silence the expression of specific genes. Synthetic RNA molecules with sequences that correspond to particular genes trigger the breakdown of the gene's mRNA.

RNA polymerase (*puh-lim'-er-ās*)
An enzyme that links together the growing chain of RNA nucleotides during transcription, using a DNA strand as a template.

RNA splicing
The removal of introns and joining of exons in eukaryotic RNA, forming an mRNA molecule with a continuous coding sequence; occurs before mRNA leaves the nucleus.

root
The underground organ of a plant. Roots anchor the plant in the soil, absorb and transport minerals and water, and store food.

rough ER (rough endoplasmic reticulum) (*reh-tik'-yuh-lum*)
A network of interconnected membranous sacs in a eukaryotic cell's cytoplasm. Rough ER membranes are studded with ribosomes that make membrane proteins and secretory proteins. The rough ER constructs membrane from phospholipids and proteins.

roundworm
An animal characterized by a cylindrical, wormlike body form and a complete digestive tract; also called a nematode.

rule of multiplication
A rule stating that the probability of a compound event is the product of the separate probabilities of the independent events.

S

saturated
Pertaining to fats and fatty acids whose hydrocarbon chains contain the maximum number of hydrogens and therefore have no double covalent bonds. Because of their straight, flat shape, saturated fats and fatty acids tend to be solid at room temperature.

savanna
A terrestrial biome dominated by grasses and scattered trees. The temperature is warm year-round. Frequent fires and seasonal drought are significant abiotic factors.

science
Any method of learning about the natural world that follows the scientific method.

seaweed
A large, multicellular marine alga.

secondary consumer
An organism that eats primary consumers.

secondary succession
A type of ecological succession that occurs where a disturbance has destroyed an existing biological community but left the soil intact. *See also* primary succession.

sedentarian
A member of a major annelid lineage that includes earthworms, leeches, and many tube-building marine worms.

seed
A plant embryo packaged with a food supply within a protective covering.

sepal (*sē'-pul*)
A modified leaf of a flowering plant. A whorl of sepals encloses and protects the flower bud before it opens.

sex chromosome
A chromosome that determines whether an individual is male or female; in mammals, for example, the X or Y chromosome.

sex-linked gene
A gene located on a sex chromosome.

sexual dimorphism
Marked differences between the secondary sexual traits of males and females.

sexual reproduction
The creation of genetically distinct offspring by the fusion of two haploid sex cells (gametes: sperm and egg), forming a diploid zygote.

sexual selection
A form of natural selection in which individuals with certain traits are more likely than other individuals to obtain mates.

shoot
The aerial organ of a plant, consisting of stem and leaves. Leaves are the main photosynthetic structures of most plants.

short tandem repeat (STR)
DNA consisting of tandem (in a row) repeats of a short sequence of nucleotides.

signal transduction pathway
A series of molecular changes that converts a signal received on a target cell's surface to a specific response inside the cell.

silencer
A eukaryotic DNA sequence that inhibits the start of gene transcription; may act analogously to an enhancer, binding a repressor.

single-blind experiment
A scientific experiment in which some information is withheld from the test subject.

sister chromatid (*krō'-muh-tid*)
One of the two identical parts of a duplicated chromosome. While joined, two sister chromatids make up one chromosome; chromatids are eventually separated during mitosis or meiosis II.

slime mold
A multicellular protist related to amoebas.

smooth ER (smooth endoplasmic reticulum) (*reh-tik'-yuh-lum*)
A network of interconnected membranous tubules in a eukaryotic cell's cytoplasm. Smooth ER lacks ribosomes. Enzymes embedded in the smooth ER membrane function in the synthesis of certain kinds of molecules, such as lipids.

solute (*sol'-yūt*)
A substance that is dissolved in a liquid (which is called the solvent) to form a solution.

solution
A liquid consisting of a homogeneous mixture of two or more substances: a dissolving agent, the solvent, and a substance that is dissolved, the solute.

solvent
The dissolving agent in a solution. Water is the most versatile known solvent.

somatic cell (*sō-mat'-ik*)
Any cell in a multicellular organism except a sperm or egg cell or a cell that develops into a sperm or egg; a body cell.

speciation (*spē'-sē-ā'-shun*)
An evolutionary process in which one species splits into two or more species.

species
A group of populations the members of which have the potential to interbreed and produce viable, fertile offspring. *See also* biological species concept.

species diversity
The variety of species that make up a biological community; the number and relative abundance of species in a biological community.

species richness

The total number of different species in a community; one component of species diversity.

sponge

An aquatic stationary animal characterized by a highly porous body, choanocytes (specialized cells used for suspension feeding), and no tissues.

spore

(1) In plants and algae, a haploid cell that can develop into a multicellular haploid individual, the gametophyte, without fusing with another cell. (2) In fungi, a haploid cell that germinates to produce a mycelium.

sporophyte (*spōr´-uh-fīt*)

The multicellular diploid form in the life cycle of organisms undergoing alternation of generations; results from a union of gametes and meiotically produces haploid spores that grow into the gametophyte generation.

stabilizing selection

Natural selection that favors intermediate variants by acting against extreme phenotypes.

stamen (*stā´-men*)

A pollen-producing part of a flower, consisting of a stalk (filament) and an anther.

starch

A storage polysaccharide found in the roots of plants and certain other cells; a polymer of glucose.

start codon (*kō´-don*)

On mRNA, the specific three-nucleotide sequence (AUG) to which an initiator tRNA molecule binds, starting translation of genetic information.

steroid (*stir´-oyd*)

A type of lipid with a carbon skeleton in the form of four fused rings: three 6-sided rings and one 5-sided ring. Examples are cholesterol, testosterone, and estrogen.

stigma (*stig´-muh*)

(plural, **stigmata**) The sticky tip of a flower's carpel that traps pollen.

stoma (*stō´-muh*)

(plural, **stomata**) A pore surrounded by guard cells in the epidermis of a leaf. When stomata are open, CO_2 enters the leaf, and water and O_2 exit. A plant conserves water when its stomata are closed.

stop codon (*kō´-don*)

In mRNA, one of three triplets (UAG, UAA, UGA) that signal gene translation to stop.

STR analysis

A method of DNA profiling that compares the lengths of STR sequences at specific sites in the genome.

stroma (*strō´-muh*)

A thick fluid enclosed by the inner membrane of a chloroplast. Sugars are made in the stroma by the enzymes of the Calvin cycle.

substrate

(1) A specific substance (reactant) on which an enzyme acts. Each enzyme recognizes only the specific substrate of the reaction it catalyzes. (2) A surface in or on which an organism lives.

sugar-phosphate backbone

The alternating chain of sugar and phosphate to which DNA and RNA nitrogenous bases are attached.

survivorship curve

A plot of the number of individuals that are still alive at each age in the maximum life span; one way to represent the age-specific death rate.

sustainability

The goal of developing, managing, and conserving Earth's resources in ways that meet the needs of people today without compromising the ability of future generations to meet their needs.

sustainable development

Development that meets the needs of people today without limiting the ability of future generations to meet their needs.

sustainable resource management

Management practices that allow use of a natural resource without damaging it.

swim bladder

A gas-filled internal sac that helps bony fishes maintain buoyancy.

symbiosis (*sim´-bē-ō´-sis*)

An interaction between organisms of different species in which one species, the symbiont, lives in or on another species, the host.

sympatric speciation

The formation of a new species in populations that live in the same geographic area. *See also* allopatric speciation.

systematics

A discipline of biology that focuses on classifying organisms and determining their evolutionary relationships.

systems biology

An approach to studying biology that aims to model the dynamic behavior of whole biological systems based on a study of the interactions among the system's parts.

T

taiga (*tī´-guh*)

The northern coniferous forest, characterized by long, snowy winters and short, wet summers. Taiga extends across North America and Eurasia, to the southern border of the arctic tundra; it is also found just below alpine tundra on mountainsides in temperate zones.

taxonomy

The branch of biology concerned with naming and classifying the diverse forms of life.

telophase

The fourth and final stage of mitosis, during which daughter nuclei form at the two poles of a cell. Telophase usually occurs together with cytokinesis.

temperate broadleaf forest

A terrestrial biome located throughout midlatitude regions where there is sufficient moisture to support the growth of large, broadleaf deciduous trees.

temperate grassland

A terrestrial biome located in a temperate zone and characterized by low rainfall and nonwoody vegetation. Tree growth is hindered by occasional fires and periodic severe drought.

temperate rain forest

A coniferous forest of coastal North America (from Alaska to Oregon) supported by warm, moist air from the Pacific Ocean.

temperate zones

Latitudes between the tropics and the Arctic Circle in the north and the Antarctic Circle in the south; regions with milder climates than the tropics or polar regions.

terminator

A special sequence of nucleotides in DNA that marks the end of a gene. It signals RNA polymerase to release the newly made RNA molecule, which then departs from the gene.

tertiary consumer (*ter´-shē-ār-ē*)

An organism that eats secondary consumers.

testcross

The mating between an individual of unknown genotype for a particular character and an individual that is homozygous recessive for that same character.

tetrapod

A vertebrate with four limbs. Tetrapods include mammals, amphibians, and reptiles (including birds).

theory

A widely accepted explanatory idea that is broader in scope than a hypothesis, generates new hypotheses, and is supported by a large body of evidence.

therapeutic cloning

The cloning of human cells by nuclear transplantation for therapeutic purposes, such as the replacement of body cells that have been irreversibly damaged by disease or injury. *See also* nuclear transplantation; reproductive cloning.

threatened species

As defined in the U.S. Endangered Species Act, a species that is likely to become endangered in the near future throughout all or a significant portion of its range.

three-domain system

A system of taxonomic classification based on three basic groups: Bacteria, Archaea, and Eukarya.

thylakoid (*thī´-luh-koyd*)

One of a number of disk-shaped membranous sacs inside a chloroplast. Thylakoid membranes contain chlorophyll and the enzymes of the light reactions of photosynthesis. A stack of thylakoids is called a granum.

thymine (**T**) (*thī´-mēn*)

A single-ring nitrogenous base found in DNA.

trace element

An element that is essential for the survival of an organism but is needed in only minute quantities. Examples of trace elements needed by people include iron and zinc.

trait

A variant of a character found within a population, such as purple flowers in pea plants or blue eyes in people.

trans fat

An unsaturated fatty acid produced by the partial hydrogenation of vegetable oils and present in hardened vegetable oils, most margarines, many commercial baked foods, and many fried foods.

transcription
The synthesis of RNA on a DNA template.

transcription factor
In the eukaryotic cell, a protein that functions in initiating or regulating transcription. Transcription factors bind to DNA or to other proteins that bind to DNA

transfer RNA (tRNA)
A type of ribonucleic acid that functions as an interpreter in translation. Each tRNA molecule has a specific anticodon, picks up a specific amino acid, and conveys the amino acid to the appropriate codon on mRNA.

transgenic organism
An organism that contains genes from another organism, typically of another species.

translation
The synthesis of a polypeptide using the genetic information encoded in an mRNA molecule. There is a change of "language" from nucleotides to amino acids. *See also* genetic code.

triglyceride (*trī-glis'-uh-rīd*)
A dietary fat that consists of a molecule of glycerol linked to three molecules of fatty acids.

trisomy 21
See Down syndrome.

trophic structure (*trō'-fik*)
The feeding relationships among the various species in a community.

tropical forest
A terrestrial biome characterized by warm temperatures year-round.

tropics
The region between the Tropic of Cancer and the Tropic of Capricorn; latitudes between 23.5° north and south.

tumor
An abnormal mass of cells that forms within otherwise normal tissue.

tumor-suppressor gene
A gene whose product inhibits cell division, thereby preventing uncontrolled cell growth.

tundra
A terrestrial biome characterized by bitterly cold temperatures. Plant life is limited to dwarf woody shrubs, grasses, mosses, and lichens. Arctic tundra has permanently frozen subsoil (permafrost); alpine tundra, found at high elevations, lacks permafrost.

tunicate
One of a group of stationary invertebrate chordates.

U

unsaturated
Pertaining to fats and fatty acids whose hydrocarbon chains lack the maximum number of hydrogen atoms and therefore have one or more double covalent bonds. Because of their bent shape, unsaturated fats and fatty acids tend to stay liquid at room temperature.

uracil (U) (*yū'-ruh-sil*)
A single-ring nitrogenous base found in RNA.

V

vacuole (*vak'-ū-ōl*)
A membrane-enclosed sac, part of the endomembrane system of a eukaryotic cell, having diverse functions.

variable
A factor or condition of an experiment that is changed, often while keeping all other factors or conditions constant.

vascular tissue
Plant tissue consisting of cells joined into tubes that transport water and nutrients throughout the plant body. Xylem and phloem make up vascular tissue.

vector
A piece of DNA, usually a plasmid or a viral genome, that is used to move genes from one cell to another.

vertebrate (*ver'-tuh-brāt*)
A chordate animal with a backbone. Vertebrates include lampreys, cartilaginous fishes, bony fishes, amphibians, reptiles (including birds), and mammals.

vesicle
A membranous sac in the cytoplasm of a eukaryotic cell.

vestigial structure (*ve-sti'-gē-al*)
A structure of marginal, if any, importance to an organism. Vestigial structures are historical remnants of structures that had important functions in the organism's ancestors.

virus
A microscopic particle capable of infecting cells of living organisms and inserting its genetic material. Viruses have a very simple structure and are generally not considered to be alive because they do not display all of the characteristics associated with life.

W

warning coloration
The bright color pattern, often yellow, red, or orange in combination with black, of animals that have effective chemical defenses.

water vascular system
In echinoderms, a radially arranged system of water-filled canals that branch into extensions called tube feet. The system provides movement and circulates water, facilitating gas exchange and waste disposal.

wavelength
The distance between crests of adjacent waves, such as those of the electromagnetic spectrum including light.

wetland
An ecosystem intermediate between an aquatic ecosystem and a terrestrial ecosystem. Wetland soil is saturated with water permanently or periodically.

whole-genome shotgun method
A method for determining the DNA sequence of an entire genome by cutting it into small fragments, sequencing each fragment, and then placing the fragments in the proper order.

wild type
The trait most commonly found in nature.

Z

zooplankton
In aquatic environments, free-floating animals, including many microscopic ones.

zygote (*zī'-gōt*)
The fertilized egg, which is diploid, that results from the union of haploid gametes (sperm and egg) during fertilization.

Index

Page numbers followed by *f* indicate figures; *t* indicate tables; those in bold indicate defined key term.

A

Abalones, eyes of, 319–320, 319*f*
Abiotic factors, **409**
    of biosphere, 410–411, 410*f*, 411*f*
    in ecosystem ecology, 474–476
Abiotic reservoirs, 474–476
ABO blood groups, **193**, 193*f*
Abortion, spontaneous, 173
Absorption, **363**
Acclimation, **412**, 412*f*
Acetic acid, 131, 131*f*
Acetyl CoA, 131, 131*f*, 134*f*
Achondroplasia, 188*t*, 189, 189*f*, 255–256
Acidification, ocean
    carbon dioxide and, 66, 66*f*
    coral bleaching, 66
Acidity, in food digestion, 65
Acid precipitation, 406, 425
Acids
    acetic, 131, 131*f*
    amino (*see* Amino acids)
    bases, pH, and, 65–66, 66*f*
    fatty, 77–78, 77*f*
    lactic, 125, 135–136, 135*f*
    nucleic (*see* Nucleic acids)
    pyruvic (*see* Pyruvic acid)
Acquired immunodeficiency syndrome (AIDS),
        224–225, 225*f*
Actinomycetes, 334*f*
Activation energy, enzymes and, **114**, 114*f*
Activators, DNA, **235**
Active site, enzyme, **116**, 116*f*
Active transport, **120**, 120*f*
Adaptations, evolutionary. *See* Evolutionary
        adaptations
Adaptive coloration, predation and, 464–465
Adelie penguins, 407, 433, 433*f*
Adenine (A), 83, 83*f*, **207**
Adenosine, 113
Adenosine diphosphate (ADP), **113**, 113*f*
Adenosine triphosphate. *See* ATP
Adenoviruses, 220*f*
Adipose tissue, 77
ADP (adenosine diphosphate), **113**, 113*f*
Adult stem cells, therapeutic cloning using, **242**
Aerobic capacity, athletic conditioning and, 125
Aerobic processes, **128**
    anaerobic processes vs., 125, 135, 137
AFP (alpha-fetoprotein), 191
African violets, 157*f*
Age structure, population, **439**, 439*f*
    human, 452–453, 452*f*, 453*f*
Agriculture
    angiosperms in, 360
    biological control of pests in, 448–449, 448*f*, 449*f*

deforestation for, 361, 361*f*, 426, 426*f*
genetically modified organisms in, 256–257,
        256*f*, 257*f*
integrated pest management in, 450, 450*f*
in temperate grassland, 422
AIDS (acquired immunodeficiency syndrome),
        224–225, 225*f*
Albinism, 188*t*, 189
Alcohol, ethyl, 137, 137*f*
Alcoholic fermentation, 136–137, 137*f*
Algae, **341**
    biofuels from, 150, 150*f*, 344
    photosynthesis experiment with, 145, 145*f*
    plant origin from, 352, 352*f*
    seaweeds, 341, 341*f*, 344–345, 345*f*
    unicellular and colonial, 341, 344, 344*f*
Alleles, **182**
    ABO blood groups as example of codominant,
        193, 193*f*
    biological diversity and, 302
    dominant vs. recessive, 182
    gene pools of, 290–292, 291*f*
    homologous chromosomes and, 183, 183*f*
    human genetic disorders and, 188–191
    law of segregation and, 182–183, 186*f*, 196*f*
    multiple, and codominance, 193
    sickle-cell disease as example of
        pleiotropy by, 194
Allopatric speciation, 308–309, 308*f*, 309*f*
Alpha-fetoprotein (AFP), 191
Alpine bumblebee, 433, 433*f*
Alpine tundra, 424
Alternation of generations, **354**, 354*f*, 356–357,
        356*f*, 357*f*
Alternative RNA splicing, **236**, 236*f*
Altitude, effects of, on terrestrial biome distribution,
        418, 418*f*
Alzheimer's disease, 188*t*
American Pika, 411, 411*f*
Amino acids, 80–82, 80*f*
    joining, 81*f*
    as monomers of proteins, 80
    origin of, 330–331, 331*f*
    substitution of, 82, 82*f*
Amino groups, 80, 80*f*
Ammonites, 279*f*
Amniocentesis, 191, 191*f*
Amniotes, **392**
Amniotic eggs, 392, 392*f*
Amoebas, 103, 103*f*, 157*f*, **342**, 342*f*
Amoebocytes, 375, 375*f*
Amoxicillin, 95
Amphibians, **391**, 391*f*
Ampicillin, 89, 95
Anabolic steroids, athletic abuse of, 79, 79*f*
Anaerobic processes, **135**
    aerobic processes vs., 125, 135, 137

Analogy, **321**
Anaphase, mitosis, **161**, 161*f*
Anaphase I, meiosis I, 166*f*
Anaphase II, meiosis II, 167*f*
Anatomical responses to environmental conditions,
        413, 413*f*
Anecdotal evidence, 44, 44*f*
Angiosperms, **353**, 353*f*
    agriculture and, 360
    fruits of, 360, 360*f*
    life cycle of, 358–360, 358*f*, 359*f*, 360*f*
    reproduction of, 358–360, 358*f*, 359*f*, 360*f*
Animals, **372**
    appearance on Earth, 328–329,
        328*f*–329*f*
    artificial selection of, 286–287
    asexual reproduction by, 155–156, 171
    body plans of, 373–374, 374*f*, 375*f*
    breeding of, by humans, 179, 190, 190*f*
    Cambrian explosion and, 373, 373*f*
    cell cycle of (*see* Cell cycle)
    cells of, 93*f*
    chordates, 388–389, 388*f*, 389*f*
    classification of, 322–323, 322*f*
    communities of (*see* Communities)
    digestive systems of (*see* Digestive systems)
    diversity of, 370, 372–374, 372*f*, 373*f*, 374*f*, 375*f*
    early, 373, 373*f*
    endangered and threatened (*see* Endangered
        species)
    evolution of, 372–374, 372*f*, 373*f*, 374*f*, 375*f*
        (*see also* Evolution)
    genetically modified pharmaceutical, 252, 256, 256*f*
    homeotic genes of, 238, 238*f*
    humans and primates (*see* Humans; Primates)
    invertebrates (*see* Invertebrates)
    life cycle of, 372, 372*f*
    nutrition for, 372, 372*f*
    osmoregulation in cells of, 119
    phylogeny of, 374, 374*f*, 375*f*
    populations of (*see* Populations)
    prion infections of, 82, 82*f*, 226
    regeneration of body parts by, 239
    reproductive cloning of, 240–241, 240*f*, 241*f*
    sexual reproduction by, 155–156, 174 (*see also*
        Meiosis)
    urban-adapted, 323, 323*f*
    vertebrates (*see* Vertebrates)
    viruses of, 222–223, 222*f*, 223*f*
Annelids (Annelida), 379–380, 379*f*
Antarctica, polar ice biome of, 424, 424*f*
Anthers, flower, **358**, 358*f*
Anthrax, 267–268, 267*f*, 339
Anthropocene, 303, 311, 323, 323*f*
Anthropoids, 395–396, 395*f*, 396*f*
Antibiotic-resistant bacteria, 51, 51*f*
    evolution of, 299, 299*f*